European Journal of
Biochemistry

EJB Reviews
1990

Contributing Authors:

C.-M. Becker, Heidelberg
N. Berry, Kobe
H. Betz, Heidelberg
H. v. Boehmer, Basel
J. A. Campus-Ortega, Köln
K. Decker, Freiburg i. Br.
F. Di Virgilio, Padova
H. v. Döhren, Berlin
C. P. Downes, Dundee
H. Eisenberg, Rehovot
R. Huber, Martinsried
U. Hübscher, Zürich
R. E. Humbel, Zürich
E. Kellenberger, Basel
H. Kleinkauf, Berlin
E. Knust, Köln

D. Langosch, Heidelberg
R. M. Lyons, Nashville
C. H. Macphee, Dundee
S. Malcolm, London
A. McPherson, Riverside
H. L. Moses, Nashville
P. P. Müller, Bern
L. Nilsson, Stockholm
Y. Nishizuka, Kobe
O. Nygård, Stockholm
D. Pietrobon, Padova
T. Pozzan, Ferrara
J. Schell, Köln
P. Thömmes, Zürich
H. Trachsel, Bern
R. Walden, Köln

 Springer-Verlag Berlin Heidelberg GmbH

Professor Dr. P. Christen
Biochemisches Institut der Universität Zürich
Winterthurerstrasse 190
CH-8057 Zürich

Professor Dr. E. Hofmann
Institut für Biochemie der Universität Leipzig
Liebigstraße 16
O-7010 Leipzig

ISBN 978-3-540-53290-3   ISBN 978-3-642-76168-3 (eBook)
DOI 10.1007/978-3-642-76168-3

© Springer-Verlag Berlin Heidelberg 1991
Originally published by Federation of European Biochemical Societies in 1991.

Typesetting, Printing and Binding: Wiesbadener Graphische Betriebe GmbH, Wiesbaden
31/3145-543210 — Printed on acid-free paper

Preface

In the mid-1980s the European Journal of Biochemistry set out to publish review articles. The enterprise proved successful with many high-level reviews written by well-known scientists appearing in the Journal. The reviews are intended to represent emerging and rapidly growing fields of research in fundamental as well as in applied areas of biochemistry, such as medicine, biotechnology, agriculture and nutrition. Novel methodological and technological approaches which stimulate biochemical research are also included. The authors of the reviews are explicitly asked to be critical, selective, evaluative and interdisciplinarily oriented. The reviews should encourage young scientists toward independent and creative thinking, and inform active investigators about the state of the art in a given field.

The good reception of the reviews by the readers of the European Journal of Biochemistry has induced the Editorial Board and Springer-Verlag to publish them annually in a separate booklet, 'EJB Reviews', in order to further their dissemination among biochemists and scientists in related biological and medical disciplines.

P. Christen
Chairman of the
Editorial Board

E. Hofmann
Reviews Editor

Reviews published in EJB before 1990

Iron transport and storage. R. R. Crichton and M. Charloteaux-Wauters (1987) 164, 485 – 506

DNA repair in human cells. M. Schweiger, B. Auer, H. J. Burtscher, M. Hirsch-Kauffmann, H. Klocker and R. Schneider (1987) 165, 1 – 6

Signals guiding proteins to their correct locations in mitochondria. G. Schatz (1987) 165, 235 – 242

Eucaryotic primase. Y.-F. Roth (1987) 165, 473 – 481

Co-operative and allosteric enzymes: 20 years on. J. Ricard and A. Cornish-Bowden (1987) 166, 255 – 272

The mechanism of the conservation of energy of biological oxidations. E. C. Slater (1987) 166, 489 – 504

Cytochrome-c oxidase. M. Brunori, G. Antonini, F. Malatesta, P. Sarti and M. T. Wilson (1987) 169, 1 – 8

Modes of attachment of acetylcholinesterase to the surface membrane. I. Silman and A. H. Futerman (1987) 170, 11 – 22

Sialylated lactosylceramides. N. V. Prokazova, E. V. Dyatlovitskaya and L. D. Bergelson (1988) 172, 1 – 6

Direct and indirect electron transfer between electrodes and redox proteins. J. E. Frew and H. A. O. Hill (1988) 172, 261 – 269

Knowledge-based protein modelling and design. T. Blundell, D. Carney, St. Gardner, F. Hayes, B. Howlin, T. Hubbard, J. Overington, D. A. Singh, B. L. Sibanda and M. Sutcliffe (1988) 172, 513 – 520

Comparative evaluation of gene expression in archaebacteria. W. Zillig, P. Palm, W.-D. Reiter, F. Gropp, G. Pühler and H.-P. Klenk (1988) 173, 473 – 482

The regulation of the intracellular pH in cells from vertebrates. Ch. Frelin, P. Vigne, A. Ladoux and M. Lazdunski (1988) 174, 3 – 14

Ecdysteroid-regulated gene expression in Drosophila melanogaster. O. Pongs (1988) 175, 199 – 204

Import of proteins into mitochondria: a multi-step process. N. Pfanner, F.-U. Hartl and W. Neupert (1988) 175, 205 – 212

Permeation of hydrophilic molecules through the outer membrane of Gram-negative bacteria. R. Benz and K. Bauer (1988) 176, 1 – 19

Enhancer sequences and the regulation of gene transcription. M. M. Müller, Th. Gerster and W. Schaffner (1988) 176, 485 – 495

Citric-acid cycle, 50 years on modifications and an alternative pathway in anaerobic bacteria. R. K. Thauer (1988) 176, 497 – 508

Nucleic acids and nuclear magnetic resonance. F. J. M. Van De Ven and C. W. Hilbers (1988) 178, 1 – 38

Intracellular transport in interphase and mitotic yeast cells. L. T. Nevalainen and M. Makarow (1988) 178, 39 – 46

Molecular mechanism of visual transduction. M. Chabre and P. Deterre (1989) 179, 255 – 266

Structure and biological activity of basement membrane proteins. R. Timpl (1989) 180, 487 – 502

Mechanisms of flavoprotein-catalyzed reactions. S. Ghisla and V. Massey (1989) 181, 1 – 17

Nucleo-mitochondrial interactions in yeast mitochondrial biogenesis. L. A. Grivell (1989) 182, 477 – 493

NMR studies of mobility within protein structure. R. J. P. Williams (1989) 183, 479 – 497

Dehydrogenases for the synthesis of chiral compounds. W. Hummel and M.-R. Kula (1989) 184, 1 – 13

Chemical model systems for drug-metabolizing cytochrome-P-450-dependent monooxygenases. D. Mansuy, P. Battioni and J.-P. Battioni (1989) 184, 267 – 285

Growth factors as transforming proteins. C.-H. Heldin and B. Westermark (1989) 184, 487 – 496

A chromosomal basis of lymphoid malignancy in man. T. Boehm and T. H. Rabbitts (1989) 185, 1 – 17

The nucleoskeleton and the topology of transcription. P. R. Cook (1989) 185, 487 – 501

Engineering of protein bound iron-sulfur clusters. H. Beinert and M. C. Kennedy (1989) 186, 5 – 15

Contents

VIII

VIII

Eur. J. Biochem. *187*, 7—22 (1990)
© FEBS 1990

Review

Thermodynamics and the structure of biological macromolecules
Rozhinkes mit mandeln

Henryk EISENBERG

Department of Polymer Research, The Weizmann Institute of Science, Rehovot

(Received June 23, 1989) — EJB 89 0780

In this review, I will discuss the role of thermodynamics in both the determination and evaluation of the structure of biological macromolecules. The presentation relates to the historical context, state-of-the-art and projection into the future. Fundamental features relate to the effect of charge, exemplified in the study of synthetic and natural polyelectrolytes. Hydrogen bonding and water structure constitute basic aspects of the medium in which biological reactions occur. Viscosity is a classical tool to determine the shape and size of biological macromolecules. The thermodynamic analysis of multicomponent systems is essential fo the correct understanding of the behavior of biological macromolecules in solution and for the evaluation of results from powerful experimental techniques such as ultracentrifugation, light, X-ray and neutron scattering. The hydration, shape and flexibility of DNA have been studied, as well as structural transitions in nucleosomes and chromatin. A particularly rewarding field of activity is the study of unusual structural features of enzymes isolated from the extreme halophilic bacteria of the Dead Sea, which have adapted to saturated concentrations of salt. Future studies in various laboratories will concentrate on nucleic-acid—protein interactions and on the so-called 'crowding effect', distinguishing the behavior in bacteria, or other cells, from simple test-tube experiments.

When I was asked by the editors of the Journal to write a review article with the above title, my first reaction was to modify a title juxtaposing, to all appearances, two concepts diametrically distinct in their physical content. Thermodynamics, majestic in its far-reaching implications, on the one hand provides a framework of broad universality, but is usually not concerned with a detailed representation of our physical, biological world. Structure on the other hand is all details and the more details become available, the more its mission is considered accomplished. At first sight, an insuperable gulf to bridge. On further thought though, the opportunity to telescope forty odd years of personal scientific activity in trying to understand the nature of the molecules of life by pushing thermodynamics into the realm of structure, proved strong enough to take on the challenge.

Around midcentury the central role of DNA had barely been appreciated and the double helix was still to come; imaginative, daring, innovative investigations had already been initiated by Perutz and Kendrew to derive the first protein structures by X-ray crystallography utilizing the radiation, detection and computational means, available then, but most primitive by present day standards. In 1989 we celebrate 150 years since the birth of J. Willard Gibbs who correctly established the fundamental laws of thermodynamics, leading to the evaluation of the free energy, both in bulk and on surfaces, which governs all chemical reactions, folding, unfolding and stability of macromolecules, interactions between macromol-

ecules, the behavior of macromolecular assemblies and of complex biological cells.

I will, in the following, endeavor to show how, by a judicious choice of experimental tools and by a correct design of experiments and their interpretation, it was possible through the application of thermodynamics to obtain novel, timely and correct information about structure and interactions in biological macromolecular systems, which has withstood the test of time. This must be viewed in the context of the extraordinary changes and advances which we have witnessed in research in biology, though the basic concepts governing our understanding of the material phenomena have not been changed in this period of time.

1. THE EFFECT OF CHARGE: POLYELECTROLYTES

I was exposed early to the ideas of Aharon Katchalsky, my teacher, who believed that as biological macromolecules are both composed of linear chains and carry electrical charges, the study and understanding of the behavior of aqueous solutions of natural and synthetic 'polyelectrolytes' should contribute a meaningful component to the understanding of the stabilization and function of biological systems; it was realized though, that specificity would not be apparent in the absence of sequential arrangement of amino acids, or nucleotide bases. Randomly coiling linear polymer chains expand upon the addition of electrical charge (Fig. 1) and attempts were made to theoretically match the electrostatic expansion with chain elasticity, deriving from Brownian motion [1, 2]. A striking transition from molecular to macroscopic behavior could be effected by crosslinking synthetic macromolecular polyelectrolyte chains (Fig. 2), showing how

---

Correspondence to H. Eisenberg, Department of Polymer Research, The Weizmann Institute of Science, Rehovot, IL-76100, Israel

Abbreviation. hMDH, halophilic malate dehydrogenase.

Note. Rozhinkes mit mandeln: raisins with almonds, a yiddish song by A. Goldfaden.

8

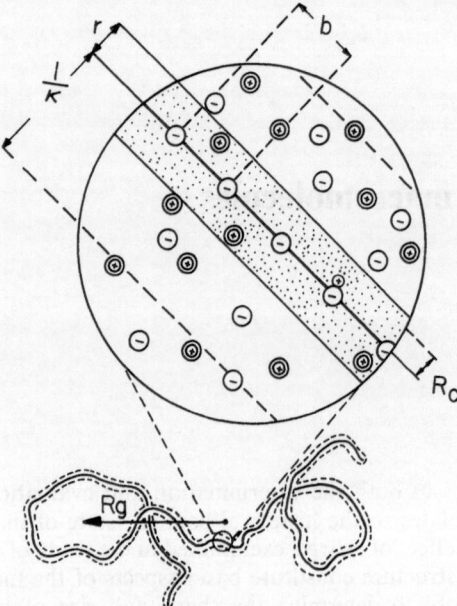

Fig. 1. *Schematic drawing of polyelectrolyte in the presence of 1:1 simple electrolyte.* Fixed negative charges ($\ominus$) are at distance b along the infinitely thin cylinder. Hydrated condensed counter-ions ($\oplus$) are within the cylinder, at radius r. Site-bound counter-ions to fixed charges are shown without hydration. The dashed line schematically indicates the Debye-Hückel reciprocal shielding length, $1/\kappa$ (whereas number of condensed counter-ions and r are independent of ionic strength, I, $1/\kappa$ increases with decreasing I); R_g denotes radius of gyration of macromolecular coil and R_c cross-section radius of gyration

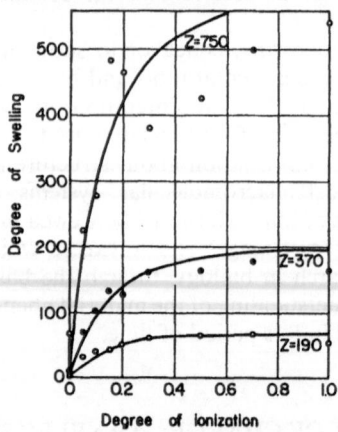

Fig. 2. *Equilibrium swelling of gels of polymethacrylic acid crosslinked with divinylbenzene in water, as a function of the degree of ionization, for different degrees of polymerization, Z, between crosslinks.* From [4] with permission

chemical energy of ionization could be transformed into work [3 – 5], an elementary, instructive, but by no means accurate simulation of muscular contraction. The early days of polyelectrolyte research and a follow-up thirty years later have been summarized [6, 7] and a recent review is available [8].

A significant step forward in polyelectrolyte research was made in the analysis of the 'cell' model of linear polyelectrolyte rods in salt-free solutions by application of the Poisson-Boltzmann equation [9, 10], later [11] extended to the more realistic situation of the charged polyelectrolyte molecule with its complement of counter-ions in the presence of added salt.

The cylindrical rod cell model is based on the assumption that over limited distances along the chain electrical charges are arranged in a linear array with well-defined charge density, allowing the calculation of the distribution of the small ions in the electrostatic field of the macromolecular rod, assumed to be of infinite length. This rather esoteric but sensible approach was restated in a novel reinterpretation of the model [12, 13], known as the ion-condensation model whose results, with few exceptions, agree with the results previously derived [14]. From Manning's basic assumption a simple picture arises [15] which has been substantiated by Record et al. [16] in their 'thermodynamic analysis of ion effects on the binding and conformational equilibria of proteins and nucleic acids, the roles of ion association or release, screening and ion effects on water activity.

The cell model introduces logarithmic potentials which persist to infinity and adequately describe the behavior of real solutions at low, experimentally accessible concentrations. Manning postulates that a critical charge density, ξ_{crit}, exists on the cylinder at which counter-ions associate with ('condense' on) the surface of the cylinder so that the net charge density is maintained at the critical value. Thus, as long as the charge density, ξ, is below the critical value, no counter-ions remain in the domain of the poly-ion at 'infinite' dilution, but when this value is exceeded a fraction of counter-ions condenses such that the value of ξ_{crit} is maintained. Ions in solution which are not condensed are distributed according to the classical Debye-Hückel theory for simple electrolytes [17]. The charge density ξ is defined by

$$\xi = e|\beta|/DkT \qquad (1)$$

where $\beta = e/b$ is the charge/unit length of cylinder (b is the distance between charged groups on the infinitely thin cylinder; in the present discussion the finite radius of the polyelectrolyte backbone is not considered), D is the relative permittivity, and ξ_{crit} is equivalent to the Bjerrum length for ion-pair formation in the classical theory of simple electrolyte solutions [17]; k is Boltzmann's constant and T is the absolute temperature. For univalent fixed charges, co-ions and counter-ions, ξ_{crit} ist taken to be unity [13, 15]. At infinite dilution the bulk relative permittivity, D, should be used. For water at 25 °C ($D = 78.5$) the critical value $\xi = 1$ corresponds to a charge spacing, b, of 0.714 nm. For double helical B-DNA, $b = 0.17$ nm, therefore the condition $\xi > 1$ (ξ is equal to 4.2) applies and condensation of ions always occurs.

What is the fraction of ions condensed? Following Manning the fraction θ of counter-ions condensed on an infinite poly-ion per poly-ion charge is

$$\theta = 1 - \xi^{-1} . \qquad (2)$$

For double helical B-DNA $\theta = 0.76$ and for single stranded DNA $\theta = 0.39$. After condensation, both forms of nucleic acids are still highly charged polyelectrolytes with an average fractional charge of ξ^{-1} charges/nucleotide unit. The screening effect of simple salt ions in solution on the interactions of these residual polyelectrolyte charges is thermodynamically equivalent to the binding of an additional fraction $(2\xi)^{-1}$ of counter-ions/poly-ion charge [18]. A thermodynamic counter-ion-binding parameter, ψ, can be defined which, in distinction to θ, includes the contribution of screening

$$\psi = \theta + (2\xi)^{-1} = 1 - (2\xi)^{-1} . \qquad (3)$$

For double helical B-DNA $\psi = 0.88$ and for single stranded DNA (or for RNA for that matter) $\psi = 0.70$. Counter-ions are thus 'released' upon DNA melting, a process enhanced by lowering and opposed by raising the ionic strength. A quantitative evaluation of the lowering of the melting temperature of DNA at low salt concentrations can thus be performed as a direct result of the lowering of ψ. Thermodynamic measurements, which are sensitive to both condensation and screening interactions of counter-ions with the polyelectrolyte, detect the ion-association parameter, ψ [16]. An understanding of ion condensation and screening is thus essential for the characterization of protein − nucleic-acid interactions and further considerations lead to the effect of the competition of single and multivalent counter-ions on the condensation and interaction process [19]. The analysis can be improved by an examination of the validity of the Poisson-Boltzman equation or the ion-condensation approach in comparison with Monte-Carlo computer simulations [20, 21], yet the major effect of charge, in particular in interactions involving nucleic acids and nucleic-acid − protein interactions, appears to be reasonably well established and a more elaborate discussion incorporating the detailed charge distribution of the DNA and the different polarizabilities of the macromolecule and solvent [22] transcends the scope of this review. The statistical mechanics of the B→Z transition of DNA and thermodynamic stability of DNA oligomers over a broad concentration range $(0.01 − 5.0 \text{ M})$ of $1:1$ electrolytes of varying counter-ion size have recently been evaluated in novel ways [23, 24].

2. A BASKET OF MIXED OBSERVATIONS ON HYDROGEN BONDING AND WATER STRUCTURE

We owe to Debye [25] the correct evaluation of activities of ions in electrolyte solutions, concepts later extended by Linderström-Lang [26] to proteins. Electrolyte conductance [17] is a good procedure for determining ionic activity and association parameters for small and large molecules. Bolaform electrolytes [27] are chain-like structures with two charged end groups. The compounds $Me_3N^+(CH_2)_2$ X $CO(CH_2)_n$ CO X $(CH_2)_2$ N^+ Me_3 show curare-like activity [28] when X = O. Physiological activity of the bis-quarternary salts is related to chain length. Succinylcholine $(n = 2)$ binds to the acetylcholine receptor, and causes depolarization of the end plate. Unlike the esters, the corresponding amides (X = NH) show no curare-like activity for $n = 0$ to $n = 6$; instead they prolong the duration of the block of neuromuscular transmission produced by certain bis-esters [28]. We could show [29], from an analysis by conductance of the ratio of the first and second counter-ion dissociation constant, for the same values of n, that the amides are hydrogen bonded and therefore the amides and esters differ in configuration. Charge separation rather than maximum chain length is the significant variable leading to curare-like activity.

Dissociation of carboxylic acids affects the stability and folding of proteins. With increasing concentration, the dissociation constants of carboxylic acids calculated from conductance measurements show pronounced deviations from the extrapolated values [30], though the limiting Debye-Hückel activity coefficients should still hold at these concentrations. Dimerization of carboxylic acids, a consequence of hydrogen bonding modulated by increasing van der Waals interactions between monomers with increasing chain length, provides a plausible explanation [31] for obtaining constant

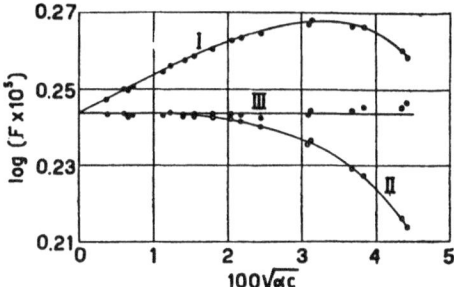

Fig. 3. *Ionization constants of acetic acid.* (I) log F = log k = log$[\alpha^2 c/(1 - \alpha)]$; classical dissociation constant. (II) log F = log k − 1.013 αc; ionization constant considering Debye-Hückel term. (III) log F = log K = log k − 1.013 $\sqrt{\alpha c}$ + log$[\sqrt{1 + 8L(1-\alpha)c} + 1]$; true ionization constant, considering both Debye-Hückel term and dimerization; c, concentration of acid (in mol/l); α, degree of ionization; L, dimerization constant (in l/mol). From [31] with permission

values of the ionization constants in the range of concentrations examined (Fig. 3).

Water, the solvent in which presently all life processes occur, is an exceptional liquid characterized by unusually high temperatures and heats of melting and of evaporation, and the curious phenomenon of maximum density in the liquid state at $4 °C$. These unusual properties, matched to a large extent by the properties of D_2O, deuterium oxide, are due to extensive hydrogen bonding, the exact extent depending on temperature and pressure. Though the structure of water has been investigated for many years, an exact interpretation of water structure and its effect on biological solutes is presently not available. In the context of our light-scattering investigation of solvents for solutions of biological macromolecules [32] we were struck by an attempt of Mysels [33] to explain the fact that the experimental Rayleigh light-scattering value of pure water is somewhat higher than the theoretical value calculated on the basis of the Einstein density-fluctuation theory, modified by the Cabannes factor for the optical anisotropy of the liquid. The additional increase was assumed by Mysels to be due to local differences in structure which occur independently of pressure fluctuations and can be rationalized on the basis of various, then current, models on water structure. Upon experimental reinvestigation of the Rayleigh ratio of water and D_2O, we found [32] that the true scattering values are significantly lower and in excellent agreement with the Einstein-Cabannes equations. Thus, no significant new information on the structure of water as a function of temperature can be derived from light-scattering experiments.

Within the context of the above investigation we became aware of interesting manifestations of water structure, which can, in principle, be derived from the refractive index, and its dependence on temperature and pressure. Measurements accurate to seven significant figures of the refractive index of water are available [34] and the influence of temperature and pressure [35] have been investigated already in the last century. On the basis of the Lorentz-Lorenz theory [36, 37] the refractive index n of pure liquids is given by

$$f(n) = (n^2 - 1)/(n^2 + 2) = P_{LL}\varrho \qquad (4)$$

where P_{LL} is the specific refraction, assumed to be independent of pressure and temperature, and ϱ the density (g/ml). In the case of most liquids, including water, Eqn (4) is however not obeyed and a number of empirical equations have been suggested, none of which enjoys universal validity for a representative class of liquids.

Close analysis [38] of refractive index data a various temperatures, wavelengths and pressures, as well as some piezo-optic measurements from our laboratory [39] yielded an empirical relationship closely describing all data

$$f(n) = A\varrho^B e^{-CT} \qquad (5)$$

where A, B and C are constants. Eqn (5) yields upon differentiation

$$(\partial \ln f / \partial P)_T = \beta_T B \qquad (6)$$

$$-(\partial \ln f / \partial T)_P = \gamma B + C \qquad (7)$$

and

$$-(\partial \ln f / \partial T)_V = C \qquad (8)$$

where β_T is the isothermal compressibility $(\partial \ln \varrho / \partial P)_T$ and γ is the volume expansion coefficient $-(\partial \ln \varrho / \partial T)_P$. Recalculated values of n are accurate to six significant decimal figures and the fact first observed by Jamin [40] that the maximum value of n for water, as is also true for D_2O, is at about 4°C below the maximum value of ϱ, is correctly predicted.

Conjecturing with respect to the molecular significance of the terms which can be isolated from the study of the dependence of n on P (or V) and T we concluded that at pressures above 200 MPa and temperatures above 100°C water may revert to the 'non-hydrogen-bonded' state, in agreement with the observation [41] that the maximum density and viscosity anomaly in water disappear at pressures above 250 MPa. Extension of refractive index and dielectric studies of water to very high pressures and temperatures might disclose information on water structure in relation to specific models. Survival of life at high pressures, as well as adaptation to other extremes such as pH, temperature, chemical environment and salt (cf. Section 6), is of great interest, though the studies involving pressure [42] have not been extended to the extreme limits suggested here. The theory for the random network, continuum model of liquid water has been redeveloped very recently [43] and should be applied to interpret the parameters derived from the study of the refractive indices.

3. VISCOSITY, AND OTHER TOOLS FOR SHAPE AND STRUCTURE STUDIES OF DOUBLE- AND SINGLE-STRANDED NUCLEIC ACIDS

Poiseuille, who gave his name to the simple capillary tube viscometer, was interested in the flow of blood [44]. Though viscosity of fluids is strongly temperature dependent, one does not even require a highly elaborate temperature stabilizer for precise measurements. Two capillary glass viscometers, one filled with solvent and the other with solute, can be run in parallel, and the difference in running time read differentially with inexpensive quartz watches available to day. Nobody should live without it. Einstein [45] derived the basic law for spheres

$$\eta_{sp} = 2.5 \, \varphi \qquad (9)$$

where the specific viscosity $\eta_{sp} = (\eta - \eta_0)/\eta_0$ is independent of the size of the spheres (η and η_0 are the viscosities of the solute and the solvent respectively and φ is the solute volume fraction). Eqn (9) has been extended to ellipsoids [46] and has found use in the analysis of the shape of proteins [47]. For non-ionic randomly coiling polymers one finds that η_{sp}/c (c is the concentration in g/ml) increases with c, due to polymer-polymer hydrodynamic interactions. This also applies to charged polymers (polyelectrolytes) in solution, at reasonable

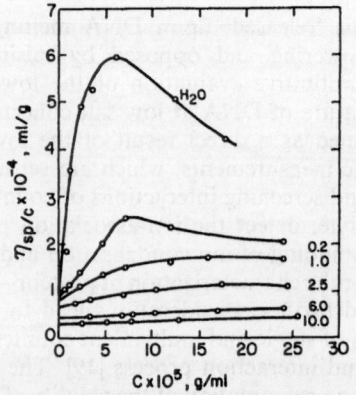

Fig. 4. *Reduced specific viscosity η_{sp}/c of aqueous solutions of a partially (61%) quaternized polyvinylpyridine as a function of polymer and NaCl concentration. From [49] with permission. The number shown are the concentrations of NaCl in mol/l*

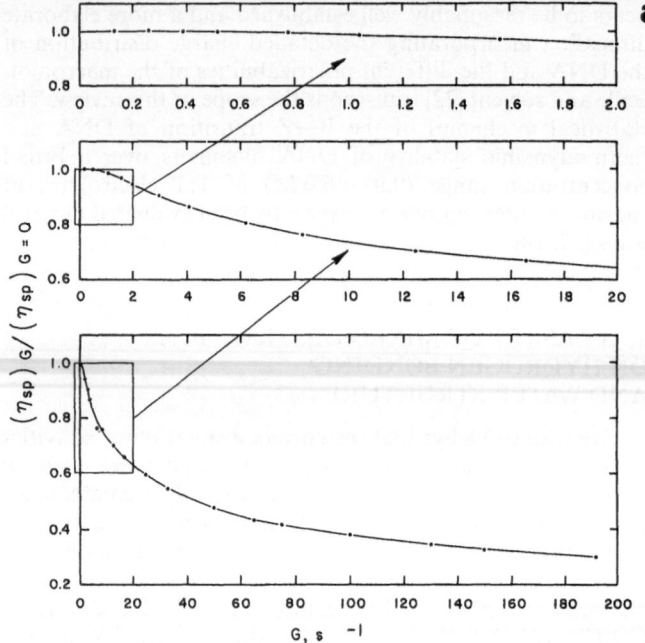

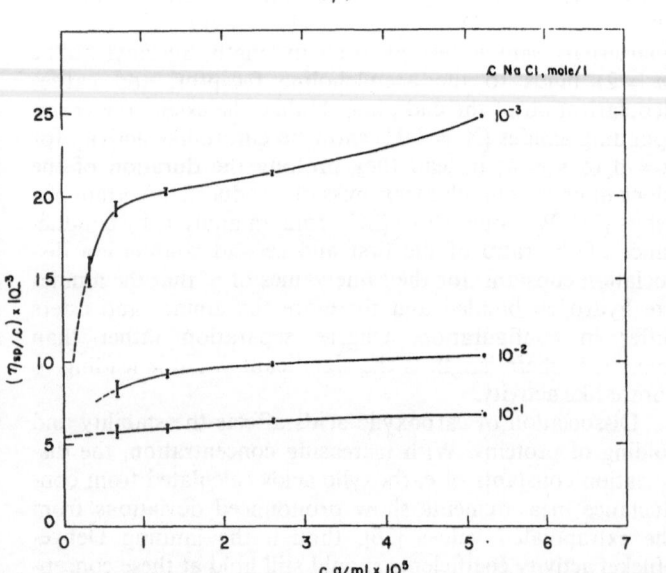

Fig. 5. *Viscosity of trout sperm DNA. (a) $(\eta_{sp})_G/(\eta_{sp})_{G=0}$ against rate of shear, G; $c_{DNA} = 5.2 \times 10^{-5}$ g/ml; $c_{NaCl} = 10^{-3}$ M; (b) η_{sp}/c at $G = 0$ as a function of c_{DNA} at various NaCl concentrations. From [52] with permission*

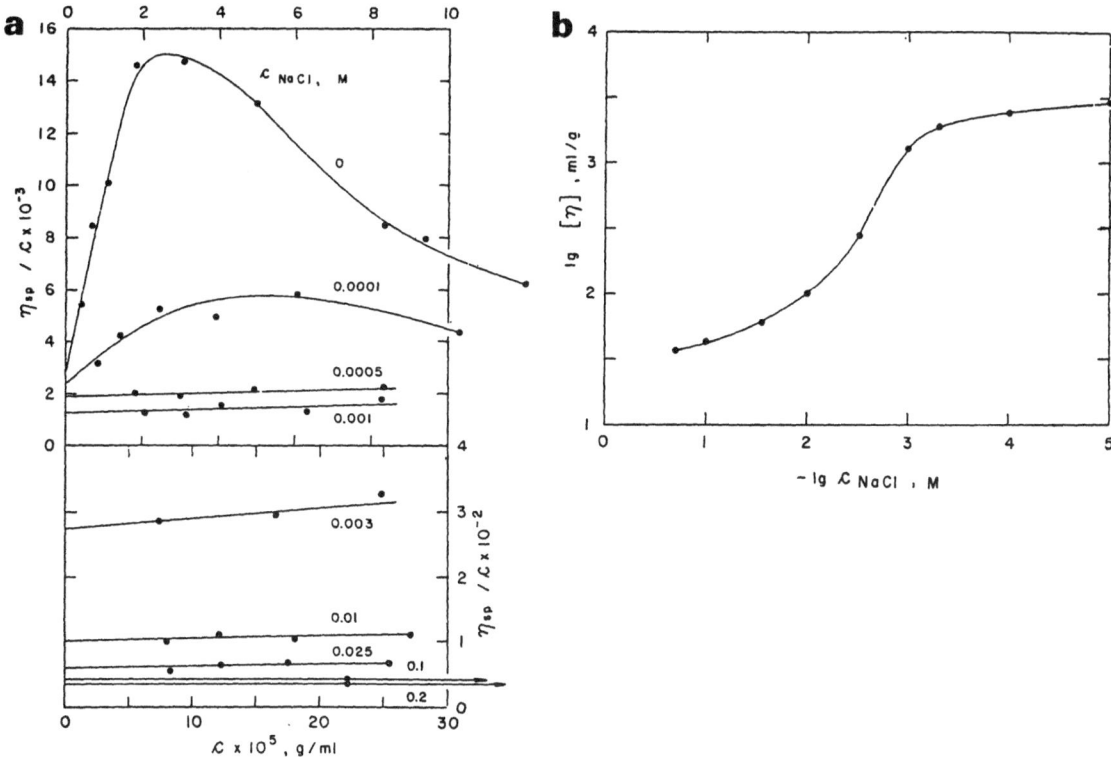

Fig. 6. *Viscosity of* E. coli *RNA solutions*. (a) η_{sp}/c at various NaCl concentrations, against RNA concentration; (b) $[\eta]$ of RNA solutions as a function of NaCl concentration. From [53] with permission

ionic strengths. A useful expression for coiling macromolecules connects $[\eta] \equiv (\eta_{sp}/c)_{c \to 0}$ with the molar mass

$$[\eta] = M^a \tag{10}$$

where $a = 0.5 - 0.8$ depending on the solvent [48]. When charges are attached to synthetic or biological macromolecular coils (polycarboxylic and sulfonic acids, nucleic acids, polypeptides, hyaluronic acid and so forth) they expand considerably due to the repulsion of the ionic charges fixed on the chains. The electrostatic force of expansion, balanced by the Brownian entropic resistance to expansion, can be evaluated on the basis of the principles described in Section 1. At very low concentrations of salt, η_{sp}/c increases dramatically with decreasing concentration c (Fig. 4) [49]. This is commonly attributed to chain expansion to fully stretched length, yet I believe [7] that polyelectrolyte molecules, though expanded, are not fully stretched, and the large increase in η_{sp}/c is mainly due to long-range electrostatic interaction between partially expanded macromolecular chains. The basic question, whether charged macromolecules expand to their fully stretched length at very low salt concentration in turn forming highly oriented complexes [50], is not of direct interest to biology in view of its remoteness from known physiological conditions (However, consider the opposite situation of very high salt concentration in the case of the extreme halophiles, discussed in Section 6. When large molecules such as DNA become oriented in flow and their solutions birefringent, then simple capillary viscometry becomes inappropriate because of the excessive rates of shear in these instruments [49]. Development of a Couette low-shear viscometer with electrostatic restoring torque, covering a broad range of rates of shear [51] led to a study [52] of the viscosity of DNA as a function of rate of shear, concentration and NaCl concentration (Fig. 5). The early measurements were best described by a Gaussian

chain model with high internal viscosity, pointing to the inherent DNA stiffness and relative insensitivity of its shape and size to salt concentration. When it then first became possible to prepare a high molecular mass ribosomal RNA preparation from *Escherichia coli* [53], it was found that the RNA preparations closely resemble coiling synthetic polyelectrolytes in their viscosity behavior and behave quite unlike DNA (Fig. 6). The viscosity behavior, birefringence of flow and potentiometric titration data were discussed in terms of a single-contractile-coil model. Similar results were obtained with RNA from rat liver [54].

Viscosity studies proved to be very useful for the study of DNA solutions and Bruno Zimm and his colleagues [55] constructed a simple rotation viscometer, which yielded the limiting low-shear viscosity, and whose use could be extended [56] to the measurement of the viscoelastic properties of solutions containing very large DNA and therefore to an estimate of the size of intact chromosomal DNA. For shorter DNA samples, capillary viscometers may still be useful and we could study, for instance, the intercalation of proflavine or other dyes into sonicated DNA samples (Fig. 7) [57]. We can appreciate the higher sensitivity exhibited by the viscosity ratio, depending on change in rotary frictional parameters only, as compared to the sedimentation ratio, involving, in addition to the less sensitive translational frictional parameters, changes in density increments (cf. Section 4). Hydrodynamic and thermodynamic studies of nucleic acids have been summarized [58].

In our search for synthetic models for the study of the influence of charge in biological macromolecules, we found [59] that aqueous solutions of polyvinylsulfonic acid,

$$-[CH_2 - CH]_n -$$
$$|$$
$$SO_3^- \; H^+,$$

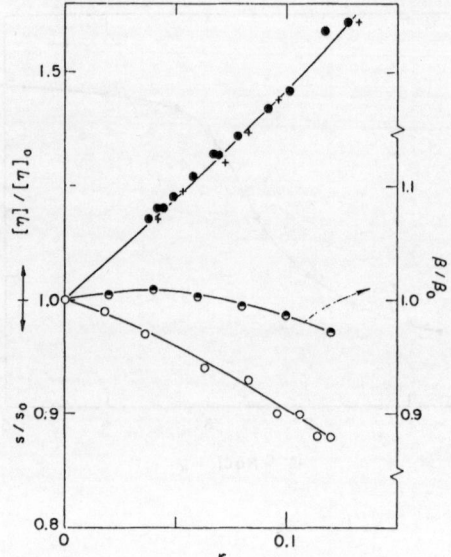

Fig. 7. *Relative change of intrinsic viscosity, sedimentation coefficient, and Mandelkern-Flory parameter β, of sonicated calf thymus DNA, with binding of proflavine. r is the number of dye molecules bound/ nucleotide.* (●) Viscosity measurements, $M = 3.95 \times 10^5$. (+) Viscosity measurements, $M = 4.95 \times 10^5$. (○) Sedimentation measurements, $M = 4.92 \times 10^5$. (◑) Mandelkern-Flory parameter β. From [57] with permission

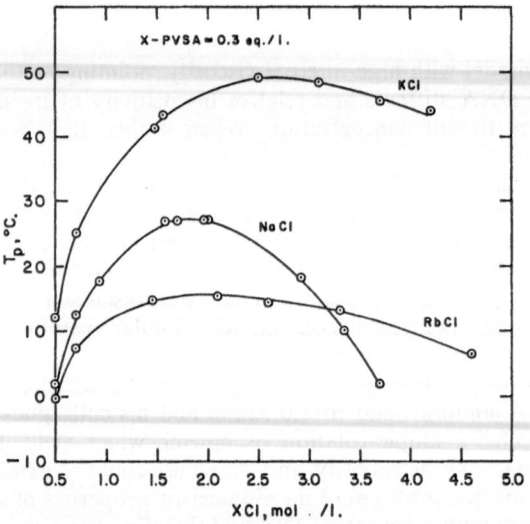

Fig. 8. *Phase-separation temperatures,* T_p, *at constant concentration of salts X-PVS of polyvinylsulfonic acid, with various electrolytes XCl (X = Na,Rb,K).* From [59] with permission

and its salts with various monovalent cations, separate into two liquid phases, in equilibrium, at high concentrations of added monovalent electrolytes. One of the two phases is very dilute and the other very concentrated in the macromolecular component. The phase separation has been studied as a function of temperature, polymer and added electrolyte concentration. The phenomenon is highly specific with respect to the monovalent electrolytes investigated (Fig. 8). In the alkaline halide series the order of specificity is NaCl < KCl > RbCl and KI > KBr > KCl; no phase separation occurs under similar conditions with HCl, LiCl, CsCl and NH_4Cl. From potentiometric and conductimetric titrations polyvinylsulfonic acid was shown to behave as a fully ionized polyelectrolyte. Specificity effects with alkali ions were shown, from viscosity and

conductivity measurements, to exist also in dilute solutions of the polymer, both in the absence and in the presence of added monovalent electrolytes; they are correlated to the order of specificity of the same cations in phase separation.

The intriguing data showing high specificity for simple monovalent cations in a polyelectrolyte of the simplest structure, in the absence of organized structure, lipids, membranes and ion channels, is highly interesting *per se*. In the present context, it led us to a method for the preparation of uniform samples of poly(A). This polynucleotide forms double helices at low pH, yet a neutral pH poly(A) forms single coils capable of base stacking at suitable temperatures and ion concentrations. We studied the conformation of poly(A) in aqueous solution at neutral pH under 'ideal' Θ (theta or 'Flory') solvent conditions [48], as a function of temperature [60]. The studies were carried out on fractions of well-defined molar mass, obtained by reversible phase separation of the polymer in NaCl solutions more concentrated than 1 M. Under such conditions, poly(A) is soluble at low and high temperatures, but only partly miscible in a range centered on 35 – 40 °C. The extent of the range varies with molar mass, polymer and salt concentrations. The upper and lower Θ temperatures for solvents in the range of NaCl concentrations 1 – 1.33 M were derived from the phase diagrams (Fig. 9).

Light-scattering experiments of poly(A) at neutral pH yielded the molar masses, radii of gyration and second virial coefficients (A_2) of several fractions at various salt concentrations, in the temperature range −2 to 62.4 °C. At a fixed salt concentration, A_2 vanished at two temperatures, the upper and lower Θ temperatures for that solvent (Fig. 10). In this way it was possible to obtain values of the radius of gyration at various Θ temperatures which were unperturbed by solvent effects, thus separating long-range, solvent-depending influences on molecular conformation, from the influences of short-range interactions due to base stacking. The sedimentation velocity and viscosity behavior of poly(A) as a function of salt concentration and temperature were also studied. In 1 M NaCl, both the sedimentation coefficient and intrinsic viscosity vary as $M^{0.5}$ at a temperature quite close to the Θ temperature.

The results show that poly(A) forms a highly extended structure at low temperatures but that this structure is rapidly disrupted as the fraction of bases in stacks falls below unity. The observations are consistent with the existence of a structure formed by a non-cooperative process. Results of this work have been subject to intensive theoretical calculations [61]. The biological significance of extended poly(A) stretches in transcribing RNA molecules was not apparent at the time our studies were undertaken.

4. MULTICOMPONENT SYSTEMS: ANALYSIS, TOOLS AND SIGNIFICANCE

It is easily appreciated that a feature common to results presented so far is the fact that all the systems investigated, as well as, for that matter, almost all biological systems, are charge-carrying multicomponent systems, containing a major solvent, water, one or more macromolecules of similar or different kind and a variety of neutral or charged low molecular mass components. It is, thus, essential, both for an understanding of the structure and function of the systems of life, and for the design and interpretation of experiments aimed at the understanding of these systems, to achieve a correct appreciation of the thermodynamics of multicomponent sys-

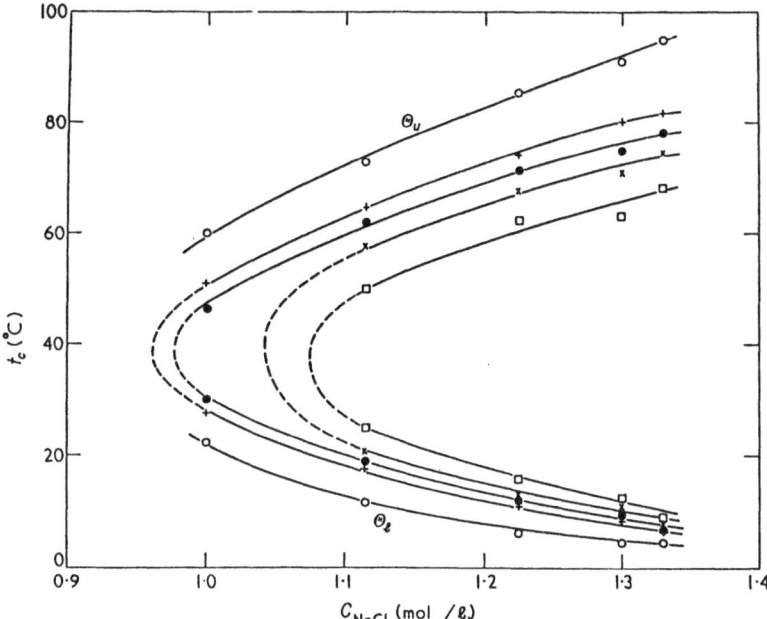

Fig. 9. *Estimated upper and lower critical consolute temperatures* t_c, *of poly(A) fractions at degrees of polymerization* Z, *against NaCl concentration.* ($\square$) $Z = 379$; ($\times$) $Z = 714$; ($\bullet$) $Z = 1123$; ($+$) $Z = 1740$. ($\bigcirc$) Upper and lower theta temperatures, Θ_u and Θ_l. From [60] with permission

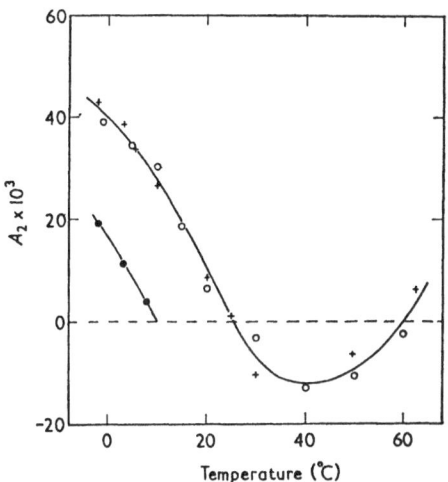

Fig. 10. *Second virial coefficients* A_2 *(from light scattering) of poly(A), as a function of temperature.* ($+$) $Z = 1740$, 1.0 M NaCl; ($\bigcirc$) $Z = 1462$, 1.0 M NaCl; ($\bullet$) $Z = 1740$, 1.3 M NaCl. From [60] with permission

tems. An extensive discussion of the thermodynamic analysis of multicomponent solutions [62] treats the general aspects of the problem, yet the major principles and applications can be grasped more easily if the discussion is reduced to an analysis of three components only [63], the major solvent (component 1), a charged macromolecular component (component 2) and a low molar mass neutral or electrolyte component 3; the negative or positive counter-ion of component 2 being identical to one of the ions composing the electrolyte component 3.

The concepts expressed in the manifestation of the thermodynamics of multicomponent systems are simple and elegant, and have been correctly understood by the great practitioners of the art, such as Zernicke, Scatchard, Debye, Hermans,

Kirkwood, Stockmayer and others. It is a total mystery why, to the present day, confusion reigns in this field and the concept persists that this is a very complex topic, which is better avoided. I believe this is to some extent due to the mistrust with which thermodynamics is looked upon by many and to the use of confusing concepts such as 'preferential hydration' which should be avoided at all costs because of the conceptual damage they are creating. In the following I will endeavor to present a clear picture of the principles and the state-of-the-art in the treatment of multicomponent systems and the rewards resulting from their application.

Consider first schematically an osmotic pressure experiment involving a neutral macromolecule in a simple solvent restricted to one compartment on one side of an osmotic membrane, (Fig. 11a). At equilibrium, μ_1, the chemical potential of component 1, is equal across the membrane and

$$\Pi/RT = c/M + A_2 c^2 + \ldots \qquad (11)$$

where Π is the osmotic pressure (in MPa), M is the molar mass (in g/mol), and A_2 is the virial coefficient (in ml·mol/g²), expressing interactions between the particles (in Θ solvents, cf. Section 3, A_2 and higher virial coefficients vanish); M and A_2 can be determined if Π is known for various values of c. It is important to note that Eqn (11) determines c/M which has the units mol/ml and thus Π/RT, in the limit $c \to 0$, yields or counts the number of moles, or particles, per unit volume. M is only obtained because we choose to express c as g (weighed in)/ml. Had c been given, for instance, in mol N (nitrogen)/ ml, M would be mol N/mol (particles). Osmotic pressure thus 'counts' particles. If we consider 'charged' particles in a salt-free system then the situation depicted in Fig. 11b arises (we have depicted a negatively charged large particle and positive counter-ions, but it could equally well be the other way round). To maintain electroneutrality, fixed charges on the particle are balanced by counter ions in solution (counter-ion condensation, cf. Section 1, reduces the effective charge). The bulk of the counter-ions cannot escape to the other side of the

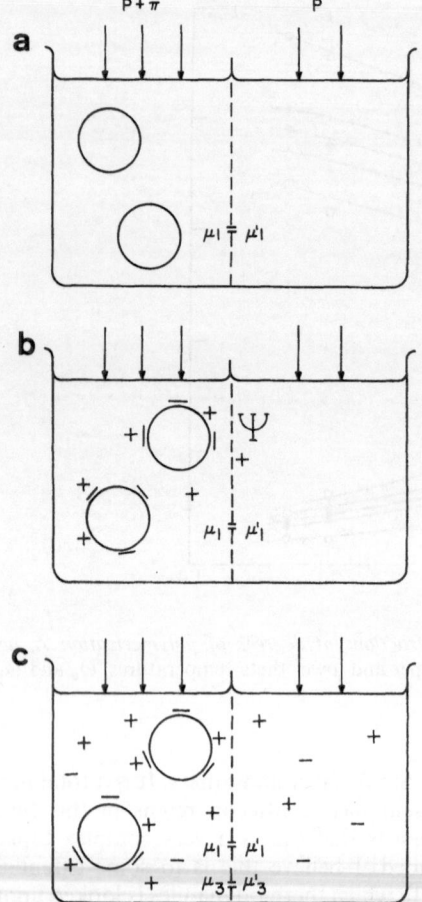

Fig. 11. *Schematic representation of osmotic-pressure-equilibrium experiments.* (a) Non-ionic particles component 2 in homogenous non-ionic solvent component 1 (not indicated). (b) Charged particles in homogenous solvent; limited escape of positive counter-ion (+) leads to membrane potential (Ψ); hydration of ions is not indicated. (c) Charged particles in homogenous solvent upon addition of low molecular mass electrolyte component 3

membrane because of electroneutrality requirements. In simple minded, but quite adequate, fashion we can now say that we still count particles, but it is a different ball game: we can almost disregard the large particle compared to the much higher number of small ions in solution and determine an approximate measure of ion condensation [15]. In more sophisticated terms we would say: electrostatic potentials are now long-ranged; they don't decrease fast enough with dilution and virial expansions are not permitted. I have already said in Section 1 that charged particles in the absence of salt are of minor interest to biology, so let us leave this situation to the physicist. Instead we add a small electrolyte, component 3 (Fig. 11c). At osmotic equilibrium not only μ_1, but also μ_3 will be equal on both sides of the osmotic membrane, as both components 1 and 3 are permeable through the membrane (component 2 is not, but as we cannot distinguish between positive ions deriving from either component 2 or 3, these may now exchange across the membrane). It can be shown that we are back now on an even keel: salt concentration is lowered on the left 'inside' compartment (this is known as Donnan 'expulsion' or 'rejection' by the charged particle) in such a precise way that Eqn (11) applies anew; the virial expansion becomes legal again and A_2 now contains additional terms relating to the ion concentration (we shall not be concerned with this in the present discussion). The number of

moles of component 2 can be determined as before. Remember, it is mol/ml which one determines, and molar mass (or 'molecular weight', a term not favored by modern nomenclature [64], however cf. [65]) may be expressed in whatever way one desires, to include counter-ions, for instance, in highly charged nucleic acids.

Similarly, references encountered in the literature to 'anhydrous' or 'hydrated' values of M are irrelevant, following the same arguments. Considerations leading to an evaluation of hydration, or interaction with component 3, will be presented below. We do not discuss here the fact that in osmotic pressure, and in related methods to be discussed below, molar mass averages of various kinds are determined in the case of heterogeneous solutes [63], and we will also, in the following and for simplicity, restrict our considerations to vanishing macromolecular concentrations.

With increasing M, Π decreases, as the number of particles decreases for a given value of c. The osmotic pressure method is therefore limited in practice to small values of M, not exceeding a few kg/mol, or kDa. A method with broad applicability is ultracentrifugation, which is close in concept to osmotic pressure. In the ultracentrifuge, particles of all kinds are pushed to the bottom of the ultracentrifuge tube (sometimes, and exceptionally, to the top if they are lighter than the medium) by the ultracentrifugal field. This leads to a concentration increase towards the bottom of the tube, in turn producing an osmotic field, countering the concentration increase. Under suitable equilibrium conditions a steady concentration distribution is reached, given by a beautifully symmetric equation, valid at each distance, r, from the center of rotation in the ultracentrifuge tube, for each component, J [62]

$$\mathrm{d}\ln c_J/\mathrm{d}r^2 = (\omega^2/2)(\mathrm{d}\Pi/\mathrm{d}c_J)^{-1}\,(\partial\varrho/\partial c_J)_\mu \tag{12}$$

where ω is the angular velocity and $(\partial\varrho/\partial c_J)_\mu$ is the increment of the solution density, ϱ, with concentration, c_J, at constant chemical potential, μ, of all components in solution. If we specialize to the macromolecular component 2, at vanishing concentration, c_2, and take the derivative of Π from Eqn (11) then Eqn (12) simplifies to

$$\mathrm{d}\ln c_2/\mathrm{d}r^2 = (\omega^2/2RT)M_2\,(\partial\varrho/\partial c_2)_\mu\,. \tag{13}$$

From the usually linear plot of $\ln c_2$ against r^2 we obtain M_2 multiplied by the buoyancy term $(\partial\varrho/\partial c_2)_\mu$. A number of observations are pertinent. Because of the logarithmic dependence any quantity proportional to c_2 may be used; the absolute concentration need not be known. In contrast to the osmotic pressure, in this case the slope, and therefore the sensitivity of the experiment, increases with increasing M_2. As we here also encounter the ratio c_2/M_2, we can express M_2 in whatever units we choose for the concentration, as the units of this ratio are mol/ml. At the position where $(\partial\varrho/\partial c_2)_\mu = 0$ a band will form in the ultracentrifuge with width inversely proportional to M_2.

For the simple, nonionic two-component system of Fig. 11a the classical result

$$(\partial\varrho/\partial c_2)_\mu = 1 - \bar{v}_2\,\varrho^0 \tag{14}$$

applies, where $\bar{v}_2$ is the partial specific volume of component 2 and ϱ^0 the solvent density. For the three-component system of Fig. 11c we have

$$(\partial\varrho/\partial c_2)_\mu = (1 - \bar{v}_2\,\varrho^0) + \xi_i(1 - \bar{v}_i\,\varrho^0) \tag{15}$$

where $i = 1.3$ and $\xi_i = (\partial w_i/\partial w_2)_\mu$ is an interaction parameter expressing change of concentration of either component 1 or

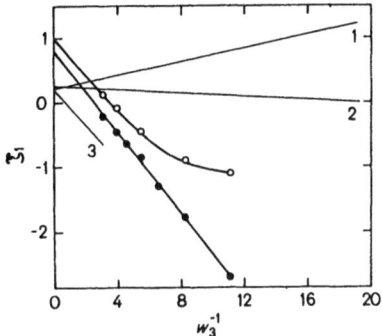

Fig. 12. *Preferential interaction parameter* ξ_1 *as a function of the reciprocal NaCl mass molality,* w_3^{-1}. ($\bullet$) *hMDH;* ($\bigcirc$) *halophilic glutamate dehydrogenase;* (1) *DNA in NaCl;* (2) *bovine serum albumin in NaCl;* (3) *bovine serum albumin in guanidine/HCl. From* [133]

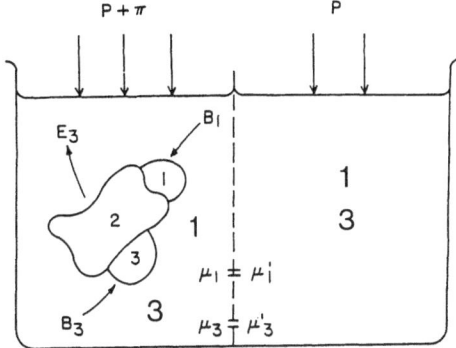

Fig. 13. *Schematic representation of binding,* B_1, *of component 1 and* B_3, *of component 3 and Donnan exclusion,* E_3 *of the latter, to the macromolecular component 2. The right-hand compartment, which does not contain component 2, is separated from the left-hand compartment by a membrane permeable to components 1 and 3;* Π *is the osmotic pressure required for complete thermodynamic equilibrium*

3 with change in concentration of component 2 (w is in g component i/g component l). From simple considerations

$$\xi_1 = -\xi_3/w_3 \qquad (16)$$

ξ_1 has been called 'preferential hydration', which is a bad term, as we shall presently see.

We can evaluate ξ_1 or ξ_3 from a knowledge of the partial specific volumes and the density increments which may be measured either directly or derived from Eqn (13) if M_2 is known. The value of $M_2 (\partial c/\partial c_2)_\mu$ can also be obtained from the Svedberg equation, which is a combination of velocity sedimentation and diffusion [66].

For reasons which will become clearer below, we plot, in Fig. 12, ξ_1 against w_3^{-1} for a number of solutions of biological macromolecules. We note that for most of these solutions ξ_1 varies strongly with w_3, and even changes sign at high salt concentration. We know, on the other hand, that in most systems which have been studied, hydration changes little or not at all with salt concentration, and negative hydration, even within the context of a term 'preferential hydration', is highly suspicious.

To achieve a meaningful analysis we recognize that ξ_1 contains a term relating to hydration but also a term relating to binding of component 3, and in some way, as already indicated by Fig. 12, we expect w_3 to play a role. We therefore calculate ξ_1 and ξ_3 on the schematic basis of a particle (Fig. 13) defined to 'bind' B_3 g component 3/g component 2,

'reject' E_3 g component 3/g component 2 by a Donnan mechanism and 'bind' B_1 g component 1/g component 2. We find

$$\xi_1 = B_1 - (B_3 - E_3)/w_3 \qquad (17a)$$

and

$$\xi_3 = (B_3 - E_3) - w_3 B_1. \qquad (17b)$$

From Eqn (17a) we now see that the slopes in Fig. 12 represent $B_3^* \equiv B_3 - E_3$, which is 'binding' of component 3 modulated by Donnan exclusion E_3, and the intercept represents hydration B_1 (Table 1). Thus bovine serum albumin binds little salt in solutions of NaCl, but considerable more in the denaturing solvent guanidine/HCl, though hydration is almost unchanged upon denaturation. In distinction to the proteins, in which the Donnan term E_3 is negligible, the slope of the curve for DNA is positive, due to the fact that in this case B_3 is negligible and the negative E_3 term takes over to yield a positive slope, which can be calculated by the methods described in Section 1 (more discussion about DNA will be given in Section 5). All the hydration values bundle closely around $0.2-0.25$ g H_2O/g component 1, except the extreme halophile enzymes glutamate and malate dehydrogenase, which will be discussed in more detail in Section 6.

Eqns (15) and (17a) or (17b) can be combined (using, $c_1 + c_3 = \varrho^0$ and $c_1\bar{v}_1 + c_3\bar{v}_3 = 1$) into

$$(\partial \varrho/\partial c_2)_\mu = (1 - \bar{v}_2 \varrho^0) + B_1(1 - \bar{v}_1 \varrho^0) + B_3^*(1 - \bar{v}_3 \varrho^0) \qquad (18)$$

or

$$(\partial \varrho/\partial c_2)_\mu = (1 + B_1 + B_3^*) - \varrho^0 (\bar{v}_2 + B_1\bar{v}_1 + B_3^*\bar{v}_3) \qquad (19)$$

where $V_t = (\bar{v}_2 + B_1\bar{v}_1 + B_3^*\bar{v}_3)$ is the total volume of the particle/g component 2. From a usually linear plot of $(\partial \varrho/\partial c_2)_\mu$ against ϱ^0, B_1 and B_3 can be calculated if the partial specific volumes are known, or assumed.

There is an additional group of powerful methods, useful in these approaches and beyond, which I will not discuss as I am afraid the editors would have rejected this review article because of undue bulk. I refer to the forward scattering of electromagnetic radiation, such as light X-rays or neutrons, all of which are based on the same principles, though they provide complementary information due to specific features [67]. The relation to osmotic pressure, in the case of radiation scattering, derives from the fact that, following the classical treatments of Einstein, for liquids, and of Debye for solutions, scattering of radiation is due to fluctuations of concentrations, which relate to $d\Pi/dc_J$, in a similar way to equilibrium sedimentation, Eqn (12). For instance, combining neutron scatter-

Table 1. *Exclusion parameters* B_1 *and* ($B_3 - E_3$) *of proteins and of DNA in salts*
These were calculated following Eqn (17a) and Fig. 12, B_1 is given in g H_2O/g protein, $B_3 - E_3$ in g NaCl/g protein. From [133]. BSA, bovine serum albumin; hGDH, halophilic glutamate dehydrogenase; Gdn, guanidine

Substance	B_1	$B_3 - E_3$
	g H_2O/g protein	g NaCl/g protein
hMDH, NaCl	0.8	0.32
hGDH, NaCl	1.0	0.29
BSA, NaCl	0.23	0.012
BSA, Gdn/HCl	0.18	0.27
DNA, NaCl	0.22	-0.052

ing and ultracentrifugation allows the determination of B_1 and B_3^*, without the rather difficult determination of $\bar{v}_2$ [68]. Additional benefits of the scattering methods reside, for instance, in the study of shape and size of solute particles by the angular dependence of scattering intensity as well in the study of contrast variation which is akin to density gradients in ultracentrifugation [69]. A good example of the application of small-angle neutron scattering of multicomponent systems is the study of the structure of phenylalanine-accepting transfer RNA and of its environment in aqueous solvents with different salts [70].

5. DNA: HYDRATION, SHAPE AND FLEXIBILITY. NUCLEOSOMES AND CHROMATIN

The density increments $(\partial\varrho/\partial c_2)_\mu$ and the partial specific volumes, $\bar{v}_2$ of DNA were determined over a wide range of salt concentration and the hydration was calculated to be 3.7 water molecules/nucleotide for NaDNA in NaCl and 5.9 water molecules/nucleotide for CsDNA in CsCl, an average of 5 water molecules/nucleotide with an uncertainty of one [71, 72]. This compares well with the value of 5 (three per phosphate, plus one each in the major and minor grooves) determined [73] from X-ray diffraction studies of a B-DNA dodecamer. Hydration is an operational concept and different values may be obtained by different experimental procedures [74]. The good agreement of our procedure with the precise location of water molecules by high-resolution X-ray diffraction provided confidence for applying the method developed to the halophilic systems to be discussed in the next section, which exhibit very unusual water and salt-binding properties in the native state.

When DNA is sedimented at high speeds in a mixed solvent containing a high proportion of a relatively heavy salt at high concentration (CsCl for instance), a well-defined density gradient, $d\varrho/dr$, forms (even in the absence of the macromolecular component) because of the redistribution of component 3 in the centrifugal field, and the macromolecular component concentrates in a narrow band, centered at some point r_b in the centrifuge cell. At equilibrium, Eqn (13) can be used to analyze the experimental results [71, 75]. The differential equation can be integrated if the dependence of $(\partial\varrho/\partial c_2)_\mu$ on r is known. If a quantity χ

$$\chi \equiv [(d/dr)\,(\partial\varrho/\partial c_2)_\mu]_{r_b} \qquad (20)$$

is defined, Eqn (13) can be integrated (for constant χ) to yield a Gaussian distribution about r_b with the standard deviation

$$\delta^2 = -RT/M_2\,\omega^2\,r_b\,\chi\,. \qquad (21)$$

Thus, if the gradient $(\partial\varrho/\partial c_2)_\mu$ can be evaluated at r_b, M_2 can be obtained unambiguously. It is possible to evaluate χ following the important observation [76] that upon isotopic substitution (^{15}N for ^{14}N) the center of the DNA band in the CsCl gradient shifts a distance Δr to the denser side of the salt gradient. Mathematical analysis [71, 75] yields

$$\chi = -(\alpha - 1)/\Delta r \qquad (22)$$

where α is the ratio of the masses of the substituted and unsubstituted DNA species. The true value of M_2 may thus be derived from Eqn (21) without any knowledge of partial volumes, interaction parameters and their dependence on pressure and r; the fact though that $(\partial\varrho/\partial c_2)_\mu$ vanishes at r_b permits the estimation of ξ_1 or ξ_3 under the conditions specified for this experiment.

Scattering of light, X-rays and neutrons are methods of broad applicability for the study of globular (proteins) and coiling (DNA and RNA) macromolecular particles and complexes (such as chromatin) [67, 69]. A combination of wavelength, λ, and scattering angle 2θ constitutes a yardstick, the absolute value of the scattering vector $q = (4\pi n/\lambda)\sin\theta$, leading to the selection of the correct method to apply. In the Guinier expansion we obtain M_2 and the radius of gyration R_g from the intercept and slope respectively of a plot of $\ln I(q)$ against q^2, and in the cross section expansion for elongated particles, the mass/unit length M/l and cross-section radius of gyration R_c from the intercept and slope respectively of a plot of $\ln qI(q)$ against q^2; $I(q)$ is the intensity of scattering. For the determination of both M and M/l, absolute calibration of scattering intensity, concentration as well as refractive index (light), electron (X-rays) or scattering length (neutrons) density increments akin to $(\partial\varrho/\partial c_2)_\mu$ are required for proper interpretation of the results; R_g and R_c determination does not require absolute calibration. Density increments in NaDNA/NaCl and CsDNA/CsCl solutions were used to evaluate the low-angle scattering of X-rays from DNA solutions [77, 78]. Scattering data [79 – 81] were found to be in reasonable agreement with the B structure of DNA in solution.

A newly proposed method [8] for obtaining partial volumes and interaction parameters by the combined use of ultracentrifugation, diffusion and neutron scattering was applied [82] to the analysis of a solution structure study of a 130-bp NaDNA fragment of known sequence, containing the strong promoter A1 of the E. coli phage T7 [83]. Thus, with a well-defined DNA fragment, and state-of-the-art experimental methods, it was possible to derive correct partial volumes and thermodynamic interaction parameters without prior assumptions or tedious procedures.

From the early days of DNA solution studies, a topic of great interest was an appreciation of DNA flexibility. The ideal structure derived from the original Watson and Crick B-DNA can be visualized as a straight cylinder or rod, yet in solution DNA bends as a result of Brownian motion. The wormlike chain of Kratky and Porod [58, 84] is characterized by a contour length L and a persistence length a. The latter increases with increasing stiffness, but is (on the basis of the model) independent of L. The relation between the radius of gyration R_g and L for worm-like linear coils without excluded volume is [85]

$$12R_g^2/L^2 = (3/2X^2)[(4X/3) - 2 + (2/X) \\ - (1 - e^{-2X})/X^2] \qquad (23)$$

where $X = L/2a$. Eqn (23) correctly yields the rigid rod relation $12R_g^2 = L^2$ in the limit $X \to 0$ and the Gaussian coil relation $6R_g^2 = L(2a)$ at large values of X. For a Gaussian coil, a is numerically equal to one-half the length of the Kuhn statistical chain element. It is thus possible, from measured vales of R_g and L (derived from properly weighted molar mass measurements and the B structure of DNA; a repeat distance of 0.34 nm/base pair, along the double helix axis) to ascribe a Kratky-Porod persistence length to DNA samples. It is also possible to express the results in terms of an elastic force constant b, $b = akT$, directly related to the pesistence length [86]. In Fig. 14 we can appreciate the dramatic transition of DNA from almost rodlike behavior, for short pieces of DNA (nucleosome core particle length DNA about 145 bp) to Gaussian coil-like behavior for linear ColE$_1$-size DNA (6646 bp), assuming reasonable values $a = 50 \pm 10$ nm.

Early work on DNA was undertaken with fragments which were polydisperse and poorly defined as a result of

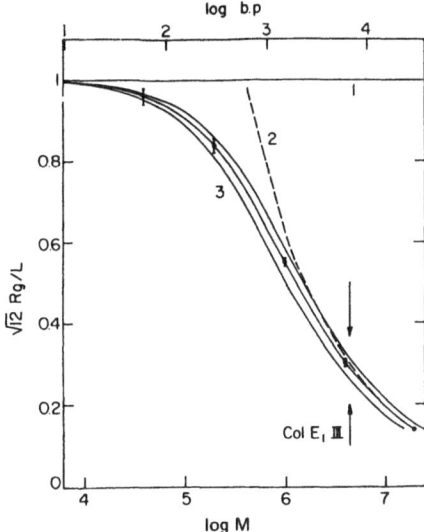

Fig. 14. *Coiling ratio* $\sqrt{12R_g^2/L}$ *of DNA chains against log* M *(or log number of base pairs, bp).* (1) Rigid rod; (2) Gaussian coil with statistical element $2a$ equal to 100 nm; (3) Kratky-Porod wormlike chains without excluded volume for (in descending order) persistence lengths a equal to 60 nm, 50 nm and 40 nm. To demonstrate the difficulty of determining DNA flexibility from the study of short pieces of DNA, experimental points have been simulated for $a = 50$ nm assuming a combined error in R_g and L of $\pm 2\%$; a repeat length 0.34 nm/bp and $M/l = 1950 \text{ g mol}^{-1} \text{ nm}^{-1}$ have been assumed for the B-form of NaDNA. From [90] with permission

preparative procedures or sonication [58]. The first monodisperse samples were first rather large size phage DNA molecules and a real breakthrough occurred with the appearance of plasmid DNAs and restriction fragments. I believe our laboratory pioneered the, now current, use for physical studies of intact supercoiled, open circular and the linear forms of ColE$_1$ plasmid DNA [87]. The value of $M_2 = (4.3 \pm 0.05) \times 10^6$ g/mol obtained from light-scattering, diffusion and sedimentation experiments before the sequence was known, corresponds closely to the value derived from the sequence 6646 bp or 4.40×10^6 g/mol, for which a definitive value has now been given [88]. We showed that the persistence length a of DNA, derived from total intensity laser-light scattering of linear ColE$_1$ DNA varies from about 68 nm in 5 mM NaCl to about 40 nm in 0.2 M NaCl, levelling off to a constant value (about 27 nm) at high NaCl $(1-4$ M) concentration [89, 90]. These observations did not quite agree with then current views on the effect of electrostatic charge and ionic conditions on DNA dimensions. The apparent diffusion constant, D_{app}, determined from laser-light-scattering autocorrelation as a function of the scattering vector q, at NaCl concentrations $0.005-4$ M, correctly yields the translational diffusion constant, D_t, at low values of q; the sedimentation coefficient, s, closely agrees with well tested empirical relations, and a combination of s and D_t, and the appropriate density increments yield correct molar masses over the whole salt concentration range. A similar limiting value of $a = 28.5$ nm was obtained for LiDNA in LiCl [91] in solutions between 3 M and 5 M LiCl in which the second virial coefficients, A_2, vanish and excluded volume corrections, which have plagued results obtained in other solvents, need not be considered. Unlike what was previously believed, it could be shown that solutions of LiDNA in LiCl are stable for reasonable periods of time between 5 M and 9 M LiCl and

the molecular parameters vary smoothly and moderately at high salt.

The field of DNA flexing and folding [92] has been dramatically changed in recent years. Following structure determinations by X-ray [93] and by NMR [94, 95] investigations and the observation of abnormal migration in agarose electrophoretic gels [96], it has been realized that specific DNA sequences may lead to naturally bent DNA for even short DNA pieces [97], by mechanisms which are today hotly debated [98]. These phenomena, which are of great biological importance in explaining facilitation of protein — nucleic-acid interactions, in the formation and stabilization of nucleosomes, for instance [99], are beyond the confines of this review, in which we have aimed to provide baseline aspects of DNA flexibility [92], and an exposition of correct procedures to be used for the characterization of deviations from these baseline properties. Other problems are the influence of polyvalent ions in promoting the 'condensation' of DNA to compact forms [100] and on the influence of polyvalent metal ions on solution properties of DNA in general [101]. Another problem of great current interest is DNA-protein interactions leading to DNA bending in a variety of biological systems [102], relating to mechanisms of control of gene regulation. Whereas, for many years Watson-Crick B-DNA was considered a boring structure, with nucleotide bases paired and surrounded by the charged backbone chain, with no visible mechanism for biological specificity, the various new exciting developments mentioned here represent the opening up of a completely new era of high interest, which would hardly have been feasible without the results which have preceded it.

Research on the very fundamental complex of DNA, histone and non-histone protein, called chromatin, would not have surged as it did without the discovery, by nuclease digestion [103], of the repeating unit, the nucleosome [104]. Nucleosomes are connected by variable stretches of linker DNA to form the basic lower order chromatin structure, at concentrations of a few mM monovalent salts, folding into the so-called higher-order structure when monovalent salt concentration is raised above 50 mM, or when a few mM divalent cations, such as Mg^{2+}, are added [105]. This constitutes a small window in the transition from active to nonactive chromatin, within the much more elaborate transition from free DNA to DNA completely folded for storage, and eventual replication and transcription, in the chromosome (Fig. 15) [106, 107]. Complex precipitation phenomena follow with further increase in various salt concentrations [108, 109], until sequential release of nonhistone and histone proteins at much higher concentration of salts.

Whereas the study of the biological implications of chromatin function has assumed a center stage role in present research [110], the physicochemical studies, in this instance, have provided some interesting observations in the dozen odd years following the first characterization of the nucleosome [111], yet no basic definitive new results have arisen from these investigations. Thus, although nucleosome core particle [112] and core histone octamer structure [113] from X-ray diffraction studies have been reported, there persist to the present day notes of criticism on solvent-dependent structural polymorphisms which have not been totally resolved [114]. The nucleosome core particle from bulk chromatin is not homogenous because of the variety of DNA chains of various sequences composing it, and presently more uniform nucleosome core particle are under investigation [115]. In our work we found that nucleosome core particles from chicken erythrocytes and calf thymus can bind tightly two or three

18

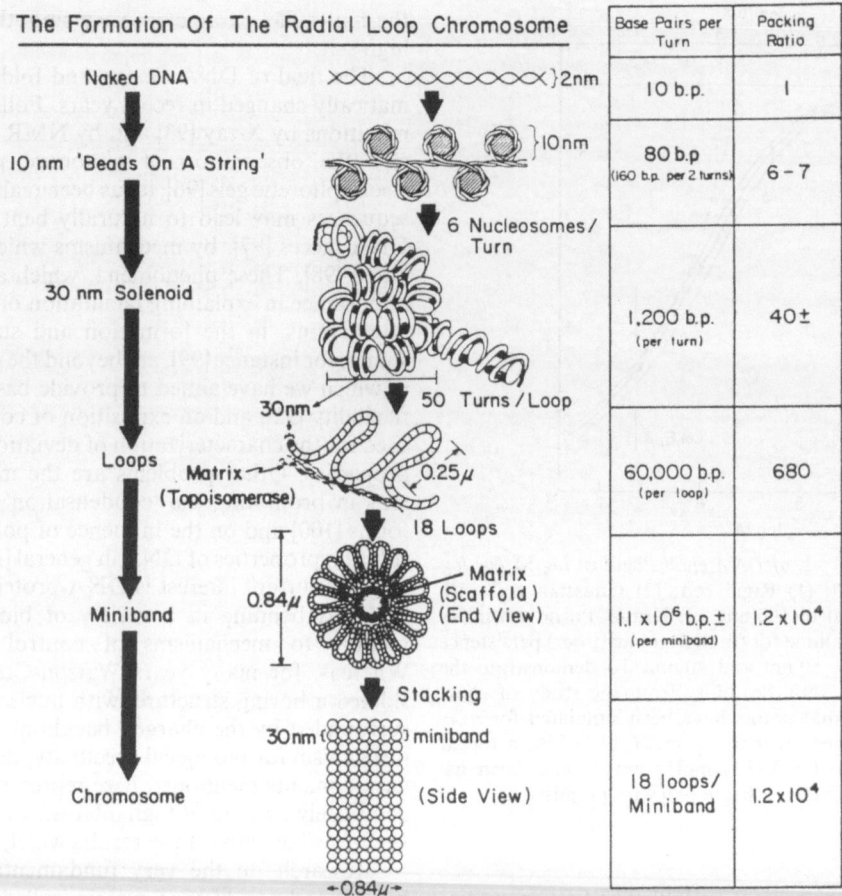

The Formation Of The Radial Loop Chromosome

	Base Pairs per Turn	Packing Ratio
Naked DNA — 2nm	10 b.p.	1
10 nm 'Beads On A String' — 10nm	80 b.p (160 b.p. per 2 turns)	6 – 7
30 nm Solenoid — 6 Nucleosomes/Turn	1,200 b.p. (per turn)	40±
Loops — 50 Turns/Loop, Matrix (Topoisomerase), 0.25μ	60,000 b.p. (per loop)	680
Miniband — 18 Loops, Matrix (Scaffold) (End View), 0.84μ	1.1 × 10⁶ b.p.± (per miniband)	1.2 × 10⁴
Chromosome — Stacking, 30nm miniband (Side View), 0.84μ	18 loops/Miniband	1.2 × 10⁴

Fig. 15. *Schematic levels of organization of DNA within a chromosome.* From [107] with permission. $0.84 \mu = 0.84 \mu m$

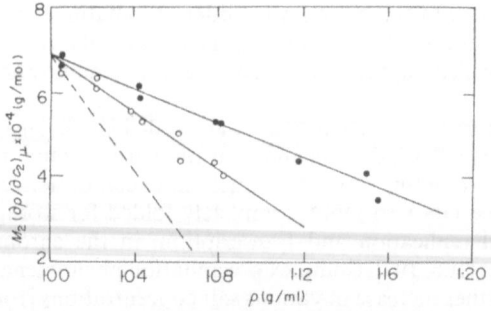

Fig. 16. *The slopes* $M_2 \; (\partial \varrho / \partial c_2)_\mu$ *from equilibrium sedimentation against medium density,* ϱ. Density-contrast variation solutes: (●) sucrose, $\xi_1 = 0.318$; (○) γ-cyclodextrin, $\xi_1 = 1.005$. The broken line is for dextran ($\xi_1 = 2.32$) from Fig. 14 in [117]. From [118] with permission

additional equivalents more of core histone octamers [116, 117] and in the range $0.1 - 0.7$ M NaCl the nucleosome core is a compact, folded structure [118]. Following the observation that the total volume ($542 \; nm^3$) of the core particle is considerably in excess of the approximate volume ($216 \; nm^3$) of its nucleic acid and protein components, we used density contrast variation in equilibrium sedimentation with a small probe (sucrose), capable of penetrating the core particle, and a larger probe (γ-cyclodextrin), incapable of doing so, to obtain particle hydration in the first instance and an estimate of the total particle volume in the second instance (Fig. 16) [118].

Much effort has been devoted in many laboratories [119] to establish valid models for the lower order, unfolded, and higher-order, folded, chromatin structure. The favorite structures were the random coil of beads on a string for the first, and the Finch and Klug solenoid for the second [120], though these have been contested on various grounds [121] and to the present day no definitive answer has been achieved [122]. The problem may be due to the fact that the structures are irregular, the linker lengths are variable, the structures are labile because of the requirement of moving with little effort from an inactive packaged to an active unfolded form and the fact that the solution methods which are applicable are low-resolution methods. Though it is possible to reconstitute nucleosome core particles with DNA and core histones, and use uniquely sequenced DNA for this purpose, it has so far not been found possible to correctly replace linker histones H1 or H5 in meaningful ways, to reform or simulate native-like chromatin, a study of which could advance our better understanding of chromatin folding.

From hydrodynamic, light and low-angle X-ray scattering studies of the folding of chicken erythrocyte chromatin fractions we concluded [108, 123] that it is feasible to identify the folding process of well-defined chicken erythrocyte chromatin fractions with a gradual compaction of a chain of freely jointed filaments or a worm-like chain, within the limits of all the experimental data obtained. I would like to emphasize that we do not claim this to be the correct folding process, we only indicate the limitations of the low-resolution methods in providing a satisfactory unique solution in this case. Thus, though this description represents a unified presentation of a

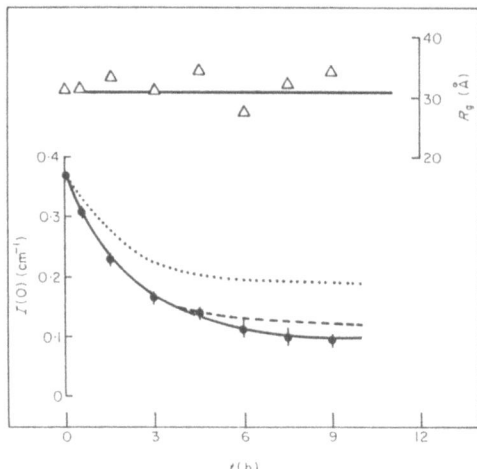

Fig. 17. *The I(0) and R$_g$ values from a Guinier extrapolation plot of the small-angle neutron-scattering intensity from a solution (5.7 g/l) hMDH in sodium phosphate (pH 7), 1 M NaCl.* Time zero is when the enzyme stock solution (4 M NaCl) was diluted to 1 M NaCl. The dotted line is calculated for no change in density increment upon denaturation; the broken and continuous lines are calculated for two models of normal hydration. From [135] with permission. 10 Å = 1 nm

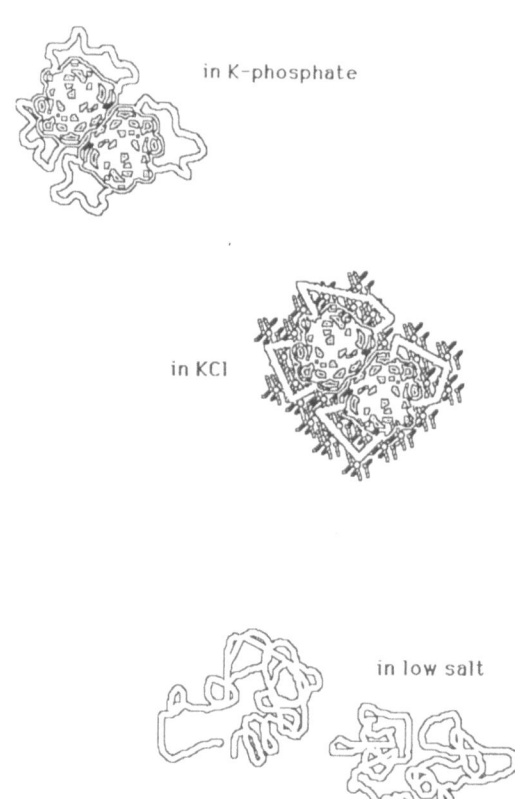

Fig. 18. *Schematic representation of hMDH solution structures.* The active structures are presumed to have two parts: a catalytically active core, conceivably similar to that in non-halophilic MDH, and protruding domains required for stabilization in KCl, NaCl or MgCl$_2$ solvents. In potassium phosphate (K-phosphate), the protein dimer is stabilized by the hydrophobicity of the core and the protruding domains are disordered. In KCl (or NaCl), the protein is stabilized by the interaction of the protruding domains in a protein-water-salt hydration network. In MgCl$_2$, a similar structure exists with the same amount of water molecules coordinated by fewer salt ions. In low salt, the protein is unfolded and its hydration is like that of non-halophilic proteins. From [137] with permission

wealth of experimental data, it by no means represents a definitive solution to an exceedingly difficult problem.

6. UNUSUAL STRUCTURAL FEATURES OF THE EXTREME HALOPHILES

The enzymes from the extreme halophilic bacteria from the Dead Sea *Halobacterium marismortui* halophilic malate dehydrogenase (hMDH) [124], halophilic glutamate dehydrogenase [125] and halophilic ferredoxin [126] have been isolated in pure active form and extensively studied and reviewed [127, 128]. Questions concerning adaptation to extreme conditions of heat, temperature, pH or solvent medium motivate active research fields [42] extending from molecular biology [129] to evolutionary aspects [130]. Thus archaebacteria have come to be recognized [131] (although this is still under debate [132]), as urkingdoms, equivalent and connecting to the eubacteria and the eukaryotes.

From an analysis of sedimentation and diffusion data of hMDH (and also of halophilic glutamate dehydrogenase) we found [133, 134] unusual values for the hydration B_1 and binding of salt $B_3^{\ddagger}$ (Table 1). The enzyme hMDH is stable at high concentrations of salt, and becomes unstable upon lowering the salt concentration below 2.5 M NaCl. It was then found, from both light scattering and sedimentation that the unusual binding values were lost upon denaturation, which involves enzyme-dimer dissociation and unfolding [134]. This result was confirmed by neutron scattering (Fig. 17) [135]. The dramatic change in hydration upon denaturation is unusual as hydration usually represents a phenomenon linked to the primary structure and is only slightly dependent upon tertiary or quarternary structure. Thus, the excess of glutamate and aspartate negative changes [127] is not sufficient to explain the high hydration in the native state and the loss of binding of water and salt upon denaturation. The huge water and salt binding was confirmed in saturated KCl solutions [136], the physiological environment of hMDH. The structure of the enzyme can also be stabilized with exceptional protein-solvent

interactions in a range of high Mg^{2+} or other 'salting-in' ions, which usually favor unfolding [137].

A novel stabilisation model resulted from the analysis of data on hMDH in salt solutions chosen for their different effects on protein structure: potassium phosphate which is a strongly 'salting-out' agent; MgCl$_2$, which is 'salting in'; NaCl and KCl which are mildly 'salting-out' [137]. Enzymatic activity and stability measurements in these various solutions were combined with neutron-scattering, ultracentrifugation and light-scattering experiments. In molar concentrations of phosphate ions, the mechanism of stabilization is similar to that in non-halophilic soluble proteins, in which the hydrophobic interaction dominates. In high concentrations of KCl, NaCl or MgCl$_2$ on the other hand, we believe a solution particle is formed in which the protein dimer interacts with a large number of salt and water molecules; under these conditions the main stabilization mechanism is the formation of hydrate bonds between the protein and hydrated salt ions, leading to protein-'connected' electrolyte concentrations highly in excess of the saturated solution electrolyte concentration (Fig. 18). These special interactions are lost upon enzyme denaturation at low salt concentration (Fig. 17).

In another recent study it has also been possible to demonstrate that hMDH can be reconstituted from guanidine/HCl solutions into the active dimer after denaturation and dissociation in various denaturants [138], confirming Anfinsen's 'thermodynamic hypothesis' of protein folding [139] in the instance even of a very unusual extreme halophilic dimer. Crystals of hMDH have been obtained in potassium phosphate [140] in which the outer loops of protein (Fig. 18) are not expected to participate in the stabilisation interaction with water and salt. Crystallization conditions in which the hydration network is likely to be maintained are being pursued. The amino acid sequence of the protein, in process of obtainment from studies of the cloned hMDH gene in our laboratories, will be explored for motifs that may give rise to potential water and salt binding domains, nucleating the stabilizing hydration network [137]. Though the stabilization mechanism proposed here appears attractive and probably true, we have no evidence so far that this represents a general scheme for protein stabilization in these extreme halophile systems and the study of a larger number of systems will be required before reaching conclusions of a more general nature.

7. PAST, PRESENT AND, ABOVE ALL, THE FUTURE

I have reviewed here scientific contributions which emphasize broadly the tremendous progress we have witnessed in the life sciences during the past forty years. At midcentury, the size, shape and mass of biopolymers and of nucleic acids were imperfectly known; the first successes in the sequencing of proteins in 1953 [141] and of nucleic acids in 1965 [142] by methods which are slow and tedious by present-day standards had not yet been achieved. A 'high molecular mass DNA' from calf thymus or salmon sperm which was used in early solution studies [58] was believed to have a molar mass of about 6×10^6 g/mol though the true value later turned out to be closer to 2×10^7 g/mol. Empirical calibration, using materials of known molar mass, in chromatography columns of various kinds, often led to incorrect results, in particular in the case of charged particles, and the exceptions most often represented the most interesting materials. It was thus of non-trivial interest and significance to develop an insight into the thermodynamics of multicomponent systems and its application to appropriate experimental methods to obtain reliable molar masses, shapes and sizes. We could show for instance [143] that proper analysis of equilibrium sedimentation measurements of rabbit muscle aldolase in guanidine/HCl solutions yielded four subunits/active enzyme, a result contested by some [144]. Bovine liver glutamate dehydrogenase was believed to have a molar mass between 2.5×10^5 g/mol and 2×10^6 g/mol and its subunit structure was unknown [145]. We could show [146] that six non-covalently bonded subunits form one active oligomer enzyme molecule of molar mass 3.1×10^5 g/mol, a conclusion in apparent opposition to the then current Monod symmetry dogma, leading one eminent referee to the comment, 'I don't believe a word the authors are saying, but I will defend to my death their right to say it'. In later contributions we could determine the correct symmetry of the active hexamer [147, 148] and show that it linearly associates in terms of a simple reversible polymerization with one equilibrium constant. The polymerization is strongly enhanced when the aqueous solutions are exposed to small amounts of toluene.

Today the broad applicability of gel electrophoresis procedures and convenient methodologies for nucleic acid and protein sequencing, the latter via the sequencing of the protein gene, have shifted focus from the exploration of properties of individual molecules to molecular complexes, such as, for instance, nucleic-acid – protein interactions in chromatin, in higher-order complexes such as the nuclear matrix, the study of transcription and replication. Our interest in understanding DNA-protein interactions in the extreme halophiles [149] has highlighted the fact that little definitive is known today about the structure of the *E. coli* bacterial chromosome. Studies indicating tremendous ionic-strength dependence of transcription [150], for instance, *in vitro*, in the test tube, may be compared with relative insensitivity to ionic strength *in vivo* [151]; furthermore, the exact nature of the ions in the *in vivo* systems may be highly significant [152]. Another aspect of importance is represented by the excluded volume, the so called 'crowding effect' [153, 154] of components which may, or may not participate in the drama of the ongoing biological reaction. Molecular biophysics is thus moving inside the biological cell, shifting methodology and emphasis, while maintaining the basic principle that thermodynamics must be obeyed, though a quantitative interpretation may be more difficult to establish in these complex systems. It remains the central challenge of science to strike out in meaningful ways into the future, the great unknown.

I would like to pay tribute to my teachers, colleagues and students, including those who are unfortunately not with us any longer, without whom this report would not have been possible. My sincere thanks to Dr Ellen Wachtel, for critical comments. I acknowledge continued support by the Israel-US Binational Foundation, the Minerva Foundation and the *Stiftung Volkswagenwerk* of the FRG, the National Institutes of Health, the National Bureau of Standards and the Office of Naval Research, of the USA. The references to the various topics touched upon here are not complete as these have not been discussed in their own right, but rather in the context of the general thrust of this review. Many more contributions may be traced through the references cited.

REFERENCES

1. Kuhn, W., Künzle, O & Katchalsky, A. (1948) *Helv. Chim. Acta 31*, 1994 – 2037.
2. Hermans, J. J. & Overbeck, J. Th. G. (1948) *Recl. Trav. Chim. Pays-Bas Belg. 67*, 761 – 776.
3. Katchalsky, A. & Eisenberg, H. (1950) *Nature 166*, 267 – 268.
4. Katchalsky, A., Lifson, S. & Eisenberg, H. (1951) *J. Polymer Sci. 7*, 571 – 574.
5. Katchalsky, A., Lifson, S. & Eisenberg, H. (1952) *J. Polymer Sci. 8*, 476.
6. Eisenberg, H. & Fuoss, R. M. (1954) in *Modern aspects of electrochemistry* (Bockris, J. O. M., ed.) vol. 1, pp. 1 – 46, Butterworths, London.
7. Eisenberg, H. (1977) *Biophys. Chem. 7*, 3 – 13.
8. Mandel, M. (1988) in *Encyclopedia of polymer science and engineering*, 2nd edn (Mark, H. F., Bikales, N. M., Overberger, C. G., Menges, G. & Kroschwitz, J. J., eds) vol. 2, pp. 739 – 829, Wiley, New York.
9. Alfrey, T. Jr, Berg, P. W. & Morawetz, H. (1951) *J. Polymer Sci. 7*, 543 – 547.
10. Fuoss, R. M., Katchalsky, A. & Lifson, S. (1951) *Proc. Natl Acad. Sci. USA 37*, 579 – 589.
11. Alexandrowicz, Z. & Katchalsky, A. (1963) *J. Polymer Sci. A1*, 3231 – 3260.
12. Oosawa, F. (1971) *Polyelectrolytes*, Dekker, New York.
13. Manning, G. S. (1972) *Annu. Rev. Phys. Chem. 23*, 117 – 140.
14. Katchalsky, A., Alexandrowicz, Z. & Kedem, O. (1966) in *Chemical physics of ionic solutions* (Conway, B. E. & Barradas, R. G., eds.) pp. 295 – 346, Wiley, New York.
15. Manning, G. S. (1978) *Q. Rev. Biophys. 11*, 179 – 245.

16. Record, M. T. Jr, Anderson, C. F. & Lohman, T. M. (1978) *Q. Rev. Biophys. 11*, 102−178.
17. Harned, H. S. & Owen, B. B. (1958) *The physical chemistry of electrolyte solutions*, Reinhold, New York.
18. Record, M. T. Jr, Lohman, T. M. & de Haseth, P. L. (1976) *J. Mol. Biol. 107*, 145−158.
19. Paulsen, M. D., Anderson, C. F. & Record, M. T. Jr (1988) *Biopolymers 27*, 1249−1265.
20. LeBret, M. & Zimm, B. H. (1984) *Biopolymers 23*, 271−285.
21. Murthy, C. S., Bacquet, R. J. & Rossky, P. J. (1985) *J. Phys. Chem. 89*, 701−710.
22. Jayaram, B., Sharp, K. A. & Honig, B. (1989) *Biopolymers 28*, 975−993.
23. Soumpasis, D.-M. (1984) *Proc. Natl Acad. Sci. USA 81*, 5116−5120.
24. Garcia, A. E. & Soumpasis, D.-M. (1989) *Proc. Natl Acad. Sci. USA 86*, 3160−3164.
25. Debye, P. & Hückel, E. (1923) *Physik, Z. 24*, 185−206.
26. Linderström-Lang, K. (1924) *C. R. Trav. Lab. Carlsberg 15*, 7.
27. Fuoss, R. M. & Edelson, D. (1950) *J. Am. Chem. Soc. 73*, 269−273.
28. Phillips, A. P. (1950) *Science 112*, 536.
29. Eisenberg, H. & Fuoss, R. M. (1953) *J. Am. Chem. Soc. 75*, 2914−2917.
30. Saxton, V. & Darken, L. S. (1940) *J. Am. Chem. Soc. 62*, 846−852.
31. Katchalsky, A., Eisenberg, H. & Lifson, S. (1951) *J. Am. Chem. Soc. 73*, 5889−5890.
32. Cohen, G. & Eisenberg, H. (1965) *J. Chem. Phys. 43*, 3881−3887.
33. Mysels, K. J. (1964) *J. Am. Chem. Soc. 86*, 3503−3505.
34. Tilton, L. W. & Taylor J. K. (1938) *J. Res. Natl Bur. Stand. 20*, 419−477.
35. Roentgen, W. C. & Zehnder, L. (1891) *Ann. Physik. Chem. Wied. 44*, 24−51.
36. Lorentz, H. A. (1880) *Ann. Physik. Chem. Wied. 9*, 641−665.
37. Lorenz, L. (1880) *Ann. Physik. Chem. Wied. 11*, 70−103.
38. Eisenberg, H. (1965) *J. Chem. Phys. 43*, 3887−3892.
39. Reisler, E. & Eisenberg, H. (1965) *J. Chem. Phys. 43*, 3875−3880.
40. Jamin, J. (1856) *Comptes Rendus 43*, 1191−1194.
41. Bridgman, P. W. (1952) *The physics of high pressure*, G. Bell and Sons, London.
42. Jaenicke, R. (1981) *Annu. Rev. Biophys. Bioeng. 10*, 1−67.
43. Henn, A. R. & Kauzmann, W. (1989) *J. Phys. Chem. 93*, 3770−3783.
44. Poiseuille, J. L. M. (1835) *Comptes Rendus 1*, 554−560.
45. Einstein, A. (1906) *Ann. Physik. 19*, 289−306.
46. Simha, R. (1940) *J. Phys. Chem. 44*, 25−34.
47. Tanford, C. (1961) *Physical chemistry of macromolecules*, Wiley, New York.
48. Flory, P. J. (1953) *Principles of polymer chemistry*, Cornell University Press Ithaca, New York.
49. Eisenberg, H. & Pouyet, J. (1954) *J. Polymer Sci. 13*, 85−91.
50. Witten, T. A. & Pincus, P. (1987) *Europhys. Lett. 3*, 315−320.
51. Eisenberg, H. & Frei, E. H. (1954) *J. Polymer Sci. 14*, 417−426.
52. Eisenberg, H. (1957) *J. Polymer Sci. 25*, 257−271.
53. Littauer, U. Z. & Eisenberg, H. (1959) *Biochim. Biophys. Acta 32*, 320−337.
54. Laskov, R., Margoliash, E., Littauer, U. Z. & Eisenberg, H. (1959) *Biochim. Biophys. Acta 33*, 247−248.
55. Zimm, B. H. & Crothers, D. M. (1962) *Proc. Natl Acad Sci. USA 48*, 905−911.
56. Klotz, L. C. & Zimm, B. H. (1972) *J. Mol. Biol. 72*, 779−800.
57. Cohen, G. & Eisenberg, H. (1969) *Biopolymers 8*, 45−55.
58. Eisenberg, H. (1974) in *Basic principles in nucleic acid chemistry* (Ts'o, P. O. P., ed.) vol. 2, pp. 171−264, Academic Press, New York.
59. Mohan, G. R. & Eisenberg, H. (1959) *J. Phys. Chem. 63*, 671−680.
60. Eisenberg, H. & Felsenfeld, G. (1967) *J. Mol. Biol. 30*, 17−37.
61a. Olson, W. K. (1975) *Biopolymers 14*, 1775−1795.
61b. Olson, W. K. (1975) *Biopolymers 14*, 1797−1810.
62. Casassa, E. F. & Eisenberg, H. (1964) *Adv. Protein Chem. 19*, 287−395.
63. Eisenberg, H. (1976) *Biological macromolecules and polyelectrolytes in solution*, Clarendon Press, Oxford.
64. *Recommendations for the presentation of thermodynamic and related data in biology* (1985) *Eur. J. Biochem. 153*, 429−434.
65. Gratzer, W. (1989) *Nature 338*, 679−680.
66. Svedberg, T. & Pedersen, K. O. (1940) *The ultracentrifuge*, Oxford University Press, London.
67. Eisenberg, H. (1981) *Q. Rev. Biophys. 14*, 537−555.
68. Zaccai, G., Wachtel, E. & Eisenberg, H. (1986) *J. Mol. Biol. 190*, 97−106.
69. Eisenberg, H. (1971) in *Procedures in nucleic acid research.* (Cantoni, G. L. & Davies, D. R., eds) vol. 2, pp. 137−175, Harper and Row, New York.
70. Zaccai, G. & Xian, S.-Y. (1988) *Biochemistry 27*, 1316−1320.
71. Cohen, G. & Eisenberg, H. (1968) *Biopolymers 6*, 1077−1100.
72. Reisler, E., Haik, Y. & Eisenberg, H. (1977) *Biochemistry 16*, 197−203.
73. Kopka, M. L., Fratini, A. V., Drew, H. R. & Dickerson, R. E. (1983) *J. Mol. Biol. 163*, 129−146.
74. Tao, N. J., Lindsay, S. M. & Rupprecht, A. (1989) *Biopolymers 28*, 1019−1030.
75. Eisenberg, H. (1967) *Biopolymers 5*, 681−683.
76. Meselson, M. & Stahl, F. W. (1958) *Proc. Natl Acad. Sci. USA 44*, 671−682.
77. Eisenberg, H. & Cohen, G. (1968) *J. Mol. Biol. 37*, 355−362.
78. Eisenberg, H. & Cohen, G. (1969) *J. Mol. Biol. 42*, 607.
79. Luzzati, V. (1963) *Prog. Nucleic Acids Res. 1*, 347−368.
80. Luzzati, V., Masson, F., Mathis, A. & Saludjian, P. (1967) *Biopolymers 5*, 491−508.
81. Bram, S. (1968) Ph. D. Thesis, University of Wisconsin.
82. Eisenberg, H. (1989) in *Landolt-Börnstein new series, biophysics − nucleic acids*, vol. 7 (1c), Springer Verlag, Berlin, in the press,
83. Lederer, H., May, R. R., Kjems, J. K., Baer, G. & Heumann, H. (1986) *Eur. J. Biochem. 161*, 191−196.
84. Kratky, O. & Porod, G. (1949) *Recl. Trav. Chim. Pays-Bas Belg. 68*, 1106−1122.
85. Benoit, H. & Doty, P. (1953) *J. Phys. Chem. 57*, 958−963.
86. Landau, L. & Lifshitz, E. (1958) *Statistical physics*, pp. 478−482, Pergamon, Oxford.
87. Voordouw, G., Kam, Z., Borochov, N. & Eisenberg, H. (1978) *Biophys. Chem. 8*, 171−189.
88. Chan, P. T., Ohmori, H., Tomizawa, J. & Lebowitz, J. (1985) *J. Biol. Chem. 260*, 8925−8935.
89. Borochov, N., Eisenberg, H. & Kam, Z. (1981) *Biopolymers 20*, 231−235.
90. Kam, Z., Borochov, N. & Eisenberg, H. (1981) *Biopolymers 20*, 2671−2690.
91. Borochov, N. & Eisenberg, H. (1984) *Biopolymers 23*, 1757−1769.
92. Eisenberg, H. (1987) *Acc. Chem. Res. 20*, 276−282.
93. Westhof, E. (1988) *Annu. Rev. Biophys. Biophys. Chem. 17*, 125−144.
94. Wüthrich, K. (1986) *NMR of proteins and nucleic acids*, Wiley, New York.
95. Van de Ven, F. J. M. & Hilbers, C. W. (1988) *Eur. J. Biochem. 178*, 1−38.
96. Wu, H.-M. & Crothers, D. M. (1984) *Nature 308*, 509−513.
97. Ulanovsky, L. E., Bodner, M., Trifonov, E. N. & Choder, M. (1986) *Proc. Natl Acad. Sci. USA 83*, 862−866.
98. Olson, W. R., Sarma, M. H., Sarma, R. H. & Sundaralingam, M. (eds) (1987) *Structure and expression*, vol. 3, Adenine Press, New York.
99. Trifonov, E. N. (1985) *CRC Crit. Rev. Biochem. 19*, 89−106.
100. Strzelecka, T. E., Davidson, M. W. & Rill, R. L. (1988) *Nature 331*, 457−460.
101. Widom, J. & Baldwin, R. L. (1983) *Biopolymers 22*, 1595−1620.
102. Gustafson, T. A., Taylor, A. & Kedes, L. (1989) *Proc. Natl Acad. Sci. USA 86*, 2162−2166.

103. Hewish, D. R. & Burgoyne, L. A. (1973) *Biochem. Biophys. Res. Commun. 52*, 504−510.
104. Kornberg, R. D. (1974) *Science 184*, 868−871.
105. Finch, J. T. & Klug, A. (1976) *Proc. Natl Acad. Sci. USA 73*, 1897−1901.
106. Felsenfeld, G. (1978) *Nature 271*, 115−122.
107. Nelson, W. G., Pienta, K. J., Barrack. E. R. & Coffey, D. S. (1986) *Annu. Rev. Biophys. Biophys. Chem. 15*, 457−475.
108. Ausio, J., Borochov, N., Seger, D. & Eisenberg, H. (1984) *J. Mol. Biol. 177*, 373−398.
109. Borochov, N., Ausio, J. & Eisenberg, H. (1984) *Nucleic Acids Res. 12*, 3089−3096.
110. Lewis, C. D., Clark, S. P., Felsenfeld, G. & Gould, H. (1988) *Genes Dev. 2*, 863−873.
111. Van Holde, K. E. (1989) *Chromatin*, Springer, Berlin.
112. Richmond, T. J., Finch, J. T., Rushton, B., Rhodes, D. & Klug, A. (1984) *Nature 311*, 532−537.
113. Burlingame, R. W., Love, W. E., Wang, B. C., Hamlin R., Xuong, N.-H. & Moudrianakis, E. N. (1985) *Science 228*, 546−553.
114. Park, K. & Fasman, G. (1987) *Biochemistry 26*, 8042−8045.
115. Richmond, T. J., Searles, M. A. & Simpson, R. T. (1988) *J. Mol. Biol. 199*, 161−170.
116. Voordouw, G. & Eisenberg, H. (1978) *Nature 273*, 446−448.
117. Eisenberg, H. & Felsenfeld, G. (1981) *J. Mol. Biol. 150*, 537−555.
118. Greulich, K. O., Ausio, J. & Eisenberg, H. (1985) *J. Mol. Biol. 186*, 167−173.
119. McGhee, J. D. & Felsenfeld, G. (1986) *Cell 44*, 375−377.
120. Widom, J. & Klug, A. (1985) *Cell 43*, 207−213.
121. Williams, S. P., Athey, B. D., Muglia, L. J., Schappe, R., Gough, A. J. & Langmore, J. P. (1986) *Biophys. J. 49*, 233−250.
122. Koch, M. H. J., Vega, M. C., Sayers, Z. & Michon, A. M. (1987) *Eur. Biophys. J. 14*, 307−319.
123. Greulich, K. O., Wachtel, E., Ausio, J., Seger, D. & Eisenberg, H. (1987) *J. Mol. Biol. 193*, 709−721.
124. Mevarech, M., Neumann, E. & Eisenberg, H. (1977) *Biochemistry 16*, 3781−3785.
125. Leicht, W., Werber, M. M. & Eisenberg, H. (1978) *Biochemistry 17*, 4004−4010.
126. Werber, M. M. & Mevarech, M. (1978) *Arch. Biochem. Biophys. 187*, 447−456.
127. Werber, M. M., Sussman, J. L. & Eisenberg, H. (1986) *FEMS Microbiol. Rev. 39*, 129−135.
128. Eisenberg, H. & Wachtel, E. (1987) *Annu. Rev. Biophys. Biophys. Chem. 16*, 69−92.
129. Dennis, P. P. (1986) *J. Bacteriol. 168*, 471−478.
130. Olsen, G. J. & Woese, C. R. (1989) *Can. J. Microbiol. 35*, 119−123.
131. Gouy, M. & Li, W. H. (1989) *Nature 339*, 145−147.
132. Cavalier-Smith, T. (1989) *Nature 339*, 100−101.
133. Pundak, S. & Eisenberg, H. (1981) *Eur. J. Biochem. 118*, 463−470.
134. Pundak, S., Aloni, H. & Eisenberg, H. (1981) *Eur. J. Biochem. 118*, 471−477.
135. Zaccai, H., Bunick, G. J. & Eisenberg, H. (1986) *J. Mol. Biol. 192*, 155−157.
136. Calmettes, P., Eisenberg, H. & Zaccai, G. (1987) *Biophys. Chem. 26*, 279−290.
137. Zaccai, G., Cendrin, F., Haik, Y., Borochov, N. & Eisenberg, H. (1989) *J. Mol. Biol. 208*, 491−500.
138. Hecht, K. & Jaenicke, R. (1989) *Biochemistry 28*, 4959−4985.
139. Epstein, C. J., Goldberger, R. F. & Anfinsen, C. B. (1963) *Cold Spring Harbor Symp. Quant. Biol. 28*, 439−449.
140. Harel, M., Shoham, M., Frolow, F., Eisenberg, H., Mevarech, M., Yonath, A. & Sussman, J. L. (1988) *J. Mol. Biol. 200*, 609−610.
141 a. Sanger, F. & Thompson, E. O. P. (1953) *Biochem. J. 53*, 353−366.
141 b. Sanger, F. & Thompson, E. O. P. (1953) *Biochem. J. 53*, 366−374.
142. Holley, R. W., Apgar, J., Everett, G. A., Madison, J. T., Marquisee, M., Merrill, J. H., Penswich, J. R. & Zamir, A. (1965) *Science 147*, 1462−1465.
143. Reisler, E. & Eisenberg, H. (1969) *Biochemistry 8*, 4572−4578.
144. Schachman, H. K. & Edelstein, S. J. (1966) *Biochemistry 5*, 2681−2705.
145. Frieden, C. (1964) *Brookhaven Symp. Biol. 17*, 98−116.
146. Eisenberg, H. & Tomkins, G. M. (1968) *J. Mol. Biol. 31*, 37−49.
147. Eisenberg, H. (1971) *Acc. Chem. Res. 4*, 379−385.
148. Eisenberg, H., Josephs, R. & Reisler, E. (1976) *Adv. Protein Chem. 30*, 101−181.
149. Mevarech, M., Hirsch-Twizer, S., Goldmann, S., Eisenberg, H., Yakobson, E. & Dennis, P. P. (1989) *J. Bacteriol. 171*, 3479−3485.
150. Record, M. T. Jr, Anderson, P. M., Mossing, M. & Roe, J.-H. (1985) *Adv. Biophys. 20*, 109−135.
151. Richey, B., Cayley, D. S., Mossing, M. C., Kolka, C., Anderson, C. F., Farrar, T. C. & Record, M. T. Jr (1987) *J. Biol. Chem. 262*, 7157−7164.
152. Leirmo, S., Harrison, C., Cayley, D. S., Burgess, R. R. & Record, M. T. Jr (1987) *Biochemistry 26*, 2095−2101.
153. Minton, A. P. (1983) *Mol. Cell. Biochem. 55*, 119−140.
154. Zimmerman, S. B. & Harrison, B. (1987) *Proc. Natl Acad. Sci. USA 84*, 1871−1875.

Eur. J. Biochem. *187*, 283–305 (1990)
© FEBS 1990

E. Antonini Plenary Lecture

A structural basis of light energy and electron transfer in biology

Robert HUBER

Max-Planck-Institut für Biochemie, Martinsried, Federal Republic of Germany

(Received September 11, 1989) — EJB 89 1099

Aspects of intramolecular light energy and electron transfer will be discussed for three protein cofactor complexes, whose three-dimensional structures have been elucidated by X-ray crystallography: components of light-harvesting cyanobacterial phycobilisomes, the purple bacterial reaction centre and the blue multi-copper oxidases. A wealth of functional data is available for these systems which allow specific correlations between structure and function, and general conclusions about light energy and electron transfer in biological materials to be made.

All life on Earth depends ultimately on the sun, whose radiant energy is captured by plants and other organisms capable of growing by photosynthesis. They use sunlight to synthesize organic substances which serve as building materials or stores of energy. This was clearly formulated by L. Boltzmann, who stated, "There exist between the sun and the earth a colossal difference in temperature... The equalization of temperature between these two bodies, a process which must occur because it is based on the law of probability will, because of the enormous distance and magnitude involved, last millions of years. The energy of the sun may, before reaching the temperature of the earth, assume improbable transition forms. It thus becomes possible to utilize the temperature drop between the sun and the earth to perform work, as is the case with the temperature drop between steam and water... To make the most use of this transition, green plants spread the enormous surface of their leaves and, in a still unknown way, force the energy of the sun to carry out chemical syntheses before it cools down to the temperature level of the Earth's surface. These chemical syntheses are to us in our laboratories complete mysteries..." [1].

Today many of these 'mysteries' have been resolved by biochemical research and the protein components and their basic catalytic functions have been defined [2].

I will focus in my lecture on Boltzmann's 'improbable transition forms', namely excited electronic states and charge-transfer states in modern terminology, the structures of biological materials involved and the interplay of cofactors (pigments and metals) and proteins. I will discuss some aspects of the photosynthetic reaction centre of *Rhodopseudomonas viridis* (see the original publications cited later and short reviews [3 – 5]) and of functionally related systems, whose structures have been studied in my laboratory: light-harvesting cyanobacterial phycobilisomes and blue oxidases. Phycobilisomes are light-harvesting organelles, peripheral to the thylakoid membrane in cyanobacteria, which carry out oxygenic photosynthesis and have photosystems I and II. A wealth of structural and functional data is available for these three systems, which make them uniquely appropriate examples for deriving general principles of light energy and electron transfer in biological materials. Indeed, there are very few systems known in sufficient detail for such purposes. The structure of the *Rhodobacter sphaeroides* reaction centre is closely related to the *Rps. viridis* reaction centre [6 – 8]. A green bacterial bacteriochlorophyll-a-containing, light-harvesting protein is well-defined in structure [9] but not in function. In the multiheme cytochromes [10, 11] the existence or significance of intramolecular electron transfer is unclear.

We strive to understand the underlying physical principles of light and electron conduction in biological materials with considerable hope for success as these processes appear to be more tractable than other biological reactions, which involve diffusive motions of substrates and products and intramolecular motions. Large-scale motions have been identified in many proteins and shown to be essential for many functions [12, 13]. Theoretical treatments of these reactions have to take flexibility and solvent into account, and become theoretically tractable only by applying the rather severe approximations of molecular dynamics [14, 15] or by limiting the system to a few active site residues, which can then be treated by quantum mechanical methods.

Light- and electron-transfer processes seem to be amenable to a more quantitative theoretical treatment. The substrates are immaterial or very small and the transfer processes on which I focus are intramolecular and far removed from solvent. Molecular motions seem to be unimportant, as shown by generally small temperature dependences. The components active in energy and electron transport are cofactors, which, in a first approximation suffice for a theoretical analysis, simplifying calculations considerably.

Copyright. The Nobel Foundation, 1989
Dedication. This review is dedicated to Christa
Correspondence to R. Huber, Max-Planck-Institut für Biochemie, D-8033 Martinsried, Federal Republic of Germany
Abbreviations. PSI and PSII, the two photosynthetic reaction centers in chloroplasts and cyanobacteria; RBP, retinol-binding protein; bilin, biliverdin IXγ; C subunit, the cytochrome *c* subunit of the reaction centre of *Rhodopseudomonas viridis*; H, L and M subunits, the other four subunits of the reaction centre of *Rhodopseudomonas viridis*; BChl and BC, bacteriochlorophyll; subscripts P, A, M and L, pair, accessory, M-subunit association and L-subunit association, respectively; P680 and P960, the primary electron donors in photosystem II and the reaction centre, respectively, of *Rhodopseudomonas viridis*; LHC, light-harvesting complex; LH_a and LH_b, light-harvesting protein pigment complexes in bacteria containing bacteriochlorophylls a and b, respectively; P* and D*, electronically excited states of P and D; Xaa, any amino acid.
Note. E. Antonini Plenary Lecture delivered at the 19th FEBS Meeting in Rome on July 2, 1989.

1. MODELS FOR ENERGY AND ELECTRON TRANSFER

To test theories developed for energy and electron transfer, appropriate model compounds are essential. Although it would be desirable, these models need not mimic the biological structures.

Förster's theory of inductive resonance [16, 17] treats the cases of strong and very weak coupling in energy transfer. Strong interactions lead to optical spectra which are very different from the component spectra. Examples include concentrated solutions of some dyes, crystalline arrays, and the pair of bacteriochlorophylls b (BCl-b or BP) discussed in 3.1. The electronic excitation is in this case delocalized over a molecular assembly. Very weak coupling produces little or no alteration of the absorption spectra but the luminescence properties may be quite different. Structurally defined models for this case are scarce. The controlled deposited dye layers of Kuhn and Frommherz [18, 19] may serve this purpose and have demonstrated the general validity of Förster's theory, but with deviations.

Synthetic models with electron transfer are abundant and have recently been supplemented by appropriately chemically modified proteins [20 – 22]. They are covered in reviews (see e.g. [23 – 30]). Fig. 1 shows essential elements of such models: donor (D; of electrons) and acceptor (A) may be connected by a bridging ligand (B) with a pendant group (P) embedded in a matrix (M).

Models with porphyrins as donors and quinones as acceptors are mimics of the reaction centre [31, 32]. Models with peptide-bridging ligands [33] merit interest especially in relation to the blue oxidases. The effect of pendant groups (P), which are not in the direct line of electron transfer [23] is noteworthy in relation to the unused electron-transfer branches in the reaction centre and the blue oxidases. It is clear, however, that the biological systems are substantially more complex than synthetic models. The protein matrix is not homogeneous and is unique in each case. Despite these shortcomings, theory and models provide the framework within which the factors controlling the transfer of excitation energy and electrons and competing processes are to be evaluated.

1.1. DETERMINANTS OF ENERGY AND ELECTRON TRANSFER

The important factors are summarized in Table 1. They may be derived from Förster's theory and forms of Marcus' theory [27] for excitation and electron transfer, respectively. These theoretical treatments may in turn be derived from classical considerations or from Fermi's 'Golden Rule' with suitable approximations (see, e.g. [34]). Excitation and electron transfer depend on the geometric relation between donors and acceptors. Excitation-energy transfer may occur over wide distances when the transition dipole moments are favourably aligned. Fast electron transfer requires sufficient electronic orbital overlap. Fast electron transfer over wide distances must therefore involve a series of closely spaced intermediate carriers with low-lying unoccupied molecular orbitals or suitable ligands bridging donor and acceptor. Bridging ligands may actively participate in the transfer process and form ligand-radical intermediates (chemical mechanism) or the electron may at no time be in a bound state of the ligands (resonance mechanism) [35]. The spectral overlap and the 'driving force', for energy and electron transfer, respectively,

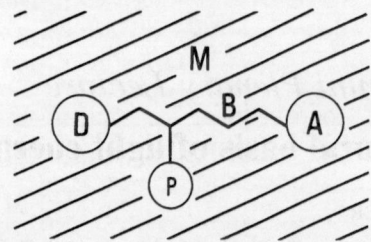

Fig. 1. *Determinants of electron-transfer models*. D, donor; A, acceptor; B, bridge; P, pendant group; M, matrix

Table 1. *Factors controlling rates of energy transfer*

Energy transfer process	Factors controlling rates
Excitation energy transfer $D^* + A \rightarrow D + A^*$ (very weak coupling)	distance and orientation (coupling of excited states); spectral overlap of emission and absorption of D and A; refractive index of medium
Electron transfer from excited state $D^* + A \rightarrow D^+ + A^-$ and from ground state $D^- + A \longrightarrow D + A^-$	distance and orientation (electronic coupling, orbital overlap); free energy change ('driving force'); reorganization in D and A; orientation polarization of medium

have obvious effects on the transfer rates and are largely determined by the chemical nature and geometry of donors and acceptors. Nuclear reorganization of donor, acceptor and the surrounding medium accompanying electron transfer is an important factor but difficult to evaluate in a complex protein system even qualitatively; we observe that the protein typically binds donors and acceptors firmly and rigidly, keeping reactant reorganization effects small. Surrounding polar groups may slow rapid electron transfer due to their reorientation. However, a polar environment also contributes to the energetics by stabilizing ion pairs ($D^+ A^-$), or lowering activation and tunneling barriers, and may increase 'driving force' and rate. Energy transfer also depends on the medium and is not favoured in media with high refractive indices.

Processes competing with productive energy and electron transfer from excited states 'lurk' everywhere (Table 2). Quite generally, they are minimized by high transfer rates and conformational rigidity of the cofactors imposed by the protein.

I will discuss these factors in relation to the biological structures later on.

2. THE ROLE OF COFACTORS

The naturally occurring amino acids are transparent to visible light and seem also to be unsuitable as single electron carriers with the exception of tyrosine. Tyrosyl radicals have been identified in PSII as Z^* and D^* intermediates, which are involved in electron transfer from the water-splitting manganese protein complex to the photooxidized $P680^+$ (for reviews, see [36, 37]). Their identification has been assisted by the observation that Tyr-L162 lies in the electron-transfer path from the cytochrome to BC_{LP} in the bacterial reaction centre

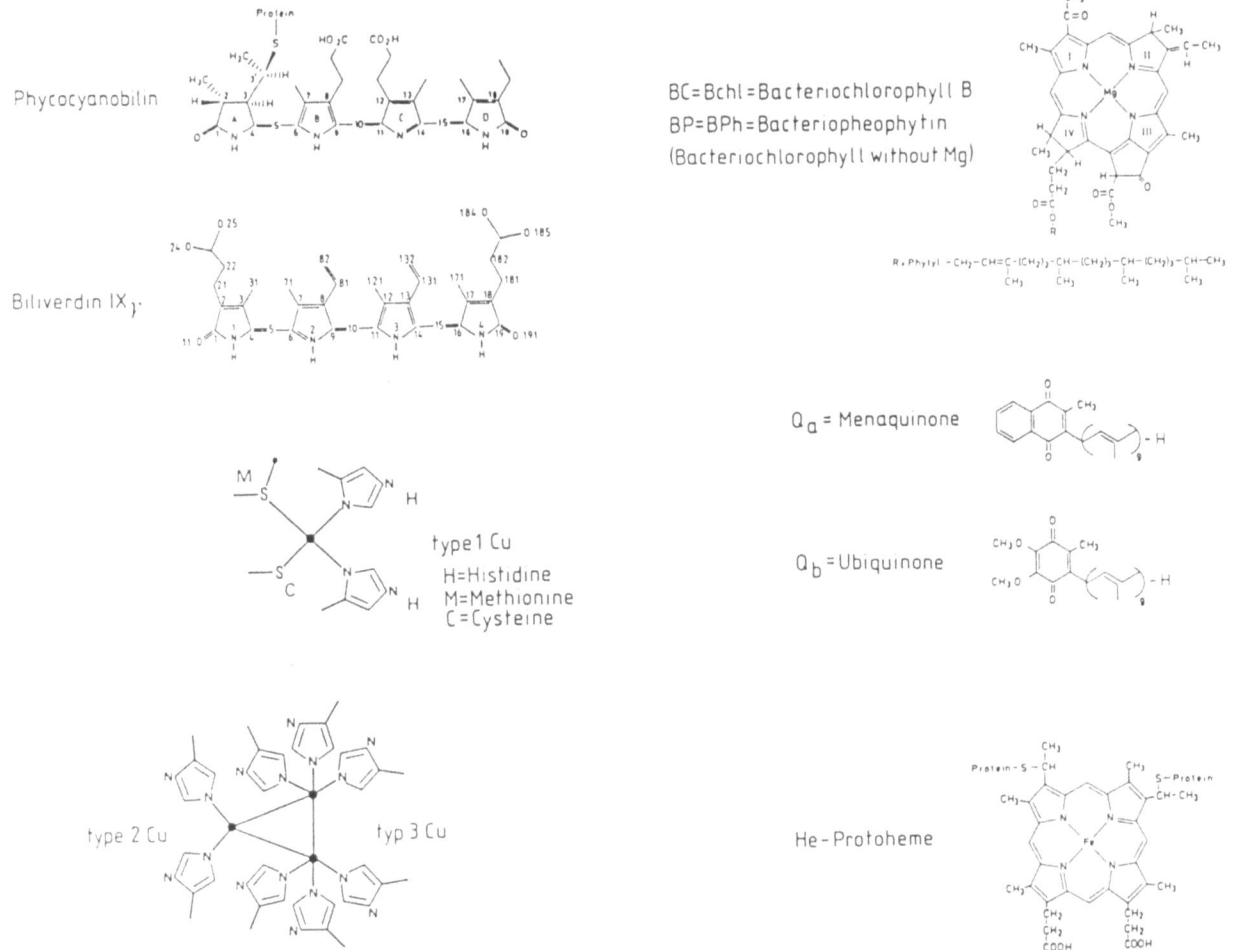

Fig. 2. *Cofactors in phycocyanin, BBP, ascorbate oxidase and the reaction centre.* Phycocyanobilins are covalently bound by thioether linkages to the protein. Biliverdin IXγ is non-covalently bound to BBP. Type-1, type-2, type-3 copper ions are linked to ascorbate oxidase by coordination to the amino acid residues indicated. Four BChl-b and two BPL-b are bound to the reaction centre. A pair of BChl-b serves as the primary electron donor, a menaquinone 9 is the primary electron acceptor (Q_A) and an ubiquinone 9 the secondary acceptor (Q_B). The four haem groups are bound by thioether linkages to the cytochrome *c*

Table 2. *Processes competing with energy transfer*

Energy transfer process	Competing processes
Excitation energy transfer $D^* + A \rightarrow D + A^*$ (very weak coupling)	non-radiative relaxation of D^* by photoisomerization and other conformational changes; excited state proton transfer; intersystem crossing; chemical reactions of D^*, A^*, D^+, A^- with the matrix; fluorescence radiation of D^*
Electron transfer from excited state $D^* + A \rightarrow D^+ + A^-$	energy transfer; as above; back reaction to ground state D, A
From ground state $D^- + A \rightarrow D + A^-$	–

[38] (see 3.2.2.2. and Fig. 10c). A tyrosyl radical is not generated in the bacterial system, because the redox potential of P960$^+$ is insufficient.

Generally, therefore, cofactors, pigments and metal ions serve as light-energy acceptors and redox-active elements in biological materials.

Fig. 2 is a gallery of the pigments and metals clusters which will be discussed further on, namely the bile pigments, phycocyanobilin and biliverdin IXγ (bilin) in the light-harvesting complexes, the BChl-b, bacteriopheophytin b (BPh-B or BP) and quinones in the purple bacterial RC, and the copper centres in the blue oxidases.

The physical/chemical properties of these cofactors determine the coarse features of the protein pigment complexes, but the protein part exerts a decisive influence on the spectral and redox properties.

3. THE ROLE OF THE PROTEIN

The role of the protein follows a hierarchy in determining the properties of the functional protein cofactor complexes shown in Table 3. These interactions are different for the various systems and shall be described separately, except point 1, as there are common features in the action of the protein as a polydentate ligand ascribed to a 'rack mechanism'.

286

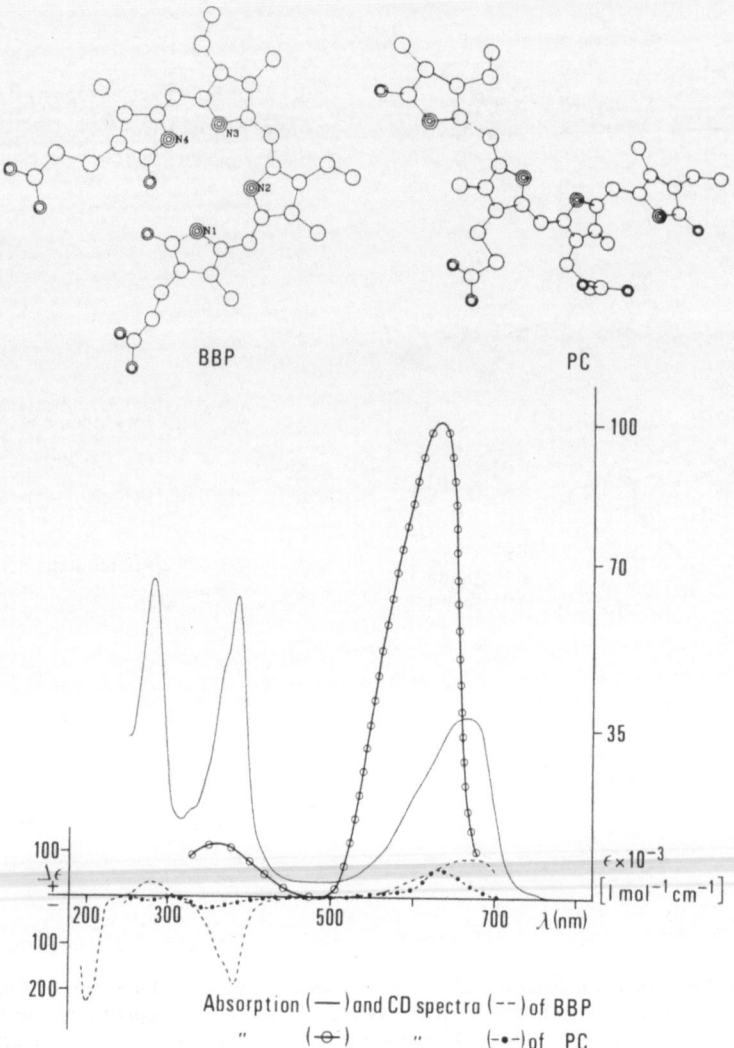

BBP   PC

Fig. 3. *Tetrapyrrole structures in phycocyanin (PC) and BBP and the associated optical and circular dichroism spectra [42, 46]*

Table 3. *Hierarchy of protein cofactor interactions*

1. Influence on configuration and conformation of the cofactors by the nature and geometry of ligands (the protein as a polydentate ligand).

2. Determination of the spatial arrangement of arrays of cofactors (the protein as a scaffold).

3. The protein as the medium.

4. Mediation of the interaction with other components in the supramolecular biological system.

3.1. THE PROTEIN AS A POLYDENTATE LIGAND

The 'rack mechanism' was introduced by Lumry and Eyring [39] and Gray and Malmström [40], to explain unusual reactivities, spectral and redox properties of amino acids and cofactors by the distortion enforced by the protein.

A comparison of isolated and protein-bound bile pigments gives a clear demonstration of this effect. Isolation bile pigments in solution and in the crystalline state prefer a macrocyclic helical geometry with configuration *ZZZ* and conformation *syn,syn,syn* and show weak absorption in the visible

range and low fluorescence quantum yield [41 – 43]. When bound as cofactors to light-harvesting phycocyanins they have strong absorption in the visible range and high fluorescence yield (Fig. 3). The auxochromic shift, essential for the light-harvesting functions, is due to a strained conformation of the chromophore, which has configuration *ZZZ* and conformation *anti,syn,anti*, stabilized by tight polar interactions with the protein [44 – 46] (Fig. 4). Particularly noteworthy is an aspartate residue (A87 in Fig. 4) bound to the central pyrrole nitrogens and conserved in all pigments sites. It influences protonation, charge and spectral properties of the tetrapyrrole systems. Tight binding is also effective against deexcitation by conformational changes. The structure shown in Fig. 3 as representative of the free pigment is in fact observed in a bilin-binding protein from insects [42, 43]. This protein serves a different function and prefers the low-energy conformer. The open-chain tetrapyrrole bilins are conformationally adaptable, a property, which makes them appropriate cofactors for different purposes.

The cyclic BChl in the reaction centre is conformationally restrained but responds to the environment by twisting and bending of the macrocycle. This may be one cause for the different electron transfer properties of the two pigment branches in the reaction centre, as will be discussed later. A

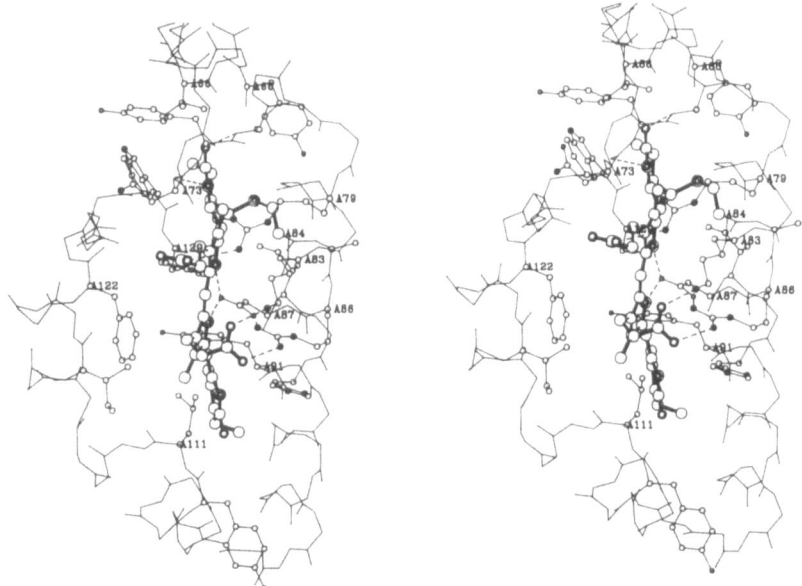

Fig. 4. *Stereo drawing of phycocyanobilin A84 (thick bonds) and its protein environment (thin bonds)*. All polar groups of the bilin exept those of the terminal D-pyrrole ring are bound by hydrogen bonds and salt links to protein groups [46]

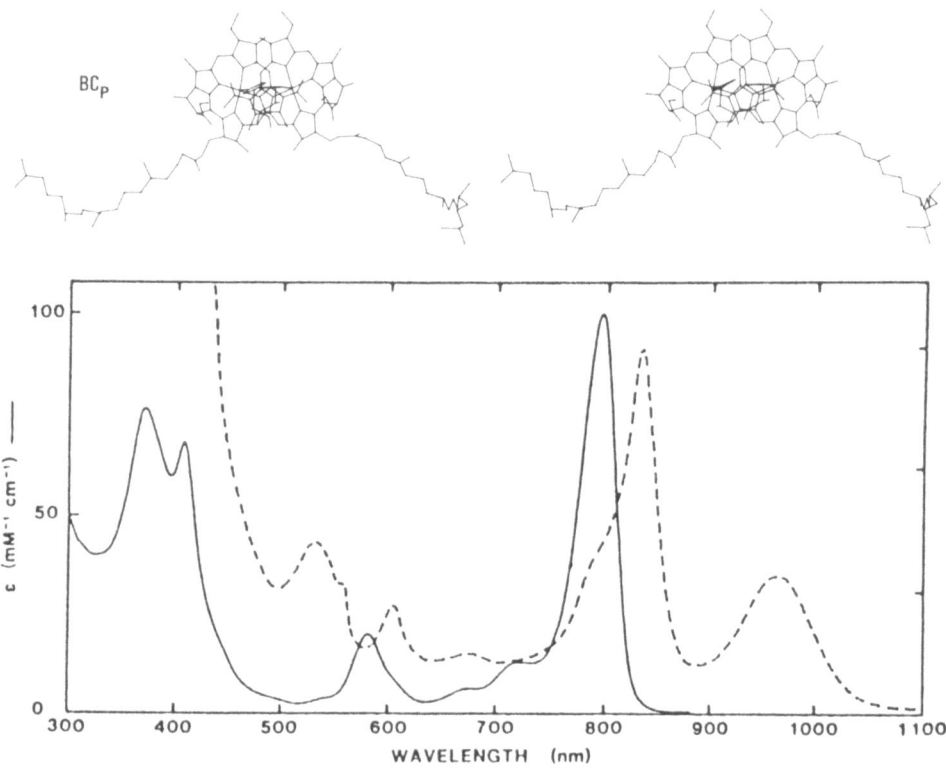

Fig. 5. *Stereo drawing of the special pair* BC_P *in the reaction centre [47] mainly responsible for the spectral alterations and the long-wavelength absorption of the reaction centre of* Rps. viridis (----) *compared with the spectra of BChl-b in ether solution* (———) *(spectra from [171])*

more profound influence of the protein on the reaction centre pigment system is seen in the absorption spectra, which differ from the composite spectra of the individual components (Fig. 5). The protein binds a pair of BChl-b (BC_P) so that the two BChl-b interact strongly between their pyrrole rings I including the acetyl substituents and the central magnesium ions [47]. Alignment of the transition dipole moments and close approach cause excitonic coupling which partially explains the long wavelength absorption band P960 [48].

The optical spectra are even more perturbed in blue copper proteins compared with cupric ions in normal tetragonal coordination (Fig. 6). The redox potential is also raised to about $300-500$ mV vs 150 mV for Cu^{2+} (aq) [49]. These effects are caused by the distorted tetrahedral coordination of the

288

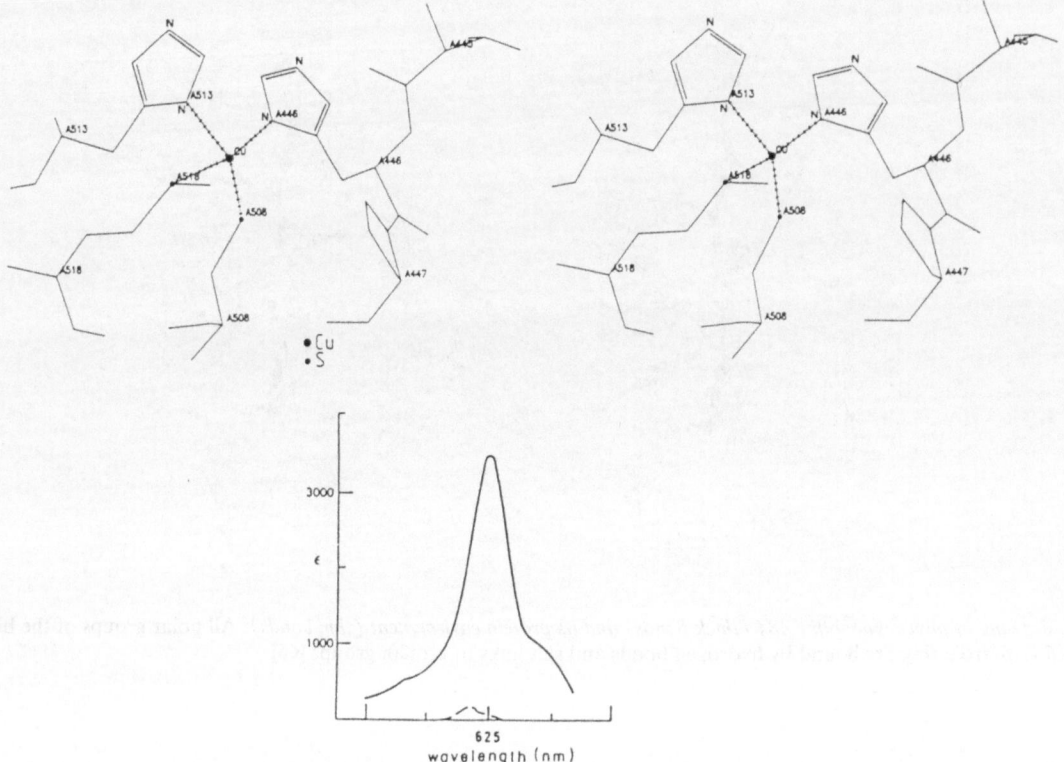

Fig. 6. *The type-1 copper and its ligands in ascorbate oxidase in stereo.* The coordination of the copper is to His-A446, His-A513, Met-A518, Cys-A508 [122]. The optical absorption spectra of 'blue' copper in copper proteins (———) are compared with normal tetragonal copper (————) (spectrum from [49])

type-1 copper (a strained conformation stabilizing the cuprous state) and a charge-transfer transition from a ligand cysteine, $S^- \rightarrow Cu^{2+}$ [40, 50].

The examples presented demonstrate the influence of the protein on the cofactors by various mechanisms, stabilization of unstable conformers and strained ligand geometries, and the generation of contacts between pigments leading to strong electronic interaction.

The fixation of the relative arrangements of systems of cofactors is the basis of the energy- and charge-transfer properties in each system.

3.2. PROTEIN AS A SCAFFOLD

3.2.1. *Light harvesting by phycobilisomes*

The limited number of pigment molecules associated with reaction centres would absorb only a small portion of incident sunlight. The reaction centres are therefore associated with the light-harvesting complex (LHC), which may be located within the photosynthetic membrane or form layers or antenna-like organelles in association with the photosynthetic membrane. Cyanobacteria have particularly intricate light-harvesting systems, the phycobilisome organelles peripheral to the thylakoid membrane. They absorb light of shorter wavelengths than do PSI and PSII, so that a wide spectral range of sunlight is used (Fig. 7). The phycobilisomes are assembled from components with finely tuned spectral properties such that the light energy is channeled along an energy gradient to PSII.

3.2.1.1. *Morphology.* Phycobilisomes consist of biliproteins and linker polypeptides. Biochemical and electron mi-

croscopy studies [51 – 54] lead to the model representative of a hemidiscoidal phycobilisomes in Fig. 7. Accordingly, phyobilisome rods are assembled in a polar way from phycoerythrin of phycoerythrocyanin and phycocyanin, which is attached to a central core of allophycocyanin. Allophycocyanin is next to the photosynthetic membrane and close to PSII (for a review see [55]). The phycocyanin component consists of α and β protein subunits, which are arranged as $(\alpha\beta)_6$ disc-like aggregates with dimensions 12 nm × 6 nm (for reviews, see [56–61]).

From crystallographic analyses, a detailed picture of phycocyanin and phycoerythrocyanin components has emerged [44–46, 62, 63]. Amino acid sequence similarity suggests that all components have similar structures.

3.2.1.2. *Structure of phycocyanin.* The phycocyanin α subunit and β subunits have 162 and 172 amino acid residues, respectively (in *Mastigocladus laminosus*). Phycocyanobilin chromophores are linked via thioether bonds to cysteine residues at positions 84 of both chains (A84, B84) and at position 155 of the β subunit (B155) [64]. Both subunits have similar structures and are folded into eight α helices (X, Y, A, B, E, F, G and H; see Fig. 13). A84 and B84 are attached to helix E, B155 to the G−H loop. α-Helices X and Y form a protruding anti-parallel pair, essential for formation of the $\alpha\beta$ unit.

The isolated protein forms $(\alpha\beta)_3$ trimers with C3 symmetry and hexamers $(\alpha\beta)_6$ as head-to-head-associated trimers with D3 symmetry (Fig. 8). The inter-trimer contact is exclusively mediated by the α subunits, which are linked by an intricate network of polar bonds. The inter-hexamer contacts within the crystal (and in the native phycobilisome rods) are made by the β subunits [46].

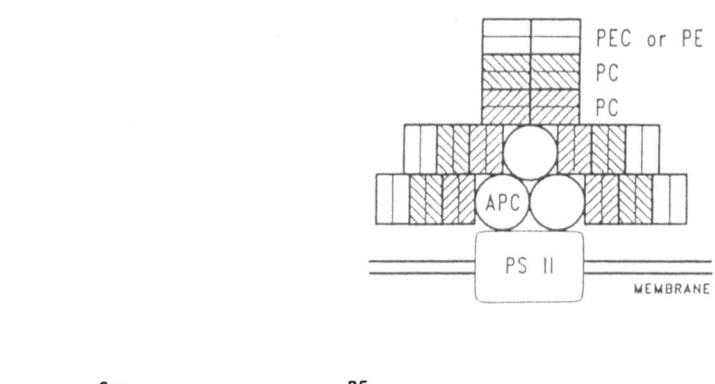

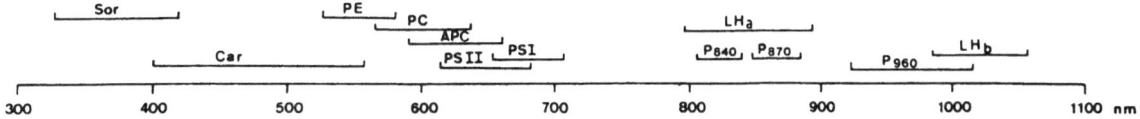

Fig. 7. *Scheme of a typical phycobilisome with the arrangements of the components and the putative spatial relationship to the thylakoid and PSII (for reviews see [54, 55]).* The component labelled PSII is thought to represent PSII and the phycobilisome attachment sites. The main absorption bands of photosynthetic protein cofactor complexes in photosynthetic organisms are also shown. The phycobilisome components absorb differently to cover a wide spectral range and permit energy flow from phycoerythrocyanin/phycoerythrin (PEC/PE) via phycocyanin (PC) and allophycoerythrin (APC) to PSII

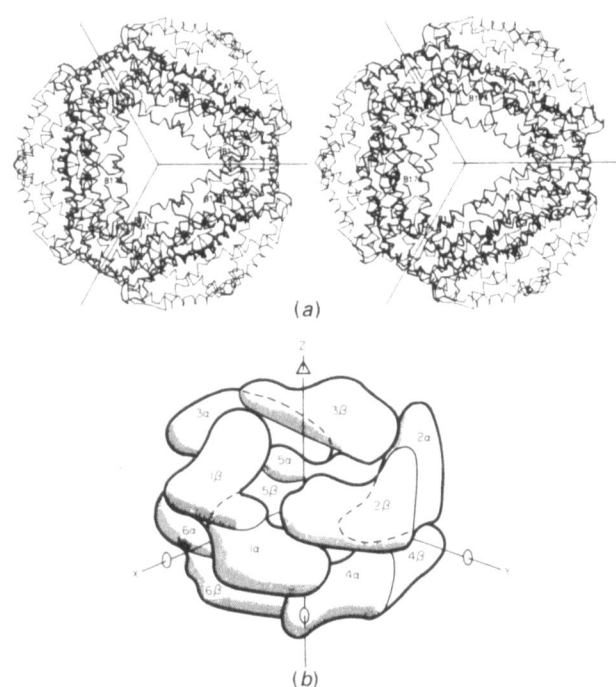

Fig. 8. *Stereo drawing of the polypeptide chain fold of an* $(\alpha\beta)_6$ *hexamer of phycocyanin seen along the disk axis (a).* The scheme (b) indicates the packing of subunits in the hexamer seen from the side

3.2.1.3. Oligomeric aggregates: spectral properties and energy transfer. The spectral properties, absorption strength and quantum yield of fluorescence of biliproteins depend on the state of aggregation. The absorption spectrum of the $\alpha\beta$ unit resembles the sum of the spectra of the constituent subunits, but the fluorescence quantum yield is somewhat higher. Upon trimer formation, the absorption is red-shifted and its strength and the quantum yield of fluorescence increased [66, 67], for a review, see [58]. In the $(\alpha\beta)_6$ linker complexes, the fluorescence is further increased and the absorption spectrum further altered [68].

These observations can be rationalized by the structure of the aggregates. Formation of $\alpha\beta$ units causes little change in the environment of the chromophores. They remain quite separat with distances > 3.6 nm (Fig. 9A). Upon trimer formation, the environment of chromophore A84 changes profoundly by approach of chromophore B84 of a related unit (Fig. 9A, a). In the hexamer (Fig. 9A, b) the A84 and B155 chromophores interact pairwise strongly across the trimer interface. Also, the molecular structures become more rigid with increasing size of the aggregates as seen in the crystals of the trimeric and hexameric aggregates [45, 46]. Rigidity hinders excitation relaxation by isomerization and thus increases the fluorescence quantum yield.

The chromophores can be divided into subsets of s(sensitizing) and f(fluorescing) chromophores [69, 70]. The s chromophores absorb at the blue edge of the absorption band and transfer the excitation energy rapidly to the f chromophores. This transfer is accompanied by depolarization [71]. Excitation at the red absorption edge (f chromophores), however, results in little depolarization, suggesting that the energy is transferred along stacks of similarly oriented f chromophores [72]. The assignment of the chromophores to s and f was made by steady-state spectroscopy on different aggregates [67], by chemical modification guided by the spatial structure [73], and conclusively by measurement of linear dichroism and polarized fluorescence in single crystals [65]. Accordingly B155 is the s, B84 the f, and A84 the intermediate chromophore.

Light energy is transferred rapidly within $50-100$ ps from the tips of the phycobilisomes to the core (for a review, see e.g. [58, 72, 74–78]). The transfer times from the periphery to the base are several orders of magnitude faster than the intrinsic fluorescence life-times of the isolated components [71, 74]. The distances between the chromophores within and between the hexamers are too large for strong (excitonic) coupling, but efficient energy transfer by inductive resonance occurs. A Förster radius of about 5 nm has been suggested by Grabowski and Gantt [79]. The relative orientations and distances of the chromophores as obtained by Schirmer et al. [46] were the basis for the calculation of the energy-transfer rates in Fig. 9A. It shows the preferred energy-transfer path-

290

ways in $\alpha\beta$ units, $(\alpha\beta)_3$ trimers, $(\alpha\beta)_6$ hexamers and stacked disks as models for native antenna rods. There is very weak coupling of the chromophores in the $\alpha\beta$ units. Some energy transfer takes place, however as indicated by steady-state polarization measurements [67, 80] probably between B155 and B84. Trimer formation generates strong coupling between A84 and B84, but B155 is integrated only weakly. In the hexamer, many additional transfer pathways are opened and B155 is efficiently coupled. Hexamers are obviously the functional units, as the energy can be distributed and concentrated on the central f chromophores, which couple the stacks of hexamers. Kinetic studies [58, 67, 72, 81] have confirmed the picture of energy transfer along the rods as a random walk (trap or diffusion limited) along a one-dimensional array of f chromophores. Sauer et al. [82] have successfully simulated the observed energy-transfer kinetics in phycocyanin aggregates on the basis of the structures using Förster's mechanism. The phycoerythrocyanin component at the tips of phycobilisome rods is extremely similar to phycocyanin [63, 83]. Its short-wavelength-absorbing chromophore A84 is located at the periphery (Fig. 10a), as are the additional chromophores in phycoerythrin, which is also a tip component (Fig. 9B).

The phycobilisome rods act as light collectors and energy concentrators from the peripheral onto the central chromophores, that is, as excitation energy funnels from the periphery to centre and from the tip to the bottom.

We may expect functional modulations by the linker polypeptides. Some of them are believed to be located in the central channel of the hexamers, where they may interact with B84.

3.2.2. Electron transfer in the reaction centre

A historical background of the development of concepts and key features of the purple bacterial reaction centre is given by Parson [83a].

3.2.2.1. *Reaction centre, composition.* The arrangement of the reaction centre in the thylakoid membranes of *Rps. viridis*, as obtained by electron microscopy, is described by Stark et al. [84].

The reaction centre of *Rps. viridis* is a complex of four protein subunits, C, L, M and H and cofactors arranged as in Fig. 10a. As shown by the amino acid sequence they consist of 336, 273, 323 and 258 residues [85–87].

The c-type cytochrome (subunit C) has four haem groups which display two redox potentials (c_{553} and c_{558}); it is covalently bound via thioether linkages and is located on the periplasmic side of the membrane. Its cofactors are four BChl-b (BC_{MP}, BC_{LP}, BC_{LA} and BC_{MA}), two BPh-b (BP_M and BP_L), one menaquinone 9 (Q_A) and a ferrous iron involved in electron transfer. A second quinone (ubiquinone 9; Q_B), which is a component of the functional complex, is partially lost during preparation and crystallization of the reaction centre. The L and M subunits are integrated into the membrane and their polypeptide chains span the membrane with five α-helices each, labelled A, B, C, D and E; they also bind BChl-b, BPh-b, menaquinone 9 and the Fe^{2+} cofactors, as well as ubiquinone 9. The H subunit is located on the cytoplasmic side of the membrane and its N-terminal α-helical segment spans the membrane.

3.2.2.2. *Chromophore arrangement and electron transfer.* The chromophores are arranged in L and M branches related by an axis of approximate twofold symmetry which meet at BC_P [47]. This axis is normal to the plane of the membrane.

While many of the optical properties of the pigment system are rather well understood on the basis of spatial structure

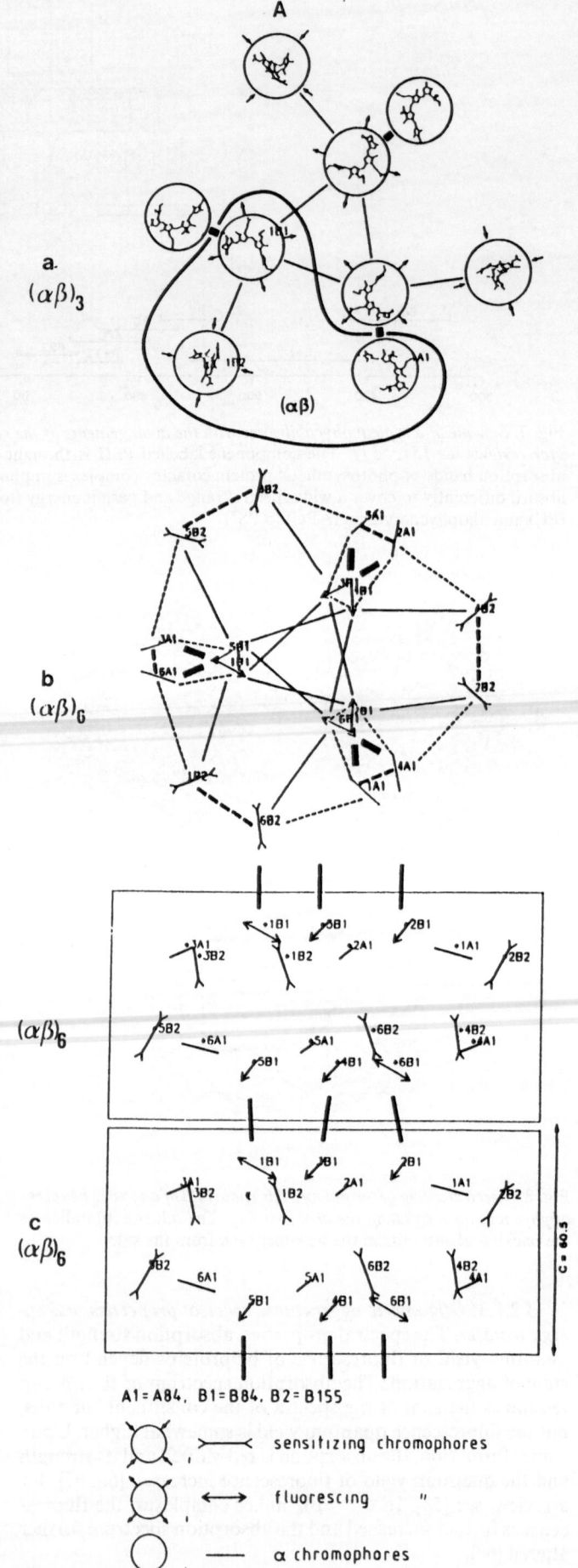

[48, 88], electron transfer is less well understood. The excited BC_P is quenched by electron transfer to BP_L in 3 ps and further on to the primary acceptor Q_A in about 200 ps, driven by the redox-potential gradient between P^*/P^+ (about -760 mV) and Q_A/Q_A^- (about -110 mV). The redox potential of BP/BP^- is intermediate with about -400 mV [88−97]. These functional data are summarized in Fig. 10a. General factors controlling the transfer rates have been summarized in Table 1 and are detailed for the reaction centre.

Fast electron transfer requires effective overlap of the molecular orbitals. The orbital interaction decreases exponentially with the edge to edge distance of donor and acceptor and is insignificant at distances larger than about 1 nm [30, 98]. In the reaction center the distance between BC_P and Q_A is far too large to allow fast, direct electron transfer; instead the electron migrates via BP_L. BP_L^- is a spectroscopically and kinetically well-defined intermediate. Although located between BC_P and BP_L, BC_{LA}^- is not an intermediate but probably involved in electron transfer by a 'super-exchange' mechanism mediating a strong quantum mechanical coupling [99] (for a review see [100]). The distance between BP_L and Q_A seems large for a fast transfer. Indeed the gap is bridged by the aromatic side chain of Trp-M250 in the L branch of the pigment system (Fig. 10b) [38, 101], which might mediate coupling via appropriate orbitals. In addition, the isoprenoid side chain of Q_A is close to BP_L. Electron transfer via long connecting chains by through-bond coupling of donor and acceptor orbitals has been observed [98, 102, 103] but there is only Van der Waals contact here.

A second important factor for electron transfer is the free energy change (ΔG), which is governed by the chemical nature of the components, by geometrical factors and by the environment (solvent polarity). It depends on the ionization potential of the donor in its excited state, the electron affinity of the acceptor, and on the coulombic interaction of the radical ion pair, which is probably small as donor and acceptor (BC_P and Q_A in the reaction centre) are far apart. The effect of the environment may be substantial by stabilizing the radical ion pair by ionic interactions and hydrogen bonds. ΔG is

a determinant of the activation energy of electron transfer. Another are nuclear rearrangements of the reactants and the environment. As the charge on donor and acceptor develops the nuclear configurations change. These changes are likely to be small in the reaction centre as the BChl-b macrocycles are relatively rigid and tightly packed in the protein and the charge is distributed over the extended aromatic electron systems. Reorientable dipolar groups (peptide groups and side chains) may contribute strongly to the energy barrier of electron transfer. A matrix with high electronic polarizability on the other hand stabilizes the developing charge in the transition state of the reaction and reduces the activation energy. An alternative picture is that the potential energy barrier to electron tunnelling is decreased. Aromatic compounds which are concentrated in the vicinity of the electron carriers in the reaction centre have these characteristics (see Trp-M250).

The electron transfer from P^* to Q_A occurs with very low activation energy [90, 93, 95, 104−108] and proceeds readily at 1 K. Thermally activated processes, nuclear motions and collisions are therefore not important for the initial very fast charge-separation steps. There is even a slight increase in rate with temperature decrease either due to a closer approach of the pigments at low temperature, or to changes of the vibrational levels which may lead to a more favourable Franck-Condon factor.

The electron transfer between primary and secondary quinone acceptors, Q_A and Q_B is rather different from the previous processes, because it is much slower (about 6 μs at pH 7, derived from [90]) and has a substantial activation energy of about 33.5 kJ mol^{-1}. In Rps. viridis, Q_A is a menaquinone 9 and Q_B a ubiquinone 9, which differ in their redox potentials in solution by about 100 mV. In other purple bacteria both Q_A and Q_B are ubiquinones. The redox-potential difference required for efficient electron transfer in these cases is generated by the asymmetric protein matrix. The protein matrix is also responsible for the quite different functional properties of Q_A and Q_B. Q_A accepts only one electron (leading to a semiquinone anion), which is transferred to Q_B before the next electron transfer can occur. Q_B, however, accepts two

B

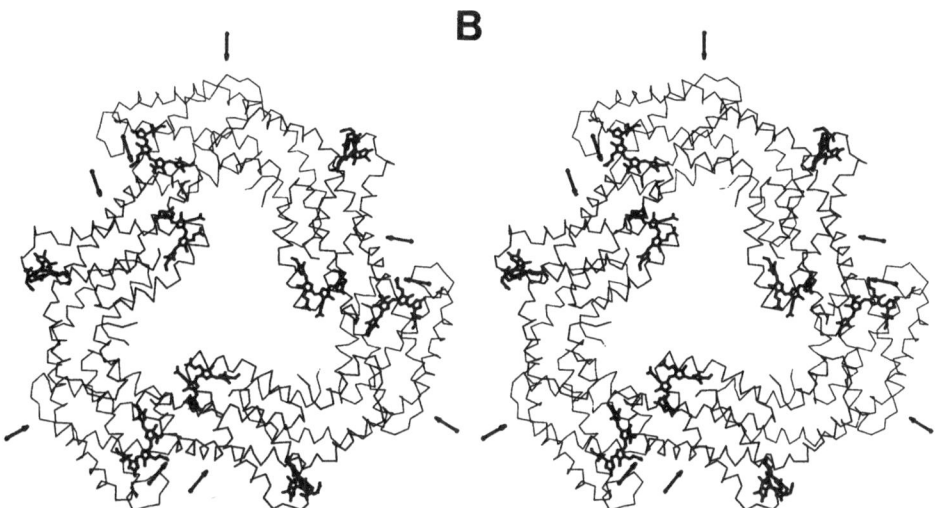

Fig. 9. *Arrangement of chromophores and preferred energy-transfer pathways in $(\alpha\beta)_3$ trimers, $(\alpha\beta)_6$ hexamers and stacked hexamers based on Table 10 in [46] (A) and model of phycoerythrin (B).* (A) For the trimer the detailed structures of the chromophores are drawn, otherwise their approximate transition dipole directions are indicated. For the trimer and hexamer the view is along the disc axis; for the stacked hexamers it is perpendicular. In the stacked hexamers only the inter-hexamer transfers are indicated. The strength of coupling is indicated by the thickness of the connecting lines. Transfer paths within and between the trimers are represented by full and broken lines, respectively. (B) Model of phycoerythrin $(\alpha\beta)_3$ on the basis of phycocyanin with the locations of the additional phycoerythrobilins indicated by arrows

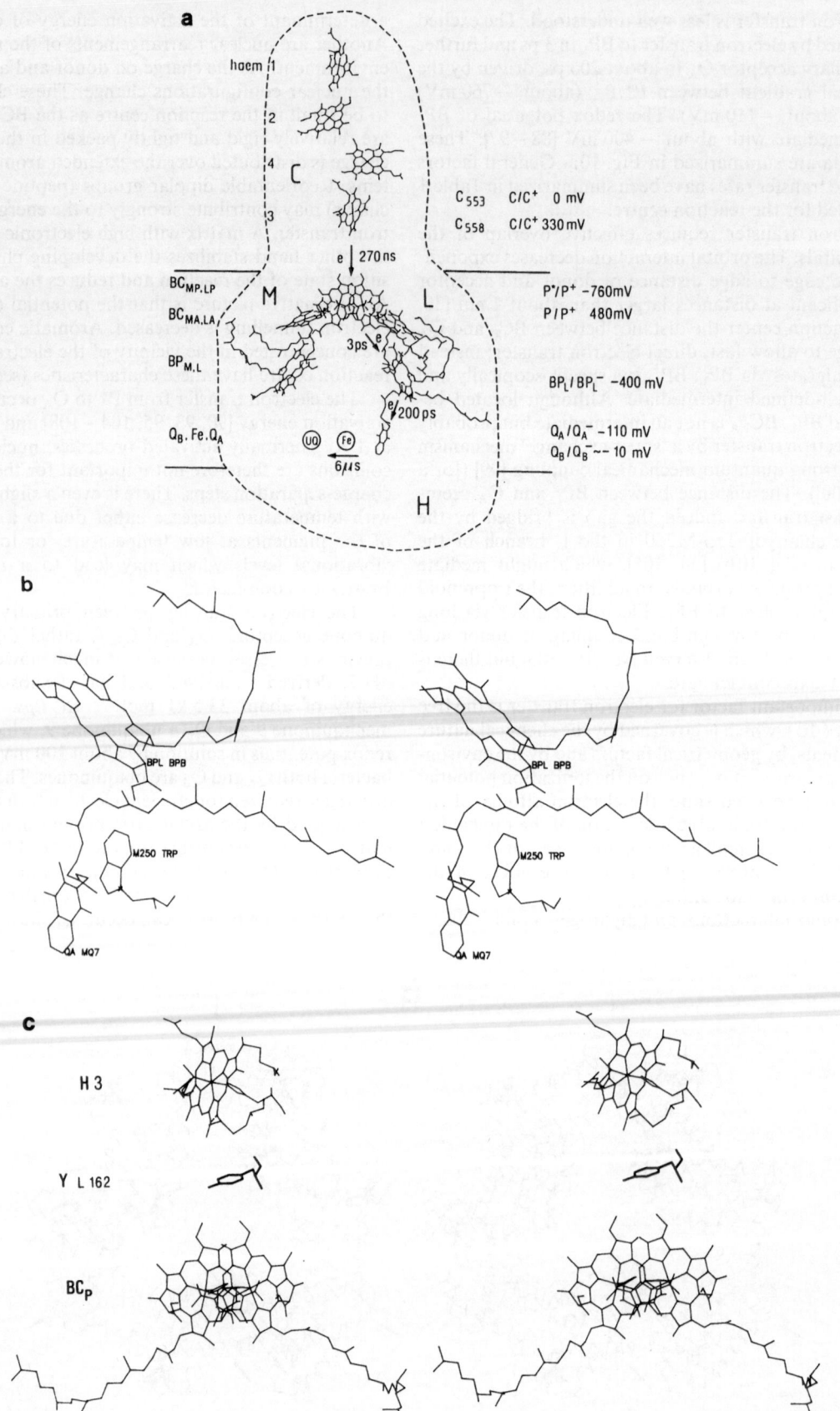

Fig. 10. *The reaction centre cofactor system.* (a) Scheme of the structure of the reaction centre cofactor system, the outline of the protein subunits (C,M,L,H), the electron transfer $t_{1/2}$, and the redox potentials of defined intermediates (for references see 3.2.2.2.). (b) Stereo drawing of the arrangement of BP_L, Trp-M250 and O_A in the L branch of the reaction centre pigment system. (c) Stereo drawing of haem 3 (H3) of the cytochrome *c*, the special pair BCp and the intercalating Tyr-L162 (Y-L162) of the L subunit [38]. The His and Met ligands to the iron of haem 3 and the His ligands to the magnesium ions of the BC_P are also shown

electrons and is protonated to form a hydroquinone, which diffuses from the reaction centre (two-electron gate [109]). Q_B is close to Glu-L212, which opens a path to the H subunit and may protonate Q_B. The path between Q_A and Q_B is very different from the environment of the primary electron-transfer components. The line connecting Q_A and Q_B (the Q_B-binding site has been inferred from the binding mode of competitive inhibitors and ubiquinone 1 in *Rps. viridis* crystals [38]) is occupied by the iron and its five-coordinating ligands, four histidines (M217, M264, L190 and L230) and a glutamic acid (M232) residue. His-M217 forms a hydrogen bond to Q_A. His-L190 is close to Q_B. Q_A and Q_B have an edge to edge distance of about 1.5 nm which might explain the slow transfer. If electron transfer and protonation are coupled, the observed pH dependence of the electron-transfer rate of Q_A to Q_B [110] could be explained and nuclear motions required for proton transfer may generate the observed activation energy barrier. The role of the charged Fe-His$_4$-Glu complex in the Q_A to Q_B electron transfer is poorly understood at present, as it occurs also in the absence of the iron [111]. Its role seems to be predominantly structural.

The cycle of electron transfer is closed by rereduction of the BC$_{P*}$ from the cytochrome bridging a distance of about 1.1 nm between pyrrole ring I of haem 3 and pyrrole ring II of BC$_{LP}$. The transfer time is 270 ns [94] (previous measurements in [112]), considerably slower than the initial processes. Tyr-L162, which is located midway (Fig. 10c) may facilitate electron transfer by mediating electronic coupling between the widely spaced donor and acceptor. The biphasic temperature dependence indicates a complex mechanism in which at high temperatures nuclear motions play a role (for reviews see [113, 114]).

The favourable rate-controlling factors discussed are a necessary but not sufficient condition for electron transfer, which competes with other quenching processes summarized in Table 2 and detailed for the reaction centre.

Energy transfer from P* back to the LHC or to other pigments may be favourable from orientation and proximity considerations but is disfavoured for energetic reasons. The special pair absorbs usually (but not in *Rps. viridis* where the maximal absorption of the reaction centre and the LHC are at 960 nm and 1.02 mm respectively, see Fig. 7) at longer wavelengths than other pigments of the photosynthetic apparatus and represents the light-energy sink. The natural radiative lifetime of the excited singlet state P* is around 20 ns [106, 115] and may serve as an estimate of the times involved in the other wasteful quenching processes. Clearly electron transfer is much faster. Non-radiative relaxation of BC$_{P*}$ by isomerizations and conformational changes is unlikely for the cyclic pigment systems tightly packed in the protein matrix.

The back reaction, $P^+Q_A^-$ to PQ_A, has a favourable driving force (Fig. 10a) and proceeds independent of temperature, but is slow and insignificant under physiological conditions (for a review [93]). The physical basis for this has yet to be explained. It may be related to a gating function of BP$_L$ by its negative redox potential compared to Q_A, to electronic properties of P^+ disfavouring charge transfer and to conformational changes induced by electron transfer.

The profound influence of the protein matrix on electron transfer in the reaction center is obvious in the observed asymmetry of electron transfer in the two branches of the BChl-b and BPh-b pigments. Only the branch more closely associated with the L subunit is active. An explanation is offered by the fact that the protein environment of both branches, although provided by homologous proteins (L and

M), is rather different in particular by the Trp-M250 located between BP$_L$ and Q_A, and the numerous differences in the Q_A- and Q_B-binding sites [38, 101]. Asymmetry is observed in the BC$_P$ due to different distortions and hydrogen bonding of the macrocycles, and in the slightly different spatial arrangements of the BC$_A$ and BP. It is suggested to facilitate electron release into the L branch [116]. The M branch may have influence as a pendant group, though.

The protein matrix also serves to dissipate the excess energy of about 650 mV [91] of the excited special pair (P*Q$_a$) over the radical ion pair $P^+Q_A^-$. These processes are probably very fast.

In summary, the very fast electron transfer from BC$_{P*}$ to Q_A occurs between closely spaced aromatic macrocycles with matched redox potentials. The protein matrix in which the pigments are tightly held is lined predominantly with apolar amino acid side chains with a high proportion of aromatic residues. The electron path is removed from bulk water.

3.2.3. *The blue oxidases*

Oxidases catalyse the reduction of dioxygen in single electron transfers from substrates. Dioxygen requires four electrons and four protons to be reduced to two water molecules. Oxidases must provide recognition sites for the two substrates, a storage site for electrons and/or means to stabilize reactive partially reduced oxygen intermediates [117–119].

The blue oxidases are classified corresponding to distinct spectroscopic properties of three types of copper which they contain: type-1 Cu^{2+} is responsible for the deep blue color of these proteins: type-2 or normal Cu^{2+} has undetectable optical absorption; type-1 and type-2 cupric ions are paramagnetic; type-3 copper has a strong absorption around 330 nm and is antiferromagnetic, indicating coupling of a pair of cupric ions. The characteristic optical and electron paramagnetic resonance spectra disappear upon reduction.

Studies of the catalytic and redox properties of the blue oxidases are well documented in several recent reviews (e.g. for laccase [119a]; for ascorbate oxidase [120]; for ceruloplasmin [121]). Basically type-1 Cu^{2+} is reduced by electron transfer from the substrate. The electron is transferred on to the type-3 and type-2 copper ions. The second substrate, dioxygen, is associated with the type-3 and/or type-2 copper ions.

3.2.3.1. Ascorbate oxidase, composition and copper arrangement. Ascorbate oxidase is a polypeptide of 553 amino acid residues folded into three tightly associated domains [122]. It is a dimer in solution, but the functional unit is the monomer. It belongs to the group of blue oxidases together with laccase and ceruloplasmin [123].

Structures of copper proteins containing only one of the different copper types are known: plastocyanin has a blue type-1 copper, which is coordinated to two histidine residues and the sulfur atoms of cysteine and methionine as a distorted tetrahedron [124]. Cu/Zn-superoxide dismutase contains a type-2 copper, which has four histidine ligands with slightly distorted quadratic coordination [125]. Hemocyanin of *Panulirus interruptus* has type-3 copper, a pair of copper ions 0.34 nm apart with six histidine ligands [126].

In domain 3 of ascorbate oxidase (see 4.4.) a copper ion is found in a strongly distorted tetrahedral (approaching trigonal pyramidal geometry) coordination by the ligands His, Cys, His and Met, as had been shown in Fig. 6. It resembles the blue type-1 copper in plastocyanin. Between domain 1 and domain 3, a trinuclear copper site is enclosed and shown in Fig. 11a. Four His-Xaa-His amino acid sequences provide the

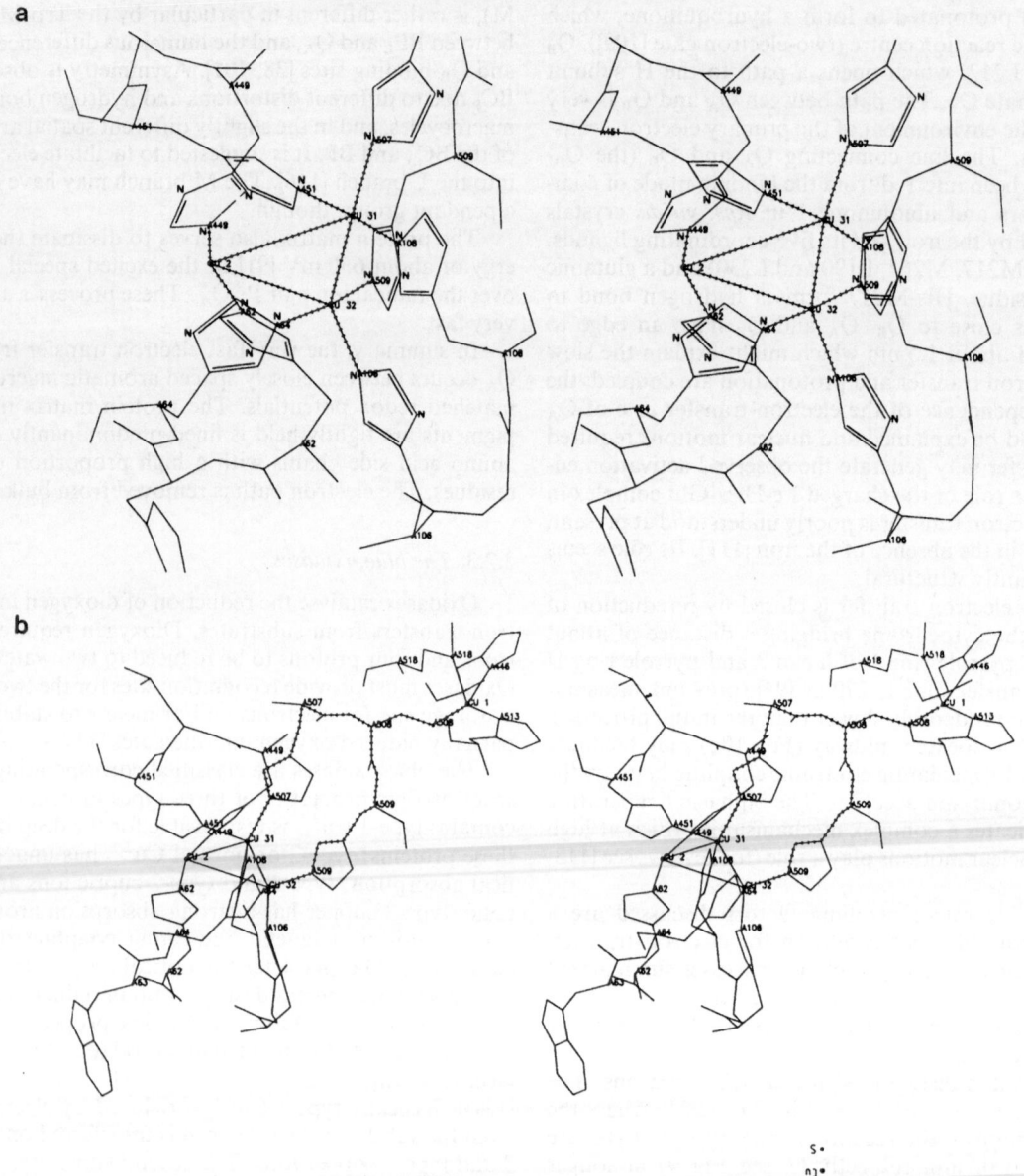

Fig. 11. *Stereo drawings of the trinuclear copper site (a) and the tridentate peptide ligand (b) in ascorbate oxidase.* (a) Coordination bonds between the copper ions and the protein residues are marked (————) [122]. (b) Stereo drawing of the tridentate peptide ligand (His507-Cys-508-His509) bridging type-1 copper (Cu1) and the trinuclear cluster (Cu31, Cu32, Cu2) [122]

eight histidine ligands. The trinuclear copper site is subdivided in a pair of coppers (Cu31, Cu32) with 2×3 histidyl (A108, A451, A507, A64, A106, A509) ligands forming a trigonal prism. It represents the type-3 copper pair, as a comparable arrangement is observed in hemocyanin. The remaining copper (Cu2) has two histidyl ligands (A62, A449). It is type-2 copper. The trinuclear copper cluster is the site where dioxygen binds, but the structural details including the presence of additional non-protein ligands require clarification. The close spatial association of the three copper ions in the cluster suggests facile electron exchange. It may function as an electron storage site and cooperative three-electron donor to dioxygen, to irreversibly break the O=O bond.

3.2.3.2. *Intramolecular electron transfer in ascorbate oxidase.* Electrons are transferred from the type-1 copper to the trinuclear site. The shortest pathway is via Cys-A508 and His-A507 or His-A509. The His-Xaa-His segment links electron donor and acceptor bridging a distance of 1.2 nm (Fig. 11 b). The cysteine sulfur and the imidazole components of the bridging ligand have low lying unoccupied molecular orbitals and may favour a chemical mechanism of electron transfer, but the intervening aliphatic and peptide chains are unlikely to form transient radicals and may participate by resonance. The optical absorption of the blue copper assigned to a cysteine $S^- \rightarrow Cu^{2+}$ charge-transfer transition supports the suggested electron pathway.

The putative electron path branches at the Cα atom of Cys-A508. Model compounds have shown inequivalence and faster transfer in the N-C direction of amide-linking groups [32]. This may apply also to the blue oxidases and cause preferred transfer to A507.

The redox potential differences between the type-1 copper and the type-3 copper are -40 mV in ascorbate oxidase. Unfortunately there are no direct measurements of the intra-

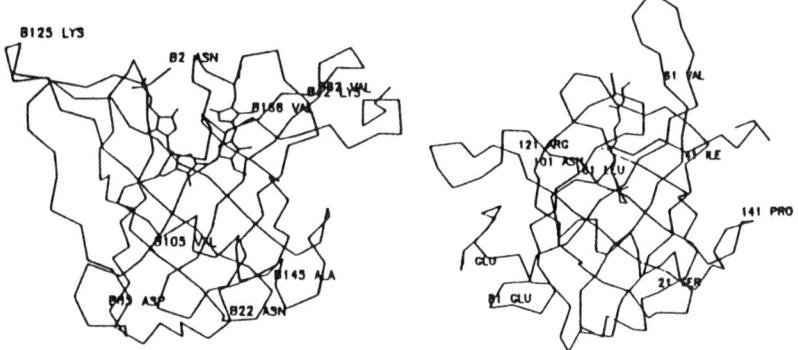

Fig. 12. *Comparison of the polypeptide chain folds of BBP and RBP with bound cofactors*

molecular electron transfer rates available. The turnover number serves as a lower limit and is 7.5×10^3 s^{-1} in ascorbate oxidase [127, 128] indicating a quite rapid transfer despite the long distance and small driving force. The electron pathway is intramolecular and removed from bulk water.

The characteristic distribution of redox centres as mononuclear and trinuclear sites in the blue oxidases may be found also in the most complex oxidase, cytochrome oxidase (see the hypothetical model of Holm et al. [129] and in the water-splitting manganese-protein complex of PSII, which carries out the reverse reaction of the oxidases. For its Mn$_4$ cofactor, either two binuclear or a tetranuclear metal centre is favoured [130], but mononuclear and trinuclear arrangements can not be excluded.

3.3. THE PROTEIN AS A MEDIUM

The boundary between the action of the protein as ligand and as medium is fluid. The protein medium is microscopically extremely complex in structure, polarity and polarisability, which may influence energy and electron transfer. There is no obvious common structural scheme in the protein systems discussed except a high proportion of aromatic residues (particularly tryptophans) bordering the electron-transfer paths in the reaction centre and ascorbate oxidase, and their wide separation from bulk water by internal location within the protein and the hydrocarbon bilayer (in the reaction centre). These effects have been mentioned in 1.3. and 3.2.2.2.

4. STRUCTURAL RELATIONSHIPS AND INTERNAL REPEATS

All four protein systems mentioned show internal repetition of structural motifs or similarities to other proteins of known folding patterns. This is quite a common phenomenon and not confined to energy- and electron-transfer proteins. It is also not uncommon that these relationships often remained undetected on the basis of the amino acid sequences, ultimately a reflection of our ignorance about the sequence/structure relationships. An analysis of structural relationships will shed light on evolution and function of the protein systems and is thus appropriate here.

4.1. RETINOL AND BILIN-BINDING PROTEINS (BBP)

The simplest case is shown in Fig. 12, where BBP [43] is compared to retinol-binding protein (RBP) [131]. The structural similarity is obvious for the bottom of the β-barrel structure, while the upper part which is involved in binding of the pigments, biliverdin and retinol, differs greatly. The molecule is apparently divided in framework and hypervariable segments which determine binding specificity in analogy to the immunoglobulins [132]. The relationship suggests carrier functions for BBP as for RBP, although it serves also for pigmentation in butterflies.

4.2. PHYCOCYANIN

The phycocyanins consist of two polypeptide chains, α and β, which are clearly related in structure (Fig. 13) and originate probably from a common precursor.

The α subunit is shorter in the G-H turn and lacks the s chromophore, B155 (see 3.2.1.3.). The loss or acquisition of chromophores during evolution may be less important than differentiation of the α and β subunits, which occupy non-equivalent positions in the $(\alpha\beta)_3$ trimer, so that the homologous chromophores A84 and B84 are not equivalent, with B84 lying on the inner wall of the disk. In addition the α and β subunits play very different roles in the formation of the $(\alpha\beta)_6$ hexamer, as had been shown in Fig. 8. Symmetrical precursor hexamers might have existed and could have formed stacks, but would lack the differentiation of the chromophores, in particular the inequivalence and close interaction of A84 and B84 in the trimer. Functional improvement has probably driven divergent evolution of the α and β subunits.

A most surprising similarity was discovered between the phycocyanin subunits and the globins shown in Fig. 13. The globular helical assemblies, A–H, show similar topology. The N-terminal X, Y α-helices forming a U-shaped extension in phycocyanin is essential for formation of the $\alpha\beta$ substructure. The amino acid sequence comparison after structural superposition reveals some similarity suggesting divergent evolution of phycobiliproteins and globins [46], however, what function a precursor of light-harvesting and oxygen-binding proteins might have had remains mysterious.

4.3. REACTION CENTRE

The reaction centre lacks symmetry across the membrane plane, not surprising for a complex, which catalyses a vectorial process across the membrane. However, there is quasi-symmetry relating the L and M subunits, and the pigment system. Structural and amino acid sequence similarities between the L and M subunits suggest a common evolutionary origin. This relationship is extended to the PSII components D1 and D2 on the basis of sequence similarity and conservation of

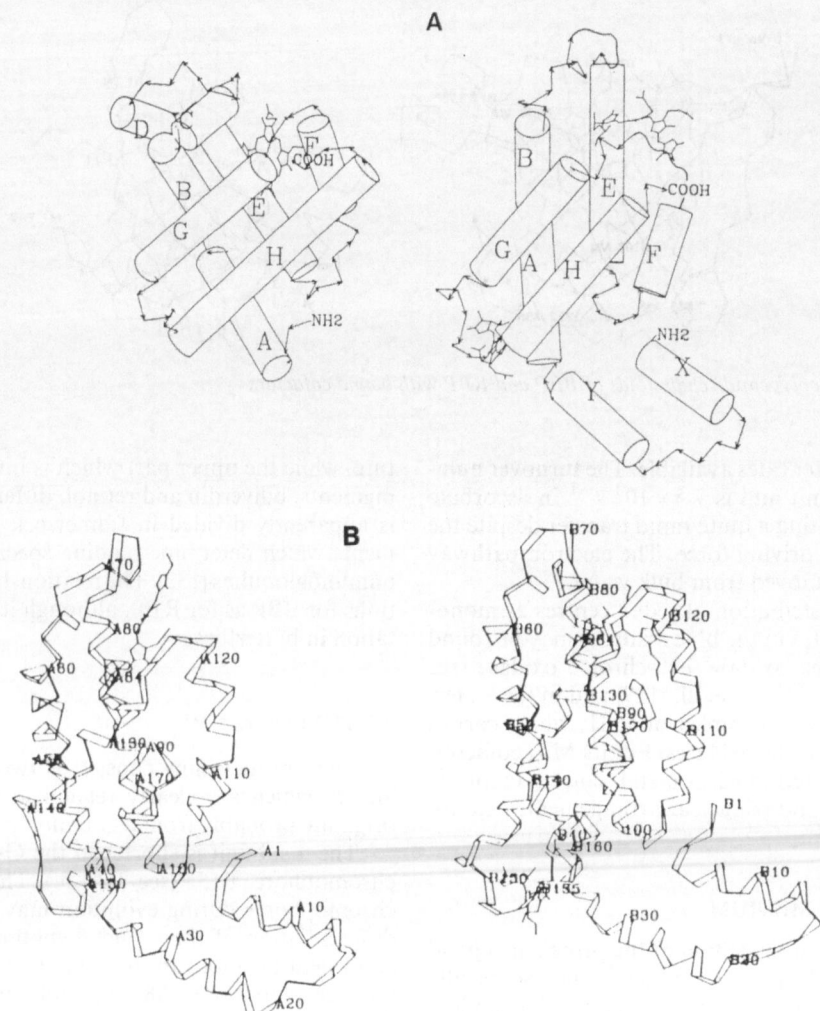

Fig. 13. *Polypeptide chain folds of the α and β subunits of phycocyanin [46] (b, left and right) and comparison of the arrangements of α-helices in myoglobin and phycocyanin (a, left and right)*

residues involved in cofactor binding (for reviews, see [133, 134]). The putative precursor was a symmetrical dimer with identical electron-transfer pathways. The interaction with the H subunit introduces asymmetry, particularly noteworthy at the N-terminal transmembrane α-helix of the H subunit (H), which is close to the E transmembrane α-helix of the M subunit and the L branch of the pigment system and Q_A (Fig. 14). The improvement of the interaction with the H subunit, which appears to play a role in the electron transfer from Q_A to Q_B, and in protonation of Q_B, might have driven divergent evolution of the L and M subunits at the expense of the inactivation of the M pigment branch. However, electron transfer from BC_P to Q_A is extremely fast and is not rate-limiting for the overall reaction. The evolutionary conservation of the M branch of pigments may be of functional significance in light harvesting and electron transfer as a pendant group. There are also structural reasons, as its deletion would generate void space.

The cytochrome subunit adds to the asymmetry of the L-M complex and shows itself an internal duplication [38]. All four heme groups are associated with a helix-turn-helix motif, but the turns are short for haem groups 1 and 3, and long for 2 and 4.

4.4. BLUE OXIDASES

Gene multiplication and divergent evolution is most evident in the blue oxidase, ascorbate oxidase. Fig. 15 shows the polypeptide chain of 553 amino acid residues folded into three closely associated domains of similar topology [122]. Although nearly twice as large, they resemble the simple, small copper protein, plastocyanin [124] (Fig. 16). The blue oxidase domains I and III enclose the trinuclear copper cluster in a quasi-symmetrical fashion, but only domain III contains the type-1 copper, the electron donor to the trinuclear site. A potential electron transfer pathway in domain I is not realized, reminiscent of the M branch of pigments in the reaction centre. Similar to the H subunit in the reaction centre, the linking domain II introduces asymmetry in ascorbate oxidase, which might have driven evolutionary divergence of domains I and III.

The proteins plastocyanin, ascorbate oxidase, laccase and ceruloplasmin are members of a family of copper proteins as indicated by structural relations and sequence similarity [122, 135–137]. They provide a record from which an evolutionary tree may be proposed (Fig. 17). The simplest molecule is plastocyanin containing only a type-1 copper. A dimer of

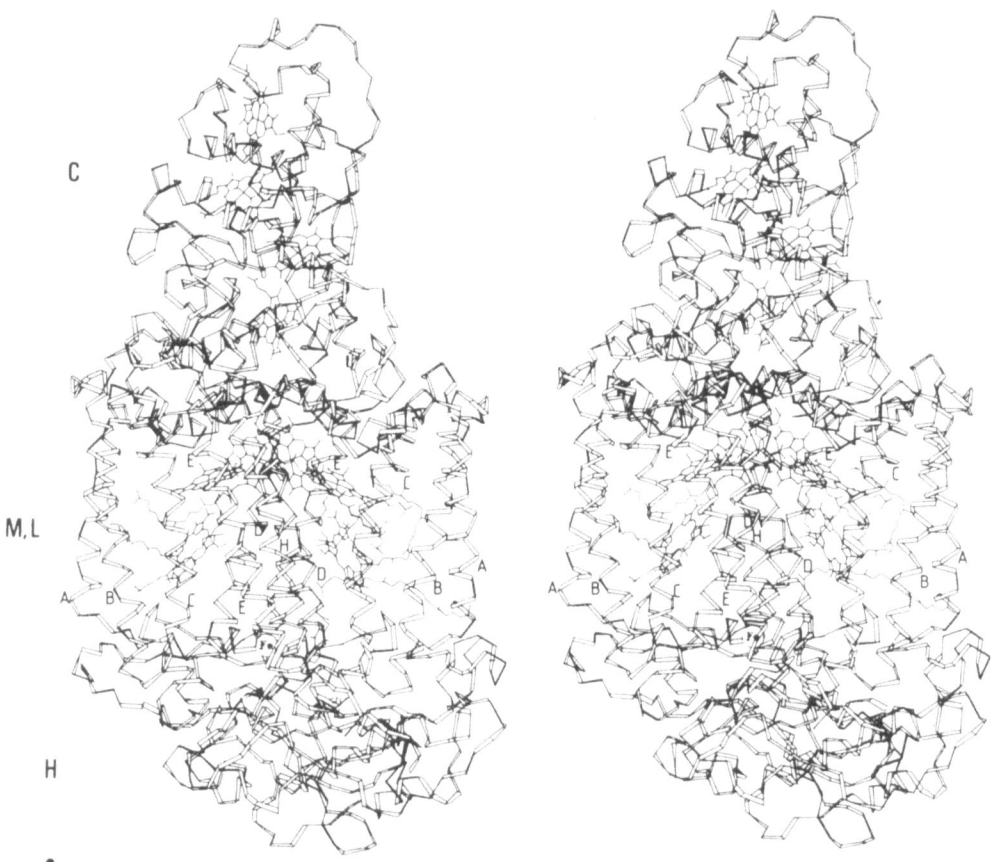

Fig. 14. *Stereo drawing of the polypeptide fold of the reaction centre subunits and the cofactor system.* The membrane-spanning α-helices of the L and M subunits (A,B,C,D,E in sequential and A,B,C,E,D in spatial order) and the H subunit (H) are labelled [38]

plastocyanin-like molecules could provide the 2 × 4 histidyl ligands for the trinuclear copper cluster, representing a symmetrical oxidase. From this hypothetical precursor the modern blue oxidases and ceruloplasmin might have evolved following different paths of gene (domain) insertion and loss or aquisition of coppers. In both the arrangement of the N- and C-terminal domains, which contain the functional copper cluster has been preserved. Recombinant DNA technology has the tools to reconstruct the hypothetical precursor oxidase. This is under investigation.

5. IMPLICATIONS FROM THE STRUCTURE OF THE REACTION CENTRE FOR MEMBRANE PROTEINS IN GENERAL

The structures of water soluble proteins show a seemingly unlimited diversity, although they are built from only a few defined secondary structural elements as helices, β-sheets and turns, and despite their construction from domains and recurring structural motifs. The proteins discussed provide ample evidence. That there seems to be a limited set of basic folds may be related to the evolution of proteins from a basis set of structures and/or to constraints by protein stability and rates of folding. These basic folding motifs do not represent rigid building blocks, however, but adapt to sequence changes and respond to the environment and association with other structural elements. Adaptability and plasticity (which is not to be confused with flexibility) is related to the fact that the entire protein and solvent system must attain the global energy minimum, not its individual components. Water is a good

hydrogen-bond donor and acceptor and is thus able to saturate polar-surface-exposed peptide groups nearly as well as intraprotein hydrogen bonds do (except for entropic effects).

Membrane proteins face the inert hydrocarbon part of the phospholipid bilayer and must satisfy their hydrogen bonds intramolecularly. Only two secondary structures form closed hydrogen bonding arrangements of their main chains, which satisfy this condition, namely the helix and the β-barrel. For assemblies of α-helices, packing rules have been derived which predict certain preferred angles between the helix axes, although with a broad distribution. Similarly, the arrangement of strands in β-sheets and β-barrels follows defined rules [138].

5.1. STRUCTURE OF THE MEMBRANE-ASSOCIATED PARTS OF THE REACTION CENTRE

The structure of the reaction centre may support some conclusions about membrane proteins in general, of which the reaction centre structure was the first to be determined at atomic resolution after the low-resolution structure of bacteriorhodopsin which has some common features [139]. The reaction centre has 11 transmembrane α-helices, which consist of 26 residues (H subunit) or 24 – 30 residues (L and M subunits), appropriate lengths to span the membrane. The amino acid sequences of these segments are devoid of charged residues (Fig. 18). Few charged residues occur close to the ends of the α-helices. Glycine residues initiate and terminate almost all α-helical segments, both the transmembrane and

298

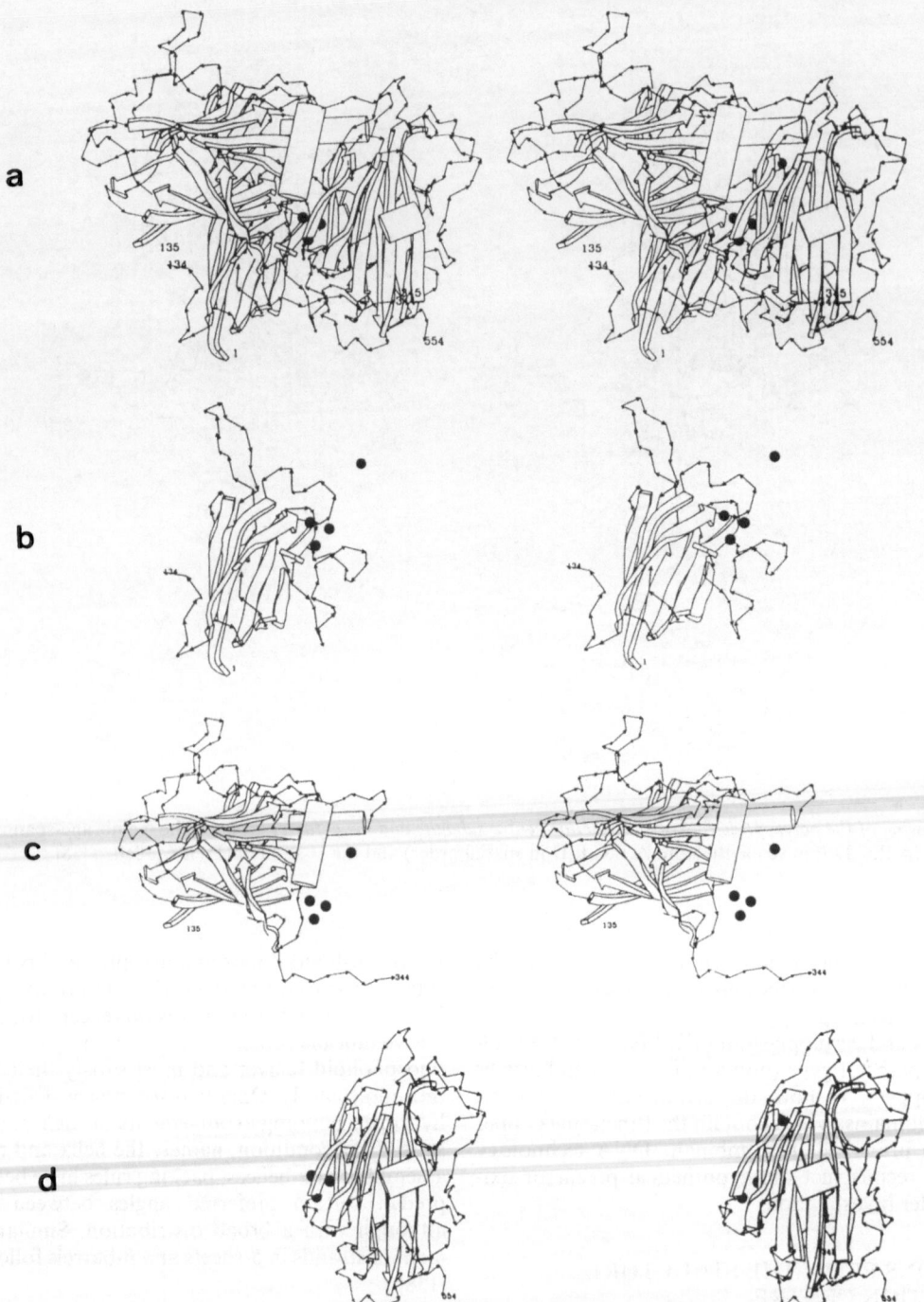

Fig. 15. *Stereo drawing of the polypeptide chain folds of ascorbate oxidase and explosion view of its three domains from top to bottom [122].* β-Strands are indicated as arrows and α-helices as cylinders (produced by the plot program of Lesk and Hardman [172]

the connecting α-helices. It is well known from soluble proteins that glycine residues are abundant in turns and often associated with flexible regions of proteins [12]. They may be important for the insertion into the membrane by allowing rearrangements. The angles between the axes of the contiguous α-helices of the L and M complex are inclined by 20–30°, a preferred angular range for the packing of the α-helices in soluble proteins. They have features in common with buried α-helices in large globular proteins, which are also characterized by the absence of charged residues and the preference of glycines and prolines at the termini [140, 141].

In addition the D and E α-helices of the L and M subunits (Fig. 18) find counterparts in soluble proteins. They are associated around the local diad axis and form the centre of the L-M module, which binds the iron and the BC_P. The four D and E α-helices of the L and M subunits are arranged as a bundle tied together by the iron ion and splay out towards the cytoplasmic side to accommodate the large special pair. This motif is quite common in soluble electron-transfer proteins [142]. I will resume this discussion later and suggest appropriate substructures of soluble proteins as models for pore forming membrane proteins.

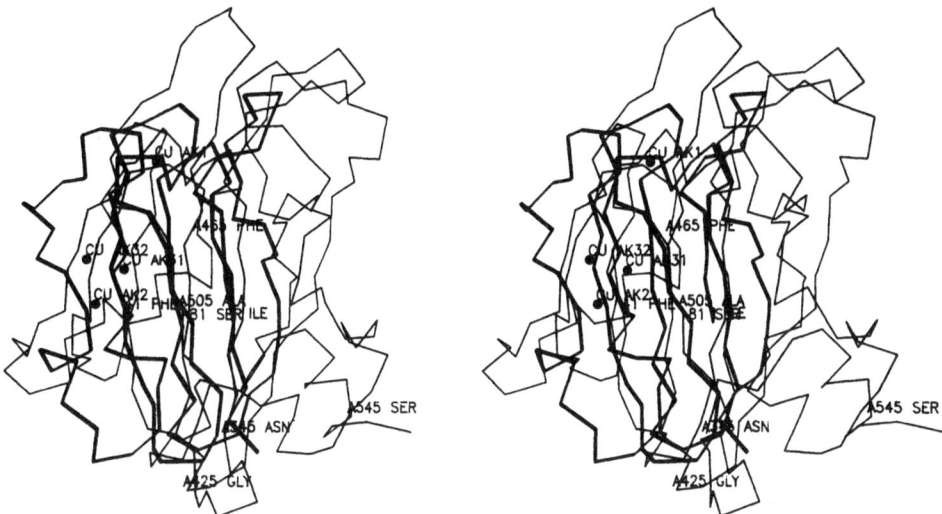

Fig. 16. *Stereo drawing and superposition of domain III of ascorbate oxidase (thin lines) and phycocyanin (thick lines).* The trinuclear copper site in ascorbate oxidase is buried between domain I (not shown) and domain III

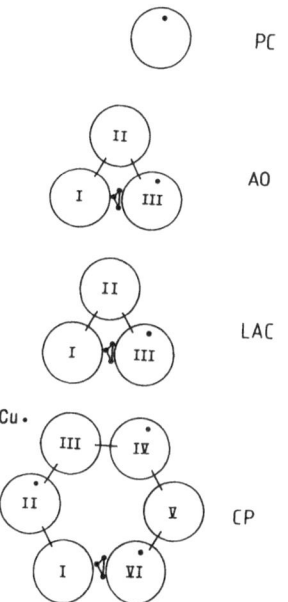

Fig. 17. *Homologous domains in plastocyanin (PC), ascorbate oxidase (AO), laccase (LAC) and ceruloplasmin (CP).* The mononuclear and trinuclear copper sites are indicated

5.2. MEMBRANE INSERTION

The structure of the reaction centre is similarly important for our views of the mechanism of integration of membrane proteins into the phospholipid bilayer. The reaction centre is composed of components which are quite differently arranged with respect to the membrane. The C subunit is located on the periplasmic side. The H subunit is folded into two parts: a globular part located on the cytoplasmic side and a transmembrane α-helix. The L and M subunits are incorporated into the phospholipid bilayer. Consequently, subunit C has to be completely translocated across the membrane from its intracellular site of synthesis. In the H, L and M subunits, the transmembrane α-helices are embedded in the bilayer. Only the N-terminal segment of H, and the C-termini and connecting segments of the α-helices located at the periplasmic side of L and M (A-B, C-D), require transfer.

It is interesting to note that only the cytochrome gene possesses a prokaryotic signal sequence, as indicated by the sequence of the gene [87]. Transfer of the large hydrophilic C subunit may require a complex translocation system and a signal sequence, while H, L and M may spontaneously insert into the bilayer due to the affinity of the contiguous hydrophobic segments with the phospholipids (for a review of this and related problems see [143]). A 'simple' dissolution still requires transfer across the membrane of those charged residues which are located at the periplasmic side [38, 101]. The increasingly favourable protein-lipid interaction which develops with insertion may assist in this process. M and L have considerably more charged residues at the cytoplasmic side (41) than at the periplasmic side (24), providing a lower activation energy barrier for correct insertion. The net charge distribution of the L-M complex is asymmetric with six positive charges at the cytoplasmic side and eight negative charges at the periplasmic side. As the intracellular membrane potential is negative, the observed orientation of the L-M complex is energetically favoured (Fig. 18).

The H subunit has a very polar amino acid sequence at the C-terminus of the transmembrane α-helix with a stretch of seven consecutive charged residues (H33 – H39) [38, 85] which may efficiently stop membrane insertion. Similarly there are 3 – 11 charged residues in each of the connecting segments of the α-helices at the cytoplasmic side of the L and M subunits, which might stop the transfer of α-helices or α-helical pairs [144]. As an alternative to sequential insertion the L and M subunits may be inserted into the membrane as assembled protein pigment complexes, because they adhere tightly by protein-protein and protein-cofactor interactions.

5.3. MODELS OF PORE-FORMING PROTEINS

It is not obvious whether the structural principles observed in the reaction centre apply also to 'pore'- or 'channel'-forming α-helical proteins. These could, in principle, elaborate quite complex structures within the aqueous channel [145], but available evidence at low resolution for gap-junction proteins [146] indicates, in this specific case, a simple hexameric arrangement of membrane-spanning amphiphilic α-helices, whose polar sides face the aqueous channel.

Guided by the observation that rules for structure and packing of α-helices derived for soluble proteins apply also to

300

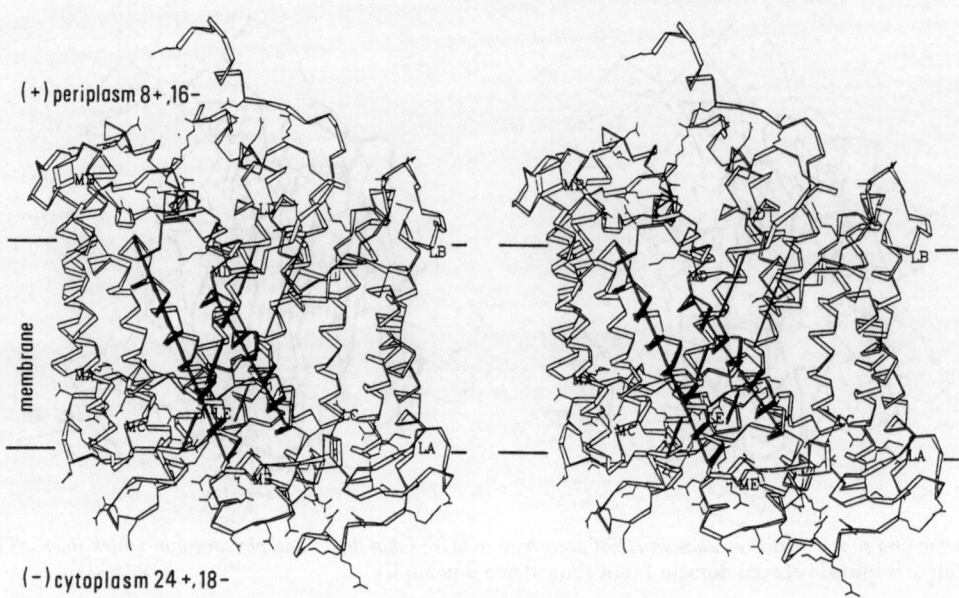

Fig. 18. *Stereo drawing of the polypeptide chains of the L and M subunits of the reaction centre in ribbon representation.* The N-terminal residues of the membrane-spanning α-helices are labelled (including the prefix M and L) and the tetrahelical motif of the D and E α-helices is marked by shading and lines. The side-chains of charged residues are drawn. Asp, Glu and carboxy termini as negatively charged, and Lys, Arg and amino termini are counted as positively charged and added for the cytoplasmic and periplasmic sides [5, 38]

the reaction centre, we may derive models for membrane-pore-forming proteins from appropriate soluble protein substructures. The pentahelical pore seen at high resolution in the icosaedral multi-enzyme complex riboflavin synthase seems to be a suitable model [147] (Fig. 19). Five amphiphilic α-helices of 23 residues each are nearly perpendicular to the capsid surface. The coiled coil of α-helices has a right-handed twist and forms a pore for the putative import of substrates and export of products. They pack with their apolar sides against the central four-stranded β-sheet of the protein, which mimics the hydrocarbon part of a phospholipid bilayer and project charged residues into the aqueous channel.

Similar modelling of membrane protein structures may be extended to another class of membrane proteins which have β-structures spanning the outer membrane, the bacterial porins [148]. In soluble proteins β-barrels observed have 4−8 or more strands. The lower limit is determined by the distortion of regular hydrogen bonds. An upper limit may be given by the possible sizes of stable protein domains. A four-stranded β-barrel with four parallel strands, duplicated head to head with symmetry D4, is seen in the ovomucoid octamer [149]. The β-strands lean against the hydrophobic core of the molecule and project their (short) polar residues into the channel (which is extremely narrow).

6. SOME THOUGHTS ON THE FUTURE OF PROTEIN CRYSTALLOGRAPHY

Thirty years after the elucidation of the first protein crystal structures by Perutz and Kendrew, and after steady development, protein crystallography is undergoing a revolution. Recent technical and methodical developments enable us to analyse large functional protein complexes like the reaction centre [7, 38], large virus structures (to mention only [150−152]), protein-DNA complexes (to mention only [153]) and multi-enzyme complexes like riboflavin synthase [147].

The significance of these studies for understanding biological functions is obvious and has excited the interest of the scientific community in general.

In addition, it was recognized that detailed structural information is a prerequisite for rational design of drugs and proteins. For an illustration, I chose human leucocyte elastase which is an important pathogenic agent. On the basis of its three-dimensional structure (Fig. 20) [154] and the criteria of optimal stereochemical fit potent, inhibitors are now being synthesized or natural inhibitors modified by use of recombinant DNA technology in many scientific and commercial institutions. Other, equally important proteins are similarly studied. This field especially benefits from the facile molecular modelling software (e.g. FRODO [155]) and a standard and depository of structural data, the Protein Data Bank [156].

Success and the new technical and methodological developments spur protein crystallography's progress. These new developments are indeed remarkable. Area detectors for automatic recording of diffraction intensities have been designed. Brilliant X-ray sources (synchrotrons) are available for very fast measurements and now permit use of very small crystals or radiation-sensitive materials. Their polychromatic radiation is used to obtain diffraction data sets within milliseconds by Laue techniques [157] and their tunability allows the optimal use of anomalous dispersion effects [158, 159].

Refinement methods including crystallographic and conformational energy terms provide improved protein models. Methods which allow the analysis of large protein complexes with internal-symmetry-averaging procedures were developed [160] leading from blurred to remarkably clear pictures. *A priori* information of a relationship to known proteins can be used to great advantage, as it is possible to solve an unknown crystal structure using a known model of a variant structure by a method discovered and named the *Faltmolekül* method by my teacher, W. Hoppe. It has become a very powerful tool in protein crystallography.

With this last paragraph I wish to pay tribute to W. Hoppe who in 1957 laid the foundation to Patterson search methods by discovering that the Patterson function (the Fourier transform of the diffraction intensities) of molecular crystals can be decomposed into sums of intramolecular and intermolecular

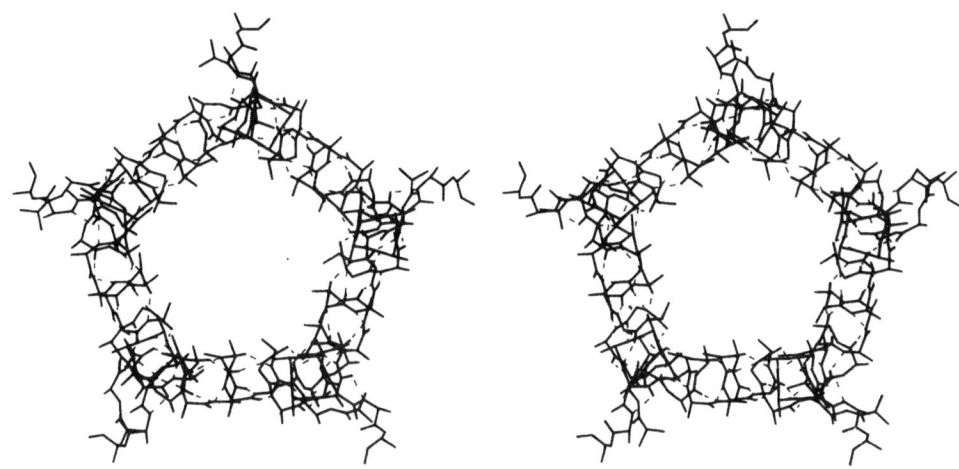

Fig. 19. *Pentahelical pore in heavy riboflavin synthase in stereo [147]*

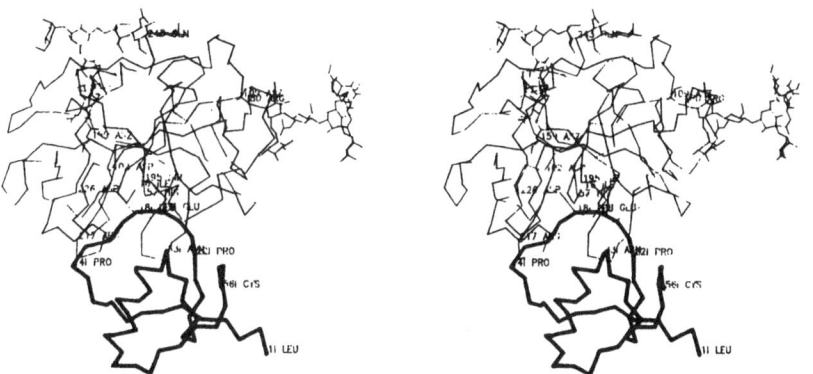

Fig. 20. *Stereo drawing of the complex between human leucocyte elastase (thin lines) and turkey ovomucoid inhibitor (thick lines) [154]*

vector sets [161] from which orientation and translation of the molecules can be derived when their approximate structure is known (Fig. 21). Hoppe's method was profoundly elaborated, computerized and reformulated [161a, 162, 163]. It provided a short-cut to the crystal structure of the reaction centre of *Rhodobacter sphaeroides* which was solved on the basis of the molecular structure of the reaction centre of *Rps. viridis* and subsequently refined [6, 7]. The molecular architectures are very similar although the *Rb. sphaeroides* reaction centre lacks the permanently bound cytochrome. The structure solution was independently confirmed using similar methods by [5]. With the *Faltmolekül* method the orientation and location of a molecule in a crystal cell can be determined. The detailed molecular structure and its deviations from the parent model have to be worked out by crystallographic refinement, to which W. Steigemann and J. Deisenhofer (in his thesis work) laid a foundation in my laboratory [164, 165].

Recently, NMR techniques (nuclear magnetic resonance) have demonstrated their capability to determine three-dimensional structure of small proteins in solution. In one case a detailed comparison between crystal and solution structure has shown very good correspondence [166, 167] but future developments will be needed to extend the power of the method to larger protein structures.

Protein crystallography is the only tool to unravel in detail the architecture of the large protein complexes described here and will continue in the foreseeable future to be the only experimental method that provides atomic resolution data on atom-atom and molecule-molecule interactions. It is the

successful analytical method E. Fischer addressed in his 9th Faraday Lecture by pointing out that 'the precise nature of the assimilation process will only be accomplished when biological research, aided by improved analytical methods, has succeeded in following the changes which take place in the actual chlorophyll granules' [168]. Yet an ultimate goal for which we all struggle is the solution of the folding problem. The growing number of known protein structures and the design of single-residue variants by recombinant DNA technology, and their analysis by protein crystallography has brought us nearer to this goal. We are able to study contributions of individual residues to rates of folding, structure, stability and function. Also, theoretical analysis of protein structures has progressed (to mention only Levitt and Sharon [169]) but a clue to the code relating sequence and structure is not in sight [170]. Like Carl von Linne, who 250 years ago created a system of plants on the basis of morphology (Genera plantarum, Leiden 1737), we classify proteins on their shapes and structures. Whether this may lead to a solution of the folding problem is unclear, but it is certain that the end of protein crystallography will only come through protein crystallography.

J. Deisenhofer's and my interest in structural studies of the photosynthetic reaction centre of *Rps. viridis* was raised by the establishment of D. Oesterhelt's department in Martinsried in 1980; he brought with him H. Michel, with whom a fruitful collaboration on the analysis of the crystal structure of this large protein complex began. Later, other members of my group, O. Epp and K. Miki, became involved. We had been studying enzymes, proteases and their

302

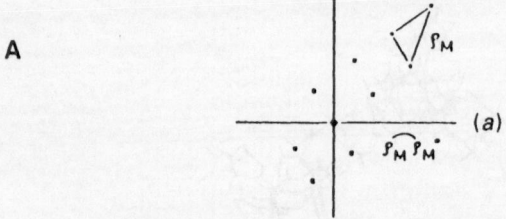

A

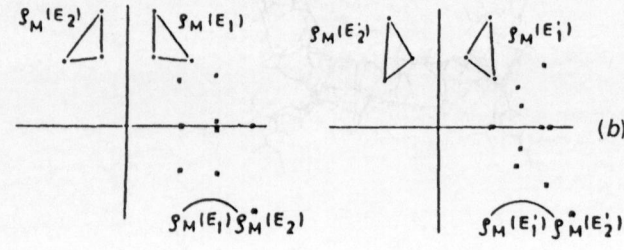

(a)

(b)

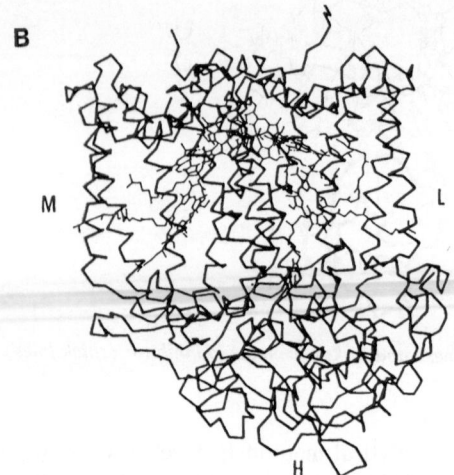

B

Fig. 21. Faltmolekül *construct (A) and the subunits and cofactors of the reaction centre (B)*. (A) $\varrho_M\varrho_{M*}$ (a) and $\varrho_M(E1)\varrho_{M*}(E2)$ (b) are the intramolecular and intermolecular vector sets of a triangular structure, ϱ_M, respectively. Their sum represents the Patterson function. The intramolecular vector set can be constructed from the molecular structure, is located at the origin and permits the determination of the orientation. From the intermolecular vector set the translation component relative to the mirror line can be derived. In (b) the intermolecular vector sets corresponding to two different orientations of ϱ_M are shown [173]. (B) Drawing of the main chain of the M, L and H subunits, and the cofactors which served as search model to solve the phase problem for the *Rb. sphaeroides* reactive centre crystal structure. For the calculation all homologous main chain and side chain atoms were included [6]

natural inhibitors, immunoglobulins and had developed methods to improve data collection, electron-density map interpretation and crystallographic refinement. The tools were available to attack a problem which was and still is the largest asymmetric protein analysed at atomic resolution today.

The 'heureka' moment of protein crystallography is at the very end when one sees for the first time a new macromolecule with the eyes of a discoverer of unknown territories. To reach this moment much sometimes tedious work has to be done with the ever present possibility of failure. I am deeply grateful to my collaborators, those who are with me and those who had left, for their dedicated and patient work over many years. I mention by name those involved in the studies of the light-harvesting cyanobacterial proteins and

the blue oxidases: W. Bode, M. Duerring, R. Ladenstein, A. Messerschmidt and T. Schirmer. These projects were collaborative undertakings with biochemists in Switzerland (H. Zuber and W. Sidler), USA (M. L. Hackert) and Italy (M. Bolognesi, A. Marchesini and A. Finazzi-Agro).

Scientific work needs a stimulating environment which was provided at the *Max-Planck-Institut für Biochemie* and it needs steady financial support, which has been provided by the *Max-Planck-Gesellschaft* and the *Deutsche Forschungsgemeinschaft*.

I thank R. Engh, S. Knof, R. Ladenstein, M. Duerring and E. Meyer for their helpful comments on this manuscript.

This lecture is published with the kind permission of the Nobel foundation.

REFERENCES

1. Boltzmann, L. (1886) *Der zweite Hauptsatz der mechanischen Wärmetheorie* (essay in Populäre Schriften) L. Boltzmann-Gesamtausgabe Bd. 7 (1919) pp. 25–46, Akad. Druck- und Verlagsanstalt Vieweg, Wiesbaden. [English translation by Arnon, D. L. (1961) in *Light and life*] (McElroy, W. D. & Glass, B. eds) pp. 489–569, The Johns Hopkins Press, Baltimore.
2. Calvin, M. & Bassham, J. A. (1962) in *The photosynthesis of carbon compounds*, pp. 1–127, Benjamin, New York.
3. Deisenhofer, J., Michel, H. & Huber, R. (1985a) *Trends Biochem. Sci. 10*, 243–248.
4. Deisenhofer, J., Huber, R. & Michel, H. (1986) *Nachrichtenbl. Chem. Tech. Lab. 34*, 416–422.
5. Deisenhofer, J., Huber, R. & Michel, H. (1989) in *Prediction of protein structure and the principles of protein conformation* (Fasman, G. D., ed.) Plenum Publishing Corp., New York, 99–116.
6. Allen, J. P., Feher, G., Yeates, T. O., Rees, D. C., Deisenhofer, J., Michel, H. & Huber, R. (1986) *Proc. Natl Acad. Sci. USA 83*, 8589–8593.
7. Allen, J. P., Feher, G., Yeates, T. O., Kemiya, H. & Rees, D. C. (1987) *Proc. Natl Acad. Sci. USA 84*, 6162–6166.
8. Chang, C.-H., Tiede, D., Tang, J., Smith, U., Norris, J. & Schiffer, M. (1986) *FEBS Lett. 205*, 82–86.
9. Tronrud, D. E., Schmid, M. F. & Matthews, B. W. (1986) *J. Mol. Biol. 188*, 443–454.
10. Pierrot, M., Haser, R., Frey, M., Payan, F. & Astier, J. P. (1982) *J. Biol. Chem. 257*, 14341–14348.
11. Higuchi, Y., Kusunoki, M., Matsuura, Y., Yasuoka, N. & Kakudo, M. (1984) *J. Mol. Biol. 172*, 109–139.
12. Bennett, W. S. & Huber, R. (1984) *CRC Crit. Rev. Biochem. 15*, 291–384.
13. Huber, R. (1988) *Angew. Chem. Int. Ed. Engl. 27*, 79–88.
14. Karplus, M. & McCammon, J. A. (1981) *CRC Crit. Rev. Biochem. 9*, 293–349.
15. Burkert, U. & Allinger, N. L. (1982) *Molecular mechanics*, American Chemical Society.
16. Förster, T. (1948) *Ann. Physik 2*, 55–75.
17. Förster, T. (1967) in *Comprehensive biochemistry* (Florkin, M. & Stotz, E. H., eds) vol. 22, pp. 61–80, Elsevier, Amsterdam.
18. Kuhn, H. (1970) *J. Chem. Phys. 53*, 101–108.
19. Frommherz, P. & Reinbold, G. (1988) *Thin Solid Films 160*, 347–353.
20. Mayo, S. L., Ellis, W. R., Crutchley, R. J. & Gray, H. B. (1986) *Science 233*, 948–952.
21. Gray, H. B. (1986) *Chem. Soc. Rev. 15*, 17–30.
22. McGoutry, J. L., Peterson-Kennedy, S. E., Ruo, W. Y. & Hoffman, B. M. (1987) *Biochemistry 26*, 8302–8312.
23. Taube, H. & Gould, E. S. (1969) *Acc. Chem. Res. 2*, 321–329.
24. Hopfield, J. J. (1974) *Proc. Natl Acad. Sci. USA 71*, 3640–3644.
25. Cramer, W. A. & Crofts, A. R. (1982) in *Electron and proton transport in photosynthesis: energy conversion by plants and bacteria*, vol. 1, pp. 387, Academic Press Inc.
26. Eberson, L. (1982) *Adv. Phys. Org. Chem. 18*, 79–185.

27. Marcus, R. A. & Sutin, N. (1985) *Biochim. Biophys. Acta 811*, 265−322.
28. Mikkelsen, K. V. & Ratner, M. A. (1987) *Chem. Rev. 87*, 113−153.
29. Kebarle, P. & Chowdhury, S. (1987) *Chem. Rev. 87*, 513−534.
30. McLendon, G. (1988) *Acc. Chem. Res. 21*, 160−167.
31. Gust, D., Moore, T. A., Lidell, P. A., Nemeth, G. A., Makings, L. R., Moore, A. L., Barrett, D., Pessiki, P. J., Bensasson, R. V., Rougée, M., Chachaty, C., De Schryver, F. C., Van der Anweraer, M., Holzwarth, A. R. & Connolly, J. S. (1987) *J. Am. Chem. Soc. 109*, 846−856.
32. Schmidt, J. A., McIntosh, A. R., Weedon, A. C., Bolton, J. R., Connolly, J. S., Hurley, J. K. & Wasielewski, M. R. (1988) *J. Am. Chem. Soc. 110*, 1733−1740.
33. Isied, S., Vassilian, A., Magnuson, R. & Schwarz, H. (1985) *J. Am. Chem. Soc. 107*, 7432−7438.
34. Barltrop, J. A. & Coyle, J. D. (1978) *Principles of photochemistry*, John Wiley and Sons, Chichester.
35. Hains, A. (1975) *Acc. Chem. Res. 8*, 264−272.
36. Barber, J. (1987) *Trends Biochem. Sci. 12*, 321−326.
37. Prince, R. C. (1988) *Trends Biochem. Sci. 13*, 286−288.
38. Deisenhofer, J., Epp, O., Miki, K., Huber, R. & Michel, H. (1985) *Nature 318*, 618−624.
39. Lumry, R. & Eyring, H. (1954) *J. Phys. Chem. 58*, 110−112.
40. Gray, H. B. & Malmström, B. G. (1983) *Comments Inorg. Chem. 2*, 203−209.
41. Scharnagl, C., Köst-Reyes, E., Schneider, S., Köst, H.-P. & Scheer, H. (1983) *Z. Naturforsch. 38c*, 951−959.
42. Huber, R., Schneider, M., Epp, O., Mayr, I., Messerschmidt, A., Pflugrath, J. & Kayser, H. (1987a) *J. Mol. Biol. 195*, 423−434.
43. Huber, R., Schneider, M., Mayr, I., Müller, R., Deutzmann, R., Suter, F., Zuber, H., Falk, H. & Kayser, H. (1987b) *J. Mol. Biol. 198*, 499−513.
44. Schirmer, T., Bode, W., Huber, R., Sidler, W. & Zuber, H. (1985) *J. Mol. Biol. 184*, 257−277.
45. Schirmer, T., Huber, R., Schneider, M., Bode, W., Miller, M. & Hackert, M. L. (1986) *J. Mol. Biol. 188*, 651−676.
46. Schirmer, T., Bode, W. & Huber, R. (1987) *J. Mol. Biol. 196*, 677−695.
47. Deisenhofer, J., Epp, O., Miki, K., Huber, R. & Michel, H. (1984) *J. Mol. Biol. 180*, 385−398.
48. Knapp, E. W., Fischer, S. F., Zinth, W., Sander, M., Kaiser, W., Deisenhofer, J. & Michel, H. (1985) *Proc. Natl Acad. Sci. USA 82*, 8463−8467.
49. Gray, H. B. & Solomon, E. I. (1981) in *Copper proteins* (Spiro, T. G., ed.) pp. 1−39, J. Wiley & Sons, New York.
50. Blair, D. F., Campbell, G. W., Schoonover, J. R., Chan, S. I., Gray, H. B., Malmström, B. G., Pecht, I., Swanson, B. F., Woodneff, W. H., Cho, W. K., English, A. R., Fry, A. H., Lum, V. & Norton, K. A. (1985) *J. Am. Chem. Soc. 10*, 5755−5766.
51. Gantt, E., Lipschultz, C. A. & Zilinskas, B. (1976) *Biochim. Biophys. Acta 430*, 375−388.
52. Mörschel, E., Koller, K.-P., Wehrmeyer, W. & Schneider, H. (1977) *Cytobiologie 16*, 118−129.
53. Bryant, D. A., Guglielmi, G., Tandeau de Marsac, N., Castets, A.-M. & Cohen-Bazire, G. (1979) *Arch. Microbiol. 123*, 113−127.
54. Nies, M. & Wehrmeyer, W. (1981) *Arch. Microbiol. 129*, 374−379.
55. MacColl, R. & Guard-Friar, D. (1987) in *Phycobiliproteins*, pp. 157−173, CRC Press Inc., Boca Raton, Florida.
56. Scheer, H. (1982) in *Light reaction path of photosynthesis* (Fong, F. K., ed.) pp. 7−45, Springer, Berlin.
57. Cohen-Bazire, G. & Bryant, D. A. (1982) in *The biology of cyanobacteria* (Carr, N. G. & Whitton, B., eds) pp. 143−189, Blackwell, London.
58. Glazer, A. N. (1985) *Annu. Rev. Biophys. Biophys. Chem. 19P*, 47−77.
59. Zilinskas, B. A. & Greenwald, L. S. (1986) *Photosynth. Res. 10*, 7−35.
60. Zuber, H. (1985) *Photochem. Photobiol. 42*, 821−844.
61. Zuber, H. (1986) *Trends Biochem. Sci. 11*, 414−419.
62. Duerring, M., Huber, R. & Bode, W. (1988) *FEBS Lett. 236*, 167−170.
63. Duerring, M., Bode, W., Huber, R., Ruembeli, R. & Zuber, H. (1989) *J. Mol. Biol.*, in the press.
64. Frank, G., Sidler, W., Widmer, H. & Zuber, H. (1978) *Hoppe-Seyler's Z. Physiol. Chem. 359*, 1491−1507.
65. Schirmer, T. & Vincent, M. G. (1987) *Biochem. Biophys. Acta 893*, 379−385.
66. Glazer, A. N., Fang, S. & Brown, D. M. (1973) *J. Biol. Chem. 16*, 5679−5685.
67. Mimuro, M., Flüglistaller, P., Rümbeli, R. & Zuber, H. (1986) *Biochim. Biophys. Acta 848*, 155−166.
68. Lundell, D. J., Williams, R. C. & Glazer, A. N. (1981) *J. Biol. Chem. 256*, 3580−3592.
69. Teale, F. W. J. & Dale, R. E. (1970) *Biochem. J. 116*, 161−169.
70. Zickendraht-Wendelstadt, B., Friedrich, J. & Rüdiger, W. (1980) *Photochem. Photobiol. 31*, 367−376.
71. Hefferle, P., Nies, M., Wehrmeyer, W. & Schneider, S. (1983) *Photobiochem. Photobiophys. 5*, 41−51.
72. Gillbro, T., Sandström, Å, Sundström, V., Wendler, J. & Holzwarth, A. R. (1985) *Biochim. Biophys. Acta 808*, 52−65.
73. Siebzehnrübl, S., Fischer, R. & Scheer, H. (1987) *Z. Naturforsch. 42c*, 258−262.
74. Porter, G., Tredwell, C. J., Searle, G. F. W. & Barber, J. (1978) *Biochim. Biophys. Acta 501*, 232−245.
75. Searle, G. F. W., Barber, J., Porter, G. & Tredwell, C. J. (1978) *Biochem. Biophys. Acta 501*, 246−256.
76. Wendler, J., Holzwarth, A. R. & Wehrmeyer, W. (1984) *Biochem. Biophys. Acta 765*, 58−67.
77. Yamazaki, I., Mimuro, M., Murao, T., Yamazaki, T., Yoshihara, K. & Fujita, Y. (1984) *Photochem. Photobiol. 39*, 233−240.
78. Holzwarth, A. R. (1986) *Photochem. Photobiol. 43*, 707−725.
79. Grabowski, J. & Gantt, E. (1978) *Photochem. Photobiol. 28*, 39−45.
80. Switalski, S. C. & Sauer, J. (1984) *Photochem. Photobiol. 40*, 423−427.
81. Holzwarth, A. R. (1985) in *Antennas and reaction centers of photosynthetic bacteria* (Michel-Beyerle, M. E., ed.) pp. 45−52, Springer-Verlag, Berlin, Heidelberg, New York.
82. Sauer, K., Scheer, H. & Sauer, P. (1987) *Photochem. Photobiol. 46*, 427−440.
83. Duerring, M. (1989) Thesis, Technical University, München.
83a. Parson, W. W. (1978) in *The photosynthetic bacteria* (Clayton, R. K. & Sistrom, W. R., eds) pp. 317−322, Plenum Press, New York, London.
84. Stark, N., Kuhlbrandt, W., Wildhaber, I., Wehrli, E. & Mühlethaler, K. (1984) *EMBO J. 3*, 777−783.
85. Michel, H., Weyer, K. A., Gruenberg, H. & Lottspeich, F. (1985) *EMBO J. 4*, 1667−1672.
86. Michel, H., Weyer, K. A., Gruenberg, H., Dunger, I., Oesterhelt, D. & Lottspeich, F. (1986a) *EMBO J. 5*, 1149−1158.
87. Weyer, K. A., Lottspeich, F., Gruenberg, H., Lang, F., Oesterhelt, D. & Michel, H. (1987) *EMBO J. 6*, 2197−2202.
88. Breton, J. (1985) *Biochim. Biophys. Acta 810*, 235−245.
89. Cogdell, R. J. & Crofts, A. R. (1972) *FEBS Lett. 27*, 176−178.
90. Carithers, R. P. & Parson, W. W. (1975) *Biochim. Biophys. Acta 387*, 194−211.
91. Prince, R. C., Leigh, J. S. & Dutton, P. L. (1976) *Biochim. Biophys. Acta 440*, 622−636.
92. Netzel, T. L., Rentzepis, P. M., Tiede, D. M., Prince, R. C. & Dutton, P. L. (1977) *Biochim. Biophys. Acta 460*, 467−479.
93. Bolton, J. R. (1978) in *The photosynthetic bacteria* (Clayton, R. K. & Sistrom, W. R., eds) pp. 419−442, Plenum Press, New York, London.
94. Holten, D., Windsor, M. W., Parson, W. W. & Thornber, J. P. (1978) *Biochim. Biophys. Acta 501*, 112−126.
95. Woodbury, N. W., Becker, M., Middendorf, D. & Parson, W. W. (1985) *Biochemistry 24*, 7516−7521.

304

96. Breton, J., Farkas, D. L. & Parson, W. W. (1985) *Biochim. Biophys. Acta 808*, 421—427.

97. Breton, J., Martin, J.-L., Migus, A., Antonetti, A. & Orszag, A. (1986) *Proc. Natl Acad. Sci. USA 83*, 5121—5125.

98. Kavarnos, G. J. & Turro, N. J. (1986) *Chem. Rev. 86*, 401—449.

99. Fleming, G. R., Marti, J. L. & Breton, J. (1988) *Nature 333*, 190—192.

100. Barber, J. (1988) *Nature 333*, 114.

101. Michel, H., Epp, O. & Deisenhofer, J. (1986b) *EMBO J. 5*, 2445—2451.

102. Pasman, P., Rob, F., Verhoeven, J. W. (1982) *J. Am. Chem. Soc. 104*, 5127—5133.

103. Moore, T. A., Gust, D., Mathis, P., Bialocq, J.-C., Chachaty, C., Bensasson, R. V., Land, E. J., Doizi, D., Liddell, P. A., Lehman, W. R., Nemeth, G. A. & Moore, A. L. (1984) *Nature 307*, 630—632.

104. Arnold, W. & Clayton, R. D. (1960) *Proc. Natl Acad. Sci. USA 46*, 769—776.

105. Parson, W. W. (1974) *Annu. Rev. Microbiol. 28*, 41—59.

106. Parson, W. W. & Cogdell, R. J. (1975) *Biochim. Biophys. Acta 416*, 105—149.

107. Kirmaier, Ch., Holten, D. & Parson, W. W. (1985) *Biochim. Biophys. Acta 810*, 33—48.

108. Kirmaier, Ch., Holton, D. & Parson, W. W. (1985) *Biochim. Biophys. Acta 810*, 49—61.

109. Wraight, C. A. (1982) in *Function of quinones in energy conserving systems* (Trumpower, B. L., ed.) pp. 181—197, Academic Press, London.

110. Kleinfeld, D., Okamura, M. Y. & Feher, G. (1985) *Biochim. Biophys. Acta 809*, 291—310.

111. Debus, R. J., Feher, G. & Okamura, M. Y. (1986) *Biochemistry 25*, 2276—2287.

112. Case, G. D., Parson, W. W. & Thornber, J. P. (1970) *Biochim. Biophys. Acta 223*, 122—128.

113. Dutton, P. L. & Prince, R. C. (1978) in *The photosynthetic bacteria* (Clayton, R. K. & Sistrom, W. R., eds) pp. 525—565, Plenum Press, New York, London.

114. DeVault, D. & Chance, B. (1966) *Biophys. J. 6*, 825—847.

115. Slooten, L. (1972) *Biochim. Biophys. Acta 256*, 452—466.

116. Michel-Beyerle, M. E., Plato, M., Deisenhofer, J., Michel, H., Bixon, M. & Jortner, J. (1988) *Biochem. Biophys. Acta 932*, 52—70.

117. Malmström, B. G. (1978) in *New trends bio-inorganic chemistry*, pp. 59—77, Academic Press, London.

118. Malmström, B. G. (1982) *Annu. Rev. Biochem. 51*, 21—59.

119. Farver, O. & Pecht, I. (1984) in *Copper proteins and copper enzymes* (Lontie, R., ed.) vol. 1, pp. 183—214, CRC Press Inc., Boca Raton, Florida.

119a. Reinhammar, B. (1984) in *Copper proteins and copper enzymes* (Lontie, L., ed.) vol. 3, pp. 1—35, CRC Press Inc., Boca Raton, Florida.

120. Mondovì, B. & Avigliano, L. (1984) in *Copper proteins and copper enzymes* (Lontie, L., ed.) vol. 3, pp. 101—118, CRC Press Inc., Boca Raton, Florida.

121. Rydén, L. (1984) in *Copper proteins and copper enzymes* (Lontie, L. ed.) vol. 3, pp. 34—100, CRC Press Inc., Boca Raton, Florida.

122. Messerschmidt, A., Rossi, A., Ladenstein, R., Huber, R., Bolognesi, M., Gatti, G., Marchesini, A., Petruzzelli, T. & Finazzi-Agrò, A. (1989) *J. Mol. Biol. 206*, 513—529.

123. Malkin, R. & Malmström, B. G. (1970) *Adv. Enzymol. 33*, 177—243.

124. Guss, J. M. & Freeman, H. C. (1983) *J. Mol. Biol. 169*, 521—563.

125. Richardson, J. S., Thomas, K. A., Rubin, B. H. & Richardson, D. C. (1975) *Proc. Natl Acad. Sci. USA 72*, 1349—1353.

126. Gaykema, W. P. J., Hol, W. G. J., Verijken, J. M., Soeter, N. M., Bak, H. J. & Beintema, J. J. (1984) *Nature 309*, 23—29.

127. Dawson, C. R. (1966) in *The biochemistry of copper* (Peisach, J., Aison, P. & Blumberg, W. E., eds) pp. 305—337, Academic Press, New York.

128. Gerwin, B., Burstein, S. R. & Westley, J. (1974) *J. Biol. Chem. 249*, 2005—2008.

129. Holm, L., Saraste, M. & Wikström, M. (1987) *EMBO J. 6*, 2819—2823.

130. Babcock, G. T. (1987) *Oxygen-evolving process in photosynthesis* (Amesz, J., ed.) Elsevier Science Publishers, Amsterdam.

131. Newcomer, M. E., Jones, T. A., Åqvist, J., Sundelin, J., Eriksson, U., Rask, I. & Peterson, P. A. (1984) *EMBO J. 3*, 1451—1454.

132. Huber, R. (1984) *Behring Inst. Mitt. 76*, 1—14.

133. Trebst, A. (1986) *Z. Naturforsch. 41c*, 240—245.

134. Michel, H. & Deisenhofer, J. (1988) *Biochemistry 27*, 1—7.

135. Ohkawa, J., Okada, N., Shinmyo, A. & Takano, M. (1989) *Proc. Natl Acad. Sci. USA 86*, 1239—1243.

136. Germann, U. A., Müller, G., Hunziker, P. E. & Lerch, K. (1988) *J. Biol. Chem. 263*, 885—896.

137. Takahashi, N., Ortel, T. L. & Putnam, F. W. (1984) *Proc. Natl Acad. Sci. 81*, 390—394.

138. Chothia, C. (1984) *Annu. Rev. Biochem. 53*, 537—572.

139. Henderson, R. & Unwin, P. N. T. (1975) *Nature 257*, 28—32.

140. Loebermann, H., Tokuoka, R., Deisenhofer, J. & Huber, R. (1984) *J. Mol. Biol. 177*, 531—556.

141. Remington, S., Wiegand, G. & Huber, R. (1982) *J. Mol. Biol. 158*, 111—152.

142. Weber, P. C. & Salemme, F. R. (1980) *Nature 287*, 82—84.

143. Rapoport, T. A. (1986) *CRC Crit. Rev. Biochem. 20*, 73—137.

144. Engelman, D. M., Steitz, T. A. & Goldman, A. (1986) *Annu. Rev. Biophys. Biophys. Chem. 15*, 321—353.

145. Lodish, H. F. (1988) *Trends Biochem. Sci. 13*, 332—334.

146. Milks, L. C., Kumar, N. M., Houghten, R., Unwin, N., Gilula, N. B. (1988) *EMBO J. 7*, 2967—2975.

147. Ladenstein, R., Schneider, M., Huber, R., Bartunik, H.-D., Wilson, K., Schott, K. & Bacher, A. (1988) *J. Mol. Biol. 203*, 1045—1070.

148. Kleffel, B., Garavito, R. M., Baumeister, N. & Rosenbusch, J. P. (1985) *EMBO J. 4*, 1589—1592.

149. Weber, E., Papamokos, E., Bode, W., Huber, R., Kato, F. & Laskowski, M. (1981) *J. Mol. Biol. 149*, 109—123.

150. Harrison, S. C., Olson, A. J., Schutt, C. E., Winkler, F. K. & Bricogne, G. (1978) *Nature 276*, 368—373.

151. Rossmann, M. G., Arnold, E., Erickson, J. W., Frankenberger, E. A., Griffith, J. P., Hecht, H.-J., Johnson, J. E., Kamer, G., Luo, M., Mosser, A. G., Rueckert, R., Sherry, B. & Vriand, G. (1985) *Nature 317*, 145—153.

152. Hogle, J. M., Chow, M. & Filman, D. J. (1985) *Science 229*, 1358—1365.

153. Ollis, D., Brick, P., Hamlin, R., Xuong, N. G. & Steitz, T. A. (1985) *Nature 313*, 762—766.

154. Bode, W., Wei, A., Huber, R., Meyer, E., Travis, P. & Neumann, S. (1986) *EMBO J. 5*, 2453—2458.

155. Jones, A. T. (1978) *J. Appl. Crystallogr. 11*, 268—272.

156. Bernstein, F. C., Koetzle, T. F., Williams, G. J. B., Meyer, E. F. Jr., Brice, M. D., Rodgers, J. R., Kennard, O., Shimonouchi, T. & Tasumi, M. (1977) *J. Mol. Biol. 112*, 535—542.

157. Hajdu, J., Acharya, K. R., Stuart, D. A., Barford, D. & Johnson, L. (1988) *Trends Biochem. Sci. 13*, 104—109.

158. Hendrickson, W. A., Smith, J. L., Phizackerly, R. P. & Merritt, E. A. (1988) *Proteins 4*, 77—88.

159. Guss, J. M., Merritt, E. A., Phizackerly, R. P., Hedman, B., Murata, M., Hodgson, K. O. & Freeman, H. C. (1988) *Science 241*, 806—811.

160. Bricogne, G. (1976) *Acta Crystallogr. A32*, 832—847.

161. Hoppe, W. (1957) *Acta Crystallogr. 10*, 750—751.

161a. Rossmann, M. G. & Blow, D. M. (1962) *Acta Crystallogr. 15*, 24—31.

162. Huber, R. (1965) *Acta Crystallogr. 19*, 353—356.

163. Crowther, R. A. & Blow, D. M. (1967) *Acta Crystallogr. 23*, 544—548.

164. Huber, R., Kukla, D., Bode, W., Schwager, P., Bartels, K., Deisenhofer, J. & Steigemann, W. (1974) *J. Mol. Biol. 89*, 70—101.

165. Deisenhofer, J. & Steigemann, W. (1975) *Acta Crystallogr. B31*, 238−280.

166. Kline, A. D., Braun, W. & Wüthrich, K. (1986) *J. Mol. Biol. 189*, 377−382.

167. Pflugrath, J. W., Wiegand, W., Huber, R. & Vertesy, L. (1986) *J. Mol. Biol. 189*, 383−386.

168. Fischer, E. (1907) *J. Chem. Soc. 91*, 1749−1765.

169. Levitt, M. & Sharon, R. (1988) *Proc. Natl Acad. Sci. USA 85*, 7557−7561.

170. Jaenicke, R. (1988) in *Mosbacher Kolloquium* (Winnacker, E.-L. & Huber, R., eds) vol. 39, pp. 16−36, Springer-Verlag, Berlin, Heidelberg.

171. Parson, W. W., Scherz, A. & Warshel, A. (1985) in *Antennas and reaction centers of photosynthetic bacteria* (Michel-Beyerle, M. E., ed.) pp. 122−130, Springer-Verlag, Berlin, Heidelberg, New York.

172. Lesk, A. M. & Hardman, K. D. (1982) *Science 216*, 539−540.

173. Huber, R. (1985) in *Molecular replacement. Proceedings of the daresbury study weekend* (Machin, P., ed.) pp. 58−61, Daresbury Laboratory.

Eur. J. Biochem. *187*, 467–473 (1990)
© FEBS 1990

Review

Transforming growth factors and the regulation of cell proliferation

Russette M. LYONS and Harold L. MOSES

Department of Cell Biology, Vanderbilt School of Medicine, Nashville, USA

(Received July 17, 1989) — EJB 89 0877

Mechanisms of growth regulation by polypeptides include endocrine, paracrine and autocrine systems. Factors, such as hormones, which enter the circulatory system and effect target cells or tissues distant from their site of synthesis, are endocrine in nature. It has been postulated that some growth factors may function in an endocrine manner. Insulin-like growth factor, which is synthesized in the liver and enters the blood stream to affect distal tissues, is an example of endocrine action of a specific growth factor [1, 2]. The paracrine mechanism of action suggests that growth factors synthesized by one cell type diffuse through the extracellular spaces to act on neighboring target cells. Finally, autocrine growth regulation may occur when cells are able to both synthesize and respond to a specific growth signal. The autocrine model of growth regulation was originated as a concept to explain the abnormal growth of cancer. This model hypothesizes that affected cells over-produce a stimulatory growth factor to which they respond, thus resulting in sustained, self-stimulated proliferation [3]. Although initially postulated to explain the unregulated growth of cancer cells, the autocrine mechanism of growth regulation, in addition to paracrine and endocrine systems, is likely to play an important role in normal cellular processes, including development, differentiation and proliferation [4, 5]. Thus local concentrations of growth factors and numbers of specific cell-surface receptors, will affect the biological response of cells to a particular growth factor. It is apparent, therefore, that regulated growth requires the complex interaction of many growth signals and that this interaction may be influenced by the synthesis and delivery of growth factors to target tissues. Of particular interest in the study of the regulation of cell proliferation are the transforming growth factors (TGF), for which paracrine and autocrine mechanisms have been postulated.

Transforming growth factor (TGF) was a term originally applied to describe an activity produced by a murine sarcoma virus-transformed cell line [6]. Medium conditioned by this cell line was able to induce the soft agar growth of normally anchorage-dependent cells, a hallmark of *in vitro* transfor-

mation. This activity has been accepted as the operational definition of TGF. Continued investigation of this activity led to the discovery that two distinct classes of TGF existed; these were termed transforming growth factor type-α (TGFα) and transforming growth factor type-β (TGFβ). Although given similar names, TGFα and TGFβ are distinct molecules in structure and function.

TGFα

Transforming growth factor type-α is a small polypeptide (50 amino acids) which has been shown to have marked structural homology to epidermal growth factor (EGF) due to conservation of cysteine residues [7]. Cloning of the human TGFα cDNA indicated that TGFα is synthesized as a 160-amino-acid precursor [8]. Subsequent complementary DNA cloning revealed that rat TGFα was synthesized as a 159-amino-acid precursor [9] and that TGFα is a highly conserved molecule. The TGFα precursor, designated pro-TGFα, contains a 23-amino-acid hydrophobic signal sequence followed by a domain which includes the mature growth factor sequence as well as potential glycosylation sites, a second hydrophobic domain, and finally a cysteine-rich domain at the carboxyl terminus. The second hydrophobic domain has been shown to span the plasma membrane [10–12]. Mature TFGα is released from the extracellular domain of pro-TGFα through proteolytic cleavage (Fig. 1).

Various larger forms of TGFα have been reported and appear to be due to incomplete cleavage and/or heterogenous glycosylation of the outer domain [13, 14]. It has been postulated that the TGFα precursor may possess biological activity and site-directed mutagenesis studies have been useful in confirming this hypothesis. Membrane-imbedded pro-TGFα was able to induce tyrosine autophosphorylation of EGF receptors on indicator cells [15, 16]. The role of the TGFα precursor *in vivo* remains uncertain, although it has been suggested that biologically active pro-TGFα may provide a highly localized and/or prolonged autocrine growth signal. The possibility that integral membrane pro-TGFα provides an accessible reservoir of TGFα for rapid paracrine action also exists.

TGFα interacts with the EGF receptor and is believed to mediate all of its biological effects through the EGF receptor. Evidence for a distinct TGFα receptor does not exist, although the possibility of distinct ligand binding domains for TGFα and EGF on this common receptor has not been excluded. The effects of TGFα in a variety of *in vitro* assays using cultured cells are essentially identical to those of EGF [17]. In contrast, some *in vivo* and organ culture studies indicate that TGFα may be more potent than EGF; TGFα is a more potent

Correspondence to H. L. Moses, Department of Cell Biology, Vanderbilt School of Medicine, Nashville, TN 37232, USA

Abbreviations. TGF, transforming growth factor; EGF, epidermal growth factor; PDGF, platelet-derived growth factor; IGF, insulin-like growth factor; IL-2, interleukin-2; BMP, bone morphogenetic protein; PAI-1, plasminogen activator inhibitor type 1; TIMP, tissue-specific inhibitor of metalloproteinase; SSC, squamous cell carcinoma; MK, mouse keratinocyte; KC, K-*ras*-transformed mouse keratinocyte; CHO, Chinese hamster ovary; NGF, nerve growth factor; MSA, multiplication stimulating activity.

468

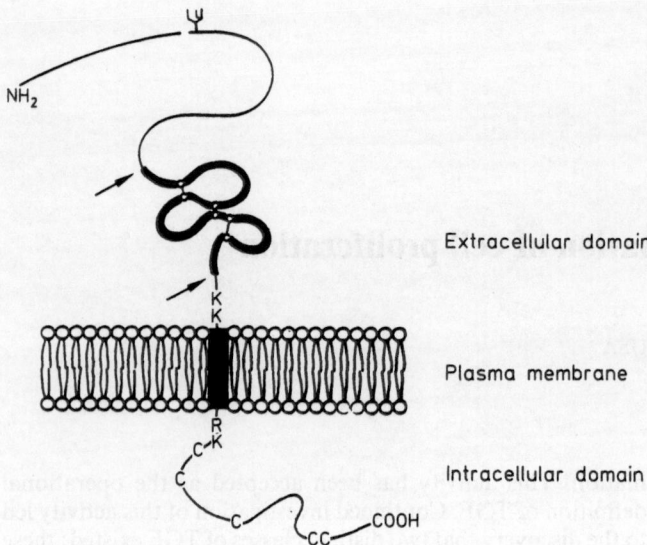

Fig. 1. *Model of pro-TGFα as a transmembrane protein.* The 23-amino-acid NH₂-terminal hydrophobic signal sequence is shown as already cleaved from the remainder of the molecule. The 50-amino-acid mature TGFα (heavy line), with three proposed cysteine disulfide bonds, is flanked by proteolytic cleavage sites (arrows). Sequences which are not contained within mature TGFα include the hydrophobic transmembrane domain (solid box) which is bordered by pairs of basic amino acids (KK and RK) and the COOH-terminal cytoplasmic domain which is rich in cysteine residues (C). An N-linked glycosylation site near the amino terminus is indicated at the top of the figure. (Reprinted with permission from [11])

stimulator of Ca^{2+} release than EGF in bone organ cultures, and at low concentrations, TGFα induces neovascularization while EGF does not (for review see [18]). Qualitative differences in biological activities of TGFα and EGF, however, have not been reported.

TGFα was initially isolated from the culture medium of sarcoma virus-transformed cells while being undetectable from the non-transformed parental cells [6]. Other transformed cell lines were also shown to produce TGFα, including cultured human carcinoma cells [19]. Continued studies of this growth factor led to the discovery that TGFα was present in embryonic tissues as well [20]. The inability to demonstrate the production of TGFα by normal adult tissues or non-transformed cells led to the popular belief that TGFα was an embryonic growth factor which may be inappropriately expressed in transformed cells. The discovery of TGFα and its seemingly aberrant expression by transformed cells led to the autocrine hypothesis as an explanation for the excessive growth that has been described for many neoplastic cells [3].

Recent investigations, however, indicate that normal adult cells do express and synthesize TGFα. Normal mouse keratinocytes are absolutely dependent upon EGF/TGFα for continued proliferation in culture [21]. Expression of TGFα mRNA and the production of TGFα protein by cultured mouse keratinocytes was shown by Coffey et al. [22]. Moreover, the expression of TGFα by these cells was dependent on the presence of EGF during culture; both EGF and TGFα were able to stimulate significant increases in TGFα mRNA levels as compared with non-treated keratinocytes. This autoinduction of TGFα message has been proposed as a mechanism of signal amplification which may be essential for the propagation of proliferative responses throughout an organ or tissue.

The discovery that normal adult cells both produce and respond to TGFα provided strong evidence that, not only transformed cell, but normal cell proliferation as well, may be regulated in an autocrine manner. Concurrent investigations of other growth factors, including platelet-derived growth factor (PDGF) [23], insulin-like growth factor-1 (IGF-1) [24] and interleukin-2 (IL-2) [25] have shown that autocrine stimulation of proliferation should be considered a general phenomenon of most cells. Thus, the autocrine hypothesis is no longer regarded as an abnormal process occurring only in transformed cells leading to excessive proliferation; rather it is believed to be an important function of the regulation of normal cell growth.

While autocrine stimulation by TGFα has been shown to be a normal phenomenon, the possibility that uncontrolled production of TGFα may play a role in neoplastic transformation has not been excluded. Alternative explanations for excessive growth of transformed cells include changes in growth factor-receptor number and/or affinity and alterations in post-receptor signalling pathways.

TGFβ

Transforming growth factor type-β is distinctly dissimilar from TGFα, both in structure and biological function. TGFβ is a dimeric polypeptide comprised of identical 112-amino-acid subunits whose association is maintained through disulfide linkages. The 25-kDa TGFβ dimer is the biologically active molecule and disruption of the dimeric nature of this growth factor with reducing agents results in a total loss of activity [26]. Cloning and sequencing of the TGFβ gene from several sources [27, 28] revealed that TGFβ is synthesized as a larger molecule of 390 amino acids. This pre-pro-TGFβ contains a typical hydrophobic signal sequence of 29 amino acids which is cleaved, resulting in pro-TGFβ. The carboxyl-terminal 112-amino-acid sequence, which represents monomeric TGFβ, is cleaved from the larger precursor at a dibasic cleavage site. In addition, several glycosylation sites have been identified within the amino-terminal portion of the TGFβ precursor (Fig. 2). While the order of these events is uncertain, it has become apparent that the synthesis of TGFβ is itself a complex issue involving several processing events [29 – 32].

It is now known that TGFβ belongs to a much larger family of closely related genes and polypeptides. The growth inhibitor originally purified from human platelets has been designated TGFβ1. A protein with striking similarity to TGFβ1 was first purified from porcine platelets [33] and bovine bone [34] and is termed TGFβ2. The binding characteristics and biological activities of TGFβ1 and TGFβ2 are very similar [33] (and R. M. Lyons, unpublished observations). The African green monkey (BSC-1) cell-secreted growth inhibitor described by Holley et al. [35] has since been shown to be essentially identical to human TGFβ2 [36]. cDNA cloning from human [37, 38], porcine [37] and chicken [39] libraries have identified a gene, TGFβ3, which shares similarity with TGFβ1 and TGFβ2. Expression and production of human recombinant TGFβ3 indicates that this protein has a similar range of *in vitro* biological activities as that of TGFβ1 and TGFβ2 [40]. TGFβ4 has been cloned from a chicken chondrocyte library [41]. The TGFβ genes indicate that there is a high degree of homology and conservation, however, as TGFβ4 protein has not been purified, the biological activities of this molecule are not known. The similarity in gene structure, however, suggests that many of the protein features

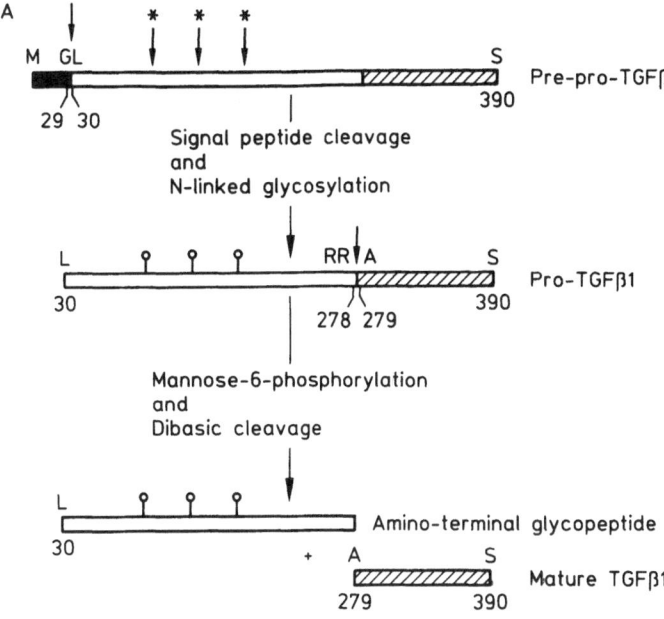

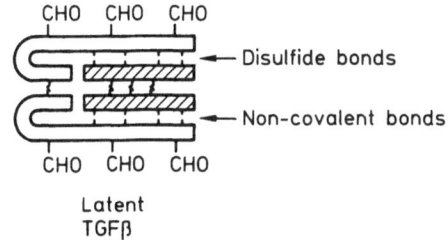

Fig. 2. *Model of synthesis and structure of latent TGFβ1.* (A) Proposed processing events of the TGFβ1 precursor. The 29-amino-acid signal sequence (solid box) is cleaved at the Gly-Leu peptide bond. Core glycosylation units ( ) are added to pro-TGFβ1 at each of the *N*-glycoslyation sites ( ). Glycoslyated pro-TGFβ1 is further processed to yield a phosphorylated glycoprotein containing complex sialated oligosaccharides. Proteolytic cleavage at the dibasic cleavage site (residue 278) separates mature TGFβ1 (crosshatched box) from the amino-terminal glycopeptide (open box), which is termed the latency associated peptide (LAP); these polypeptides remain non-covalently associated to form the latent TGFβ1 complex. Disulfide bond formation and dimerization occur at some point during synthesis or transit. The order of these processing steps, and the tertiary structure of the TGFβ1 molecules represented, are unknown. (Reprinted with permission from [30].) (B) Model of cell-secreted latent TGFβ1. The 25-kDa TGFβ1 homodimer (crosshatched box) remains noncovalently associated with the amino-terminal glycopeptide (open box) after proteolytic cleavage at the dibasic cleavage site. Events which disrupt this complex, such as changes in ionic strength or alterations in confromation, may destabilize the latent complex and allow the release of mature, biologically active TGFβ1

described earlier for TGFβ1 will be present in these members of the TGFβ family as well. The vast majority of studies performed to date have employed TGFβ1 due to the lack of availability of TGFβ3 or TGFβ4 protein from natural sources and the striking similarity in structure and function of TGFβ2. Thus, unless otherwise stated, the term TGFβ will refer to TGFβ1 for the remainder of this discussion.

Similarity of TGFβ to several other molecules or genes has been reported by various investigators. Proteins which share significant structural similarity to the TGFβ molecule include Müllerian inhibiting substance [42], the inhibins and their α subunits, the activins [43] and bone morphogenetic proteins (BMPs) [44]. Genes with sequence similarity with the TGFβ gene are the *Drosophila* decapentaplegic gene complex [45], the *Xenopus* Vg-1 gene [46] and the mouse Vgr-1 gene [47]. Thus, it is clear that TGFβ belongs to a large family of related molecules. While the function of many of these molecules is unknown, many have been shown to be important regulators of cell growth and developmental processes.

Specific high-affinity TGFβ receptors have been described for a wide variety of cells, both normal and transformed [48–50]. Dissociation constants range from 25 pM to 140 pM and receptor numbers range over 2000–40000/cell, depending on the cell line [51]. Affinity cross-linking studies have shown that TGFβ interacts specifically with three cell-surface molecules [33]. While the contribution of each of these cell-surface-binding molecules to the overall transduction of the TGFβ signal is unclear, several TGFβ-resistant cell lines have been established to investigate this problem. The studies of Boyd and Massagué [52] and Segarini et al. [53] indicate that the low-molecular-mass receptor (type I) is required for TGFβ sensitivity; whether the other TGFβ receptors (type II and type III) are required for TGFβ responsiveness is not known as mutants lacking these receptors have not been identified. Interestingly, there have been no reports of kinase activity, intracellular Ca^{2+} fluxes or inositol phosphate involvement associated with the TGFβ receptor. However, Howe et al. [54] report that a GTP-binding protein may be involved in TGFβ signal transduction. These investigators have demonstrated an increase in GTP[γS] binding and GTPase activity in mouse fibroblasts treated with TGFβ. Furthermore, these investigators have mutagenized mink lung epithelial cells and established TGFβ-resistant lines. Interestingly, TGFβ did not induce an increase in GTP[γS] binding or an increase in GTPase activity in these TGFβ-resistant mutants. In addition, at least one mutant line lacked the TGFβ type-I receptor (unpublished results), suggesting that the coupling of a G protein to the type-I receptor is required for transduction of the TGFβ signal.

In addition to alterations in the TGFβ receptor types expressed on the cell surface, changes in the number of TGFβ receptors present may be involved in determining cell responsiveness to this growth factor. An increase in the number of high-affinity TGFβ receptors on T cells following activation has been demonstrated [55]. This increase in TGFβ receptor number was accompanied by an increase in the expression and production of TGFβ. Furthermore, treatment of T cells with TGFβ was shown to inhibit the IL-2-induced up-regulation of IL-2 receptors. These results are particularly interesting in view of earlier reports concerning IL-2 receptor modulation. Expression of the IL-2 receptor and synthesis and secretion of IL-2 occur when T cells are presented with antigen [56, 57]. Once a critical number of high-affinity IL-2 receptors have specifically bound ligand, DNA synthesis and cell division occur [58]. Thus, T cell responses to positive (IL-2) and negative (TGFβ) signals may be dependent on the number of specific receptors expressed and the number of receptors may be modulated by the presence of the growth factors themselves.

The biological activities which have been described for TGFβ are both numerous and diverse. TGFβ was originally shown to stimulate the growth of cells in soft agar [59] and was later shown to stimulate the monolayer growth of cells of mesenchymal origin [60]. The mitogenic effect of TGFβ in monolayer, however, displayed altered kinetics with respect to other described mitogens. Rather than DNA synthesis occurring at 24 h, stimulation of DNA synthesis was delayed until 36 h following TGFβ addition. It was also shown that TGFβ caused an induction of the c-*sis* proto-oncogene and the synthesis of PDGF-like material [61]. Thus, it is likely that the mitogenic effect of TGFβ on fibroblasts is not a direct effect, but rather due to c-*sis* induction.

Postlewaite et al. [62] have shown that TGFβ is a potent chemotactic agent for fibroblasts; it appears to be chemotactic for macrophages as well [63]. There is evidence that TGFβ may induce terminal differentiation of bronchial epithelial cells [64]; however, contrasting reports indicate that it inhibits the differentiation of adipocytes [65] and myoblasts [66]. Whether the role of TGFβ in differentiation is tissue- or cell-specific remains unclear.

TGFβ has a pronounced effect on extracellular matrix production. Different investigators have shown that it causes an increase in collagen, fibronectin and proteoglycan expression [67, 68]; an increase in integrin expression [69]; a decrease in the synthesis of proteases which degrade extracellular matrix components, such as collagenase [70] and transin [71]; and an increase in expression of protease inhibitors, such as plasminogen activator inhibitor type 1 (PAI-1) [72] and tissue-specific inhibitor of metalloprotease (TIMP) [70]. The concerted effect of TGFβ is thus increased production and deposition of extracellular matrix components [73] and the stimulation of connective tissue formation [74]. These activities ascribed to TGFβ suggest that a major physiological role of this growth factor may be in wound healing. This hypothesis has been strengthened by studies of wounds *in vivo* and the ability of TGFβ to accelerate healing [75]. In addition, platelets are a major source of TGFβ, thus the involvement of TGFβ in wound repair has gained significant support.

In addition to the above-mentioned biological activities, TGFβ has been described as a potent growth inhibitor of a variety of cell types, including epithelial, endothelial, lymphoid and myeloid cells [76−78]. Inhibition of proliferation by TGFβ is reversible and thus not the result of cytotoxicity [79]. It has been proposed that autocrine inhibition may be an important aspect of growth regulation in some cells [35, 77]. Interestingly, most epithelial cells lack PDGF receptors and therefore TGFβ is not expected to act as an indirect mitogen for epithelial cells as it does for fibroblastic cells. This has provided a useful system in which to study the inhibitory actions of TGFβ. While TGFβ is a potent growth inhibitor of all normal and many neoplastic epithelial cells, TGFβ has no effect on the clonal growth of the squamous cell carcinoma cell line, SSC-25 [79]. Subsequent studies on several squamous carcinoma cell lines have shown that loss of the normal inhibitory response is common in this type of cancer cell [60, 80, 81].

Mouse keratinocytes (MK), a clonal epithelial cell line, have been used to study the inhibitory effects of TGFβ on cell proliferation and gene expression. Half-maximal inhibition of DNA synthesis was observed at a concentration of 80 pM TGFβ [82]. Genes which are presumably involved in early events leading to cell division have also been examined in this system. Within 1 h of exposure of rapidly growing MK cells to TGFβ, there is a pronounced decrease in c-*myc* expression [83]. The inhibition of gene expression was specific, as β-actin levels increased and c-*fos* expression was not affected. These data suggest that the selective control of gene expression by

TGFβ may be indicative of a mechanism of inhibition of cellular proliferation by TGFβ.

A line of MK cells which was derived by Kirsten murine sarcoma virus transformation of the parental cells has been established and designated KC cells [21]. This cell line has been used to study the role of TGFβ on the growth regulation of transformed cells. Interestingly, these K-*ras* transformed keratinocytes are inhibited by TGFβ; however, variant cells which have lost the TGFβ inhibitory response arise at a relatively high frequency [84]. Affinity crosslinking studies indicate that the resistant variants may have an alteration in the low-molecular-mass TGFβ receptor (type I). Thus, the loss of TGFβ responsiveness in this system may be due to the inability of the altered type-I receptor to transduce the inhibitory signal rather than to a complete loss of the type-I receptor. These resistant variants may prove useful in future studies aimed at defining functional domains of the type-I TGFβ receptor.

It has become increasingly clear that the regulation of TGFβ expression and synthesis, and other growth factors as well, will be crucial to the maintenance of appropriate growth control. TGFβ1 expression appears to be tissue-specific, with highest mRNA levels present in spleen, lung and placenta of the adult mouse, while the expression of TGFβ2 and TGFβ3 show different patterns of expression [85, 86]. Although detectable in most cultured cells, the level of expression of TGFβ1, TGFβ2 and TGFβ3 mRNA also varies across cell lines and cell types [37]. In addition, TGFβ1 is able to regulate the expression of its own mRNA in a manner similar to that previously described for TGFα. A 2−3-fold amplification of TGFβ1 mRNA by TGFβ1 has been reported for many normal and transformed cells in culture [87] and as much as a 20−25-fold increase in the mouse fibroblast cell line, AKR-2B (unpublished results). The mechanism of amplified TGFβ expression in response to exogenous TGFβ is both increased transcriptional activity and mRNA stabilization (unpublished results).

The similarity of TGFβ to polypeptides which have been shown to be involved in embryonic development has generated interest in determining whether TGFβ itself may play a role in early development. Based on the localization of TGFβ1 protein by immunohistochemical techniques to mouse mesenchymal tissues [88] and localization of TGFβ1 mRNA by *in situ* hybridization to fetal bone, liver megakaryocytes and the overlying epithelia of these mesenchymal tissues [89, 90], it has been suggested that TGFβ may act in both a paracrine and autocrine manner during development. TGFβ2 mRNA has been localized by *in situ* hybridization to mesenchymal tissues with an adjacent epithelium [91]. These patterns of protein and mRNA localization suggest a role for TGFβ1 and TGFβ2 in epithelial/mesenchymal interactions.

Another potential site for the regulation of TGFβ action is the growth factor itself. Many investigators have shown that TGFβ is secreted by cells in culture, and released by platelets, in an inactive, or latent, form [92, 93]. Latent TGFβ cannot interact with cell-surface TGFβ receptors and therefore elicits none of the aforementioned activities of TGFβ. The activation of latent TGFβ is thus likely to be a key regulatory event. Acidification of cell-conditioned medium, or acid extraction of tissue, has become a standard method of activation. On the basis of this information, it has been suggested that TGFβ is associated with other polypeptides in a latent complex. The transfection of Chinese hamster ovary (CHO) cells with the full length TGFβ1 gene and subsequent amplification of TGFβ1 protein synthesis provided valuable information with respect to the polypeptide composition of

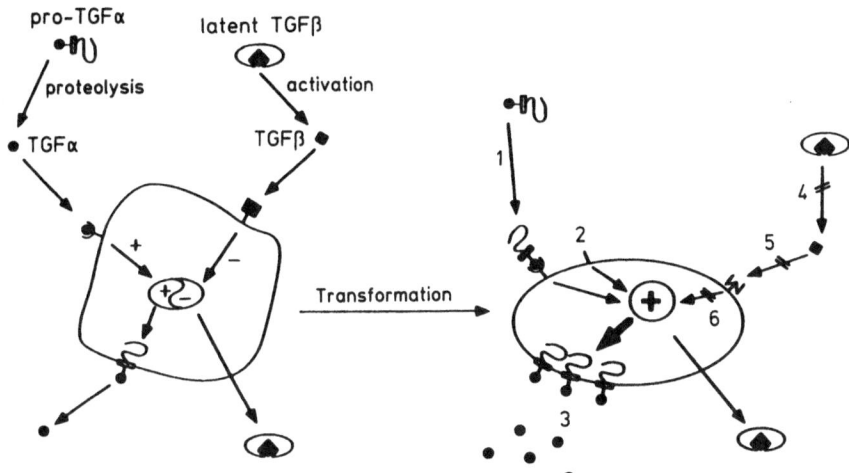

Fig. 3. *Potential positive/negative growth control of epithelial cells.* TGFα provides a growth stimulatory signal and TGFβ provides a growth inhibitory signal resulting in normal growth regulation. Neoplastic transformation may result from aberrations at one or more of the following points: (1) accumulation of pro-TGFα embedded in plasma membranes of adjacent cells providing a prolonged and localized positive signal; (2) alterations in the EGF/TGFα receptor leading to constitutive activation of positive post-receptor signal-transduction pathways; (3) amplification of TGFα mRNA levels resulting in an enhanced positive signal; (4) loss of the ability to activate latent TGFβ; (5) loss of TGFβ receptors; (6) changes in negative post-receptor signal-transduction pathways. Alterations such as those indicated above may contribute to transformation of epithelial cells due to an imbalance in positive and negative growth signals

latent TGFβ [29]. The 25-kDa mature TGFβ remains in a noncovalent association with the amino-terminal glycopeptide portion of the TGFβ precursor resulting in a 110-kDa latent complex (unpublished results). While acidification is capable of disruption of these noncovalent interactions, thus releasing active TGFβ, protease activation has been achieved as well. Specifically, plasmin, a serine protease, appears to activate latent TGFβ1 by cleavage within the amino-terminal glycopeptide [94] (and unpublished results). This may result in conformational changes which destabilize the latent complex and ultimately release mature TGFβ. That plasmin may be a physiological activator of TGFβ is supported by the observations of Sato and Rifkin [95]. They observed that the *in vivo* activation of TGFβ during co-culture of bovine endothelial cells and pericytes could be blocked by specific inhibitors of plasmin action. While plasmin is an attractive candidate as an *in vivo* activator of latent TGFβ, this mechanism of activation may be restricted to sites in which plasminogen is available, such as wound repair. Other mechanisms may be operating in the absence of plasminogen, including alternative proteases which may provide tissue specificity for TGFβ activation.

Interestingly, Miyazano et al. [96] have shown that alterations of glycosylation within the amino-terminal glycopeptide can lead to activation of latent TGFβ. Thus, the possibility of deglycosylation as a mechanism of latent TGFβ activation must be considered. In general, activation of latent TGFβ appears to be the result of any event which leads to an unstable conformation of the latent TGFβ complex, thus enabling the release of mature TGFβ.

Regulation of growth factor action at a post-translational level, as suggested for TGFβ, may be a common feature of growth factors which function in an endocrine or paracrine manner. The secretion of biologically active factors at sites other than the target cell, tissue or organ could lead to inappropriate responses. Mechanisms for the release of active molecules from inactive complexes could potentially provide both growth factor and tissue specificity. Biologically inactive growth factor complexes have been reported for several factors including EGF [97], nerve growth factor (NGF) [98],

insulin-like growth factor-1 (IGF-1) [99] and multiplication stimulating activity (MSA; rat IGF-2) [100]. These inactive complexes and the mechanisms for the release of active growth factors are distinct and may contribute to response specificity. While the mechanism of release of an active factor may be specific for a particular molecule, these data suggest that, in general, inactive growth factor complexes, and therefore mechanisms of activation, play an important role in the regulation of growth factor action.

SUMMARY

The number of different growth regulatory molecules which have been isolated and characterized is continuing to increase. As more information is obtained, it has become apparent that the cooperative actions of many factors with distinct activities is necessary for appropriate proliferative responses. An interplay of both growth stimulatory and growth inhibitory factors is essential for normal growth. Of crucial importance, therefore, is the appropriate regulation of growth factors. Unregulated expression, synthesis, post-translational processing or activation of either positive or negative growth signals may contribute to neoplastic transformation (Fig. 3). Altered responses to normally positive or negative signals by transformed cells have been demonstrated by several investigators [64, 79, 84]. While altered growth factor responses in transformed cells are well documented, the mechanisms responsible for the loss of growth control are poorly understood and are likely to be both complex and numerous. Continued efforts to dissect and comprehend fully growth factor action on normal cells will be necessary before an understanding of neoplastic transformation can be achieved.

REFERENCES

1. McConaghey, P. & Sledge, C. B. (1970) *Nature 225*, 1249–1250.
2. Schalch, D. S., Heinrich, U. E., Draznin, B., Johnson, C. J. & Miller, L. L. (1979) *Endocrinology 104*, 1143–1151.

472

3. Sporn, M. B. & Todaro, G. J. (1980) *N. Eng. J. Med. 303*, 878–880.

4. Barnard, J. A., Bascon, C. C., Lyons, R. M., Sipes, N. J. & Moses, H. L. (1988) *Am. J. Med. Sci. 31*, 159–163.

5. Bascom, C. C., Sipes, N. J., Coffey, R. J. & Moses, H. L. (1989) *J. Cell. Biochem. 39*, 25–32.

6. DeLarco, J. & Todaro, G. J. (1978) *Proc. Natl Acad. Sci. USA 75*, 4001–4005

7. Marquardt, H., Hunkapiller, M. W., Hood, L. E., Twardzik, D. R., DeLarco, J. E., Stephenson, J. R. & Todaro, G. J. (1984) *Proc. Natl Acad. Sci. USA 80*, 4684–4688.

8. Derynck, R., Roberts, A. B., Winkler, M. E., Chen, E. Y. & Goeddel, D. V. (1984) *Cell 38*, 287–297.

9. Lee, D. C, Rose, T. M., Webb, N. R. & Todaro, G. J. (1985) *Nature 313*, 489–491.

10. Texidó, J., Gilmore, R., Lee, D. C. & Massagué, J. (1987) *Nature 326*, 883–885.

11. Bringman, T. S., Lindquist, P. B. & Derynck, R. (1987) *Cell 48*, 429–440.

12. Gentry, L. E., Twardzik, D. R., Lim, G. J., Ranchalis, J. E. & Lee, D. C. (1987) *Mol. Cell. Biol. 7*, 1585–1591.

13. Texidó, J. & Massagué, J. (1988) *J. Biol. Chem. 263*, 3924–3929.

14. Luetteke, N. C., Michalopoulos, G. K., Texidó, J., Gilmore, R., Massagué, J. & Lee, D. C. (1988) *Biochemistry 27*, 6487–6494.

15. Wong, S. T., Winchell, L. F., McCune, B. K., Earp, H. S., Texidó, J., Massagué, J., Herman, B. & Lee, D. C. (1989) *Cell 56*, 495–506.

16. Brachmann, R., Lindquist, P. B., Nagashima, M., Kohr, W., Lipari, T., Napier, M. & Derynck, R. (1989) *Cell 56*, 691–700.

17. Anzano, M. A., Roberts, A. B., Smith, J. M., Sporn, M. B. & DeLarco, J. E. (1983) *Proc. Natl Acad. Sci. USA 80*, 6264–6268.

18. Derynck, R. (1986) in *Oncogenes and growth control* (Kahn, P. & Graf, T., eds) pp. 58–63, Springer-Verlag, Berlin.

19. Todaro, G. J., Fryling, C. & DeLarco, J. E. (1980) *Proc. Natl Acad. Sci. USA 77*, 5258–5262.

20. Twardzik, D. R., Sherwin, S. A., Ranchalis, J. E. & Todaro, G. J. (1982) *J. Natl Cancer Inst. 69*, 793–798.

21. Weissman, B. E. & Aaronson, S. A. (1983) *Cell 32*, 599–606.

22. Coffey, R. J. Jr, Derynck, R., Wilcox, J. N., Bringman, T. S., Goustin, A. S., Moses, H. L. & Pittlekow, M. R. (1987) *Nature 328*, 817–820.

23. Betsholtz, C., Westermark, B., Ek, B. & Heldin, C.-H. (1984) *Cell 39*, 447–457.

24. Clemmons, D. R. & Van Wyk, J. J. (1985) *J. Clin. Invest. 75*, 1914–1918.

25. Smith, K.A. (1982) *Immunobiology 161*, 157–173.

26. Assoian, R. K., Komoriya, A., Meyers, C. A., Miller, D. M. & Sporn, M. B. (1983) *J. Biol. Chem. 258*, 7155–7160.

27. Derynck, R., Jarrett, J. A., Chen, E. Y., Eaton, D. H., Bell, J. R., Assoian, R. K., Roberts, A. B., Sporn, M. B. & Goeddel, D. V. (1984) *Nature 316*, 701–705.

28. Sharples, K., Plowman, G. D., Rose, T. M., Twardzik, D. R. & Purchio, A. F. (1987) *DNA 6*, 239–244.

29. Gentry, L. E., Webb, N. R., Lim, G. J., Brunner, A. M., Ranchalis, J. E., Twardzik, D. R., Lioubin, M. N., Marquardt, H. & Purchio, A. F. (1987) *Mol. Cell. Biol. 7*, 3418–3427.

30. Gentry, L. E., Lioubin, M. N., Purchio, A. F. & Marquardt, H. (1988) *Mol. Cell. Biol. 8*, 4162–4168.

31. Purchio, A. F., Cooper, J. A., Brunner, A., Lioubin, M. N., Gentry, L. E., Kovacina, K. S., Roth, R. A. & Marqurdt, H. (1988) *J. Biol. Chem. 263*, 14211–14215.

32. Brunner, A. M., Gentry, L. E., Cooper, J. A. & Purchio, A. F. (1988) *Mol. Cell. Biol. 8*, 2229–2232.

33. Cheifetz, S., Weatherbee, J. A., Tsang, M. L.-S., Anderson, J. K., Mole, J. E., Lucas, R. & Massagué, J. (1987) *Cell 48*, 409–415.

34. Seyedin, S. M., Thompson, A. Y., Bentz, H., Rosen, D. M., McPherson, J. M., Conti, A., Siegel, N. R., Galluppi, G. R. & Piez, K. A. (1986) *J. Biol. Chem. 261*, 5693–5695.

35. Holley, R. W., Armour, R. & Baldwin, J. H. (1978) *Proc. Natl Acad. Sci. USA 75*, 1864–1866.

36. Hanks, S. K., Armour, R., Baldwin, J. H., Maldonado, F., Spiess, J. & Holley, R. W. (1988) *Proc. Natl Acad. Sci. USA 85*, 79–82.

37. Derynck, R., Lindquist, P. B., Lee, A., Wem, D., Tamm, J., Graycar, J. L., Rhee, L., Mason, A. J., Miller, D. A., Coffey, R. J. Jr, Moses, H. L. & Chen, E. Y. (1988) *EMBO J. 7*, 3737–3743.

38. ten Dijke, P., Hansen, P., Iwata, K. K., Peiler, C. & Foulkes, J. G. (1988) *Proc. Natl Acad. Sci. USA 85*, 4715–4719.

39. Jakowlew, S. B., Kondaiah, P., Dillard, P. J., Sporn, M. B. & Roberts, A. B. (1988) *Mol. Endocrinol. 2*, 747–755.

40. Graycar, J. L., Miller, D. A., Arrick, B. A., Lyons, R. M., Moses, H. L. & Derynck, R. (1989) *Mol. Endocrinol.*, in the press.

41. Jakowlew, S. B., Dillard, P. J., Sporn, M. B. & Roberts, A. B. (1988) *Mol. Endocrinol. 2*, 1186–1195.

42. Cate, R. L., Mattalianok, R. J., Hession, C., Tizard, R., Farber, N. M., Cheung, A., Ninfa, E. G., Frey, A. Z., Gash, D. J., Chow, E. P., Fisher, R. A., Bertonis, J. M., Torres, G., Wallner, B. P., Ramachandran, K. L., Ragin, R. C., Managanaro, T. F., MacLauchlin, D. T. & Donahoe, P. K. (1986) *Cell 45*, 685–698.

43. Mason, A. J., Hayflick, J. S., Ling, N., Esch, F., Ueno, N., Ying, Y., Guillemin, R., Niall, H. & Seeburg, P. H. (1985) *Nature 318*, 659–663.

44. Wozney, J. M., Rosen, V., Celeste, A. J., Mitsock, L. M., Whitter, M. J., Kriz, R. W., Herwick, R. M. & Wang, E. A. (1988) *Science 242*, 1528–1534.

45. Padgett, R. W., St. Johnston, R. D. & Gelbart, W. M. (1987) *Nature 325*, 81–84.

46. Weeks, D. L. & Melton, D. A. (1987) *Cell 51*, 861–867.

47. Lyons, K., Graycar, J. L., Lee, A., Hashimi, S., Lindquist, P. B., Chen, E. Y., Hogan, B. L. M. & Derynck, R. (1989) *Proc. Natl Acad. Sci. USA 86*, 4554–4558.

48. Tucker, R. F., Branum, E. L., Shipley, G. D., Ryan, R. J. & Moses, H. L. (1984) *Proc. Natl Acad. Sci. USA 81*, 6757–6761.

49. Frolik, C. A., Wakefield, L. M., Smith, D. M. & Sporn, M. B. (1984) *J. Biol. Chem. 259*, 10995–11000.

50. Massagué, J. (1985) *J. Biol. Chem. 260*, 7059–7066.

51. Wakefield, L. M., Smith, D. M., Masui, T., Harris, C. C. & Sporn, M. B. (1987) *J. Cell Biol. 105*, 965–975.

52. Boyd, F. T. & Massagué, J. (1989) *J. Biol. Chem. 263*, 2808–2816.

53. Segarini, P. R., Rosen, D. M. & Seyedin, S. M. (1989) *Mol. Endocrinol. 3*, 261–272.

54. Howe, P. H. & Leof, E. B. (1989) *Biochem. J. 261*, 879–886.

55. Kehrl, J. H., Wakefield, L. M., Roberts, A. B., Jakowlew, S., Alvarez-Mon, M., Derynck, R., Sporn, M. B. & Fauci, A. S. (1986) *J. Exp. Med. 163*, 1037–1050.

56. Smith, K. A. (1980) *Immunol. Rev. 51*, 337–357.

57. Robb, R. J., Munck, A. & Smith, K. A. (1981) *J. Exp. Med. 154*, 1455–1474.

58. Smith, K. A. & Cantrell, D. A. (1985) *Proc. Natl Acad. Sci. USA 82*, 864–868.

59. Moses, H. L., Branum, E. L., Proper, J. A. & Robinson, R. A. (1981) *Cancer Res. 41*, 2842–2848.

60. Shipley, G. D., Tucker, R. F. & Moses, H. L. (1985) *Proc. Natl Acad. Sci. USA 82*, 4147–4151.

61. Leof, E. B., Proper, J. A., Goustin, A. S., Shipley, G. D., DiCorletto, P. E. & Moses, H. L. (1986) *Proc. Natl Acad. Sci. USA 83*, 2453–2457.

62. Postlewaite, A. E., Keski-Oja, J., Moses, H. L. & Kang, A. H. (1987) *J. Exp. Med. 165*, 251–256.

63. Wahl, S. M., Hunt, D. A., Wakefield, L. M., McCartney-Francis, N., Wahl, L. M., Roberts, A. B. & Sporn, M. B. (1987) *Proc. Natl Acad. Sci. USA 84*, 5788–5792.

64. Masui, T., Wakefield, L. M., Lechner, J. F., LaVeck, M. A., Sporn, M. B. & Harris, C. C. (1986) *Proc. Natl Acad. Sci. USA 83*, 2438–2442.

65. Ignotz, R. A. & Massagué, J. (1985) *Proc. Natl Acad. Sci. USA 82*, 8530–8534.

66. Massagué, J., Cheifetz, S., Endo, T. & Nadal-Ginard, B. (1986) *Proc. Natl Acad. Sci. USA 83*, 8206–8210.
67. Ignotz, R. A., Endo, R. & Massagué, J. (1987) *J. Biol. Chem. 262*, 6443–6446.
68. Raghow, R., Postlewaite, A. E., Keski-Oja, J., Moses, H. L. & Kang, A. H. (1987) *J. Clin. Invest. 79*, 1285–1288.
69. Ignotz, R. A. & Massagué, J. (1987) *Cell 51*, 189–197.
70. Edwards, D. R., Murphy, G., Reynolds, J. J., Whitham, S. E., Docherty, J., Angel, P. & Heath, J. K. (1987) *EMBO J. 6*, 1889–1904.
71. Matrisian, L. M., Leroy, P., Ruhlmann, C., Gesnel, M. & Breathnach, R. (1986) *Mol. Cell. Biol. 6*, 1679–1686.
72. Laiho, M., Saksela, O., Andreasen, P. A. & Keski-Oja, J. (1986) *J. Cell Biol. 103*, 2403–2410.
73. Ignotz, R. A. & Massagué, J. (1986) *J. Biol. Chem. 261*, 4337–4345.
74. Roberts, A. B., Sporn, M. B., Assoian, R. K., Smith, J. M., Roche, N. S., Wakefield, L. M., Heine, U. I., Liotta, L. A., Falanga, V., Kehrl, J. H. & Fauci, A. S. (1986) *Proc. Natl Acad. Sci. USA 83*, 4167–4171.
75. Mustoe, T. A., Pierce, G. F., Thomason, A., Gramates, P., Sporn, M. B. & Deuel, T. F. (1987) *Science 237*, 1333–1336.
76. Tucker, R. F., Shipley, G. D., Moses, H. L. & Holley, R. W. (1984) *Science 226*, 705–707.
77. Moses, H. L., Tucker, R. F., Leof, E. B., Coffey, R. J. Jr, Halper, J. & Shipley, G. D. (1985) *Cancer Cells 3*, 65–71.
78. Moses, H. L. & Leof, E. B. (1986) in *Oncogenes and growth control* (Kahn, P. & Graf, T., eds) pp. 51–57, Springer-Verlag, Berlin.
79. Shipley, G. D., Pittelkow, M. R., Wille, J. J., Scott, R. E. & Moses, H. L. (1986) *Cancer Res. 46*, 2068–2071.
80. Jetten, A. M., Shirley, J. E. & Stoner, G. (1986) *Exp. Cell Res. 167*, 539–549.
81. Hebert, C. D. & Birnbaum, L. S. (1989) *Cancer Res. 49*, 3196–3202.
82. Coffey, R. J. Jr, Sipes, N. J., Bascom, C. C., Graves-Deal, R., Pennington, C. Y., Weissman, B. E. & Moses, H. L. (1988) *Cancer Res. 48*, 1596–1602.
83. Coffey, R. J. Jr, Bascom, C. C., Sipes, N. J., Graves-Deal, R., Weissman, B. E. & Moses, H. L. (1988) *Mol. Cell. Biol. 8*, 3088–3093.
84. Sipes, N. J., Lyons, R. M. & Moses, H. L. (1990) *Mol. Carcinogenesis*, in the press.
85. Miller, D. A., Lee, A., Pelton, R. W., Chen, E. Y., Moses, H. L. & Derynck, R. (1989) *Mol. Endocrinol. 3*, 1108–1114.
86. Miller, D. A., Lee, A., Chen, E.Y., Moses, H. L. & Derynck, R. (1989) *Mol. Endocrinol.*, in the press.
87. Van Obberghen-Schilling, E., Roche, N. S., Flanders, K. C., Sporn, M. B. & Roberts, A. B. (1988) *J. Biol. Chem. 263*, 7741–7746.
88. Heine, U. I., Munoz, E. F., Flanders, K. C., Ellingsworth, L. R., Lam, H.-Y. P., Thompson, N. L., Roberts, A. B. & Sporn, M. B. (1987) *J. Cell Biol. 105*, 2861–2876.
89. Lehnert, S. A. & Akhurst, R. J. (1988) *Development 104*, 263–273.
90. Wilcox, J. N. & Derynck, R. (1988) *Mol. Cell. Biol. 8*, 3415–3422.
91. Pelton, R. W., Nomura, S., Moses, H. L. & Hogan, B. L. M. (1989) *Development 106*, 759–767.
92. Lawrence, D. A., Pircher, R., Kryceve-Martinerie, C. & Jullien, P. (1984) *J. Cell. Physiol. 121*, 184–188.
93. Pircher, R., Jullien, P. & Lawrence, D. A. (1986) *Biochem. Biophys. Res. Commun. 136*, 30–37.
94. Lyons, R. M., Keski-Oja, J. & Moses, H. L. (1988) *J. Cell Biol. 106*, 1659–1665.
95. Sato, Y. & Rifkin, D. B. (1989) *J. Cell Biol. 109*, 309–315.
96. Miyazano, K. & Heldin, C.-H. (1989) *Nature 338*, 158–160.
97. Taylor, J. M., Cohen, S. & Mitchell, W. M. (1970) *Proc. Natl Acad. Sci. USA 67*, 164–171.
98. Stach, R. W. & Shooter, E. M. (1980) *J. Neurochem. 34*, 1499–1505.
99. Hintz, R. L. & Liu, F. (1977) *J. Clin. Endocrinol. Metabol. 45*, 988–995.
100. Moses, A. C., Nissley, S. P., Passamani, J. & White, R. M. (1979) *Endocrinology 104*, 536–546.

Eur. J. Biochem. *189*, 1 – 23 (1990)
© FEBS 1990

Review

Current approaches to macromolecular crystallization

Alexander McPHERSON

Department of Biochemistry, University of California at Riverside, USA

(Received September 18, 1989) – EJB 89 1133

Given our current expertise, and the certain future developments in genetically altering organisms to produce proteins of modified structure and function, the concept of protein engineering is nearing reality. Similarly, our ability to describe and utilize protein structure and to define interactions with ligands has made possible the rational design of new drugs and pharmacological agents. Even in the absence of any intention toward applied use or value, the correlation of regulation, mechanism, and function of proteins with their detailed molecular structure has now become a primary concern of modern biochemistry and molecular biology.

At the present time, there are numerous physical-chemical approaches that yield information regarding macromolecular structure. Some of these methods, such as NMR and molecular dynamics, are becoming increasingly valuable in defining detailed protein structure, particularly for lower-molecular-mass proteins. There is, however, only one general technique that yields a detailed and precise description, in useful mathematical terms, of a macromolecule's structure, a description that can serve as a basis for drug design, and an intelligent guide for protein engineering. The method is X-ray diffraction analysis of single crystals of proteins, nucleic acids, and their complexes with one another and with conventional small molecules. Some inspirational examples of representative crystals are shown in Fig. 1 – 3.

In the past 20 years, the practice of X-ray crystallography has made enormous strides. Nearly all of the critical and time-consuming components of the technique have been improved, accelerated, and refined. X-ray crystallography today is not simply an awesome method used by physical chemists to reveal the vast beauty of macromolecular architecture; it is a practical, reliable, and relatively rapid means to obtain straightforward answers to perplexing questions.

X-ray diffraction data that once required years to obtain, can now be collected in a matter of weeks, even days in some cases. Computers of extraordinary speed and capacity are now common tools as are computer graphics systems of a versatility and cleverness that would have been unimaginable only a few years ago. Software, too, exists that is sophisticated yet friendly, flexible yet reliable, and readily available to anyone in need of it. The question, then, is where does the problem lie? What prevents the full utilization and exploitation of this enormously powerful approach.

The answer, of course, is that for application of the method to a particular macromolecule, the protein or nucleic acid must first be crystallized. Not only must crystals be grown, but they must be good quality crystals, crystals suitable for a high-resolution X-ray diffraction analysis. '*Aye, there's the rub*' as Hamlet might say, for in general, this is not an easy task. While some proteins may be trivially simple to crystallize, many others, invariably those of greatest personal interest, are elusive and stubborn [2].

The reason that the crystallization step has become the primary obstacle to expanded structural knowledge is the necessarily empirical nature of the methods employed to overcome it [3 – 6]. Macromolecules are extremely complex physical-chemical systems whose properties vary as a function of many environmental influences such as temperature, pH, ionic strength, contaminants and solvent composition to name only a few. They are structurally dynamic, microheterogeneous, aggregating systems, and they change conformation in the presence of ligands (for a survey of protein structure and function, see [7 – 9]). Superimposed on this is the poor state of our current understanding of macromolecular crystallization phenomena and the forces that promote and maintain protein and nucleic acid crystals.

As a substitute for the precise and reasoned approaches that we commonly apply to scientific problems, we are forced, for the time being at least, to employ a strictly empirical methodology. Macromolecular crystallization is, thus, a matter of searching, as systematically as possible, the ranges of the individual parameters that impact upon crystal formation, finding a set or multiple sets of these factors that yield some kind of crystals, and then optimizing the variable sets to obtain the best possible crystals for X-ray analysis. This is done, most simply, by conducting a long series, or establishing a vast array, of crystallization trials, evaluating the results, and using information obtained to improve matters in successive rounds of trials. Because the number of variables is so large, and their ranges so broad, intelligence and intuition in designing and evaluating the individual and collective trials becomes essential.

Crystals grow from supersaturated solutions

In a saturated solution, including one saturated with respect to protein, two states exist in equilibrium, the solid phase, and one consisting of molecules free in solution. At saturation, no net increase in the proportion of solid phase can accrue since it would be counterbalanced by an equivalent dissolution. Thus, crystals do not grow from a saturated solution. The system must be in a non-equilibrium, or supersaturated, state to provide the thermodynamic driving force for crystallization.

Correspondence to A. McPherson, Department of Biochemistry, University of California at Riverside, Riverside, California 92521-0129,USA

1

2

3

When the objective is to grow crystals of any compound, a solution of the molecule must by some means be transformed or brought into the supersaturated state whereby its return to equilibrium forces exclusion of solute molecules into the solid state, the crystal. If, from a saturated solution, for example, solvent is gradually withdrawn by evaporation, temperature is lowered or raised appropriately, or some other property of the system is altered, then the solubility limit will be exceeded and the solution will become supersaturated. If a solid phase is present, or introduced, then strict saturation will be reestablished as molecules leave the solvent to join the solid phase.

If no solid is present, as conditions are changed, then solute will not immediately partition into two phases, and the solution will remain in the supersaturated state. The solid state does not necessarily develop spontaneously as the saturation limit is exceeded because energy, analogous to the activation energy of a chemical reaction, is required to create the second phase, the stable nucleus of a crystal or a precipitate. Thus, a kinetic or energy barrier allows conditions to proceed further and further from equilibrium, into the zone of supersaturation. On a phase diagram [10, 11], like that seen in Fig. 4, the line indicative of saturation is also a boundary that marks the requirement for energy-requiring events to occur in order for a second phase to be established, the formation of the nucleus of a crystal or the nonspecific aggregate that characterizes a precipitate [12].

Once a stable nucleus has formed in a supersaturated solution, it will continue to grow until the system regains equilibrium. While non-equilibrium forces prevail and some degree of supersaturation exists to drive events, a crystal will grow or precipitate continue to form.

It is important to understand the significance of the term 'stable nucleus'. Many aggregates or nuclei spontaneously form once supersaturation is achieved, but most are, in general, not 'stable'. Instead of continuing to develop, they redissolve as rapidly as they form and their constituent molecules return to solution. A 'stable nucleus' is a molecular aggregate of such size and physical coherence that it will enlist new molecules into its growing surfaces faster than others are lost to solution; that is, it will continue to grow so long as the system is supersaturated.

In classical theories describing crystal growth of conventional molecules (see [13−16]), the region of supersaturation that pertains above saturation is further divided into what are termed the metastable region and the labile region [10−13], as shown in Fig. 4. By definition, stable nuclei cannot form in the metastable region just beyond saturation. If, however, a stable nucleus or solid is already present in the metastable region, then it can and will continue to grow. The labile region of greater supersaturation is discriminated from the metastable in that stable nuclei can spontaneously form. Further, because they are stable they will accumulate molecules and thus deplete the liquid phase until the system reenters the metastable, and ultimately, the saturated state.

An important point, shown graphically in Fig. 4, is that there are two regions above saturation, one of which can support crystal growth but not formation of stable nuclei, and the other which can yield nuclei as well as support growth. Now the rate of crystal growth is some function of the distance of the solution from the equilibrium position at saturation. Thus a nucleus that forms far from equilibrium and well into the labile region will grow very rapidly at first and, as the solution is depleted and moves back toward the metastable state, it will grow slower and slower. The nearer the system is to the metastable state when a stable nucleus first forms, then the slower it will proceed to mature.

It might appear that the best approach for obtaining crystals is to press the system as far into the labile region, supersaturation, as possible. There, the probability of nuclei formation is greatest, the speed of growth is greatest, and the likelihood of crystals is maximized. As the labile region is penetrated further, however, the probability of spontaneous and uncontrolled nucleation is also enhanced. Thus crystallization from solutions in the labile region far from the metastable state frequently results in extensive and uncontrolled 'showers' of crystals. By virtue of their number, none is favored and, in general, none will grow to a size suitable for X-ray diffraction studies. In addition, when crystallization is initiated from a point of high supersaturation, then initial growth is extremely rapid. Rapid growth is frequently associated with the occurrence of flaws and dislocations. Hence crystals produced from extremely saturated solutions tend to be numerous, small, and afflicted with growth defects.

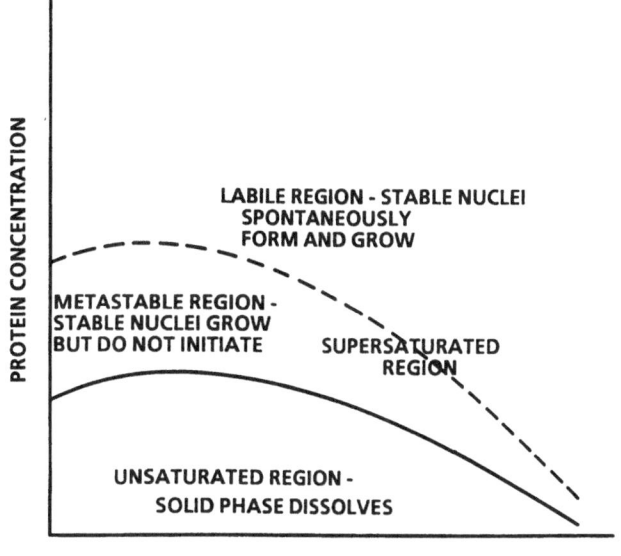

Fig. 4. *A phase diagram for a hypothetical protein showing its solubility as a function of precipitant concentration.* The solid line represents the maximum solubility or saturation curve, for the protein. Note that the protein is less soluble at very low and very high concentrations of the precipitant, corresponding to the 'salting in' and 'salting out' regions. The supersaturated region lies above the maximum solubility curve and is, in turn, demarcated by a boundary discriminating the metastable region of supersaturation from the labile region. In the labile region, crystal nuclei can both spontaneously form and grow, while in the metastable region they can only grow

Fig. 1. *Photomicrograph of crystals of the heme-containing enzyme catalase from beef liver.* One of the earliest enzymes crystallized [1], it provides a good model for studies on protein crystal growth

Fig. 2. *Orthorhombic crystals of the major seed-storage protein from the jack bean (canavalin).* These crystals can grow to sizes of several millimeters on an edge. This protein is now being studied for its crystallization behavior in microgravity

Fig. 3. *Tetragonal crystals of hen egg lysozyme, one of the easiest proteins known to crystallize.* It has provided a source for many studies on the mechanisms of protein crystal growth

4

5

6

Fig. 5. *Crystals of the protein concanavalin B, first crystallized by J. B. Sumner in 1919*. These crystals are unusually stable and resistant to physical stress

Fig. 6. *Large hexagonal plates of the plant satellite tobacco mosaic virus, a protein/nucleic acid particle of over 1 MDa*. In spite of its great size, the virus crystallizes by procedures identical to those used for many proteins

In terms of the phase diagram, ideal crystal growth would begin with nuclei formed in the labile region but just beyond the metastable. There, growth would occur slowly, the solution, by depletion, would return to the metastable state where no more stable nuclei could form, and the few nuclei that had established themselves would continue to grow to maturity at a pace free of defect formation. Thus in growing crystals for X-ray diffraction analysis, one attempts, by either dehydration or alteration of physical conditions, to transport the solution into a labile, supersaturated state, but one as close as possible to the metastable phase.

Why crystals grow

The natural inclination of any system proceeding toward equilibrium is to maximize the extent of disorder, or entropy, by freeing individual constituents from physical and chemical constraint. At the same time, there is a thermodynamic requirement to minimize the free energy (or Gibbs energy) of the system. This is achieved by the formation of chemical bonds and interactions which generally provide negative free energy. Clearly the assembly of molecules into a fixed lattice severely reduces their mobility and freedom, yet crystals do form and grow.

It follows, then, that crystal nucleation and growth must be dominated by non-covalent chemical and physical bonds arising in the crystalline state that either cannot be formed in solution or are stronger than those that can. These bonds are, in fact, what hold crystals together. They are the energetically favorable intermolecular interactions that drive crystal growth in spite of the resistance to molecular constraint. From this it is clear that if one wishes to enhance the likelihood of crystal

Fig. 7. *Electron micrographs of (left) a microcrystal of pig pancreas α-amylase and (right) a microcrystal of orthorhombic canavalin.* The microcrystals are negatively stained with uranyl acetate so that the light-colored areas represent protein while the darker areas reflect the presence of the stain. On the left the light-colored oval units are composed of the two molecules of α-amylase of M_r 50000 that comprise the asymmetric unit of these crystals. Note the extensive order that remains in the crystals and the clarity of the protein molecules even after dehydration and heavy metal staining. Note also the high proportion, nearly 50%, of the crystal that would be occupied by solvent, here replaced by the uranyl acetate

nuclei formation and growth, then one must do whatever is possible to ensure the greatest number of most stable interactions between the solute molecules in the solid state.

One may ask why molecules should arrange themselves into perfectly ordered and periodic crystal lattices, exemplified by those in Fig. 5, when they could equally well form random and disordered aggregates which we commonly refer to as precipitate. The answer is the same as for why solute molecules leave the solution phase at all: to form the greatest number of most stable bonds, to minimize the free energy, or free enthalpy, of the system. While precipitates represent, in general, a low-energy state for solute in equilibrium with a solution phase, crystals not precipitates are the states of lowest free energy.

A frequently noted phenomenon has been the formation of precipitate followed by its slow dissolution concomitant with the formation and growth of crystals. The converse is not observed. This is one empirical demonstration that crystals represent more favorable energy states.

Proteins present special problems for crystallographers

In principle, the crystallization of a protein, nucleic acid, or virus (like that shown in Fig. 6) is little different than the crystallization of conventional small molecules. Crystallization requires the gradual creation of a supersaturated solution of the macromolecule followed by spontaneous formation of crystal growth centers or nuclei. Once growth has commenced, emphasis shifts to maintenance of virtually invariant conditions so as to sustain continued, ordered addition of single molecules, or perhaps ordered aggregates, to surfaces of the developing crystal.

The perplexing difficulties that arise in the crystallization of macromolecules in comparison with conventional small

molecules stem from the greater complexity, lability and dynamic properties of proteins and nucleic acids. The description offered above of labile and metastable regions of supersaturation are still applicable to macromolecules, but it must now be borne in mind that as conditions are adjusted to transport the solution away from equilibrium by alteration of its physical and chemical properties, the very nature of the solute molecules is changing as well. As temperature, pH, pressure or solvation are changed, so may be the conformation, charge state or size of the solute macromolecules.

In addition, proteins and nucleic acids are very sensitive to their environment and if exposed to sufficiently severe conditions may denature, degrade or randomize in a manner that ultimately precludes any hope of their forming crystals. They must be constantly maintained in a thoroughly hydrated state at or near physiological pH and temperature. Thus common methods for the crystallization of conventional molecules such as evaporation of solvent, dramatic temperature variation, or addition of strong organic solvents are unsuitable and destructive. They must be supplanted with more gentle and restricted techniques.

Properties of macromolecular crystals

Macromolecular crystals are composed of approximately 50% solvent on average, though this may vary over 25 – 90% depending on the particular macromolecule [17]. The protein or nucleic acid occupies the remaining volume so that the entire crystal is in many ways an ordered gel with extensive interstitial spaces through which solvent and other small molecules may freely diffuse. This is seen quite dramatically in electron micrographs of small protein crystals such as those in Fig. 7.

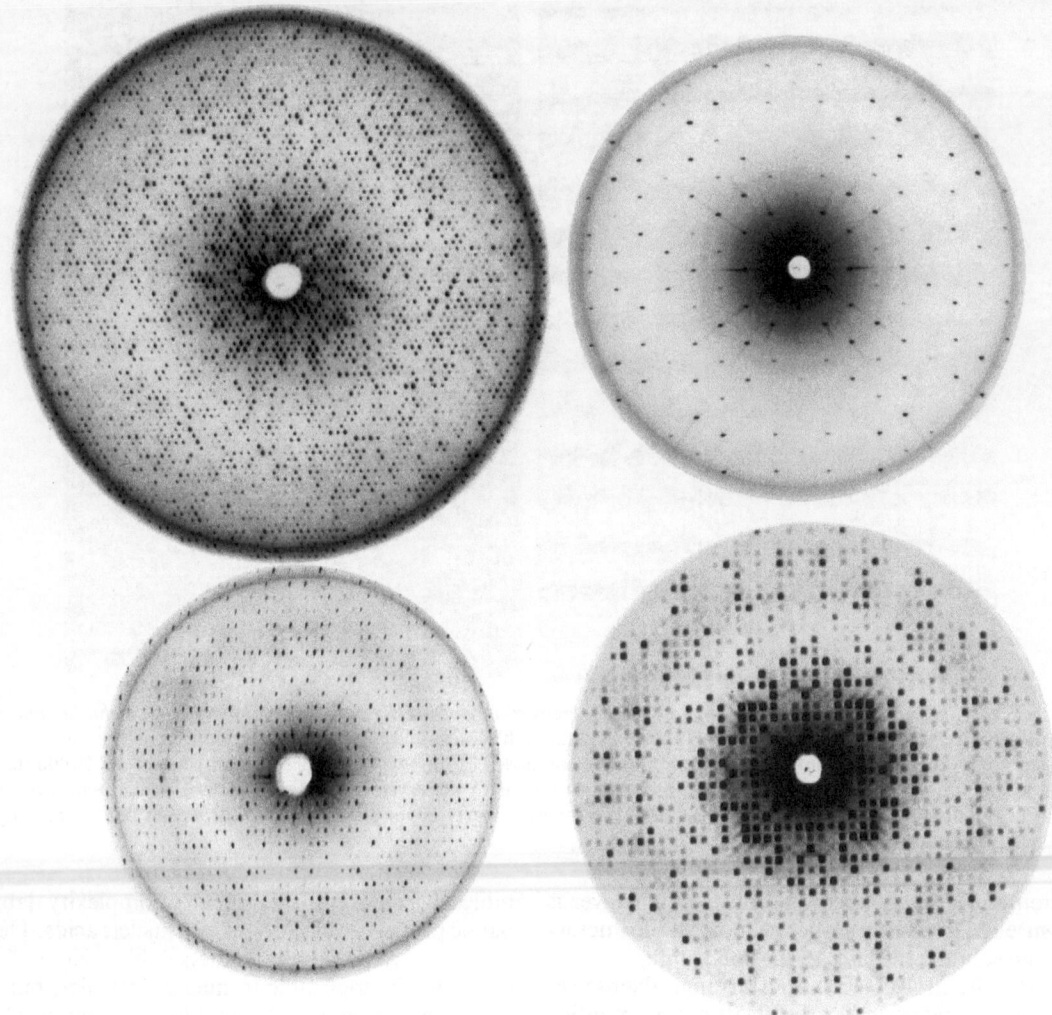

Fig. 8. *Four examples of good X-ray diffraction photographs obtained from different protein crystals.* Upper left, a hexagonal crystal of canavalin showing sixfold symmetry; upper right, a monoclinic crystal of the gene-5 DNA-unwinding protein with mm symmetry; lower left, an orthorhombic crystal of the complex between RNase A and the oligonucleotide $(dA)_4$; lower right, a tetragonal crystal of dogfish lactate dehydrogenase showing its characteristic fourfold symmetry. All of these diffraction patterns extend to a high level of resolution, and all have provided a basis for the structure determination of their constituent macromolecules

In proportion to molecular mass, the number of bonds (salt bridges, hydrogen bonds, hydrophobic interactions) that a conventional molecule forms in a crystal with its neighbors far exceeds the very few exhibited by crystalline macromolecules. Since these contacts provide the lattice interactions that maintain the integrity of the crystal, this largely explains the difference in properties between crystals of salts or small molecules, and macromolecules, as well as why it is so difficult to grow protein and nucleic acid crystals.

Because proteins are sensitive and labile macromolecules that readily loose their native structures, the only conditions that can support crystal growth are those that cause little or no perturbation of the molecular properties. Thus protein crystals, maintained within a narrow range of pH, temperature and ionic strength, must be grown from a solution to which they are tolerant. This is called the mother liquor. Because complete hydration is essential for the maintenance of structure, protein crystals are always, even during data collection, bathed in the mother liquor.

Although morphologically indistinguishable, there are important practical differences between crystals of low-molecular-mass compounds and crystals of proteins and nucleic acids. Crystals of small molecules exhibit firm lattice forces, are highly ordered, generally physically hard and brittle, easy to manipulate, usually can be exposed to air, have strong optical properties, and diffract X-rays intensely. Macromolecular crystals are by comparison usually more limited in size, are very soft and crush easily, disintegrate if allowed to dehydrate, exhibit weak optical properties and diffract X-rays poorly. Macromolecular crystals are temperature-sensitive and undergo extensive damage after prolonged exposure to radiation. In general, many crystals must be analyzed for a structure determination to be successful (for reviews of crystal structure analysis by X-ray diffraction, see [4, 18–21]), although the advent of area detectors and high intensity X-ray sources has greatly lessened this constraint in recent times.

The extent of the diffraction pattern from a crystal is directly correlated with its degree of internal order. The more extensive the pattern, or the higher the resolution to which it extends, the more uniform are the molecules in the crystal and the more precise is their periodic arrangement. The level of detail to which atomic positions can be determined by a crystal structure analysis corresponds closely with the degree of crystalline order. While conventional molecular crystals often diffract almost to their theoretical limit of resolution, protein crystals by comparison are characterized by diffraction patterns of limited extent. Some better examples of diffraction patterns from protein cyrstals are shown in Fig. 8.

The liquid channels and solvent cavities that characterize macromolecular crystals are primarily responsible for the limited resolution of the diffraction patterns. Because of the relatively large spaces between adjacent molecules and the consequent weak lattice forces, every molecule in the crystal may not occupy exactly equivalent orientations and positions in the crystal but they may vary slightly from lattice point to lattice point. Furthermore, because of their structural complexity and their potential for conformational dynamics, protein molecules in a particular crystal may exhibit slight variations in the course of their polypeptide chains or the dispositions of side groups.

Although the presence of extensive solvent regions is a major contributor to the poor quality of protein crystals, it is also responsible for their value to biochemists. Because of the very high solvent content, the individual macromolecules in protein crystals are surrounded by hydration layers that maintain their structure virtually unchanged from that found in bulk solvent. As a consequence, ligand binding, enzymatic and spectroscopic characteristics and other biochemical features are essentially the same as for the native molecule in solution. In addition, the size of the solvent channels is such that conventional chemical compounds, which may be ions, ligands, substrates, coenzymes, inhibitors, drugs or other effector molecules, may be freely diffused into and out of the crystals. Crystalline enzymes, though immobilized, are completely accessible for experimentation through alteration of the surrounding mother liquor. Thus, a protein crystal can serve as a veritable ligand binding laboratory [21].

Crystallization strategy

The strategy employed to bring about crystallization is to guide the system very slowly toward a state of reduced solubility by modifying the properties of the solvent. This is accomplished by increasing the concentration of precipitating agents or by altering some physical property, such as pH. In this way, a limited degree of supersaturation is achieved. Whatever the procedure used, no effort must be spared in refining the parameters of the system, solvent and solute, to encourage and promote specific bonding interactions between molecules and to stabilize them once they have formed. This latter aspect of the problem generally depends on the chemical and physical properties of the particular protein or nucleic acid being crystallized.

In very concentrated solutions the macromolecules may aggregate as an amorphous precipitate. This result is to be avoided if possible and is indicative that supersaturation has proceeded too extensively or too swiftly. One must endeavor to approach very slowly the point of inadequate solvation and thereby allow the macromolecules sufficient opportunity to order themselves in a crystalline lattice.

The classical procedure for inducing proteins to separate from solution and produce a solid phase is to gradually increase the level of saturation of a salt. Traditionally the salt has been ammonium sulfate, but others are also in common use. Most frequently the protein separates as a precipitate, but with appropriate care, manipulation of salt concentration can be used to grow protein crystals. At the present time, in fact, this approach has probably yielded more varieties of protein crystals than any other [22].

For a specific protein, the precipitation points or solubility minima are usually critically dependent on the pH, temperature, the chemical composition of the precipitant, and the properties of both the protein and the solvent (for more extensive discussions, see [23–27]). At very low ionic strength a phenomenon known as 'salting-in' occurs in which the solubility of the protein increases as the ionic strength increases from zero (see the solubility curve in Fig. 4 for example). The physical effect that diminishes solubility at very low ionic strength is the removal of ions essential for satisfying the electrostatic requirements of the protein molecules. As these ions are removed, and in this region of low ionic strength cations are most important [23, 24], the protein molecules seek to balance their electrostatic requirements through interactions among themselves. Thus they tend to aggregate and separate from solution. Alternatively, one may say that the chemical activity of the protein is reduced at very low ionic strength.

The salting-in effect, when applied in the direction of reduced ionic strength, can itself be used as a crystallization tool. In practice, one extensively dialyzes a protein that is soluble at moderate ionic strength against distilled water. Many proteins such as catalase, concanavalin B, and a host of immunoglobulins and seed proteins have been crystallized by this means [3, 4].

As ionic strength is increased the solution again reaches a point where the solute molecules begin to separate from solvent and preferentially form self interactions that result in crystals or precipitate. The explanation for this 'salting out' phenomenon is that the salt ions and macromolecules compete for the attention of the solvent molecules, that is, water. Both the salt ions and the protein molecules require hydration layers to maintain their solubility. When competition between ions and proteins becomes sufficiently intense, the protein molecules begin to self associate in order to satisfy, by intermolecular interactions, their electrostatic requirements. Thus dehydration, or the elimination and perturbation of solvent layers around protein molecules, induces insolubility.

Just as proteins may be driven from solution at constant pH and temperature by the addition or removal of salt [26], they can similarly be crystallized or precipitated at constant ionic strength by changes in pH or temperature. This is because the electrostatic character of the macromolecule, its surface features, or its conformation may change as a function of pH, temperature and other variables as well [23]. By virtue of its ability to inhabit a range of states, proteins may exhibit a number of different solubility minima as a function of the variables, and each of these minima may afford the opportunity for crystal formation. Thus, we may distinguish the separation of protein from solution according to methods based on variation of precipitant concentration at constant pH and temperature from those based on alteration of pH, temperature or some other variable at constant precipitant concentration. The principles described here for salting-out with a true salt are not appreciably different if precipitating agents such as poly(ethylene glycol) are used instead. In practice, proteins may equally well be crystallized from solution by increasing the poly(ethylene glycol) concentration at constant pH and temperature, or at constant poly(ethylene glycol) concentration by variation of pH or temperature [5, 28].

The most common approach to crystallizing macromolecules, be they proteins or nucleic acids, is to alter gradually the characteristics of a highly concentrated protein solution to achieve a condition of limited supersaturation. As discussed above, this may be achieved by modifying some physical property such as pH or temperature, or through equilibration with precipitating agents. The precipitating agent may be a salt such as ammonium sulfate, an organic solvent such as ethanol or methylpentanediol, or a highly soluble synthetic polymer such as poly(ethylene glycol). The three types of precipitants act by slightly different mechanisms, though all share some common properties.

In highly concentrated salt solutions competition for water exists between the salt ions and the polyionic protein molecules. The degree of competition will depend on the surface charge distribution of the protein as well. This is a function primarily of pH. Because protein molecules must bind water to remain solvated, when deprived of sufficient water by ionic competition, they are compelled to associate with other protein molecules. Aggregates may be random in nature and lead to linear and branched oligomers, and eventually to precipitate. When the process proceeds in an orderly fashion and specific chemical interactions are used in a repetitive and periodic manner to give three-dimensional aggregates, then the nuclei of crystals will form and grow.

The removal of available solvent by addition of precipitant is in principle no different than the crystallization of sea salt from tidal pools as the heat of the sun slowly drives the evaporation of water. It is a form of dehydration but without physical removal of water.

A similar effect may be achieved as well by the slow addition to the mother liquor of certain organic solvents such as ethanol or methylpentanediol. The only essential requirement for the precipitant is that at the specific temperature and pH of the experiment, the additive does not adversely effect the structure and integrity of the protein. This is often a very stringent requirement and deserves more than a little consideration. The organic solvent competes to some extent like salt for water molecules, but it also reduces the dielectric screening capacity of the intervening solvent. Reduction of the bulk dielectric increases the effective strength of the electrostatic forces that allow one protein molecule to be attracted to another.

Polymers such as poly(ethylene glycol) also serve to dehydrate proteins in solution as do salts, and they alter somewhat the dielectric properties in a manner similar to organic solvents. They produce, however, an additional important effect. Poly(ethylene glycol) perturbs the natural structure of the solvent and creates a more complex network having both water and itself as structural elements. A consequence of this restructuring of solvent is that macromolecules, particularly proteins, tend to be excluded and phase separation is promoted [29, 30].

Crystallization of macromolecules may also be accomplished by increasing the concentration of a precipitating agent to a point just below supersaturation and then adjusting the pH or temperature to reduce the solubility of the protein. Modification of pH can be accomplished very well with the vapor diffusion technique, which is described below, when volatile acids and bases such as acetic acid and ammonium hydroxide are used. This process is analogous to saturating boiling water with sugar and then cooling it to produce rock candy.

Creating the supersaturated state

Crystallization of a novel protein using any of the precipitation methods is unpredictable as a rule. Every macromolecule is unique in its physical and chemical properties because every amino acid or nucleotide sequence produces a unique three-dimensional structure having distinctive surface characteristics. Thus, lessons learned by investigation of one protein are only marginally applicable to others. This is compounded by the behavior of macromolecules which is complex owing to the variety of molecular masses and shapes, aggregate states, and polyvalent surface features that change with pH and temperature, and to their dynamic properties [7].

Because of the intricacy of the interactions between solute and solvent, and the shifting character of the protein, the methods of crystallization must usually be applied over a broad set of conditions with the objective of discovering the particular minimum (or minima) that yield crystals. In practice, one determines the precipitation points of the protein at sequential pH values with a given precipitant, repeats the procedure at different temperatures, and then examines the effects of different precipitating agents.

There are a number of devices, procedures and methods for bringing about the supersaturation of a protein solution, generally by the slow increase in concentration of some precipitant such as salt or poly(ethylene glycol). Many of these same approaches can be used as well for salting-in, modification of pH and the introduction of ligands that might alter protein solubility. These techniques have been reviewed elsewhere [3 – 6, 20, 31] and will not be dealt with exhaustively here. Only three of these, microdialysis, free interface diffusion, and vapor equilibration, will be described as examples of the best methods in current use. A drawing summarizing these techniques is seen in Fig. 9.

Dialysis is familiar to nearly all biochemists as a means of changing some properties of a protein-containing solution. The macromolecule solution is maintained inside a membrane casing or container having a semi-permeable membrane partition. The membrane allows, through its pores, the passage of small molecules and ions, but the pore size excludes passage of the much larger protein molecules. The vessel or dialysis tube containing the protein is submerged in a larger volume of liquid having the desired solution properties of pH, ionic strength, ligands, etc. With successive changes of the exterior solution and concomitant equilibration of small molecules and ions across the semipermeable membrane, the protein solution gradually acquires the desired properties of the exterior fluid.

Exactly this same procedure, in some manifestation or other, can and has been used to crystallize a number of proteins on a bulk scale [32, 33]. It is generally applicable on a large scale, however, only when substantial amounts of the protein are available. It has the advantage that by liquid-liquid diffusion through a semi-permeable membrane, a protein solution can be exposed to a continuum of potential crystal-producing conditions without actually altering directly the mother liquor. Diffusion through the membrane is slow and controlled. Because the rate of change of substituents in the mother liquor is proportional to the gradient of concentrations across the membrane, the nearer the system approaches equilibrium, the more slowly it changes.

This method has been adapted to much smaller amounts of protein by crystallographers who now use almost exclusively microtechniques involving no more than 5 – 50 μl protein solution in each trial. First described by Zeppenzauer and Zeppenzauer [34, 35] and subsequently modified and refined by numerous others, the method confines a protein solution to the interior of a glass capillary, or the microcavity of a small plexiglass button. The cavity of the button or the ends of the microcapillary tube are then closed off by a semipermeable dialysis membrane. The whole arrangement,

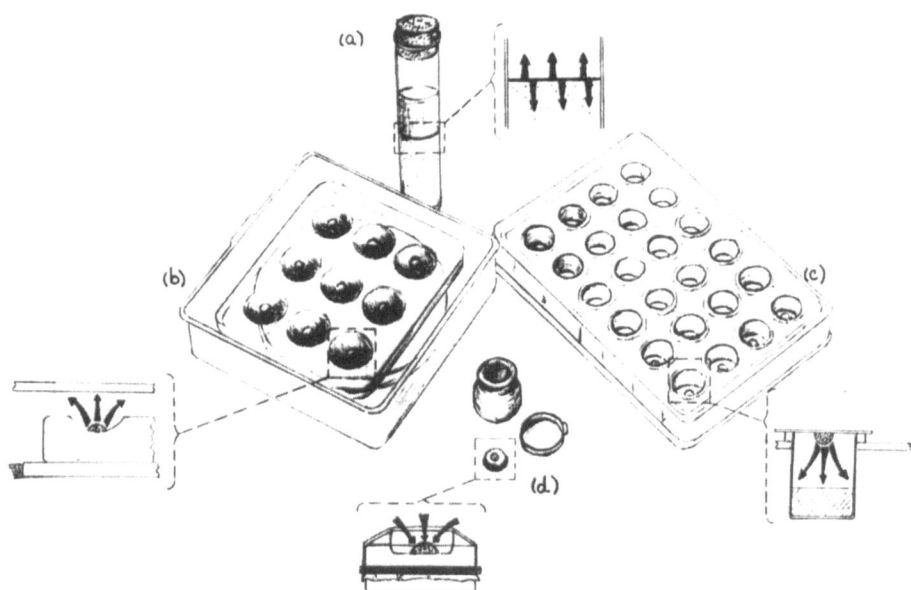

Fig. 9. *A drawing of an array of the most common microtechniques currently in use for the crystallization of macromolecules.* (a) The free interface diffusion technique; (b and c) two useful vapor diffusion methods using sitting drops on glass depression plates and hanging drops in tissue-culture plates; (d) a liquid dialysis button and a small vial which serves as the exterior liquid reservoir. All can be used with a variety of conditions and precipitating agents and each allows gradual equilibration of the protein and precipitating solutions to attain supersaturation

charged with protein solution, is then submerged in a much larger volume of an exterior liquid and the whole system kept within a closed vessel such as a test tube or vial.

If the exterior solution is at an ionic strength or pH that causes the mother liquor to become supersaturated, crystals may grow. If not, the exterior solution may be exchanged for another and the experiment continued.

The dialysis buttons, seen in Fig. 10, are particularly ingenious. Not only are they compact and easy to examine, but they have a shallow groove about their waist. After a section of wet dialysis membrane is placed over the mother-liquor-filled cavity, it can be held firmly and precisely in place by simply slipping a common rubber O ring over the top of the button and seating it in the groove.

These buttons, available from Cambridge Repetition Parts (Cambridge, UK), are now in wide use, and have proven themselves quite successful. Their cavities range in size from 5 µl to 50 µl and they can be reused many times.

A modification of the liquid-liquid diffusion method is the free interface diffusion technique [36, 37]. Here, the membrane is dispensed with completely and the mother liquor is simply layered upon a second precipitating solution in a glass tube or capillary. In some applications, the bottom solution is first frozen before the second is layered to ensure a sharp demarcation between the two.

In the free interface diffusion method, direct diffusive and convective mixing at the interface generates concentration gradients that produce regions of local supersaturation. These can, in turn, yield nuclei that may grow to a size useful for diffraction analysis. Modifications of this technique are currently being planned for experiments in zero gravity aboard the space shuttle. In zero gravity, where only diffusive interchange occurs and where stable concentration gradients of precipitant and protein can be established and maintained, the method may prove to be even more successful than on earth.

Currently, the most widely used method for bringing about supersaturation in microdrops of protein mother liquid is

vapor diffusion [3, 4, 31, 38]. This approach also exhibits a diversity and may be divided into those procedures that use a 'sitting drop' and those employing a 'hanging drop.' In any form, the method relies on the transport of either water or some volatile agent between a microdrop of mother liquor, generally 5 – 25 µl volume, and a much larger reservoir solution of 0.75 – 25 ml volume. Through the vapor phase, the droplet and reservoir come to equilibrium, and because the reservoir is of such larger volume, the final equilibration conditions are essentially those of the initial reservoir state. A variety of devices currently in use for protein crystal growth by vapor diffusion are shown in Fig. 11.

Through the vapor phase, then, water is removed slowly from the droplet of mother liquor, its pH may be changed, or volatile solvents such as ethanol may be gradually introduced. As with the liquid-liquid dialysis and diffusion methods, the procedure may be carried out at a number of different temperatures to gain advantage of that parameter as well.

According to a popular procedure, droplets of 10 – 20 µl are placed in the nine wells of depression spot plates (Corning Glass no. 7220). The samples are then sealed in transparent containers, such as Pyrex dishes or plastic boxes, which hold, in addition, reservoirs of 20 – 50 ml of the precipitating solution. The plates bearing the protein or nucleic acid samples are held off the bottom of the reservoir by the inverted half of a disposable Petri dish. Through the vapor phase, the concentration of salt or organic solvent in the reservoir equilibrates with that in the sample. In the case of salt precipitation, the droplet of mother liquor must initially contain a level of precipitant lower than the reservoir, and equilibration proceeds by distillation of water out of the droplet and into the reservoir. This holds true for nonvolatile organic solvents, such as methylpentanediol and for poly(ethylene glycol) as well. In the case of volatile precipitants, none need be added initially to the microdroplet, as distillation and equilibration proceed in the opposite direction.

This method has the advantage that it requires only small amounts of material and is ideal for screening a large number

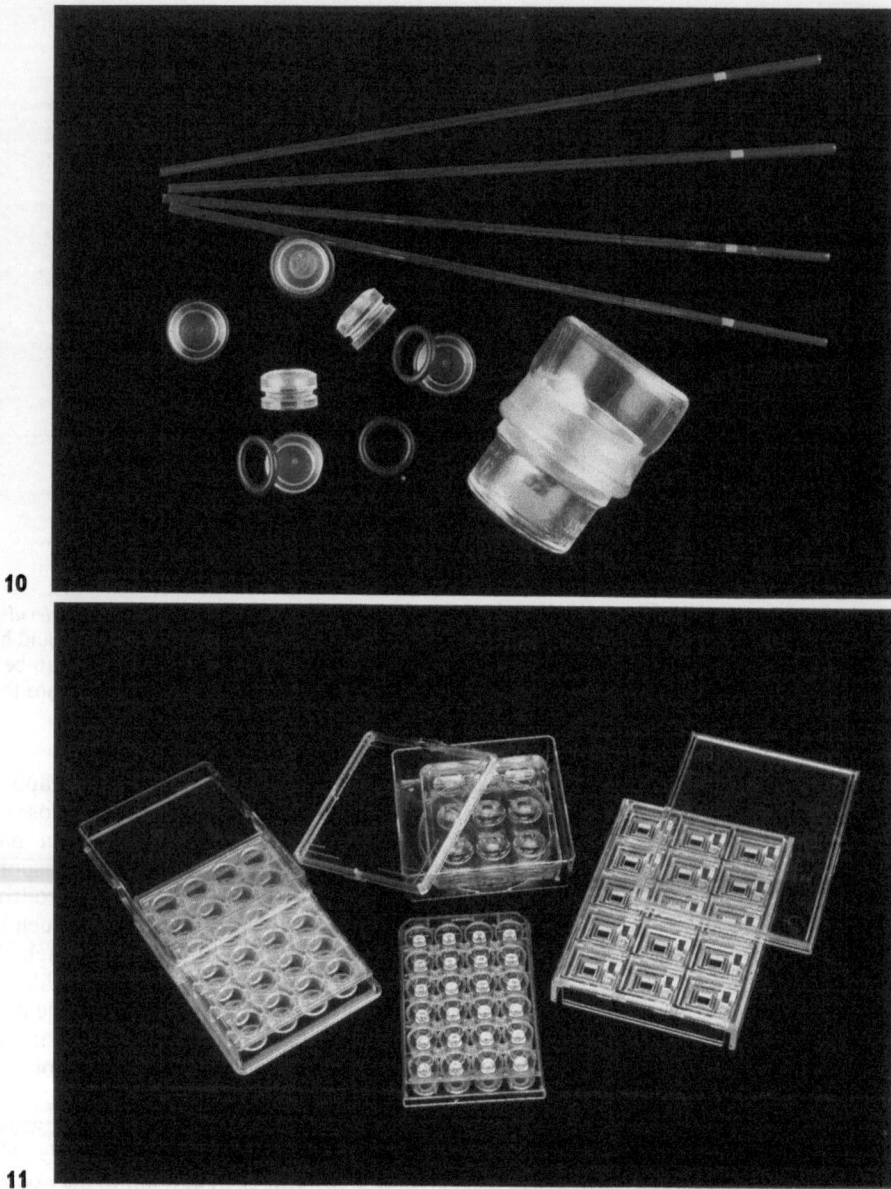

10

11

Fig. 10. *Plexiglass dialysis buttons with the O rings than maintain the dialysis membrane in place, and a small weighing bottle that serves to hold the exterior reservoir and the button.* Some microcapillaries are seen at top in which free interface diffusion can be carried out on a microscale

Fig. 11. *The four varieties of crystallization apparatus or plates now in common use.* Left the Linbro tissue-culture plate appropriated for use in hanging-drop experiments; top, the traditional sitting-drop apparatus consisting of a Corning glass depression plate in a plastic box; bottom, the Cryschem multiwell vapor diffusion plate; right, the plate from FLO Labs for both sitting and sandwich drops. The Linbro and FLO Lab plates are used in conjunction with glass cover slips while the Cryschem plate is covered with transparent plastic tape

of conditions. The major disadvantage is that all samples in a single box must be equilibrated against the same reservoir solution. It does, however, permit some flexibility in varying conditions once the samples have been dispensed, by modification of the concentration or pH of precipitants in the reservoir. When clear plastic boxes are used, large numbers of samples can be quickly inspected for crystals under a dissecting microscope and conveniently stored.

The disadvantage of identical reservoir conditions for all samples in a single box has been overcome to a great extent with the introduction of two plastic plates specifically designed for protein crystallization. One of these, sponsored by the American Crystallographic Association and manufactured by FLO Labs, Inc., is a plastic plate having accommodation for 15 protein samples. Each chamber has a separate reservoir

compartment and the mother liquor microdroplet may be either suspended from the underside of a glass cover slip as in the 'hanging drop' method or sandwiched between two glass cover slips. Sealing of the chambers from air requires silicone grease or oil between cover slips and the plastic rims of the chambers. With these plates, the optical properties are very good but equilibration tends to be slow.

A second crystallization plate [39], seen in Fig. 11, is produced by Cryschem Inc. (Riverside, CA). With these plates, the drop is sitting on the top of a clear support post that protrudes upward from a circular reservoir cavity containing the precipitating solution. The chambers can be rapidly and conveniently sealed from air by clear plastic tape pressed onto the upper surface of the plate after the reservoirs have been filled and the drops of mother liquor dispensed. Equilibration,

Table 1. *Precipitants used in macromolecular crystallization*

Salts	Organic solvents	Polymers
1. Ammonium or sodium sulfate	1. Ethanol	1. Poly(ethylene glycol)
2. Lithium sulfate	2. Isopropanol	1000, 3350, 6000, 20000
3. Lithium chloride	3. 1,3-Propanediol	2. Jeffamine T
4. Sodium or ammonium citrate	4. 2-Methyl-2,4-pentanediol	3. Polyamine
5. Sodium or potassium phosphate	5. Dioxane	
6. Sodium or potassium or ammonium chloride	6. Acetone	
7. Sodium or ammonium acetate	7. Butanol	
8. Magnesium or calcium sulfate	8. Acetonitrile	
9. Cetyltrimethyl ammonium salts	9. Dimethyl sulfoxide	
10. Calcium chloride	10. 2,5-Hexanediol	
11. Ammonium nitrate	11. Methanol	
12. Sodium formate	12. 1,3-Butyrolactone	
	13. Poly(ethylene glycol) 400	

as with the other plate, is through the vapor phase. While the optical properties are somewhat less favorable with the Cryschem devices, they are inexpensive, convenient, compact and can be rapidly utilized in vast screens of crystallization conditions.

The 'hanging drop' procedure also uses vapor phase equilibration but, with this approach, a microdroplet of mother liquor (as small as 5 µl) is suspended from the underside of a microscope cover slip, which is then placed over a small well containing 1 ml of the precipitating solution. The wells are most conveniently supplied by disposable plastic tissue culture plates (Linbro model FB−16−24-TC) that have 24 wells with rims that permit sealing by application of silicone vacuum grease or oil around the circumference. These plates provide the further advantages that they can be swiftly and easily examined under a dissecting microscope and they allow compact storage. The hanging drop technique can be used both for the optimization of conditions and for the growth of large single crystals.

While the principle of equilibration with both the 'sitting drop' and the 'hanging drop' are essentially the same, they frequently do not give the same results even though the reservoir solutions and protein solutions are identical. Presumably because of the differences in the apparatus used to achieve equilibration, the path to equilibrium is different even though the end point may be the same. In some cases there are striking differences in the degree of reproducibility, final crystal size, morphology, required time, or degree of order. These observations illustrate the important point that the pathway leading to supersaturation, the kinetics of the process, may be as important as the final point achieved.

As noted earlier, one of the most powerful techniques for producing a supersaturated protein solution is adjustment of the pH to values where the protein is substantially less soluble. This may be done in the presence of a variety of precipitants so that a spectrum of possibilities can be created whereby crystals might form. The gradual alteration of pH is particularly useful because it may be accomplished by a variety of gentle approaches that do not otherwise perturb the system or introduce unwanted effects.

Although microdialysis is probably equally suitable, more success has been achieved with the vapor diffusion method using 'sitting' microdroplets on spot plates or in one of the plastic plates available for protein crystallization. The ambient salt, effector, or buffer conditions are established prior to dispensing the microdroplets in the depressions on the plate. The pH is then slowly raised or lowered by adding a small

amount of volatile acid or base to the reservoir. Diffusion of the acid or base then occurs from reservoir to sample, just as for a volatile precipitant.

If the pH is to be raised, for example, a small drop of concentrated ammonium hydroxide can be added to the reservoir; a drop of acetic acid may be used to lower it. The pH can also be gradually lowered over a period of days by simply placing a tiny chip of solid CO_2 in the reservoir. The liberated CO_2 diffuses and dissolves in the mother liquor to form weak carbonic acid.

When a specific pH end point is required, the mother liquor may be buffered with suitable compounds at that point and then moved significantly away by addition of acid or base. The microdroplets of mother liquor may then be returned to the buffer point by addition of an appropriate volatile acid or base to the reservoir.

As with pH, proteins may vary in solubility as a function of temperature, and some are quite sensitive. One can take advantage of this property with both bulk and microtechniques [40−42]. Many of the earliest examples of protein crystallization were based on the formation of concentrated solutions at elevated temperatures followed by slow cooling. Osborne in 1892 [43] reported the crystallization of numerous plant seed globulins by cooling relatively crude extracts from 60°C to room temperature in the presence of varying concentrations of sodium chloride. These same procedures were followed by Bailey in 1942 [32, 33] and Vickery et al. in 1941 [44] to crystallize other proteins. More recent examples are those of glucagon [45], which is crystallized by dissolving the protein at 60°C in appropriate buffers and cooling slowly to room temperature, insulin [46] and deoxyribonuclease [47].

If temperature change is an important consideration or the primary means for inducing crystal formation, its rate may be manipulated to some extent by enclosing the sample at elevated temperature in a Dewar flask or insulated container and then placing the container at the desired final temperature. The use of thermal insulation in this regard has been reported for insulin and has been used as well for the crystallization of numerous conventional small molecules of biological interest.

Most protein and nucleic acids are conformationally flexible or exist in several conformational equilibrium states. In addition, they may assume a substantially different conformation when they have bound coenzyme, substrate, or other ligand. Frequently a protein with bound effector may exhibit appreciably different solubility properties than the native protein. In addition, if many conformational states are available, the presence of effector may be used to select for only one of

12

these, thereby engendering a degree of conformity of structure and system microhomogeneity that would otherwise be absent.

The effect of ligands can be employed to induce supersaturation and crystallization in those cases where its binding to the protein produces solubility differences under a given set of ambient conditions. The effector may be slowly and gently combined with the protein, for example by dialysis, so that the resulting complex is at a supersaturating level.

The addition of ligands, substrates, and other small molecules has seen widespread use in protein crystallography, since it provides useful alternatives if the apoenzyme itself cannot be crystallized.

Precipitating agents

Protein precipitants fall into four broad categories: (a) salts, (b) organic solvents, (c) long-chain polymers and (d) low-molecular-mass polymers and non-volatile organic compounds. The first two classes are typified by ammonium sulfate and ethyl alcohol respectively, and higher polymers such as poly(ethylene glycol) 4000 are characteristic of the third. In the fourth category we might place compounds such as methylpentanediol and low-molecular-mass poly(ethylene glycol). Common members of the four groups are presented in Table 1.

As already described, salts exert their effect by dehydrating proteins through competition for water molecules. Their ability to do this is proportional to the square of the valences of the ionic species composing the salt [23, 26]. Thus, multivalent ions, particularly anions are the most efficient precipitants. Sulfates, phosphates and citrates have traditionally been employed with success.

One might think there would be little variation between different salts so long as their ionic valences were the same, or that there would be little variation with two different sulfates such as Li_2SO_4 and $(NH_4)_2SO_4$. This, however, is often not the case. In addition to salting out, which is a general dehydration effect or lowering of the chemical activity of water, there are also specific protein − ion interactions that may have other consequences. This is particularly true because of the unique polyvalent character of individual proteins, their structural complexity, and the intimate dependence of their physical properties on environmental conditions and interacting molecules. It is never sufficient, therefore, when attempting to crystallize a protein to examine only one or two salts and ignore a broader range. Changes in salt can sometimes produce crystals of varied quality, morphology, and in some cases diffraction properties.

It is usually not possible to predict the degree of saturation or molarity of a salt required for the crystallization of a particular protein without some prior knowledge of its behavior. In general, however, it is a concentration just a small percentage less than that which yields an amorphous precipitate, and this can be determined for a macromolecule under a given set of conditions using only minute amounts of material.

To determine approximately the precipitation point with a particular agent, a 10-μl droplet of a 5 − 15 mg/ml protein solution can be placed in a well of a depression slide and observed under a low-power light microscope as increasing amounts of saturated salt solution or organic solvent (in 1-μl or 2-μl increments) are added. If the well is sealed between additions with a cover slip, the increases can be made over a period of many hours. Indeed, the droplet should be allowed to equilibrate for 10 − 30 min after each addition, and longer in the neighborhood of the precipitation point. With larger amounts of material the sample may be dialyzed in standard 6.35-mm (0.25 in) celluloid tubing or in a microdialysis button against a salt solution that is incremented over a period of time until precipitation occurs.

In general, the most common organic solvents utilized have been ethanol, acetone, butanols and a few other common laboratory reagents [48, 49]. It might be noted here that organic solvents have been of more general use for the crystallization of nucleic acids, particularly tRNA and the duplex oligonucleotides. There they have been the primary means for crystal growth. This in part stems from the greater tolerance of polynucleotides to organic solvents and their polyanionic surfaces which appear to be even more sensitive to dielectric effects than are proteins.

The only general rules are that organic solvents should be used at a low temperature, at or below 0°C, and they should be added very slowly and with good mixing. Since most are volatile, vapor diffusion techniques are equally suitable for both bulk or micro amounts. Ionic strength should, in general, be maintained as low as possible and whatever means are available should be taken to protect against denaturation.

Poly(ethylene glycol) is a polymer produced in various lengths, containing from several to many hundred monomers. It exhibits as its most conspicuous feature a regular alteration of ether oxygens and terminal glycols. In addition to its volume exclusion property, it shares some characteristics with salts that compete for water and produce dehydration, and with organic solvents which reduce the dielectric properties of the medium.

Aside from its general applicability and utility in obtaining crystals for diffraction analysis [28], poly(ethylene glycol) also has the advantage that it is most effective at minimal ionic strength and provides a low-electron-density medium. The first feature is important because it provides for higher ligand binding affinities than does a high-ionic-strength medium such as concentrated salt. As a consequence there is greater ease in obtaining isomorphous heavy-atom derivatives and in forming protein − ligand complexes for study by difference Fourier techniques. The second characteristic, a low-electron-density medium, implies a generally lower background or noise level for protein structures derived by X-ray diffraction and presumably, therefore, a more ready interpretation.

A number of protein structures have now been solved using crystals grown from poly(ethylene glycol). These confirm that the protein molecules are in as native a condition in this medium as in those traditionally used. This is perhaps even more so, since the larger-molecular-mass poly(ethylene glycols) probably do not even enter the crystals and therefore do not directly contact the interior molecules. In addition, it appears that crystals of many proteins when grown from poly(ethylene glycol) are essentially isomorphous with, and exhibit the same unit cell symmetry and dimensions as, those grown by other means.

Poly(ethylene glycol) is produced in a variety of polymer ranges. The low-molecular-mass species are oily liquids while those of M_r above 1000, at room temperature, exist as either waxy solids or powders. The size specified by the manufacturer is the mean M_r of the polymeric molecules, and the distribution about that mean may vary appreciably. Poly(ethylene glycol) in its commercial form does contain contaminants; this is particularly true of the high-molecular-mass forms such as those of M_r 15000 or 20000. These may be removed by simple purification procedures [50] or, in the case of the 20000-

M_r form by dialysis in low-pass dialysis or collodian tubes. There have been reports that repurified poly(ethylene glycol) has proved more effective [51]. Certainly the contaminants could be disadvantageous for some proteins.

All of the poly(ethylene glycol) sizes from M_r 400 to 20 000 have provided protein crystals, but the most useful are those in the range 2000 – 6000. Occasionally, however, a protein can not be easily crystallized using this range but yields in the presence of polymer with M_r 400 or 20000. The sizes are generally not completely interchangeable for a given protein even within the mid-range, some producing the best-formed and largest crystals only at, say, M_r 4000 and less perfect examples at other M_r. This is a parameter which is best optimized by empirical means along with concentration and temperature.

A distinct advantage of poly(ethylene glycol) over other agents is that most proteins (but not all) crystallize within a fairly narrow range of poly(ethylene glycol) concentration, this being about 4 – 18%. In addition, the exact poly(ethylene glycol) concentration at which crystals form is rather insensitive and if one is within 2 – 3% of the optimal value some success will be achieved. With most crystallizations from high-ionic-strength solutions or from organic solvents, one must be within 1 – 2% of an optimum lying anywhere between 10 – 85% saturation. The advantage of poly(ethylene glycol) is that, when conducting a series of initial trials to determine what conditions will give crystals, one can use a fairly coarse selection of concentrations and over a rather narrow total range. This means fewer trials with a corresponding reduction in the amount of protein expended. Thus, it is well suited for particularly precious proteins of very limited availability.

The time required for crystal growth with poly(ethylene glycol) as the precipitant is also generally shorter than with ammonium sulfate or methylpentanediol but occasionally longer than required by volatile organic solvents such as ethanol. Although equilibration times will depend on the differential between starting and target concentrations, if this is no more than 3 – 4%, then crystallization may occur within a few hours or a few days. It seldom requires more than two weeks. Thus evaluation of results can be made without undue demands on patience. It should be noted, however, that protein-poly(ethylene glycol) solutions are excellent media for microbes, particularly molds, and if crystallization is being attempted at room temperature or over extended periods of time, then some retardant such as azide (commonly 0.1%) must be included in the protein solutions.

Since poly(ethylene glycol) solutions are not volatile, this precipitant must be used like salt and equilibrated with the protein by dialysis, slow mixing, or vapor equilibration. This latter approach, utilizing either 15-μl hanging drops over 0.5-ml reservoirs or multi-depression glass plates in sealed chambers, has proved the most popular. When the reservoir concentration is in the range 5 – 12%, the protein solution to be equilibrated should be at an initial concentration of about half of that, which is conveniently obtained by adding an equal volume of the reservoir to that of the protein solution. When the target poly(ethylene glycol) concentration is much higher than 12%, it is advisable to start the protein equilibrating at no more than 4 – 5% below the final value. This reduces unnecessary time lags during which the protein might denature.

Crystallization of proteins with poly(ethylene glycol) has proved most successful when the ionic strength is low and difficult when high. Good buffer conditions in the neutral range are, for example, 10 – 20 mM Tris or cacodylate buffer.

Table 2. *Factors that do or could affect protein crystal growth*

1. pH and buffer
2. Ionic strength
3. Temperature and temperature fluctuations
4. Concentration and nature of precipitant
5. Concentration of macromolecule
6. Purity of macromolecules (see Table 3 regarding Microheterogeneity)
7. Additives, effectors and ligands
8. Organism source of macromolecule
9. Substrates, coenzymes, inhibitors
10. Reducing or oxidizing environment
11. Metal and other specific ions
12. Rate of equilibration and rate of growth
13. Surfactants or detergents
14. Gravity, convection and sedimentation
15. Vibrations and sound
16. Volume of crystallization sample
17. Presence of amorphous or particulate material
18. Surfaces of crystallization vessels
19. Proteolysis
20. Contamination by microbes
21. Pressure
22. Electric and magnetic fields
23. Handling by investigator and cleanliness
24. Viscosity of mother liquor
25. Heterogeneous or expitaxial nucleating agents

If crystallization proceeds too rapidly, addition of some neutral salt may be used to slow growth and better effect crystal form. Poly(ethylene glycol) is useful over the entire pH range and over a broad temperature range and shows no anomalous effects in response to either.

Factors influencing protein crystal growth

Table 2 lists physical, chemical and biological variables that may influence to a greater or lesser extent the crystallization of proteins. The difficulty in properly arriving at a just assignment of importance for each factor is substantial for several reasons. Every protein is different in its properties and, surprisingly perhaps, this applies even to proteins that differ by no more than one or just a few amino acids. There are even cases where the identical protein prepared by different procedures or at different times may show significant variations. In addition, each factor may differ considerably in importance for individual proteins. α-Amylase and catalase, for example, are clearly sensitive to temperature change, while ovalbumin and ferritin show little, if any, variation in crystallization properties as a function of that variable.

Because each protein is unique, there are few means available to predict in advance the specific values of a variable, or sets of conditions that might be most profitably explored. Finally, the various parameters under one's control are not independent of one another and their interrelations may be complex and difficult to discern. It is, therefore, not easy to elaborate rational guidelines relating to physical factors or ingredients in the mother liquor that can increase the probability of success in crystallizing a particular protein. The specific components and conditions must be carefully deduced and refined for each individual.

As already noted, temperature may be of great importance or it may have little bearing at all. In general, it is wise to duplicate all crystallization trials and conduct parallel investigations at 4°C and at 25°C. Even if no crystals are observed at either temperature, differences in the solubility behavior of the protein with different precipitants and with various effector molecules may give some indication as to whether temperature is likely to play an important role. If crystals are observed to grow at one temperature and not, under otherwise identical conditions, at the other, then further refinement of this variable is necessary. This is accomplished by conducting the trials under the previously successful conditions over a range of temperatures centered on the one that initially yielded crystals.

The only rules with regard to temperature seem to be that proteins in a high salt solution are usually more soluble at cold than warmer temperatures. Proteins, however, generally precipitate or crystallize from a lower concentration of poly(ethylene glycol), methylpentanediol or organic solvent at cold than at warmer temperature. One must remember, however, that diffusion rates are less and equilibration occurs more slowly at cold than higher temperature, so that the times required for precipitation or crystal formation may be longer at colder temperatures.

After precipitant concentration, the next most important variable in protein crystal growth appears to be pH. This follows since the charge character of a protein and all of its attendant physical and chemical consequences are intimately dependent on the ionization state of the amino acids or chemical groups that comprise the macromolecule. Not only does the net charge on the protein change with pH, but the distribution of those charges, the dipole moment of the protein, its conformation, and in many cases its aggregation state. Thus, an investigation of the behavior of a specific protein as a function of pH is perhaps the single most essential analysis that should be carried out in attempting to crystallize the macromolecule.

As with temperature, the procedure is to first conduct multiple crystallization trials at coarse intervals over a broad pH range and then repeat the trials over a finer matrix of values in the neighborhoods of those that initially showed promise. The only limitations on the breadth of the initial range screened are the points at which the protein begins to show indications of denaturation. In refining the pH for optimal growth, it should be recalled that the difference between amorphous precipitate, microcrystals, and large single crystals may be only a ΔpH of less than 0.5 [34, 35].

In addition to adjusting pH for the optimization of crystal size, it is sometimes also useful to explore variation of pH as a means of altering the habit or morphology of a crystalline protein. This is occasionally necessary if the initial crystal form is not amenable to analysis because it grows as fine needles or flat, thin plates or demonstrates some other unfavorable tendency such as striation or twinning.

There have been virtually no systematic studies of such factors as pressure, sound, vibrations, electrical and magnetic fields, or viscosity on the rate of growth or final quality of protein crystals. Similarly, studies are only now being undertaken to evaluate the effects of gravity, convection and fluid flow on protein crystal growth, final size, and perfection [52–54]. Thus it is not possible at this time to evaluate their influence definitively.

Some useful considerations

The earliest investigators of protein crystals noted that the concentration of protein in the mother liquor should be as high as possible, 10–100 mg/ml. This is particularly true if one is attempting to grow crystals of a protein for the first time. The probability of obtaining crystals is certainly enhanced by increasing the concentration of protein. Concentration alone is sometimes sufficient to drive the system into a state of supersaturation and into the labile region where stable nuclei can form. This may not, however, be the best approach in growing large, perfect crystals once optimal conditions for all other parameters have been established.

Once conditions for nucleation and growth have been identified and the investigation of variables more or less complete, the concentration of the protein should be gradually reduced in increments to moderate the growth of the crystals. As a general rule, the largest and most perfect crystals result when the rate of accretion of molecules is slow and orderly. Reduction of concentration is an effective means for controlling this.

The time required for the appearance and growth of protein crystals is quite variable and may range from a few hours in the best of cases to several months in others. Because no truly systematic investigations have been carried out, how rapidly crystals grow once visible nuclei have formed remains in question. The rate of growth may not be reflected at all in the total amount of time required to obtain crystals adequate for analysis. This includes the time required for solvent equilibration to be achieved, for crystal nuclei to form, and for full growth to occur.

When one is screening variables to establish optimal parameters, then the practical objective is to promote crystallization at the greatest possible speed to expedite determination of most probable conditions. When optimizing and refining crystallization parameters, time itself becomes an important parameter and long periods of slow growth are generally desirable.

One caution is in order. If it is observed that a long period elapses without the formation of crystals and then, well beyond the time required for solvent equilibration to have occurred, crystals begin to appear, then some possible causes should be explored. One likelihood is that the protein has, over the long time period, undergone some physical or chemical change. It may have undergone limited proteolysis, lost a coenzyme or metal ion, or undergone a slow conformational change. By forcing this same event to occur before the crystallization trials are carried out the time required for growth may be substantially reduced. Another possibility is that the apparatus in which the crystallization experiments were carried out was leaking and that very slow evaporation occurred. Thus the final concentration of precipitant may have been appreciably higher than believed. A final possibility is change in the ambient temperature. This is particularly likely when crystallization is being carried out at room temperature and heating or air conditioning systems are switched on and off as the seasons change.

The most intriguing questions with regard to optimizing crystallization conditions concern what additional components or compounds should comprise the mother liquor in addition to solvent, protein and precipitating agent. The most probable effectors are those which maintain the protein in a single, homogeneous, and invariant state. Reducing agents such as glutathione or 2-mercaptoethanol are useful to secure sulfhydryl groups and prevent oxidation. EDTA and EGTA are good if one wishes to protect the protein from heavy or transition metal ions or the alkaline earths. Inclusion of these components may be particularly desirable when crystallization requires a long period of time to reach completion.

When crystallization is carried out at room temperature in poly(ethylene glycol) or low-ionic-strength solutions, then attention must be given to preventing the growth of microbes.

These generally secrete proteolytic enzymes that may have serious effects on the integrity of the protein under study. Inclusion of sodium azide or thymol at low levels may be necessary to discourage invasive bacteria and fungi.

Substrates, coenzymes and inhibitors often serve to fix an enzyme in a more compact and stable form. Thus a greater degree of structural homogeneity may be imparted to a population of macromolecules and a reduced level of dynamic behavior achieved by complexing the protein with a natural ligand before attempting its crystallization.

In some cases an apoprotein and its ligand complexes may be significantly different in their physical behavior and can, in terms of crystallization, be treated as almost entirely separate problems. This may permit a second or third opportunity for growing crystals if the native apoprotein appears refractile. Thus, it is worthwhile, when determining or searching for crystallization conditions, to explore complexes of the macromolecule with substrates, coenzymes, analogues and inhibitors very early. In many ways, such complexes are inherently more interesting in a biochemical sense than the apoprotein when the structure is ultimately determined.

It should be pointed out that, just as natural substrates or inhibitors are often useful, they also can have the opposite effect of obstructing crystal formation. In such cases, care must be taken to eliminate them from the mother liquor and from the purified protein before crystallization is attempted. This is exemplified by many sugar binding proteins such as lectins. Concanavalin A and *Abrus precatorius* lectin can be crystallized only with great difficulty or not at all when glucosamine or galactose, respectively, are present. Pig pancreas α-amylase can also be crystallized only after residual oligosaccharides are removed from the preparation.

Finally, it should be noted that the use of inhibitors or other ligands may sometimes be invoked to obtain a crystal form different from that grown from the native protein. When crystals of the apoprotein are poorly suited for analysis, this may provide an alternative approach.

It was noted that microbial growth frequently results in proteolysis of protein samples, something to be avoided. This, however, is not always the case. It has been shown in a number of instances [55−59] that limited and controlled proteolytic cleavage of a protein can render it crystallizable while in the native state it was not. In other cases [60], limited proteolysis results in a change of crystal form to a more suitable and useful habit. It should be emphasized that these represent examples of controlled proteolysis where the end product is an essentially homogeneous population of molecules, albeit cleaved molecules.

Proteases, it seems, occasionally trim off loose ends or degrade macromolecules to stable, compact domains. These abbreviated proteins are, as a result, more invariant, less conformationally flexible and they often form crystals more readily than the native precursor. Although one might prefer the intact protein, a partially degraded form sometimes exhibits the activity and physical properties that are of primary interest. If a molecule can undergo limited digestion, this form should also be included in the crystallization strategy.

Various metal ions have been observed to induce or contribute to the crystallization of proteins and nucleic acids. In some instances these ions were essential for activity and it was, therefore, reasonable to expect that they might aid in maintaining certain structural features of the molecule. In other cases, however, metal ions, particularly divalent metal ions of the transition series, were found that stimulated crystal growth but played no known role in the macromolecules'

activity. One of the oldest examples of an animal protein being crystallized is horse spleen ferritin that forms perfect octahedra when a solution containing the protein is exposed to concentrations of Cd^{2+} ions [61]. α-Lactalbumin was similarly shown to crystallize in the presence of this ion [62] and several varieties of α-amylase crystallize spontaneously when presented with Ca^{2+} ions [63, 64]. Metal ions should be included for investigation in that class of additives which for any reason might tend to stabilize or engender conformity by specific interaction with the macromolecule.

Typical trial arrays

It is sometimes useful for those of limited experience with protein crystallization to have a flow chart or plan in advance to guide their first efforts. Similarly, it is often helpful to have a few simple objectives firmly in mind, to know where to begin. Presented in Figs 12 and 13 are general schemes for conducting crystallization trials on a protein that has not previously been crystallized. In Figs 14 and 15 are 'details' from those schemes, elaborated to show what several initial trial elements, or arrays, might typically be like.

Initially, the parameters that one wishes to establish as rapidly as possible are optimal concentration for each precipitant used, optimal pH for solubilization and crystallization, and the effect of temperature. The two precipitants that should be examined first are ammonium sulfate and poly(ethylene glycol) 4000 as representatives of salts and poly(ethylene glycol), the two major classes of precipitants in use. If quantity of protein permits than the additional two classes of organic solvents and short chain alcohols should be investigated as well. The best representatives of these latter groups are ethanol and methylpentanediol, respectively; suggestions for their use are shown in Fig. 16.

Initially, a pH range of 3.5−9.0 should be explored in ΔpH intervals of 0.5 but the range should be extended, abridged, or modified in appropriate cases. Generally, it is sufficient to set up two parallel sets of trials and maintain one set at 4°C and the other at 25°C. This will provide an indication of the possible influence and value of temperature as a variable.

If crystals of any sort are obtained in the first round of trials, then the coarse matrix of conditions is more finely sampled, evaluated, and in successive rounds the growth of the crystals optimized. If no crystals are obtained, ligand complexes or alternative forms of the protein are explored. If this fails, then effectors such as metal ions and detergents are introduced, and so on.

A major consideration in screening crystallization conditions is a reduction in the number of trials that must be carried out. Even in those happy cases where the quantity of protein is not a limitation, reduction of trials means less time and effort. Thus, one seeks to avoid conditions that are certain to be unprofitable. For example, if the protein is observed to precipitate rapidly at salt concentrations greater than 50% saturation, or at pH below 5.0, or at 4°C, then clearly the trials lying beyond those limits or at that temperature can be eliminated.

The entire strategy of crystallizing proteins is often a process of picking out those areas of variable space that have some chance of yielding success and intuiting those likely to produce failure. A major difficulty in this pursuit is that only a narrow range of conclusions are possible from each crystallization trial. The mother liquor (a) contains precipitate, (b) it is clear, (c) large crystals are present, or (d) microcrystals are present. This makes it rather difficult to know how close

16

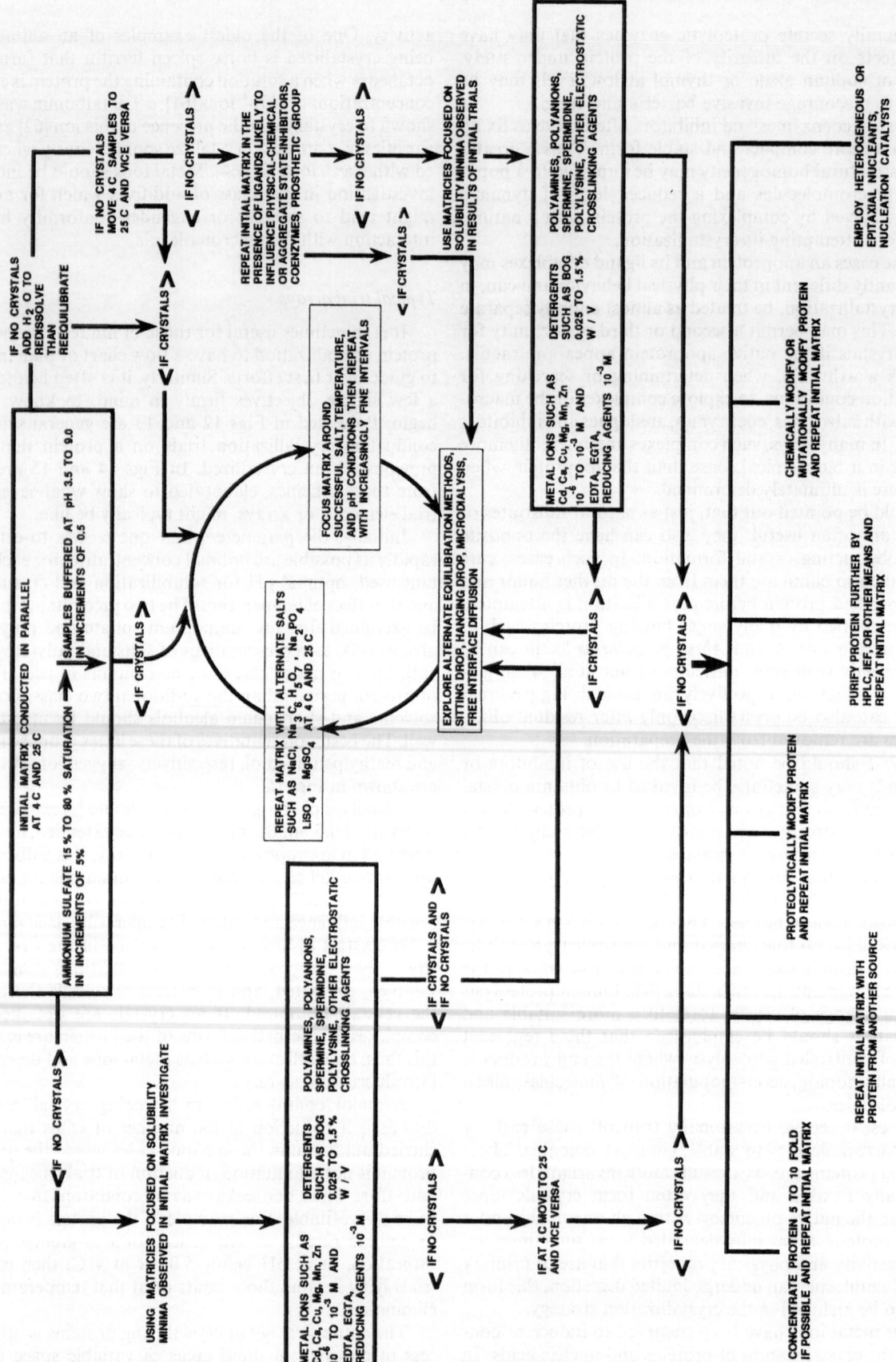

Fig. 12. *Investigation of crystallization conditions using salt*. This is a flow diagram used by the author showing the succession of variables and procedures investigated, and the order in which they are explored. Progression through the network hopefully leads to the crystallization of a protein or nucleic acid and the optimization of its growth. Other such plans of action could undoubtedly be drawn and every laboratory has its favorite variations and additions, but this diagram should serve as a guide for new experimenters. The assumption in this scheme is that ammonium sulfate will be the initial precipitate used and all other experiments will employ that, or some comparable salt. See Fig. 14 for a detailed outline of the starting matrix

17

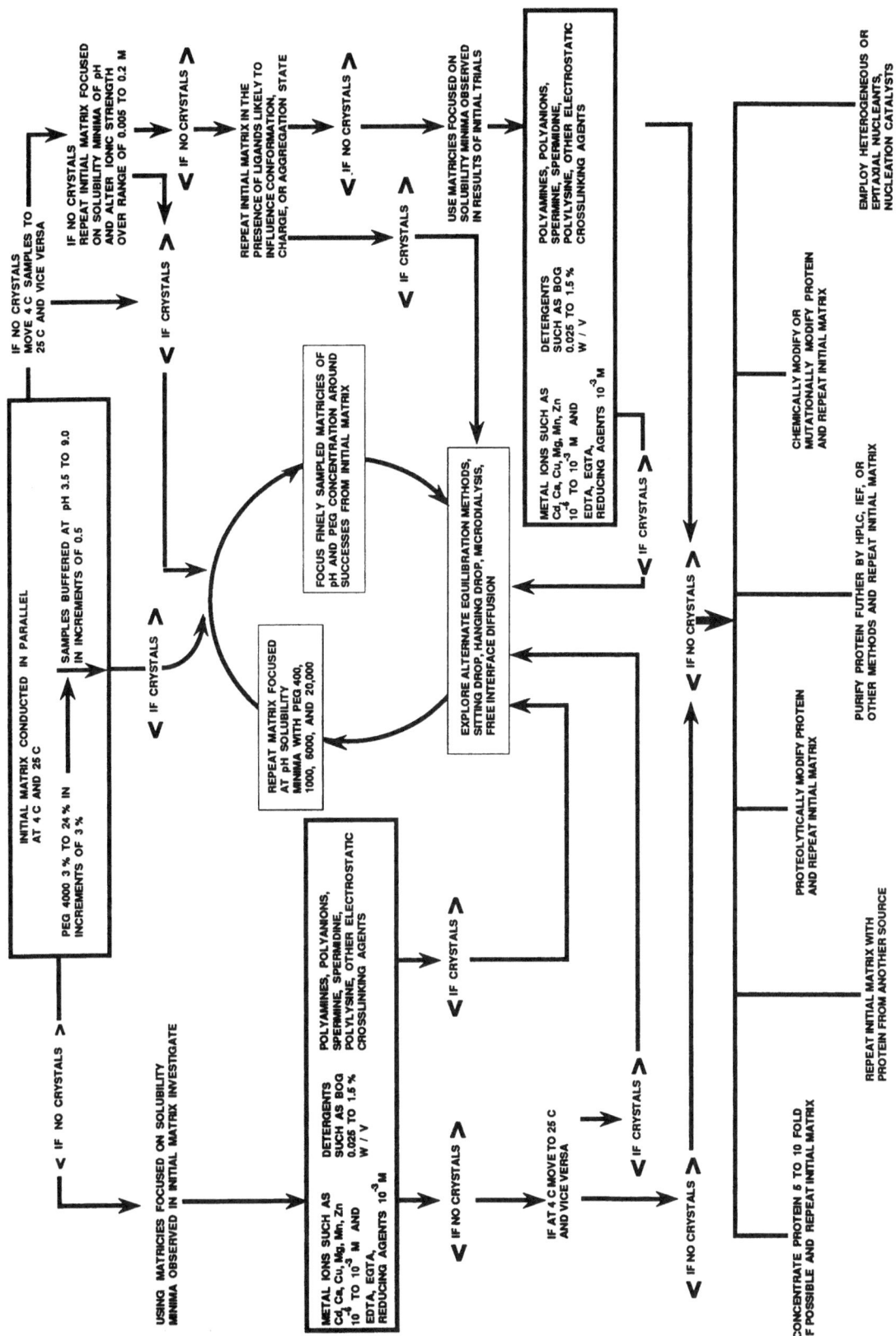

Fig. 13. *Investigation of crystallization conditions using poly(ethylene glycol)*. This is a second flow diagram, corresponding to that using salt as a precipitant (Fig. 12), but here based on poly(ethylene glycol) 4000 (PEG 4000) as the initial precipitating agent. While similar to the scheme of Fig. 12, it contains some important differences. In general, when one is attempting to grow crystals of a particular protein, the salt and poly(ethylene glycol) schemes are carried out in parallel. Often protein is limiting, and the investigator must choose between several options and decide when and how to abbreviate a specific trial matrix based on his biochemical understanding of the protein. Interpreting the results of the trials is a skill that must be developed. See Fig. 15 for a detailed outline of the starting matrix

18

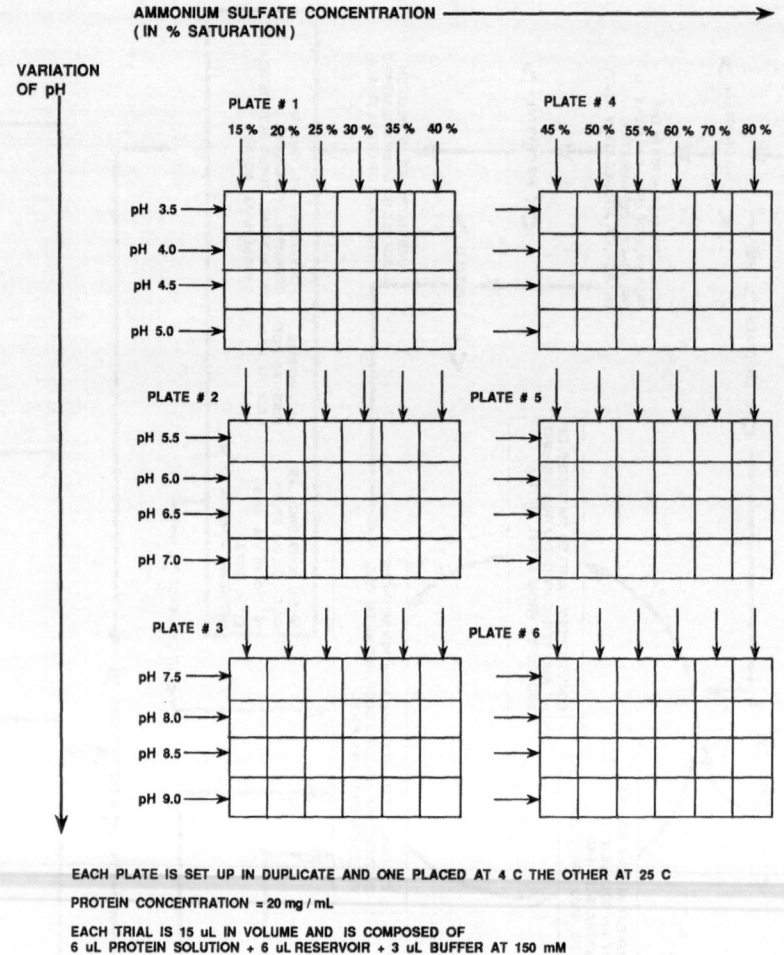

AMMONIUM SULFATE CONCENTRATION
(IN % SATURATION)

VARIATION
OF pH

PLATE # 1
15 % 20 % 25 % 30 % 35 % 40 %

pH 3.5
pH 4.0
pH 4.5
pH 5.0

PLATE # 4
45 % 50 % 55 % 60 % 70 % 80 %

PLATE # 2
pH 5.5
pH 6.0
pH 6.5
pH 7.0

PLATE # 5

PLATE # 3
pH 7.5
pH 8.0
pH 8.5
pH 9.0

PLATE # 6

EACH PLATE IS SET UP IN DUPLICATE AND ONE PLACED AT 4 C THE OTHER AT 25 C

PROTEIN CONCENTRATION = 20 mg / mL

EACH TRIAL IS 15 uL IN VOLUME AND IS COMPOSED OF
6 uL PROTEIN SOLUTION + 6 uL RESERVOIR + 3 uL BUFFER AT 150 mM

Fig. 14. *Initial salt matrix*. A detailed presentation of an initial screening matrix using ammonium sulfate as the precipitating agent and vapor diffusion by either the 'hanging drop' in a 24-well Linbro plate or the 'sitting drop' in a Cryschem 24-well vapor diffusion plate. This initial matrix investigates the effect of salt concentration and pH, generally the most important parameters, on precipitation and crystallization behavior. Equivalent matricies should be investigated at 4° and 22°C to evaluate the effect of temperature

a trial is to success unless crystals are actually present. Nevertheless, systematic approaches to the interpretation of crystallization trials are under development, and have proven useful in a number of cases [22, 65, 66].

Careful examination of precipitates formed in the mother liquor are frequently of some value. Granular precipitate, for example, sometimes is actually microcrystalline when examined under a high-power microscope; a globular or oil-like precipitate often indicates hydrophobic aggregation and suggests the use of detergents; a light, fluffy precipitate is generally a strong negative; a clear trial means a higher precipitation level is needed or another pH, and so on.

Timing is also important, and when one is carrying out initial trials it is good to examine the crystallization samples frequently, every 12 – 24 h for the first few days. In this way, conditions that cause very rapid precipitation or crystal growth can be identified. Once optimal crystal growth conditions have been precisely defined, then that is the time to lay the trials down like fine wine, in a cool, quiet place.

It is also wise to pay attention to what might be considered trivial matters. Be certain that the workplace is clean to minimize dust and microbes in the samples. When making a microdroplet, see that it is as hemispherical as possible and does not spread on the glass or plastic to yield a large surface/ volume ratio. Microfilter protein samples, work quickly to

avoid evaporation, do not carry on philosophical conversations while dispensing ingredients. Be alert for unusual events that may later explain anomalous results. Be patient.

The importance of protein purity and homogeneity

With regard the rate of growth of protein crystals, there are two important effects to consider: the transport of molecules to the face of a growing nucleus or crystal, and the frequency with which the molecules orient and attach themselves to the growing surface. Crystal growth rates can therefore be considered in terms of transport kinetics and attachment kinetics. For protein crystals which grow relatively slowly, transport kinetics, dependent primarily on physical forces and movements in the solution phase, is almost certainly the less important of the two, although as seen in Fig. 17, its effects are sometimes evident. There is not much doubt that the predominant limitation on the rate at which protein crystals nucleate and grow is, at least over most of the period of growth, a function of the rate of attachment.

The capture of molecules by a growing crystal surface requires, as in any multi-component chemical reaction, first, that the molecules to be incorporated have the correct orientation when they approach the crystal surface and, second, that they be in the proper chemical state to form interactions

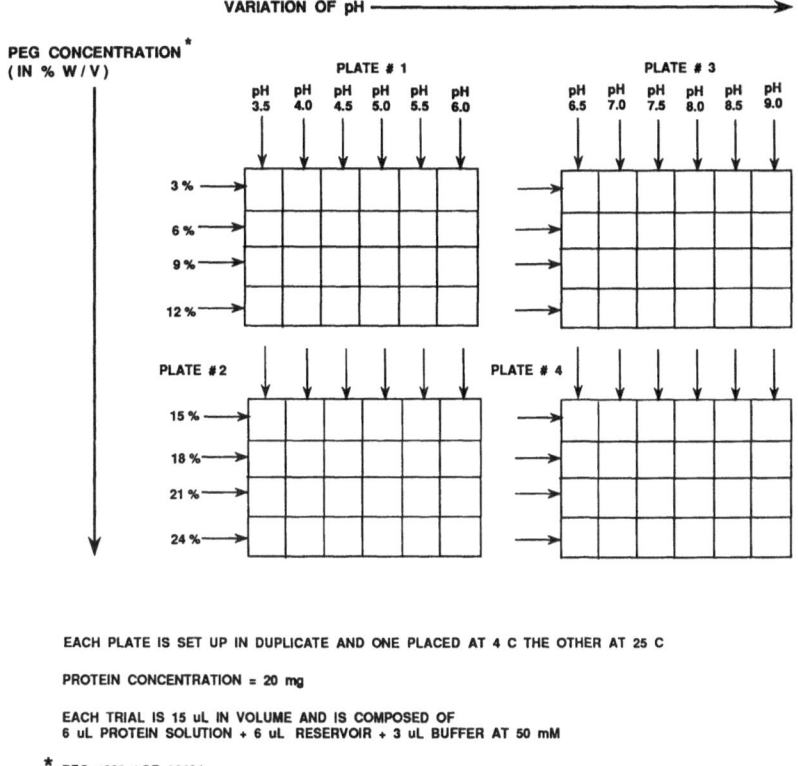

Fig. 15. *Initial poly(ethylene glycol) matrix*. A detailed plan for an initial screening matrix using poly(ethylene glycol) (PEG) the precipitating agent, but otherwise corresponding to the salt-based matrix seen in Fig. 14. Again, the matricies should be reviewed at 4° and 22°C to evaluate temperature effects, and if ligand complexes are available, these provide the basis for additional starting matricies. Either 'hanging drop' or 'sitting drop' procedures may be used in the 24-well Linbro or Cryschem plates

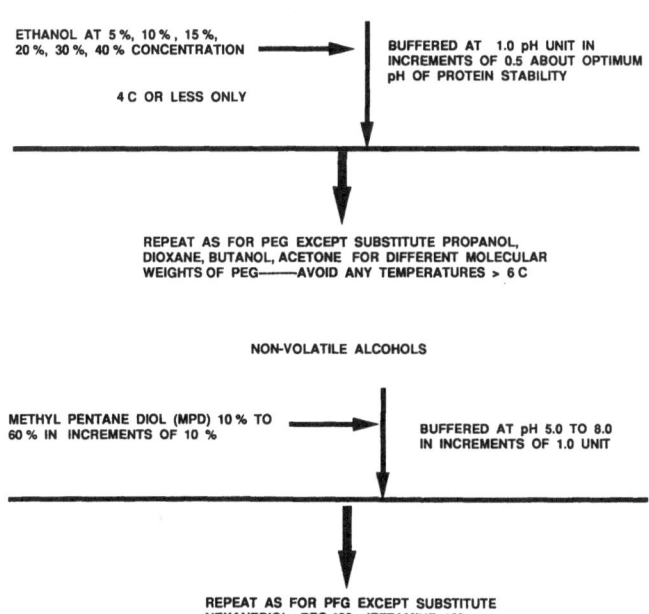

Fig. 16. *Volatile organic precipitants*. If the amount of protein is not limiting, if no success is attained with the salt or poly(ethylene glycol) approaches seen in Fig. 12 and 13, or if biochemical evidence suggests, then two additional approaches can be considered. Fundamentally, these are the same as the procedures and trials using poly(ethylene glycol) but begin and use in one case volatile organic solvents, and in the second case non-volatile reagents. These latter schemes may be particularly appropriate when the target molecule is a nucleic acid, though they have also worked well with many proteins

essential for coupling to a set of neighbors. Although there may be some things we can do to improve the statistical probability of proper orientation, there is not likely to be very much. On the other hand, we may have many opportunities to effect the frequency of attachment by enhancing the number and strength of the interactions between molecules in the lattice. We do this, for example, by optimizing the charge state of the proteins by adjusting pH, providing electrostatic crossbridges, or by minimizing the dielectric shielding between potential bonding partners by adding organic solvents such as ethanol.

Electrostatic crosslinking of protein molecules in the crystal lattice may be produced by a number of agents. This is an area of macromolecular crystallization that has been little investigated, but which the literature suggests might profitably be undertaken. We know, for example, that metal ions such as Cd^{2+} and Ca^{2+} can bridge and stabilize intermolecular contacts in crystals. This is undoubtedly the effect that causes Cd^{2+} to promote the crystallization of ferritin [61] and β-lactalbumin [67] or Ca^{2+} the crystallization of α-amylase [68]. In a similar fashion, polyamines such as spermine and spermidine have been widely used in the crystallization of nucleic acids [4], and short negatively charged oligonucleotides such as $(dA)_4$ and $(dT)_4$ were shown to be useful in promoting crystal growth of the positively charged RNase protein [69].

Certainly one major means of promoting periodic bond formation is to ensure that the population of molecules to be crystallized is as homogeneous as possible. As suggested by Table 3, this is not always straightforward. It means not only that contaminating proteins of unwanted species be elimin-

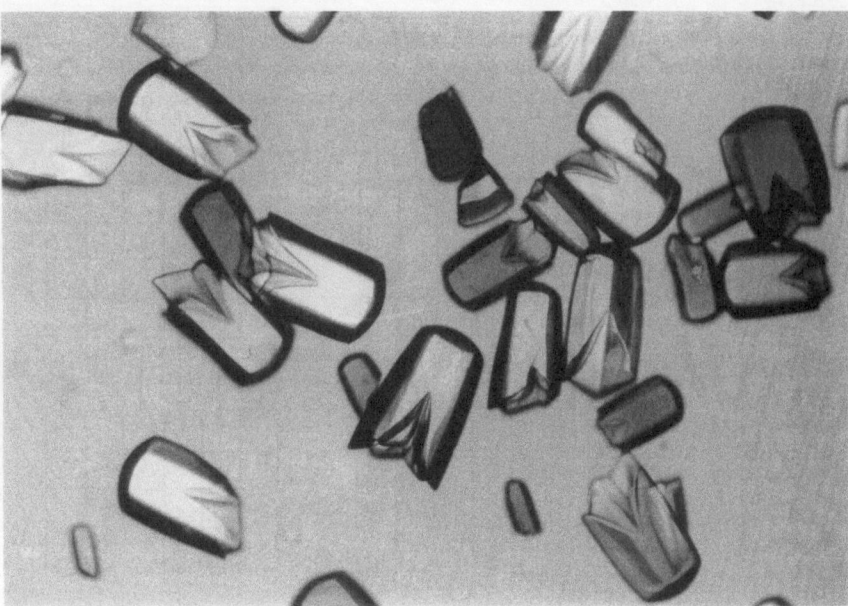

Fig. 17. *Hexagonal crystals of canavalin*. These crystals show the effects of asymmetric, rapid growth. One end of the hexagonal prisms is flat and represents the starting point of growth, the opposite end shows a deep cusp in the center arising from solute depletion during rapid growth

ated, but that within a target population all individuals assume absolute physical and chemical conformity. Because crystals have as their essential elements perfect symmetry and periodic translational relationships between molecules in the lattice, then nonuniform protein units cannot properly enter the crystal. They will not bear a proper correspondence to their neighbors. Thus, imperfect molecules will serve as inhibitors of crystal growth and bear a generally negative effect on the attachment rate. Should they enter the lattice in spite of their peculiarities, they will introduce imperfections which, by accumulation, will ultimately produce defects, dislocations, and probably termination of crystal growth.

For proteins difficult to crystallize, it is essential to take all possible measures to purify the protein free of contaminants and to do whatever is necessary to engender a state of maximum structural and chemical homogeneity. Frequently, we are misled by our standard analytical approaches, such as PAGE or IEF, into believing that a specific protein preparation is completely homogeneous. This is often illustrated for us by distinctive differences in the crystallizability of several preparations even when all analyses indicate they are identical. These imperceptible differences may be due to various degrees of microheterogeneity within preparations that lie at the margin of our ability to detect them. Table 3 lists a number of possible causes for microheterogeneity. Others could undoubtedly be added.

The pronounced effects of microheterogeneity on protein crystallization have recently received much more attention from investigators than previously. Giege et al. [70] have discussed this point in detail and provided broad evidence that purification plays a crucial role in successful crystal growth. Bott et al. [71] similarly showed the pronounced beneficial effects of isoelectric focusing on an otherwise 'pure' protein.

There are occasions when even the most intense efforts to crystallize a specific protein fail in spite of the best efforts at ultra-purification and elimination of microheterogeneity. When this occurs, an alternative is to turn to a different source of the protein. Often only very small variations in amino acid sequence, as found for example between different species of organisms, is enough to produce dramatic differences in the

Table 3. *Sources of microheterogeneity*

1. Presence, absence, or variation in a bound prosthetic group, substrate, coenzyme, inhibitor, or metal ion
2. Variation in the length or composition of the carbohydrate moiety of a glycoprotein
3. Proteolytic modification of the protein during the course of isolation or crystallization
4. Oxidation of sulfhydryl groups during isolation
5. Reaction with heavy metal ions during isolation or storage
6. Presence, absence, or variation in post-translational side-chain modifications such as methylation, amidiation, and phosphorylation
7. Microheterogeneity in the amino or carboxy terminus or modification of termini
8. Variation in the aggregation or oligomer state of the protein due to association/dissociation
9. Conformational instability due to the dynamic nature of the molecule
10. Microheterogeneity due to the contribution of multiple but nonidentical genes to the coding of the protein, isozymes
11. Partial denaturation of sample
12. Genetically different animals, plants or microorganisms that make up the source of protein preparations
13. Bound lipid, nucleic acid or carbohydrate material, or substances such as detergents used in the isolation

crystallization behavior of a protein. Thus if the protein from one source proves intractable, consider another.

It might be noted that proteins manufactured in bacteria by recombinant DNA techniques appear to be especially favorable for crystallization. There are now numerous reports of such crystalline proteins in the literature. Proteins produced in this way are apparently less subject to post translational modifications and many of the other sources of microheterogeneity that characterize naturally occurring proteins. Because this technique also provides a means of amplifying the available quantity of otherwise scarce or rare proteins, it will undoubtedly play an important role in future protein crystallization strategies.

The utility of mild detergents in the crystallization of membrane proteins is now well known, and is discussed in detail elsewhere [72], but it is useful to point out here that detergents may be of value in the crystallization of otherwise soluble proteins as well [73]. Many protein molecules, particularly when they are highly concentrated and in the presence of precipitating agents such as poly(ethylene glycol) or methylpentanediol, tend to form transient and sometimes metastable non specific aggregates. The existence of a spectrum of varying sizes, shapes and charges presents problems not appreciably different from the crystallization of a protein from a heterogeneous mixture or an impure solution composed of dissimilar macromolecules. An objective in crystallizing proteins is to limit the formation of nonuniform states and reduce the population to a set of standard individuals that can form identical interactions with one another.

Indeed, evidence from inelastic light scattering experiments [65, 74, 75] suggest that the formation of nonspecific or disordered aggregates, particularly linear aggregates, may be a major obstacle to the appearance of crystals. Conditions that tend to produce a preponderance of such aggregates, therefore, are to be avoided in favor of those yielding ordered three dimensional arrangements. Many laboratories are currently investigating and developing methods to predict, even prior to the observation of microscopic crystals, which conditions favor the latter over the former.

Non-specific aggregation is primarily a consequence of hydrophobic interactions between molecules. These place few geometrical constraints on the orientations and bonding patterns between molecules that make up an oligomer. Hydrophobic contacts make proteins adhere to one another in a more or less random fashion. Hydrogen bonds and arrays of electrostatic interactions on the other hand generally demand geometrical complementarity between the protein carriers in order to form. They thereby force macromolecules to orient themselves in specific ways with respect to one another. Thus, another objective in obtaining crystals of a protein is to discourage hydrophobic interactions and to encourage those having an electrostatic basis.

A means for limiting nonspecific aggregation is the inclusion of mild, usually nonionic, detergents in the crystallization mother liquor. McPherson et al. [73] have shown that for a fairly wide range of proteins the neutral detergent octyl β-glucoside was a positive factor in obtaining crystals useful for diffraction analysis. In addition, it was demonstrated that other detergents also exhibit helpful properties in altering crystal morphology, decreasing microcrystal formation, or improving growth patterns.

Because the key to crystallizing a macromolecule successfully often lies in the procedure, means, or solvent used to solubilize it, some careful consideration should be given to this initial step. This is particularly true of membrane, lipophilic, or other proteins which, for one reason or another, are only marginally soluble in water solutions. In addition to mild detergents, for example, there are a number of chaotropic agents that can also be employed for the solubilization of proteins. These include such compounds as urea, guanidinium hydrochloride, and relatively innocuous anions such as SCN^-, ClO_4^-, I^-, Br^- and NO_3^- [76]. These compounds, even at relatively low concentrations, may serve to increase dramatically the solubility of a protein under conditions where it would otherwise be insoluble. Gradual withdrawal of the chaotrop, for example by dialysis, could then serve as a mechanism for the crystallization of the macromolecule.

Seeding

Often it is desirable to reproduce crystals of a protein previously grown where either the formation of nuclei is limiting, or spontaneous nucleation occurs at such a profound level of supersaturation that poor growth patterns result. In such cases it is desirable to induce growth in a directed fashion at low levels of supersaturation. This can sometimes be accomplished by seeding a metastable, supersaturated protein solution with crystals from earlier trials. The seeding techniques fall into two categories, those employing microcrystals as seeds and those using larger macroseeds. In both methods, the fresh solution to be seeded should be only slightly supersaturated so that controlled, slow growth will occur. The two approaches have been described in some detail by Fitzgerald [77] and by Thaller et al. [78, 79] respectively.

In the method of seeding with microcrystals, the danger is that too many nuclei will be introduced into the fresh supersaturated solution and masses of crystals will result, none of which are suitable for diffraction analysis. To overcome this, a stock solution of microcrystals is serially diluted over a very broad range. Some dilution sample in the series will, on average, have no more than one microseed per microliter. Others will have severalfold more or none at all. Then 1 µl of each sample in the series is added to fresh protein-crystallization trials under what are perceived to be the optimal conditions for growth to occur. This empirical test should, ideally, identify the correct sample to use for seeding by yielding only one or a small number of single crystals when crystal growth is completed. Seeding solutions containing too many seeds will yield additional showers of microcrystals and seeding solutions containing too low a concentration of seeds will produce nothing at all. The optimal seeding concentration as determined by the test can then be used to seed many additional samples.

The second approach to seeding involves crystals large enough to be manipulated and transferred under a microscope. Again the most important consideration is to eliminate spurious nucleation by transfer of too many seeds. Even if a single large crystal is employed, microcrystals adhering to its surface may be carried across to the fresh solution. To avoid this, it is recommended that the macroseed be thoroughly washed by passing it through a series of intermediate transfer solutions. In so doing, not only are microcrystals removed, but if the wash solutions are chosen properly, some limited dissolution of the seed may take place. This has the effect of freshening the seed crystal surfaces and promoting new growth once it is introduced into the fresh protein solution. Again, the new solution must be at least saturated with respect to protein but not extremely so in order to ensure slow and proper growth.

Seeding is frequently a useful technique for promoting nucleation of protein crystals, or initiating nucleation and growth at a lower level of supersaturation than might otherwise spontaneously occur. This can only be done, however, where crystals, even poor crystals, of the protein under investigation have previously been obtained and can be manipulated to serve as seeds. A common problem in macromolecular crystallization is inducing crystals to grow that have never previously been observed. This reflects, of course, the salient fact that the formation of stable nuclei of protein crystals is most often the single major obstacle to obtaining any crystals at all. In those cases where the immediate problem is simply growing crystals, any crystals, then attention must be focused on the nucleation problem, and any approach that might help promote nucleation should be considered.

22

One such technique, borrowed in part from classical small molecule crystal growth methodology, is the use of heterogeneous or epitaxial nucleants. In principle, this means the induction of growth of crystals of one substance on crystal faces of another. The classical example is galium arsenide crystals that nucleate and grow from the faces of crystals of silicon.

Because protein molecules possess chemical groups, both charged and neutral, that often readily interact with small molecules, membranes, or other surfaces, the possibility presents itself that the faces of natural and synthetic minerals might help order protein molecules at their surfaces and thereby induce the formation of ordered two dimensional arrays of the macromolecules. This ordering might occur by mechanical means due to steps and dislocations on the crystal faces or by chemical means derived from a complementarity between groups on the mineral and the protein. Such cooperation between mineral faces and nascent protein crystals might be particularly favored when the lattice dimensions of the protein unit cell are integral multiples of natural spacings in the mineral crystal.

Recently, McPherson and Schlicta [80] have shown in a series of experiments using 50 different water insoluble minerals and five different proteins that both heterogeneous nucleation and epitaxial growth of protein crystals from mineral faces do indeed occur. For each of the five proteins, certain specific sets of minerals were empirically identified that promoted nucleation and growth at earlier times and lower levels of supersaturation than occurred through spontaneous events.

A second approach to enhancing the formation of crystal nuclei has been described by Ray [81]. He introduced microdroplets of various concentrations of poly(ethylene glycol) into protein solutions that were also sufficiently high in salt concentration (approximately 50% saturated with ammonium sulfate) to support crystal growth once stable nuclei were formed. He was able to show that protein left the salt-dominated phase of the mixture and concentrated itself in the poly(ethylene glycol)-rich microdroplets, sometimes reaching effective concentrations in these droplets of several hundred milligrams/milliliter. By light microscopy techniques it was demonstrated that crystal nuclei appeared first at the surface of the droplets and then proceeded to grow into the supersaturated salt solution that surrounded them, finally reaching a terminal size appropriate for X-ray analysis. In the absence of the droplets, no crystals were ever observed to form.

These experiments are encouraging in that other, perhaps even more effective, heterogeneous precipitant/solvent systems might be found that will assist in the enhancement of crystal nucleation by what Ray refers to as 'crystallization catalysts'.

A final thought

A last word of advice regarding success. Once crystals are obtained, then that should not signal the end of the chase. Better crystals for analysis, larger crystals, a more favorable crystallographic symmetry or unit cell or crystals that diffract to a higher level of resolution might all be obtained by continued examination of conditions. The ability of a specific protein to form isomorphous heavy atom derivatives and ligand complexes is often very much dependent on the crystal lattice interactions. Thus, the search for improvements should go forth in parallel as the X-ray analysis commences.

I would like to acknowledge the capable assistance of Ms Josephine Cheung and many useful conversations with Stan Koszelak, Dave Martin, Bob Cudney, John Day, David Birdsall and Ping Ko. The writing of this review was supported in part by a grant from the National Science Foundation of the United States.

REFERENCES

1. Sumner, J. B. & Somers, G. F. (1943) *The enzymes*, Academic Press, New York.
2. McPherson, A. (1989) *Sci. Am. 260*, 62 – 69.
3. McPherson, A. (1976) *Methods Biochem. Anal. 23*, 249 – 345.
4. McPherson, A. (1982) *The preparation and analysis of protein crystals*, John Wiley and Sons, New York.
5. McPherson, A. (1985) *Methods Enzymol. 114*, 112 – 120.
6. McPherson, A. (1990) *Crit. Rev. Biochem.*, in the press.
7. Creighton, T. E. (1984) *Proteins: structures and molecular properties*, Freeman Co., San Francisco.
8. Schulz, G. E. & Schirmer, R. H. (1979) *Principles of protein structure*, Springer-Verlag, New York.
9. Richardson, J. (1981) *Adv. Protein Chem. 34*, 167 – 339.
10. Miers, H. A. & Isaac, F. (1906) *J. Chem. Soc. 89*, 413.
11. Miers, H. A. & Isaac, F. (1907) *Proc. R. Soc. Lond. A79*, 322.
12. Petrov, T. G., Trevius, E. B. & Kasatkin, A. P. (1969) *Growing crystals from solution*, Plenum, New York.
13. Buckley, H. E. (1951) *Crystal growth*, Ch. 1, pp. 1-23, John Wiley and Sons, London.
14. Feigelson, R. S. (1988) *J. Cryst. Growth 90*, 1 – 13.
15. Boistelle, R. & Astier, J. P. (1988) *J. Cryst. Growth 90*, 14 – 30.
16. Rosenberger, F. (1986) *J. Cryst. Growth 76*, 618.
17. Matthews, B. W. (1968) *J. Mol. Biol. 33*, 491 – 497.
18. Glusker, J. P. & Trueblood, K. N. (1972) *Crystal structure analysis: a primer*, Oxford University Press, Oxford.
19. Stout, G. H. & Jensen, L. H. (1968) *X-ray structure determination*, MacMillan Company, New York.
20. Blundell, T. L. & Johnson, L. N. (1976) *Protein crystallography*, Academic Press, New York.
21. McPherson, A. (1987) in *Crystallography reviews* (Moore, M., ed.) vol. 1, 191 – 250, Gordon and Breach, London.
22. Gilliland, G. L. (1988) *J. Cryst. Growth 90*, 51 – 59.
23. Cohn, E. J. & Ferry, J. D. (1950) in *Proteins, amino acids and peptides* (Cohn, E. J. & Edsall, J. T., eds) Reinhold, New York.
24. Czok, R. & Bücher, Th. (1960) *Adv. Protein Chem. 15*, 315 – 415.
25. Northrop, J. H., Kunitz, M. & Herriott, R. M. (1948) *Crystalline enzymes*, Columbia University Press, New York.
26. Hofmeister, T. (1887) *Arch. Exp. Pathol. Pharmakol. 24*, 274.
27. Timasheff, S. N. & Arakawa, T. (1988) *J. Cryst. Growth 90*, 39 – 46.
28. McPherson, A. (1976) *J. Biol. Chem. 251*, 6300 – 6303.
29. Hermans, J. (1982) *J. Chem. Phys. 77*, 2193.
30. Lee, J. C. and Lee, L. L. Y. (1981) *J. Biol. Chem. 256*, 625 – 631.
31. Hirs, C. H. W., Timasheff, S. N. & Wyckoff, H. (eds) (1986) *Methods Enzymol. 114*.
32. Bailey, K. (1942) *Trans. Faraday Soc. 38*, 186.
33. Bailey, K. (1940) *Nature 145*, 934.
34. Zeppenzauer, M., Eklund, H. & Zeppenzauer, E. (1968) *Arch. Biochem. Biophys. 126*, 564.
35. Zeppenzauer, M. (1971) *Methods Enzymol. 22*, 253.
36. Salemme, F. R. (1972) *Arch. Biochem. Biophys. 151*, 533.
37. Weber, B. A. & Goodkin, P. E. (1970) *Arch. Biochem. Biophys. 141*, 489 – 498.
38. Hampel, A., Labanauskas, M., Conners, P. G., Kirkegard, L., RajBhandary, U. L., Sigler, P. B. & Bock, R. M. (1968) *Science 162*, 1384.
39. Morris, D., Kim, C. Y. & McPherson, A. (1989) *Biotechniques 7*, no. 5.
40. McPherson, A. (1985) *Methods Enzymol. 114*, 121 – 124.
41. Rosenberger, F. & Meehan, E. J. (1988) *J. Cryst. Growth 90*, 74 – 78.
42. Jacoby, W. B. (1968) *Anal. Biochem. 26*, 295.

43. Osborne, T. B. (1924) *The vegetable proteins*, 2nd edn, Longmans Green, London.
44. Vickery, H. B., Smith, E. L., Hubbell, R. B. & Nolan, L. S. (1941) *J. Biol. Chem. 140*, 613 – 624.
45. King, M. V. (1965) *J. Mol. Biol. 11*, 549.
46. Baker, E. N. & Dodson, G. (1970) *J. Mol. Biol. 54*, 605 – 609.
47. Kunitz, M. (1952) *J. Gen. Physiol. 35*, 423.
48. King, M. V., Bello, J., Pagnatano, E. H. & Harker, D. (1962) *Acta Crystallogr. 15*, 144.
49. King, M. V., Magdoff, B. S., Adelman, M. B., & Harker, D. (1956) *Acta Crystallogr. 9*, 460.
50. Ray, W. J. & Puvathingal, J. N. (1985) *Anal. Biochem. 146*, 307.
51. Jurnak, F. (1985) *J. Mol. Biol. 185*, 215 – 217.
52. DeLucas, L. J., Suddath, F. L., Snyder, R., Naumann, R., Broom, M. B., Pusey, M., Yost, V., Herren, B., Carter, D., Nelson, B., Meehan, E. J., McPherson, A. & Bugg, C. E. (1986) *J. Cryst. Growth 76*, 681 – 693.
53. Pusey, M. L. & Nauman, R. (1986) *J. Cryst. Growth 76*, 593.
54. Pusey, M., Witherow, W. K. & Naumann, R. (1988) *J. Cryst. Growth 90*, 105 – 111.
55. Wyckoff, H. W., Tsernoglou, D., Hanson, A. W., Knox, J. R., Lee, B. & Richards, F. M. (1970) *J. Biol. Chem. 245*, 305.
56. McPherson, A. & Spencer, R. (1975) *Arch. Biochem. Biophys. 169*, 650 – 661.
57. Waller, J. P., Risler, J. L., Monteilhet, C. & Zelwer, C. (1971) *FEBS Lett. 16*, 186 – 188.
58. Sorensen, S. P. L. & Hoyrup, M. (1915 – 1917) *C. R. Trav. Lab. Carlesberg 12*, 12.
59. Solomon, A., McLaughlin, C. L., Wei, C. H. & Einstein, J. R. (1970) *J. Biol. Chem. 245*, 5289 – 5291.
60. Jurnak, F. A., McPherson, A., Wang, A. H. J. & Rich, A. (1980) *J. Biol. Chem. 255*, 6751 – 6757.
61. Granick, S. (1941) *J. Biol. Chem. 146*, 451.
62. Green, D. W. & Aschaffenburg, R. (1959) *J. Mol. Biol. 1*, 54.
63. McPherson, A. & Rich, A. (1972) *Biochim. Biophys. Acta 285*, 493 – 497.
64. McPherson, A. & Rich, A. (1973) *J. Ultrastruct. Res. 44*, 75 – 84.
65. Carter, C. W., Baldwin, E. T. & Frick, L. (1988) *J. Cryst. Growth 90*, 60 – 73.
66. Cox, M. J. & Weber, P. C. (1988) *J. Cryst. Growth 90*, 318 – 324.
67. Aschaffenburg, R., Green, D. W. & Simmons, R. M. (1965) *J. Mol. Biol. 13*, 194 – 201.
68. McPherson, A. & Rich, A. (1972) *Biochim. Biophys. Acta 285*, 493 – 497.
69. Brayer, G. D. & McPherson, A. (1981) *J. Biol. Chem. 257*, 3359 – 3361.
70. Giege, R., Dock, A. C., Kern, D., Lorber, B., Thierry, J. C. & Moras, D. (1986) *J. Cryst. Growth 76*, 554.
71. Bott, R. R., Navia, M. A. & Smith, J. L. (1982) *J. Biol. Chem. 257*, 9883.
72. Michael, H. (ed.) (1990) *Crit. Rev. Biochem.*, in the press.
73. McPherson, A., Koszelak, S., Axelrod, H., Day, J., Williams, R., Robinson, L., McGrath, M. & Cascio, D. (1986) *J. Biol. Chem. 261*, 1969 – 1975.
74. Kam, Z., Shore, H. B. & Feher, G. (1978) *J. Mol. Biol. 123*, 539.
75. Kadima, W., McPherson, A., Dunn, M. F. & Jurnak, F. A. (1990) *Biophys. J. 57*, 125 – 132
76. Hatefi, Y. & Hanstein, W. G. (1969) *Proc. Natl Acad. Sci. USA 62*, 1129 – 1136.
77. Fitzgerald, P. M. D. & Madsen, N. B. J. (1987) *J. Crystal Growth 76*, 600.
78. Thaller, C., Eicher, G., Weaver, L. H., Wilson, E., Karlsson, R., Jansonius J. N. (1985) *Methods Enzymol. 115*, 132 – 135.
79. Thaller, C., Weaver, L. H., Eichele, G., Wilson, E., Karlsson, R. & Jansonius, J. N. (1981) *J. Mol. Biol. 147*, 465.
80. McPherson, A. & Shlichta P. (1988) *Science 2398*, 385 – 387.
81. Ray, W. & Bracker, C. Jr (1986) *J. Crystal Growth 76*, 562 – 576.

Eur. J. Biochem. *189*, 205−214 (1990)
© FEBS 1990

Review

Protein kinase C and T cell activation

Nicola BERRY and Yasutomi NISHIZUKA

Department of Biochemistry, Kobe University School of Medicine, Kobe, Japan

(Received November 27, 1989) − EJB 89 1420

Understanding the intracellular mechanisms by which binding of ligands, such as hormones and growth factors, to their specific receptors elicits the appropriate cellular response has long been a topic of great interest. Considerable excitement was generated when it was recognised that several receptor-ligand interactions operate via the hydrolysis of inositol phospholipids. This yields, at least, two 'second messengers', namely, inositol 1,4,5-trisphosphate [Ins(1,4,5)P_3], which causes the release of Ca^{2+} from intracellular stores, and 1,2-diacylglycerol (ac$_2$Gro), which activates the serine/threonine-specific enzyme, protein kinase C(PKC), reviewed in [1] and [2]. The pertinent question that follows is, how do PKC activation and elevation of the intracellular Ca^{2+} concentration evoke cell responses? In this review, attention has been focussed on PKC, and the consequences of its activation in resting human T cells. Evidence that PKC activity is, at least partially, responsible for activation of resting human T cells will be examined, and some of the more recent research investigating how PKC activation elicits this cell response will be described.

ACTIVATION OF RESTING HUMAN T CELLS

T cell activation

Before beginning to discuss the signal transduction mechanisms involved, the phenomenon which, in this review, constitutes 'T cell activation' should be described. Appropriate stimulation of T cells results in the transcription of over fifty genes, leading to the expression of a variety of molecules (reviewed in [3]). These include interleukin-2 (IL-2) and a 55-kDa polypeptide which combines with a constitutively expressed 70−75-kDa polypeptide [4, 5], to form the high-affinity interleukin receptor (IL-2R; reviewed in [6]). Expression of the 55-kDa polypeptide (IL-2Rα) and secretion of IL-2 is an identifiable activation state in which T cells are irreversibly committed to proliferation, for autocrine binding of IL-2 to the high-affinity IL-2R triggers DNA synthesis and completion of the mitotic cycle [7, 8]. Activation can be determined, therefore, by IL-2Rα expression and secretion of IL-2, or, since these molecules are controlled at the level of transcription [9, 10], by appearance of the mRNA for IL-2Rα or IL-2. Proliferation can be used as an indirect measurement of expression of the high-affinity IL-2R and secretion of IL-2.

Correspondence to N. Berry, Department of Biochemistry, Kobe University School of Medicine, Chuo-ku, Kobe, Japan, 650

Abbreviations. Ins(1,4,5)P_3, inositol 1,4,5-trisphosphate; ac$_2$Gro, 1,2-diacylglycerol; PKC, protein kinase C; IL, interleukin; IL-2R, interleukin-2 receptor; IL-2Rα, 55-kDa subunit of the IL-2R; Ti, T cell receptor; LFA-3, lymphocyte function-associated antigen-3; TPA, 12-*O*-tetradecanoylphorbol-13-acetate; Oco$_2$Gro, 1,2-dioctanoylglycerol; PDBt, 4β-phorbol-12,13-dibutyrate.

Note. The nomenclature of T cell surface molecules (CD2, CD3, CD4, CD8) is according to the style proposed by E. L. Reinherz [151].

There is substantial evidence that PKC activity is involved in T cell activation and this has been documented in several reviews [11 − 14], but a number of recent advances in this field have defined more fully the role of PKC in this process. One of these advances is the recognition that investigations of activation requirements should be performed on populations of highly purified resting T cells, stringently depleted of macrophages and pre-activated T cells, which markedly affect the cell response [12, 15, 16]. An analysis of the activation requirements of such T cell populations shows that at least three signals are required for optimal activation, an 'antigenic' and two types of 'accessory' signals (Table 1). It is well recognised that transcription of the IL-2Rα and IL-2 genes are differentially regulated, a matter which will be discussed in greater detail in the last section of this review, but transcription of each of these genes requires all three activation signals [17]. In two studies [16, 18] the cells were induced to respond to exogenous IL-2 by fewer signals but IL-2 responsiveness does not correlate with IL-2Rα expression [16] and the mechanism and significance of this activation state is not clear. It should be noted that even highly purified resting T cells are composed of T cell subsets which are heterogenous with respect to morphology and function (not all T cells, for example, secrete IL-2) [20]. The current information on the signal-transduction pathways involved in activation of separate T cell subsets is, however, too limited for a detailed discussion.

The antigenic signal

Physiologically, T cells are activated when antigen, appropriately presented by an accessory cell, is specifically recognised by the T cell receptor (Ti) (reviewed in [21]). This receptor is composed of a covalently linked heterodimer, usually

Table 1. Analysis of the activation requirements of populations of highly purified, resting T cells
Cell response, as determined by proliferation, or, where stated, by IL-2 secretion; nd, not done; −, no response; + + +, very good response; +, poor response

Purified, resting human T cells	Reference	Cell response to			
		antigenic signal soluble anti-Ti/ CD3 antibody	soluble anti-Ti/ CD3 antibody + IL-1	Sepharose-linked anti-Ti/CD3 antibody	Sepharose-linked anti-Ti/CD3 antibody + IL-1
	18	−	−	+	+ + +
	17	−	−	−	+ + +
	19	−	−	−	+ + +
	16	−	nd	−	+ + +
	12	−	nd	−	+ + +
Jurkat T cell line					
	15	nd	nd	− (IL-2)	+ + + (IL-2)

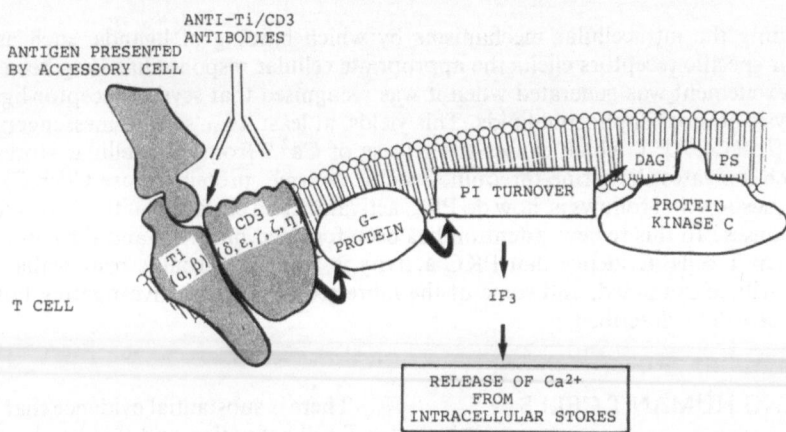

Fig. 1. A schematic diagram of the signal-transduction mechanism operated by the antigenic signal. PI, phosphatidylinositol; PS, phosphatidylserine; DAG, 1,2-diacylglycerol; G-protein, GTP-binding protein; IP₃, inositol 1,4,5-trisphosphate

containing subunits α and β, although in some cells, particularly immature T cells, a different set of subunits, γ and δ, have been recognised. These molecules are members of the 'immunoglobulin family' and are clonally defined, such that each T cell recognises a distinct antigenic epitope and, theoretically, a mature T cell population is capable of responding to a full repertoire of antigens. Transduction of the signal of antigen binding into intracellular events is probably not carried out by Ti but by a multicomponent, invariant, transmembrane complex, CD3, which is non-covalently associated with Ti (see [22]). CD3, in mice, is composed of five subunits: γ, δ, ε, ζ and η, with ζ expressed either as a homodimer or as a heterodimer with η. Homologues of γ, δ, ε and ζ have been identified in human T cells, and it is likely that η also exists (see [23] and references therein). Investigations of the signal-transduction mechanisms leading to T cell activation have been facilitated by the finding that the antigenic signal can be mimicked by antibodies directed against either Ti or CD3 (Ti/CD3), resulting in polyclonal activation of T cells [24, 25].

Evidence that the antigenic signal results in PKC activation is summarised in Fig. 1. The majority of evidence is based on experiments using cells of the human leukemia T cell line, Jurkat, which, although they must be defective at some point in their control of proliferation, resemble resting T cells both in morphology and activation requirements [15] (see Table 1), and have the advantage of being available in the large quantities necessary for such experiments. An antigenic

signal delivered by appropriately presented antigen to T cell clones or by anti-Ti/CD3 antibodies to Jurkat cells triggers hydrolysis of inositol phospholipids (PtdIns turnover), probably via a GTP-binding protein [26–29]. Ti/CD3 mutants of a murine hybridoma lacking the η subunit are unable to generate hydrolysis of inositol phospholipids effectively in response to treatment with anti-Ti/CD3 antibodies, suggesting that the η subunit is responsible for the coupling of ligand binding to this signal-transduction pathway [30].

The consequences of inositol phospholipid hydrolysis are generation of $Ins(1,4,5)P_3$, which releases Ca^{2+} from intracellular stores, and ac₂Gro, which activates PKC. Accordingly, treatment of Jurkat cells or purified resting T cells with anti-Ti/CD3 antibodies induces an immediate elevation in intracellular Ca^{2+} concentrations due to release of Ca^{2+} from intracellular stores, followed by a slightly lower, but prolonged, elevation in intracellular Ca^{2+} levels derived from extracellular sources [26, 31] (reviewed in [12]), which is the result of activation of a voltage-insensitive plasma membrane Ca^{2+} channel (reviewed in [32]). In both Jurkat cells and highly purified resting T cells, anti-Ti/CD3 antibodies result in translocation of PKC from the soluble to the particulate fraction of the cells, indicating that PKC has been activated [19, 33]. In the latter study, however, even though PKC is activated, there is apparently no elevation in intracellular Ca^{2+} concentrations. This is obviously at variance with the studies described above, the reason for which is not clear.

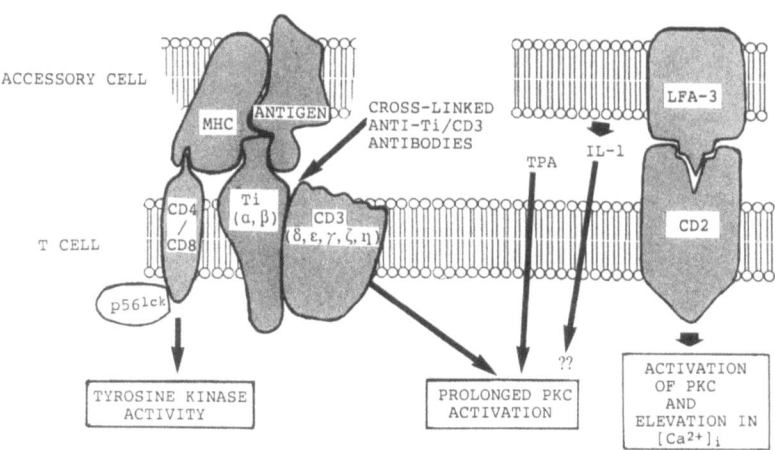

Fig. 2. A schematic diagram of the signal-transduction mechanisms operated by accessory signals. MHC, major histocompatablity complex

It is possible that a tyrosine kinase is also activated by the antigenic signal, for treatment of murine T cell clones (which can be considered as pre-activated cells), or a population of T cells mixed with accessory cells, with either antigen or anti-Ti/CD3 antibodies results in serine phosphorylation of the δ and γ subunits of CD3 and tyrosine phosphorylation of the ζ subunit [34 – 36]. The former is a consequence of activation of PKC and the latter of activation of a tyrosine kinase. Using the criteria adopted in this review, however, activation of a tyrosine kinase cannot be classified as an antigenic signal until it has been demonstrated in a population of purified resting T cells. Indeed, it will be shown later that the ζ subunit can be phosphorylated by the tyrosine kinase, $p56^{lck}$, which is thought to deliver an accessory signal via the CD4/CD8 membrane antigens. T cells contain several tyrosine kinases (see [37]), however, and clarification of the involvement of tyrosine kinases in T cell activation remains to be resolved. Whatever the case, since the antigenic signal is not sufficient for T cell activation, the signal transduction systems activated via accessory signals must be considered.

The accessory signal via CD2

The possible signal-transduction mechanisms by which the various accessory signals operate are summarised in Fig. 2. Accessory cells, such as macrophages, endothelial cells and fibroblasts, are capable of delivering a signal which is not antigen dependent but requires membrane interactions with T cells [38]. It is possible that this accessory signal is generated via the CD2 antigen, a 50-kDa transmembrane molecule expressed by T cells [39], whose physiological ligand is the lymphocyte function-associated antigen-3 (LFA-3) [40, 41]. LFA-3 is expressed on a variety of cell types, including, of significance, macrophages, endothelial cells and fibroblasts [42].

Monoclonal antibodies have been used to show that the CD2 molecule expresses several epitopes, which can be divided into two categories [see 12]. The first group, the 'high-density' epitopes are expressed at high levels on both resting and activated T cells. Monoclonal antibodies against these epitopes induce a conformational change in the CD2 molecule such that the second group of epitopes, the 'low-density' epitopes, which are cryptic on resting T cells, are exposed. Monoclonal antibodies against the latter epitopes stimulate signals leading to T cell proliferation and a combination of monoclonal antibodies against the high-density and low-den-

sity epitopes is capable, therefore, of inducing IL-2-dependent proliferation [43 – 46]. Alternatively, purified LFA-3 can synergise with certain anti-CD2 antibodies to stimulate proliferation [47, 48]. In certain T cell populations that do not express the Ti/CD3 complex, namely immature T cells and natural killer cells, the CD2 antigen is able to function independently. In mature T cells, however, activation via CD2 is dependent on surface expression of the Ti/CD3 complex [48 – 50]. Stimulation of this complex induces exposure of the low-density CD2 epitopes [46, 51] and T cells can proliferate, therefore, in response to a combination of either antigen or anti-Ti/CD3 antibodies, i.e. an antigenic signal and LFA-3 [52, 53].

Although the mechanism of signal transduction via the CD2 antigen has not been firmly established, the general consensus is that, like the Ti/CD3 complex, it operates via PtdIns turnover. Mitogenic combinations of either anti-CD2 antibodies or of an anti-CD2 antibody and LFA-3 induce PtdIns turnover in Jurkat cells and Jurkat cell clones, which, as expected, is accompanied by increases in the intracellular Ca^{2+} concentration [48, 54]. As with signal transduction via the Ti/CD3 complex, the Ca^{2+} is derived partially from intracellular stores and partially from extracellular sources [49, 54], although one group do not detect the former [51]. As stimulation of cell responses via the CD2 antigen requires exposure of a low-density epitope, there is a time lag of a few minutes while this occurs, before the elevation in Ca^{2+} levels is observed [51]. Stimulation of the CD2 antigen also results in activation of PKC, as determined both by translocation of PKC from the soluble to the particulate fraction of the cells [55] and by phosphorylation of the γ subunit of CD3 [50], which occurs as a result of PKC activation [56]. As with stimulation of cell responses, signal transduction requires the cell surface expression and function of the Ti/CD3 complex [48, 51] and stimulation of both the CD2 antigen and the Ti/CD3 complex results in a greater increase in the intracellular Ca^{2+} concentration than when the Ti/CD3 complex alone is stimulated [57].

The accessory signal via CD4/CD8

There is evidence of another accessory signal that is dependent on membrane interactions of T cells and accessory cells, indicated by the discovery that a *src*-related tyrosine kinase, $p56^{lck}$, is associated with the CD4/CD8 T cell membrane antigens [58, 59]. The physiological ligands of the CD4/CD8 antigens are the class I and class II major histocompatability antigens, which are the molecules with which antigen is presented on the surface of accessory cells (reviewed in [60]). It is conceivable, then, that when antigen is recognised by the Ti receptor, interactions between the major histocompatability antigen and CD4/CD8 could activate the tyrosine kinase and generate an accessory signal. Indeed, stimulation of the CD4 antigen by cross-linked antibodies is capable of activating p56[lck], resulting in phosphorylation of a tyrosine residue on the ζ subunit of CD3 [61]. Although the effect of p56[lck] activity on T cell responses has not been firmly established, a recent review argues persuasively in favour of the view that it enhances the T-cell-activation response [62].

In an experimental system in which purified resting T cells are activated by anti-Ti/CD3 antibodies, membrane interactions between T cells and accessory cells can be replaced by cross-linking the antibodies [17, 18]. Treatment of Jurkat cells with Sepharose-linked anti-Ti/CD3 antibodies prolongs the activation of PKC, which remains translocated to the membrane for greater than 2 h, compared to 10 min when soluble antibodies are used [19]. It appears, therefore, that the accessory signal resulting from membrane interactions, whether generated via the CD2 antigen or by Sepharose-linked antibodies, serves, at least partially, to intensify the signal generated through the Ti/CD3 complex.

The accessory signal delivered by IL-1

Activation via the CD2 antigen or by Sepharose-linked anti-Ti/CD3 antibodies does not evoke a significant cell response until a second type of accessory signal is provided in the form of interleukin-1 (IL-1), a lymphokine secreted by activated accessory cells [46] (see Table 1). Alternatively, this stimulus can be provided by antibodies against the T cell surface antigens CD28 (Tp44) or Tp67, although the physiological ligands for these molecules have not been identified [63, 64]. The signal transduction pathway generated by IL-1 is far from clear, for there are a number of reports, each suggesting a different mechanism. One point on which there is general agreement is that IL-1 does not cause PtdIns turnover, elevation in intracellular Ca^{2+} levels nor PKC activation [65–67]. Despite this, two groups have proposed mechanisms in which IL-1 prolongs or intensifies PKC activation. One demonstrated that IL-1 generates ac_2Gro from phosphatidylcholine, i.e. in the absence of PtdIns turnover [68], a mechanism that has been proposed as stimulating prolonged PKC activation in several cell types [69]. Another showed that IL-1 can reverse an inhibition of phosphatidylserine synthesis caused by anti-Ti/CD3 or anti-CD2 antibodies, and suggested that, since phosphatidylserine is required for activation of PKC, this would effectively intensify PKC activation [70]. These reports remain unsubstantiated, however, and the latter suggestion needs to be verified by evidence that activation of PKC is limited by the inhibition of phosphatidylserine synthesis observed during early T cell activation. The signals generated by antibodies to the Tp67 antigen have not yet been established and although anti-CD28 antibodies at high concentrations induce PtdIns turnover, at the lower concentrations sufficient for accessory signals, the mechanism is not clear [63].

Accessory signals can be replaced by phorbol esters

The possibility that accessory signals operate, at least partially, by prolonging or intensifying the effects of PtdIns turn-

over, particularly PKC activation, is confirmed by the demonstration that, in Jurkat cells and normal resting T cells, accessory signals can be replaced by treatment with the β-phorbol ester, 12-O-tetradecanoylphorbol-13-acetate (TPA) [18, 71]. PKC is the cellular receptor for β-phorbol esters, and, consequently, TPA treatment directly activates PKC [72]. Treatment with α-phorbol esters, such as 4α-phorbol-12,13-didecanoate, which do not activate PKC, is not able to replace accessory signals [73]. Furthermore, accessory signals can also be replaced by treatment of Jurkat cells with NaF which generates inositol monophosphate and ac_2Gro [but not Ins-$(1,4,5)P_3$ or elevation in the intracellular Ca^{2+} concentration], although the mechanism by which this occurs is not clear [74].

A PARADOX IN THE INVOLVEMENT OF PKC ACTIVITY IN T CELL ACTIVATION

The evidence outlined in the previous section reveals a paradox central to the issue of PKC and T cell activation, for, whilst stimulation of the Ti/CD3 complex activates PKC sufficiently for phosphorylation of certain target proteins, accessory signals are required to prolong or intensify PKC activation before the T cells become activated. There are three main explanations for this paradox: firstly, in the past few years it has been demonstrated that PKC is not a single homologous enzyme, but exists in the form of several subspecies which may be differentially activated and evoke specific cell responses; secondly, transient PKC activation by the antigenic signal and prolonged activation by a combination of antigenic and accessory signals may have qualitatively different effects on the cell response; finally, accessory signals may evoke other signal-transduction mechanisms which are required for T cell activation as well as PKC activation. Each of these possibilities will be examined.

Are PKC subspecies differentially activated by antigen or accessory signals?

Molecular cloning studies to establish the structure of PKC have shown that there are several subspecies of this enzyme, which can be divided into two groups on the basis of their structure (for a review see [75]). All of the subspecies are composed of a regulatory domain and a protein kinase domain, but the first group, α, βI, βII and γ, have common amino acid sequences in four conserved regions, C1–C4 (C1 and C2 constituting the regulatory domain; C3 and C4, the protein kinase domain), and different sequences in five variable regions, V1–V5, whilst the second group, δ, ε and ζ, do not have the C2 region (see Fig. 3).

It is possible to isolate α, β and γ subspecies from a mixture of PKC subspecies by hydroxyapatite column chromatography, the elution point of each subspecies being identified by comparison with the elution profiles of PKC purified from COS7 cells transfected by either α, β or γ cDNA [76, 77]. βI and βII PKC are derived from a single gene by alternative splicing [78], and, differing only in approximately 50 amino acids in the V5 region, cannot be separated by chromatographic techniques. Isolation of the first group of PKC subspecies, along with Northern blot analysis with specific oligonucleotide probes, or immunocytochemical staining with antisera specific for each of the subspecies, including βI and βII, has allowed a comparison of their biochemical characteristics and tissue localization. The second group of subspecies was identified more recently and their characterisation is more

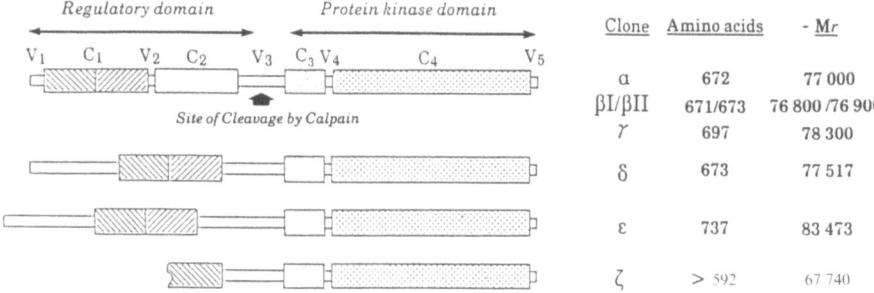

Fig. 3. A schematic representation of the structures of PKC subspecies

limited. Nevertheless, it is clear that each of the PKC sub-species differ with respect to their activation requirements and regional localization implying that each subspecies plays a specific, distinct role in signal transduction. The general characteristics of the subspecies have been reviewed recently [75], and only those pertaining to T cell activation will be described here.

In our experiments, the PKC activity obtained from purified resting T cells, when applied to a hydroxyapatite column, is resolved into two peaks, containing approximately equal amounts of PKC activity, and corresponding to rat brain type II (β) and type III (α) PKC [79]. This was confirmed by immunoblotting and by immunocytochemical staining with subspecies-specific antisera, which also revealed that the type II enzyme is composed of comparatively low amounts of βI PKC and high amounts of βII PKC. Another report has been published showing that hydroxyapatite chromatography of the PKC from each of a T-cell-enriched mononucelar cell population, Jurkat cells, and T cells from a patient with pro-lymphocytic leukemia can be resolved into two peaks, corre-sponding to rat brain type I (γ) and type II (β) PKC [80]. This group have confirmed that the second peak is β PKC, but were unable to confirm that the first peak is γ PKC. Even allowing for the difference in cell source, it is more likely that T cells express β and α, but not γ PKC because, to date, γ PKC has been detected only in central nervous tissue [76, 77, 81–83]. Also, both chromatographic isolation of the PKC subspecies and in situ hybridisation histochemistry have shown that human leukemia-lymphoma cell lines, and tissues which contain T cells, such as spleen and thymus express only type II and type III PKC [82–86] (please note that these authors use a different nomenclature for the subspecies). Of the second group of subspecies δ, ε and ζ, it has been demon-strated that ε is not expressed in either spleen or thymus tissue [87], but analysis of δ and ζ awaits the development of techniques and reagents to identify these subspecies in tissues.

As with the PKC subspecies in rat brain, the type II and type III PKC isolated from resting human T cells show differ-ential activation requirements in that, in the presence of phospholipid and ac$_2$Gro, type II, but not type III PKC, can be significantly activated in the nominal absence of Ca^{2+} [79]. The significance of this is not so much that it defines the specific activation requirements of each subspecies, for in experimental assay systems using mixed micelles instead of the vesicles used in our experiments, rat brain PKC does not show this difference in sensitivity to Ca^{2+} [88]. Rather it is an indication that there are circumstances in which the subspecies can be differentially activated.

To determine whether α, βI and βII PKC are differentially activated by antigen and accessory signals, T cells were treated with ionomycin and either synthetic forms of ac$_2$Gro, such as

1,2-dioctanoylglycerol (Oco$_2$Gro), or TPA [89]. The effect of synthetic ac$_2$Gro on T cell activation is somewhat contro-versial, for a combination of Oco$_2$Gro and ionomycin is capable of stimulating significant levels of IL-2Rα expression and proliferation in some cases, [19, 90] but not in others [65, 89]. Presumably the cell responses depend on both the composition of the T cell population and the culture con-ditions. Using human T cells stringently depleted of accessory cells and pre-activated T cells, and incubated in autologous plasma to avoid the stimulatory effects of serum, synthetic Oco$_2$Gro, in synergy with ionomycin, deliver an antigenic signal, whilst TPA and ionomycin replace the requirement for both the antigenic and the accessory signal. Direct activation of PKC in this way eliminates the possible involvement of other signal-transduction mechanisms operated via T cell sur-face molecules.

Immunocytochemical staining with subspecies-specific antisera showed that treatment of T cells with either synthetic ac$_2$Gro or TPA for 10 min, with and without ionomycin, results in a redistribution of α, βI and βII PKC from a gener-ally diffuse cytosolic distribution to a discrete focal area towards the periphery of the cell [89]. The only observable difference between the effects of TPA and synthetic ac$_2$Gro on these cells is not related to PKC subspecies, but that TPA causes a redistribution of PKC for more than 6 h, whilst synthetic ac$_2$Gro cause only a transient redistribution, maxi-mal after 10 min. These results suggest that the PKC sub-species are not differentially activated by antigenic and acces-sory signals, although this needs to be confirmed by criteria other than immunostaining. The possible involvement of δ and ζ PKC also remains to be investigated.

Is prolonged PKC activity required for T cell activation?

The hypothesis that accessory signals function, at least partially, by prolonging PKC activation is supported by sev-eral recent studies in which PKC was directly activated for different periods of time by either phorbol esters or synthetic ac$_2$Gro. Phorbol esters cause a prolonged translocation of PKC form the soluble to the particulate fraction of treated cells [91]. In both purified, resting T cells and Jurkat cells, treatment with TPA causes maximal PKC translocation be-tween 3 min and 20 min, after which the level of translocation remains constant for at least 4 h [19, 92]. Control of the duration of PKC activation can be achieved by use of the more hydrophilic phorbol ester, 4β-phorbol-12,13-dibutyrate (PDBt), which, unlike TPA, can be washed out of the culture effectively (see [93]). Activation of a population of T cells contaminated by approximately 5% macrophages and pre-activated T cells, is dependent on the duration of treatment with PDBt [9, 93]. It is important to note that T cell activation requires prolonged elevation of the intracellular Ca^{2+} concentration by ionomycin as well as prolonged PKC activation by PDBt. Both IL-2Rα expression and IL-2 secretion increase from negligible to maximal levels with increased exposure to both PDBt and ionomycin between 15 min and 4–6 h. Similar results have been reported using purified resting T cells and employing an activation system in which an antigenic signal is provided by anti-Ti/CD3 antibodies and an accessory signal by incubation with PDBt for different lengths of time [94]. This report showed that initiation of significant IL-2Rα expression and IL-2 secretion requires prolonged PKC activation for 4 h, but in this study the level of these responses continues to increase with increased exposure to PDBt for up to 18 h.

A requirement for prolonged PKC activation is also shown by treating T cells with synthetic, membrane permeable ac_2Gro, which activate PKC in a more physiological manner, for, like the ac_2Gro generated by the hydrolysis of inositol phospholipids, they are rapidly converted into the respective phosphatidic acid by diacylglycerol kinase and, consequently, only activate PKC transiently [95]. Oco_2Gro induces maximal PKC translocation after 10 min [19] or 20 min [92] followed by a gradual return to control values. In these two studies, the duration of PKC activation by Oco_2Gro is markedly different (40 min [19] as compared to 4 h [92]), probably due to the methods of PKC extraction and assay employed (see [92]). Increasing the duration of PKC activation by treating the cells with multiple doses of Oco_2Gro and ionomycin activates the cells and the levels of IL-2Rα expression and proliferation increase in proportion to the number of doses delivered [92]. Furthermore, an additional dose delivered after 15 min, whilst the PKC is still maximally activated by the initial dose, increases the responses only slightly, and the greatest increase is obtained when the additional dose is delivered after 4 h, when the PKC activity has returned to control levels. Once again, multiple doses of Oco_2Gro or ionomycin alone are not capable of activating the T cells, which requires co-stimulation with ionomycin [92].

Analysis of the presence of the mRNA for IL-2Rα and IL-2 confirmed that prolonged co-stimulation with Oco_2Gro and ionomycin is required for maximal transcription of these genes [9, 94]. Furthermore, IL-2 mRNA levels decline to control levels within 2 h following removal of the Oco_2Gro and ionomycin, whilst IL-2Rα mRNA levels remain constant. Prolonged PKC activation (as well as prolonged elevation of the intracellular Ca^{2+} concentration) is required, therefore, both for initiation of IL-2Rα expression and IL-2 secretion and for maintenance of IL-2 secretion.

Are other signal transduction mechanisms operated by accessory signals?

Multiple doses of Oco_2Gro, in synergy with ionomycin stimulate approximately 50% of the magnitude of the responses obtained with TPA and ionomycin [92]. Although this could be due to TPA exerting non-physiological effects, it is also possible that another signal transduction mechanism, in addition to prolonged PKC activation and elevation of intracellular Ca^{2+} levels, is required for maximal T cell activation. Treatment of T cells with an antigenic signal, either in the presence of accessory cells, or in combination with TPA, results in secretion of IL-2 and is accompanied by phosphorylation of the p56[lck] tyrosine kinase on serine residues [96]. This phosphorylation event is not prevented by inhibitors of PKC and is possibly the result, therefore, of activity of another protein kinase. The identity of this kinase is not clear because the PKC inhibitors, at the concentrations used, also collectively inhibited cAMP-dependent protein kinase, cGMP-dependent protein kinase, and calmodulin-dependent protein kinases. Another study [97] also showed that treatment of T cells with TPA activates a kinase other than PKC, but in this study the kinase activity was prevented by an inhibitor of calmodulin-dependent protein kinase, although its identity was not confirmed. It is possible that such a kinase may increase the magnitude of T cell activation by its phosphorylation of p56[lck], whose involvement in T cell activation has been discussed earlier in this review. Also, the possibility that the activity of the unknown TPA-activated kinase may stimulate other, as yet unrecognised, pathways should be considered.

PKC-ACTIVATED INTRACELLULAR EVENTS THAT RESULT IN T CELL ACTIVATION

The intracellular pathways by which PKC activation exerts its effects, although widely investigated, remain largely unelucidated. One of the reasons for this is that PKC activation is involved in a diverse range of cell responses (for a review see [2]). Some responses are a direct result of phosphorylation by PKC, such as the decreased affinity of the epidermal growth factor receptor for epidermal growth factor [98] and, in T cells, decreased surface expression of the Ti/CD3 complex [99] whilst other responses may be the indirect result of mechanisms initiated by PKC activity. For example, the activity of many ion channels can be regulated by PKC phosphorylation, thus altering the intracellular environment, and consequently, cell responses [100]. Some cell responses, such as exocytosis and release reactions, require only transient PKC activation and the physiological signal can be replaced by a single dose of synthetic ac_2Gro [98, 101–104]. Others, such as the differentiation of a granulocyte macrophage precursor cell line [105] and mitogenesis of Swiss 3T3 cells [106], as well as the activation of T cells, require prolonged PKC activation. Recognition that T cell activation requires prolonged PKC activation contributes to clarification of the intracellular pathways involved in this process.

Does T cell activation require down-regulation of PKC?

It is possible that prolonged activation of PKC alters the level of activity of the enzyme itself, for, in some cell types, prolonged activation of PKC by high concentrations of TPA results in a time-dependent decrease of PKC activity from the cell, probably as a result of proteolysis by the calcium-dependent protease, calpain [107–110]. Calpain catalyses the cleavage of each of α, β and γ PKC subspecies in the V3 region resulting in the generation of two major PKC fragments, the regulatory domain and the protein kinase domain [110] (see Fig. 3). The protein kinase domain is catalytically active *in vitro* in the absence of phospholipid and calcium. The physiological significance of this cleavage is not clear, as it may represent either the generation of a constitutively active PKC fragment called M-kinase [108, 109, 111] or the removal of all PKC activity from the cell [107]. In the latter case it is possible that transient activation of PKC exerts an inhibitory effect on certain cell responses which is removed when PKC is down-regulated. With respect to this, it is interesting that, in the LBRM-331A5 T cell line, transient activation of PKC by treatment with synthetic ac_2Gro inhibits mitogen-induced hydrolysis of inositol phospholipids as well as the consequent elevation in the intracellular Ca^{2+} concentration [65, 112]. This inhibitory activity has also been noted in other cell systems [113, 114]. Down-regulation of PKC in LBRM-331A5 cells by overnight treatment with relatively high concentrations of TPA (1 µM) resulted in the secretion of significantly more IL-2 in response to mitogen stimulation than in control cells. In contrast to this study, a different T cell line, Ar-5, is prevented from responding to antigen or anti-Ti/CD3 antibodies by depletion of PKC [115]. Also, prolonged activation of PKC by lower concentrations of phorbol esters, that still elicit maximal T cell responses, results only in partial down-regulation. In one study [116], treatment of lymphocytes with 100 nM TPA caused a 33–80% decrease in cellular PKC levels, in another [117], 70 nM PDBt caused a 75–80% decrease and in a third [92], 10 nM TPA, caused only a 5–30% decrease.

Partial down-regulation of total PKC activity could be due to selective down-regulation of certain PKC subspecies, particularly as α, β and γ PKC subspecies are differentially susceptible to cleavage by calpain [110] and treatment of the pre-B, pre-T leukemia cell line, KM3, with TPA causes the disappearance of the intact subspecies at different rates [118]. In T cells, however, immunohistochemical staining with antisera specific for α, βI and βII PKC, after 24 h of treatment with 10 nM TPA, showed that all three subspecies are still present [89]. The expression of either of δ and ζ PKC subspecies, and their activity and possible down-regulation, remain to be resolved, but the above results suggest that the different cell responses to prolonged and transient PKC activation are not a result of selective down-regulation of PKC subspecies. In confirmation of this, prolonged activation of PKC by multiple doses of Oco_2Gro does not cause significant down-regulation of PKC (N. Berry, unpublished observations). It appears, therefore, that prolonged PKC activity is required to exert a positive effect on gene transcription via protein phosphorylation.

Possible targets of PKC in T cells

A variety of proteins in T cells have been reported as being substrates for PKC and in several cases the phosphorylation is associated with T cell activation [119–131]. The most well-characterised of the phosphorylated proteins are the γ subunit of the CD3 complex [56, 126], the CD4 and CD8 antigens [128–131] and the CD45 antigen [132]. Phosphorylation of CD3 and CD4 is accompanied by decreased expression of these molecules on the cell surface [99, 128, 130], which, in the case of CD3, is due to an increased rate of internalisation [133]. The functional significance of this is not clear, but it has been suggested that it results in decreased responsiveness of the cell to the ligand [99]. If this were the case, activation of PKC would have a dual effect on T cells, a positive stimulatory effect on the transcription of the IL-2Rα and IL-2 genes, and a negative feedback role, regulating activation of the cell via cell surface molecules. Such a 'dual effect' of PKC activation has been noted in other cell types (see [75]). Phosphorylation of the CD45 antigen may be functionally significant because it is a tyrosine phosphatase [134] which is capable of regulating signal transduction and, consequently, cell responses, via both the Ti/CD3 complex and the CD2 antigen [135]. Despite this, no link between phosphorylation of any of the proteins mentioned above and regulation of gene transcription has been identified.

Control of transcription of the IL2-Rα and IL-2 genes

A more profitable approach has been to identify the regions of the IL-2Rα and IL-2 genes that control transcription and then to investigate the signal-transduction pathways that might regulate them. Transcription of IL-2Rα is controlled by an enhancer region lying 5′ to the transcription-initiation site. This enhancer region contains two functional sequences, one of which binds the NFκB transcription factor, or an NFκB-like factor named HIVEN 86A [136, 137]. The presence of this and another, as yet uncharacterised, sequence are critical for transcription [137]. The IL-2 gene is also under control of an enhancer region upstream of the transcription initiation site, but this contains the consensus binding sites for four transcription factors, namely, NFAT-1, NFIL-2A, AP-1 and NFκB which, apparently, act in concert to induce transcription [138] (reviewed in [3]). PKC activation does not stimulate binding of NFAT-1 and NFIL-2A to their respective functional sites, and so transcription of the IL-2 gene may be controlled by other signal-transduction pathways as well as PKC. This is borne out by reports in which secretion of IL-2 is more stringently controlled than expression of IL-2Rα. In mixed populations of T cells and accessory cells, IL-2Rα is expressed in response to PKC activity alone [139, 140], whilst in a 98% pure T cell population maximal expression requires low concentrations of calcium ionophores [93, 141]. In both mixed and purified T cell populations, however, appearance of IL-2 mRNA and secretion of IL-2 requires a higher threshold concentration of the ionophore [139, 141], (reviewed in [12]). It is possible, therefore, that activation of either one or both of NFAT-1 and NFIL-2A is controlled by a Ca^{2+}-dependent signal-transduction mechanism. Likewise, the tyrosine kinase activity and the unknown TPA-activated kinase described above may contribute to T cell activation via these regulatory regions.

The requirement for PKC activity in transcription of the IL-2Rα and IL-2 genes becomes apparent when activation of the AP-1 and NFκB consensus binding sites is examined. AP-1 binds to a 9-bp motif that is present in the enhancer regions of several genes in a variety of cell types. Transcription of each of these genes is initiated by treatment of the cells with TPA, and tandem-linkage of synthetic AP-1-binding sites to heterologous genes confers TPA inducibility [142, 143]. AP-1 consists of a complex of c-jun and c-fos or fos-related antigens, which are induced in stimulated cells (see [144]). The c-fos-oncogene is induced in cells by treatment with growth factors or with TPA [see 144]. Accordingly, activation of T cells with mitogens or with phorbol esters and ionomycin cause an early and transient appearance of c-fos mRNA followed by c-fos protein [145, 146]. The induction of c-jun in T cells has not been reported. An interesting scenario can be envisioned, therefore, in which activation of PKC firstly stimulates transcription of the c-fos, and possibly c-jun, proto-oncogenes, which then induce transcription of the IL-2Rα and IL-2 genes. Furthermore, both c-fos and c-jun are phosphorylated on serine and threonine residues, which may regulate the interaction of AP-1 with its binding site. Although it has been demonstrated that the phosphorylation of c-fos product is due to the activity of cAMP-dependent kinase [147], it is possible that phosphorylation by PKC may also be directly or indirectly involved. It should be noted, however, that treatment of T cells with the protein synthesis inhibitor cycloheximide does not abolish the transcription of either the IL-2Rα or IL-2 genes, suggesting that induction of c-fos or c-jun is not a prerequisite for T cell activation [145].

Like AP-1, the NFκB transcription factor is involved in the regulation of transcription of a variety of genes in several cell types. In a recent review [148], a model for the mechanism of regulation of gene transcription by NFκB was proposed based on the results from several recent studies on this transcription factor. In resting T cells NFκB is complexed to a protein, inhibitor-κB, that inhibits its DNA-binding activity. Activation of the cell causes modulation of inhibitor-κB such that NFκB is released and translocates to the nucleus. As treatment of the cells with phorbol esters induces the appearance of NFκB DNA-binding activity and its translocation to the nucleus, the modulation of inhibitor-κB may take the form of phosphorylation by PKC. This hypothesis is supported by experiments in which treatment of cytosol with PKC *in vitro* releases NFκB from its inhibitor but there is no evidence as to whether PKC phosphorylates inhibitor-κB directly, or whether it acts via intermediate proteins. In T cells, NFκB DNA-binding activity is induced by treatment with anti-Ti/CD3 antibodies [149, 150]. Since the phosphorylation of intracellular proteins is finely regulated by the activity of kinases and phosphatases (see [37, 93]), prolonged PKC activity may be required for maintaining the phosphorylation of inhibitor-κB for a critical length of time, whilst NFκB stimulates initiation of gene transcription.

CONCLUSION

In conclusion, a review of the literature on the intracellular mechanisms involved in T cell activation reveals that this process is probably the result of interaction of a number of signal-transduction pathways. One of these is PKC activation, which is required for a prolonged length of time, brought about, physiologically, by a combination of antigenic and accessory signals. This may be required for maintenance of phosphorylation of a protein whose activity is critical in regulation of transcription of the IL-2Rα and IL-2 genes, such as a transcription factor. There is less evidence that the other transduction mechanisms mentioned, such as tyrosine kinases and Ca^{2+}-dependent kinases play a crucial role in T cell activation. Recently, however, considerable interest has centred on cross-talk between a variety of different kinases and it is likely that T cell activation results form dynamic inter-reactions between PKC and other transduction pathways.

We would like to thank Dr. M. S. Shearman for carefully checking this manuscript. N. B. is a recipient of a research fellowship from the Japan Society for the Promotion of Science. Work from this laboratory cited in this review was supported in part by research grants from the Scientific Research Fund of the Ministry of Education, Science and Culture, Japan, Muscular Dystrophy Association, USA, Yamanouchi Foundation for Research on Metabolic Disorders, Juvenile Diabetes Foundation International, USA, Merck Sharp & Dohme Research Laboratories, Biotechnology Laboratories of Takeda Chemical Industries, Ajinomoto Central Research Laboratories, Meiji Institute of Health Sciences and New Lead Research Laboratories of Sankyo Co.

REFERENCES

1. Berridge, M (1989) *Nature 341*, 197–205.
2. Nishizuka, Y. (1986) *Science 233*, 253–392.
3. Crabtree, G. R. (1989) *Science 243*, 355–361.
4. Nishi, M., Ishida, Y. & Honjo, T. (1988) *Nature 331*, 267–269.
5. Dukovich, M., Wano, Y., Thuy, L-t-B., Katz, P., Cullan, B. R., Kehrl, J. H. & Greene, W. C. (1987) *Nature 327*, 518–522.
6. Wang, H.-M. & Smith, K. A. (1987) *J. Exp. Med. 166*, 1055–1069.
7. Meuer, S. C., Hussey, R. E., Cantrell, D. A., Hodgdon, J. C., Schlossman, S. F., Smith, K. A. & Reinherz, E. L. (1984) *Proc. Natl Acad. Sci. USA 81, 1509–1513.*
8. Cantrell, D. A. & Smith, K. A. (1984) *Science 224*, 1312–1316.
9. Kumagai, N., Benedict, S. H., Mills, G. B. & Gelfand, E. W. (1987) *J. Immunol. 139*, 1393–1399.
10. Kronke, M., Leonard, W. J., Depper, J. M. & Greene, W. C. (1985) *J. Exp. Med. 161*, 1593–1598.
11. Isakov, N., Mally, M. I., Scholz, W. & Altman, A. (1987) *Immunol. Rev. 95*, 89–111.
12. Linch, D. C., Wallace, D. L. & O'Flynn, K. (1987) *Immunol. Rev. 95*, 137–159.
13. Nel, A. E., Wooten, M. W. & Galbraith, R. M. (1987) *Clin. Immunol. Immunopathol. 44*, 167–186.
14. Weiss, A. & Imboden, J. B. (1987) *Adv. Immunol. 41*, 1–38.
15. Manger, B., Weiss, A., Weyand, C., Goronzy, J. & Stobo, J. D. (1985) *J. Immunol. 135*, 3669–3673.
16. Meuer, S. C. & Meyer zum Buschenfelde, K.-H. (1986) *J. Immunol. 136*, 4106–4112.
17. Williams, J. M., Deloria, D., Hansen, J. A., Dinarello, C. A., Loertsher, R., Shapiro, H. M. & Strom, T. B. (1985) *J. Immunol. 135*, 2249–2255.
18. Palacios, R. (1985) *Eur. J. Immunol. 15*, 645–651.
19. Manger, B., Weiss, A., Imboden, J., Laing, T. & Stobo, J. D. (1987) *J. Immunol. 139*, 2755–2760.
20. Mosmann, T. R., Cherwinski, H., Bond, M. W., Giedlin, M. A. & Coffman, R. L. (1986) *J. Immunol. 136*, 2348–2357.
21. Marrack, P. & Kappler, J. (1987) *Science 238*, 1073–1079.
22. Weiss, A., Imboden, J., Hardy, K., Manger, B., Terhorst, C. & Stobo, J. (1986) *Annu. Rev. Immunol. 4*, 593–619.
23. Baniyash, M., Garcia Morales, P., Bonifacino, J. S., Samelson, L. E. & Klausner, R. D. (1988) *J. Biol. Chem. 263*, 9874–9878.
24. van Wauwe, J. P., de Mey, J. R. & Goossens, J. G. (1980) *J. Immunol. 124*, 2708–2713.
25. Meuer, S. C., Hodgdon, J. C., Hussey, R. E., Protentis, J. P., Schlossmann, S. F. & Reinherz, E. L. (1983) *J. Exp. Med. 158*, 988–993.
26. Imboden, J. B. & Stobo, J. D. (1985) *J. Exp. Med. 161*, 446–456.
27. Imboden, J., Weyand, C. & Goronzy, J. (1987) *J. Immunol. 138*, 1322–1324.
28. Mire-Sluis, A. R., Hoffbrand, A. V. & Wickremashinghe, R. G. (1987) *Biochem. Biophys. Res. Commun. 148*, 1223–1231.
29. Imboden, J. B., Shoback, D. M., Pattison, G. & Stobo, J. D. (1986) *Proc. Natl Acad. Sci. 83*, 5672–5677.
30. Mercep, M., Bonifacino, J., Garcia-Morales, P., Samelson, L. E., Klausner, R. D. & Ashwell, J. D. (1988) *Science 242*, 571–574.
31. Wacholtz, M. C. & Lipsky, P. E. (1987) *Fed. Proc. 46*, 463.
32. Gardner, P. (1989) *Cell 59*, 15–20.
33. Melton, L. B., Lakkis, F., Moscovitch-Lopatin, M., Smith, B. R., Williams, J. M., Rosoff, P. M. & Strom, T. B. (1989) *Clin. Immunol. Immunopathol. 50*, 171–183.
34. Patel, M. D., Samelson, L. E. & Klausner, R. D. (1987) *J. Biol. Chem. 262*, 5831–5838.
35. Samelson, L. E., Patel, M. D., Weissman, A. M., Harford, J. B. & Klausner, R. D. (1986) *Cell 46*, 1083–1090.
36. Weissman, A. M., Ross, P., Luong, E. T., Garcia-Morales, P., Jelachich, M. L., Biddison, W.-E., Klausner, R. D. & Samelson, L. E. (1988) *J. Immunol. 141*, 3532–3536.
37. Alexander, D. R. & Cantrell, D. A. (1989) *Immunol. Today 10*, 200–205.
38. Roska, A. K. & Lipsky, P. E. (1985) *J. Immunol. 135*, 2953–2961.
39. Sewell, W. A., Brown, M. H., Dunne, J., Owen, M. J. & Crumpton, M. J. (1986) *Proc. Natl Acad. Sci. 83*, 8718–8722.
40. Selvaraj, P., Plunkett, M. L., Dustin, M., Sanders, M. E., Shaw, S. & Springer, T. A. (1987) *Nature 326*, 400–403.
41. Dustin, M. L., Sanders, M. E., Shaw, S. & Springer, T. A. (1987) *J. Exp. Med. 165*, 677–692.

42. Krensky, A. M., Sanchez-Madrid, F., Robbins, E., Nagy, J. A., Springer, T. A. & Burakoff, S. J. (1983) *J. Immunol. 131*, 611 – 616.
43. Meuer, S. C., Hussey, R. E., Fabbi, M., Fox, D., Acuto, O., Fitzgerald, K. A., Hodgdon, J. C., Protentis, J. P., Schlossmann, S. F. & Reinherz, E. L. (1984) *Cell 36*, 897 – 906.
44. Brottier, P., Boumsell, L., Gelin, C. & Bernard, A. (1985) *J. Immunol. 135*, 1624 – 1631.
45. Huet, S., Wakasuge, H., Sterkers, G. & Gilmour, J. (1986) *J. Immunol. 137*, 1420 – 1428.
46. Yang, S. Y., Chouaib, S. & Dupont, B. (1986) *J. Immunol. 137*, 1097 – 1100.
47. Hunig, T., Tiefenthaler, G., Meyer zum Buschenfelde, K.-H. & Meuer, S. C. (1987) *Nature 326*, 298 – 301.
48. Bockenstedt, L. K., Goldsmith, M. A., Dustin, M., Olive, D., Springer, T. A. & Weiss, A. (1988) *J. Immunol. 141*, 1904 – 1911.
49. June, C. M., Ledbetter, J. A., Rabinovitch, P. S., Martin, P. J., Beatty, P. G. & Hansen, J. A. (1986) *J. Clin. Invest. 77*, 1224 – 1232.
50. Breitmayer, J. B., Daley, J. F., Levine, H. B. & Schlossman, S. F. (1987) *J. Immunol. 139*, 2899 – 2905.
51. Alcover, A., Weiss, M. J., Daley, J. F. & Reinherz, E. L. (1986) *Proc. Natl Acad. Sci. USA 83*, 2614 – 2618.
52. Bierer, B. E., Peterson, A., Gorga, J. C., Herrmam, S. H. & Burakoff, S. J. (1988) *J. Exp. Med. 168*, 1145 – 1156.
53. Bierer, B. E., Barbosa, J., Herrmam, S. & Burakoff, S. J. (1988) *J. Immunol. 140*, 3358 – 3363.
54. Panteleo, G., Olive, D., Poggi, A., Kozumbo, W. F., Moretta, L. & Moretta, A. (1987) *Eur. J. Immunol. 17*, 55 – 60.
55. Bagnasco, M., Nunes, J., Lopez, M., Cerdan, C., Pierres, A., Mawas, C. & Olive, D. (1989) *Eur. J. Immunol. 19*, 823 – 827.
56. Alexander, D. R., Hexham, J. M. & Crumpton, M. J. (1988) *Biochem. J. 256*, 885 – 892.
57. Suthanithiran, M. (1988) *Cell. Immunol. 112*, 112 – 122.
58. Veillette, A., Bookman, M. A., Horak, E. M. & Bolen, J. B. (1988) *Cell 55*, 301 – 308.
59. Rudd, C. E., Trevillyan, J. M., Dasgupta, J. D., Wong, L. L. & Schlossman, S. F. (1988) *Proc. Natl Acad. Sci. USA 85*, 5190 – 5194.
60. Bierer, B. A., Sleckman, B. P., Ratnofsky, S. E. & Burakoff, S. J. (1989) *Annu. Rev. Immunol. 7*, 579 – 599.
61. Veillette, A., Bookman, M. A., Horak, E. M., Samelson, L. E. & Bolen, J. B. (1989) *Nature 338*, 257 – 259.
62. Mustelin, T. & Altman, A. (1989) *Immunol. Today 10*, 189 – 192.
63. Weiss, A., Manger, B. & Imboden, J. (1986) *J. Immunol. 137*, 819 – 825.
64. Ledbetter, J. A., Martin, P. J., Spooner, C. E., Wofsy, D., Tsu, T. T., Beatty, P. G. & Gladstone, P. (1985) *J. Immunol. 135*, 2331 – 2336.
65. Abraham, R. T., Ho, S. N., Barna, T. J. & McKean, D. J. (1987) *J. Biol. Chem. 262*, 2719 – 2728.
66. Avissar, S., Stenzel, K. H. & Novogrodzky, A. (1985) *Cell. Immunol. 96*, 4462 – 4471.
67. Mukaida, N., Yagisawa, H., Kawai, T. & Kasahara, T. (1988) *Biochem. Biophys. Res. Commun. 154*, 187 – 193.
68. Rosoff, P. M., Savage, N. & Dinarello, C. A. (1988) *Cell 54*, 73 – 81.
69. Exton, J. H. (1989) *J. Biol. Chem. 264*, 21689 – 21698.
70. Didier, M., Aussel, C., Pelassey, C. & Fehlmann, M. (1988) *J. Immunol. 141*, 3078 – 3080.
71. Weiss, A., Wiskocil, R. L. & Stobo, J. D. (1984) *J. Immunol. 133*, 123 – 128.
72. Castagna, M., Takai, Y., Kaibuchi, K., Sano, K., Kikkawa, U. & Nishizuka, Y. (1982) *J. Biol. Chem. 257*, 7847 – 7851.
73. Isakov, N. & Altman, A. (1987) *J. Immunol. 138*, 3100 – 3107.
74. Aussel, C., Didier, M., Peyron, J.-F., Pelassy, C., Ferrua, B. & Fehlmann, M. (1988) *J. Immunol. 140*, 215 – 220.
75. Nishizuka, Y. (1988) *Nature 334*, 661 – 665.
76. Kikkawa, U., Ono, Y., Ogita, K., Fujii, T., Asaoka, Y., Sekiguchi, K., Kosaka, Y., Igarashi, K. & Nishizuka, Y. (1987) *FEBS Lett. 217*, 227 – 231.
77. Huang, F. L., Yoshida, Y., Nakabayashi, H. & Huang, K.-P. (1987) *J. Biol. Chem. 262*, 15714 – 15720.
78. Ono, Y., Kikkawa, U., Ogita, K., Fujii, T., Kurokawa, T., Asaoka, Y., Sekiguchi, K., Ase, K., Igarashi, K. & Nishizuka, Y. (1987) *Science 236*, 1116 – 1120.
79. Shearman, M. S., Berry, N., Oda, T., Ase, K., Kikkawa, U. & Nishizuka, Y. (1988) *FEBS Lett. 234*, 387 – 391.
80. Beyers, A. D., Hanekom, C., Rheeder, A., Strachan, A. F., Wooten, M. W. & Nel, A. E. (1988) *J. Immunol. 141*, 3463 – 3470.
81. Shearman, M. S., Naor, Z., Kikkawa, U. & Nishizuka, Y. (1987) *Biochem. Biophys. Res. Commun. 147*, 911 – 919.
82. Kosaka, Y., Ogita, K., Ase, K., Nomura, H., Kikkawa, U. & Nishizuka, Y. (1988) *Biochem. Biophys. Res. Commun. 151*, 973 – 981.
83. Farago, A., Gyongyi, F., Meszaros, G., Buday, L., Antoni, F. & Seprodi, J. (1989) *FEBS Lett. 243*, 328 – 332.
84. Sawamura, S., Ase, K., Berry, N., Kikkawa, U., McCaffrey, P. G., Minowada, J. & Nishizuka, Y. (1989) *FEBS Lett. 247*, 353 – 357.
85. Ohno, S., Kawasaki, H., Imajoh, S., Suzuki, K., Inagaki, M., Yokokura, H., Sakoh, T. & Hidaka, H. (1987) *Nature 325*, 161 – 166.
86. Brandt, S. J., Niedel, J. E., Bell, R. M. & Young III, W. S. (1987) *Cell 49*, 57 – 63.
87. Schaap, D., Parker, P. J., Bristol, A., Kriz, R. & Knopf, J. (1989) *FEBS Lett. 243*, 352 – 357.
88. Huang, K.-P., Huang, F. L., Nakabayashi, H. & Yoshida, Y. (1988) *J. Biol. Chem. 263*, 14839 – 14845.
89. Berry, N., Ase, K., Kikkawa, U., Kishimoto, A. & Nishizuka, Y. (1989) *J. Immunol. 143*, 1407 – 1413.
90. Subramaniam, P., Sehajpal, P., Murthi, V. K., Stenzel, K. H. & Suthanthiran, M. (1988) *Cell. Immunol. 116*, 439 – 449.
91. Kraft, A. S. & Anderson, W. B. (1983) *Nature 301*, 621 – 623.
92. Berry, N., Ase, K. & Nishizuka, Y. (1990) *Proc. Natl Acad. Sci. USA*, in the press.
93. McCrady, C. W., Ely, C. M., Westin, E. & Carchman, R. A. (1988) *J. Biol. Chem. 263*, 18537 – 18544.
94. Davis, L. S. & Lipsky, P. E. (1989) *Cell. Immunol. 118*, 208 – 221.
95. Kaibuchi, K., Takai, Y., Sawamura, M., Hoshijima, M., Fijikura, T. & Nishizuka, Y. (1983) *J. Biol. Chem. 258*, 6701 – 6704.
96. Marth, J. D., Lewis, D. B., Cooke, M. P., Mellins, E. D., Gearn, M. E., Samelson, L. E., Wilson, C. B., Miller, A. D. & Perlmutter, R. M. (1989) *J. Immunol. 142*, 2430 – 2437.
97. Jung, L. K. L., Bjorndahl, J. M. & Fu, S. M. (1988) *Cell. Immunol. 117*, 352 – 359.
98. McCaffrey, P. G., Friedman, B. A. & Rosner, M. (1984) *J. Biol. Chem. 259*, 12502 – 12507.
99. Cantrell, D. A., Davies, A. A. & Crumpton, M. J. (1985) *Proc. Natl Acad. Sci. USA 82*, 8158 – 8162.
100. Shearman, M. S., Sekiguchi, K. & Nishizuka, Y. (1990) *Pharmacol. Rev. 41*, 211 – 237.
101. Rink, T. J., Sanchez, A. & Hallam, T. J. (1983) *Nature 305*, 317 – 319.
102. Katakami, Y., Kaibuchi, K., Sawamura, M., Takai, Y. & Nishizuka, Y. (1984) *Biochem. Biophys. Res. Commun. 121*, 573 – 578.
103. Fujita, I., Irita, K., Takeshige, K. & Hinakami, S. (1984) *Biochem. Biophys. Res. Commun. 120*, 318 – 324.
104. Sasakawa, N., Ishii, K., Yamamoto, S. & Kato, R. (1985) *Biochem. Biophys. Res. Commun. 238*, 913 – 920.
105. Ebeling, J. G., Vandenbark, G. R., Kuhn, L. J., Ganong, B. R., Bell, R. M. & Niedel, J. E. (1985) *Proc. Natl Acad. Sci. USA 82*, 815 – 819.
106. Davis, R. J., Ganong, B. R., Bell, R. M. & Czech, M. P. (1985) *J. Biol. Chem. 260*, 1562 – 1566.
107. Rodriguez-Pena, A. & Rozengurt, E. (1984) *Biochem. Biophys. Res. Commun. 120*, 1053 – 1059.
108. Melloni, E., Pontremoli, S., Michetti, M., Sacco, O., Sparatore, B. & Horecker, B. L. (1986) *J. Biol. Chem. 261*, 4104 – 4105.
109. Chida, K., Kato, N. & Kuroki, T. (1986) *J. Biol. Chem. 261*, 13013 – 13018.

214

110. Kishimoto, A., Mikawa, K., Hashimoto, K., Yasuda, I., Tanaka, S., Tominaga, M., Kuroda, T. & Nishizuka, Y. (1989) *J. Biol. Chem. 264*, 4088−4092.

111. Tauber, A. I., Cox, J. A., Curnutte, J. T., Carrol, P. M., Nakakuma, H., Warren, B., Gilbert, H. & Blumberg, P. M. (1989) *Biochem. Biophys. Res. Commun. 158*, 884−890.

112. Mills, G. B., May, C., Hill, M., Ebanks, R., Roifman, C., Mellors, A. & Gelfand, E. W. (1989) *J. Immunol. 142*, 1995−2003.

113. Rittenhouse, S. E. & Sasson, J. P. (1985) *J. Biol. Chem. 260*, 89657−8660.

114. Mizuguchi, J., Beaven, M. A., Hu, L. J. & Paul, W. E. (1986) *Proc. Natl Acad. Sci. USA 83*, 4474−4478.

115. Valge, V. E., Wong, J. G. P., Datlof, B. M., Sinskey, A. J. & Rao, A. (1988) *Cell 55*, 101−112.

116. Grove, D. S. & Mastro, A. M. (1988) *Biochem. Biophys. Res. Commun. 151*, 94−99.

117. Cantrell, D. A., Verbi, W., Davies, A., Parker, P. & Crumpton, M. J. (1988) *Eur. J. Immunol. 18*, 1391−1396.

118. Ase, K., Berry, N., Kikkawa, U., Kishimoto, A. & Nishizuka, Y. (1988) *FEBS Lett. 236*, 396−400.

119. Avissar, S., Stenzel, K. H. & Novogrodsky, A. (1985) *Cell. Immunol. 96*, 462−471.

120. Friedrich, B., Noreus, K., Cantrell, D. A. & Gullberg, M. (1988) *Immunobiology 176*, 465−478.

121. Peyron, J.-F., Pont, S., Pierres, M. & Fehlmann, M. (1988) *Eur. J. Immunol. 18*, 1139−1142.

122. Swift, A. M., Davidson, S. O. & Berger, A. E. (1988) *J. Biol. Chem. 263*, 2389−2396.

123. Samstag, Y., Emmrich, F. & Staehlin, T. (1988) *Proc. Natl Acad. Sci. USA 85*, 9689−9693.

124. Kaibuchi, K., Takai, Y. & Nishizuka, Y. (1985) *J. Biol. Chem. 260*, 1366−1369.

125. Hirata, F., Matsuda, K., Notsu, Y., Hattori, T. & del Carmine R. (1984) *Proc. Natl Acad. Sci. USA 81*, 4717−4721.

126. Cantrell, D., Davies, A. A., Londei, M., Feldman, M. & Crumpton, M. J. (1987) *Nature 325*, 540−542.

127. Chatila, T. A. & Geha, R. S. (1988) *J. Immunol. 140*, 4308−4314.

128. Acres, R. B., Conlon, P. J., Mochizuki, D. Y. & Gallis, B. (1986) *J. Biol. Chem. 261*, 16210−16214.

129. Acres, R. B., Conlon, P. J., Mochizuki, D. Y. & Gallis, B. (1987) *J. Immunol. 139*, 2268−2274.

130. Blue, M.-L., Hafler, D. A., Craig, K. A., Levine, H. & Schlossman, S. (1987) *J. Immunol. 139*, 3949−3954.

131. Hoxie, J. A., Matthews, D. M., Callahan, K. J., Cassel, D. L. & Cooper, R. A. (1986) *J. Immunol. 137*, 1194−1201.

132. Autero, M. & Ghamberg, C. C. (1987) *Eur. J. Immunol. 17*, 1503−1506.

133. Minami, Y., Samelson, L. E. & Klausner, R. D. (1987) *J. Biol. Chem. 262*, 13342−13347.

134. Tonks, N. K., Charbonneau, H., Diltz, C. D., Fischer, E. H. & Walsh, K. A. (1988) *Biochemistry 27*, 8695−8701.

135. Ledbetter, J. A., Tonks, N. K., Fischer, E. H. & Clark, E. A. (1988) *Proc. Natl Acad. Sci. USA 85*, 8628−8632.

136. Bohnlein, E., Lowenthal, J. W., Siekewitz, M., Ballard, D. W., Franza, B. R. & Greene, W. C. (1988) *Cell 53*, 827−836.

137. Cross, S. L., Halden, N. F., Lenardo, M. J. & Leonard, W. J. (1989) *Science 244*, 466−469.

138. Hoyos, B., Ballard, D. W., Bohnlein, E., Siekevitz, M. & Greene, W. C. (1989) *Science 244*, 457−460.

139. Mills, G. B., Cheung, R. K., Grinstein, S. & Gelfand, E. W. (1985) *J. Immunol. 134*, 1640−1643.

140. Depper, J. M., Leonard, W. J., Kronke, M., Noguchi, P. D., Cunningham, R. E., Waldmann, T. A. & Greene, W. C. (1984) *J. Immunol. 133*, 3054−3061.

141. Chopra, R. K., Nagel, J. E., Chrest, F. J., Boto, W. M., Pyle, R. S., Dorsey, B., McCoy, M., Holbrook, N. & Adler, W. H. (1987) *Clin. Exp. Immunol. 69*, 433−440.

142. Angel, P., Imagawa, M., Chiu, R., Stein, B., Imbra, R. J., Rahmsdorf, H. J., Jonat, C., Herrlich, P. & Karin, M. (1987) *Cell 49*, 729−739.

143. Lee, W., Mitchell, P. & Tjian, P. (1987) *Cell 49*, 741−752.

144. Curran, T. (1988) in *The oncogene handbook* (Reddy, E. P., Skalka, A. M. & Curran, T., eds.) pp. 307−325, Elsevier Science Publishers, Amsterdam.

145. Reed, J. C., Alpers, J. C., Nowell, P. C., Hoover, R. G. (1986) *Proc. Natl Acad. Sci. USA 83*, 3982−3986.

146. Pompidou, A., Corral, M., Michel, P., Defer, N., Kruh, J. & Curran, T. (1987) *Biochem. Biophys. Res. Commun. 148*, 435−442.

147. Curran, T., Gordon, M. B., Rubino, K. L. & Sambucetti, L. C. (1987) *Oncogene 2*, 79−84.

148. Lenardo, M. J. & Baltimore, D. (1989) *Cell 58*, 227−229.

149. Sen, R. & Baltimore, D. (1986) *Cell 47*, 921−928.

150. Tong-Starksen, S. E., Luciw, P. A. & Peterlin, B. M. (1989) *J. Immunol. 142*, 702−707.

151. Reinherz, E. L. (1987) *Nature 325*, 660−663.

Eur. J. Biochem. *190*, 1−10 (1990)
© FEBS 1990

Review

Molecular analysis of a cellular decision during embryonic development of *Drosophila melanogaster*: epidermogenesis or neurogenesis

José A. CAMPOS-ORTEGA and Elisabeth KNUST

Institut für Entwicklungsphysiologie, Universität zu Köln, Federal Republic of Germany

(Received November 21, 1989) − EJB 89 1392

In *Drosophila melanogaster*, the neuroblasts (neural progenitor cells) develop from a special region of the ectoderm, called the neuroectoderm. During early embryonic development, the neuroblasts separate from the remaining cells of the neuroectoderm, which develop as epidermoblasts (epidermal progenitor cells). The separation of these two cell types is the result of cellular interactions. The available data indicate that a signal chain formed by the products of several identified genes regulates the cell's decision to enter either neurogenesis or epidermogenesis. Various kinds of data, in particular from cell transplantation studies and from genetic and molecular analyses, suggest that the proteins encoded by the genes *Notch* and *Delta* interact at the membrane of the neuroectodermal cells to provide a regulatory signal. This signal is thought to lead, on the one hand, to epidermal development through the action of the genes of the *Enhancer of split* complex, a gene complex that encodes several functions related to the transduction and further processing of the signal, including the genetic regulation in the receiving cell; on the other hand, the signal is thought to lead to neural development through the participation of the genes of the *achaete-scute* complex and *daughterless*, which are members of a family of DNA-binding regulatory proteins and of the gene *vnd* whose molecular nature is still unknown.

The origin of cell diversity is one of the major problems in developmental biology. Although this problem has received a great deal of attention during the last several years, the mechanisms controlling the direction and timing of determination and cellular differentiation are still largely unknown. Nevertheless, considerable progress has been made in our understanding of the problem. For instance, results from a variety of studies show that many steps in the development from the zygote to the differentiated multicellular organism depend on interactions between neighboring cells. In recent years, several examples of cellular interactions have been shown to be mediated by proteins similar to growth factors and their receptors [1−3]. We will show below that such molecules are also involved in the development of the neural primordium in the fly *Drosophila melanogaster*.

In *Drosophila*, neurogenesis is initiated by the separation of neural progenitor cells (neuroblasts) from the neurogenic region of the ectoderm. The neuroblasts move inward from the surface of the embryo where they build up the primordium of the central nervous system. The remaining cells of the neuroectoderm, with which the presumptive neuroblasts are intermingled prior to their segregation, take on a different developmental fate: they develop as epidermal progenitor cells (epidermoblasts) to give rise to the epidermal sheath of ventral and cephalic regions [4, 5]. It seems, thus, that each cell within the neuroectoderm has to choose between a neural and an epidermal developmental fate; these two are the only possible fates of the neuroectodermal cells and they are mutually exclusive. Observations from embryology [4] point to a peculiar behavior of the *Drosophila* neuroectodermal cells. They suggest that all of these cells initially acquire the capability to develop as neuroblasts: in the parlance of developmental biology, neurogenesis is the primary fate of the neuroectodermal cells. Results from laser ablations in grasshoppers [6] and cell transplantations in *Drosophila* [7, 8] indicate that interactions among the neuroectodermal cells are essential to deflect a substantial proportion of these cells from the primary neural into the secondary epidermal fate and, thus, make possible the proper segregation of the two cell lineages. The intention of this review is to summarize recent results from the analysis of the processes leading to the divergence of neural and epidermal cell lineages in *Drosophila* and, in particular, to discuss these results with respect to the possible functions of the proteins involved. The developmental process under discussion is highly complex and presents a variety of different facets, which have been investigated by using various approaches and techniques, e.g. from the fields of classical and experimental embryology, cell biology, and transmission

---

Correspondence to J. A. Campos-Ortega, Institut für Entwicklungsphysiologie, Universität zu Köln, Gyrhofstrasse 17, D-5000 Köln 41, Federal Republic of Germany

Abbreviations. amx, almondex; *AS-C*, the *achaete-scute* complex; *bib, big brain*; EGF, epidermal growth factor; *Dl, Delta*; *E(spl)C*, the *Enhancer of split* complex; *E(spl)^D*, *Enhancer of split* dominant allele; HLH, helix-loop-helix motif; *mam, master mind*; *neu, neuralised*; *vnd, ventral nervous system condensation defective.*

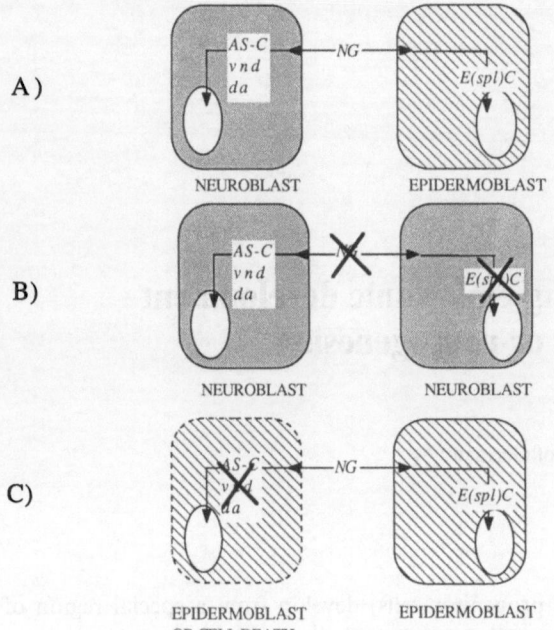

A)

B)

C)

Fig. 1. *Interactions between neuroblasts and epidermoblasts*. (A) Two interacting cells of the neuroectoderm in the wild type. A large number of genes, i.e. the neurogenic genes (*NG*), including the genes of the *enhancer of split* complex (*E(spl)C*), the genes of the *achaete-scute* complex (*AS-C*), *ventral nervous system condensation-defective* (*vnd*) and *daughterless*, encode the proteins of a regulatory signal chain which allows the cells to develop either as neuroblasts or as epidermoblasts. The functions encoded by the genes of *AS-C*, *vnd* and *daughterless* are required to regulate the genetic activity of the neuroblasts, those encoding the various genes of *E(spl)C* to regulate the genetic activity of the epidermoblasts. (B) Mutation of one of the neurogenic genes results in the development of all neuroectodermal cells as neuroblasts. (C) Mutation in the genes of *AS-C*, *vnd* or *daughterless* results in either the development of additional neuroectodermal cells as epidermoblasts, at the expense of neuroblasts, or to cell death

and molecular genetics. Therefore, we shall deal below with data from these various approaches.

The segregation of neuroblasts and epidermoblasts is controlled by two groups of genes with apparently complementary functions

In *Drosophila*, the correct separation of neural and epidermal progenitor cells depends on the action of two groups of genes, the so-called neurogenic genes [9, 10], on the one hand, and genes of the subdivision 1B of the X chromosome, i.e. the various members of the *achaete-scute* complex (*AS-C*) and the locus of *ventral nervous system condensation defective* (*vnd*), as well as the second chromosomal *daughterless* gene, on the other hand (Fig. 1). The evidence to support the participation of these genes in the process under discussion derives from the phenotypes of their mutations. In the wild type, the neuroectoderm contains approximately 2000 cells, 500 of which choose the neural pathway, whereas the remaining 1500 take on an epidermal fate and develop as epidermoblasts; loss of function of any of the neurogenic genes causes all the neuroectodermal cell to develop as neuroblasts, i.e. these mutants initiate neurogenesis with 2000 instead of the normal 500 neuroblasts [11]. Consequently, the mutant embryos die with a highly hyperplasic nervous system and lack ventral and cephalic epidermis (Fig. 2C). Apparently, thus, the functions

of the neurogenic genes are required to suppress the neural fate in 1500 neuroectodermal cells and to allow those cells to develop as epidermoblasts [10]. Upon mutation, the suppressive effect normally exerted by the neurogenic genes on neurogenesis is relieved, or eliminated, this leading to realization of the primary fate, that is, the development of all the neuroectodermal cells as neuroblasts.

The phenotype of loss-of-function mutations in the other genes is the opposite of that of neurogenic mutations: embryos lacking the genes of *AS-C*, *vnd* or *daughterless* initiate neurogenesis with less than the normal complement of 500 neuroblasts: depending on the mutant, 20–30% of all neuroblasts are missing (F. Jiménez & J.A. Campos-Ortega, and M. Brand & J.A. Campos-Ortega, unpublished results). In addition, during later stages, large numbers of cells degenerate within the neural primordium of all these mutants [12, 13]. Consequently, the fully differentiated mutant embryos die with a highly hypoplasic central nervous system [12, 14, 15] (cf. Fig. 2D). These observations can be interpreted to mean that the genes of the subdivision 1B (i.e. those of *AS-C* and *vnd*) and *daughterless* are required for the development of the normal 500 neuroblasts. Since the populations of neuroblasts affected by mutations in any one of these genes do not seem to overlap significantly, *AS-C*, *vnd* or *daughterless* may each control the development of particular groups of neuroblasts.

This observation demonstrates a major difference between the phenotypes of mutations in the neurogenic genes and in the subdivision 1B-*daughterless* genes (Fig. 2). Neurogenic mutations exhibit global effects, that is to say, the complete elimination of the function of any of them leads to the misrouting into neurogenesis of all of the neuroectodermal cells; in addition, the defect is identical in each case, irrespective of the gene considered. As we will discuss below, this is due to the fact that the neurogenic genes (with the exception of *big brain*) are interrelated, serving a common functional pathway; thus, perturbing one of the functions results in the disruption of the whole system. Contrasting this situation, mutation in any of the subdivision 1B-*daughterless* genes causes only partial effects, i.e. roughly speaking, only some of the neuroblasts are missing in each case and the affected neuroblasts are different depending on which gene is mutated. This difference in the intensity of the effects on neurogenesis of mutations is striking. We shall see below that it reflects the pattern of expression of the corresponding genes.

The neurogenic genes, on the one hand, and the subdivision 1B-*daughterless* genes, on the other, have been described as two different groups of genes. Such a distinction is justified by the phenotype of their mutations: loss-of-function from any of the neurogenic genes causes neural hyperplasia, and from any of the subdivision 1B-*daughterless* genes causes neural hypoplasia. However, this distinction is in fact an artificial one for, as we shall discuss in the following, the products of all of these genes are apparently engaged in the formation of a complex genetic network to contribute to the same process: the separation of neural and epidermal cell progenitors.

The neurogenic genes encode different steps of a signal chain

We have mentioned that interactions between the neuroectodermal cells are required for the proper segregation of the neural and epidermal progenitors. We assume that the products of the neurogenic genes contribute to a cell communication process. Indeed, evidence from a genetic analysis suggests that the functions of six of the neurogenic genes, i.e. *Notch*, *almondex*, *master mind*, *Delta*, *neuralized* and the genes

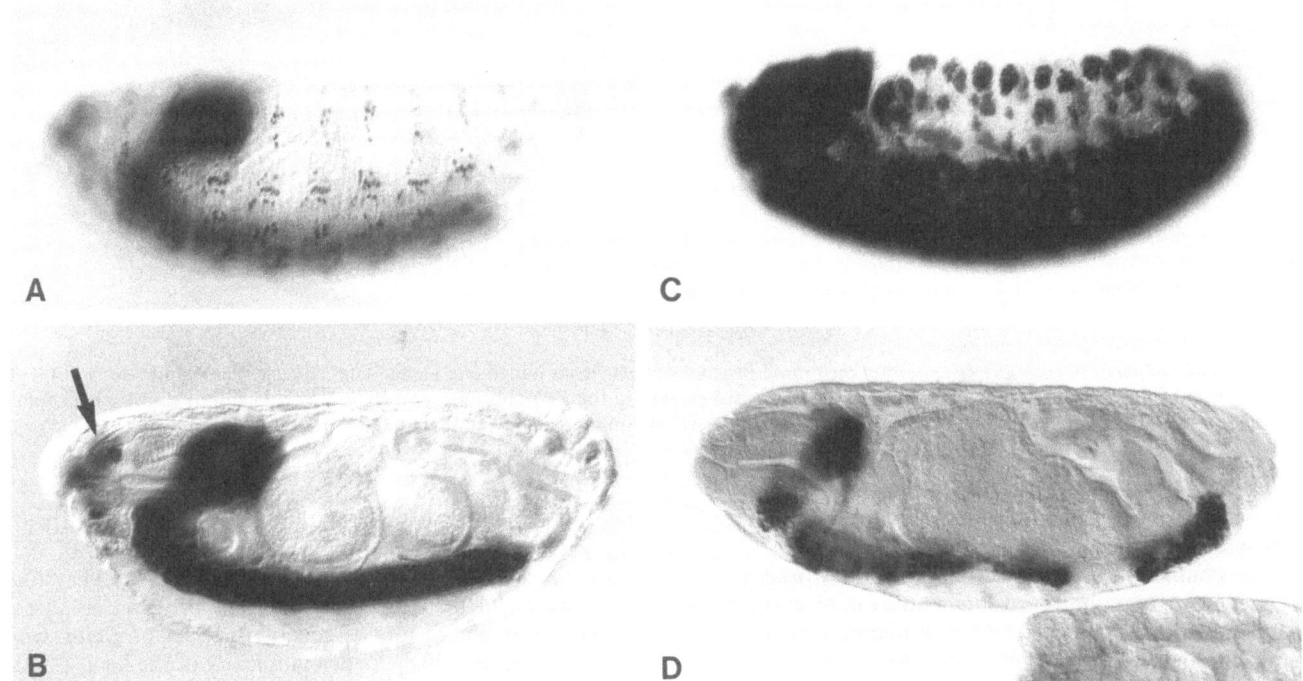

Fig. 2. *Wild-type and mutant embryos.* Lateral (A) and median (B) planes of focus through the same wild-type embryo; C is a neurogenic mutant; D is a *daughterless* mutant. All three embryos have been stained with a neural specific antibody (44c11, kindly provided by Y. N. Jan, San Francisco) that recognizes an antigen present in all neuronal nuclei of the embryo; therefore only the nuclei of neurones are stained. A shows the exquisite pattern of some of the sensory neurones, B the central nervous system. The arrow points to one of several sensory organs that are visible on this plane of focus. Notice the conspicuous (central and peripheral) neural hyperplasia of the neurogenic mutant shown in C, and the neural hypoplasia of the *daughterless* mutant shown in D. Notice that the cephalic sensory organs (arrow in A), as well as all other sensory organs, are missing

of the *Enhancer of split* complex [*E(spl)C*], inasmuch as their participation in the segregation of neuroblasts from epidermoblasts is concerned, form the links in a chain of epistatic relationships [16]; that is to say, the function of each of the genes is dependent on that of another member of the group and consequently the function of the entire chain will be interrupted if any of the links is missing (Fig. 3). Within this chain, *almondex* acts as the first link and the genes of *E(spl)C* as the last one; *big brain* seems to act independently of the others. We should perhaps mention that epistatic, and hypostatic, relationships between two genes can be derived from the results of genomic combinations and their effect on phenotypic traits. If an increased wild-type dosage of a gene, for example *Notch*, modifies the phenotype caused by the loss-of-function of another gene, for example *neuralised*, it is very likely that the former gene is regulated by the latter [16]. Such considerations allow one to propose formal schemes like the one shown in Fig. 2. Its molecular basis has still to be established, but such schemes are of great use in the design of further experiments.

In addition, results of cell transplantations support the notion that mutations in the different neurogenic genes affect specific parts of the signal chain [17]. Mutations in *Notch*, *almondex*, *big brain*, *master mind*, *Delta* or *neuralised* behave as if the gene products were required to produce and/or send a regulatory signal that leads to epidermogenesis (Fig. 3). When neuroectodermal cells lacking any of these gene products are transplanted into the neurogenic ectoderm of a wild-type embryo, the mutant cells do not autonomously express their phenotypes; rather they develop as epidermoblasts with the same frequency as wild-type cells. This result suggests that the transplanted mutant cells are still capable of receiving the

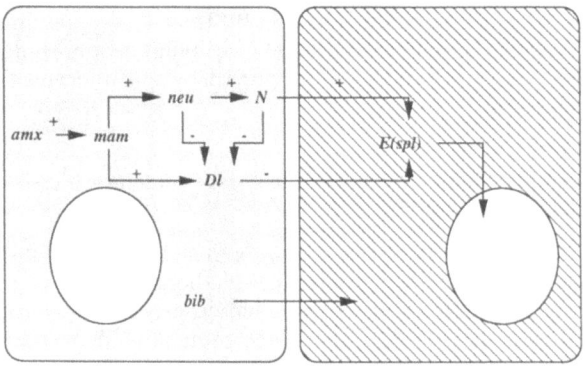

NEUROECTODERM

Fig. 3. *The network of relationships of the neurogenic genes.* This diagram is based on formal arguments derived from transmission genetics and gives no indication concerning the molecular level at which the proposed interactions take place. Positive or negative signs reflect the kind of functional influences, i.e. repression or activation, assumed to be exerted by one gene product upon the next one. The data suggest that *E(spl)* provides an epidermalizing signal to the neuroectodermal cells. Although *bib* takes part in the process of neuroblast segregation, its function is apparently independent of the other six genes. (From [16], modified)

regulatory signal from the neighboring wild-type cells and processing it to their own genomes. In contrast, neuro-ectodermal cells lacking the functions of *E(spl)C* never adopt an epidermal fate following transplantation into the wild-type

4

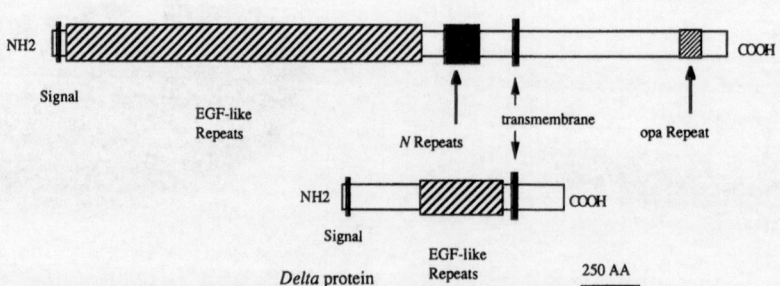

Fig. 4. *A comparison of the primary structures of the proteins encoded by the genes* Notch *and* Delta. The *Notch* protein comprises 36 EGF-like repeats. *N* and opa are other repetitive sequences which are not present in the *Delta* protein. The *Delta* protein exhibits a single hydrophobic transmembrane domain and a stretch of nine EGF-like repeats. AA = amino acids (from [23] and [27], modified)

neurogenic region, suggesting that this locus is required to receive and/or process the signal.

The chain of epistatic interrelationships worked out by genetic means can be placed in a functional context, and correlated with the results of the transplantation experiments using mutant cells. Accordingly, we postulate that the function of the genes *Notch*, *almondex*, *big brain*, *master mind*, *Delta* and *neuralised* is required in the cells sending the signal, both to make and to transmit it; contrarily, the function of *E(spl)C* is required in cells to receive and/or process the signal.

We should mention that the evidence for the non-autonomy of *Notch* gene expression, that is found under the conditions of the transplantation experiment, is still controversial. Under different experimental conditions from those of the transplantation, cells lacking the *Notch* gene are not able to take on the epidermal fate [18, 19] (and P. Hoppe and R. Greenspan, unpublished results). This point is important, for it may indicate that the *Notch* protein acts as a receptor (see Discussion).

Notch and *Delta* encode transmembrane proteins with EGF-like repeats

The *Notch* locus has a size of about 40 kbp (i.e. 40000 base pairs) and encodes a transcript of about 10.2 kb which is developmentally regulated, but ubiquitously expressed during neurogenesis [20–22]. The *Notch* protein, as deduced from sequences of genomic and cDNA clones [23, 24], comprises 2703 amino acids and shows features typical of a transmembrane protein: two hydrophobic domains, an amino-terminal signal peptide and a membrane-spanning domain (Fig. 4). Indeed, antibodies raised against different parts of the *Notch* protein confirm its location in the cell membrane [25, 26]. A striking feature of the extracellular domain is a tandem array of 36 cysteine-rich motifs each of about 40 amino acids with striking similarity to the epidermal growth factor (EGF) and other proteins of mammals. In addition, there are three other cysteine-rich repeats, called *Notch* repeats. The putative cytoplasmic domain of about 1000 amino acids contains a putative nucleotide phosphate binding site.

By mapping restriction site polymorphisms, the *Delta* locus has been localized within about 25 kbp of genomic DNA that exhibits a high degree of transcriptional complexity. Three major developmentally regulated and largely overlapping RNAs of 5.4, 4.6 and 3.7 kb have been identified as products of the *Delta* gene [27, 28] (and M. Haenlin, B. Kramatschek & J.A. Campos-Ortega, unpublished results);

two minor RNAs, of 3.8 and 2.8 kb are also encoded within the same genomic region, but their relationships of *Delta* function are not understood. Nonetheless, it is very likely that the *Delta* locus encodes different proteins, as already suggested by the results of genetic analyses of *Delta*, which point to considerable genetic complexity of the locus [29, 30]. In contrast to *Notch*, which is ubiquitously transcribed during the segregation of the neuroblasts [22], the transcription of *Delta* is spatially regulated and the distribution of the RNA is compatible with a function of *Delta* as the source of the epidermalizing signal [27]. *Delta* RNA is expressed in all regions with neurogenic abilities (e. g. the neurogenic ectoderm, the primordia of the optic lobes and of the sensory organs, etc.). In these regions, *Delta* is initially transcribed in all cells, but after the segregation of the progenitors for the two cell lineages, i.e. neural and epidermal, *Delta* RNA can only be detected in the cells which have adopted the neural fate.

Sequencing of cDNA clones complementary to the largest *Delta* RNA revealed a structure of the putative *Delta* protein that is similar to the one proposed for the *Notch* protein: a transmembrane protein, with a hydrophobic signal sequence and a membrane-spanning domain [27, 28]. The extracellular domain contains nine EGF-like repeats, arranged in tandem like those in the *Notch* protein (Fig. 4).

Physical interactions between the *Delta* and *Notch* proteins?

As mentioned above, the proteins encoded by *Delta* and *Notch* seem to act at the signal source side of the postulated regulatory chain; there is evidence that both are membrane-bound proteins with EGF-like repeats in their putative extracellular domains. These EGF-like repeats may well represent essential parts of the signal, by being involved in protein–protein interactions. Several examples are known of proteins, mainly from mammals, that contain EGF-like repeats. They are either secreted (e. g. EGF [31], blood-clotting factors [32]), membrane-bound (e. g. low-density-lipoprotein receptor [33]), or components of the extracellular matrix, like laminin [34]. The participation of EGF-like repeats in specific ligand–receptor interactions has been demonstrated for the binding of EGF and laminin to their corresponding receptors [36, 37]. In the case of *Notch* and *Delta*, genetic mosaic analysis indicates that their products cannot diffuse over long distances. Furthermore, the results from transplanting single mutant cells into wild-type hosts lead to the conclusion that the membrane-bound proteins are involved in cellular communication. Therefore, it is tempting to speculate that this process is mediated by the EGF-like repeats acting between adjacent cells.

The view that the EGF-like repeats of *Notch* mediate protein–protein interactions is further supported by the data obtained from sequencing some point mutations in the *Notch* locus, namely *split* and several *Abruptex* alleles. The putative *split* and *Abruptex* proteins each differ from the *Notch* wild-type protein by just single amino acid exchanges in one of the 36 EGF-like repeats [22, 38]. The phenotypes associated with the various alleles whose sequences were determined suggest that the EGF-like repeats of the extracellular domain, although very similar in structure, may perform different functions. One possibility could be that they interact with different proteins at different times in development.

Finally, direct physical interrelationships between the *Delta* and *Notch* proteins are supported by the existence of allele-specific interactions between both genes. Particular alleles of *Delta* are known which suppress the phenotype of the *split* mutation [39]. A functional complex formed by the direct association of the *Delta* and *Notch* proteins would explain why the embryonic phenotype of *Notch* mutations is restricted to defects in neurogenesis, although the protein is ubiquitously expressed [25, 26]. In that case, the phenotype of embryos lacking the *Notch* protein would actually reflect the perturbance of the *Delta* protein.

Genetic evidence for two gene complexes regulating the differentiation of epidermoblasts and neuroblasts

Genetic (and molecular data to be discussed further below) have shown that two groups of functionally related genes (gene complexes, using the terminology of transmission genetics), the *Enhancer of split* complex *E(spl)C* [40–42] and the *achaete-scute* complex *AS-C* [43–50], each consisting of several functionally and structurally related genes, appear to act on either end of the signal chain. That is to say, we assume that these genes regulate the genetic activity in the epidermoblasts and neuroblasts, respectively. The precise composition of *E(spl)C* is not yet definitely established, but the available evidence points to at least five different genes; this problem will be discussed further below.

Several related functions are encoded by the *Enhancer of split* gene complex

Two different aspects of the *Enhancer of split* complex have to be distinguished. One aspect concerns the gene that gave the locus its name and the function of this gene in compound eye development; the other concerns the participation of the complex in the process of neuroblast segregation. The gene that gave the *Enhancer of split* complex its name was defined by a mutation with dominant expression [called *E(spl)^D*] that enhances the phenotype of *split*, in particular the effects of the *split* mutation on compound eye development [51]. This observation is important since *split* is a recessive allele of *Notch*, and the relationships of *split* and *E(spl)^D* are allele-specific. Thus, we assume that the proteins encoded by the two genes interact. The molecular analysis has unequivocally shown that *E(spl)^D* corresponds to a mutation in the transcription unit *m8* (see below; Fig. 5). That is to say, since the mutant *m8* protein causes the enhancement of *split*, we may conclude for the wild-type that the protein encoded by *m8* interacts with the *Notch* protein, either directly or indirectly.

On the other hand, the current genetic and molecular evidence indicates that *E(spl)C* participates in the early neuro-epidermal lineage segregation with several related, partially redundant functions [40–42]. The most significant observation with respect to functional redundancy among the members of *E(spl)C* derives from genetic analysis. It has been found that only large deletions of the locus lead to the severe phenotypes characteristic of neurogenic mutations, whereas 'point' lethal mutations, or chromosomal breaks that apparently affect single functions within *E(spl)C*, cause only a weak neurogenic phenotype with incomplete penetrance, or even no neurogenic phenotype at all. That is to say, although individual genetic functions of the region are essential for viability, several of them must be simultaneously affected in order to disturb the neurogenic function of the locus [41]. Among these functions, those encoded by the genes *m8* and *m9–m10* seem to play essential roles [40, 42, 52, 53] (and unpublished observations from our laboratory).

E(spl)C extends over at least 34 kbp of genomic DNA containing 10 major transcription units and 11 transcripts, named *m1–m11* (see Fig. 5). Most of these transcripts are expressed during the first half of embryonic development, i.e. during early neurogenesis. Other evidence, e.g. the temporal and spatial expression pattern, the presence of aberrant transcripts in mutant embryos, and results of P-element-mediated germ-line transformation experiments, indicate that the transcripts *m3, m4, m5, m7, m8, m9* and *m10* encode proteins which participate in functions of *E(spl)C* related to the divergence of neural and epidermal lineages [40, 42, 52].

A modification of the *m8* transcript causes the enhancement of the *split* phenotype in *E(spl)^D* mutants

Three major restriction site polymorphisms have been detected in the DNA of the *E(spl)^D* mutant by Southern blot analysis [40]. One of these polymorphisms is caused by the insertion of a middle repetitive element of about 5 kbp in the *m9–m10* transcription unit; the other two correspond to small deletions in the *m8* transcription unit (see Fig. 5). The more distal deletion causes a reduction in the size of the *m8* RNA (1.0 kb in the wild-type vs about 0.7 kb in the mutant) which, in addition, is more abundantly transcribed in *E(spl)^D* embryos. In contrast, the insertion in the *m9–m10* transcription unit does not change the size of these two RNAs, presumably because the insertion is located in an intron. Comparison of the sequence of the genomic DNA of the mutant *m8* region with the corresponding sequence of the wild-type showed that the larger of the two deletions eliminates the last 56 amino acids in the putative *m8* protein; the *m8* protein of the *E(spl)^D* mutant is further modified by the addition of nine amino acids to its truncated carboxy-terminus [42].

In order to find out whether the molecular defects of the *m8* transcript have any causal relation to the phenotypic traits associated with the *E(spl)^D* mutation, P-element-mediated transformation experiments have been carried out using mutant and wild-type genes [42]. These types of experiments, in which a transposable element is used as a vector, allow one to establish the function of cloned DNA fragments upon their reintroduction into the germ line of mutant or wild-type flies. In this particular case, a genomic fragment containing the entire coding and flanking regions of the *m8* transcription unit (named *m8^D*, see Fig. 4), was isolated from the mutant DNA and cloned in a transformation vector. Transgenic flies were created, which carry one or two copies of the mutant *m8^D* gene, in addition to two copies of the wild-type *m8* gene. In combination with *split*, the transforming *m8^D* gene brings about the same enhancement of the *spl* phenotype as the

6

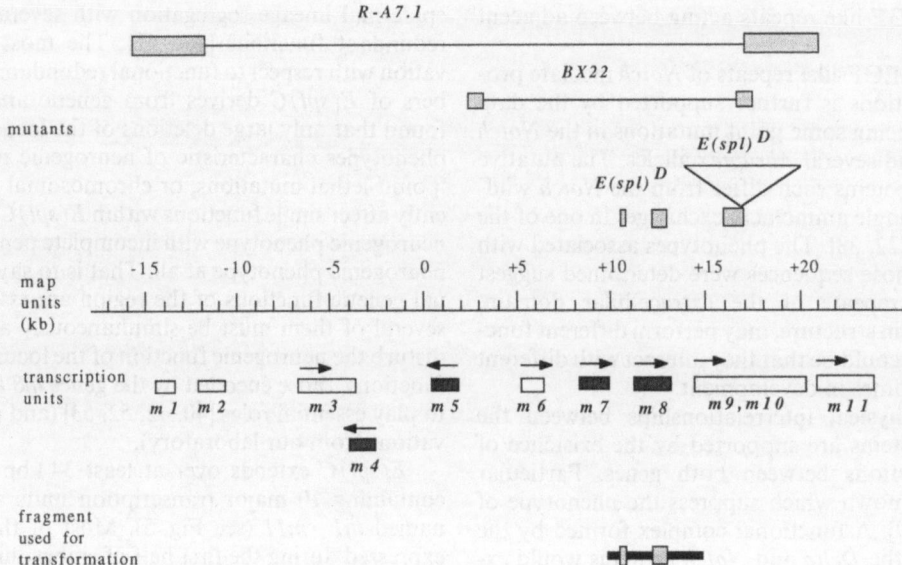

Fig. 5. *Molecular map of the* E(spl) *gene complex.* (Modified from [40].) Several mutants [40, 52] have been mapped to the 34 – 36 kb genomic DNA deleted in *Df(3R)E(spl)*$^{R-A7.1}$, the deletion that defines the minimum extent of the locus; transcription units and direction of transcription are indicated beneath the physical map. Below is shown the *E(spl)*^{D} fragment including the *m8* transcription unit of the mutant (deletions are marked stippled) that causes enhancement of the *spl* phenotype used for transformation experiments [42]

E(spl)^D mutation itself. This experiment directly demonstrates that the transcription unit *m8* and the gene *E(spl)* are identical. Similar germ line transformation experiments using the mutant *m9 – m10* transcription unit, which carries a 5-kbp middle repetitive DNA fragment inserted in an intron, show that this gene does not modify the expression of *split*, demonstrating that *m9 – m10* is not *E(spl)*.

The transcripts *m4, m5, m7* and *m8*: a molecular basis for functional redundancy?

Besides being the molecular counterpart of the gene *E(spl)*, *m8* is one of four (*m4, m5, m7* and *m8*) RNAs that are likely to provide the molecular basis for the functional redundancy found in the genetic analysis [41]. During early stages of embryogenesis, these four transcripts exhibit nearly identical spatial distributions, as shown by *in situ* hybridization experiments to embryonic tissue sections [40]; this pattern is partially complementary to that of *Delta* [27] and compatible with the function of *E(spl)C*, which is assumed to be required in the epidermal progenitor cells to permit their further development.

Sequencing of genomic and cDNA clones comprising the coding region of *m4, m5, m7* and *m8* has shown that the protein encoded by *m4* is rather acidic, whereas those of *m5, m7* and *m8* are fairly basic. In addition, the latter three proteins show a high degree of sequence similarity to each other, with three conserved domains. The amino-terminal domain exhibits sequence similarity to a region comprising a helix-loop-helix (HLH) motif [54] conserved in several different proteins (cf. Fig. 6), among them being members of the *myc* family [55 – 58], several proteins involved in muscle development (*MyoD1, myogenin*, and others [59 – 62]) and two immunglobulin enhancer binding proteins [63]. Interestingly, the same HLH motif has been found in other proteins of *Drosophila*, like *twist* [64] and *hairy* [65] and, more important for our present purposes, in the proteins encoded by the transcripts *T2, T3, T5* and *T1a* of *AS-C* [46, 47] and in the *daughterless* [66] protein (see below).

The transcripts *m9* and *m10*

Both RNAs are transcribed from the same transcription unit, and *m9* is the only one of the 11 transcripts mapped in this region predominantly expressed during oogenesis [40, 52]. Since genetic data indicate a strong component of maternal expression with respect to the neurogenic function of *E(spl)C*, this finding suggests that protein(s) encoded by the transcription unit *m9 – m10* are related to neurogenesis. In addition, P-element-mediated transformation experiments have demonstrated that a genomic fragment comprising the whole coding region of *m9* and *m10* can rescue the lethality of *E(spl)* mutations that cause weak neurogenic phenotypes with incomplete penetrance. However, neither the neurogenic phenotype of the 34-kbp deletion which eliminates the 11 transcripts, nor that of a smaller deletion [*Df(3R)E(spl)*^{BX22} 14 kbp, eliminating transcripts *m6* to *m9 – m10*] can be completely rescued by this fragment [52] (and H. Schrons, E. Knust and J.A. Campos-Ortega, unpublished). Since the neurogenic defect caused by these deletions cannot be restored to the wild type by the addition of just the functions encoded by *m9 – m10*, the phenotype must be the result of the simultaneous lack of *m9 – m10* and some of the other genes contained in the region.

The two RNAs differ in the length of their 3′ untranslated regions; the amino acid sequence of the putative protein encoded by these RNAs has been deduced from several cDNAs. Interestingly, it contains a repeated motif which, among others, is similar to the β-subunit of transduction, a G-protein involved in phototransduction [53]. This structure is compatible with a role for *m9 – m10* in the processing of the epidermalizing signal, in good agreement with the inferences drawn from the genetic and transplantation studies discussed above.

AS-C and *daughterless* are members of a gene family encoding putative regulatory proteins

The *achaete-scute* complex *AS-C* includes four genes *achaete, scute, lethal of scute* and *asense* [50], the names being

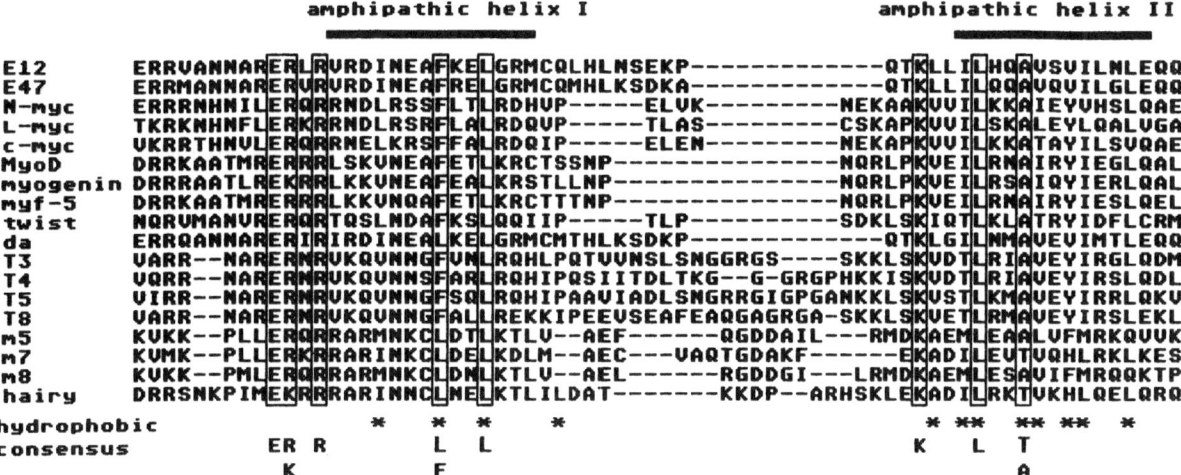

Fig. 6. *The conserved region, including the helix-loop-helix (HLH) motif, found in many different proteins of mammals and insects.* The consensus sequence is indicated below. Bars indicate the extent of the two putative helices, separated by a putative loop region of variable extent. A basic region is found at the amino-terminal side of the proteins. Amino acids conserved in all proteins are boxed. Refer to text for further details

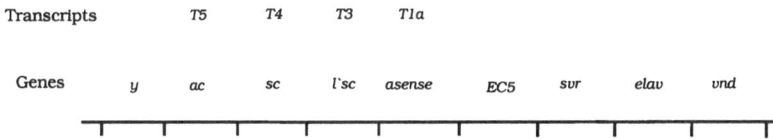

Fig. 7. *Genetic and transcriptional map of the subdivision 1B.* The map includes data from [43–45, 48, 50] and K. White (personal communication)

derived from the phenotypic effects of their mutations on bristle development and viability. The genomic DNA of *AS-C* has been cloned and characterized by Modolell and colleagues [44, 45, 48]. The complex is contained within approximately 85–90 kb of genomic DNA. The region contains a large number of transcription units, four of which, *T5*, *T4*, *T3* and *T1a* (corresponding to *T8* in [47]) have been identified as corresponding to *achaete*, *scute*, *lethal of scute* and *asense* functions (Fig. 7) [45, 48].

Sequence analyses have shown that the proteins encoded by the four *AS-C* genes *T3*, *T4*, *T5* and *T1a* (*T8*) are similar to each other, particularly in two domains; a C-terminal acidic domain of 15 amino acids, which may be involved in transcription activation, and an N-terminal basic domain, which exhibits substantial similarity to the HLH motif mentioned above while discussing the *E(spl)C* transcripts [46–48] (Fig. 6).

The *daughterless* locus has recently been cloned and found to contain a single transcription unit with two overlapping RNAs of 3.2 and 3.7 kb [66, 69]. The conceptual translation of the corresponding cDNA sequences uncovers the same conserved HLH motif present in the proteins encoded by the *AS-C* transcripts *T3*, *T4*, *T5* and *T1a*, and in the proteins encoded by the *E(spl)C* transcripts *m5*, *m7* and *m8*, among other genes (Fig. 6).

The helix-loop-helix motif
is sufficient for dimer formation and DNA binding

The presence of the HLH motif in several developmentally regulated proteins is striking. Murre et al. [63], who have found it in two proteins (E_{12} and E_{47}) that bind to the κE2 DNA motif in the immunglobulin kappa-chain enhancer, showed that the HLH motif is sufficient to form dimers and bind to specific DNA sequences. Moreover, it has been shown that members of the HLH protein family, including E_{12}, E_{47}, *MyoD*, *T3* and *daughterless*, are able to form homo- and hetero-dimers, which specifically bind to DNA *in vitro* with high affinity [54]. This is a particularly important finding which strongly supports the contention that all these proteins function *in vivo* as transcriptional regulators. A high degree of specificity and complexity in the regulatory functions of the corresponding genes may be achieved through the combination of different proteins to form heterodimers. Furthermore, the formation of heterodimers between the various proteins encoded by *E(spl)C* would also explain the functional redundancy found in the genetic analysis of the gene complex. This possibility is being tested experimentally at present.

Transcription of *AS-C* genes
suggests a function in neuroblast commitment

Interesting results have been obtained from the study the spatial distribution of the *T3*, *T4* and *T5* transcripts on sections of staged embryos using *in situ* hybridization [67, 68]. The distribution is very similar for all three transcripts and shows a high degree of correlation with the processes of neuroblast segregation, and development of sensory organs and stomatogastric nervous system, i.e. those processes in which the functions of the genes are known to be required. During early neurogenesis, the three transcripts are expressed in partially overlapping clusters of cells within the neuroectoderm [67]. Since their domains of expression overlap partially, some neuroblasts may contain all three RNAs, and their products, whereas other neuroblasts contain only one or two of them. This pattern of transcription suggests a role for the *AS-C* genes in neuroblast commitment. In addition, the correspondence between deletion of some *AS-C* genes and defects in particular subsets of sensory organs [49, 50] suggest specific roles for the *AS-C* genes during development of central and peripheral neural progenitor cells. Cabrera et al. [67] have proposed that the *AS-C* gene products provide the neuroblasts with specific identities, based on the particular combination of the products of the *AS-C* genes that are expressed in each neuroectodermal cell.

Interactions between the neurogenic genes and the genes of *AS-C* and *daughterless* complete the regulatory chain

Genetic interactions between some of the *AS-C* genes and *daughterless* have recently been described [70], suggesting that these genes are involved in the same function. Thus, the recent finding [54] that the *T3* and *daughterless* proteins may form DNA binding heterodimers corroborates observations from genetic analysis. On the other hand, evidence has been obtained for interactions of the neurogenic genes with the *AS-C* genes and with *daughterless* [13]. Double mutants for neurogenic genes and *AS-C* or *daughterless* show that the severity of the phenotype of homozygous neurogenic mutations is considerably reduced by the simultaneous presence of a mutation of *AS-C* or of *daughterless* in homo- or hemizygosity in the genome. Although such interactions deduced from phenotypic changes are difficult to interpret without corresponding biochemical data, they are certainly a manifestation of functional interrelationships between the genes involved. Phenotypic interactions may, for example, reflect regulatory interactions at the level of transcription; or, they may also reflect interactions at the protein level, e. g. heterodimer formation between the HLH motifs encoded by members of *E(spl)C*, *AS-C* and *daughterless*, similar to those between *T3* and *daughterless*, mentioned above [54].

This problem has been approached and partially solved by studying the distribution of RNA from the *AS-C* genes in neurogenic mutants (Fig. 8). The results suggest that at least some of the interactions between neurogenic and *AS-C* genes are likely to involve an influence on the transcription of these genes. Changes in the pattern of transcription of the genes *lethal of scute* and *achaete* (*T3* and *T5*) have been observed in embryos carrying any of several neurogenic mutations [13]. In these embryos, the early expression of *T3* and *T5* is indistinguishable from the wild type. However, at the stage when the neuroblasts normally segregate, RNA from *T3* and *T5* is found in more cells than in the wild type. Hence, transcriptional interactions seem to operate, or at least to become evident, at the time when the segregation of lineages is taking place. In the wild type, a restriction of *T5* transcription occurs from an initial group of about nine ectodermal cells to one or a few neuroblasts, as they segregate from the epidermoblasts [67]. In neurogenic mutants, this restriction fails to occur; moreover, the total number of *T5* transcribing cells per cluster is larger than nine. In contrast to this finding, no significant modification of the pattern of transcription of neurogenic genes are observed in embryos lacking most of the *AS-C* genes [13]. Therefore, the polarity of the functional relationships between the two gene groups is likely to be from the neurogenic to the *AS-C* genes, and not vice versa.

These observations on the transcription of *AS-C* genes in neurogenic mutants suggest that the cellular interactions mediated by the neurogenic genes are responsible for the refinement of the territories of *T3 − T5* expression in the wild type and that the neurogenic genes exert this function by directly or indirectly suppressing the transcription (or the accumulation) of RNA from these two genes in some of the neuroectodermal cells. This means that from among approximately nine neuroectodermal cells which in a given region transcribe one of the *AS-C* genes, only one or two will finally become neuroblasts. Since these are the only cells of the group in which RNA from the corresponding gene is still present, a causal relationship between the presence of RNA of the *AS-C* genes in a given cell and its development as a neuroblast is very likely. This has important consequences and it would explain why all neuroectodermal cells of neurogenic mutants take on the neural fate.

Conclusion and open questions

The final picture arising from reviewing the current data points to cells interacting by direct physical contact as a prerequisite for them to take on a given developmental fate. A main conclusion to be drawn from the current data, albeit still tentative, because of being based on circumstantial evidence, is that the proteins encoded by the neurogenic genes, the genes of the *AS-C* and *vnd*, and *daughterless* are functionally interrelated, forming a regulatory network that permits the cells to take on the neural or the epidermal fate. Cellular interactions seem to occur at the cell membrane, and are necessary for the regulatory signal to be passed from one cell to another. We suggest that, at the cell's membrane, the cellular interactions are mediated by the *Delta* and *Notch* proteins, probably at the EGF-like repeats present in the extracellular domains of both proteins. Since we postulate a signal chain, a very appealing possibility is that the interactions between both proteins represent relationships between ligand and receptor. On the basis of its expression pattern, we propose that *Delta* acts as the signal itself.

Unfortunately, the question of the nature of the receptor molecule for the postulated signal is still open. Transplantation of mutant cells into the neuroectoderm of the wild type has shown that only the *E(spl)C* behaves as if it encodes functions related to the reception of the signal. However, none of the products from genes of *E(spl)C* whose structure has been determined so far resembles a receptor-like protein. The structures of the *E(spl)C* proteins already known are compatible, however, with functions necessary for the transduction of a signal from the membrane (the G-protein-like molecule encoded by *m9 − m10* [53]) and for transcriptional regulation (the proteins encoded by *m5*, *m7*, and *m8*, with the DNA-binding HLH motif [42]). Thus, until the composition of the entire *E(spl)C* complex and the molecular nature of its constituents are completely known, it will be impossible to decide whether or not the complex encodes the postulated receptor protein. Nevertheless, we assume that the proteins encoded by these latter genes, and probably others, are responsible for the regulation of the specific genetic activities of the neuroectodermal cells that enable them to develop as epidermoblasts.

The results of transplantations of mutant cells into the neuroectoderm of the wild-type argues against the notion that *Notch* acts as a receptor. However, we have already mentioned that the implications of this evidence are still uncertain since under different experimental conditions *Notch* cells behave differently. Thus, we must await additional experiments to resolve this issue.

The molecular structure of the *AS-C* and *daughterless* proteins, together with the phenotypes of their mutants, immediately suggest that these proteins carry out the regulatory

Pattern of transcription of T5

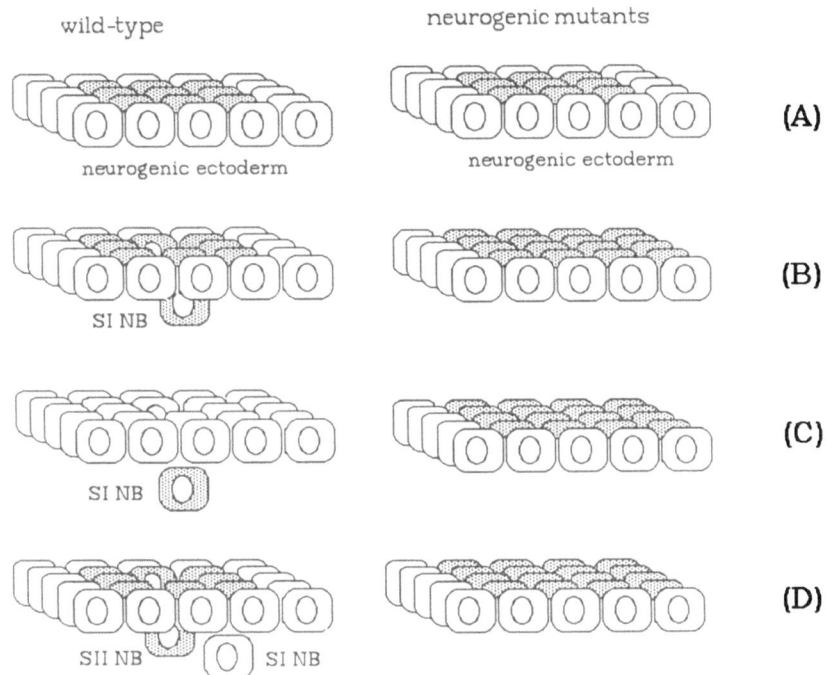

wild-type   neurogenic mutants

neurogenic ectoderm   neurogenic ectoderm   (A)

SI NB   (B)

SI NB   (C)

SII NB   SI NB   (D)

Fig. 8. *Spatial distribution of* T5 *transcripts in wild-type [67, 68] and neurogenic mutants [13].* An array of 25 neuroectodermal cells is shown at different stages in both cases. In (A), before the segregation of the neuroblasts, a group of approximately nine of these cells transcribe the T5 RNA; in (B), when the segregation of the SI neuroblasts (SI NB) takes place [4], one of them segregates as an SI NB and continues transcribing T5 for some time, whereas the other cells of the group no longer contain T5 RNA (C). At the end of this stage (D), transcription of T5 reappears in a cluster of nine cells of which one will segregate as an SII NB, which continues transcribing T5. In the neurogenic mutants (right side), the number of cells transcribing T5 is comparable to the wild type in young stages, but it increases at the time when SI NB segregation normally occurs; since all neuroectodermal cells develop as NBs, no segregation of lineages occurs in the neurogenic mutants. Transcription of T5 is therefore continued for some time in a large group of neuroectodermal cells (from [13])

functions necessary for development of the neuroblasts. DNA-binding properties have been demonstrated recently for T3 and *daughterless* [54] and the same can be assumed for T4, T5 and T1a. Since the HLH motif permits the formation of heterodimers, which have been found to bind to DNA with higher affinity than the corresponding homodimers [54], combinations between these proteins are possible.

We would like to thank Dr Paul Hardy for constructive criticisms on the manuscript. The research reported here was supported with several grants from the *Deutsche Forschungsgemeinschaft* (DFG).

REFERENCES

1. Weeks, D. L. & Melton, D. A. (1987) *Cell 51*, 861−867.
2. Kimelman, D. & Kirschner, M. (1987) *Cell 51*, 869−877.
3. Padgett, R. W., St. Johnson, R. D. & Gelbart, W. M. (1987) *Nature 325*, 81−84.
4. Hartenstein, V. & Campos-Ortega, J. A. (1984) *Roux's Arch. Dev. Biol. 193*, 308−325.
5. Technau, G. M. & Campos-Ortega, J. A. (1985) *Roux's Arch. Dev. Biol. 194*, 196−212.
6. Doe, C. Q. & Goodman, C. S. (1985) *Dev. Biol. 111*, 206−219.
7. Technau, G. M. & Campos-Ortega, J. A. (1986) *Roux's Arch. Dev. Biol. 195*, 445−454.
8. Technau, G. M., Becker, T. & Campos-Ortega, J. A. (1988) *Roux's Arch. Dev. Biol. 197*, 413−418.
9. Poulson, D. F. (1937) *Proc. Natl Acad. Sci. USA 23*, 133−137.
10. Lehmann, R., Jimenez, F., Dietrich, U. & Campos-Ortega, J. A. (1983) *Roux's Arch. Dev. Biol. 192*, 62−74.
11. Campos-Ortega, J. A. (1988) *Trends Neurosci. 11*, 400−405.
12. Jimenez, F. & Campos-Ortega, J. A. (1979) *Nature 282*, 310−312.
13. Brand, M. & Campos-Ortega, J. A. (1988) *Roux's Arch. Dev. Biol. 197*, 457−470.
14. Jimenez, F. & Campos-Ortega, J. A. (1987) *J. Neurogen. 4*, 179−200.
15. White, K. (1980) *Dev. Biol. 80*, 322−344.
16. de la Concha, A., Dietrich, U., Weigel, D. & Campos-Ortega, J. A. (1988) *Genetics 118*, 499−508.
17. Technau, G. M. & Campos-Ortega, J. A. (1987) *Proc. Natl Acad. Sci. USA 84*, 4500−4504.
18. Dietrich, U. & Campos-Ortega, J. A. (1984) *J. Neurogen. 1*, 315−332.
19. Hoppe, P. E. & Greenspan, R. J. (1986) *Cell 46*, 773−783.
20. Artavanis-Tsakonas, S., Muskavitch, M. A. T. & Yedvobnick, B. (1983) *Proc. Natl Acad. Sci. USA 80*, 1977−1981.
21. Kidd, S., Lockett, T. J. & Young, M. W. (1983) *Cell 34*, 421−433.
22. Hartley, D. A., Xu, T. & Artavanis-Tsakonas, S. (1987) *EMBO J. 6*, 3407−3417.
23. Wharton, K. A., Johansen, K. M., Xu, T. & Artavanis-Tsakonas, S. (1985) *Cell 43*, 567−581.
24. Kidd, S., Kelley, M. R. & Young, M. W. (1986) *Mol. Cell. Biol. 6*, 3094−3108.
25. Kidd, S., Baylies, M. K., Gasic, G. P. & Young, M. W. (1989) *Genes Dev. 3*, 1113−1129.
26. Johansen, K. M., Fehon, R. G. & Artavanis-Tsakonas, S. (1990) *J. Cell Biol.*, in the press.
27. Vässin, H., Bremer, K. A., Knust, E. & Campos-Ortega, J. A. (1987) *EMBO J. 6*, 3431−3440.

10

28. Kopczynski, C. C., Alton, A. K., Fechtel, K., Kooh, P. J. & Muskavitch, M. A. T. (1988) *Genes Dev. 2*, 1723—1735.
29. Vässin, H. & Campos-Ortega, J. A. (1987) *Genetics 116*, 433—445.
30. Alton, A. K., Fechtel, K., Terry, A. L., Meikle, S. B. & Muskavitch, M. A. T. (1988) *Genetics 118*, 235—245.
31. Carpenter, G. & Cohen, S. (1979) *Annu. Rev. Biochem. 48*, 193—216.
32. Furie, B. & Furie, B. C. (1988) *Cell 53*, 505—518.
33. Rusell, D. W., Schneider, W. J., Yamamoto, T., Luskey, K. L., Brown, M. S. & Goldstein, J. L. (1984) *Cell 37*, 577—585.
34. Sasaki, M., Kato, S., Kohno, K., Martin, G. R. & Yamada, Y. (1987) *Proc. Natl Acad. Sci. USA 84*, 935—939.
35. Komoriya, A., Hortsch, M., Meyers, C., Smith, M., Kanety, H. & Schlesinger, J. (1984) *Proc. Natl Acad. Sci. USA 81*, 1351—1355.
36. Heath, W. F. & Merrifield, R. B. (1986) *Proc. Natl Acad. Sci. USA 83*, 6367—6371.
37. Graf, J., Iwamoto, Y., Sasaki, M., Martin, G. R., Kleinman, H. K., Robey, F. A. & Yamada, Y. (1987) *Cell 48*, 989—996.
38. Kelley, M. R., Kidd, S., Deutsch, W. A. & Young, M. W. (1987) *Cell 51*, 539—548.
39. Brand, M. & Campos-Ortega, J. A. (1990) *Roux's Arch. Dev. Biol. 198*, 275—285.
40. Knust, E., Tietze, K. & Campos-Ortega, J. A. (1987) *EMBO J. 6*, 4113—4123.
41. Ziemer, A., Tietze, K., Knust, E. & Campos-Ortega, J. A. (1988) *Genetics 119*, 63—74.
42. Klämbt, C., Knust, E., Tietze, K. & Campos-Ortega, J. A. (1989) *EMBO J. 8*, 203—210.
43. Garcia-Bellido, A. (1979) *Genetics 91*, 491—520.
44. Carramolino, L., Ruiz-Gomez, M., Guerrero, M. C., Campuzano, S. & Modolell, J. (1982) *EMBO J. 1*, 1185—1191.
45. Campuzano, S., Carramolino, L., Cabrera, C. V., Ruiz-Gomez, M., Villares, R., Boronat, A. & Modolell, J. (1985) *Cell 40*, 327—338.
46. Villares, R. & Cabrera, C. V. (1987) *Cell 50*, 415—424.
47. Alonso, M. C. & Cabrera, C. V. (1988) *EMBO J. 7*, 2585—2591.
48. González, F., Romani, S., Cubas, P., Modolell, J. & Campuzano, S. (1990) *EMBO J.*, in the press.
49. Dambly-Chaudière, C. & Ghysen, A. (1987) *Genes Dev. 1*, 297—306.
50. Ghysen, A. & Dambly-Chaudière, C. (1988) *Genes Dev. 2*, 495—501.
51. Knust, E., Bremer, K. A., Vässin, H., Ziemer, A., Tepaß, U. & Campos-Ortega, J. A. (1987) *Dev. Biol. 122*, 262—273.
52. Preiss, A., Hartley, D. A. & Artavanis-Tsakonas, S. (1988) *EMBO J. 7*, 3917—3928.
53. Hartley, D. A., Preiss, A. & Artavanis-Tsakonas, S. (1988) *Cell 55*, 785—795.
54. Murre, C., Schonleber McCaw, P., Vaessin, H., Caudy, M., Jan, L. Y., Jan, Y. N., Cabrera, C. V., Buskin, J. N., Hauschka, S. D., Lassar, A. B., Weintraub, H. & Baltimore, D. (1989) *Cell 58*, 537—544.
55. Watt, R., Stanton, L. W., Marcu, K. B., Gallo, R. C., Croce, C. M. & Rovera, G. (1983) *Nature 303*, 725—728.
56. Kohl, N. E., Legouy, E., DePinho, R. A., Nisen, P. D., Smith, R. K., Gee, C. E. & Alt, F. W. (1986) *Nature 319*, 73—77.
57. Stone, J., DeLange, T., Ramsey, G., Jakobovits, E., Bishop, J. M., Varmus, H. & Lee, W. (1987) *Mol. Cell Biol. 7*, 1697—1709.
58. Legouy, E., DePinho, R., Zimmerman, K., Collum, R., Yancopoulos, G., Mitsock, L., Kriz, R. & Alt, F. W. (1987) *EMBO J. 6*, 3359—3366.
59. Tapscott, S. J., Davis, R. L., Thayer, M. J., Cheng, P.-F., Weintraub, H. & Lassar, A. B. (1988) *Science 242*, 405—411.
60. Wright, W. E., Sassoon, D. A. & Lin, V. K. (1989) *Cell 56*, 607—617.
61. Edmondson, D. G. & Olson, E. N. (1989) *Genes Dev. 3*, 628—640.
62. Braun, T., Buschhausen-Denker, G., Bober, E., Tannich, E. & Arnold, H. H. (1989) *EMBO J. 8*, 701—709.
63. Murre, C., Schonleber McCaw, P. & Baltimore, D. (1989) *Cell 56*, 777—783.
64. Thisse, B., Stoetzel, C., Gorostiza-Thisse, C. & Perrin-Schmitt, F. (1988) *EMBO J. 7*, 2175—2183.
65. Rushlow, C. A., Hogan, A., Pinchin, S. M., Howe, K. M., Lardelli, M. & Ish-Horowicz, D. (1989) *EMBO J. 8*, 3095—3103.
66. Caudy, M., Vässin, H., Brand, M., Tuma, R., Jan, L. Y. & Jan, Y. N. (1988) *Cell 55*, 1061—1067.
67. Cabrera, C. V., Martinez-Arias, A. & Bate, M. (1987) *Cell 50*, 425—433.
68. Romani, S., Campuzano, S. & Modolell, J. (1987) *EMBO J. 6*, 2085—2092.
69. Cronmiller, C., Schedl, P. & Cline, T. W. (1988) *Genes Dev. 2*, 1666—1676.
70. Dambly-Chaudière, C., Ghysen, A., Jan, L. Y. & Jan, Y. N. (1988) *Roux's Arch. Dev. Biol. 197*, 419—423.

Eur. J. Biochem. *190*, 233—248 (1990)
© FEBS 1990

Review

Form determination of the heads of bacteriophages

Edward KELLENBERGER

Department of Microbiology, Biozentrum, University of Basel, Switzerland

(Received July 3, 1989/February 23, 1990) — EJB 89 0853

The shape of the DNA-containing heads of many bacteriophages is not only determined by the properties of the protein subunits which build the shell (capsid) but also by the scaffolding core which is a transient structure of the prohead. The form-determining properties of the scaffolding proteins have been characterized by genetic methods based on conditional mutants and site-directed mutagenesis. The mechanism of form determination has been studied by *in vitro* assembly experiments. The theoretical background is discussed and different models for mechanisms of form determination are considered. Definitive decisions about the validity of a model is still limited by the difficulty of obtaining unambiguous answers on the stoichiometry and the fine structure of the scaffold because of their high instability.

The conservation of the specific shape of any living species is the very basic observation that gave rise to the notion of genetics. The mechanisms which underlay form determination still remain among the fundamental unsolved problems of biology although the descriptive part, the genesis of shape (morphogenesis) from the fertilized egg to the fully grown organism, is known in detail for most multicellular organisms. In recent decades one has also learned that the different steps of the morphogenesis of multicellular organisms must be determined through the successive switching on and off of groups of genes. In consequence, today's research efforts are concentrated on the mechanisms of the genetic control of gene expression. Only once these are understood and the newly expressed genes identified, can the functions of the corresponding gene products, the proteins, be studied. The phenomena of cellular recognition and of contact inhibition of growth must be understood before one is even able to formulate plausible hypotheses for the form determination of multicellular organs and organisms. How, for example, does a regenerating, partly ectomized liver, know how to stop cellular proliferation when the correct shape is again achieved?

The situation is much better, although quite different, for multi-molecular structures represented by elements of the cytoskeleton or by other cellular organelles and, above all, by viruses [1—3].

Everybody is more or less familiar with what is called self-assembly of a multimolecular structure. It is represented, for instance, by some cases of virus capsids, where the shape and binding properties of the protein subunit are such that they can assemble only into one type of hollow sphere. Such cases are considered as form determination of the first order. This situation is somewhat trivial; indeed, for the large majority of viruses, form determination is not so simple: inner scaffolds are transiently needed to help in imposing any form to the capsid. The form of the capsid is therefore determined by the information contributed by both the properties of the subunits

of the capsid and by those of the protein subunits that compose the scaffold. Since the information for the form is now carried jointly by several genes, they are called form determinations of higher orders. The present paper concentrates on such a specific, well studied case: the head of bacteriophage T_{even}.

Bacteriophages have provided the first experimental systems by which the genetic approach has been exploited to the full. By introducing conditional lethal mutants, the functions of nearly all of the genes of some bacteriophages could be described in genetic terms. The suppressor-sensitive mutants (sus [4], amber [5, 6]) were particularly useful because, in the non-permissive host, they produce only a fragment of the 'gene product'. The nonpermissive host lacks a tRNA able to read UAG ('amber') or UAA ('ochre'). Comparisons of proteins and fragments produced in the permissive and nonpermissive hosts, respectively, by electrophoresis permits correlation of proteins with genes. This part of the work is particularly easy when the infecting phage arrests completely the expression of the host genome, as is the case with the T_{even} coliphages. Radioactive markers (^{35}S- or ^{14}C-labeled amino acids) added at the time of infection only label the newly synthesized, phage-related proteins; these are identified on an autoradiograph of an electrophoretic gel. This method demonstrated that only a few groups of commonly controlled genes exist: all of the 'late genes' are turned-on at the same time; they are all concerned with the assembly of the virus particles. Products of late genes continue to be synthesized at various gene-specific rates until the cell lyses. With the additional help of electron microscopy and centrifugation, it was possible to show that phage morphogenesis within individual infected cells is completely asynchronous; initiations occur randomly in time, and the assembly and maturation of an individual T_{even} particle only takes 7—15 min. When certain genes that mediate cellular lysis and liberation of phage are mutated, this asynchronous synthesis of T_{even} phage particles can continue for 90—120 min! It is thus obvious that, in contrast to multicellular assemblies, this morphogenesis is not controlled at the level of gene expression but is mediated by sequential protein—protein interactions. The control is exerted by 'actuation' as will be explained below.

Correspondence to E. Kellenberger, Department of Microbiology, Biozentrum, University of Basel, Klingelbergstrasse 70, CH-4056 Basel, Switzerland

234

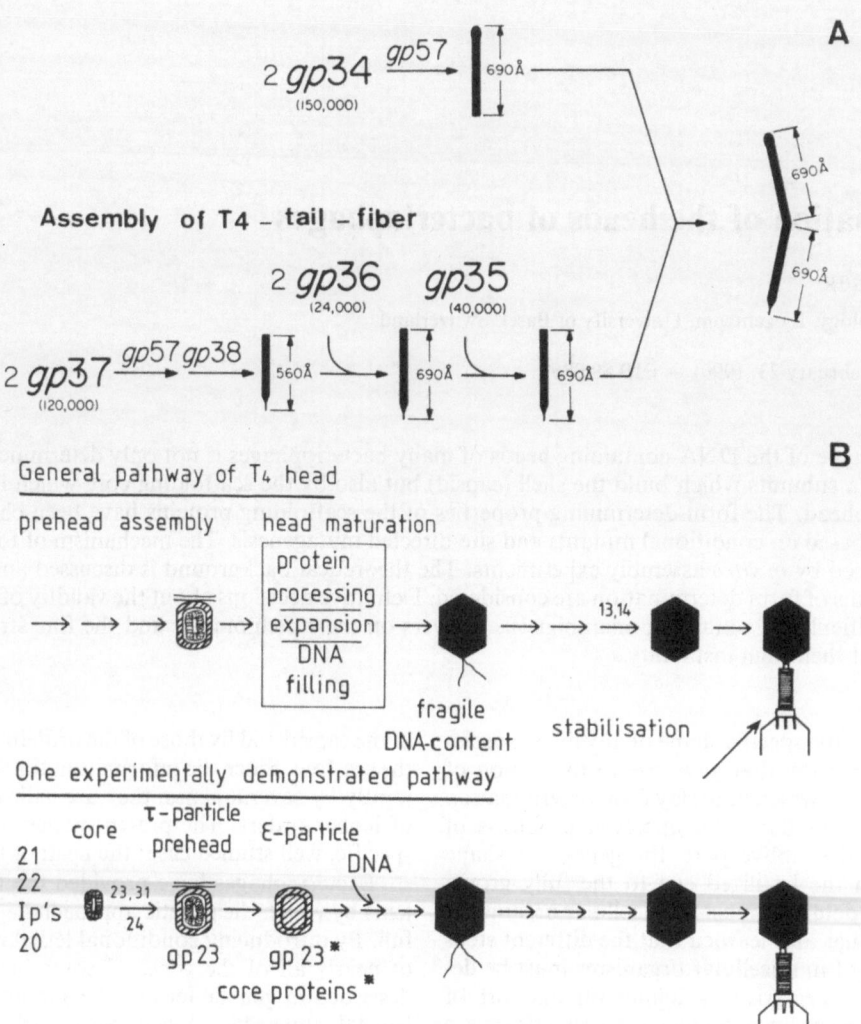

Fig. 1. *Morphogenesis of T4 substructures.* (A) The assembly of tail fibers with the successive additions of gene products (gp). The temporary sequence is not determined on the level of gene expression but is fixed by the conformation of the assembling fiber. The newly added gene product is actuated, i.e. put into action (slightly modified after Wood [86]). 690 Å ≡ 69 nm. (B) Pathway of T4 head maturation. This pathway is more complicated, because it cannot be described by simple successive additions of gene products. Some of the maturation steps occur intrinsically, as interactions between proteins that are already present within the particle. The upper pathway accounts for all the experimental facts reported and which were obtained under various experimental conditions and/or with different mutants. The lower pathway accounts for results under specific conditions and their general validity is not yet established. Completely finished cores without shells are observed *in vivo* and *in vitro.* Some evidence is available for an *in vivo* maturation of cores into proheads [61]. The existence of the ε particle is demonstrated both in the case of mutants 16 and 17 and in presence of 9-aminoacridine [70]. In the latter case, but not the former, its capacity to mature *in vivo* was shown. Numbers indicate those of the genes involved. IP's are the gene products which lead, after cleavage, to three different internal proteins of this phage. The asterisk denotes that, of these marked proteins, 5−20% have been amputated by partial proteolysis (T4ppase is T4-coded protease [20]). (From Schärli and Kellenberger [71])

By analyzing the conditional lethal mutants, the assembly-maturation pathways of many bacteriophages became established [2] (see Fig. 1 for an example). The lack of a functioning gene product normally leads to an accumulation of the precursor particle with which the concerned gene product was predestined to interact. Sometimes, however, the accumulated precursors do not remain unaltered but rather undergo abortive modifications. Only in the most simple systems are the precursor particles, observed by electron microscopy as part of an asynchronously produced population, true precursors. The possibility of abortive pathways must always be considered. This can be advantageous in that abortive products can explain many fundamental properties of the form-determining mechanisms.

All the DNA-containing bacteriophages studied showed a prohead as precursor which then becomes filled with DNA and matures by further steps into a stable head. The prohead features all the parameters which characterize the final shape of the head, although, in many cases, it undergoes a regular enlargement prior or parallel to packaging and maturation. The finished head is then fitted with a tail or another smaller device needed for infecting bacteria. Such a device is required to push the DNA-injection apparatus through the tough, outer layers of the cell envelopes.

In contrast to multicellular organisms where some small variations of shape necessarily occur between individuals of the same species, this is not the case with viruses. All observations made converge to the surprising fact that the number of subunits used to make the rod-like tail of a bacteriophage or the shell of a near-globular virus or bacteriophage head is precisely determined. The number is that given by defined geometric bodies. Variants which do occur are frequently non-

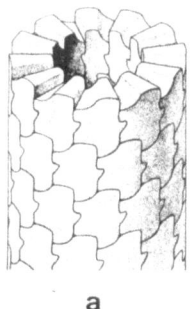 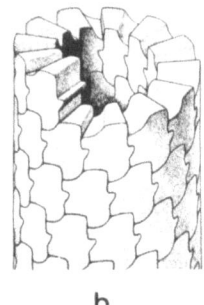 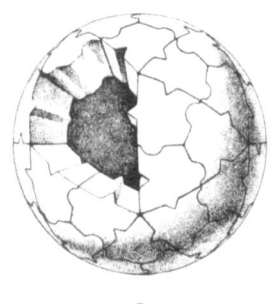

a   b   c

Fig. 2. *Cylindrical and spherical assemblies made of identical subunits.* The reality is simplified by assuming only a purely geometric fit, as in the case of a puzzle. (a, b) Examples of two conformations: (a) stacked disc and (b) helical, as they occur, for instance, with tobacco mosaic virus (TMV). In the sphere (c) the 60 subunits have perfect icosahedral symmetry. In each of the 12 vertices of the generating icosahedron 5 subunits are placed with fivefold rotational symmetry. (From Kellenberger [87])

viable structures. Either viable or inactive, they still observe the geometric rules which, for closed shells, are possible only with discrete, defined numbers of subunits. In this paper, the mechanisms determining the form of near-globular phage heads, including width and length, are discussed. Space does not allow the beautiful example of length determination, as represented by phage tails, to be considered. This appears to be determined by a type of ruler made of a fibrous protein, its length given by the length of the gene coding for this protein. Such a correlation between the lengths of a gene and of the tail has been found for phage lambda [7]. A final proof has recently been provided by engineering the gene [7]. For the tail of phage T_{even}, the presence of a central protein fiber is suggested by electron microscopy [8].

From a large number of observations which are too abundant to be described here, certain fundamental concepts emerged which are axiomatic because they are difficult to demonstrate directly with the experimental tools presently available. They are, however, necessary to understand the phenomena in a broader, holistic context.

FUNDAMENTAL CONCEPTS

Shape determining properties of the protein subunits

Hollow cylinders and hollow spheres can be built from identical subunits whose shape is fully form-determining for the assembly product. Or, expressed in other words, the shape of the subunits is such that only a cylinder or a sphere of given diameter can be built with them. Fig. 2 shows this situation, recognized by early mathematicians long before its application to viruses by Crick and Watson [9]. The sphere shows an icosahedral arrangement of exactly 60 subunits. Five subunits together form a capsomer with a local fivefold symmetry axis which passes through the center. These capsomeres correspond to the vertices of an icosahedron folded from paper (Fig. 3). Other symmetries are theoretically possible, but these do not occur for reasons discussed below.

The building principles for viruses had been enunciated masterfully by Caspar and Klug [10]. They introduced the fundamental concept of quasi-equivalence. By folding a two-dimensional lattice of hexagonal symmetry (p6) it is easy to build larger icosahedra (Fig. 3). However, the number of subunits is discrete (60, 180, 420 etc.). When examining these paper models, one sees that the local curvature of the lattice is not equal on the whole surface. When replacing the flat, two-dimensional subunit of the paper model with a more realistic three-dimensional subunit one realizes that only the

icosahedron with 60 subunits can become transformed into a sphere (with identical curvature). In the model with 60 sub-units all of the subunits are related to each other by rotational movements (symmetry operations) around the center and the local fivefold axes, but in the larger bodies which are made of many more subunits these same symmetry relations are not fulfilled. The subunits are only in quasi-equivalent positions. When this fundamental concept was presented, the deviation from equivalence was described as deviations of the bonding angles. We will see below that these deviations must now be envisaged rather as elastic deformations of the subunits.

In their classic paper, Caspar and Klug [10] stated that the 'amount' of required quasi-equivalence increases with the size, i.e. with the number of identical subunits involved. They considered the possibility that the largest shells might need additional information through scaffolds. In other words, the form and size of shells might be determined fully by shape-specifying subunits in the simplest cases, but additional information might be needed for larger structures [10, 11].

Clearly, the purely geometric models cannot be applied to real proteins which do not have the edgy, precisely defined shape needed for such a geometric fit. The specific binding properties must be added in order to obtain a full shape determination. Protein — protein interactions in supramolecular assemblies only very rarely involve covalent links of the disulfide type. Mostly they are based on weak interactions, particularly on hydrophobic bonding [12, 13]. Some ten weak interactions are needed to give as strong a bond as would be achieved covalently. The ten interactions therefore concern an area that comprises at least ten amino-acid residues. That this is so is indicated by the fact that a large majority of virus shells are only dissociated into individual polypeptides by chemical agents which, at the same time, also denature the subunits. In short, the strength of a protein — protein interaction is approximatively proportional to the area of interaction.

Thus quasi-equivalence must be imagined as quasi-equivalent deformations of the protein subunits to maintain the areas of interaction.

The intrinsic strain of protein subunits and the notion of intrinsic curvature

Let us make the following *Gedankenexperiment.* A protein might have evolved to form two different interaction areas (or sites) by which it can polymerize into a linear polymer chain. It is obvious that, in general cases, such a chain would form a regular helix, without contact between the windings. If we want to build a cylinder, we would have to compress the

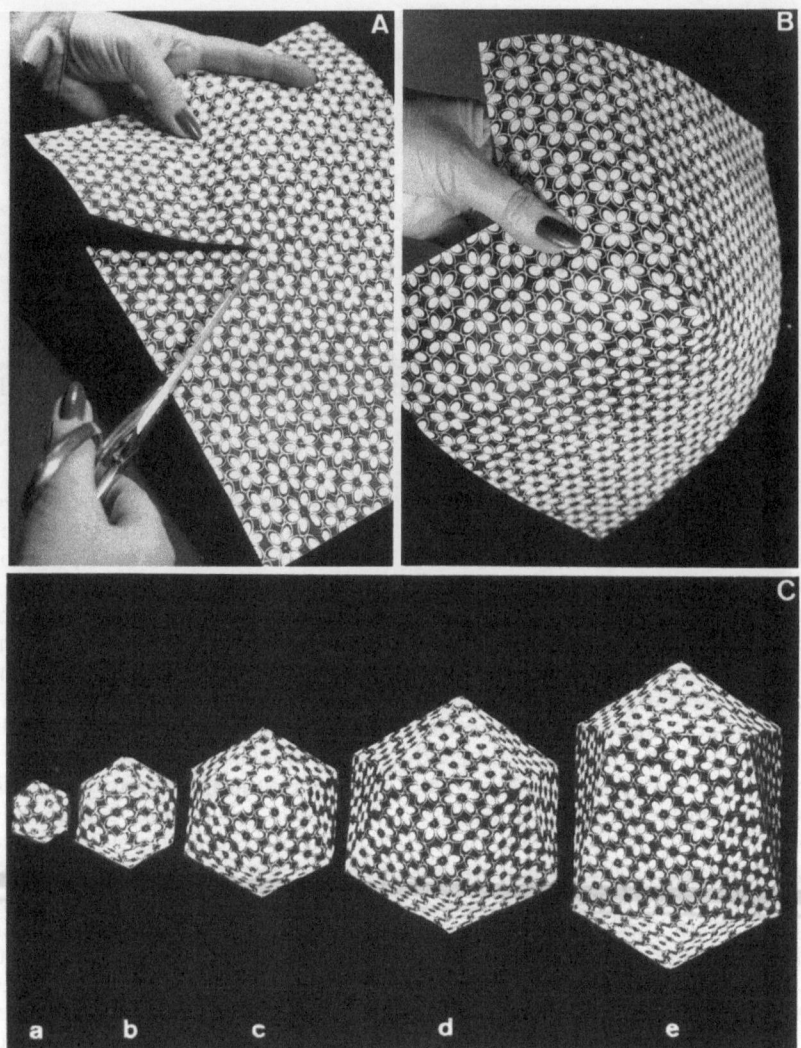

Fig. 3. *Icosahedral bodies folded from a sheet of subunits arranged in p6 symmetry.* (A) Out of the p6 net, a 60° sector is cut out. The sector has its origin in one of the six coordinated groups of subunits. After removal of the sector, the edges can be joined. A cone results with a pentamer at its top (vertex). On the surface the net is again continuous. The curvature, however, decreases with locations further away from the vertex. By successive cuttings of 60° sectors geometric bodies with icosahedral properties can be made as shown in C and Fig. 7B. Models of icosahedral bodies. (a) This corresponds to Fig. 2c with 60 subunits ($T = 1$). The number of subunits is calculated from the triangulation number T as $N = 60T$, or $N = 30 (T + Q)$. (b, c, d) Models with $T = 3$, 7 and 13, respectively. They require quasi-equivalence of the subunits as explained in the text. The subunits frequently form groups ('capsomers') of six when situated on the faces but of only five in the vertices. (e) The prolate derivative of d. The 'equatorial band' of the icosahedron is enlarged by a number of subunits that is expressed by Q. The head of bacteriophage lambda is an icosahedron with $T = 7$, that of phage T4 a prolate icosahedron of $T = 13$ and $Q = 21$ [77]. (Models by courtesy of Dr M. Wurtz)

helix to bring the neighboring windings into contact (Fig. 4a and b). The compression must act against a spring force; additional interactions are needed to hold the windings together. One can imagine another situation in which the compression is replaced by extension. Then the interaction between the windings can be very weak. Obviously, evolution will tend to optimize shape and energy interactions to obtain a strongly bonded cylinder with a minimal expenditure of energy. It is essential, however, to be aware of how the interactions counterbalance insufficiencies in the molecular shape. One should also keep in mind that protein molecules are never symmetric.

Instead of compressing a relatively open helix into a cylinder, the same result might be achieved by several parallel helices. In such a multistranded helical cylinder the 'lateral'

interactions between the strands do not need to be very strong (Fig. 4c).

In all these cases a minimum of two sets of specific interactions are needed per subunit for producing a stable cylinder. These are characterized by prominent areas of contact between neighbouring subunits. A line following each type of these pairs of contacts describes a set of different generating helices (Fig. 4b and c). In the closest packing, three sets of interactions are possible; accordingly, three types of generating helices are defined. In a particular case of multistranded helices one of the generating helices might degenerate into independent circles; we then speak of the stacked disc situation which is frequently encountered.

In all cases, even in that of the multistrand model, an intrinsic strain is also likely to be involved for statistical re-

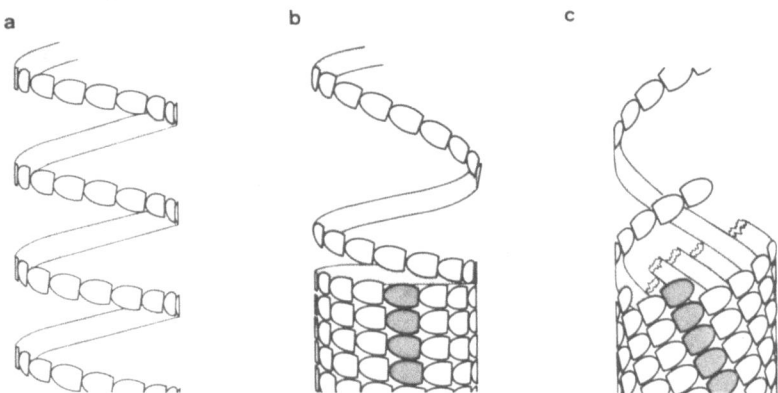

Fig. 4. *The helices of cylindrical assemblies.* The general form of a linear assembly of subunits is an open helix (a). In this case only two complementary protein−protein interaction sites are involved. If a second, sufficiently strong set is available, the helix is compressed into a cylinder (b). Several helices might also generate a multistranded helix (c). In all cases of cylinders at least two sets of interaction sites are needed. With two sets of interactions one can always distinguish two sets of generating helices, according to the choice of one or the other set of interactions. If there are more sets of interaction, the number of different choices for generating helices increases accordingly. In (b) and (c) only two sets of interactions (= sites of contact) have been chosen. Of the additional set of helices one is indicated in grey. The three figures are strongly schematized, the depth of the subunit is not considered

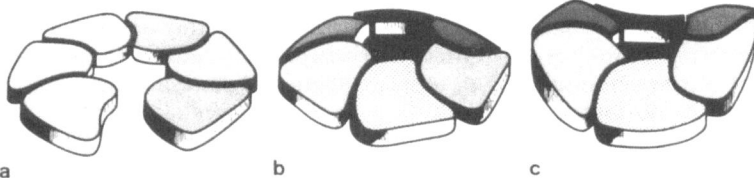

Fig. 5. *The different possibilities for the form of a capsomer.* In globular capsids of icosahedral symmetry we find on the facets and over the 'edges' six-coordinated groups of protein subunits (hexamers) and in the vertices five-coordinated groups (pentamers). Frequently the binding forces in between hexamers are stronger than of subunits within the hexamer such that trimeric or even dimeric clustering is predominant. As in cylinders, at least two sets of interactions are needed. According to their strength, dissociation goes either over pentons and hexons or over trimers and dimers. In the figure, hexamers are considered. It is again assumed that evolution has not led to a 'perfectly flat' hexamer; if the dimensions of the subunits are too small, a cup-like deformation is obtained (b); if they are slightly too large, a saddle-like hexamer is obtained (c). While assemblies from the cup-like hexamers (b) have no preference for a cylindrical bending, this is not the case with the saddle-like hexamer (c) which prefers bending in one direction. This is a possible explanation of the intrinsic curvature reflected in tubular assemblies by a preference of diameter (see text)

asons. It would be rare indeed that the resulting curvature be such as to terminate the cylinder into a continuous, axially symmetric structure without a small additional effort of bending which is contributed again by the lateral interactions.

These models suggest that the intrinsic curvature of the protein subunits determines the diameter of the helix and, with additional slight strains, that of the continuous compact cylinder. In order to maintain a cylinder in a stable state, at least two sets of specific interactions must exist.

The hollow sphere requires similar strains, although the situation is much more difficult to describe. To illustrate the concepts, we will consider hexons and pentons, the six- or five-coordinated groups of subunits around local sixfold or fivefold axes. They form rings, as shown in Fig. 5. If the subunits are slightly too small (or if one is lacking), we obtain a cup-shaped ring; it has an intrinsic, symmetric curvature (Fig. 5b). When the penton is made of the same subunits as the hexon, it is clear why the pentons are in the vertices of large icosahedra. If the subunits are slightly too large, then the hexon reaches a curious saddle-like form (Fig. 5c). When such hexons have to associate, they cannot do so without strain. This strain might be substantial if the association product is a flat sheet, but less so when it forms a cylinder. Thus, head-shell subunits of a few bacteriophages might form tubular variants as soon as the action of the scaffold is lacking, as shown below. Again, as in the case of cylindrical and globular assemblies, the subunits need at least two sets of interactions.

The order of morphopoiesis in form determination and the role of scaffolds

If a protein subunit is fully shape-determining, such an assembly is considered to be of the first order. The information contained in one gene suffices to determine a supramolecular structure made of one species of protein. The order of a morphopoiesis is described by the number of genes that contribute information toward the determination of a given form [11]. If a subunit is made of two nonidentical polypeptide chains, then we have a morphopoiesis of the second order, although the dimer is still fully shape-determining. The notion of the order of morphopoiesis becomes really meaningful for those cases where gene products are only transiently acting as scaffolds and where these are essential to achieve a certain form.

From the considerations on the strain in the subunits which increases, for instance, with the size of a virus shell, it is plausible from theoretical grounds that such shells might need more information than that provided by the subunit alone. It is plausible that such a need might increase further for geometric bodies which have only an axial symmetry instead of the central symmetry. Such is the case with prolate, i.e. elongated, phage heads. Prolate heads are found rather frequently in nature; two of them, T_{even} coliphages and the *Bacillus subtilis* phage $\phi 29$, have been subjects of particularly intensive research (for references see [1] and [3]). Phage T4 was the first for which a form-giving function of the internal core observed with phage-related particles was postulated [11, 14]. Later, nearly all DNA-containing phages which mature via a prohead have been shown to need a scaffolding core at the prohead stage, without which abnormal capsids result [15 – 17]. The shapes of these phage heads depend therefore not only on the information contributed by the subunit(s) of the shell, but also on that contributed by the scaffold. The relative amount of information is variable according to the species and difficult to assess quantitatively.

The concept of the temporal control of protein – protein interactions through sequentially appearing interaction sites [18]; the principle of protein actuation

During the assembly and maturation pathways of the sub-assemblies (e.g. tail, head, tail fiber for T_{even}) of bacteriophages, all proteins involved are present and synthesized continuously. They form a mixture of subunits of all the different species of proteins ('the soup') that finally form the mature particle. They do not, however, assemble at once: they are called into action (actuated) whenever the maturing particle is ready to interact with the next gene product. The different protein – protein interactions needed are displayed in a sequence. Each maturing particle decides individually about the protein species needed next and selects it out of the soup of subunits. This sequence of actuations of gene products is depicted in the pathways such as the one exemplified in Fig. 1. Many more have been described in detail by Casjens [2]. These pathways are frequently misunderstood by believing that the sequence is under transcriptional control, in that the gene of the needed protein is expressed only at the exact time it is needed. That this is not the case has already been discussed and is demonstrated by the complete asynchrony of the particle production within one single cell. The new notion of actuation or actuational control therefore simply defines these facts.

Before we can discuss examples of mechanisms of actuation, we have to recall the theories of initiation (or nucleation) of multimolecular assemblies made of one single species of subunits, as is represented, for instance, by the well studied case of actin. The present understanding is based on probability considerations.

When two subunits encounter each other they might interact. The interaction with only one set of complementary sites being reversible, the chance to become separated again is rather large. If a third subunit happens to join before they separate, then additional interactions with additional pairs of sites now stabilize the trimeric structure, such that the probability to separate is drastically reduced. Each supplementary subunit attaches by involving two sets of interactions; consequently, they are also stably bound. The step to reach a trimer is called the initiation or nucleation, while that of the addition of further subunits is the elongation.

In multimolecular assemblies, made of several different species of proteins, as are represented by our example of bacteriophages, we have also to assume an initiation mediated by one species of protein. The rate at which this species is synthesized might thus be the rate-limiting step of particle formation. Once an initiator is formed, new sites of interaction become revealed which now actuate the addition of subunits of another species.

The subunits of one type after another would, in this way, each become actuated in turn. An example of such a straightforward sequential actuation is given by the tail fibers

of bacteriophage T4 (Fig. 1 A). The pathways of tails mostly follow similar straightforward patterns, although with a few complications [19]. In the pathway of the head, further very interesting complications are observed (Fig. 1 B). Once the prohead is assembled, apparently by the sequential addition of proteins, a series of events follow which could be described as an intrinsic maturation. A series of proteolytic modifications of nearly all the proteins occurs which seem to be all produced by a phage-coded protease, the T4ppase, which is derived by a cascade of proteolytic cleavages from the product of gene 21, which is situated within the prohead [20]. How far host proteases might also be involved in the first steps is still not established. The modifications are rather profound; they had been studied both structurally and biochemically with the capsid. The prohead shell, when transformed into the capsid, has its subunit mass reduced by some 20% but, despite this, the capsid is enlarged and chemically strongly stabilized (for references and further discussion see [20]). The scaffolding core of the prohead becomes digested and most of it is eliminated during the ensuing process of DNA packaging.

The processes that underlie the maturation of a particle are probably combinations of activations and actuations. Bonds established during assembly of proheads, for instance those between the scaffolding core and the shell, are opened while others are enforced or modified. Whether the straightforward cases of sequential actuations of proteins are considered as additional bonds or just changes of bonds within a maturing particle, one has to assume conformational shifts of the constituting protein subunits such that new interaction sites become available or disappear. The formation of sites by the juxtaposition of two subunits as a consequence of polymerization has also to be considered.

It is not within the scope of this paper to discuss further these interesting mechanisms of sequential interactions and reactions by successively induced conformational changes. It is necessary, however, to retain for further discussions about form determination the insight that the essential parameters of the form of the bacteriophage head have already been fixed in the prohead. What happens later is only an enlargement by the regular expansion of the surface lattice of the prohead shell [20, 21]. In order to understand more about the mechanisms of form determination, we should know more about the assembly of the prohead and, because of its fundamental influence in form determination, particularly the assembly of the scaffolding core. Unfortunately, we have only very few solid data either on important structural details of the prohead or about its precise pathway of assembly.

THE EXPERIMENTAL RESULTS

The experimental systems

Bacteriophages with their host, the bacterial cell, are particularly suited for experiments both *in vivo* and *in vitro*. Experiments conducted *in vitro* can mostly not be performed as a perfect, faithful simulation of what happens *in vivo*. Some of the parameters of the intracellular environment are, in most cases, not yet known and therefore cannot be simulated correctly *in vitro*; they are discussed below. Bacteriophages present probably the best known experimental system which allows both types of experiments to be done with about equal precision.

For the time being, it is the only experimental system in which the conditional lethal amber mutations can be used to

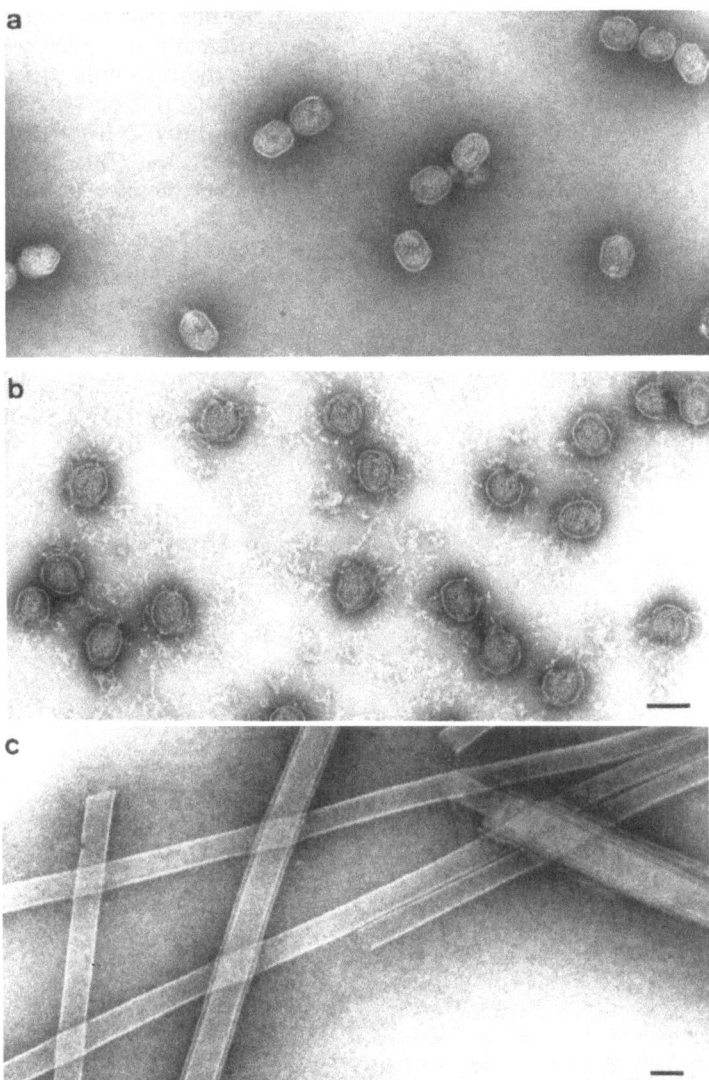

Fig. 6. *Proheads or τ particles and the tubular variant of its shell, the polyhead.* (a, b) Native and *in vitro* reassembled preheads of phage T4. These micrographs (obtained by Dr R. Van Driel and E. Couture [41] represent native proheads: (a) after isolation and purification, (b) after reassembly. The small aggregates visible on (b) are the proteins which did not reassemble into a correctly organized structure. Some of these non-structured aggregates always occur. Both figures are negatively stained. The bar represents 100 nm. (c) Tubular variants (polyheads) of the shell of proheads, here without core. They are composed only of gp23. (Negatively stained; micrograph by Dr R. van Driel and E. Couture.) The bar represents 100 nm

Table 1. *Location and function of T4-head-related gene products*

Protein	Location and function
gp23	shell of the prohead, after proteolytic modification forms the capsid (gp23*)
gp 24	in the vertices of the capsid as gp24* (still to be confirmed), responsible for osmotic shock resistance
gp21	zymogen of the T4ppase, probably in the center of the scaffold
gp22	main protein of the scaffold (60%)
IPIII	dispensable protein of the scaffold
IPII	dispensable protein of the scaffold
IPI	dispensable protein of the scaffold
gp67	in the scaffold
gp68	in the scaffold
gp20	in the portal vertex of the phage as a dodecamer; connects head to tail

the full extent of their advantages, which lie in the fact that individual gene functions can be eliminated without affecting all the others. The resulting phenotypes are therefore not the result of a missense, as is usually the case with mutants, but that of a nonsense. With bacteriophages, the study of form determination by genetic means has therefore provided information far ahead of that obtained by all other approaches.

The results obtained with bacteriophage T4, in particular, will now be described mainly because, due to the prolate shape of its head, it furnishes an adequate system to study the genetic basis of form determination. Phages with isometric heads show, as a rule, only size variants. In these phages the role of form determination of the scaffolding core, relative to that of the shell, is thus much reduced, although it is not nil. Indeed, in most phages extensively studied, the absence of scaffolding core proteins lead to particles which no longer have a defined form (references in last section of Experimental Results).

Results of the genetic approach

The system of conditional lethal amber mutants allowed a thorough investigation of the functions of the head-related genes of phage T4 (Table 1) [5]. In the permissive host the amber (am) mutation leads to a normally active protein, while in the nonpermissive host only a fragment of the gene product is synthesized which, as a rule, is nonfunctional. Very often this fragment is digested by cellular proteases. When infecting a nonpermissive cell with an amber mutant phage, all genes, except the one with the mutation, produce active proteins at their normal rates. What we observe as the phenotype is therefore produced in the absence of a functional product of one given gene (the one which carries the amber mutation). This is very different from 'conventional' mutants, where the complete protein is synthesized except that, compared to the wild type, one amino acid has been changed into another as a consequence of a point mutation. As a result, the properties of the protein might occasionally be changed as well, and, in consequence, alter the phenotype.

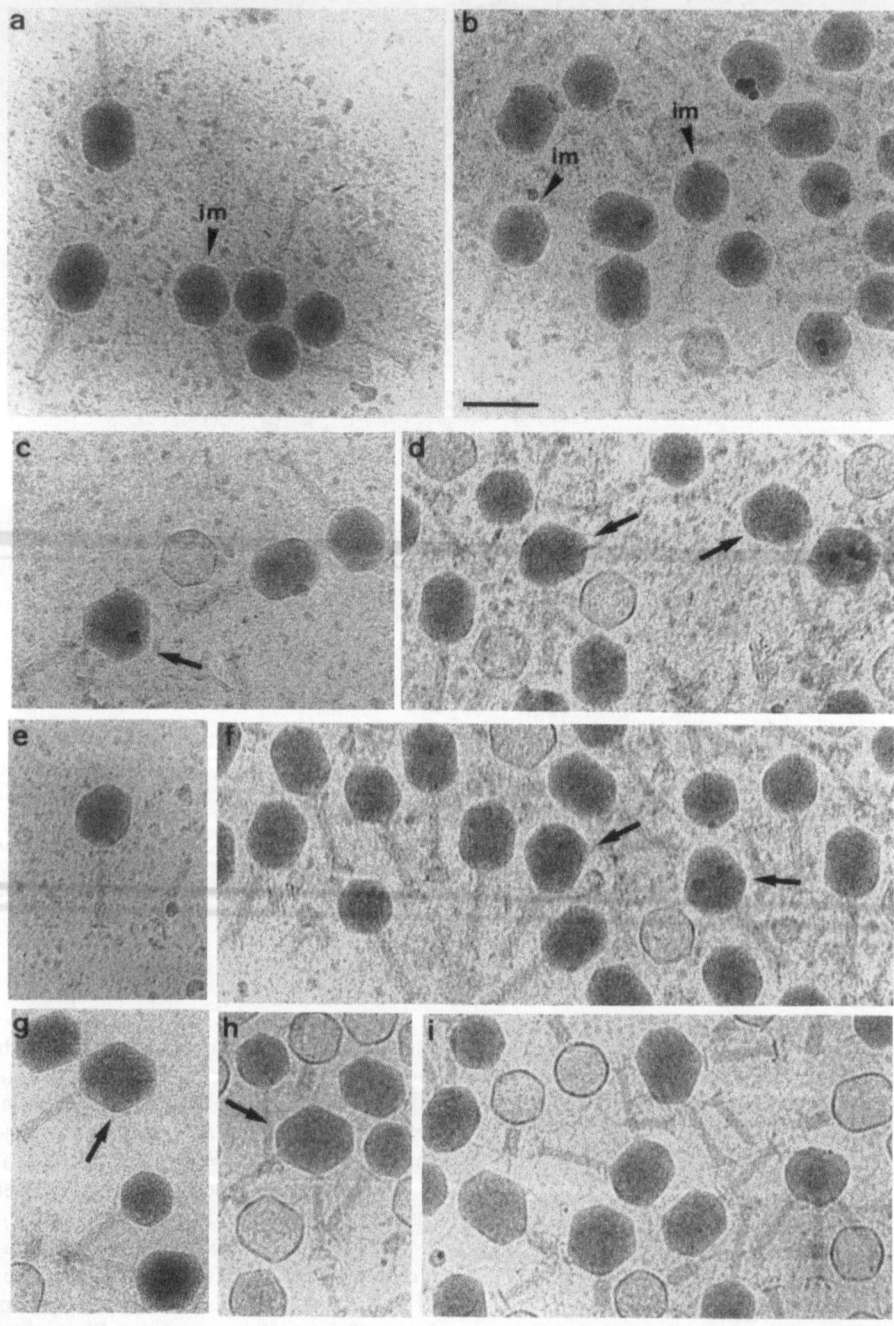

Fig. 7A. *Electron micrographs of form variants of mature T4 heads.* The particles were observed in a thin layer of ice. The material was rapidly frozen and observed when the water was still present and in the form of ice. (a–f) From a crude lysate of 68dela mutant of phage T4; in (a, b) members of the three length classes of isometric, intermediate and prolate phages are clearly distinguishable. The arrowheads labeled 'im' point to intermediate-length particles. The arrow in c indicates a biprolate particle; arrows in d, e and f show intermediate-length particles with a protrusion on one side and different tail attachment vertices. (g, h and k) Electron micrographs of mature phages obtained from 22ts*A447* at intermediate temperatures. The arrows point to particles wider than normal. The bars indicate 100 nm. (All micrographs by J. Lepault, and published by Keller et al. [39])

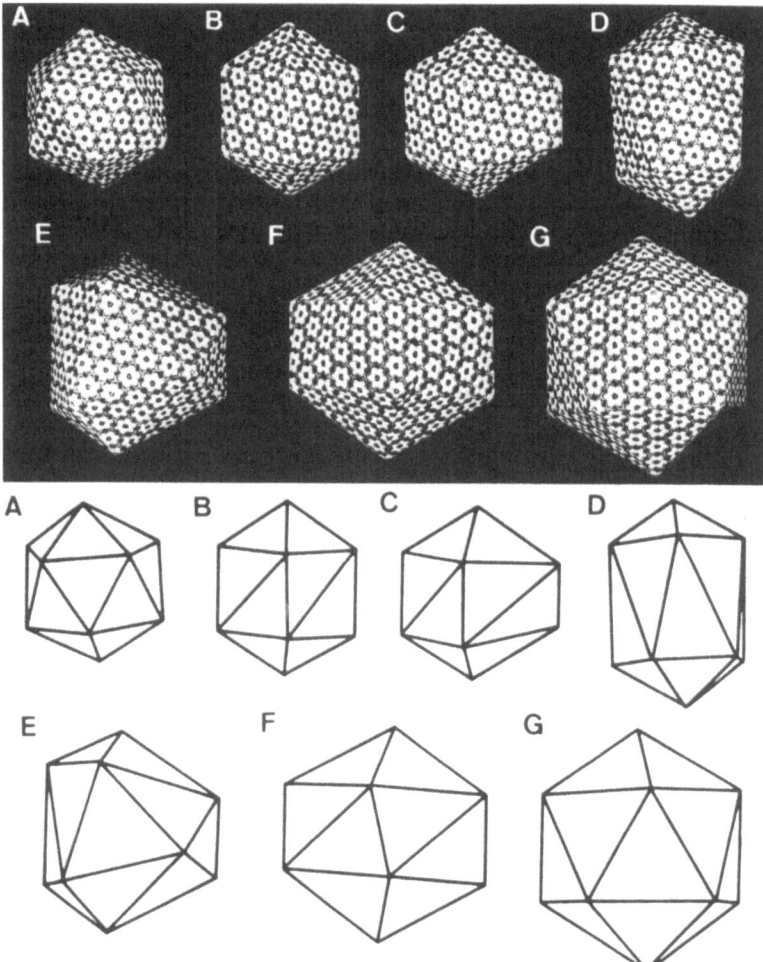

Fig. 7B. *Paper models and schematic drawings of aberrant head structures.* These are based on electron micrographs such as those shown in Fig. 7A made by folding a hexagonal plane net. All these polyhedra are based on the principles which direct the construction of icosahedral viruses (see text). (Models courtesy of M. Wurtz, published by Keller et al. [39])

Temperature-sensitive mutants (ts) of phages are selected as conditional lethals. They are point mutations that lead to a changed amino acid; at the permissive temperature (30 °C) the function of the protein is normal while at the non-permissive temperature its function is either abolished completely or so strongly altered that no mature phages are produced.

Different mutants in gene 23, of which the product is the main component of the shell, were used for studying the action of its gene product. With am mutants in gene 23, no particles are produced at all, except a few 'naked' scaffolding cores [22]. A ts mutant leads to abnormal ('crummy') proheads which no longer mature [23].

With 'conventional' point mutants in gene 23, slightly modified proteins are made which are able to build normal particles in addition to a strongly increased number of head-length variants [24, 25]. With a few of these mutants one finds up to 30% isometric variants; with others about 5–10% of the phages are isometrics or giants with variable lengths. These phages are infective, although the isometrics contain 30% less DNA; they are not able to infect singly, but, because of the cyclic permutation in the T4 genome, several together can do so.

A somewhat puzzling observation was an increased number of giants with a ts mutant in gene 24 grown at intermediate temperature. The same production of giants was achieved by gene dosage, i.e. an infection by a mixture of wild-type phage with another am mutant in gene 24 [26]. It is thought that gp24 is located in the vertices of the phage [29]. Interestingly, this gene is also responsible for the sensitivity towards osmotic shock [27].

Am mutants in the genes coding for the scaffolding core show various morphological effects; these are most dramatic with mutations in gene 20 where gp23, the head shell protein, assembles into long tubes (polyheads) about the length of the cell (Fig. 6). The width and pitch of single-layered polyheads, or the first layer of multilayered polyheads, are slightly variable [28, 29] but only slightly less than that of the prohead [29]. The shells of both polyheads and proheads are made of uncleaved gp23. They are fundamentally distinct by having a unique pitch for the prohead and a different, slightly variable one, for the single-layered polyheads [29]. The first layer of multilayered polyheads corresponds to the single-layered one, while the next layers are obviously larger and have accordingly different pitches [28]. Polyheads are also produced with mutants in some other genes [26]. Even in wild-type infection, they appear long after and apart from normal phages.

In vivo, polyheads contain a core which, with some preparation procedures for electron microscopy, might not be preserved or might be altered or even disappear. Purified gp23 assembles *in vitro* into polyheads without cores as easily as *in vivo* [30]. We have to conclude that the assembly of polyheads

Table 2. *Form variation in T4 as the result of the absence of a gene function or as a consequence of a mutated but active gene product*

	Protein	Variant observed	Relative amount
Inactive gene product (*am*)	gp20	tubular assembly forms of gp23 ('polyheads')	alone
	gp22	multilayered polyheads	alone
	gp24	maturable proheads and polyheads	about equal
	gp21	τ particles (abortively modified proheads)	alone
	gp 67	width variants (monsters)	with about 50%
	gp68	and isometrics	of active phages
Mutationally modified gp	gp 23	isometrics and/or giants of various lengths	variable portions
	gp23(ts)	abnormal proheads (crummy proheads)	alone
	gp22(ts)	width and length variants (monsters)	together with active phages
Gene dosage	gp24	giants of variable lengths	less than 10% among normal phages

Table 3. *Assembly products* in vitro

Gene products made available	Product	References
gp23	polyheads (tubes)	Van Driel [45] Caldentey & Kellenberger [30]
Scaffolding proteins alone	scaffolding cores	Van Driel & Couture [42]
Scaffolding proteins with gp23	τ particles	Van Driel & Couture [41]
gp22, gp67, gp68 + IP mixture	cylindrical scaffolding cores	Caldentey et al. [46]
gp22, gp67, gp68 + IP mixture + native connectors	'Kato particles'	Kato, unpublished (see text)
gp20 with minor proteins	symmetrical bodies	Driedonks et al. [31]
gp20 with minor proteins	dodecamers, morphologically identical to native connectors	Caldentey (in our lab), unpublished

is a first-order morphopoiesis and is not dependent on the presence of a core. The intrinsic curvature of the capsomer, composed of six protein subunits, determines the only slightly variable width and pitch of the tubes.

The gp20 forms a dodecamer of rotational symmetry [31]. It is the major part of the 'connector' located at the portal vertex of the shell, at which the tail, later, becomes attached. Several observations suggest a role of this connector in initiating the assembly of the scaffolding core [20].

With am mutations in gene 20 no such connectors are formed and therefore normal proheads cannot be initiated. The abundant gp23 eventually starts to form polyheads.

Am mutants in gene 22 (Table 2) produce multi-layered polyheads, which, upon lysis of the cell, easily separate into individual tubes [28]. The innermost tube has a diameter which is a little smaller than that of the prohead [29]. The lattice pitch of the succeeding tubes is variable so as to allow snug layering [28]. In all of these tubular variants, as in the shells of the prohead, gp23 is regularly arranged in a two-dimensional lattice which has been the subject of numerous studies (for references see [20, 21, 32–34]).

Point mutants in gene 22 (e.g. ts mutants at intermediate temperatures) produce, besides isometric particles, also variants with altered widths [35]. Two proteins of 13 kDa and 17 kDa are known as components of proheads [23]. No spontaneous am mutants were found in the genes coding for these two proteins, but am and other mutants were constructed by genetic engineering [36b, 37]. On the map these two new genes, 67 and 68, are located between genes 20 and 21. Mutants of genes 67 and 68 obtained by site-directed mutagenesis produce a very high proportion of isometric particles and 'monsters' [38, 39]. The monsters are characterized by a larger width (Fig. 7A). Isometric particles are found with two sizes: one ('short heads') corresponds to the width of normal phage while that of the other corresponds to the length of the prolate head [39]. Between the isometric, short head and the normal prolate

head, one, or possibly even two, intermediates exist [40]. Most gratifying is the observation that these variant particles can all be explained as geometric bodies which are derived from the icosahedron and produced as foldings of the same p6 lattice of the major subunit (Fig. 7B).

The products of genes 67 and 68 are both part of the scaffolding core (Table 1). Unfortunately, their precise location in the scaffold is still unknown, as is the determinate structure of scaffolds themselves. Because of the instability of scaffolds, electron microscopy has not yet provided micrographs which are reproducible and interpretable (Figs 6, 8 and 11).

In vitro *assemblies*

Using the approach described in the preceding section, the protein composition of proheads and heads was analysed, the proteins were identified genetically (as being produced by a given gene) and the function of each gene and thus of each protein was described by its expression in the phenotype (Table 3). The way was opened to isolate and purify the various gene products and to proceed to *in vitro* assembly experiments. These should provide a deeper understanding of the form-giving mechanisms which are based on mutual protein–protein interactions.

In the first set of experiments the scaffold proteins were isolated from proheads and separated from the gp23 of the shell. Preparations of purified protein gp23 was derived from polyheads [41–44]. Polyheads (Fig. 6c) assemble from gp23 *in vitro* at increased temperature or increased salt concentration [30, 45], conditions which also promote many other protein assemblies [12]. In these assemblies the protein subunits are highly ordered, while in solution they are random. Order is thus apparently increased by higher temperatures. This observation seems to contradict the laws of entropy. The contradiction was resolved by the discovery that, in such processes, the entropy increase is provided by the decrease of order which accompanies the 'melting' of the water of the hydration shell of the proteins (for references see [12]). The interactions of gp23 in the polyheads are thus predominantly hydrophobic in nature. In similar experiments the scaffold proteins produced clear 'naked cores' [42] (Fig. 9a). When the

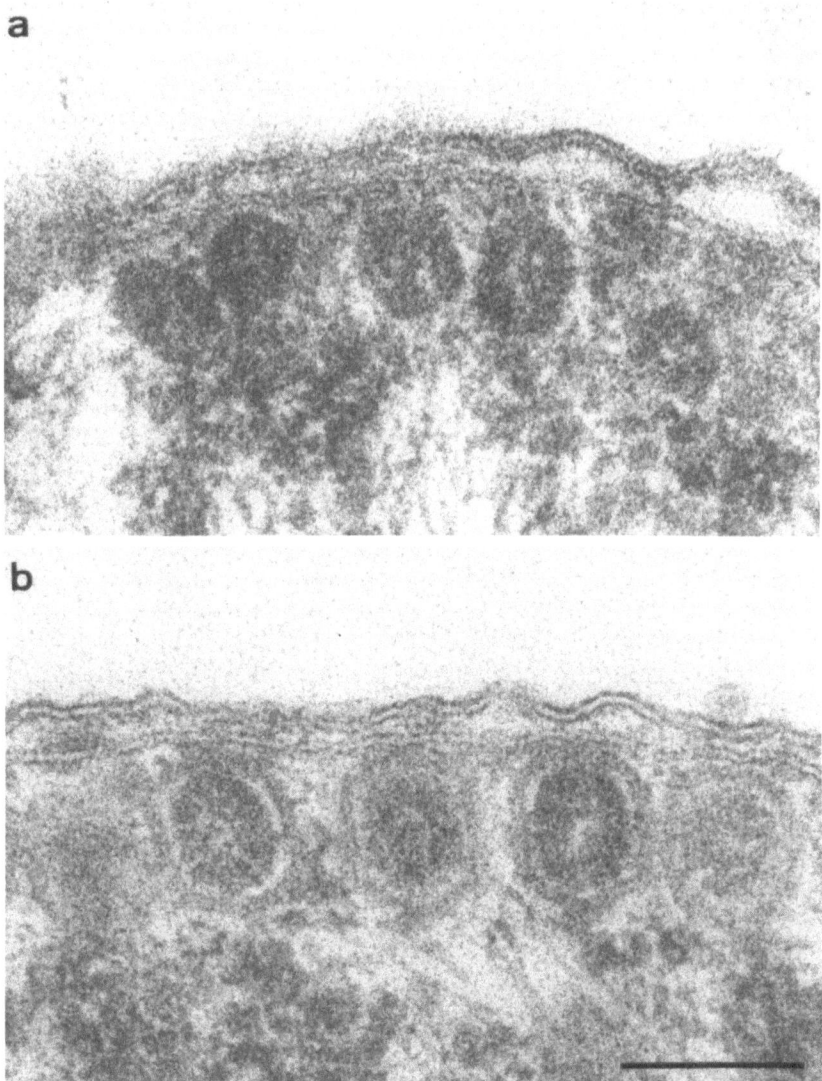

a

b

Fig. 8. *Naked scaffolding cores of T4 compared with proheads.* (a, b) Electron micrographs of thin sections of mutant-infected *E. coli* cells [22]. In (b) typical proheads attached to the plasma membrane can be seen. In (a) are naked scaffolding cores as obtained in relatively small amounts when gp23 is not available. Some of the differences of aspect are the consequence of the level of sectioning; the sections are so thin that a particle appears in two or three consecutive sections. (For proheads prepared by cryo-methods see also Fig. 11.) Cores were also isolated and analysed for the component major proteins [88]. (Micrographs made in our laboratory by Marlies Maeder.) The bar represents 100 nm

subunits were all mixed together, morphologically correct-looking proheads were produced [41] (Fig. 6b).

For subsequent experiments, the scaffolding proteins were purified individually, except for the internal proteins (IPI, IPII and IPIII) which were still used as a mixture of all three. At first, gp22, gp67, gp68 and the internal proteins were mixed [46]; they produced long, cylindrical 'polycores' (Fig. 10). All proteins had to be added together, except the mixture of IPs which could be added later. A few particles were found which resembled isometric cores. Disregarding these few isometric particles, this finding suggests that the input proteins only contributed to the morphopoietic information for the width, but not for the length. gp20 and gp21 were virtually absent from the mixtures (only present in extremely small amounts, as 'contaminants'). Comparing these results with the first set of experiments in which 'normal' cores were produced, gp20 and gp21 emerge as candidates for the length determination of the scaffold. Adding gp20 that was produced from the cloned gene failed to provide cores of limited length (unpublished data of our laboratory).

Although connectors were assembled from the cloned proteins which were morphologically identical to 'native' connectors [31], they were not morphopoietically active. They disassembled under the conditions needed for the assembly of the scaffold proteins. 'Native' connectors, isolated from infected cells, contained additional minor proteins (H. Kato et al., unpublished results). When added to the scaffolding proteins (which alone made only polycores) globular particles were now produced (H. Kato and Ch. Baschong-Prescianotto, unpublished results; Fig. 9b). Although of a comparable size, these isometric particles differed in morphology from the previously observed 'naked' scaffolding cores (Fig. 9a). Nothing is known as yet about their composition.

This morphological difference can be interpreted in two ways: either the particles isolated by Kato et al. are precursors to the other scaffolding cores or they are aberrant assemblies with an increased number of connectors. The Kato particles might be related to those native scaffolds which finally, *in vivo*, give rise to multi-tailed particles [39, 47]. In Kato's interpretation, gp20 would be in all vertices initially and then be re-

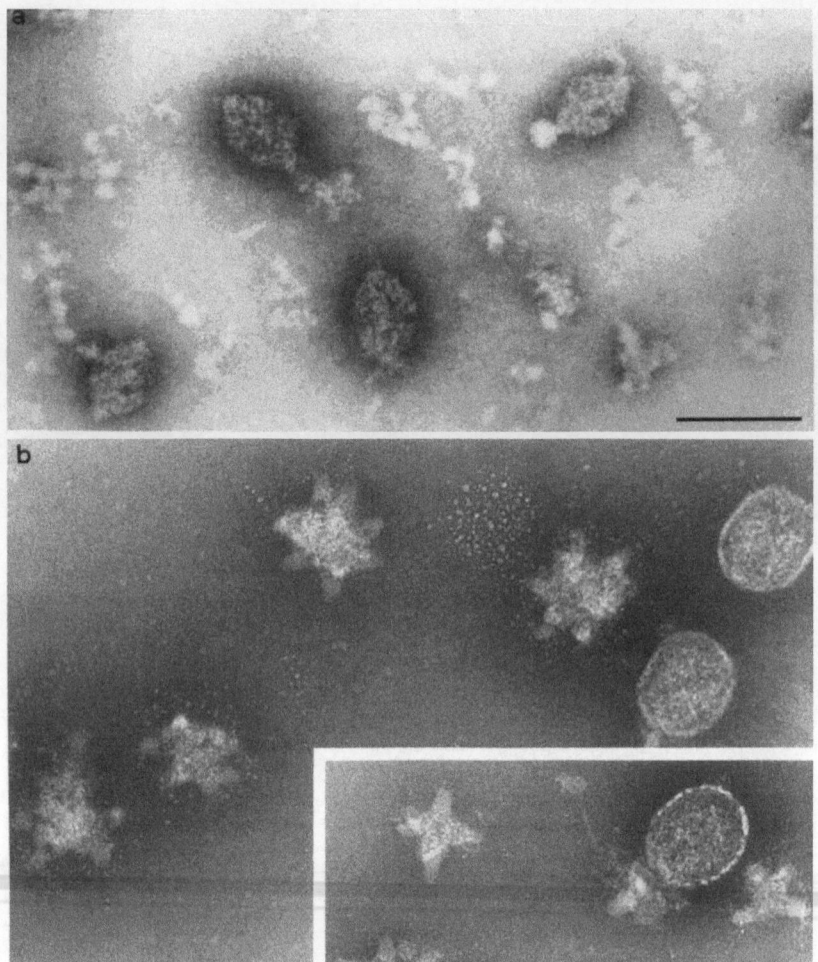

Fig. 9. In vitro *assembled scaffolding cores obtained from various mixtures of core proteins*. Prepared for electron microscopy by negative stain. (a) Cores obtained from a crude mixture without gp23. This mixture also contains gp21 and minor core proteins. (From Van Driel and Couture [42] and Van Driel [44].) (b) Particles obtained through *in vitro* assembly from a mixture of individually purified gp22, gp67, gp68 and a mixture of IPI, IPII and IPIII, to which native 'connectors' (containing gp20 and minor proteins), was added. Three proheads are also shown; they were added for the purpose of comparison. It is hypothesized by Kato that the spires (gp20) of the star-like particles could determine the vertices of the prohead. (Experiments in our laboratory by Dr H. Kato and C. Baschong-Prescianotto.) Negative staining. The bar represents 100 nm. (Micrographs by C. Baschong-Prescianotto)

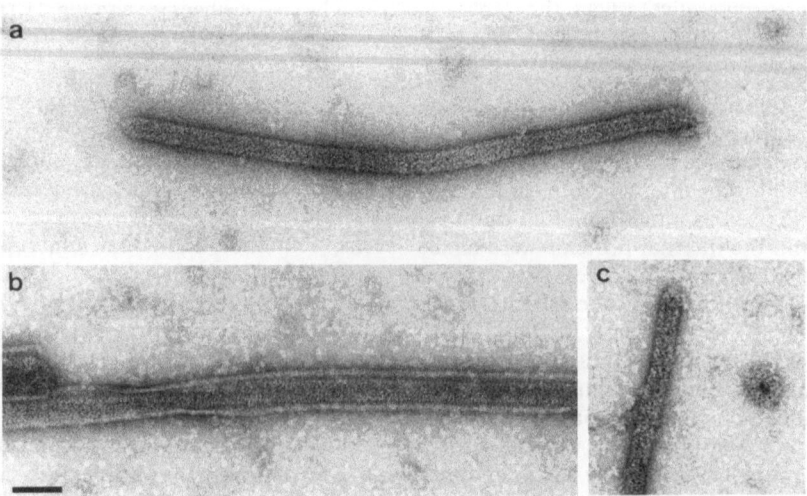

Fig. 10. *Polycores from in vitro assembly of purified proteins*. Particles reassembled from a mixture of individually purified gp22, gp67, gp68 and of a mixture of the IPs as in Fig. 9b, but without added connectors. The reassembled particles were fixed with 2% formaldehyde and negatively stained. A well-preserved polycore is shown in (a). Note the presence of an isometric core-like particle in (c) with a distinctly less dense central region. Micrographs of reassemblies from mixtures containing, in addition, the shell protein gp23 are shown in (b). The bar represents 75 nm in all micrographs. (Micrographs from J. Caldentey [46])

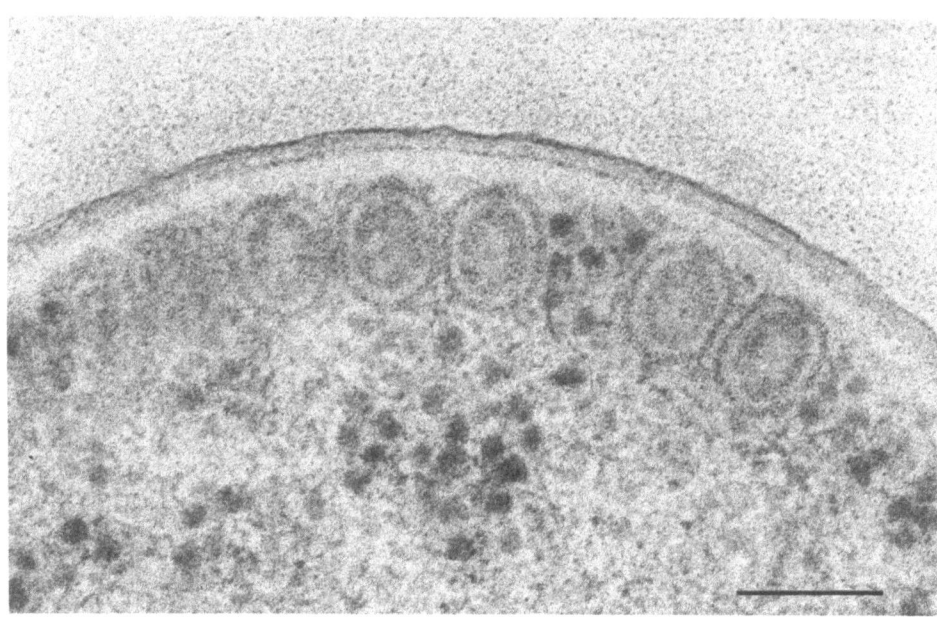

Fig. 11. *Proheads T4 after cryofixation-freeze-substitution.* The same prohead-related τ particles as in Fig. 8a but now prepared with a fixation and embedding procedure which is reputed to preserve structures best. The infected bacteria in a thin layer are very rapidly frozen at liquid helium temperatures and then the ice substituted by acetone containing 2.5% OsO_4 kept at $-80\,°C$. From there the cells are transferred into epon and polymerized at $60\,°C$. The detailed procedure is described by Hobot et al. [90]. The better preservation of the cells is indicated by the absence of wrinkling of the cell envelopes, in contrast to that seen in Fig. 8. The periplasm is also regularly spaced. The cytoplasmic membrane to which the proheads are attached is not visible in this procedure, probably because lipids, which cannot be fixed by aldehydes, are extracted. Only the proteins cross-linked to other proteins remain. The prohead itself, after normal on-section staining with uranyl acetate and lead citrate, looks similar to the conventional micrograph (Fig. 8a). It is, however, not reproducible even with this technique. The unstained space between the core and the shell is puzzling. It is difficult to envisage this as a protein-free space, because then the scaffolding action of the core is not understandable without complicated additional hypothesis. It is more likely that matter is present, but not stainable. A lack of staining ability was also observed on the hydrophobic part of a membrane protein [89]. The T4 mutant used here is a multiple amber mutant in genes 10 (baseplate), 18 (tail sheath) and 21. Electron microscopy by W. Villiger of our laboratory. The bar represents 100 nm

placed by gp24. Much more experimental work is needed to clarify this picture.

Most needed are reliable data on the structure of the scaffold. At present, not even the stoichiometry of T4 proheads can be established with sufficient reproducibility; it has been investigated only with prohead-related τ particles. These accumulate, for instance, with amber mutants in gene 21. Morphologically nearly indistinguishable particles accumulate with ts mutants in gene 21; they seem to differ by the presence or absence of matter (gp21) in the center of the core [48]. The latter particles are not capable of maturation *in vivo* after a temperature shift [49]. Am mutants in gene 24 lead *in vivo* to both τ particles and polyheads. After a temperature shift, a large proportion of these τ particles are capable of maturation [50, 51]; they are genuine proheads. It should be remembered here that the τ particles of T4 are the only known proheads which are firmly attached to the plasma membrane. A reproducible method is not yet available for their isolation and people working on isolated 'proheads' never agree about the best procedure. The yield of isolated particles varies every time, although electron microscopy of thin sections has shown that the number of τ particles that accumulate in mutants of gene 21 corresponds to that of phage obtained with wild-type phage (with H. Wunderli and M. van der Broek, unpublished data). It appears that the particles become mechanically altered through the procedures of isolation, necessary to tear them off the plasma membrane. Equally unsatisfactory results are obtained by electron microscopy of isolated particles observed as frozen-hydrated specimens (with J. Dubochet, EMBL, unpublished data).

Even observation of intracellular τ particles in thin sections of cryofixed, cryosubstituted and embedded infected cells does not provide absolutely regular morphology (Fig. 11). The conventionally fixed and embedded infected cells apparently provide more regular results! It appears as if the procedures involved initiate a degradative process that results in a more homogeneous, although abortive, product. That this is a plausible explanation is justified by a parallel observation of other head-related particles which, during processing of sections or upon isolation, expand [52, 53].

The packing density of the scaffold proteins, calculated from the stoichiometry, does not correspond to observations by electron microscopy. In samples prepared by negative staining, as well as in those maintained frozen in a layer of ice ('frozen-hydrated'), the concentration of matter within the particles appears much lower than expected. Other proheads (lambda, $\phi29$) exhibit similarly low packing densities of the scaffold proteins (with J. Dubochet, EMBL and D. L. Anderson, University of Minnesota, unpublished results). On thin sections from cryoprocessed infected cells, the protein packing density of the cores of these proheads is only slightly below that of their shells, in agreement with predictions from stoichiometry.

Determination of the nearest neighbors should, in principle, be possible by chemical crosslinking studies. Such experiments are very difficult because of the high number of identical proteins [54].

Efforts are being made to crystallize proheads for diffraction studies using phage $\phi29$ (D. Anderson, personal communication). It is quite obvious that knowledge of the scaffold structure is indispensable for a full understanding of the mechanisms of form determination.

Additional studies on the form determination with other phages

Other extensively studied phages have isometric heads, but even here the form-determining role of the scaffold has been demonstrated clearly. With these phages the number of different scaffold proteins is small, one for phage P22 [13] and two for phage lambda [55]. With both phages the absence of scaffold protein(s) leads to abnormal particles, either smaller ones with phage P22 [15] or to monsters with phage lambda [56] (see also [57]). An interesting size variation is observed with phages P2/P4: The genome of P4 does not encode all functions needed for particle assembly. A helper phage, for instance P2, has to provide them. P2 furnishes the capsid protein, which P4 lacks. P4 assembles it into a smaller head than does P2. Despite a few genes being known to be involved in this size determination, the underlying mechanism is not yet understood [57a].

As we have seen above, for T_{even} the head-tail connector (gp20 + minor proteins) seems to act first as initiator of prohead assembly. Its lack leads in vivo to the tubular variants of the prohead, the polyheads. It is very interesting to note that, in phage P22, the connector (gp1 + minor proteins [58]) seems not to play a role in prohead assembly either in vivo [59] or in vitro [60]. This might be correlated also with the attachment of proheads to the cytoplasmic membrane of the host, which is observed with T4 but not with P22 nor with lambda. In phage P22 the in vitro assembly of proheads seems to proceed like copolymerisation of scaffold and shell proteins [60]. For T_{even}, the fact that scaffolds can be observed 'naked' in vivo [22] and produced in vitro without the coat protein [42] has given rise to the hypothesis that scaffolding cores might act as precursors for proheads. On the one hand, some in vivo experimental results can be interpreted in this way [61]. On the other hand, the observation of Van Driel [44] that in vitro assembled scaffolding cores cannot become coated later by adding gp23 is in favor of a simultaneous assembly of shell and scaffold with T_{even}.

For the phages lambda and P22 a capsid expansion step has been demonstrated as with T_{even}; it occurs sometime during maturation and DNA-packaging. The bacteriophage $\phi29$ is too small to measure expansion with sufficient precision. Other phages, like T3/7, T5 and other B. subtilis phages (for references see [3]), also appear to have a capid expansion step like T_{even}, lambda and P22 mentioned above.

DISCUSSION AND PERSPECTIVES

The basic concepts for the genetic analysis of form determination in phage T4 were worked out as early as 1966 [11, 14]. It was essential for further work to demonstrate the relationship of the τ particle to the maturable prohead [50, 51, 62]. Progress at that time was slowed by the broad acceptance of our proposal that a naked DNA condensate might act as a precursor [63], although shortly afterwards we showed that our data did not exclude the possibility that the DNA was enveloped by a thin protein shell to form a fragile head [64].

The suggestion of a precursor made only of proteins met with such resistance that we removed it from our paper on petite lambda, the head-related, smaller particle that contained no DNA [65]. Attempts with Karamata, Hofschneider and then later with Granboulan to demonstrate the precursor nature of the petite lambda, all failed and were not published. Then the Hohns [66] took up the problem in our laboratory

in parallel with Kaiser et al. [67] in Stanford, and the precursor nature of the petite lambda was finally demonstrated successfully by the detour of in vitro DNA packaging.

With T4 the experimental results of the first attempts to show a DNA-free precursor [68] were as equally misleading as our postulated naked condensates because the presumed precursor was the capsid of artificially emptied particles [69, 70].

The fragility of precursors and intermediates remains a barrier to a rapid experimental progress. The τ particle (an abortive relative of the prohead) and, later, a very unstable intermediate, the ε particle [71], were observed in an only minimally altered state in thin sections obtained by modern cryoprocedures. In such micrographs the scaffolding core of the prohead and of the τ particle (Fig. 11) is still present in the ε particle, although in a heavily modified form.

The structure of the scaffolds observed by electron microscopy show substantial variations according to the procedures employed and even within the same cell. Without knowing the precise structure of the scaffold, we can only speculate about the mechanisms by which the length and width of a particle are determined. Some speculations now follow.

The length determination of cylindrical structures is obviously less complicated than that of a non-isometric globular structure and has been discussed many times [72–74]. In the case of the phage tail, the simplest model became true: a filamentous protein, of which the length is determined by the length of the gene, acts as a ruler around which the cylinder assembles and becomes limited in length [7]. For the head, an attempt was made to reduce the form determination to that of length only, as shown below. This is a dangerous simplification since it is known from strong experimental data that, for the prolate T4 phage head, not only the length but also the width is genetically determined [38].

Two mechanisms had been proposed to explain length variations of the head. First, the Vernier model which is based on the existence of two concentric structures of distinct periodicity along one of the dimensions, the length. The linear growth would be arrested when the two periods happen to be in register ($na = mb$, where a and b are the two periods, n and m two different integers). This model was proposed by Anderson and Stephens [75] for the tail where the tail tube and the tail sheath were proposed to represent the concentric structures. As already mentioned, this model was not confirmed for phage tails. Instead, the model of the ruler, as discussed above, has found good evidence [7, 8]. For proheads of prolate phage particles, the scaffolding core and the shell could possibly act as the concentric structures for a Vernier mechanism. Laemmli and his group [35] had some experimental data that might be explained with this model, although the authors did not claim to provide proof. An intriguing set of experiments has been provided by Lane et al. [76], as we will discuss now. The icosahedral viruses are described by the triangulation number T [10]. The prolate derivative is described by an additional number Q, which fixes the cylindrical part between the two caps [77, 78]. For T_{even} phages, $T = 13$. For the isometric particle, by definition, $Q = 13$ also. The elongated (prolate) wild-type particle has $Q = 22$ [77]. Considering crystallography, Q could change by steps of 1.

For the T_{even} phages, the length of DNA that is packaged in the head is determined by the size of the capsid; we speak of the 'head-full' mechanism [79]. By analysing the length of the DNA, one should find increments corresponding to the unit increase of Q. Surprisingly, the experimental measurements showed that the increments corresponded to $\Delta Q = 4$

[76]. The authors interpreted $\Delta Q = 4$ by assuming a Vernier model: the two concentric structures would be in register only when $\Delta Q = 4$. Another explanation of these findings is based on geometry and model building. Only $Q = 13, 17, 21$, etc., lead to heads with an approximately cylindrical part between the two pyramidal caps. For values of Q in between, these cylindrical parts degenerate into fivefold prisms, which require increased quasi-equivalence. The building of such particles would be energetically unfavorable.

Another model is based on the equilibrium of two rates of assembly in one of the dimensions, again of two concentric structures [20, 80]. Several experimental findings based on 'gene dosage' favored this model so that it is widely accepted. The relative amounts of gp23 (shell) and gp22 (scaffolding core) were varied by multiple infections of a mixture of am mutants of these genes with their wild type. The amount of protein is then directly proportional to the relative number of wild-type genes present (hence the 'gene dosage'). The results showed that the ratio of the occurrence of genes 22/23 has a strong influence on the length of the head. Later, the discovery of a polar effect of am mutants in gene 22 on the expression of gene 23 [81], cast doubt on the validity of the experiments with gene dosage of these two genes. Because of the polarity, the biochemically determined gp22 and gp23 did not follow the predictions of gene dosage [81].

The unambiguous demonstration of scaffold-dependent width variants [38] has made all of the models mentioned above less attractive because they reduce the form determination to a linear model, while reality shows size variations in three dimensions. It is highly unlikely that all three dimensions are individually controlled by three different Vernier measurements.

The instability of the precursors must be overcome in some way in order to elucidate the structure of the scaffold. In addition, in order to isolate native structures and assemble them efficiently *in vitro*, more needs to be known about the intracellular 'environment'. For instance, we know that K^+ rather than Na^+ is the predominant intracellular ion. Its counterions, however, are not Cl^- but organic anions; in *Escherichia coli* and other bacteria, these are glutamate [82]. The intracellular K^+ concentration ranges over $100 - 500$ mM, at outside osmotic pressures up to 350 mOs. At these osmolarities, *E. coli* still grows normally. Multiplication of the intracellular phage T4 is also not affected significantly by these changes [83], while the concentration of potassium or sodium chloride is crucial for *in vitro* assembly. This is true also for temperatures between $25 - 40\,°C$ which have virtually no influence on the *in vivo* situation (lower temperatures delay the onset of expression of the late proteins but do not significantly change the rate of assembly; unpublished results with E. Stauffer). The effect of temperature on *in vitro* assembly is well known for all entropy-driven protein assemblies (as mentioned above). Equally important is the intracellular concentration of Mg^{2+}, as demonstrated for the *in vivo* packaging of phage T7 DNA [84]. Hence the current use of chelators for divalent cations must be avoided.

To learn more about the structure of scaffolds, it would probably be wise to start with phages simpler than T_{even} i.e. those which have scaffolds of only one or two protein species. We have already mentioned the *B. subtilis* phage $\phi 29$, the proheads of which have been crystallized but not yet investigated by X-ray crystallography (D. L. Anderson, University of Minnesota, personal communication). It is likely that substantial progress will be made with this approach.

In spite of all these remaining difficulties, it is rewarding to see how far the genetic approach has advanced our knowledge. For T4 we can already clearly state the form-determining functions of each gene product. The morphopoiesis is at least of the fifth order, including the effect of the connector (made of gp20) as putative initiator of head assembly. Similarly, the scaffold has been found to have a strong modulating influence on the width, counteracting the (weak) intrinsic curvature of the shell subunit gp23. More than ten point mutants in gene 23 affect only the shell length, presumably by influencing the rate of assembly. None has, as yet, been found which changes the radius of curvature. The predominant action of the scaffold for determining the shell width is generally accepted, having already been strongly suggested by the existence of multilayered polyheads [28, 85]. Each subsequent layer is assembled with the previous layer acting as a scaffold, regardless of the lesser curvature imposed.

Phages are, without question, a most suitable experimental system for studying the mechanism of form determination of macromolecular assemblies. Intracellular organelles of eucaryotes are examples of such assemblies. As a warning I reiterate that phage morphogenesis and morphopoiesis should not be mistaken as models for embryogenesis, where the sequences of pathways are controlled at the transcriptional level, quite in contrast to phage particle formation, where the control is exclusively at the level of actuation, which governs the successive protein − protein interactions.

I am most grateful to Dwight L. Anderson who critically read the manuscript and to Fred Eiserling who communicated new results obtained with Lane and Serwer prior to publication. My thanks go obviously also to the numerous colleagues and collaborators whose results and micrographs I was able to quote. To R. A. Driedonks I am particularly indebted because we started model-building of capsomers and intrinsic curvature (Fig. 5) together. M. Wurtz was a great help in preparing models and photographs thereof. Elvira Amstutz patiently did all the typewriting and editing, Marlies Zoller and Hedy Frevel the excellent photographic work. Experiments of our own laboratory were supported by the Kanton of Basel-Stadt and the Swiss National Science Foundation.

REFERENCES

1. Mathews, C. K., Kutter, E. M., Mosig, G. & Berget, P. B. (eds) (1983) *Bacteriophage T4*, American Society for Microbiology, Washington DC.
2. Casjens, S. (1985) *Virus structure and assembly*, Jones & Bartlett, Boston MA.
3. Calendar, R. (1988) *The bacteriophages*, vols I & II, Plenum Press, New York.
4. Campbell, A. (1961) *Virology 14*, 22−32.
5. Epstein, R. H., Bolle, A., Steinberg, C. M., Kellenberger, E., Boy de la Tour, M. E., Chevalley, R., Edgar, R. S., Susman, M., Denhardt, G. H. & Lielausis, A. (1963) *Cold Spring Harbor Symp. Quant. Biol. 28*, 375−394.
6. Edgar, R. S., Denhardt, G. H. & Epstein, R. T. (1964) *Genetics 49*, 635−648.
7. Katsura, I. (1987) *Nature 327*, 73−75.
8. Duda, R. L., Wall, J. S., Hainfeld, J. F., Sweet, R. M. & Eiserling, F. A. (1985) *Proc. Natl Acad. Sci. USA 82*, 5550−5554.
9. Crick, F. H. C. & Watson, J. D. (1956) *Nature 177*, 473−475.
10. Caspar, D. L. O. & Klug, A. (1962) *Cold Spring Harbor Symp. Quant. Biol. 27*, 1−23.
11. Kellenberger, E. (1966) in *Principles of biomolecular organisation* (Wolstenholme, G. E. W. & O'Connor, M., eds) pp. 192−228, Churchill, London.
12. Lauffer, M. A. (1975) *Entropy-driven processes in biology*, Springer-Verlag, New York.

248

13. Tanford, Ch. (1980) *The hydrophobic effect*, 2nd edn, Wiley-Interscience, New York.
14. Kellenberger, E. (1969) in *Symmetry and function of biological systems at the macromolecular level* (Engström, A. & Strandberg, B., eds) pp. 349–366, Almquist & Wiksell, Stockholm.
15. Earnshaw, W. & King, J. (1978) *J. Mol. Biol. 126*, 721–747.
16. Hohn, T., Flick, H. & Hohn, B. (1975) *J. Mol. Biol. 98*, 107–120.
17. Hagen, E. W., Reilly, B. E., Tosi, M. E. & Anderson, D. L. (1976) *J. Virol. 19*, 501–517.
18. Kellenberger, E. (1972) in *Polymerization in Biological Systems* (Wolstenholme, G. E. W. & O'Connor, M., eds) pp. 189–206, Churchill, London.
19. Berget, P. B. & King, J. (1983) in *Bacteriophage T4* (Mathews, Ch. K., Kutter, E. M., Mosig, G. & Berget, P. B., eds) pp. 246–258, American Society of Microbiology, Washington DC.
20. Black, L. W. & Showe, M. K. (1983) in *Bacteriophage T4* (Mathews, Ch. K., Kutter, E. M., Mosig, G. & Berget, P. B., eds) pp. 219–245, American Society of Microbiology, Washington DC.
21. Kellenberger, E. (1978) in *Electron microscopy*, 9th Int. Congr., vol. III (Sturgess, J. M., ed.) pp. 441–449, Microscopical Society of Canada, Toronto.
22. Traub, F. & Maeder, M. (1984) *J. Virol. 49*, 892–901.
23. Onorato, L., Stirmer, B. & Showe, M. K. (1978) *J. Virol. 27*, 409–426.
24. Doermann, A. H., Eiserling, F. D. & Boehner, L. (1973) *J. Virol. 12*, 374–385.
25. Doermann, A. H. & Pao, A. (1987) *J. Virol. 61*, 2835–2842.
26. Bijlenga, R., Aebi, U. & Kellenberger, E. (1976) *J. Mol. Biol. 103*, 469–498.
27. Leibo, S. P., Kellenberger, E., Kellenberger, C., Frey, T. G. & Steinberg, C. M. (1979) *J. Virol. 30*, 327–338.
28. Yanagida, M., Boy de la Tour, E., Alff-Steinberger, C. & Kellenberger, E. (1970) *J. Mol. Biol. 50*, 35–58.
29. Steven, A. C., Aebi, U. & Showe, M. K. (1976) *J. Mol. Biol. 102*, 373–407.
30. Caldentey, J. & Kellenberger, E. (1986) *J. Mol. Biol. 188*, 39–48.
31. Driedonks, R. A., Engel, A., ten Heggeler, B. & Van Driel, R. (1981) *J. Mol. Biol. 152*, 641–662.
32. Kistler, J., Showe, M. K., Carrascosa, J. L., Stirmer, B. & Onorato-Showe, L. (1982) *J. Mol. Biol. 162*, 607–622.
33. Buhle, E. L. Jr & Aebi, U. (1984) *J. Ultrastruct. Res. 89*, 165–178.
34. Steven, A. C., Couture, E., Aebi, U. & Showe, M. K. (1976) *J. Mol. Biol. 106*, 187–221.
35. Paulson, J. R., Lazaroff, S. & Laemmli, U. K. (1976) *J. Mol. Biol. 103*, 155–174.
36a. Völker, T. A., Gafner, J., Bickle, T. A. & Showe, M. K. (1982) *J. Mol. Biol. 161*, 479–489.
36b. Völker, T. A., Kuhn, A., Showe, M. K. & Bickle, T. A. (1982) *J. Mol. Biol. 161*, 491–504.
37. Keller, B., Sengstag, C., Kellenberger, E. & Bickle, T. A. (1984) *J. Mol. Biol. 179*, 415–430.
38. Keller, B., Maeder, M., Becker-Laburte, C., Kellenberger, E. & Bickle, T. A. (1986) *J. Mol. Biol. 190*, 83–95.
39. Keller, B., Dubochet, J., Adrian, M., Maeder, M., Wurtz, M. & Kellenberger, E. (1988) *J. Virol. 62*, 2960–2969.
40. Mosig, G., Caringan, J. R., Bibring, J. B., Cole, R., Bock, H.-G. O. & Bock, S. (1972) *J. Virol. 9*, 857–871.
41. Van Driel, R. & Couture, E. (1978) *J. Mol. Biol. 123*, 115–128.
42. Van Driel, R. & Couture, E. (1978) *J. Mol. Biol. 123*, 713–719.
43. Van Driel, R. (1980) *J. Mol. Biol. 138*, 27–42.
44. Van Driel, R. (1980) in *Electron microscopy at molecular dimension* (Baumeister, W. & Vogell, W., eds) pp. 129–136, Springer-Verlag, Berlin, Heidelberg.
45. Van Driel, R. (1977) *J. Mol. Biol. 114*, 61–78.
46. Caldentey, J., Lepault, J. & Kellenberger, E. (1987) *J. Mol. Biol. 195*, 637–647.
47. Boy de la Tour, E. & Kellenberger, E. (1965) *Virology 27*, 222–225.
48. Van Driel, R., Traub, F. & Showe, M. K. (1980) *J. Virol. 36*, 220–223.
49. Laemmli, U. K. & Johnson, R. A. (1973) *J. Mol. Biol. 80*, 601–611.
50. Bijlenga, R. K. L., Scraba, D. & Kellenberger, E. (1973) *Virology 56*, 250–267.
51. Bijlenga, R. K. L., van den Broek, R. & Kellenberger, E. (1974) *J. Supramol. Struct. 2*, 45–59.
52. Carrascosa, J. L. (1978) *J. Virol. 26*, 420–428.
53. Carrascosa, J. L. & Kellenberger, E. (1978) *J. Virol. 253*, 831–844.
54. Granboulan, P. (1980) *J. Ultrastruct. Res. 70*, 336–346.
55. Hohn, T. & Katsura, I. (1977) *Curr. Top. Microbiol. Immunol. 78*, 69–110.
56. Ray, P., Murialdo, H. (1975) *Virology 64*, 247–263.
57. Hohn, Th. & Katsura, I. (1977) *Current Top. Microbiol. Immunol. 78*, 69–110.
57a. Calendar, R. (1988) *The bacteriophages*, vol. II, pp. 303–306, Plenum Press, New York.
58. Bazinet, Ch., Benbasat, J., King, J., Carazo, J. M. & Carrascosa, J. L. (1988) *Biochemistry 27*, 1849–1856.
59. Bazinet, Ch. & King, J. (1988) *J. Mol. Biol. 202*, 77–86.
60. Prevelidge, P. E., Thomas, D. & King, J. (1988) *J. Mol. Biol. 202*, 743–757.
61. Kuhn, A., Keller, B., Maeder, M. & Traub, F. (1987) *J. Virol. 61*, 113–118.
62. Simon, L. D. (1972) *Proc. Natl Acad. Sci. USA 69*, 907–911.
63. Kellenberger, E., Séchaud, J. & Ryter, A. (1959) *Virology 8*, 478–498.
64. Kellenberger, E. (1961) *Adv. Virus Res. 8*, 1–61.
65. Karamata, D., Kellenberger, E., Kellenberger, G. & Terzi, M. (1962) *Pathol. Microbiol. 25*, 575–585.
66. Hohn, B., Wurtz, M., Klein, B., Lustig, A. & Hohn, T. (1974) *J. Supramol. Struct. 2*, 302–317.
67. Kaiser, A. D., Syvanen, M. & Masuda, T. (1974) *J. Supramol. Struct. 2*, 318–328.
68. Luftig, R. B., Wood, W. B. & Okinaka, R. (1971) *J. Mol. Biol. 57*, 555–573.
69. Hamilton, D. L. & Luftig, R. B. (1976) *J. Virol. 17*, 550–567.
70. Wunderli, H., van den Broek, J. & Kellenberger, E. (1977) *J. Supramol. Struct. 7*, 135–161.
71. Schärli, C. & Kellenberger, E. (1980) *J. Virol. 33*, 830–844.
72. Kellenberger, E. (1972) in *Polymerization in Biological Systems* (Wolstenholme, G. E. W. & O'Connor, M., eds) pp. 295–298, Churchill, London.
73. Kellenberger, E. (1976) *Phil. Trans. R. Soc. Lond. 276*, 27–28.
74. Kellenberger, E. (1984) *Helv. Phys. Acta 57*, 188–201.
75. Anderson, T. F. & Stephens, R. (1964) *Virology 23*, 113–116.
76. Lane, T., Serwer, P., Hayes, S. J. & Eiserling, F. (1990) *Virology 174*, in the press.
77. Baschong, W., Aebi, U., Baschong-Prescianotto, C., Dubochet, J., Landmann, L., Kellenberger, E. & Wurtz, M. (1988) *J. Ultrastruct. Mol. Struct. Res. 99*, 189–202.
78. Moody, M. F. (1965) *Virology 26*, 567–576.
79. Streisinger, G., Emrich, J. & Stahl, M. M. (1967) *Proc. Natl. Acad. Sci. USA 57*, 292–295.
80. Showe, M. K. & Onorato, L. (1978) *Proc. Natl Acad. Sci. USA 75*, 4165–4169.
81. Grütter, T. (1983) Ph.D. Thesis, Microbiology, University of Basel.
82. Leirmo, S., Harrison, C., Cayley, D. S., Burgess, R. R. & Record, M. T. Jr (1987) *Biochemistry 26*, 2095–2101.
83. Kuhn, A. & Kellenberger, E. (1985) *J. Bacteriol. 163*, 906–912.
84. Kuhn, A., Jütte, H. & Kellenberger, E. (1983) *J. Virol. 47*, 540–552.
85. Kellenberger, E., Eiserling, F. A. & Boy de la Tour, E. (1968) *J. Ultrastruct. Res. 21*, 335–360.
86. Wood, W. B. & Crowther, R. A. (1983) in *Bacteriophage T4* (Mathews, Ch. M., Kutter, E. M., Mosig, G. & Berget, P. B., eds) pp. 259–269, American Society of Microbiology, Washington, DC.
87. Kellenberger, E. (1966) *Sci. Am. 215*, 32–40.
88. Traub, F., Keller, B., Kuhn, A. & Maeder, M. (1984) *J. Virol. 49*, 902–908.
89. Garavito, R. M., Carlemalm, E., Colliex, C. & Villiger, W. (1982) *J. Ultrastruct. Res. 80*, 344–353.
90. Hobot, J. A., Villiger, W., Escaig, J., Maeder, M., Ryter, A. & Kellenberger, E. (1985) *J. Bacteriol. 162*, 960–971.

Eur. J. Biochem. *192*, 815 (1990)
© FEBS 1990

Corrections and Addition

Volume 190, No. 2

Form determination of the heads of bacteriophages, by Eduard Kellenberger:

Page 240, legend to Fig. 7A, line 6: *for* J. Lepault *read* J. Lepault and M. Adrian.

Page 242, Table 3, line 16: *for* gp20 with minor proteins *read* native connectors gp20 with minor proteins.

Page 242, Table 3, line 16: *for* symmetrical bodies *read* dodecamers.

Page 242, Table 3, line 18: *for* gp20 with minor proteins *read* gp20 from cloned gene in *E. coli.*

Page 242, Table 3, line 18: *for* (in our lab.), unpublished *read* et al. [46].

Page 243, column 1, second line from bottom: *for* unpublished data of our laboratory *read* [46].

Page 245, column 2, fourth line from bottom: *insert* J. Caldentey was able to use the technique of nearest neighbours rather successfully for the study of T4 proheads. In his thesis for the Swiss certificate of molecular biology he presented data showing cross-links between gp23 and gp24, confirming both to be integral parts of the shell. He also showed that pip (precursor of internal peptides), which was later identified as gp67 [36b], is part of the core together with gp22 and the IPs.

Eur. J. Biochem. *190*, 445–462 (1990)
© FEBS 1990

Review

Insulin-like growth factors I and II

René E. HUMBEL

Biochemisches Institut, Universität Zürich, Switzerland

(Received December 27, 1989) — EJB 89 1540

Insulin-like growth factors (IGF) I and II are two peptides which have been isolated originally from a Cohn fraction of human serum [1]. The designation IGF has been proposed [2] to stress the fact that they showed mitogenic effects at concentrations in the nanomolar range *in vitro*, that they also exhibited insulin-like effects in adipose and muscle tissue and that their structure was homologous to that of (pro)insulin [3–5].

IGF I and II are also known as somatomedins. The original observation was made by Salmon and Daughaday [6] that pituitary growth hormone (GH) *in vitro* did not correct the defect in the synthesis of matrix proteins of hypophysectomized rats, whereas the serum of GH-treated hypophysectomized rats did. This has led to the so-called somatomedin hypothesis [7] which states that GH acts on skeletal tissues by inducing the formation of a growth factor (= somatomedin) circulating in the blood and acting on the peripheral tissue. The subsequent isolation and amino acid sequence determination of somatomedin C [8] and A [9] has shown their identity with IGF I. The use of the two designations somatomedin and IGF I as synonyms is now generally accepted [10]. However to equate IGF II with somatomedin is, at least at the present time, not justified because the formation of IGF II is under less stringent control of GH than that of IGF I, and because the physiological role of IGF II is far from clear.

The literature on IGF has in recent years become so vast that the attempt to cover the entire field risks exceeding the competence of a single author and the length constraints of review papers in this journal. By necessity, therefore, this review can not be comprehensive. Certain topics will be treated in more detail than others, reflecting the partiality of the author. Particularly, a certain bias towards the significance of IGF in man and against the application of IGF in animal husbandry could not be overcome. Some excellent reviews with emphasis on particular topics have been published recently [11–19].

Correspondence to R. E. Humbel, Biochemisches Institut, Universität Zürich-Irchel, Winterthurerstrasse 190, CH-8057 Zürich, Switzerland

Abbreviations. IGF I, insulin-like growth factor I; IGF II, insulin-like growth factor II; rhIGF, recombinant human IGF; GH, growth hormone; GRF, GH-releasing factor; EGF, epidermal growth factor; PDGF, platelet-derived growth factor.

CURSORY OVERVIEW

This chapter will serve as a mini-review of the field of IGF, presenting the most prominent biochemical and clinical aspects of IGF, in which only review articles will be cited.

IGF I and II are single-chain polypeptides of about 7.5 kDa which occur in blood plasma at concentrations of 20–80 nM and at lower concentrations in most if not all tissues of the body. IGF circulating in the blood is probably produced by the liver, whereas tissue IGF is produced to a greater part locally. The significance of endocrine vs paracrine/autocrine actions is at present not clear. The expression of IGF I and, to a lesser degree, of IGF II is under the control of GH. IGF I acts as feedback inhibitor at the pituitary level of the GH release and/or GH gene transcription. IGF I can mimic most, but probably not all, effects of GH [14, 15].

The single-chain peptides consist of the peptide domains, B, C, A and D, in which domains A and B are structural homologs of the insulin A and B chains (see Fig. 1). Domain C is analogous to the connecting (C) peptide in proinsulin, whereas the D domain is not found in insulin [20]. The amino acid sequences of six species (human, bovine, porcine, ovine, rat and mouse) have so far been determined. IGF I in all these species consists of 70 amino acid residues while IGF II has 67 [15].

The three-dimensional structure of the IGFs has not yet been determined but models, obtained by computer-assisted molecular modelling analogously to the structure of insulin, have been built [20]. Predictions of the binding sites of the IGFs to receptors and IGF binding proteins have been made which can now be tested using site-directed mutagenesis.

Radioimmunoassays, and to a lesser degree radioreceptor assays, are currently being used to determine concentrations of IGF I and II in serum [15]. In adult man, normal concentrations of IGF I and II are about 200 and 700 ng/ml, respectively. At birth, the concentrations of IGF I are about half the adult values. Just before puberty, IGF I reaches adult levels. In puberty, IGF I rises 2–3-fold, whereas IGF II does not change significantly. In persons with untreated GH deficiency (hypopituitary dwarfs), concentrations of IGF I are reduced to about 10 ng/ml and those of IGF II to about 200 ng/ml. Comparisons between growth rates of normal versus GH-deficient children suggest that IGF I increases growth in later childhood and that it is particularly responsible for the

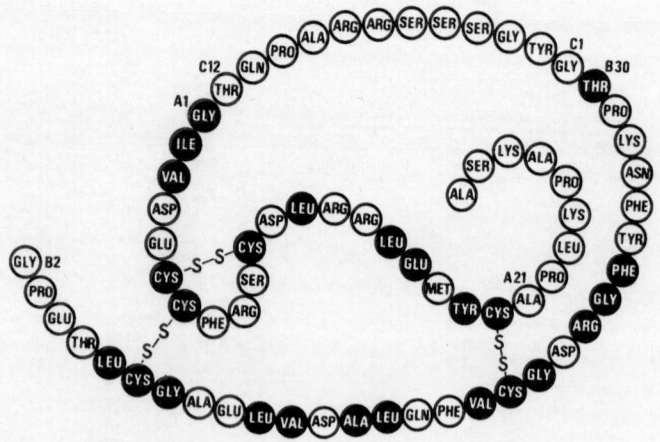

Fig. 1. *Amino acid sequence of human IGF I*. Identical positions to those in human insulin are in black. The numbering (B2, B30, C1, C12, A1, A21) corresponds to the numbering residues in proinsulin. (From [20] with permission)

pubertal growth spurt. Conversely, patients with GH-secreting tumors (gigantism or acromegaly, depending on time of onset) show increased concentrations of IGF I (600 – 1000 ng/ml), but normal or subnormal levels of IGF II. The determination of the serum concentrations of IGF I in such patients serves as an adjunct for the correct diagnosis. There are, however, other clinical conditions with decreased concentrations of IGF I such as cirrhosis of the liver, diabetes, primary hypothyroidism and undernutrition [15, 21]. No IGF-producing tumors leading to elevated serum IGF have so far been described.

Serum IGF is for the most part bound to specific IGF binding proteins. These binding proteins prolong the half-life of IGF, they are responsible for avoiding hypoglycemia, and are reported to exert stimulatory and inhibitory effects on IGF action [12, 21]. There are at least three different binding proteins which have been cloned recently. In the serum of normal individuals, the greater part of the IGFs is bound to a 150-kDa binding protein. This binding protein is also under control of GH [12].

IGF I and II show *in vitro* and *in vivo* insulin-like metabolic effects (such as on glucose transport and on blood glucose), but only at relatively high concentrations. The presence of IGF binding proteins in plasma prevents hypoglycemia under physiological conditions [21]. The biologically relevant effects of IGF I at nanomolar concentrations are the stimulation of cell proliferation and, at least in certain tissues, cell differentiation [21, 22]. The region-specific expression and relative abundance of IGF I and II in the central nervous system has led to speculations on their roles in the development and/or function of the brain [11]. *In vitro*, IGF II mimicks all effects of IGF I [21]. In rats, IGF II is considered by some investigators to be a fetal growth factor [22]. Evidence in man does not support this view [21]. The most honest statement is, therefore, that the biological role of IGF II is not clear and that even its biological relevance is under dispute [17, 19, 21].

Half-maximal effects on cell proliferation *in vitro* occur at concentrations of about 1 nM. These mitogenic effects are mediated by a specific IGF receptor, the type 1 IGF receptor, which is a receptor kinase and which is homologous to the insulin receptor. Signal transduction at the type 1 IGF receptor involves tyrosine autophosphorylation and phosphorylation of other proteins [13, 23]. The type 2 IGF receptor, which

shows no structural similarity to the type 1 receptor, has a somewhat greater affinity to IGF II than to IGF I [13, 16, 23]. The recent identification of the type 2 receptor with the mannose-6-phosphate receptor has so far not contributed to our understanding of the biological role of this receptor nor to that of IGF II [17].

Recently, recombinant human IGF I and II (rhIGF) have become available. Phase I trials are under way to test clinical applications. Besides the treatment of a rare form of GH-resistant dwarfism (Laron dwarfism), treatment of diabetics (in conjunction with insulin) and application for the acceleration of wound healing will be tried [21].

IGF I AND II GENES AND mRNAs

IGF I and II are each the product of a single gene, localized in man on the long arm of chromosome 12 [24, 25] and on the short arm of chromosome 11, respectively [24, 25]. The IGF II gene is contiguous with the insulin gene [26]. The gene for tyrosine hydroxylase is also contiguous with the insulin gene [27], the order being 5′ – tyrosine-hydroxylase – insulin – IGF II – 3′.

The human IGF I gene contains five exons and is at least 90 kb long [18, 28 – 30]. Fig. 2 shows a map of the human IGF I gene and of the mRNAs. Rotwein [31] has shown that there are two different mRNAs (IGF Ia and Ib) transcribed from the single gene, exon 4 being used for transcription of the C-terminal part in IGF Ib, exon 5 in IGF Ia (Fig. 2). Some controversy exists over the length of the prepeptide, since there are three codons for methionine (Met) in the open reading frame 5′ to the codon for the first residue of the peptide. Homology with the preproinsulin gene favors the starting Met at position -25, considerations of consensus sequences suggest Met-22, whereas *in vitro*, translation is initiated at Met-48 [32].

The product of translation is the IGF precursor consisting of a signal peptide (of 48, 25 or 22 residues), of the four domains (BCAD) of the processed growth factor with 70 residues, and of a propeptide region at the C-terminus (E peptide) which contains 35 (Ia) and 77 (Ib) residues, respectively [26, 28 – 30, 33, 34].

In man, the use of exons 4 and 5 is mutually exclusive, while in the rat and mouse, both exons 4 and 5 are present in the same mRNA. The expression of two mRNAs (Ia and Ib) in rat and mouse is due to an insertion in the case of Ib of a 'mini-exon' of 52 bp, thus leading to a frameshift [35]. Consequently, the C-terminal sequences of mouse rat Ib prepeptides are totally different from human IGF Ib.

The human IGF II gene spans about 30 kb and consists of three promotors, five non-coding exons (exons 1 – 4 and 4_B) and three protein-encoding exons [15, 18, 26, 28, 30, 36 – 42]. The last exon also contains a 3′-untranslated region of about 3.8 kb. Fig. 3 shows a map of the human IGF II gene and of the mRNAs. The IGF II precursor consists of a signal peptide of 24 residues, the 67 amino acid residues of the growth factor, and of an E peptide 89 residues long. Gray et al. [39] and Irminger et al. [40] have found that the 5′ ends of IGF II cDNAs isolated from adult liver differ radically from those of other tissues. Evidence for similar complexity in the transcription of the rat IGF II gene has been found earlier by Soares et al. [43, 44]. But it is well to remember that the various transcripts vary in their untranslated and/or E-peptide regions only, so that they all code for the same BCAD domains of the peptide finally processed. For further information on this topic, the reader is referred to the excellent recent review by Sussenbach [18].

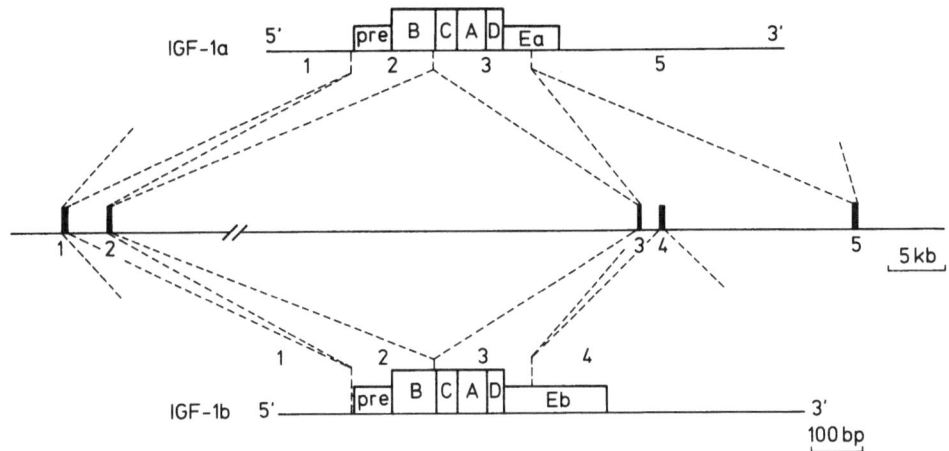

Fig. 2. *Map of the human IGF I gene and mRNAs*. The coding regions are indicated by blocks. (From [18] with permission)

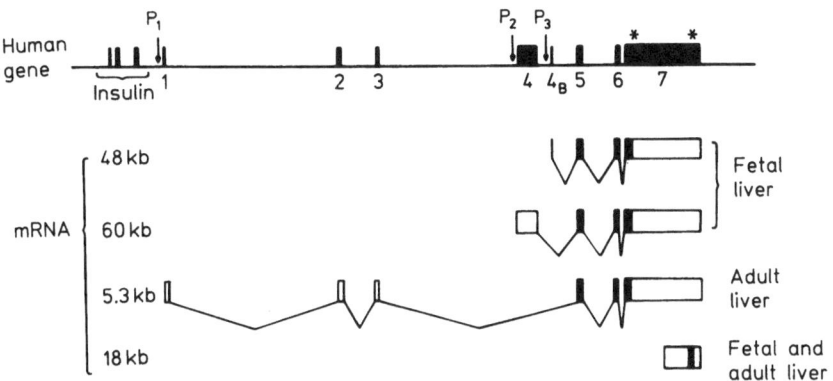

Fig. 3. *Map of the human IGF II gene and mRNAs*. In the mRNAs the coding regions are indicated as black boxes. Asterisks indicate polyadenylation sites. (From [18] with permission)

The expression of the various transcripts is tissue- and development-specific [39 – 43, 45 – 47] (see also below). The occurrence of IGF variants will be discussed in the following chapter on IGF structures.

STRUCTURES OF THE IGFs

Amino acid sequences

The first determination of the amino acid sequences of IGF I and II was made in 1978 [3, 4] on the peptides isolated from 1 kg of an acetone powder of a Cohn fraction (precipitate B) of human serum [1, 2]. Meanwhile, more efficient isolation procedures have been developed [48 – 55] and amino acid sequences of IGFs from several other species have been determined [56 – 61] and deduced from IGF cDNAs [33, 35, 38, 43, 62 – 64]. Fig. 4 lists the primary structures of IGF I and II from the six species known today. IGF I in all six species consists of 70 amino acid residues, IGF II of 67, all being grouped into domains A and B (similar to insulin), C (analogous to the connecting peptide of proinsulin) and D (not present in insulins). Chicken IGF I and II have only been partially sequenced [65].

The three intrachain disulfide bridges in IGF I and II have been predicted to be located in analogous positions to those in (pro)insulin [3]. Only recently, this hypothesis has been

verified by elaborate degradation and mass spectrometry techniques on recombinant human IGF I [66]. Similarly, the SS bridges of natural rat IGF II and of a recombinant human IGF II have been determined by fast-atom-bombardment mass spectrometry and found to be as predicted [67].

In addition to the 'classical' IGF I and II with 70 and 67 residues, several variants of different length have been described. Truncated IGF I, lacking the first three residues at the amino terminus, has been found in fetal and adult human brain [68, 69], in bovine colostrum [57], and in porcine uterus [70]. In all likelihood, it is the product of differential processing of prepro-IGF. The biological potency of this truncated form has been reported to be 1.4 – 10 times higher than that of the full-length form [57, 70, 71]. This finding is rather puzzling, since the amino terminus has been assumed to be highly flexible and not involved in receptor binding [5]. Evidence is, however, accumulating that reduced binding of the truncated form to IGF-binding proteins may be responsible for the increased biological potency [72 – 76]. A shortened variant of IGF II lacks the amino-terminal alanine residue [4], again probably a product of differential processing.

In human spinal fluid, a 9-kDa form of IGF II has been found [77]. In human brain, several higher-molecular-mass forms [78] and in human serum, a 15-kDa form of IGF II have been described [79]. So far, the structures of these forms have not been determined. Most likely, they represent partially unprocessed forms of pro-IGF-II containing all or part of the

IGF I

HUMAN	GPETLCGAELVDALQFVCGDRGFYFNKPT	GYGSSSRRAPQT	GIVDECCFRSCDLRRLEMYCA	PLKPAKSA
BOVINE	-----------------------------	------------	---------------------	--------
PORCINE	-----------------------------	------------	---------------------	--------
OVINE	-----------------------------	------------	---------------------	--------
RAT	------------------P----------	-----I------	---------------------	-T------
MOUSE	------------------P----------	-----I------	---------------------	-T-A----

IGF II

HUMAN	AYRPSETLCGGELVDTLQFVCGDRGFYF	SRPASRVSRRSR	GIVEECCFRSCDLALLRTYCA	TPAKSE
BOVINE	----------------------------	---S--IN----	---------------------	------
PORCINE	----------------------------	-------N----	---------------------	------
OVINE	----------------------------	---S--IN----	---------------------	A-----
RAT	----------------------S-----	---S--AN----	---------------------	------
MOUSE	--G-G-----------------S-----	---S--AN----	---------------------	------
DOMAINS	B	C	A	D

Fig. 4. *Amino acid sequences of IGF I and II*. Amino acid residues are given in single-letter code. Dashes signify residues identical to the top (human) sequence. The peptide domains are indicated in the bottom line. Data for IGF I from [3, 33] (human), [56] (bovine), [59, 62] (porcine), [58] (ovine), [60] and [63] (rat), [35] (mouse). Data for IGF II from [4, 36, 37] (human), [57] (bovine), [59] (porcine), [58] (ovine), [38, 43, 61] (rat), [64] (mouse)

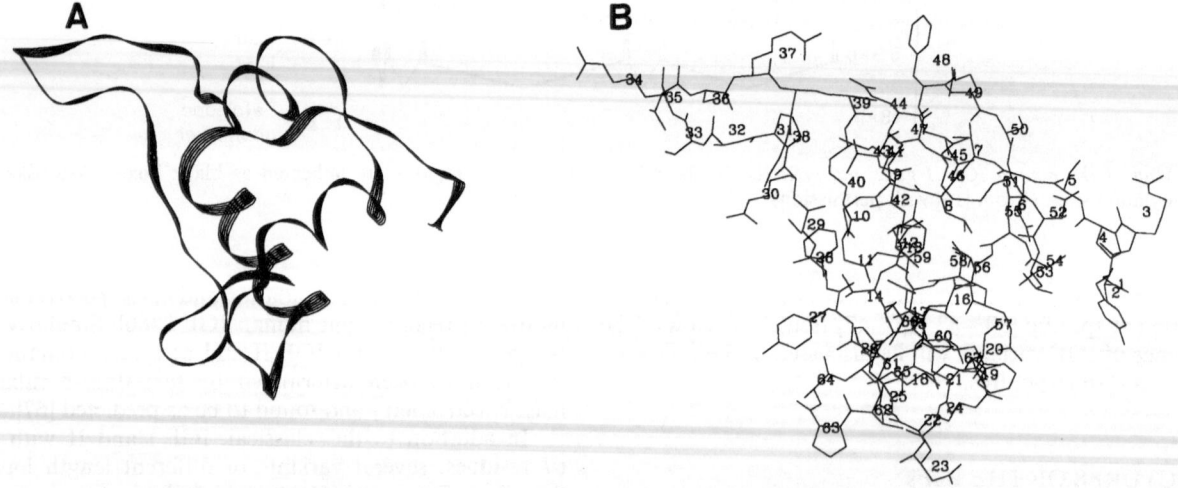

Fig. 5. *View of a computer-generated model of IGF II according to the model atomic coordinates [5] as deposited at the Brookhaven data bank.* (A) Ribbon model of the backbone structure. (B) Skeletal model with all amino acid side chains numbered

E domain. A C-terminal extension of 21 residues (part of the E domain) has been determined in a 9-kDa variant of IGF II from human serum which, in addition to the extension, also contains a substitution of Ser33 by the tripeptide Cys-Gly-Asp [80]. A second variant with an insert of the tetrapeptide Arg-Leu-Pro-Gly substituting for Ser29 was observed in a human IGF II cDNA [36]. Since available evidence excludes the presence of more than one gene for IGF II, alternative splicing [36, 80] or allelic variation [80] have been offered as explanations for these inserts. In view of the heterogeneity of IGF II preparations from serum [80, 81] upon separation by isoelectric or chromato-focusing, there may be even more variants, not counting deamidated forms. Since most forms occur also in individual sera [81], allelic variation is an unlikely explanation.

Models for the three-dimensional structures

An alignment of the amino acid sequences of IGF I and II and of insulin shows the conservation of all residues which are invariant in all insulins with only one exception (Asn A21 becomes Ala in IGFs) [5, 82]. Residues contributing to the hydrophobic core are strictly conserved, as are structurally important Cys and Gly residues. This indicates that the folding of the IGFs may be very similar to that of insulin, and allowed the modelling of the three-dimensional structure of the IGFs by molecular graphics based on the known structure of insulin [5, 82]. In the absence of X-ray crystallographic data and with only preliminary studies with two-dimensional nuclear magnetic resonance, these model studies have been further refined [83]. The models can be used to make predictions about binding regions to antibodies, IGF binding proteins and IGF receptors which can be tested experimentally (see below). A view of such a model of IGF II is given in Fig. 5.

The structural similarities of insulin and the IGFs are such that they allow crossreaction with the heterologous receptors. On the other hand, insulin cannot interact with IGF binding proteins. Testing synthetic insulin/IGF hybrids for recognition by the insulin and IGF receptors and the IGF binding proteins has been a first approach to define these interaction sites [84–88]. More detailed information has been obtained by substituting insulin B25 Phe by Tyr, Ala, Ser or Leu [89], substitutions which have been modelled after an insulin mutation found in a patient [90]. Replacement of insulin B25 Phe by Tyr increased, replacement by Ala, Ser or Leu decreased, the affinity to type 1 IGF receptor. The corresponding residue in IGF I and II is Tyr which had been proposed before to be part of the type 1 receptor binding site [83, 91].

Recently, the expression of IGF I mutants prepared by site-directed mutagenesis has become feasible, a technique both more flexible and more efficient than chemical synthesis and semisynthesis. If the sequence Phe23-Tyr24-Phe25 (corresponding to positions B24 to B26 in insulin) is replaced by the insulin sequence Phe-Phe-Tyr, the analog is almost equipotent with IGF I at the insulin receptor and at the type 1 and 2 IGF receptors [92]. This is a rather unexpected finding in view of the increased affinity of the B25 Tyr insulin analog to type 1 receptor [89]. The presence of an aromatic hydroxyl group at position 25 rather than 24 seems to be equally effective. When, however, the single Tyr24 was substituted by Leu or Ser, the affinity to type 1 receptor decreased 32-fold and 16-fold, respectively [92]. On the other hand, replacement of Phe23 by Ser led to a drastic decrease of affinities to all three receptors and to IGF binding proteins [92]. This finding was interpreted as a destabilization of the tertiary structure by the loss of aromaticity [92]. A decreased affinity of IGF I to the type 2 receptor without altering the affinity to the type 1 receptor has been obtained with the mutant Thr49-Ser50-Ile51 (wild-type: Phe-Arg-Ser), while the substitution of Arg55-Arg56 by Tyr-Gln increased the affinity to the type 2 receptor, again without altering the affinity to the type 1 receptor [93]. The same group also studied a mutant with an altered B region (Gln3-Ala4, Tyr15-Leu16) whose affinity to IGF binding proteins was reduced by two orders of magnitude [74, 94]. The main conclusions of these studies are as follows.

a) The N-terminal part of IGF is a recognition site for the IGF binding proteins; either deletion of the first three residues [72–76] or substitution of residues 3, 4, 15 and 16 [74, 94] decreases the affinity drastically.

b) The recognition site for the type 1 receptor is similarly located to the one for the insulin receptor. Definitely involved are the aromatic residues at positions 23–25 [92]. Further studies are clearly required for a complete characterization.

c) Part of the A region (at least positions 49–51) seems to be involved in binding to the type 2 receptor [93].

IGF DISTRIBUTION IN TISSUES AND BODY FLUIDS

The realization that IGF is also produced in tissues other than liver [95] has challenged the original somatomedin hypothesis [6, 7, 14]. But it is now generally accepted that IGF I and II act both as endocrine hormones via the blood and as paracrine and autocrine growth factors locally. GH stimulates the biosynthesis of IGF I in liver and in the other organs and tissues [96]. Rat fibroblasts in culture mimick the developmental switch from IGF II to IGF I production observed at birth in rat serum [97]. This has led to the rash and generalized conclusion that IGF II represents the fetal growth factor, whereas IGF I is the corresponding growth factor after birth. However, studies in other species [98, 99] including man, have shown that the rat may be quite unique in not maintaining high concentrations of IGF II postnatally.

The distribution of IGFs in tissues has been determined both by measuring mRNAs and by assaying IGFs by radioimmunoassays [100, 101]. As far as one can judge, concentrations of mRNAs and peptides correlate well except in certain tumors [102]. Highly elevated IGF II transcripts have been reported in Wilms' tumors [103]. This could be confirmed [102], but the concentrations of IGF II were found to be in the same range as in the adjacent normal kidney tissue. Since the putative Wilms' tumor gene (Wg) is on chromosome 11 near the IGF II gene, these findings have led to speculations that the enhanced expression of IGF II may be involved in the pathogenesis of Wilms' tumor [103, 104]. However, elevated IGF II expression is not always found in these tumors [105]. Conflicting results have been reported on another type of tumor, pheochromocytoma. In three such human tumors, IGF II concentrations were about 20 times higher than in normal adrenal medulla, whereas the corresponding mRNA was not increased [102]. In another publication, IGF II mRNAs were found to be increased in 8 out of 8 pheochromocytomas [106]. Furthermore, the IGF type 1 receptor-specific antibody αIR3 was found to inhibit proliferation of a human neuroblastoma cell line which synthesized large amounts of IGF II mRNAs and of IGF II [106]. These authors conclude that IGF II mediates the autonomous growth of this cell line. It is, however, highly unlikely that autocrine growth stimulation by IGF II is of general importance in tumorigenesis.

Of particular interest are findings on IGF in the central nervous system. In pooled human cerebrospinal fluid, about 50 ng/ml IGF II, but only about 2 ng/ml IGF I have been found [77]. Almost half of the immunoreactive IGF II is present as a 9-kDa form. Cerebrospinal fluid IGF I and II are apparently not regulated by GH [107]. In human brain, various amounts of IGF II including larger molecular sizes have been found in different regions, but again very little of IGF I [78]. But surprisingly, appreciable amounts of a 'truncated' IGF I, lacking the three amino-terminal amino acid residues, have been demonstrated in both fetal and adult human brain [68, 69]. Adult rat brain, in contrast to other rat tissues, expresses IGF II [108], particularly in the choroid plexus and the leptomeninges [109, 110]. Both IGF I and II seem to be differentially expressed in different regions in the developing and adult rat brain [111]. All these findings, together with the different distribution patterns of type 1 and 2 IGF receptors in rat brain [112] and some clinical observations [113, 114], suggest that IGF I and II may play an important role in development and/or function of the central nervous system [115].

IGF in serum

Nowhere else in the body are concentrations of IGF as high as in blood. Standard procedures for their determination have been the radioimmunoassay for IGF I [116] and the radioreceptor assay for IGF II [117]. But reliable separate measurements of serum IGF I and II have been notoriously difficult. The main reasons are as follows.

a) Antisera raised against the IGFs available at the time (IGF I from human serum, IGF II from rats) crossreacted only partially in a heterologous assay.

b) IGFs used as immunogens and as antigens were cross-contaminated (IGF I with IGF II and vice versa).

c) In different laboratories, different standard preparations of varying purity were used.

d) Over 95% of serum IGF is bound to binding proteins which interfere with the antigen — antibody reaction by competing with the antibody. Complete removal of binding proteins from IGF is therefore a prerequisite for the radioimmunoassay of IGFs. Among the many proposed procedures to separate IGFs from binding proteins, acidification with acetic acid and gel filtration in 1.0 M acetic acid [118] seems to yield the most reliable results. A more detailed discussion of the various assay procedures is to be found elsewhere [15].

With the advent of commercially available recombinant IGF I and II, the first three difficulties can be overcome. Also, the use of C_{18} cartridges (Sep Pak) to separate IGFs from binding proteins [119] has gained wide acceptance, because it lends itself to the simultaneous treatment of a large number of serum samples. Several commercially available radioimmunoassay kits are being developed or are already on the market.

Despite these difficulties, several valuable studies on serum concentrations of IGF I and II in man have been published before the availability of recombinant IGFs. Furlanetto et al. [120] were the first to make the still valid observation that in acromegalics, serum concentrations of IGF I are about 4 times higher and in hypopituitary children about 20 times lower than in normals (Fig. 6). Concentrations of IGF I in normal adult man are about 200 ng/ml serum [118]. Thus, the concentrations of IGF I generally reflect the concentrations of GH [116, 118, 121, 122], in accord with the cardinal tenet of the somatomedin hypothesis [7]. IGF II, in contrast, is not further elevated in conditions of elevated GH nor drastically reduced in patients with GH deficiency [118, 122] (Fig. 6).

In the rat [123] and in the lamb [124], fetal serum IGF I is much lower than in adult serum, whereas high concentrations of IGF II in the fetus fall dramatically just before birth. In the guinea pig, serum IGF I and II at midgestation are almost as high as in adult serum. Furthermore, IGF II rises to very high levels shortly before birth and falls then back to adult levels [125]. The situation in the rat and in the lamb together with earlier observations on the developmental switch from IGF II to IGF I in cultures of rat fibroblasts [97] suggested that IGF II is primarily a fetal somatomedin and IGF I the postnatal somatomedin. However, findings in the guinea pig [125] and in man [126, 127] do not support this as a general concept. In man, both IGF I and II are low in fetal serum and IGF II persists at high levels into adulthood. The same concentrations of IGF I and II as in normals are found in anencephalic fetuses [127] which excludes pituitary GH as a regulator of IGF of the fetus. Furthermore, a significant correlation was found between IGF I, but not IGF II, and fetal weight, placental weight and fetal length [127].

During puberty, there is a 2 – 3-fold rise of serum IGF I in boys and girls [128, 129], whereas no significant changes of IGF II occur [118]. This rise in IGF I is probably due to an increase in the amplitude of the pulsatile GH secretion [130] and, directly or indirectly, to the increase in sex steroids [131]. Adult serum concentrations remain thereafter quite constant.

Hormonal regulation of IGF serum concentrations

As described above, serum concentrations of IGF reflect the GH status. Also, treatment of GH-deficient children [132] or of hyophysectomized rats [133] with GH causes a rise of serum IGF I. The synthesis and secretion of IGF I in liver, quantitatively the most important source of serum IGF [97, 134, 135], were reduced in livers of hypophysectomized rats, but restored after treatment of the hypophysectomized animals with GH [134]. In rat liver and pancreas, the concentration of IGF I mRNAs was 8 – 10-fold lower in hypophysectomized rats than in normal rats; within 4 h after injecting GH into the hypophysectomized animals the IGF I mRNA concentrations were restored to normal [136]. In transgenic pigs carrying a metallothioneine promoter fused to a GH gene, serum concentrations of IGF I were raised 3-fold compared to controls [137].

Since IGF I is a hormone under GH control, one would expect that IGF I might act as a feedback inhibitor of GH release. In a monolayer pituitary cell culture system, IGF I and, although less potently, IGF II inhibited the GH release stimulated by a synthetic GH releasing factor (GRF) [138]. Similarly, IGF I was shown to suppress the spontaneous release of GH in rats [139]. IGF I was found to reduce GH mRNA levels in pituitary cells, an effect due to decreased gene transcription as shown by nuclear run-off experiments [140].

Somatostatin has long been known to inhibit GH release at the pituitary level. In a recent publication, somatostatin was shown to inhibit both basal and GRF-induced GH release, whereas IGF I decreased GH gene transcription [141]. In immunodeficient rats bearing an IGF II secreting tumor, IGF II serum concentrations were sevenfold higher, yet basal and GRF-stimulated GH release was unchanged [142]. This lack of feedback inhibition by IGF II further corroborates our contention that IGF II is not a somatomedin which is under control and controlling GH.

In the same paper [142], the observation was made that 7-fold higher concentrations of IGF II led to a lowering of IGF I. Since liver IGF I mRNAs were not reduced, high levels of IGF II probably do not reduce IGF I gene transcription. The reverse situation has been found after infusion of recombinant IGF I into two healthy subjects, where serum IGF II was depressed [143]. It is possible that the high doses of IGF I displaced IGF II from the IGF binding proteins with consecutive rapid degradation [143].

Disorders with decreased IGF serum concentrations

Whereas acromegaly (see above) is the only known disorder with elevated serum IGF (IGF I), disorders with decreased serum IGF are more numerous.

In an intersting study on pygmies, serum IGF I was found to be low, while IGF II and GH were normal [144]. It was speculated that the deficiency of IGF I might be the cause of the short stature of pygmies. In a later study [145], prepubertal pygmy children and controls did not differ in linear growth or in serum IGF I and II concentrations. However, in pygmy adolescent boys, the mean IGF I concentrations were only one third of control adolescents. It was concluded that the failing burst of IGF I during puberty is responsible for the lack of the pubertal growth spurt and this in turn for the short stature of adult pygmies. Furthermore, testosterone seems not to accelerate growth without a concomitant increase of IGF I, at least in pygmies.

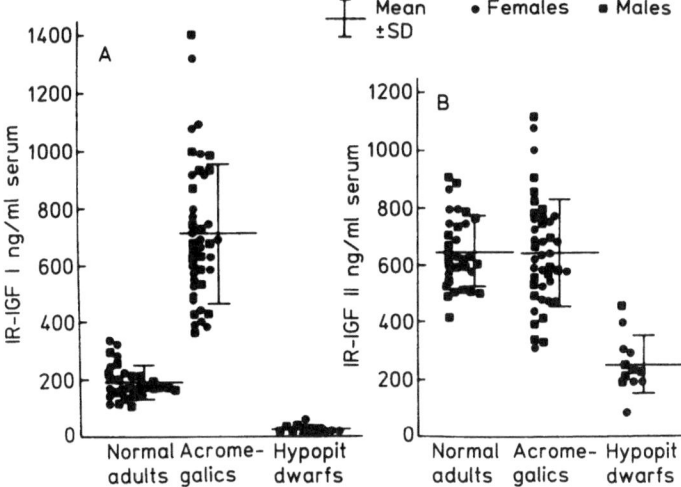

Fig. 6. *Serum concentrations of IGF I (A) and II (B) determined by radioimmunoassay in normal adults, acromegalic patients and patients with isolated GH deficiency.* Bars indicate the mean ± SD. (From [118] with permission)

Another, but rare form of dwarfism is the one first described by Laron [146]. The clinical picture resembles that of patients with GH deficiency with their proportionate dwarfism and low IGF I and II concentrations [147] but, distinct from GH-deficient patients, Laron patients have elevated concentrations of GH [146]. Their refractoriness to treatment with GH suggested that the cause of the disease is a defect of the GH receptor. Preliminary evidence for this is the failure of liver membrane preparations of such patients to bind GH [148]. Similarly, the GH binding protein normally found was lacking in the sera of three Laron patients [149]. Since available data suggest that this GH binding protein represents a truncated shed GH receptor, the lack of the circulating GH binding protein in Laron patients supports the assumption that this disorder stems from a GH receptor defect. Moreover, the determination of GH binding proteins in sera of patients with short stature might be a simple test for Laron dwarfism [149]. Very recently, a partial deletion of the GH receptor gene in two out of nine Laron patients has been directly established [150]. The finding that tissues of Laron patients are sensitive to IGF I [151] gives hope that such patients may be treated successfully with recombinant IGF I.

GH or GH receptor deficiencies are not the only disorders with decreased serum IGF concentrations. Malnutrition, severe insulin deficiency, liver disease and primary hypothyroidism all lead to decreased IGF I concentrations [152]. It must, however, be kept in mind that the relative importance of circulating (endocrine) IGF versus locally secreted (para- and autocrine) IGF is by no means clarified [14, 15] and that the activity of IGF at the receptor level is probably not only determined by its concentration, but also by the number of receptors and by the presence of IGF binding proteins.

IGF binding proteins

Plasma forms the largest IGF-containing pool in the body. In fact, the concentrations of IGF I and II combined are about 750 µg/l or 100 nM, which is about 1000 times more than the concentration of insulin. Since IGF has a hypoglycemic potency of about 5% of that of insulin, this IGF concentration would suffice to provoke hypoglycemia [153], unless the

greater part of IGF were hindered to bind to the insulin/IGF receptors. This is exactly one effect of the IGF binding proteins [154], another is the prolongation of the half-life of IGF [155]. There are, however, also reports on IGF binding proteins which enhance the biological response to IGF [156–158]. One reason for the conflicting results may be the heterogeneity of the IGF binding proteins as used *in vitro* [157] and secreted and circulating *in vivo* [12].

IGF binding proteins bind IGF I and II with high affinity and specificity; insulin and other tested proteins are not bound. In serum, more than 95% of the IGFs are bound to two main protein fractions of about 150 and 50 kDa [159]. The 150-kDa species in rat and human serum was shown to be GH-dependent, the 50-kDa species varies inversely with the GH status [160, 161]. Many different IGF-binding proteins from serum and cell supernatants have been described over the years [162, 163] which can now be grouped into three different classes, called IGF binding proteins 1, 2 and 3 [164].

The cDNA for an IGF binding protein previously found in amniotic fluid, placental membranes and Hep G2 cells has been cloned and sequenced recently [165–168]. The corresponding 25-kDa protein is expressed in decidua, secretory endometrium, liver and in Hep G2 cells [168] and is identical with the previously described placental protein 12 (PP12) [169]. This IGF binding protein, whose concentrations are inversely related to those of GH, is now called IGF binding protein 1 [164].

IGF binding protein 2 is present in conditioned medium of rat BRL-3A cells [170]. Its cDNA has been cloned recently from both a BRL-3A [171] and a human fetal liver cDNA library [172]. The cDNA encodes a 31-kDa protein [172]. IGF binding proteins 1 and 2 are homologous proteins with a sequence identity of somewhat less than 40% [172]. Both proteins contain a cysteine-rich amino-terminus, regions of clustered Pro, Glu, Ser and Thr residues (so-called PEST regions) as well as an Arg-Gly-Asp sequence (RGD motif) [168, 172]. The PEST regions are characteristic for proteins of short half-lives, the RGD motif may function as a cell recognition signal (see [168, 172] for further references).

The 150-kDa IGF binding protein consists of an acid-labile 85-kDa subunit and a 53-kDa acid-stable binding subunit (IGF binding protein 3) bound to IGF I or II [173]. A human cDNA for IGF binding protein 3 has been cloned and sequenced [174]. It is this 150-kDa complex which contains most of the IGF in the circulation and which is GH-dependent. IGF binding protein 3 is lacking in hypophysectomized rats; it can be induced by infusion of IGF I [175]. However, the induced binding protein 3 needs the presence of GH to form the 150-kDa complex. The authors speculate that whereas IGF induces the binding protein 3, GH is needed for the formation of the acid-labile subunit [175].

IGF binding protein 1 has been shown to have a higher affinity for IGF I than for IGF II [176], whereas IGF binding protein 2 preferentially binds IGF II [172]. IGF binding protein 3 appears to bind both IGFs with equal affinity [173]. Binding protein 1 seems to occur mostly in amniotic fluid, binding protein 2 in fetal serum and binding protein 3 in adult serum. Further speculations about their role and interplay in the regulation of IGF have to await further experiments with purified or recombinant proteins.

Very recently, two other IGF binding proteins have been described: one, in conditioned medium of human bone cells, has a limited amino-terminal sequence identity with the other binding proteins and seems to inhibit IGF [177]; the other was found to be the preponderant binding protein in cerebrospinal fluid and to be specific for IGF II [178].

A large binding protein previously observed in fetal rat serum [179] was later identified with circulating type 2 IGF receptor [180]. The serum receptor is 10 kDa smaller than the membrane-bound receptor [181] and is thus probably produced by shedding. This shed receptor has been shown to be truncated in the carboxyl-terminal domain [182]. It is assumed that it does not function as an IGF binding protein, but rather as a carrier of mannose-6-phosphate containing ligands [182].

IGF RECEPTORS

IGF I and insulin are structural homologs which both elicit the same two types of biological responses, i.e. long-term effects on cell proliferation and short-term metabolic effects such as glucose transport, but with differing potencies. This fact has been explained by the presence of an insulin and an IGF receptor, each of which can crossreact at lower affinity with the heterologous ligand [183]. Indeed, competitive binding and crosslinking experiments have provided evidence for the presence of separate receptors (see [23] for review). On the other hand, more recent data have shown that both the insulin and the type 1 IGF receptor can mediate metabolic and mitogenic responses to either ligand (see below). The functions of IGF II and of the type 2 IGF receptor remain mysterious.

Structures of the IGF receptors

Both the insulin and the type 1 IGF receptor are glycosylated heterotetramers, composed of two α subunits (135 kDa) and two β subunits (90 kDa) linked by disulfide bonds [184–186]. The α subunits contain the extracellular ligand binding site, the β subunits a transmembrane domain and the intracellular tyrosine autophosphorylation site [187–189]. In human and rat brain, the size of the α subunits of the type 1 IGF receptor has been shown to be about 115 kDa [190, 191] rather than 135 kDa as in other organs. The type 2 IGF receptor is structurally distinct from the type 1 receptor: it is a monomer (250 kDa) without autokinase activity [184, 186, 192]. Although the type 2 receptor is devoid of kinase activity, it can serve as a substrate for a membrane-associated tyrosine kinase [192].

The cDNAs for the human insulin receptor [193, 194] and the human type 1 IGF receptor [195] have been cloned and sequenced and found to be very similar regarding overall structure, subunit size and deduced amino acid sequence (Fig. 7). The most pronounced similarity is in the tyrosine kinase domain (84% identity). The type 1 IGF receptor is synthesized as a precursor of 1367 amino acid residues, a 30-residue signal is removed during translocation into the endoplasmic reticulum and an RKRR sequence is cleaved to yield the 135-kDa and 90-kDa glycosylated subunits [195]. The mass of the unglycosylated precursor sequence can be calculated to be 152 kDa, whereas bigger forms of 190 and 210 kDa are assumed to represent the high-mannose and the fully glycosylated precursor, respectively [196].

Recently, a human cDNA for the type 2 IGF receptor has been found to be very similar to the bovine cation-independent mannose-6-phosphate receptor [197], thus to a receptor which is believed to be involved in the transport of lysosomal enzymes. The predicted molecular mass of the coded protein is 270 kDa after removal of the 40-residue leader peptide, the molecular mass of the mature glycosylated receptor is esti-mated to be about 300 kDa. Of the receptor sequence, 92% is oriented extracellularly, the extracellular domain consisting mostly of 15 cysteine-containing repeat sequences. A 43-residue insert in repeat 13 is similar to the type II region of fibronectin (Fig. 7). In a later report [198], the cDNA sequence of the human mannose-6-phosphate receptor was found to be 99.8% identical to that of human type 2 IGF receptor. This identity has so far rather amplified the puzzle [17] than solved it (see below). The original proposal that the IGF II/mannose-6-phosphate receptor is a multifunctional receptor with at least two distinct binding sites, one for IGF II and one for mannose-6-phosphate-labelled proteins [199], has since been confirmed [199–202] (see below).

The single genes for the human insulin receptor and the human type 1 and 2 IGF receptors all map to different chromosomes, i.e. to the short arm of chromosome 19 [203], to the long arm of chromosome 15 [195] and to the long arm of chromosome 6 [204], respectively. Furthermore, the human gene for the structurally unrelated 46-kDa cation-dependent mannose-6-phosphate receptor has been assigned to chromosome 12 [205].

In the liver of the stingray, a cartilaginous fish, a high-affinity receptor has been described, which is apparently a homodimer with a monomer size of 210 kDa [206]. The ligand binding and tyrosine kinase region are both contained on the same subunit. Both insulin and IGF I stimulate tyrosine phosphorylation. This receptor may represent an evolutionary relic of the state before divergence of insulin and IGFs.

Ligand binding of the IGF receptors

On most cells, type 1 and type 2 IGF receptors coexist and, on many cells, insulin receptors are there additionally. In binding or crosslinking experiments, radiolabelled IGF I or II is distributed between the different receptor types according to their concentrations and affinities. Exact estimations of the affinities of the three ligands to the three receptors are thus not feasible by this procedure. In addition, the preparations of IGF I and II used were usually not free from cross-contaminants. However, most workers in the field agree that the following relative binding affinities are found in most cells (see [13, 16, 23] for reviews): insulin receptor, insulin ≫ IGF II > IGF I; type 1 IGF receptor, IGF I > IGF II ≫ insulin; type 2 IGF receptor, IGF II > IGF I (insulin does not bind).

Soon after the identification of the type 2 IGF receptor with the cation-independent mannose-6-phosphate receptor [197, 198], this receptor was found to bind both ligands, IGF II and mannose 6-phosphate [199, 201, 202, 207] at different sites [208]. Some controversy exists whether binding of the two ligands occurs independently [199] or whether a bound ligand alters the binding affinity of the other [201, 207, 209]. Perhaps mannose 6-phosphate can displace from the receptor endogenous lysosomal proteins which interfere with binding of IGF II [210]. The transforming growth factor β1 precursor contains mannose 6-phosphate and can bind to the type 2 receptor [211].

An interesting observation has been made on mannose-6-phosphate receptors of chicken [212, 213] and frogs [213]. Human, rat or chicken IGF II failed to bind to the purified liver mannose-6-phosphate receptor which suggests the high-affinity binding sites for IGF II on the mannose-6-phosphate receptor have emerged late in evolution [213]. Since IGF II circulates in chickens at concentrations similar to mammalian species [13] and since IGF II has many and pronounced effects in chick fibroblasts despite an early recognized lack of type 2

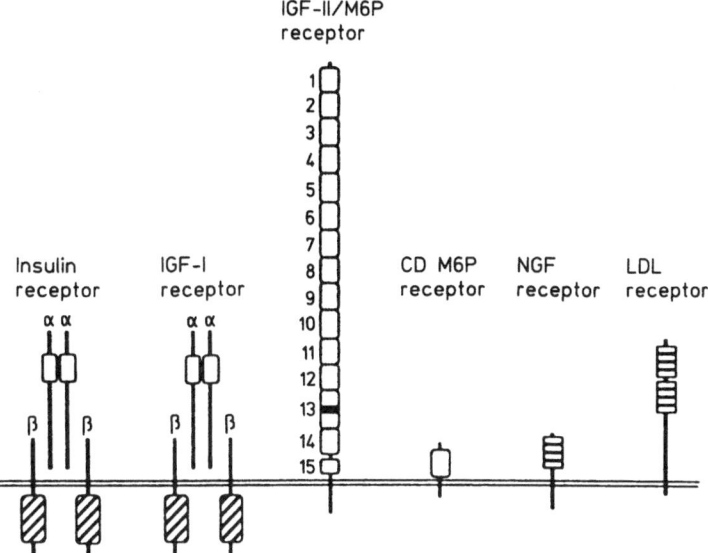

Fig. 7. *Schematic comparison of the IGF II/mannose-6-phosphate receptor structure with related receptors.* (Abbreviations: CD M6P receptor, cation-dependent mannose-6-phosphate receptor; NGF, nerve growth factor; LDL, low-density lipoprotein.) Tyrosine kinase domains are indicated by boxes with diagonal lines. The numbered boxes of the IGF II/Man6P receptor indicate extracellular cysteine-rich regions and repeat sequences. (From [197] with permission)

IGF receptors [23], it is probable that IGF II acts in chick fibroblasts via the type 1 IGF receptor [212].

The tools are now available to determine precise affinity constants by using recombinant IGFs as ligands and cells overexpressing one type of receptor, provided that expression of the endogenous receptors is sufficiently low. In Chinese hamster ovary cells overexpressing the transfected cDNA for type 1 IGF receptor, the three ligands were bound with the following affinites: IGF I, K_d 1.5 nM; IGF II, $K_d \approx$ 3 nM; insulin, $K_d \approx$ 100 nM [214].

SIGNAL TRANSDUCTION AND BIOLOGICAL RESPONSE

Insulin and type 1 IGF receptor

The fact that insulin at supraphysiological concentrations is mitogenic and that IGF at supraphysiological concentrations shows insulin-like metabolic effects has been interpreted to mean that metabolic effects are mediated via the insulin receptor and mitogenic effects via the type 1 IGF receptor, while the biological significance of the type 2 receptor remained unclear. Experiments with antireceptor antibodies seemed at first to confirm this assumption [215]. Later, the monoclonal antibody αIR3 [216], which is directed against the type 1 IGF receptor, was widely used. This antibody abolished mitogenic effects of low concentrations of IGF I and II [217], suggesting that the mitogenic effect of IGF II is mediated via the type 1 receptor. A growing number of papers appeared which presented evidence that insulin at low concentrations can elicit mitogenic effects via its own receptor, while at high concentrations the greater part of the mitogenic effect of insulin is mediated via the type 1 IGF receptor [218 – 221]. Likewise, the use of antibodies against the insulin receptor suggested that IGF I can have metabolic effects via its own receptor [221 – 223].

Experiments employing cells overexpressing insulin or type 1 IGF receptors have shown more directly that both receptors can mediate metabolic as well as mitogenic responses to both insulin and IGF I [214, 224, 225]. However, *in vivo*, insulin primarily regulates metabolic and IGF mitogenic responses. If the two receptors use both overlapping receptor and postreceptor signalling pathways, what are then the mechanisms responsible for the different main actions of insulin and IGF I *in vivo*? It appears now 'that the different physiological roles of these two hormones are determined by the distribution of the two receptors on different cells and/or the pharmacodynamics of the two hormones' [214]. A partial modification of this concept has been proposed on the basis of results obtained with overexpression of hybrid receptors consisting of extracellular insulin domains and intracellular type 1 IGF domains [226]. The two wild-type and the chimeric receptors stimulated glucose transport with similar efficiencies, whereas receptors with a type 1 IGF cytoplasmic domain were 10 times more active than the insulin receptor in stimulating DNA synthesis [26]. Whether the results obtained in cells overexpressing these receptors represent the situation under physiological conditions is is too early to tell.

Several reports have described findings which suggest that other type 1 IGF receptors might exist: β subunit doublets found by SDS electrophoresis [214, 227], an atypical insulin receptor from human placenta with increased affinities to IGF [228, 229], a monoclonal antibody (5D9) which recognizes some, but not all type 1 IGF receptors [230], the presence of hybrid receptors in NIH-3T3 and in HepG2 cells [231], a fetal type 1 IGF receptor with a 105-kDa β subunit [232] and a protein sequence deduced from a nucleotide sequence of human and guinea pig genomic DNA which appears to represent a receptor distinct from, but similar to, the insulin and the type 1 IGF receptor [233]. In addition, on purified placental type 1 IGF receptors, two distinct binding sites have been described: one with highest affinity to IGF I which can be blocked with αIR3, and one with highest affinity to IGF II which is not blocked by αIR3 [234]. Some of these observations may explain some of the discrepancies in the literature on receptor affinities, but they are at present hard to fit into a plausible scheme.

High occupancy of the receptors leads to down-regulation (for a review, see [23]). Once a ligand is bound to either the insulin or the type 1 IGF receptor, signal transduction involves tyrosine autophosphorylation of the β subunits [22, 188] and phosphorylation of tyrosines of several other proteins [214, 235]. In addition, phosphorylation also occurs at serine and threonine residues of the β subunit and several cellular serine kinases are activated (see [13] for review). Receptor aggregation has been implicated in the signalling pathway [236] which might explain the ability of αIR3 to induce biological responses in cells with a high number of type 1 receptors [214]. Interestingly, αIR3 did not lead to tyrosine phosphorylation of either the β subunit or of other major substrates [214]. It is speculated that receptor cross-phosphorylation can occur, not only between identical β subunits in the heterotetramer, but also between β subunits of the insulin and the type 1 IGF receptor [237]. Whether this cross-phosphorylation involves the formation of disulfide-linked heteroreceptor complexes [13] is at present not known.

In rat hepatoma cells [238] and in cells expressing high levels of the insulin receptor [239], insulin was found to induce c-*myc* and c-*fos*. Also, the effect of IGF I on cells expressing high levels of normal p21 H-*ras* had similar effects as the expression of mutant H-*ras* [240] which was interpreted as a functional linkage between insulin/IGF I and normal p21 H-*ras*. But it has to be stressed that beyond kinase activation, the signal transduction pathway is poorly understood.

Type 2 IGF receptor

The physiological relevance of the type 2 IGF receptor has been questioned [241]. There are reports on biological effects of IGF II via the type 2 receptor (see [17] for review); however, these results are difficult to interpret. The type 2 receptor can be grouped into the class of membrane receptors involved in nutrient transport such as the low-density lipoprotein receptor (see Fig. 7) that contrast with those of the growth factor type such as the insulin and the type 1 IGF receptor. Typical for the former type of receptors is the short cytoplasmic domain, the lack of tyrosine autokinase activity, the principal occurrence in intracellular membranes and a ligand-independent membrane recycling [242]. That IGF II might associate with type 2 receptors not for a growth signal but for degradation has been shown in experiments using anti-(type 2 receptor) antibodies which inhibited IGF II degradation by more than 90% [243]. Further support for the concept of a non-growth receptor comes from the fact, already mentioned, that mannose-6-phosphate receptors of chickens and frogs fail to bind IGF II [212, 213].

The failure to signal a growth effect does of course not exclude a relevant function in the action of IGF II. It has long been known that insulin increases binding of IGF II to the type 2 receptor in fat and hepatoma cells [244, 245]. Apparently, insulin induces a redistribution of type 2 receptors from intracellular to the plasma membranes qualitatively similar to its effects on the glucose transporter [245, 246]. A similar increase of mannose-6-phosphate binding sites at the surface of human skin fibroblasts by mannose-6-phosphate, IGFs, insulin and epidermal growth factor has been described [247] which is associated with a decreased phosphorylation of the receptor in the plasma membrane [248]. Binding of either mannose 6-phosphate or IGF II to the type 2 receptor initiates a receptor redistribution, but via different signal transducing pathways [247], whereby the one elicited by mannose 6-phosphate seems to be linked to G proteins [247]. In another report, IGF II was also reported to elicit a signal linked to G proteins [249].

Despite the encouraging progress, we are far from a unified and clear picture of the physiological roles of IGF II and of the type 2 receptor. Further advances can be expected from cells overexpressing these receptors. A beginning has been made by transfecting Chinese hamster ovary cells with a type 2 receptor cDNA [250]. In these cells, insulin and IGF I were found to inhibit protein catabolism. It is speculated that an insulin-induced translocation of the receptor to the cell membrane 'could disrupt the flow of lysosomal enzymes and thereby account for the ability of insulin to inhibit protein catabolism' [250].

BIOLOGICAL EFFECTS OF IGF

Challenges to the somatomedin hypothesis

The original somatomedin hypothesis of Daughaday [6, 7] stated that GH acts on peripheral tissues not directly but by the intermediary of a plasma growth factor, somatomedin, which is produced under the influence of GH. This hypothesis has been seriously challenged. It was observed that GH administered locally and unilaterally into epiphyseal growth plates led to significant bone growth only on the injected side [251] which seemed to implicate a local and direct action of GH without the intermediary of plasma IGF. When, however, IGF antiserum was injected concurrently, the GH effect was abolished [252]. Later experiments showed that at the growth plate, GH stimulates the expression of IGF I mRNA [253]. In an *in vitro* study with osteoblast-like cells, the GH effect on proliferation could be inhibited by IGF I antiserum in a dose-dependent manner. Yet, a partial additive effect of GH and IGF I pointed to an additional effect of GH not mediated by IGF [254]. A local biosynthesis of IGF I under control of GH is consistent with the proposal that IGF acts not exclusively in an endocrine but also in a paracrine and autocrine fashion [95, 255]. In a careful analysis of data from the literature, Daughaday concluded that local action of GH on cartilage and bone does exist, but that the systemic action of GH via IGF I is predominant in the regulation of skeletal growth [14].

The so-called dual effector theory of GH action [256] was based on observations of the conversion of 3T3-F442A pre-adipocytes into differentiated adipocytes under the influence of GH and stated that GH induces the differentiation step, whereupon IGF I is able to promote clonal expansion of the differentiated cells [256]. A variation of this theory assigned to GH the effect of priming such cells and insulin to convert the primed cells into adipocytes [257]. In another line of pre-adipocytes (ob1771), GH was shown to induce expression of the IGF I gene during, but not before, differentiation [258]. In a well documented paper studying the preadipocyte cell line 3T3-L1, the authors found that IGF I played a central role in activating expression of an adipocyte differentiation program and that these effects were completely independent of GH [259]. The reasons for the discrepant results are not clear.

All these studies suggest that the original somatomedin hypothesis may need to be modified, because IGF I has been found in certain systems not to be able to replace GH completely and because GH may be essential for inducing IGF I also locally. I propose the following modification of the original hypothesis: the growth promoting activiy of GH is due, on the one hand, to direct effects on the periphery enabling cells to produce and to respond to IGF I and, on the other hand, to indirect effects, mostly on liver, to increase serum concentrations of IGF I; IGF I can mimick most, but not all effects of GH. The relative contributions of circulating IGF, locally produced IGF and directly acting GH are at present disputed [14]. The history of science teaches us that this description of an apparently complex process in all probability represents no more than the current state of ignorance.

EFFECTS *in vitro*

Metabolic and mitogenic effects

In early publications on IGFs (see [14, 20, 22] for a historical overview), the sulfation- and thymidine-uptake-stimulating activities in cartilage and the insulin-like activity on adipose tissue have been the most studied *in vitro* effects. Later, many more cell types have been found to respond to nanomolar concentrations of IGF, mostly by increased DNA synthesis with or without subsequent cell proliferation. Indeed, in nearly all tissues from fetal and adult animals, receptors for IGF are present [23] which may explain the target cell diversity of IGFs.

In vitro effects of IGFs are now commonly classified into short-term insulin-like metabolic effects such as stimulation of glucose uptake and glycogen and lipid synthesis in adipose tissue, and into long-term mitogenic effects such as stimulation of protein, RNA and DNA synthesis and cell proliferation. It is as well to remember that generally insulin, IGF I and IGF II can substitute for each other in promoting all these effects, albeit with differing potencies. Insulin is more potent in stimulating metabolic parameters by a factor of $5-10$ (muscle) to about 100 (adipose tissue), whereas IGFs are more potent in stimulating mitogenic effects by an average factor of around 100, but with wide fluctuations depending on the cell type [21, 22]. IGF I and II were often found to be about equipotent in stimulating both metabolic and mitogenic parameters [23].

Usually, the addition of IGFs as the only growth factors to a chemically defined medium cannot reproduce the potent mitogenic effects of serum [21, 22]. In studies on BALB/c 3T3 cells, the consecutive action of platelet-derived growth factor (PDGF), IGF and/or epidermal growth factor (EGF) was shown to be necessary for a complete traverse of the cell cycle [260]. Quiescent cells arrested in G_0/G_1 phase, when shortly exposed to PDGF, were made 'competent' to respond to 'progression' factors such as IGF and EGF. This model was refined by demonstrating restriction points in G_1 controlled by specific growth factors [261]. EGF in combination with suboptimal amounts of IGF allowed PDGF-treated cells to traverse the first 6 h of G_1 to a distinct growth-arrest point termed the V point [261]; IGF I at optimal concentrations mediated traverse of the remaining 6 h of G_1 and entry into the S phase [262]. Traverse of S, G_2 and M was found to be independent of added growth factors [263]. Exponentially growing (not growth-arrested) BALB/c 3T3 cells require IGF only, suggesting that the 6-h G_1 phase of exponentially growing cells is equivalent to the V to S traverse [264]. The mechanism by which IGF mediates late G_1 traverse and entry into S phase is not known. Preliminary data on concomitant protein modifications have been published [265]. In another study, IGF I was shown to regulate the expression of about 30 genes [266]. The greater part of these genes were constitutively transcribed and IGF I seemed to increase the stability of the transcripts [266].

Whether BALB/c 3T3 cells are a valid model for the effects of IGF on all cells is by no means certain. Indeed, experiments on subconfluent, serum-deprived human diploid fibroblasts using flow cytometric analysis yielded somewhat different results [267]. These authors found that EGF could regulate the proportion of cells capable of entering the cell cycle from the quiescent state as well as PDGF, whereas IGF I only regulated the rate of exit from G_1 into S phase without affecting the cycling fraction [267]. It cannot be excluded that growth factors may also stimulate secretion of other growth factors, as has been shown for PDGF which stimulated IGF I secretion [268]. Furthermore, growth factors may modulate receptors. This has been demonstrated for PDGF which down-regulates EGF receptors [269].

The number of reports on stimulation of wound healing by IGF does not reflect the great commercial interest and therapeutic value such effects, if well documented, would have. Only recently, a combination of PDGF and IGF I applied topically was found to stimulate the healing of skin wounds in pigs [270].

Effects on differentiation

Evidence is accumulating that IGFs can also have distinct effects on the differentiation of cells of mesodermal origin [21]. In a cell culture system, chick cells from the body walls were shown to differentiate into contracting myotubes without concomitant drastic increase of the cell number [271]. When osteoblast-like cells of newborn rats were cultivated for 6 days in the presence of IGF I, the cell number increased only slightly, whereas alkaline phosphatase activity, an indicator of differentiation into mature osteoblasts, increased considerably. PDGF, on the other hand, depressed alkaline phosphatase and increased cell number and led to cell dedifferentiation [272]. In fetal rat calvariae, IGF I stimulated bone matrix formation in excess of its effect on bone cell replication [273]. In a serum-free cell system, an effect of IGF I on erythroid colony formation was observed [274]. The same effect on erythropoiesis could be detected *in vivo* (see below). Finally, several papers have reported on the effects of IGF on cytodifferentiation of the ovarian granulosa cell and on androgen production in ovarian thecal-interstitial cells [275, 276]. Effects on differentiation may be as important as effects on cell proliferation under physiological conditions.

Effects on the nervous system

The role IGF might play in the development and function of the nervous system has aroused considerable interest in recent years [11]. The discovery in the central nervous system of significant amounts of IGF [68, 69, 77, 78, 277], of differential expression of IGF I and II which is developmentally regulated [111], of IGF receptors [190, 278], and of release of immunoreactive IGF in brain explants [279] has initiated the IGFs to the habitués of neurobiology.

IGF I was found to be a potent inducer of oligodendrocyte development in cerebral cells obtained from 1-day-old rats [280]. In a serum-free medium, 100 ng/ml of IGF I increased the oligodendrocyte numbers 60-fold, whereas IGF II produced less than a 2-fold increase.

IGF I and/or II have been found to be a necessary constituent for maintaining in serum-free medium fetal rat brain neurons [281] and human fetal brain cells [282]. IGF II was reported to promote neurite outgrowth in cultured chick dorsal root ganglionic sensory neurons [283] and in sympathetic and sensory neurons [284]. In contrast to nerve growth factor (NGF), IGF I stimulated DNA synthesis in tyrosine-hydroxylase-expressing rat sympathetic presumptive neuroblasts [285].

Effects of IGFs on release of acetylcholine [286], catecholamine [287] and neuropeptides [282] and on triggering differentiation of catecholaminergic precursors [288] have been reported. Human brain cell cultures contained more choline acetyltransferase activity in the presence of IGF II, whereas IGF I was without effect [282].

IGF has been even implicated with the complex process of synaptogenesis. A correlation between IGF II expression and turnover of neuromuscular synapses has been claimed in fetal rat hind limb muscles [289]. After transection of the sciatic nerve, IGF II mRNA was up-regulated [289]. This last finding may be complemented by the observation that IGF I stimulates the regeneration of crushed rat sciatic nerves when administered by miniosmotic pumps during 3 – 4 days [290].

A word of caution may be in order here. Some of the cited publications in this new field lack the necessary care in the design of the experiments and in the interpretation of the results. One way of dealing with such publications would be to neglect them in a review such as this one. Since, however, the habit is growing of allowing wishful thinking be the selection criterion for publications to be cited irrespective of their qualities, I feel obliged to expose rather than to drop such papers. I now give examples from just two of the references cited above. I do not understand how it can be stated that 'insulin and IGF II concentrations as low as 10 pM were effective' in a medium with 10% serum which contains approximately 10 nM IGF II! Or the claim of 'a striking correlation' between two parameters, one measured by quantification of Northern blots, the other cited from the literature; if one looks up the cited publication, one learns that the cited results were admittedly not clear-cut.

EFFECTS *in vivo*

Metabolic effects

When a bolus of 40 µg recombinant human IGF I (rhIGF I) was injected into rats (120 – 140 g) intravenously, a transient hypoglycemia with a maximum at 15 min was observed [155]. Even more pronounced was the fall of the blood sugar, when a bolus of 100 µg/kg body mass was injected into healthy human volunteers [291]. Obviously, acute injections lead temporarily to such high concentrations of IGF I that they exceed the binding capacity of the IGF binding proteins. Under physiological conditions, free IGF never rises to such concentrations as seen after bolus injections [19]. Similar hypoglycemic effects were seen after injection of rhIGF I into Laron patients [292].

A very interesting recent paper reports on effects of rhIGF I when infused subcutaneously for 6 days into normal human subjects [143]. Whereas GH caused hyperinsulinism, IGF I led to decreased insulin secretion. IGF I furthermore increased the glomerular filtration rate, whereas fluid retention was seen in the GH-treated subject. During infusion of IGF I, the concentrations of GH and of IGF II fell, the former probably being due to feed-back inhibition, the latter possibly to its displacement from IGF binding proteins with subsequent rapid degradation [143].

Growth effects

The first demonstration of a growth effect of IGF *in vivo* succeeded in 1982: IGF I isolated from human serum was administered over 6 days subcutaneously via mini-pumps into hypophysectomized rats [293]. Due to the slow, continuous infusion, no hypoglycemia occurred. But there was an increase of body weight, of thymidine incorporation into costal cartilage and of the tibial epiphyseal width over controls comparable to that obtained with GH. Similar results were obtained in normal rats [294], in Snell dwarf mice [295] and again in hypophysectomized rats, but this time comparing the effects of IGF I with those of IGF II [296]. IGF II was found to be less potent than IGF I.

All these results were either hailed as proof for the original somatomedin hypothesis or critized for using suboptimal doses of GH which could simulate an equipotent effect of IGF. In similar experiments in hypophysectomized rats using recombinant methionyl IGF I, only slight effects on the epiphyseal width at the highest doses could be observed [297]. A further criticism referred to the short duration of the experiments which might be crucial in the light of the dual effector theory of GH action (see above). The original experiments [293] were therefore repeated with higher doses of both IGF I and of GH and with infusions for 18 days [298]. Again, body weights gain, tibial epiphyseal width and epiphyseal growth rate were stimulated to the same extent by either GH or IGF I. The accumulated longitudinal bone length was slightly less with IGF I. However, distinct effects on organ weights were observed: in contrast to IGF I, GH increased the weights of two examined muscles; IGF I, but not GH, increased the weights of the kidney, spleen and thymus and decreased the weight of the fat pads [298]. In conclusion, IGF I administered systemically can mimic the effects of GH on differentiation and proliferation of epiphyseal cartilage. On the other hand, the dose/response curves on epiphyseal width and the influence on some organ weights were clearly different between GH- and IGF-treated animals. Similar differential effects on organ size and insulin concentrations of GH and IGF I were obtained in transgenic mice expressing chimeric genes composed of the coding sequence of either bovine GH, human GRF or human IGF I fused to mouse metallothioneine I promoter [299].

It is possible that the differences in the effects of the two hormones GH and IGF I can be attributed to a different presentation of IGF to different cells due to the difference in the IGF binding protein pattern in hypophysectomized GH-treated versus IGF-treated animals [298]. Therefore, these *in vivo* experiments can hardly settle the quarrel about whether or not GH has other effects in addition to the one to stimulate the release of IGF I.

To study effects of high concentrations of IGF II, an animal model has been developed transplanting rat IGF II secreting cells into nude rats and mice [300]. Neither body weight nor tail length were different from control animals. However, serum concentrations of IGF I were found to be significantly lowered. This last effect may be due to the same mechanisms as the lowering of IGF II after infusion of IGF I [143]. In contrast, hypophysectomized rats subcutaneously infused with rhIGF II during 7 days showed increases in body weight gain, tibial epiphyseal cartilage width and femur hydroxyproline concentrations, as well as a decrease in serum urea nitrogen [301]. The effects of GH were similar. These apparently discrepant results may reside in the different GH status of the two animal models: in the former study [300], the animals disposed of normal levels of GH and of the sum of IGF I and II, so that no further increase of body mass or growth rate could be observed by substituting IGF II for IGF I; in the latter study [301], the animals lacked GH and

IGF I, so that the infused IGF II could, at least partially, elicit a growth response, probably via the type 1 IGF receptor.

With the aim of investigating the role of IGF I and II in fetal development, 10-day-old rat embryos and 16-day-old fetal rat paws were transplanted under the kidney capsule of syngeneic hosts and the effects of infusions into the kidney artery with IGF I or II or IGF I antiserum studied [302]. Significant growth retardation of the embryos could be observed after 9-days infusion with the antiserum. Infusion with IGF for 6 days stimulated the growth of the embryos with differing potencies: purified rat IGF II > rhIGF II > rhIGF I.

Again in hypophysectomized rats, infusion of rhIGF I led to increases of ^{59}Fe incorporation into erythrocytes and to a rise in reticulocytes and serum erythropoietin [303]. Since the rise in reticulocytes preceded the one of erythropoietin, the action of IGF I on erythropoiesis was postulated to be twofold: a direct, erythropoietin-like effect and an indirect one via stimulation of erythropoietin.

Hopes that IGFs might prove useful in wound healing [21] have so far not been nourished by many data [270, 290].

EXPRESSION OF RECOMBINANT IGF

A major obstacle in IGF research, the scant availability of the pure peptides, has now been overcome with the advent of recombinant human IGFs (rhIGF). Seven years after the publication of the amino acid sequences of IGFs [3, 4] and two years after the isolation of a human IGF I cDNA [33], rIGF I has been expressed successfully [304 – 307].

The first rIGF I was expressed as a fusion protein in *Escherichia coli* which was cleaved by cyanogen bromide [305]. Since IGF I contains a methionine residue at position 59, Met59 was substituted by Thr. This [Thr^{59}]IGF I was found to be approximately equipotent to natural IGF I isolated from serum [304] and was made commercially available.

One of the difficulties in bacterial systems is the accumulation of the expressed protein as inclusion bodies in a denatured form which is difficult to renature in good yield [305, 308]. To avoid denatured precipitates, expression with fused leader sequences has been tried to direct the protein into the periplasmic space [308, 309]. Only part of the expressed protein was soluble, correctly processed and secreted into the medium [309, 310]. Another approach was gene fusion with another sequence coding for an exported protein [306, 311 – 313]. The difficulty here was to achieve efficient cleavage while sparing the desired product. Particularly attractive is the fusion with protein A [306, 311], because the fusion protein can be easily isolated on a IgG affinity column.

Yeast, a eucaryotic organism, allows the expression of a pre-pro-sequence with proper processing into the final product and its secretion into the medium [91 – 93, 314, 315]. Interestingly, a glycosylated form of rhIGF I was found in addition to the unglycosylated native form [315]. The glycosylated analog was shown to contain two mannose residues O-linked to Thr29 and to be biologically fully active [315]. A mammalian system using transfected mouse fibroblasts has been described with a fused GH signal peptide [316]; however, improper cleavage left an additional alanine residue at the amino-terminus [317].

Expression of rhIGF II is apparently more cumbersome. For unknown reasons, it tends to be more denatured and insoluble in procaryotic systems [318], more difficult to renature and prone to proteolytic degradation. The expression of rhIGF II was described in *E. coli* as a fusion protein [319] and as a double-fusion protein [320] for protection against proteolysis.

OUTLOOK

Superimposed on the exponentially increasing output of publications in the field of the IGFs, two jumps can be observed: the first after the publication of the amino acid sequences and the second after the isolation of the cDNAs. A third jump can be expected in the near future due to the production of now sizable quantities of rhIGF I and, after some delay, of rhIGF II. This will certainly facilitate worldwide efforts to study further the mode of action of the IGFs and their clinical applications.

So far, our knowledge is still scanty and, as we penetrate deeper, digging will get harder. The interplay of the IGFs with each other, with other hormones, with their binding proteins and receptors is apparently more complex than most of us imagined. Particularly, the physiological roles of IGF II and of the enigmatic type 2 IGF receptor should be a challenge to our inquiring minds. This view is in contrast to statements sometimes heard (although not published) characterizing IGF II as a functionless evolutionary relic and the type 2 IGF receptor as a garbage receptor. In my mind, nature is a sphinx, presenting us with IGF II as a riddle. We better be as clever as was Oedipus, although I would abhor the sphinx committing suicide afterwards [321].

The authors own work has been supported by the University of Zürich, by the Swiss National Science Foundation and by the Hartmann Müller Foundation. A long and fruitful collaboration with the group of E. R. Froesch is gratefully acknowledged. I also wish to thank J. Zapf and my present and former collaborators D. Bürgisser, G. Haselbacher, A.-M. Honegger, C. Lüthi, B. Roth, J. Zarn and P. Zumstein for criticism and suggestions.

REFERENCES

1. Rinderknecht, E. & Humbel, R. E. (1976) *Proc. Natl Acad. Sci. USA 73*, 2365 – 2369.
2. Rinderknecht, E. & Humbel, R. E. (1976) *Proc. Natl Acad. Sci USA 73*, 4379 – 4381.
3. Rinderknecht, E. & Humbel, R. E. (1978) *J. Biol. Chem. 253*, 2769 – 2776.
4. Rinderknecht, E. & Humbel, R. E. (1978) *FEBS Lett. 89*, 283 – 286.
5. Blundell, T. L., Bedarkar, S., Rinderknecht, E. & Humbel, R. E. (1978) *Proc. Natl Acad. Sci. USA 75*, 180 – 184.
6. Salmon, W. D. Jr & Daughaday, W. H. (1957) *J. Lab. Clin. Med. 49*, 825.
7. Daughaday, W. H., Hall, K., Raben, M. S., Salmon, W. D. Jr, Van den Brande, J. L. & Van Wyk, J. J. (1972) *Nature 235*, 107.
8. Klapper, D. G., Svoboda, M. E. & Van Wyk, J. J. (1983) *Endocrinology 112*, 2215 – 2217.
9. Enberg, G., Carlquist, M. Jörnvall, H. & Hall, K. (1984) *Eur. J. Biochem. 143*, 117 – 124.
10. Daughaday, W. H., Hall, K., Salmon, W. D. Jr, Van den Brande, J. L. & Van Wyk, J. J. (1987) *J. Clin. Endocrinol. Metab. 65*, 1075 – 1076.
11. Baskin, D. G., Wilcox, B. J., Figlewicz, D. P. & Dorsa, D. M. (1988) *Trends Neurosci. 11*, 107 – 111.
12. Baxter, R. C. & Martin, J. L. (1989) *Progr. Growth Factor Res. 1*, 49 – 68.
13. Czech, M. P. (1989) *Cell 59*, 235 – 238.
14. Daughaday, W. H. (1989) *Perspect. Biol. Med. 32*, 194 – 211.

458

15. Daughaday, W. H. & Rotwein, P. (1989) *Endocrine Rev. 10*, 68–91.
16. Rosenfeld, R. G. (1989) in *Advances in growth hormone and growth factor research* (Müller, E. E., Cocchi, D. & Locatelli, V., eds) pp. 133–143, Pythagora Press, Rome and Springer-Verlag, Berlin.
17. Roth, R. A. (1988) *Science 239*, 1269–1271.
18. Sussenbach, J. S. (1989) *Progr. Growth Factor Res. 1*, 33–48.
19. Zapf, J., Guler, H. P., Schmid, Ch., Kurtz, A. & Froesch, E. R. (1989) in *Advances in growth hormone and growth factor research* (Müller, E. E., Cocchi, D. & Locatelli, V., eds) pp. 145–162, Pythagora Press, Rome and Springer-Verlag, Berlin.
20. Humbel, R. E. (1984) in *Hormonal proteins and peptides*, vol. XII (Li, C. H., ed.) pp. 57–79, Academic Press, Orlando FL.
21. Froesch, E. R., Schmid, C., Schwander, J. & Zapf, J. (1985) *Annu. Rev. Physiol. 47*, 443–467.
22. Van Wyk, J. J. (1984) in *Hormonal proteins and peptides*, vol. XII (Li, C. H., ed.) pp. 81–125, Academic Press, Orlando FL.
23. Rechler, M. M. & Nissley, S. P. (1985) *Annu. Rev. Physiol. 47*, 425–442.
24. Brissenden, J. E., Ullrich, A. & Francke, U. (1984) *Nature 310*, 781–784.
25. Tricoli, J. V., Rall, L. B., Scott, J., Bell, G. I. & Shows, T. B. (1984) *Nature 310*, 784–786.
26. Bell, G. I., Gerhard, D. S., Fong, N. M., Sanchez-Pescador, R. & Rall, L.B. (1985) *Proc. Natl Acad. Sci. USA 82*, 6450–6454.
27. O'Malley, K. L. & Rotwein, P. (1988) *Nucleic Acids Res. 16*, 4437–4446.
28. Ullrich, A., Berman, C. H., Dull, T. J., Gray, A. & Lee, J. M. (1984) *EMBO J. 3*, 361–364.
29. Rotwein, P., Pollock, K. M., Didier, D. K. & Krivi, G. G. (1986) *J. Biol. Chem. 261*, 4828–4832.
30. de Pagter-Holthuizen, P., van Schaik, F. M. A., Verduijn, G. M., van Ommen, G. J. B., Bouma, B. N., Jansen, M. & Sussenbach, J. S. (1986) *FEBS Lett. 195*, 179–184.
31. Rotwein, P. (1986) *Proc. Natl Acad. Sci. USA 83*, 77–81.
32. Rotwein, P., Folz, R. J. & Gordon, J. I. (1987) *J. Biol. Chem. 262*, 11807–11812.
33. Jansen, M., van Schaik, F. M. A., Ricker, A. T., Bullock, B., Woods, D. E., Gabbay, K. H., Nussbaum, A. L., Sussenbach, J. S. & Van den Brande, J. L. (1983) *Nature 306*, 609–611.
34. Le Bouc, Y., Dreyer, D., Jaeger, F., Binoux, M. & Sondermeyer, P. (1986) *FEBS Lett. 196*, 108–112.
35. Bell, G. I., Stempien, M. M., Fong, N. M. & Rall, L. B. (1986) *Nucleic Acids Res. 14*, 7873–7882.
36. Jansen, M., van Schaik, F. M. A, von Tol, H., Van den Brande, J. L. & Sussenbach, J. S. (1985) *FEBS Lett. 179*, 243–246.
37. Bell, G. I., Merryweather, J. P., Sanchez-Pescador, R., Stempien, M. M., Priestley, L., Scott, J. & Rall, L. B. (1984) *Nature 310*, 775–777.
38. Dull, T. J., Gray, A., Hayflick, J. S. & Ullrich, A. (1984) *Nature 310*, 777–781.
39. Gray, A., Tam, A. W., Dull, T. J., Hayflick, J., Pintar, J., Cavenee, W. K., Koufos, A. & Ullrich, A. (1987) *DNA 6*, 283–295.
40. Irminger, J. C., Rosen, K. M., Humbel, R. E. & Villa-Komaroff, L. (1987) *Proc. Natl Acad Sci. USA 84*, 6330–6334.
41. de Pagter-Holthuizen, P., Jansen, M., van Schaik, F. M. A., van der Kammen, R., Oosterwijk, C., Van den Brande, J. L. & Sussenbach, J. S. (1987) *FEBS Lett. 214*, 259–264.
42. de Pagter-Holthuizen, P., Jansen, M., van der Kammen, R. A., van Schaik, F. M. A. & Sussenbach, J. S. (1988) *Biochim. Biophys. Acta 950*, 282–295.
43. Soares, M. B., Ishii, D. N. & Efstratiadis, A. (1985) *Nucleic Acids Res. 13*, 1119–1134.
44. Soares, M. B., Turken, A., Ishii, D., Mills, L., Episkopou, V., Cottier, S., Zeitlin, S. & Efstratiadis, A. (1986) *J. Mol. Biol. 192*, 737–752.
45. Brown, A. L., Graham, D. E., Nissley, S. P., Hill, D. J., Strain, A. J. & Rechler, M. M. (1986) *J. Biol. Chem. 261*, 13144–13150.
46. Rotwein, P., Pollock, K. M., Watson, M. & Milbrandt, J. D. (1987) *Endocrinology 121*, 2141–2144.
47. Schofield, P. N. & Tate, V. E. (1987) *Development 101*, 793–803.
48. Baxter, R. C. & Brown, A. S. (1982) *Clin. Chem. 28*, 485–487.
49. Zumstein, P. P. & Humbel, R. E. (1985) *Methods Enzymol. 109*, 782–798.
50. Chernausek, S. D., Chatelain, P. G., Svoboda, M. E., Underwood, L. E. & Van Wyk, J. J. (1985) *Biochem. Biophys. Res. Commun. 126*, 282–288.
51. Petrides, P. E., Hintz, R. L., Bohlen, P. & Shively, J. E. (1986) *Endocrinology 118*, 2034–2038.
52. Povoa, G., Wennberg, G. & Hall, K. (1986) *Biochem. Biophys. Res. Commun. 136*, 253–259.
53. Greenstein, L. A., Gaynes, L. A., Romanus, J. A., Lee, L., Rechler, M. M. & Nissley, S. P. (1987) *Methods Enzymol. 146*, 259–269.
54. Tally, M., Florell, K. & Enberg, G. (1988) *Biosci. Rep. 8*, 293–297.
55. Hey, A. W., Browne, C. A., Simpson, R. J. & Thorburn, G. D. (1989) *Biochim. Biophys. Acta 997*, 27–35.
56. Honegger, A. & Humbel, R. E. (1986) *J. Biol. Chem. 261*, 569–575.
57. Francis, G. L., Upton, F. M., Ballard, F. J. & McNeil, K. A. (1988) *Biochem. J. 251*, 95–103.
58. Francis, G. L., McNeil, K. A., Wallace, J. D., Ballard, F. J. & Ownes, Ph. C. (1989) *Endocrinology 124*, 1173–1183.
59. Francis, G. L., Owens, P. C., McNeil, K. A., Wallace, J. C. & Ballard, F. J. (1989) *J. Endocrinol. 122*, 681–687.
60. Tamura, K., Kobayashi, M., Ishii, Y., Tamura, T., Hashimoto, K., Nakamura, S., Niwa, M. & Zapf, J. (1989) *J. Biol. Chem. 264*, 5616–5621.
61. Marquardt, H., Todaro, G. J., Henderson, L. E. & Oroszlan, S. (1981) *J. Biol. Chem. 256*, 6859–6865.
62. Tavakkol, A., Simmen, F. A. & Simmen, R. C. (1988) *Mol. Endocrinol. 2*, 674–681.
63. Shimatsu, A. & Rotwein, P. (1987) *J. Biol. Chem. 262*, 7894–7900.
64. Stempien, M. M., Fong, N. M., Rall, L. B. & Bell, G. I. (1986) *DNA 5*, 357–361.
65. Dawe, S. R., Francis, G. L., McNamara, P. J., Wallace, J. C. & Ballard, F. J. (1988) *J. Endocrinol. 117*, 173–181.
66. Raschdorf, F., Dahinden, R., Maerki, W., Richter, W. J. & Merryweather, J. P. (1988) *Biomed. Mass. Spectrom. 16*, 3–8.
67. Smith, M. C., Cook, J. A., Furman, Th. C. & Occolowitz, J. L. (1989) *J. Biol. Chem. 264*, 9314–9321.
68. Sara, V. R., Carlsson-Skwirut, C., Andersson, C., Hall, E., Sjögren, B., Holmgren, A. & Jörnvall, H. (1986) *Proc. Natl Acad. Sci. USA 83*, 4904–4907.
69. Carlsson-Skwirut, C., Jörnvall, H., Holmgren, A., Andersson, C., Bergman, T., Lundquist, G., Sjögren, B. & Sara, V. R. (1986) *FEBS Lett. 201*, 46–50.
70. Ogasawara, M., Karey, K. P., Marquardt, H. & Sirbasku, D. A. (1989) *Biochemistry 28*, 2710–2721.
71. Carlsson-Skwirut, C., Lake, M., Hartmanis, M., Hall, K. & Sara, V. R. (1989) *Biochim. Biophys. Acta 1011*, 192–197.
72. Bagley, Ch. J., May, B. L., Szabo, L., McNamara, P. J., Ross, M., Francis, G. L., Ballard, F. J. & Wallace, J. C. (1989) *Biochem. J. 259*, 665–671.
73. Ballard, F. J., Francis, G. L. Ross, M., Bagley, Ch. J., May, B. & Wallace, J. C. (1987) *Biochem. Biophys. Res. Commun. 149*, 398–404.
74. Cascieri, M. A., Hayes, N. S. & Bayne, M. L. (1989) *J. Cell. Physiol. 139*, 181–188.
75. Szabo, L., Mottershead, D. G., Ballard, F. J. & Wallace, J. C. (1988) *Biochem. Biophys. Res. Commun. 151*, 207–214.
76. Ross, M., Francis, G. L., Szabo, L., Wallace, J. C. & Ballard, F. J. (1989) *Biochem. J. 258*, 267–272.
77. Haselbacher, G. K. & Humbel, R. E. (1982) *Endocrinology 110*, 1822–1824.
78. Haselbacher, G. K., Schwab, M. E., Pasi, A. & Humbel, R. E. (1985) *Proc. Natl Acad. Sci. USA 82*, 2153–2157.

79. Gowan, L. K., Hampton, B., Hill, D. J., Schlueter, R. J. & Perdue, J. F. (1987) *Endocrinology 121*, 449−458.
80. Zumstein, P. P., Lüthi, C. & Humbel, R. E. (1985) *Proc. Natl Acad. Sci. USA 82*, 3169−3172.
81. Blum, W. F., Ranke, M. B., Lechner, B. & Bierich, R. B. (1987) *Acta Endocrinol. 116*, 445−451.
82. Blundell, T. L., Bedarkar, S. & Humbel, R. E. (1983) *Fed. Proc. 42*, 2592−2597.
83. Dafgård, E., Bajaj, M., Honegger, A. M., Pitts, J., Wood, S. & Blundell, T. (1985) *J. Cell. Sci. (Suppl.) 3*, 53−64.
84. King, G. L., Kahn, C. R., Samuels, B., Danho, W., Bullesbach, E. E. & Gattner, H. G. (1982) *J. Biol. Chem. 257*, 10869−10873.
85. De Vroede, M. A., Rechler, M. M., Nissley, S. P. Joshi, S., Burke, G. Th. & Katsoyannis, P. G. (1985) *Proc. Natl Acad. Sci. USA 82*, 3010−3014.
86. Tseng, L. Y.-H., Schwartz, G. P., Sheikh, M., Chen, Z. Z., Joshi, S., Wang, J.-F., Nissley, S. P., Burke, G. Th., Katsoyannis, P. G. & Rechler, M. M. (1987) *Biochem. Biophys. Res. Commun. 149*, 672−679.
87. Chen, Z. Z., Schwartz, G. P., Zong, L., Burke, G. Th, Chanley, J. D. & Katsoyannis, P. G. (1988) *Biochemistry 27*, 6105−6111.
88. Joshi, S., Burke, G. T. & Katsoyannis, P. G. (1985) *Biochemistry 24*, 4208−4212.
89. Cara, J. F., Nakagawa, S. H. & Tager, H. S. (1988) *Endocrinology 122*, 2881−2887.
90. Tager, H., Given, B., Baldwin, D., Mako, M., Markese, J., Rubenstein, A., Olefsky, J., Kobayashi, M., Kolterman, O. & Poucher, R. (1979) *Nature 281*, 122−125.
91. Maly, P. & Lüthi, Ch. (1988) *J. Biol. Chem. 263*, 7068−7072.
92. Cascieri, M. A., Chicchi, G. G., Applebaum, J., Hayes, N. S., Green, B. G. & Bayne, M. L. (1988) *Biochemistry 27*, 3229−3233.
93. Cascieri, M. A., Chicchi, G. G., Applebaum, J., Green, B. G., Hayes, N. S. & Bayne, M. L. (1989) *J. Biol. Chem. 264*, 2199−2202.
94. Bayne, M. L., Applebaum, J., Chicchi, G. G., Hayes, N. S., Green, B. G. & Cascieri, M. A. (1988) *J. Biol. Chem. 263*, 6233−6239.
95. D'Ercole, A. J., Applewhite, G. T. & Underwood, L. E. (1980) *Dev. Biol. 75*, 315−328.
96. D'Ercole, A. J., Stiles, A. D. & Underwood, L. E. (1984) *Proc. Natl Acad. Sci. USA 81*, 935−939.
97. Adams, S. O., Nissley, S. P., Handwerger, S. & Rechler, M. M. (1983) *Nature 302*, 150−153.
98. Daughaday, W. W., Yanow, C. E. & Kapadia, M. (1986) *Endocrinology 119*, 490−494.
99. Ashton, I. K., Zapf, J., Einschenk, I. & MacKenzie, I. Z. (1985) *Acta Endocrinol. 110*, 558−563.
100. Han, V. K. M., D'Ercole, A. J. & Lund, P. K. (1987) *Science 236*, 193−197.
101. Han, V. K. M., Lund, P. K., Lee, D. C. & D'Ercole, A. J. (1988) *J. Clin. Endocrinol. Metab. 66*, 422−429.
102. Haselbacher, G. K., Irminger, J.-C., Zapf, J., Ziegler, W. H. & Humbel, R. E. (1987) *Proc. Natl Acad. Sci. USA 84*, 1104−1106.
103. Reeve, A. E., Eccles, M. R., Wilkins, R. J., Bell, G. I. & Millow, L. J. (1985) *Nature 317*, 258−262.
104. Irminger, J. C., Schoenle, E. J., Briner, J. & Humbel, R. E. (1989) *Eur. J. Pediatr. 148*, 620−623.
105. Little, M. H., Ablett, G. & Smith, P. J. (1987) *Carcinogenesis 8*, 865−868.
106. El-Badry, O. M., Romanus, J. A. Helman, L. J., Cooper, M. J., Rechler, M. M. & Israel, M. A. (1989) *J. Clin. Invest. 84*, 829−839.
107. Backstrom, M., Hall, K. & Sara, V. R. (1984) *Acta Endocrinol. 197*, 171−178.
108. Murphy, L. J., Bell, G. I & Friesen, H. G. (1987) *Endocrinology 120*, 1279−1282.
109. Stylianopoulou, F., Herbert, J., Soares, M. B. & Efstratiadis, A. (1988) *Proc. Natl Acad. Sci. USA 85*, 141−145.
110. Hynes, M. A., Brooks, P. J., Van Wyk, J. J. & Lund, P. K. (1988) *Mol. Endocrinol. 2*, 47−54.
111. Rotwein, P., Burgess, S. K., Milbrandt, J. D. & Krause, J. E. (1988) *Proc. Natl Acad. Sci. USA 85*, 265−269.
112. Lesniak, M. A., Hill, J. M. Kiess, W., Rojeski, M., Pert, C. B. & Roth, J. (1988) *Endocrinology 123*, 2089−2099.
113. Laron, Z. & Galatzer, A. (1985) *Brain Dev. 7*, 559−567.
114. Schoenle, E. J., Haselbacher, G. K., Briner, J., Janzer, R. C., Gammeltoft, S., Humbel, R. E. & Prader, A. (1986) *J. Pediatr. 108*, 737−740.
115. Sara, V. R. & Carlsson-Skwirut, C. (1988) *Progr. Brain Res. 73*, 87−99.
116. Furlanetto, R. W. & Marino, J. M. (1987) *Methods Enzymol. 146*, 216−226.
117. Daughaday, W. H. (1987) *Methods Enzymol. 146*, 248−259.
118. Zapf, J., Walter, H. & Froesch, E. R. (1981) *J. Clin. Invest. 68*, 1321−1330.
119. Kao, P. C., Tateishi, K., Abboud, C. F., Zimmerman, D., Randall, R. V. & Li, C. H. (1988) *Ann. Clin. Lab. Sci. 18*, 120−130.
120. Furlanetto, R. W., Underwood, L., Van Wyk, J. J. & D'Ercole, A. J. (1977) *J. Clin. Invest. 60*, 648−657.
121. Clemmons, D. R., Van Wyk, J. J., Ridgway, E. C., Kliman, B., Kjellberg, R. N. & Underwood, L. E. (1979) *N. Engl. J. Med. 301*, 1138−1142.
122. Rosenfeld, R. G., Wilson, D. M., Lee, P. D. K. & Hintz, R. L. (1986) *J. Pediatr. 109*, 428−433.
123. Moses, A. C., Nissley, S. P., Short, P. A., Rechler, M. M., White, R. M., Knight, A. B. & Higa, O. Z. (1980) *Proc. Natl Acad. Sci. USA 77*, 3649−3653.
124. Gluckman, P. D. & Butler, J. H. (1983) *J. Endocrinol. 99*, 223−232.
125. Daughaday, W. H., Yanow, C. E. & Kapadia, M. (1986) *Endocrinology 119*, 490−494.
126. Bennett, A., Wilson, D. M., Liu, F., Nagashima, R., Rosenfeld, R. G. & Hintz, R. L. (1983) *J. Clin. Endocrinol. Metab. 57* 609−611.
127. Ashton, I. K., Zapf, J., Einschenk, I. & MacKenzie, I. Z. (1985) *Acta Endocrinol. 110*, 558−563.
128. Luna, A. M., Wilson, D. M., Wibbelsman, D. M., Brown, R. C., Nagashima, R. J. Hintz , R. L. & Rosenfeld, R. G. (1983) *J. Clin. Endocrinol. Metab. 57*, 268−271.
129. Silbergeld, A., Litwin, A., Bruchis, S., Varsana, I. & Laron, Z. (1986) *Clin. Endocrinol. (Oxf) 25*, 67−74.
130. Mauras, N. E., Blizzard, R., Link, K., Johnson, M. L., Rogol, A. D. & Veldhuis, J. D. (1987) *J. Clin. Endocrinol. Metab. 64*, 596−601.
131. Harris, D. A., Van Vliet, G., Egli, C. A., Grumbach, M. M., Kaplan, S. L., Styne, D. M. & Vainsel, M. (1985) *J. Clin. Endocrinol. Metab. 61*, 152−159.
132. Kemp, S. F., Rosenfeld, R. G., Liu, F., Gaspich, S. & Hintz, R. L. (1981) *J. Clin. Endocrinol. Metab. 52*, 616−621.
133. Kaufmann, U., Zapf, J. & Froesch, E. R. (1978) *Acta Endocrinol. 87*, 716−727.
134. Schwander, J., Hauri, C., Zapf, J. & Froesch, E. R. (1983) *Endocrinology 113*, 297−305.
135. Scott, C. D., Martin, J. L. & Baxter, R. C. (1985) *Endocrinology 116*, 1094−1101.
136. Hynes, M. A., Van Wyk, J. J., Brooks, P. J., D'Ercole, A. J., Jansen, M. & Lund, P. K. (1987) *Mol. Endocrinol. 1*, 233−242.
137. Miller, K. F., Bolt, D. J., Pursel, V. G., Hammer, R. E., Pinkert, C. A., Palmiter, R. D. & Brinster, R. L. (1989) *J. Endocrinol. 120*, 481−488.
138. Brazeau, P., Guillemin, R., Lang, N., Van Wyk, J. & Humbel, R. E. (1982) *C. R. Acad. Sci. Paris 295*, 651−654.
139. Abe, H., Molitch, M. E., Van Wyk, J. J. & Underwood, L. E. (1983) *Endocrinology 113*, 1319−1324.
140. Yamashita, S. & Melmed, S. (1987) *J. Clin. Invest. 79*, 449−452.
141. Namba, H., Morita, S. & Melmed, S. (1989) *Endocrinology 124*, 1794−1799.

460

142. Wilson, D. M., Perkins, S. N., Thomas, J. A., Seelig, S., Berry, S. A., Hamm, T. E., Hoffman, A. R., Hintz, R. L. & Rosenfeld, R. G. (1989) *Metabolism 38*, 57–62.

143. Guler, H. P., Schmid, C., Zapf, J. & Froesch, E. R. (1989) *Proc. Natl Acad. Sci. USA 86*, 2868–2872.

144. Merimee, T. J., Zapf, J. & Froesch, E. R. (1981) *N. Engl. J. Med. 305*, 965–968.

145. Merimee, T. J., Zapf, J., Hewlett, B. & Cavalli-Sforza, L. L. (1987) *N. Engl. J. Med. 316*, 906–911.

146. Laron, Z. (1974) *Isr. J. Med. Sci. 10*, 1247–1253.

147. Zapf, J., Morell, B., Walter, H., Laron, Z. & Froesch, E. R. (1980) *Acta Endocrinol. 95*, 505–517.

148. Eshet, R., Laron, Z., Pertzelan, A., Arnon, R. & Dintzman, M. (1984) *Isr. J. Med. Sci. 20*, 8–11.

149. Daughaday, W. H. & Trivedi, B. (1987) *Proc. Natl Acad. Sci USA 84*, 4636–4640.

150. Godowski, P. J., Leung, D. W., Meacham, L. R., Galgani, J. P., Hellmiss, R., Keret, R., Rotwein, P. S., Parks, J. S., Laron, Z. & Wood, W. I. (1989) *Proc. Natl Acad. Sci. USA 86*, 8083–8087.

151. Geffner, M. E., Golde, D. W., Lippe, B. M., Kaplan, S. A., Bersch, N. & Li, C. H. (1987) *J. Clin. Endocrinol. Metab. 64*, 1042–1046.

152. Zapf, J. & Froesch, E. R. (1986) *Hormone Res. 24*, 160–165.

153. Baxter, R. C. (1988) *Comp. Biochem. Physiol. 91B*, 229–235.

154. Zapf, J., Schoenle, E., Jagers, E., Sand, I. & Froesch, E. R. (1979) *J. Clin. Invest. 63*, 1077–1084.

155. Zapf, J., Hauri, C., Waldvogel, M. & Froesch, E. R. (1986) *J. Clin. Invest. 77*, 1768–1775.

156. Clemmons, D. R., Elgin, R. G., Han, V. K., Casella, S. J., D'Ercole, A. J. & Van Wyk, J. J. (1986) *J. Clin. Invest. 77*, 1548–1556.

157. Elgin, R. G., Busby, W. H. & Clemmons, D. R. (1987) *Proc. Natl Acad. Sci. USA 84*, 3254–3258.

158. Blum, W. F., Jenne, E. W., Reppin, F., Kietzmann, K., Ranke, M. B. & Bierich, J. R. (1989) *Endocrinology 125*, 766–772.

159. Zapf, J., Waldvogel, M. & Froesch, E. R. (1975) *Arch. Biochem. Biophys. 168*, 638–645.

160. Moses, A. C., Nissley, S. P., Choen, K. L. & Rechler, M. M. (1976) *Nature 263*, 137–140.

161. Hintz, R. L., Liu, F., Rosenfeld, R. G. & Kemp, S. F. (1981) *J. Clin. Endocrinol. Metab. 53*, 100–104.

162. Hardouin, S., Hossenlopp, P., Segovia, B., Seurin, D., Portolan, G., Lassarre, C. & Binoux, M. (1987) *Eur. J. Biochem. 170*, 121–132.

163. Hossenlopp, P., Seurin, D., Segovia, B., Portolan, G. & Binoux, M. (1987) *Eur. J. Biochem. 170*, 133–142.

164. Ballard, J., Baxter, R., Binoux, M., Clemmons, D., Drop, S., Hall, K., Hintz, R., Rechler, M., Rutanen, E. & Schwander, J. (1989) *Acta Endocrinol. 121*, 751–752.

165. Brinkman, A., Groffen, C., Kortleve, D. J., Geurts van Kessel, A. & Drop, S. L. S. (1988) *EMBO J. 7*, 2417–2423.

166. Lee, Y.-L., Hintz, R. L., James, P. M., Lee, P. D. K., Shively, J. E. & Powell, D. R. (1988) *Mol. Endocrinol. 2*, 404–411.

167. Brewer, M. T., Stetler, G. L., Squires, C. H., Thompson, R. C., Busby, W. H. & Clemmons, D. R. (1988) *Biochem. Biophys. Res. Commun. 152*, 1289–1297.

168. Julkunen, M., Koistinen, R., Aalto-Setala, K., Seppälä, M., Janne, O. A. & Kontula, K. (1988) *FEBS Lett. 236*, 295–302.

169. Koistinen, R., Kalkkinen, N., Huhtala, M.-L., Seppälä, M., Bohn H. & Rutanen, E.-M. (1986) *Endocrinology 118*, 1375–1378.

170. Mottola, C., MacDonald, R. G., Brackett, J. L., Mole, J. E., Anderson, J. K. & Czech, M. P. (1986) *J. Biol. Chem. 261*, 11180–11188.

171. Brown, A. L., Chiarotti, L., Orlowsky, C. C., Mehlman, T., Burgess, W. H., Ackerman, E. J., Bruni, C. B. & Rechler, M. M. (1989) *J. Biol. Chem. 264*, 5148–5154.

172. Binkert, C., Landwehr, J., Mary, J.-L., Schwander, J. & Heinrich, G. (1989) *EMBO J. 8*, 2497–2502.

173. Baxter, R. C. & Martin, J. L. (1989) *Proc. Natl Acad. Sci. USA 86*, 6898–6902.

174. Wood, W. I., Cachianes, G., Henzel, W. J., Winslow, G. A., Spencer, S. A., Hellmiss, R., Martin, J. L. & Baxter, R. C. (1989) *Mol. Endocrinol. 2*, 1176–1185.

175. Zapf, J., Hauri, C., Waldvogel, M., Futo, E., Häsler, H., Binz, K., Guler, H. P., Schmid, C. & Froesch, E. R. (1989) *Proc. Natl Acad. Sci. USA 86*, 3813–3817.

176. Baxter, R. C., Martin, J. L. & Wood, M. H. (1987) *J. Clin. Endocrinol. Metab. 65*, 423–431.

177. Mohan, S., Bautista, C. M., Wergedal, J. & Baylink, D. J. (1989) *Proc. Natl Acad. Sci. USA 86*, 8338–8342.

178. Roghani, M. R., Hossenlopp, P., Lepage, P., Balland, A. & Binoux, M. (1989) *FEBS Lett. 255*, 253–258.

179. White, R. M., Nissley, S. P. Short, P. A., Rechler, M. M. & Fennoy, I. (1982) *J. Clin. Invest. 69*, 1239–1252.

180. Kiess, W., Greenstein, L. A., White, R. M., Lee, L., Rechler, M. M. & Nissley, S. P. (1987) *Proc. Natl Acad. Sci. USA 84*, 7720–7724.

181. Causin, C., Waheed, A., Braulke, T., Junghans, U., Maly, P., Humbel, R. E. & von Figura, K. (1988) *Biochem. J. 252*, 795–799.

182. MacDonald, R. G., Tepper, M. A., Clairmont, K. B., Perregaux, S. B. & Czech, M. P. (1989) *J. Biol. Chem. 264*, 3256–3261.

183. Megyesi, K., Kahn, C. R., Roth, J., Froesch, E. R., Humbel, R. E., Zapf, J. & Neville, D. M. Jr (1974) *Biochem. Biophys. Res. Commun. 57*, 307–315.

184. Kasuga, M., Van Obberghen, E., Nissley, S. P. & Rechler, M. M. (1981) *J. Biol. Chem. 256*, 5305–5308.

185. Chernausek, S. D., Jacobs, S. & Van Wyk, J. J. (1981) *Biochemistry 20*, 7345–7350.

186. Massague, J. & Czech, M. P. (1982) *J. Biol. Chem. 257*, 5038–5045.

187. Kasuga, M., Karlsson, F. A. & Kahn, C. R. (1982) *Science 215*, 185–186.

188. Jacobs, S., Kull, F. C. Jr, Earp, H. S., Svoboda, M. E., Van Wyk, J. J. & Cuatrecasas, P. (1983) *J. Biol. Chem. 258*, 9581–9584.

189. Rubin, J. B., Shia, M. A. & Pilch, P. F. (1983) *Nature 305*, 438–440.

190. Gammeltoft, S., Haselbacher, G. K., Humbel, R. E., Fehlmann, M. & Van Obberghen, E. (1985) *EMBO J. 4*, 3407–3412.

191. Rosenfeld, R. G., Pham, H., Keller, B. T., Borchardt, R. T. & Pardridge, W. M. (1987) *Biochem. Biophys. Res. Commun. 149*, 159–166.

192. Corvera, S., Whitehead, R. E., Mottola, C. & Czech, M. P. (1986) *J. Biol. Chem. 261*, 7675–7679.

193. Ullrich, A., Bell, J. R., Chen, E. J., Herrera, R., Petruzzelli, L. M., Dull, T. J., Gray, A., Coussens, L., Liao, Y.-C., Tsubokawa, M., Mason, A., Seeburg, P. H., Grunfeld, C., Rosen, O. M. & Ramachandran, J. (1985) *Nature 313*, 756–761.

194. Ebina, Y., Ellis, L., Jarnagin, K., Edery, M., Graf, L., Clauser, E., Ou, J.-H., Masiarz, F., Kan, Y. W., Goldfine, I. D., Roth, R. A & Rutter, W. J. (1985) *Cell 40*, 747–758.

195. Ullrich, A., Gray, A., Tam, A. W., Yang-Feng, T., Tsubokawa, M., Collins, C., Henzel, W., Le Bon, T., Kathuria, S., Chen, E., Jacobs, S., Francke, U., Ramachandran, J. & Fujita-Yamaguchi, Y. (1986) *EMBO J. 5*, 2503–2512.

196. Jacobs, S., Kull, F. C. & Cuatrecasas, P. (1983) *Proc. Natl Acad. Sci. USA 80*, 1228–1231.

197. Morgan, D. O., Edman, J. C., Standring, D. N., Fried, V. A., Smith, M. C., Roth, R. A. & Rutter, W. J. (1987) *Nature 329*, 301–307.

198. Oshima, A., Nolan, C. M., Kyle, J. W., Grubb, J. H. & Sly, W. S. (1988) *J. Biol. Chem. 263*, 2553–2562.

199. Tong, P. Y., Tollefsen, S. E. & Kornfeld, S. (1988) *J. Biol. Chem. 263*, 2585–2588.

200. Kiess, W., Blickenstaff, G. D., Sklar, M. M., Thomas, C. L., Nissley, S. P. & Sahagian, G. G. (1988) *J. Biol. Chem. 263*, 9339–9344.

201. Roth, R. A., Stover, C., Hari, J., Morgan D. O., Smith, M. C., Sara, V. & Fried, V. A. (1987) *Biochem. Biophys. Res. Commun. 149*, 600–606.

202. Braulke, T., Causin, C., Waheed, A., Junghans, U., Hasilik, A., Maly, P., Humbel, R. E. & von Figura, K. (1988) *Biochem. Biophys. Res. Commun. 150*, 1287–1293.
203. Yang-Feng, T. L., Francke, U. & Ullrich, A. (1985) *Science 228*, 728–731.
204. Laureys, G., Barton, D. E., Ullrich, A. & Francke, U. (1988) *Genomics 3*, 224–229.
205. Pohlmann, R., Nagel, G., Schmidt, G., Stein, M., Lorkowski, G., Krentler, C., Cully, J., Meyer, H. E., Grzeschik, K.-H., Mersmann, G., Hasilik, A. & von Figura K. (1987) *Proc. Natl Acad. Sci. USA 84*, 5575–5579.
206. Stuart, C. A. (1988) *J. Biol. Chem. 263*, 7881–7886.
207. MacDonald, R. G., Pfeffer, S. R., Coussens, L., Tepper, M. A., Brocklebank, C. M., Mole, J. E., Anderson, J. K., Chen, E., Czech, M. P. & Ullrich, A. (1988) *Science 239*, 1134–1137.
208. Waheed, A., Braulke, T., Junghans, U. & von Figura, K. (1988) *Biochem. Biophys. Res. Commun. 152*, 1248–1254.
209. Kiess, W., Thomas, C. L., Greenstein, L. A., Lee, L., Sklar, M. M., Rechler, M. M., Sahagian, G. G. & Nissley, S. P. (1989) *J. Biol. Chem. 264*, 4710–4714.
210. Polychronakos, C., Guyda, H. J. & Posner, B. I. (1988) *Biochem. Biophys. Res. Commun. 157*, 632–638.
211. Purchio, A. F., Cooper, J. A., Brunner, A. M., Lioubin, M. N., Gentry, L. E., Kovacina, K. S., Roth, R. A. & Marquardt, H. (1988) *J. Biol. Chem. 263*, 14211–14215.
212. Canfield, W. M. & Kornfeld, S. (1989) *J. Biol. Chem. 264*, 7100–7103.
213. Clairmont, K. B. & Czech, M. P. (1989) *J. Biol. Chem. 264*, 16390–16392.
214. Steele-Perkins, G. Turner, J., Edman, J. C., Hari, J., Pierce, S. B., Stover, C., Rutter, W. J. & Roth, R. A. (1988) *J. Biol. Chem. 263*, 11486–11492.
215. King, G. L., Kahn, C. R., Rechler, M. M. & Nissley, S. P. (1980) *J. Clin. Invest. 66*, 130–140.
216. Kull, F. C., Jacobs, S., Su, Y.-F., Svoboda, M. E., Van Wyk, J. J. & Cuatrecasa, P. (1983) *J. Biol. Chem. 258*, 6561–6566.
217. Conover, C. A., Misra, P., Hintz, R. L. & Rosenfeld, R. G. (1986) *Biochem. Biophys. Res. Commun. 139*, 501–508.
218. Flier, J. S., Usher, P. & Moses, A. C. (1986) *Proc. Natl Acad. Sci. USA 83*, 664–668.
219. Chaiken, R. L., Moses, A. C., Usher, P. & Flier, J. S. (1986) *J. Clin. Endocrinol. Metab. 63*, 1181–1185.
220. Conover, A., Hintz, R. L. & Rosenfeld, R. G. (1989) *Horm. Metab. Res. 21*, 59–63.
221. Kadowaki, T., Koyasu, S., Nishida, E., Sakai, H., Takaku, F., Yahara, I. & Kasuga, M. (1986) *J. Biol. Chem. 261*, 16141–16147.
222. Sasaoka, T., Kobayashi, M., Takata, Y., Ishibashi, O., Iwasaki, M., Shigeta, Y., Goji, K. & Hisatomi, A. (1988) *Diabetes 37*, 1515–1523.
223. Verspohl, E. J., Maddux, B. A. & Goldfine, I. D. (1988) *J. Clin. Endocrinol. Metab. 67*, 169–174.
224. Roth, R. A., Steele-Perkins, G., Hari, J., Stover, C., Pierce, S., Turner, J., Edman, J. C. & Rutter, W. J. (1988) *Cold Spring Harbor Symp. Quant. Biol. 53*, 537–543.
225. Hofmann, C., Goldfine, I. D. & Whittaker, J. (1989) *J. Biol. Chem. 264*, 8606–8611.
226. Lammers, R., Gray, A., Schlessinger, J. & Ullrich, A. (1989) *EMBO J. 8*, 1369–1375.
227. Morgan, D. O., Jarnagin, K. & Roth, R. A. (1986) *Biochemistry 25*, 5560–5564.
228. Jonas, H. A., Newman, J. D. & Harrison, L. C. (1986) *Proc. Natl Acad. Sci. USA 83*, 4124–4128.
229. Jonas, H. A., Cox, A. J. & Harrison, L. C. (1989) *Biochem. J. 257*, 101–107.
230. Morgan, D. O. & Roth, R. A. (1986) *Biochem. Biophys. Res. Commun. 138*, 1341–1347.
231. Moxham, C. P., Duronio, V. & Jacobs, S. (1989) *J. Biol. Chem. 264*, 13238–13244.
232. Alexandrides, T. K. & Smith, R. J. (1989) *J. Biol. Chem. 264*, 12922–12930.
233. Shier, P. & Watt, V. M. (1989) *J. Biol. Chem. 264*, 14605–14608.
234. Casella, S. J., Han, V. K., D'Ercole, J., Svoboda, M. E. & Van Wyk, J. J. (1986) *J. Biol. Chem. 261*, 9268–9273.
235. Chou, C. K., Dull, T. J., Russell, D. S., Gherzi, R., Lebwohl, D., Ullrich, A. & Rosen, O. M. (1987) *J. Biol. Chem. 262*, 1842–1847.
236. Ikari, N., Yoshino, H., Moses, A. C. & Flier, J. S. (1988) *Mol. Endocrinol. 2*, 831–837.
237. Beguinot, F., Smith, R. J., Kahn, C. R., Maron, R., Moses, A. C. & White, M. F. (1988) *Biochemistry 27*, 3222–3228.
238. Taub, R., Roy, A., Dieter, R. & Koontz, J. (1987) *J. Biol. Chem. 262*, 10893–10897.
239. Stumpo, D. J., Stewart, T. N., Gilman, M. Z. & Blackshear, P. J. (1988) *J. Biol. Chem. 263*, 1611–1614.
240. Burgering, B. M. T., Snijders, A. J., Maassen, J. A., van der Eb, A. J. & Bos, J. L. (1989) *Mol. Cell. Biol. 9*, 4312–4322.
241. Mottola, C. & Czech, M. P. (1984) *J. Biol. Chem. 259*, 12705–12713.
242. Oka, Y. & Czech, M. P. (1986) *J. Biol. Chem. 261*, 9090–9093.
243. Kiess, W., Haskell, J. F., Lee, L., Greenstein, L. A., Miller, B. E., Arons, A. L., Rechler, M. M. & Nissley, S. P. (1987) *J. Biol. Chem. 262*, 12745–12751.
244. Zapf, J., Schoenle, E. & Froesch, E. R. (1978) *Eur. J. Biochem. 87*, 285–296.
245. Oppenheimer, C. L., Pessin, J. E., Massague, J., Gitomer, W. & Czech, M. P. (1983) *J. Biol. Chem. 258*, 4824–4830.
246. Appell, K. C., Simpson, I. A. & Cushman, S. W. (1988) *J. Biol. Chem. 263*, 10824–10829.
247. Braulke, T., Tippmer, S., Neher, E. & von Figura, K. (1989) *EMBO J. 8*, 681–686.
248. Corvera, S., Roach, P. J., De Paoli-Roach, A. A. & Czech, M. P. (1988) *J. Biol. Chem. 263*, 3116–3122.
249. Nishimoto, I., Murayama, Y., Katada, T., Ui, M. & Ogata, E. (1989) *J. Biol. Chem. 264*, 14029–14038.
250. Kovacina, K. S., Steele-Perkins, G. & Roth, R. A. (1989) *Mol. Endocrinol. 3*, 901–906.
251. Isaksson, O. G. P., Jansson, J. O. & Gause, I. A. M. (1982) *Science 216*, 1237–1239.
252. Schlechter, N. L., Russell, S. M., Spencer, E. M. & Nicoll, C. S. (1986) *Proc. Natl Acad. Sci. USA 83*, 2932–2934.
253. Isgaard, J., Möller, C. & Isaksson, O. G. P. (1988) *Endocrinology 122*, 1515–1520.
254. Ernst, M. & Froesch, E. R. (1988) *Biochem. Biophys. Res. Commun. 151*, 142–147.
255. Cook, J. J., Haynes, K. M. & Werther, G. A. (1988) *J. Clin. Invest. 81*, 206–212.
256. Green, H., Morikawa, M. & Nixon, T. (1985) *Differentiation 29*, 195–198.
257. Guller, S., Sonenberg, M., Wu, K.-Y., Szabo, P. & Corin, R. E. (1989) *Endocrinology 125*, 2360–2367.
258. Doglio, A., Dani, C., Fredrikson, G., Grimaldi, P. & Ailhaud, G. (1987) *EMBO J. 6*, 4011–4016.
259. Smith, P. J., Wise, L. S., Berkowitz, R., Wan, C. & Rubin, C. S. (1988) *J. Biol. Chem. 263*, 9402–9408.
260. Pledger, W. J., Stiles, C. D., Antoniades, H. N. & Scher, C. D. (1977) *Proc. Natl Acad. Sci. USA 74*, 4481–4485.
261. Pledger, W. J., Stiles, C. D., Antoniades, H. N. & Scher, C. D. (1978) *Proc. Natl Acad. Sci. USA 75*, 2839–2843.
262. Leof, E. B., Wharton, W., Van Wyk, J. J. & Pledger, W. J. (1982) *Exp. Cell Res. 141*, 107–115.
263. Wharton, W. (1983) *J. Cell. Physiol. 117*, 423–429.
264. Campisi, J. & Pardee, A. B. (1984) *Mol. Cell. Biol. 4*, 1807–1814.
265. Olashaw, N. E., Van Wyk, J. J. & Pledger, W. J. (1987) *Am. J. Physiol. 253*, C575–579.
266. Zumstein, P. & Stiles, C. D. (1987) *J. Biol. Chem. 262*, 11252–11260.
267. Chen, Y. & Rabinovich, P. S. (1989) *J. Cell. Physiol. 140*, 59–67.
268. Clemmons, D. R., Underwood, L. E. & Van Wyk, J. J. (1981) *J. Clin. Invest. 67*, 10–19.
269. Heldin, C.-H., Wasteson, A. & Westermark, B. (1982) *J. Biol. Chem. 257*, 4216–4221.

462

270. Lynch, S. E., Colvin, R. B. & Antoniades, H. N. (1989) *J. Clin. Invest. 84*, 640−646.
271. Schmid, C., Steiner, T. & Froesch, E. R. (1983) *FEBS Lett. 161*, 117−121.
272. Schmid, C., Steiner, T. & Froesch, E. R. (1984) *FEBS Lett. 173*, 48−52.
273. Hock, J. M., Centrella, M. & Canalis, E. (1988) *Endocrinology 122*, 254−260.
274. Kurtz, A., Jelkmann, W. & Bauer, C. (1982) *FEBS Lett. 149*, 105−108.
275. Adashi, E. Y., Resnick, C. E., Svoboda, M. E. & Van Wyk, J. J. (1984) *Endocrinology 115*, 1227−1229.
276. Cara, J. F. & Rosenfield, R. L. (1988) *Endocrinology 123*, 733−739.
277. Sara, V. R., Hall, K., Rodeck, C. H. & Wetterberg, L. (1981) *Proc. Natl Acad. Sci. USA 78*, 3175−3179.
278. Sara, V. R., Hall, K., von Holtz, K., Humbel, R., Sjögren, B. & Wetterberg, L. (1982) *Neurosci. Lett. 34*, 39−44.
279. Binoux, M., Hossenlopp, P., Lassarre, C. & Hardouin, N. (1981) *FEBS Lett. 118*, 1835−1842.
280. McMorris, F. A., Smith, T. M., De Salvo, S. & Furlanetto, R. W. (1986) *Proc. Natl Acad. Sci. USA 83*, 822−826.
281. Aizenman, Y. & de Vellis, J. (1987) *Brain Res. 406*, 32−42.
282. Haselbacher, G., Groscurth, P., Otten, U., Vedder, H., Lutz, U., Sonderegger, P., Bulatko, A., Greeff, N. & Humbel, R. (1989) *J. Neurosci. Methods 30*, 121−131.
283. Bothwell, M. (1982) *J. Neurosci. Res. 8*, 225−231.
284. Recio-Pinto, E., Rechler, M. M. & Ishii, D. N. (1986) *J. Neurosci. 6*, 1211−1219.
285. DiCicco-Bloom, E. & Black, I. B. (1988) *Proc. Natl Acad. Sci. USA 85*, 4066−4070.
286. Nilsson, L., Sara, V. R. & Nordberg, A. (1988) *Neurosci. Lett. 88*, 221−226.
287. Dahmer, M. K. & Perlman, R. L. (1988) *J. Neurochem. 51*, 321−323.
288. Xue, Z. G., Le Douarin, N. M. & Smith, J. (1988) *Cell. Differ. Dev. 25*, 1−10.
289. Ishii, D. N. (1989) *Proc. Natl Acad. Sci. USA 86*, 2898−2902.
290. Kanje, M., Skottner, A., Sjoberg, J. & Lundborg, G. (1989) *Brain Res. 486*, 396−398.
291. Guler, H.-P., Zapf, J. & Froesch, E. R. (1987) *N. Engl. J. Med. 317*, 137−140.
292. Laron, Z., Erster, B., Klinger, B. & Anin, S. (1988) *Lancet II*, 1170−1172.
293. Schoenle, E., Zapf, J., Humbel, R. E. & Froesch, E. R. (1982) *Nature 296*, 252−253.
294. Hizuka, N., Takano, K., Shizume, K., Asakawa, K., Miyakawa, M. & Tanaka, I. (1986) *Eur. J. Pharmacol. 125*, 143−146.
295. Van Buul-Offers, S., Ueda, I. & Van den Brande, J. L. (1986) *Pediatr. Res. 20*, 825−827.
296. Schoenle, E., Zapf, J., Hauri, C., Steiner, T. & Froesch, E. R. (1985) *Acta Endocrinol. 108*, 167−174.
297. Skottner, A., Clark, R. G., Robinson, I. C. & Fryklund, L. (1987) *J. Endocrinol. 112*, 123−132.
298. Guler, H.-P., Zapf, J., Scheiwiller, E. & Froesch, E. R. (1988) *Proc. Natl Acad. Sci. USA 85*, 4889−4893.
299. Quaife, C. J., Mathews, L. S., Pinkert, C. A., Hammer, R. E., Brinster, R. L. & Palmiter, R. D. (1989) *Endocrinology 124*, 40−48.
300. Wilson, D. M., Thomas, J. A., Hamm, T. E. Jr, Wyche, J., Hintz, R. L. & Rosenfeld, R. G. (1987) *Endocrinology 120*, 1896−1901.
301. Shaar, C. J., Tinsley, F. C., Smith, M. C., Clemens. J. A. & Neubauer, B. L. (1989) *Endocrine Res. 15*, 403−411.
302. Liu, L., Greenberg, S., Russell, S. M. & Nicoll, C. S. (1989) *Endocrinology 124*, 3077−3082.
303. Kurtz, A., Zapf, J., Eckardt, K. U., Clemons, G., Froesch, E. R. & Bauer, C. (1988) *Proc. Natl Acad. Sci. USA 85*, 7825−7829.
304. Schalch, D., Reismann, D., Emler, C., Humbel, R., Li, C. H., Peters, M. & Lau, E. (1984) *Endocrinology 115*, 2490−2492.
305. Peters, M. A., Lau, E. P., Snitman, D. L., Van Wyk, J. J., Underwood, L. E., Russell, W. E. & Svoboda, M. E. (1985) *Gene 35*, 83−89.
306. Nilsson, B., Holmgren, E., Josephson, S., Gatenbeck, S., Philipson, L. & Uhlen, M. (1985) *Nucleic Acids Res. 13*, 1151−1162.
307. Buell, G., Schulz, M.-F., Selzer, G., Chollet, A., Movva, N. R., Semon, D., Escanez, S. & Kawashima, E. (1985) *Nucleic Acids Res. 13*, 1923−1938.
308. Meng, H., Burleigh, B. D. & Kelly, G. M. (1988) *J. Chromatogr. 443*, 183−192.
309. Wong, E. Y., Seetharam, R., Kotts, C. E., Heeren, R. A., Klein, B. K., Braford, S. R., Mathis, K. J., Bishop, B. F., Siegel, N. R., Smith, C. E. & Tacon, W. C. (1988) *Gene 68*, 193−203.
310. Obukowicz, M. G., Turner, M. A., Wong, E. Y. & Tacon, W. C. (1988) *Mol. Gen. Genet. 215*, 19−25.
311. Moks, T., Abrahmsén, L., Holmgren, E., Bilich, M., Olsson, A., Uhlén, M., Pohl, G., Sterky, C., Hultberg, H., Josephson, S., Holmgren, A., Jörnvall, H. & Nilsson, B. (1987) *Biochemistry 26*, 5239−5244.
312. Nishikawa, S., Yanase, K., Tokunaga-Doi, T., Kodama, K., Gomi, H., Uesugi, S., Ohtsuka, E., Kato, Y., Suzuki, F. & Ikehara, M. (1987) *Protein Eng. 1*, 487−492.
313. Seito, Y., Yamada, H., Niwa, M. & Ueda, I. (1987) *J. Biochem. (Tokyo) 101*, 123−134.
314. Bayne, M. L., Applebaum, J., Chicchi, G. G., Hayes, N. S., Green, B. G. & Cascieri, M. A. (1988) *Gene 66*, 235−244.
315. Gellerfors, P., Axelsson, K., Helander, A., Johansson, S., Kenne, L., Lindqist, S., Pavlu, B., Skottner, A. & Fryklund, L. (1989) *J. Biol. Chem. 264*, 11444−11449.
316. Bayne, M. L., Cascieri, M. A., Kelder, B., Applebaum, G., Chicchi, G., Shapiro, J. A., Pasleau, F. & Kopchick, J. J. (1987) *Proc. Natl Acad. Sci. USA 84*, 2638−2642.
317. Cascieri, M. A., Hayes, N. S., Kelder, B., Kopchick, J. J., Chicchi, G. C., Slater, E. E. & Bayne, M. L. (1988) *Endocrinology 122*, 1314−1320.
318. Furman, T. C., Epp, J., Hsiung, H. M., Hoskins, J., Long, G. L., Mendelsohn, L. G. Schoner, B., Smith, D. P. & Smith, M. C. (1987) *Bio/technology 5*, 1047−1051.
319. Hummel, M., Herbst, H. & Stein, H. (1989) *Eur. J. Biochem. 180*, 555−561.
320. Hammarberg, B., Nygren, P.-A., Holmgren, E., Elmblad, A., Tally, M., Hellman, U., Moks, T. & Uhlén, M. (1989) *Proc. Natl Acad. Sci. USA 86*, 4367−4371.
321. Sophokles: Oidipus Tyrannos, cited in: Durant, W. (1939) *The life of greece (The story of civilisation*, Part II), pp. 393−394, Simon and Schuster, New York.

Eur. J. Biochem. *191*, 1–17 (1990)
© FEBS 1990

Review

Translational dynamics

Interactions between the translational factors, tRNA and ribosomes during eukaryotic protein synthesis

Odd NYGÅRD and Lars NILSSON

Department of Cell Biology, The Wenner-Gren Institute, University of Stockholm, Sweden

(Received December 4, 1989) — EJB 89 1455

De novo synthesis of proteins in the cell is a complex process involving multiple steps and numerous components. The basic machinery is provided by the ribosomal particles with which the translational enzymes, mRNA and aminoacyl-tRNA interact during protein synthesis. Protein synthesis in eukaryotes and prokaryotes are grossly similar but some aspects have turned out to be quite different. The analogous components are listed in Table 1. The most marked difference exists in the initiation step of protein synthesis. In the eukaryotic organism more than ten different protein initiation factors (eIF) are required in comparison to three factors in prokaryotes (Table 2). Also the initiator tRNA, Met-tRNA$_f$, differs as the methionine is formylated in prokaryotes. Furthermore, the structure of the mRNA vary between prokaryotes and eukaryotes. In the latter organisms the mRNA contain a so-called cap structure at the 5′ end and a poly(A) tail at the 3′ end which both seem to be essential for effective translation of the mRNA. An additional difference in the mRNA structure is the lack of a Shine-Dalgarno sequence in eukaryotic mRNA, indicating that the mechanism of identification of the initiation codon differs between the two systems. The elongation cycle seems to work in a highly similar way but the prokaryotic and the eukaryotic elongation factors (EF and eEF respectively) are not interchangeable. Also, termination has a similar mechanism in the two systems, but only one release factor has been found in eukaryotes compared to three in bacteria.

It has become clear during the last decade that the protein synthesis machinery is not only capable of synthesizing protein with a high degree of accuracy, but that the level of expression of mRNA can be regulation both quantitatively and qualitatively. The expression of genes to mature proteins can be regulated at several levels including regulation of the rate of translation. Most studies concerning translational regulation

Table 1. *Functionally similar translational components in eukaryotes and prokaryotes*

Component	Eukaryotes	Prokaryotes
Ribosome	80 S	70 S
Subunits	40 S and 60 S	30 S and 50 S
rRNA	18S	16 S
	28 S, 5.8 S	26 S
	5 S	5 S
Elongation factors	eEF-1α	EF-Tu
	eEF-1β/eEF-1γ	EF-Ts
	eEF-2	EF-G

have dealt with regulatory mechanisms influencing the rate of initiation but it has become increasingly evident during the last years that the elongation process is also involved in translational regulation.

Protein synthesis is one of the most studied cellular processes and a considerable amount of information has accumulated during the last 30 years. In this review we have chosen to focus on the areas of protein synthesis where recent reviews are lacking. We have also tried to emphasize the dynamic properties of the ribosome and the translational factors. Furthermore, considerable attention has been given to the role of ribosomal RNA since the latest developments have given RNA a more direct role in the function of the ribosome than previously assumed. The review deals with the protein synthesis in eukaryotes but data from the prokaryotic system are discussed when information from the eukaryotic system is unavailable or when similarities between the two systems can clarify the structure and function of the components involved.

Due to the rapid progress in crystallisation of ribosomes and the improved techniques for probing RNA structures *in situ*, it has become possible to follow the alterations in the ribosomal structure during the translation process. The latest development has shown that protein synthesis proceeds via several previously unknown intermediates. The new technology, together with the increasing interest for translational regulation in the control of gene expression, points to a highly exciting future for the field of protein synthesis.

Correspondence to O. Nygård, Department of Cell Biology, University of Stockholm, Biology E5, S-10691 Stockholm, Sweden

Abbreviations. EF, prokaryotic elongation factor; eEF, eukaryotic elongation factor; eIF, eukaryotic initiation factors; GEF, guanosine nucleotide exchange factor; GuoPP[CH$_2$]P, guanosine 5′-[β,γ-methylene]triphosphate; oGTP, oxidised GTP; RF, termination factor; SRP, signal recognition particle.

2

Table 2. *Sizes and functions of the eukaryotic initiation factors*

Initiation factor	M_r of		Function
	protein	subunits	
eIF-2	128000	55000, 38000, 35000	binding of Met-tRNA$_f$ to 40S subunits
GEF	270000	85000, 67000, 52000, 37000, 27000	nucleotide exchange on eIF-2
eIF-3	0.7×10^6	several subunits	dissociation of 80S ribosomes
eIF-4A	46000		ATP-dependent unwinding of mRNA
eIF-4B	80000		binding of mRNA to 40S subunits
eIF-4C	17000		dissociation of 80S ribosomes
eIF-4E	24000		cap recognition
eIF-4F	290000	220000, 46000, 24000	cap recognition and mRNA binding
eIF-5	58000 − 62000		GTP-dependent subunit joining
eIF-6	25000		dissociation of 80S ribosomes

The ribosome

Structure

The eukaryotic ribosome is a 4.3-MDa RNP complex with a sedimentation coefficient of approximately 80 S. The ribosome consists of two subunits with sedimentation coefficients of 40 S and 60 S with the corresponding masses of 1.4 MDa and 2.9 MDa, respectively. The 40 S particle is comprised of one molecule of 18 S rRNA and approximately 30 proteins, whereas the large subparticle is composed of one molecule each of 5 S, 5.8 S and 28 S rRNA and approximately 45 proteins [1]. The ribosomal components show a high degree of similarity between prokaryotic and eukaryotic ribosomes. The ribosomal RNAs from both lower and higher organisms contain highly conserved sequences and the proposed secondary structure of ribosomal RNAs from various organisms show an extensive similarity [2, 3].

Structural data obtained by electron microscopy show that the eukaryotic ribosome is a particle with marked similarities in the general morphology to that of the smaller prokaryotic ribosome. The structure of the 40 S ribosomal subunit can be divided into a head and a body region separated by a neck (Fig. 1). The particle has one major platform protruding from the body up to the neck region and a bill extending from the head in the opposite direction [4, 5]. The large subunit has a stalk and a central protuberance similar to that of the prokaryotic 50 S particle [4]. The folding of the ribosomal RNAs in the ribosome and the position of individual rRNA domains in the eukaryotic ribosome are not known. However, due to the extensive similarity between the predicted secondary structure of ribosomal RNA from prokaryotes and eukaryotes, it is reasonable to assume that the positioning of the corresponding regions are conserved [2, 3]. Extensive studies of the spatial arrangement of ribosomal proteins in the two subunits have been performed by chemical cross-linking [6 − 11]. The most detailed studies have been performed with the small ribosomal subunit. In this particle the position of several individual proteins is known from immunoelectron microscopy studies [12, 13] (Fig. 2). However, due to the broad distribution of antigenic determinants observed for several proteins, the protein maps obtained by cross-linking and by immunoelectron microscopy can not easily be combined. Less information is available on the location of proteins in the large

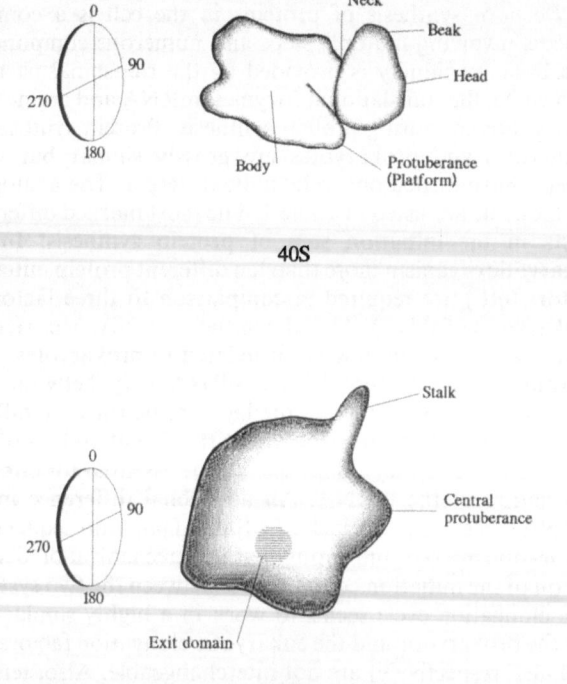

Fig. 1. *Schematic presentation of the basic morphological details of the small and the large ribosomal subunits of the eukaryotic ribosome.* The model is based on electron microscopy data [4, 12]

subunit. It is therefore at present not obvious how the protein topography obtained by cross-linking should be oriented with respect to the structure of the particle as seen by electron microscopy. One of the problems of fitting the available data into a comprehensive model similar to that available for the prokaryotic ribosome [14, 15] is the left-handed and right-handed positioning of the platform on the 40 S particle, used by various authors [4, 5, 12, 13, 16, 17]. In the schematic illustrations used here we have placed the platform in a similar direction as depicted in the prokaryotic ribosome [4, 18].

Almost all ribosomal proteins in the prokaryotic ribosome have antigenic determinants located at the surface [19 − 22].

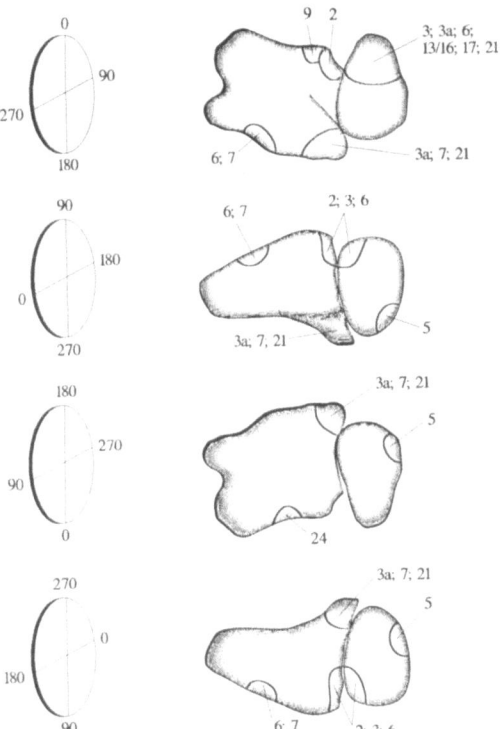

Fig. 2. *Location of proteins within the small ribosomal subunit.* The position of the antigenic determinants of the individual proteins are based on the immunoelectron microscopic data of Lutsch et al. [12, 13] with the modification that the position of the platform is assumed to be equivalent to that in the prokaryotic small ribosomal subunit [4]

Also neutron-scattering experiments and the electron distribution indicate a peripheral localization of the proteins with an rRNA core at the centre of the ribosomal particle [23 – 28]. Thus, the interface between the two ribosomal particles contains few proteins [29] and a direct contact between the two major ribosomal RNAs across the subunit interface has been observed [30]. Herr et al. suggested that complementary sequences in the ribosomal RNA of the small and large ribosomal subunits contribute to ribosomal association by direct base pairing [31]. This hypothesis is supported by the observations that mutation of G^{791} to A^{791} or G^{1416} to U^{1416} in *Escherichia coli* 16 S rRNA significantly reduces the subunit association [32, 33]. Despite the low abundance of proteins at the interface, proteins from one subunit have been linked across the interface of the ribosome to both ribosomal RNA and proteins from the complementary subunit, suggesting that some proteins are located directly at or close to the interface border [10, 34 – 36].

Ribosomal dynamics during translation

The ribosome is a highly flexible structure in which the dynamic properties are most likely a prerequisite for function. The ribosome reciprocates between the pre- and post-translocational phases of the elongation cycle and one can distinguish these two types of ribosomes based on their ability to interact with the two elongation factors since pre-translocational ribosomes preferentially bind eEF-2 and the post-translocational particles eEF-1α. The pre- and post-translocational forms of the ribosomes do not depend on the location of peptidyl-tRNA as ribosomes are able to mimic the pre- and post-translocational form even in the absence of mRNA and

peptidyl-tRNA [37]. However, both ribosomal subunits are required, suggesting that the flexibility involves an altered contact between the subunits. Prokaryotic ribosomes are present in two different structures [38] and it has been suggested that the ribosome reciprocates between an extended and a contracted conformation [39]. The differences are related to variations in the structure of the large ribosomal subparticle [40] and seem to be coupled to the structure of 23 S rRNA [41]. The two ribosomal RNA structures are interconvertible in the presence of EF-G and GTP and correlate well with the differences found in the 50 S subunit between pre- and post-translocational form [42]. These interconversions have been suggested to be the driving force in translocation [43].

Various naturally occuring toxins, such as α-sacrin, ricin and the translational inhibitor from barely seeds, have been used to characterize the ribosomal structure and function. The inhibitors block protein synthesis by modifying the 28 S rRNA. Both ricin and the barely inhibitor hydrolyse the *N*-glycosidic bond at A^{4324} whereas α-sarcin cleaves the nucleic acid at a position between G^{4325} and A^{4326} [44 – 47]. These sites are located in a region of the 28 S rRNA that is similar to a loop in the prokaryotic 23 S rRNA that interacts with EF-Tu and EF-G [48, 49]. It is therefore not surprising that these modifications affect the binding and the GTPase activity of the two elongation factors and more detailed studies have demonstrated toxin-specific differences in the functional alterations. The barley inhibitor reduces the eEF-1α-dependent hydrolysis of GTP whereas ricin drastically increases the eEF-2-dependent GTPase activity [50, 51]. The effects of these inhibitors seem therefore to alter the flexibility of the ribosomal particle, locking the ribosome in a pre- and post-translocational structure respectively. It has also been shown that the toxins alter the chemical reactivity of ribosomal proteins. Ricin treatment specifically affects the modification of protein L4 whereas α-sarcin alters the reactivity of L3 and L4 [52]. The differences found suggest that the toxins induce specific changes in the ribosomal conformation although their sites of action of the 28 S rRNA are located close to each other. Thus, it seem as if the inhibitors prevent interconversion between the ribosomal states of the elongation cycle. The differential effects of the function and structure should therefore depend on the conformation in which the ribosome was frozen. This could be achieved by a preferential ribosomal access of the individual inhibitors related to the ribosomal phase of the elongation cycle.

The flexibility of the ribosome is one of the major difficulties when analyzing the functional domains of the particle. Removal or modification of individual proteins or RNAs may have profound secondary effects. It is therefore not surprising that attempts to localize functional domains by reconstitution experiments have failed, although the technique has been successfully used in prokaryotes. Thus, the functional domains have mainly been studied by other techniques such as cross-linking with bifunctional reagents and ultraviolet irradiation, affinity labelling, immunoelectron microscopy and by accessibility to protein- and RNA-modifying reagents. The results of these investigations are summarized in Tables 3 and 4. The functional domains mapped so far seem all to be located in the area known as the interface region, suggesting that the interface is the 'active site' of the ribosome.

Ribosomal binding sites for tRNA

Decoding of the mRNA required codon – anticodon interaction on the ribosome. Cross-linking data show that the

Table 3. *Components involved in the various functional domains on the eukaryotic ribosome*

Functional domain	Protein in	
	40S subunit	60S subunit
eIF-2 binding site [53]	18S rRNA	
eIF-2 binding site [54, 55]	S3, S3a, S6, S13/16, S15/15a, S19, S24	
eIF-3 binding site [56]	18S rRNA	
eIF-3 binding site [57, 58]	S2, S3, S3a, S4, S5, S6, S7, S8, S9, S10, S11, S12, S14, S15, S16, S19, S23/24, S25, S26, S27	
eEF-1α binding site [59]	S23/24, S26	L12, L23, L39
eEF-2 binding site [60, 61]	S3/3a, S6, S7, S23/24, S11	L3, L4, L5, L8, L9, L12, L13, L21, L23, L26, L27a, L31, L32, LA33, P2
eEF-2 binding site [62]		5S rRNA
mRNA binding site [63–65]	S1, S3/3a, S6, S11	L5, L6
mRNA binding site [66, 67]	18S rRNA	L5, 5S rRNA
P-site [68–76]	S3/3a, S6, S7, S13/15	L3, L5, L6, L7, L8, L13, L21/23, L32
A-site [68, 72–74, 76–79]	S4, S6, S10, S13, S14, S15, S23, S24	L4, L6, L13/15, L18, L21, L28, L29, L30
Peptidyltransferase [78, 80–84]		L10, L13/15, L21, L23, L24, L26, L27, L28, L31, L32/33, L36
Met-tRNAf binding site [85]	S3a, S6	
Phe-tRNA binding site [86]	S2, S3a, S6/6a, S10, S13/15	

Table 4. *Cross-linking between RNA and protein within the eukaryotic ribosome*

RNA	Associated proteins in	
	40S subunit	60S subunit
18S rRNA [36, 87]	S3a, S6, S7, S8, S11, S16/18, S23/24, S25	L3, L6, L7, L8
28S rRNA [36]	S2, S3, S4, S6, S7, S13	L3, L4, L6, L7, L7a, L8, L10, L15, L19, L23/23a
3′ End of 18S rRNA [88]	S3a	
3′ End of 28S rRNA [88]		L3
5S rRNA [89, 90]		L5
5S and 5.8S rRNA [72, 73, 75]		L5, L6, L7, L8
5.8S rRNA [75]	complete subunit	
Interface [10]	SA30, S2, S3a, S4, S6	LA33, L11, L5, L24

mRNA is positioned at the interface between the two ribosomal subunits. The tRNAs in the A- and P-sites must therefore, at least partly, be bound to the ribosome in the interface region. Both proteins from the large and the small ribosomal subunit have been suggested to organise the A- and P-sites and some of these proteins have also been shown to belong to the ribosomal interface region (Table 3). Even though the tRNA molecule is rather extended and the two tRNA copies cover a substantial portion of the ribosomal interface, the large number of proteins depicted as A- and P-site proteins seems incompatible with the low protein content of the interface region. As the interface is mainly comprised of ribosomal RNA, it could be expected that tRNA molecules placed in the A- and P-sites are in contact with ribosomal RNA. In yeast, the anticodon of Phe-tRNA located in the P-site is cross-linked to the nucleotide C^{1626} in the rRNA from the small ribosomal subunit [91]. Analogous cross-linkages have also been observed in higher eukaryotes and in *E. coli* (C^{1400}) [92–94]. In *E. coli*, the cross-linked nucleotide is located on the interface side of the cleft between the body and the platform of the 30 S particle and it is reasonable to assume that the corresponding base in eukaryotes has the same position [95] (Fig. 3). Even A-site-located tRNA is cross-linked to C^{1400} in 16 S rRNA, suggesting that the anticodons of A-

and P-site-located tRNA are in close proximity [14, 96, 97]. A direct interaction between tRNA and 23 S rRNA is also possible since the acceptor arm of the tRNA in the P-site can be cross-linked to G^{1945} in the 23 S rRNA [98]. It therefore seems likely that tRNA is in close contact with ribosomal RNA at the subunit interface and that rRNA is part of the A- and P-sites.

Besides the canonical A- and P-sites, two additional ribosomal sites for tRNA, a pre-A site or recognition site (R-site) [99] and a post- P-site or exit (E) site [100] have been suggested. The R-site is proposed to be responsible for an initial selection of aminoacyl-tRNA in the ternary eEF-1α · GTP · aminoacyl-tRNA complex, but it still remains an open question whether this site is to be regarded as a physically separated site or if the initial recognition process proceeds in a modified A-site. The experimental evidence supporting the existence of an E-site is more solid [49, 101, 102]. After the translocation-mediated displacement of the deacylated tRNA from the P-site to the E-site, the tRNA exclusively interacts with the large ribosomal subparticle as demonstrated by the protective effect of tRNA on 23 S rRNA against chemical modification [49]. On the other hand, the A- and P-sites obviously involve both the small and large ribosomal subunit. In the P-site, tRNA interacts with both 16 S and 23 S rRNA [49,

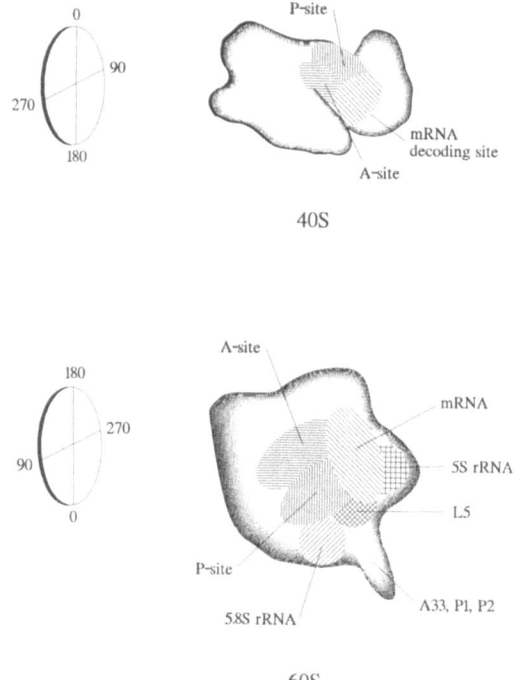

Fig. 3. *Tentative model showing the topography of the ribosomal interface.* The positions of 5 S and 5.8 S rRNA as well as of the ribosomal proteins A33, P1 and P2 are based on similarity with the prokaryotic ribosome

103, 104]. This is also the case in the A-site provided that EF-Tu has left the complex. In case where EF-Tu still remains in the complex, tRNA only interacts with the 16 S rRNA while EF-Tu itself protects regions of the 23 S rRNA.

Ribosomal interaction with initiation factors and mRNA

Initiation

The initiation process serves to decode the proper initiation signal on the mRNA thereby positioning the ribosome on the mRNA in the correct reading frame. The strategic role of this process has led to the suggestion that post-transcriptional regulation of protein synthesis in eukaryotes is primarily mediated by an extensive control of the activity of the initiation process. It is therefore not surprising that over the last 15 years most of the interest concerning eukaryotic protein synthesis has focused on the initiation step and the factors involved [105 – 108].

The first step in the initiation is the dissociation of the 80 S ribosome into its two subunits (Fig. 4). The equilibrium of this reaction is shifted towards dissociation by interaction of the initation factors eIF-3 and eIF-4C with the 40 S subunit [109] and eIF-6 with the 60 S particle [110]. In the following reaction, initiator tRNA, Met-tRNA$_f$ in a ternary complex with GTP and initiation factor eIF-2 is bound to the 40 S · eIF-3 · eIF-4C complex forming a 43 S pre-initiation complex. The ternary complex can associate with 40 S subunits lacking other bound factors, but the association is enhanced by eIF-3 and eIF-4C [109, 111 – 113]. The 40 S particle with its associated factors, GTP and Met-tRNA$_f$, is now ready to pick up the mRNA. In eukaryotes this step involves the initiation factors eIF-4A, eIF-B, eIF-4E and eIF-4F. Initiation factor eIF-4F is a multi-subunit complex consisting of eIF-4A, eIF-4E and a 220-kDa protein. eIF-4E is a cap-binding protein

that in the eIF-4F complex recognises the 5′-cap structure on the mRNA. eIF-4A seems to unfold mRNA in an ATP-driven process [108]. Also, eIF-4B is needed for the ATP-dependent unfolding of the secondary structure of the mRNA. The unwinding of the mRNA presumably generates a suitable site for the association of the 43 S pre-initiation complex. Unlike the situation in prokaryotes, the position of the proper intiation codon AUG is not identified by a base complementarity between mRNA and the 3′ end of the ribosomal RNA of the small subunit. Thus the initiation codon has to be identified by other means. eIF-4B has affinity for the AUG codon and may be responsible for identifying the initation signal. It has been proposed that the 43 S pre-initiation complex associates with the cap structure of the mRNA-factor complex and scans the mRNA from the 5′ end to the proper initiation codon [114]. Alternatively, mRNA-recognizing factors are responsible for the scanning until eIF-4B identifies the initiation codon, thereby guiding the initiating 40 S particle to the initiator codon [118]. Once the initiation signal is identified the programmed 40 S particle is ready to associate with the 60 S subunit in a programmed 80 S ribosome. This reaction requires initiation factor eIF-5 and is accompanied by hydrolysis of GTP. During this subunit joining, the factors associated with the pre-initiation complex are released from the ribosome [105 – 107, 301].

Regulation of protein synthesis at the level of initiation has been extensively studied. Many of the initiation factors undergo post-translational modifications assumed to affect their activity in the initiation process. Despite all efforts, only the phosphorylation of the α subunit of eIF-2 has been correlated to a functional alteration. The catalytic activity of phosphorylated eIF-2 is inhibited due to the sequestration of the factor in a complex with the guanosine nucleotide exchange factor (GEF). Another relatively well understood mechanism is the effect of polio virus infection on the rate of initiation. Infection leads to a proteolytic degradation of the 220-kDa subunit of eIF-4F and a subsequent loss of the ability of the factor to promote mRNA binding to the 40 S subunit. As several reviews in this field are available, we refer to these for further reading [105 – 108, 115].

Ribosomal binding domain for initiation factors

The initiation step involves mainly the small ribosomal subunit. Hence, the attachment sites for most initiation factors are generally believed to be on the 40 S particle and many of the factors are, in fact, associated with the so-called native 40 S subunits isolated from the cell, i.e. associated with the particles actively involved in the initiation process. The individual factor concentrations vary but eIF-3 is present in a stoichiometric complex with the ribosomal particle. Protein synthesis experiments based on native 40 S subunits show that eIF-2, eIF-4A and eIF-4B are present in limiting amounts on the native 40 S particles while the concentration of eIF-5 is saturating [116]. Despite the association of the initiation factors with the small subunit, little is known about the requirements for ribosome association and the structure and localization of the ribosomal binding sites for the initiation factors. The only exceptions are the binding sites for eIF-2 and eIF-3 which have been characterised by both chemical cross-linking with bifunctional reagents and electron microscopy [16, 54 – 58, 64, 85, 87, 117 – 119] (Table 3). eIF-3 is a huge multi-subunit assembly with a mass of approximately 0.7 MDa. Due to its size, eIF-3 is the only initiation factor which has been individually resolved by conventional electron

6

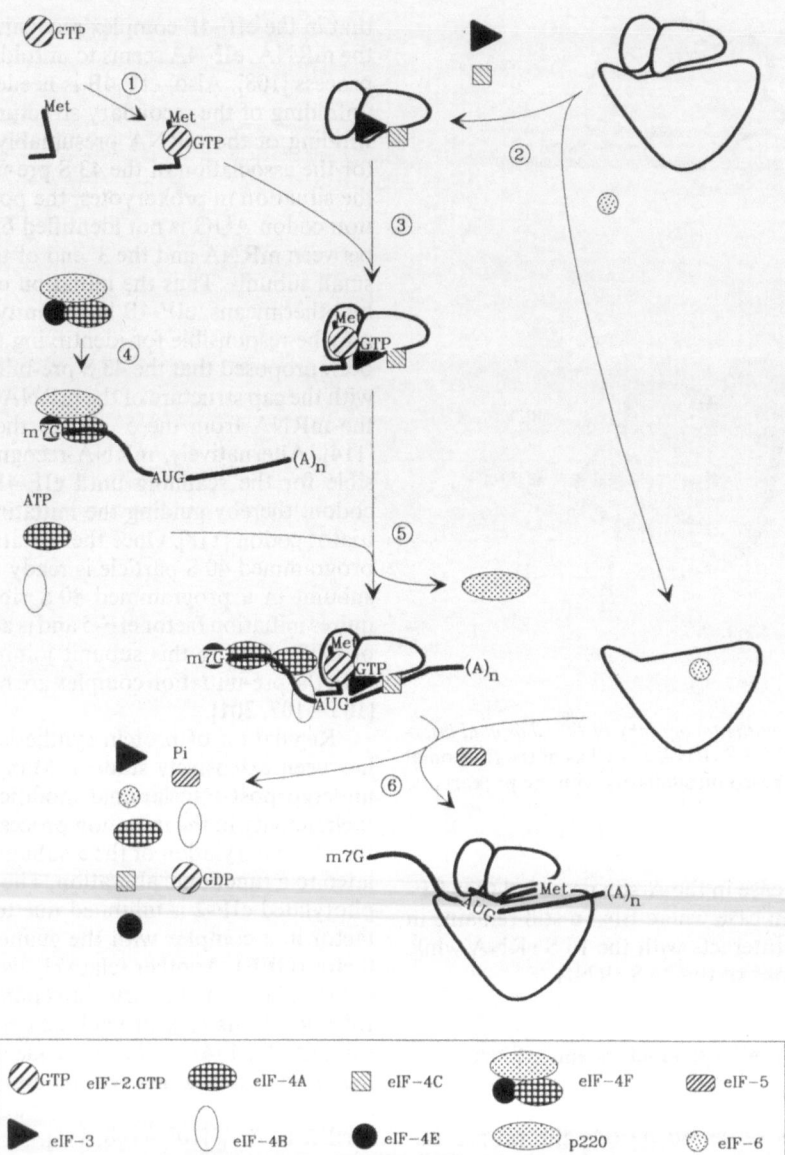

GTP eIF-2.GTP   eIF-4A   eIF-4C   eIF-4F   eIF-5

eIF-3   eIF-4B   eIF-4E   p220   eIF-6

Fig. 4. *Schematic illustration of the various steps in the eukaryotic protein synthesis initiation.* (1) eIF-2-dependent formation of a ternary complex containing initiator-tRNA. (2) Dissociation of 80 S ribosomes in the presence of eIF-3, eIF-4C and eIF-6. (3) Association of the ternary complex with the small ribosomal subunit. (4) eIF-4F-dependent identification of the cap structure on mRNA. (5) Binding of the mRNA to the small ribosomal subunit (6). Association of the 60 S subunit with the 40 S-mRNA complex in the present of eIF-5

microscopy. Electron microscopy and hydrodynamic measurements show that the factor has the shape of a triangular prism [120]. Emmanuilov et al. found eIF-3 localized at the interface of the small subunit close to the protuberance [16]. In contrast, Lutsch et al. located the factor to the outer side of the protuberance at a position in which the factor did not directly interfere with the interface region [118]. This position is also supported by cross-linking experiments. Both in native small subunits and in complexes consisting of derived 40 S particles and purified eIF-3, the factor was covalently linked to ribosomal proteins which, according to immune electron microscopy, are located at the protuberance of the particle. The schematic illustrations indicating the position of eIF-3 on the 40 S particle so far presented show the small subunit as a left-handed particle with the protuberance extending to the right. However, recent reports have placed the protuberance on the left side in analogy with the position of the platform on the prokaryotic 30 S particle [4, 5]. If we assume that

the position of the protrusion is evolutionary conserved, the binding site for eIF-3 would be as indicated in Fig. 5. Cross-linking studies using a short (0.4 nm) bifunctional reagent have also shown that the factor binding site contains 18 S rRNA. The data suggests that the binding site for eIF-3 is located close to the interface. This assumption is supported by the observation that most of the ribosomal protein cross-linked to eIF-3 are also cross-linked to 28 S rRNA in the intact 80 S ribosome. Further support for localization of the factor close to the interface comes from the observed close contact with the mRNA [64]. The subunit anti-association activity of eIF-3 may therefore, at least in part, be accounted for by steric hindrance.

eIF-2 cannot be directly visualized by electron microscopy. This disadvantage has prompted the use of antibodies directed towards individual subunits of the factor. By this technique, Bommer et al. were able to show that the factor is an elongated molecule with a binding site located at the neck region on the

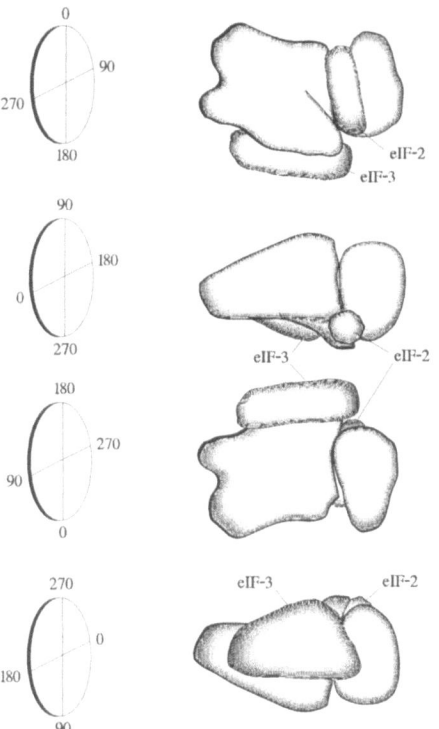

Fig. 5. *Localisation of the binding sites for eIF-2 and eIF-3 on the small ribosomal subunit.* See text for details

16 S rRNA are cross-linked to the mRNA in *E. coli* [122]. Thus, the binding domain for mRNA in the prokaryotic ribosome must contain parts of the 16 S rRNA. Similarly, 18 S rRNA has been found in the binding site for mRNA on the 40 S subunit although evidence for a direct interaction between the two RNA species is lacking. In the eukaryotic ribosome, 5 S rRNA also seems to be in close vicinity to the mRNA in the programmed ribosome. The accumulated data point to a positioning of the mRNA in the interface region between the two ribosomal subparticles in analogy to the prokaryotic ribosome where the mRNA is supposed to be located in the interface region with the codon–anticodon interaction occurring in the area around the platform [18] (Fig. 3).

Elongation

eEF-1

Binding of the cognate aminoacyl-tRNA to the ribosomes is catalysed by elongation factor eEF-1 in the presence of GTP (Fig. 6). During purification it was discovered that the eEF-1 activity occurred in multiple forms with molecular masses of 50 kDa, 140 kDa and up to more than 500 kDa [123–125]. The low-molecular-mass form contained a single 50-kDa polypeptide called eEF-1α, whereas the high-molecular-mass forms (eEF-1$_H$) contained several additional proteins named eEF-1β [126], eEF-1γ [127] and eEF-1δ (Table 5) [128]. The observed dissociation of eEF-1$_H$ to free eEF-1α upon acitvation of the cryptobiotic cysts of *Artemia salina* and during germination in wheat, suggests that the high-molecular-mass aggregates of eEF-1 are a storage form [124, 129]. The mechanism responsible for the dissociation is not unterstood at present but guanosine nucleotides and treament with protease or phospholipase lead to a disaggregation [123, 130]. eEF-1β seems to be responsible for the formation of eEF-1$_H$ and the function of eEF-1β is dependent on a cholesteryl 14-methyl-hexadecanoate [131]. The recent observation that eEF-1α contains ethanolamine and glycerol may indicate that the protein is anchored to membranes or to the cytoskeleton [132, 133]. An interesting observation is that eEF-1γ, a highly hydrophobic protein, can be co-precipitated with tubulin under nondenaturing conditions. This suggests a role for eEF-1γ in associating eEF-1$_H$ to the cytoskeleton [134].

Interaction of eEF-1α with guanosine nucleotides and aminoacyl-tRNA

eEF-1α requires GTP for its biological function. The part of the factor that is involved in guanosine nucleotide binding has not been determined, but both eEF-1α and eEF-2 have sequences similar to the region in EF-Tu and *ras* p21 constituting the guanosine-nucleotide-binding pocket (amino acids 9–28, 81–138 in eEF-1α) [135–139]. This highly conserved domain seems to be a common feature of most GTP binding proteins [140, 141]. eEF-1α and EF-Tu show a major difference in affinity for guanosine nucleotides. EF-Tu has much higher affinity for GDP than for GTP (K_d 4.9 nM and 3.6 µM respectively) [142] whereas eEF-1α has approximately the same affinity for GDP and GTP (K_d 0.5 µM and 0.2 µM respectively) [143, 144]. Despite the slightly higher affinity for GTP than for GDP, the guanosine nucleotide exchange is rate-limiting for the catalytical function of eEF-1α, indicating that both the on and off rates are slow. The rate of exchange is stimulated by a factor of 5 in the presence of eEF-1β [134,

interface side of the 40 S particle [119] (Fig. 5). Monospecific antibodies raised against some of the proteins from the area of the subunit were able to inhibit the attachment of the ternary eIF-2 · Met-tRNA$_f$ · GTP complex [55]. This location of the factors is further supported by the identification of ribosomal proteins cross-linked to eIF-2. Some of the proteins also belong to the eIF-3 binding site. As both factors bind to the 40 S subunit simultaneously, the two factors cannot have overlapping binding sites but the results suggest that the two factors are in close contact in the 40 S initiation complex. The positioning of eIF-2 in the interface area suggests that Met-tRNA$_f$, is directly located at the future P-site by the interaction of the ternary complex with the ribosome. In fact, some of the proteins cross-linked to eIF-2 have also been identified as P-site proteins (Table 3).

The localization of the binding sites for eIF-2 and eIF-3 to the region between the head and the body of the small subunit close to the protuberance is analogous to that found in prokaryotes. Proteins involved in ribosomal binding of the three initiation factors IF-1, IF-2 and IF-3 have been located to the platform region by both cross-linking and immuno-electron microscopy [4].

Ribosomal binding domain for mRNA

Several proteins have been appointed candidates for the mRNA binding site on the ribosome (Table 3). Proteins from both the small and the large ribosomal subunits have close contact with mRNA although mRNA binding is an exclusive property of the small ribosomal subunit. During the prokaryotic initiation process, a short sequence at the 5' end of the mRNA base pair with a complementary sequence located near the 3' end of the 16 S rRNA [12]. Anticodon–codon interaction seems to take place close to the platform on the small subunit as nucleotides close to C^{1400} (1394–1399) in

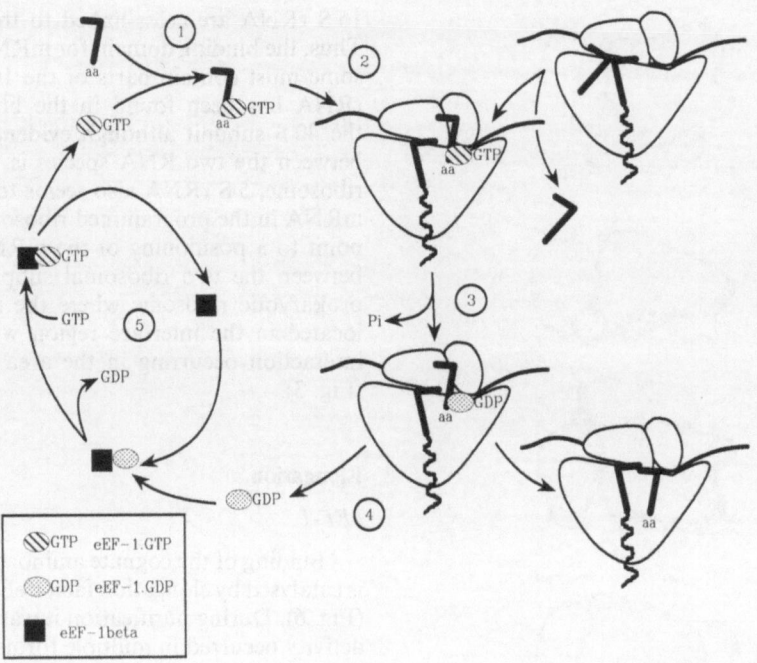

Fig. 6. *Schematic illustration of the eEF-1α cycle*. In step (1), eEF-1α · GTP associate with aminoacyl-tRNA to form a ternary complex in which the ester bond is protected by eEF-1α. (2) The ternary complex is bound to the ribosome in a modified A-site. (3) The GTP is hydrolysed during proofreading. (4) Release of the eEF-1α · GDP complex facilitates appropriate positioning of the aminoacyl-tRNA in the ribosomal A-site. (5) eEF-1β-catalysed exchange. The eEF-1β · eEF-1α · GTP intermediate is highly unstable, but can be detected at $0-2°C$ [145]

Table 5. *Physical properties of the eukaryotic factors involved in elongation and termination*

Factor	Organism	M_r	Number of amino acids	pI	Sequences known from other organisms
eEF-1α	*Artemia* [138]	50274	461	8.3 [125]	human [283], *Mucor racemosus* [284], yeast [285], tomato [286], mouse [299]
eEF-1β	*Artemia* [126]	23300	208	—	
eEF-1γ	*Artemia* [127]	49200	429	—	
eEF-2	hamster [139]	95192	857	6.5 [287]	rat [288], *Dictyostelium* [289] human [300]
eEF-3	yeast [290]	110000−125000	—	5.9	
SUF12	yeast [259]	88000	685		
RF	rabbit [265]	105000	(dimer)		*Artemia* [291]

145] and is further increased by a factor of 2 in the presence of eEF-1γ although the mechanism behind the latter stimulation is unknown.

The exchange of GDP for GTP leads to a increase in the affinity of eEF-1α for aminoacyl-tRNA [144, 146]. The region that interacts with tRNA has not been identified in either eEF-1α or EF-Tu. However, in EF-Tu His[66] and His[118] in the guanosine binding domain and Lys[208] and Lys[237] in the middle domain have been suggested to interact with the tRNA molecule [147−150]. Binding of the aminoacyl-tRNA to eEF-1α leads to a protection of the otherwise labile ester bond between the amino acid and the tRNA [143, 151, 152], thereby allowing the cell to store aminoacyl-tRNA in a protected form. This requires that the intracellular concentration of eEF-1α is stoichiometric to that of the aminoacyl-tRNA as in the case with EF-Tu [153]. eEF-1α has been shown to be a highly abundant protein although its relative concentration varies between different cell types [154]. In reticulocytes, there are approximately six copies of eEF-1α per ribosome [155], but no systematic correlation between the *in vivo* level of tRNA and eEF-1α has been made.

eEF-1α has a trypsin-sensitive arginine-glycine bond between positions 68 and 69 [140, 156, 157]. The trypsin sensitivity is low in the eEF-1 · GDP complex but increases as the GDP is replaced by GTP. The sensitivity is further incrased by both tRNA and aminoacyl-tRNA. Position 68 is situated in the GDP binding region of the factor and corresponds to the trypsin-sensitive Arg[58] in EF-Tu [158] and Arg[66] in eEF-2 [159]. The data demonstrate that the factor undergoes structural rearrangements during binding of guanosine nucleotids and aminoacyl-tRNA.

Interactions of eEF-1α with ribosomes

The interaction of the ternary eEF-1α · GTP · aminoacyl-tRNA complex with the ribosomes has not been studied in great detail. However, the available data show that the ternary complex associates only with post-translocational ribosomes having an empty A-site which ensures that the ternary complex associates with ribosomes in the correct phase of the elongation cycle [160−162]. Binding of the factor to the ribosome is not primarily dependent on a codon−anticodon interaction but on the ribosomal structure [163]. After selection of the cognate aminoacyl-tRNA, the GTP is hydrolysed to GDP and inorganic phosphate. The hydrolytic domain is located on eEF-1α and the factor alone can cleave GTP, albeit much less efficiently than in the presence of ribosomes and aminoacyl-tRNA [125, 164]. The rate of hydrolysis is further increased by the addition of mRNA [i.e. poly(U)] [163, 164], suggesting that codon−anticodon interaction is a prerequisite for the induction of the GTPase. The intrinsic discrimination based on the codon−anticodon interaction is insufficient to account for the low error rate observed in prokaryotes and eukaryotes (approximately 10^{-5} to 10^{-4}) [165−168]. The question of how such a high accuracy can be achieved has been intensively studied in the prokaryotic system and a kinetic proof-reading model has been suggested [169]. Essentially, the mechanism is based on a two-step selection process. An initial selection of the ternary complex is followed by the GTP hydrolysis and a second selection before the formation of the peptide bond. According to this model, binding of aminoacyl-tRNA should be accompanied by a more than stoichiometric hydrolysis of GTP and this has, in fact, been observed [170, 171]. Proof-reading after GTP hydrolysis depends on a delay between binding of the aminoacyl-tRNA to the A-site (i.e. when the interaction between the tRNA and eEF-1α is broken) and peptide bond formation. Molecular evidence for a sequential binding of cognate aminoacyl-tRNA to the ribosome has recently been presented [49]. In the first phase of the binding reaction, the anticodon region interacts with mRNA and 16 S rRNA on the 30 S subunit while EF-Tu interacts with 23 S rRNA. After GTP hydrolysis and release of EF-Tu · GDP, the CCA end of the tRNA interacts with 23 S rRNA in the proposed peptidyltransferase centre. The sequential binding of aminoacyl-tRNA is also supported by affinity labelling studies using a photoreactive derivative of Phe-tRNA [172]. It is obvious that the accuracy of the protein synthesis depends on a correct interaction of the components involved in the binding of aminoacyl-tRNA. This has been shown by analysis of the effect of mutations in ribosomal protein, tRNA and eEF-1α/EF-Tu on the translational accuracy, frame-shift and non-sense suppression [173−181]. The mutations in eEF-1α/EF-Tu that alter the translational accuracy were localized in regions that seem to affect the stability of the ternary complex.

Regulation of the activity of eEF-1

The rate of elongation is reduced with age whereas the rates of initiation and termination are unaffected [182, 183]. A change in eEF-1α activity following aging in rats and *Drosophila* has also been reported [182, 184−187] and the activity of the factor seems to be correlated with the extent of methylation of eEF-1. eEF-1α is methylated in Siemian-virus-40-transformed 3T3 cells to a higher extent than in the non-transformed cells [188]. It has also been shown that the methylation and activity of eEF-1α increases during germination of the fungus *Mucor*, while protein and mRNA levels remain constant [189, 190]. The individual step of the eEF-1α cycle that is affected by methylation has not been identified.

Phosphorylation of eEF-1α, eEF-1β and eEF-1γ has been reported [191, 192, 292]. The phosphorylation of eEF-1α seems to alter the ribosomal binding properties of the factor as isolated ribosomes do not contain any phosphorylated factor [191]. However, at present it is not clear whether this modification interferes directly with the ribosomal association of eEF-1α · GTP · aminoacyl-tRNA, or indirectly by altering the

nucleotide or aminoacyl-tRNA binding properties. eEF-1β is phosphorylated at Ser[89] by an endogenous kinase found in preparations of eEF-1β/γ. Phosphorylation of eEF-1β reduces the guanosine nucleotide exchange activity and thereby inhibits the regeneration of eEF-1α · GTP from eEF-1α · GDP [192]. This reduces the rate of elongation by decreasing the amount of available ternary eEF-1α · GTP · aminoacyl-tRNA complexes. eEF-1γ is phosphorylated by a cell-division-controlled protein kinase p34[cdc2] [292]. The functional role of this modification is not yet known, but the phosphorylation seems to correlate with the changes in protein synthesis activity during cell division.

The peptidyltransferase centre

After binding of aminoacyl-tRNA to the A-site, the amino acid is coupled to the nascent peptide chain in the P-site via a peptide bond. The capacity to catalyse the formation of the peptide bond is an inherent property of the large ribosomal subunit. The reaction centre is most likely small and distinct in space and the catalytic activity seems to be dependent on a correct positioning of the two tRNA molecules in the P- and A-sites. This ensures the necessary structural features for the formation of the peptide bond. A direct interaction of the two adjacent tRNA molecules has also been proposed [139, 194] and data from energy transfer experiments show that the two tRNA molecules are separated by a distance of not more than $0.2 - 1$ nm [195].

The components of the peptidyltransferase centre have as yet not been unambiguously identified. A number of ribosomal proteins from the large subunit seem to be part of the peptidyl transferase domain (Table 3); interestingly, some proteins from the small subparticle have also been suggested to be involved in the catalytic centre (Table 3). It is therefore highly likely that the peptidyltransferase is located at the interface region of the ribosome. However, it should be born in mind that most of the data supporting localization of these proteins to the active domain are indirect. Not even in the case of the well characterized prokaryotic ribosome have the active components of the peptidyltransferase centre been properly identified. Although ribosome reconstitution experiments have demonstrated an absolute requirement for a restricted number of ribosomal proteins, direct evidence for their involvement in the catalytic process is still lacking.

The little success so far in pinpointing the functional active proteins of the peptidyltransferase centre may indicate that ribosomal RNA plays a more direct role in this process. This assumption is further supported by functional studies of the prokaryotic ribosome showing that the peptidyltransferase centre involves evolutionarily conserved sequences of 23 S rRNA [196, 197].

eEF-2

After peptide bond formation, the peptidyl-tRNA is positioned in the A-site whereas the P-site is occupied by an empty tRNA. Before addition of the next amino acid, the ribosomal A-site has to be made available for interaction with eEF-1α · GTP · aminoacyl-tRNA. This is achieved by an eEF-2-catalysed translocation involving the following simultaneous actions: (a) movement of the peptidyl-tRNA from the A-site to the P-site; (b) movement of the deacylated tRNA from the P-site to the E-site and (c) sliding of the mRNA three bases forward, placing a new codon in the A-site (Fig. 7).

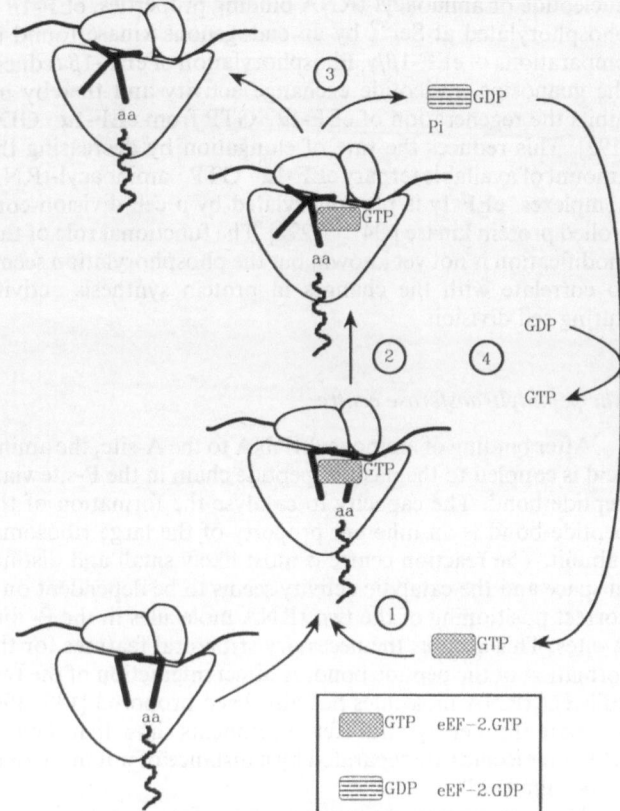

Fig. 7. *Schematic representation of the eEF-2 cycle.* (1) eEF-2 in a complex with GTP associates with pre-translocation ribosomes. (2) Translocation. (3) The release of eEF-2 is facilitated by the hydrolysis of GTP. (4) The nucleotide exchange, does not require additional factors

Interaction of eEF-2 with guanosine nucleotides

Unlike eEF-1α, eEF-2 has higher affinity for GDP than for GTP and the dissociation constants for the corresponding factor—nucleotide complexes have been determined to 0.4 μM and 2 μM respectively [198, 199]. Despite the higher affinity for GDP than for GTP, no nucleotide-exchange factor has been identified, suggesting that the factor—nucleotide complex has a fast dissociation rate. Determinations of the kinetic constants of the eEF-2 and ribosome-dependent GTPase activity also indicate that the nucleotide exchange is not limiting for the eEF-2 function (unpublished observation). The factor is capable of binding guanosine nucleotide analogues such as guanosine $5'$-[β,γ-methylene]triphosphate (GuoPP[CH$_2$]P) (K_d 20 μM), oxidized GTP (oGTP) [199], 5-p-fluorosulfonylbenzoyl guanosine [200] and dGTP [201]. The relatively high affinities observed for dGTP and oGTP show that the $2',3'$ part of the ribose ring is of little importance for nucleotide binding.

oGTP is covalently bound to an amino group between amino acids Gly[262] and Trp[343] in eEF-2 via the two aldehydes in the $2'$ and $3'$ positions of the opened ribose ring [159, 202]. The covalent complex is fully capable of binding to the ribosome but the γ-phosphate of the nucleotide cannot be hydrolysed by the factor, suggesting that the formation of a covalent bond inhibits the conformational adapation necessary for activating the hydrolytic domain [159]. Covalent attachment of oGTP also inhibits the diphtheria-toxin-catalysed ADPribosylation of eEF-2 (see below), indicating that the

structures of the two factor domains are interdependent. The importance of a structural flexibility for the factor function is also demonstrated by proteolytic degradation of the factor. eEF-2 is relatively insensitive to degradation and trypsin treatment only results in a rapid cleavage at Arg[66] but a second site at Lys[571/572] becomes exposed in the guanosine-nucleotide-containing complex, indicating that the nucleotide binding causes a structural rearrangement in this region.

Interaction of eEF-2 with ribosomes

Binding of eEF-2 · GTP to the pre-translocational ribosome leads to a translocation of peptidyl-tRNA from the A-site to the P-site, a distance of less than 3 nm [195]. Translocation as such is not dependent on the eEF-2 as the process spontaneously proceeds in the absence of factor, albeit at an extremely slow rate [18, 203, 204]. Furthermore, binding of eEF-2 in the presence of non-hydrolysable GTP analogous leads to translocation, showing that the process does not require GTP hydrolysis [205]. However, translocation is slow in the absence of hydrolysis and only proceeds to approximately 50% completion [206]. Even in the presence of eEF-2 · GTP, the translocation is close to non-detectable if the temperature is kept at 0 °C [205]. Hence, it seems as if the translocation involves a high-energy intermediate that represses spontaneous translocation. A high-energy intermediate has been postulated from theoretical considerations [207] and the group of Wintermeyer has recently confirmed this in the prokaryotic system by measuring the activation energy of the translocation [208]. The activation energy was reduced from 80 to 32 kJ/mol in the presence of EF-G, indicating that the role of the factor is to eliminate the high-energy barrier that normally reduces the rate of spontaneous translocation.

The tRNA molecule has dynamic properties of vital importance for the translocation process, as seen by the difference in structure between A- and P-site-located tRNA. In the A-site, the tRNA molecule seems to loosen the tertiary interactions of the T- and D-loops while the structure refolds after translocation of the P-site [209]. Stereochemical analysis of tRNA bound to the A- and P-sites has also shown that the angle between the two limbs of the L-shaped tRNA molecule is wider in the A-site than in the P-site [210]. As a consequence, the acceptor stem and the anticodon loop are approximately 0.5 nm further apart in the A-site than in the P-site. These structural changes indicate that the energy required for the translocation is provided by the aminoacyl-tRNA positioned in the A-site [211]. This interpretation has also been supported by kinetic analysis showing that tRNA has a considerably higher affinity for the P-site than for the A-site [102, 212] and by the factor-independent translocation [213]. The additional E-site is believed to shift the equilibrium in favour of translocation by binding the deacylated tRNA after translocation from the P-site [214]. Based on the data from chemical footprinting experiments, Moazed and Noller have proposed a two-step model for the translocation process [293]. In the first step, which occurs spontaneously after formation of the peptide bond, the acceptor end of the peptidyl-tRNA moves relative to the large ribosomal subunit forming a hybrid particle in which the acceptor end is in the P-site with the anticodon still remaining in the A-site. Simultaneously the CCA end of the deacylated tRNA in the P-site moves over to the E-site and interacts with 23 S rRNA, probably by baise-pairing [294]. In the second step, which is promoted by EF-G, the deacylated tRNA is completely transferred to the E-site and the translocation is completed by moving the anticodon end of the peptidyl-tRNA relative to the small ribosomal subunit. The mRNA molecule is a passive partner in the translocation, with the anticodon – codon interaction acting as the handle that drives the mRNA forward [215]. High salt concentrations leads to a spontaneous translocation which could be due to a loosening of the tRNA – ribosome interaction or an altered equilibrium between the ribosomal structures.

eEF-2 is dependent on guanosine nucleotides for binding to the ribosome [216]. In the presence of GDP, the factor forms a ribosomal complex with an apparent dissociation constant of 0.4 μM [37]. The corresponding dissociation constant for the GTP-containing complex can not be determined due to the rapid hydrolysis of GTP. Thus, the affinity of the triphosphate complex must be studied in the presence of nonhydrolysable GTP analogues such as $GuoPP[CH_2]P$. The binary eEF-2 · $GuoPP[CH_2]P$ complex binds to ribosomes with two distinct affinities depending on the ribosomal conformation. Pre-translocational ribosomes interact with eEF-2 · $GuoPP[CH_2]P$ with high affinity ($K_d \approx 1$ nM) whereas the affinity for post-translocational ribosomes is comparable to that of eEF-2 · GDP. Thus, high-affinity binding is dependent on a correct structure of the ribosome as well as of the factor. The high-affinity complex is also formed with empty reconstituted ribosomes showing that the high-affinity complex does not depend on the presence of peptidyl-tRNA in the A-site, but on a correct pre-translocational structure of the ribosome.

The GTPase activity observed during translocation seems to depend on an activation of a catalytic domain located on eEF-2 [200, 217, 218]. Hydrolysis occurs after translocation and eEF-2 continues to hydrolyse GTP in the presence of post-translocational ribosomes provided that eEF-1α and aminoacyl-tRNA are absent [51]. Thus, the energy released by the hydrolysis is not directly coupled to the translocation. Kinetic analysis of the GTPase has shown that the hydrolysis is triggered by ribosomes in the post-translocational state, even in the absence of tRNA and mRNA [37]. The reaction can take place in the presence of 60 S subunits alone [37, 219], showing that the ribosome domain responsible for activating the GTPase is within the 60 S subunit. However, due to the low affinity of the factor for 60 S subunits the K_m of the reaction is considerably higher than for hydrolysis in the presence of complete post-translocational ribosomes.

Reducing agents have to be present in order to retain the function of eEF-2 [220]. Alkylation of the cysteines inhibits the formation of ternary eEF-2 · GTP · ribosome complexes without an apparent loss of the guanosine nucleotide binding ability [221], indicating that the cysteine residues are essential for ribosome binding. However, as the cysteines are scattered over the whole molecule, this experiment does not show which parts of the molecule that are involved in the ribosome interaction. More promising is the finding that ADPribosylation at position 715 leads to a reduced ability of the factor to bind ribosomes [222], suggesting that this region of the factor is involved in the ribosomal binding. This view is further supported by results from site-directed mutagenesis [223, 224]. Furthermore, regions 411 – 430, 501 – 584, 661 – 740 and 741 – 841 in eEF-2 are similar in parts to EF-G, suggesting that the C-terminal region is involved in the factor function during ribosomal binding and translocation [139, 225].

Cleavage at Arg[66] causes an inhibition of the binding of eEF-2 to pre-translocational ribosomes, but does not affect the post-translocational binding or the GTPase activity, suggesting that this region could also be a part of the contact region with the ribosome [159]. This view is supported by the observation that this cleavage site is protected when eEF-2 is bound to the ribosome (unpublished observation). The sequences 54 – 78 in eEF-2 and 57 – 81 in eEF-1α are similar to the prokaryotic elongation factors [138, 139, 225, 226], indicating that this region is a common feature of the elongation factors.

Regulation of the activity of eEF-2

eEF-2 is ADPribosylated in the presence of diphtheria toxin and NAD^+ due to the occurrence of a unique post-translationally modified amino acid 2-[3-carboxyamido-3(trimethylammonio)propyl]histidine, called diphtamide, at position 715 [227 – 230]. The diphtamide residue as such is not essential for the function of eEF-2 in protein synthesis as mutated cells with eEF-2 lacking the modified histidine are vital, suggesting that the role of the modification is to allow the cell to regulate the rate of elongation by ADPribosylation [229]. This assumption is supported by recent reports describing the existence of an endogenous ADPribosyltransferase activated upon serum deprivation. Suggests that ADPribosylation is involved in a negative regulation of the protein synthesis during nutritional down-shifts [231].

The ADPribosylation leads to a total inhibition of protein synthesis and a reduction in the amount of ribosome-associated eEF-2. This effect has been suggested to be due to a decreased nucleotide exchange on eEF-2, a reduced affinity of eEF-2 for RNA and an interference with the ribosome – factor interaction [22, 232, 233]. Kinetic analysis using the eEF-2-catalysed ribosome-dependent hydrolysis of GTP shows that the GDP/GTP exchange was not rate-limiting for the function of eEF-2, suggesting that the protein synthesis inhibition seen after ADPribosylation was not due to a lack of functional eEF-2 · GTP complexes. Instead the kinetc data show that ADPribosylated eEF-2 has reduced affinity for the pre-translocation type of ribosome and a decreased ability to catalyse the hydrolysis of GTP [302].

It was recently found that eEF-2 is phosphorylated by a Ca^{2+}/calmodulin-dependent protein kinase [234 – 237]. The kinase incorporates approximately one phosphate residue/molecule of eEF-2 to a threonine located at the N-terminus of the protein between amino acids Ala[51] and Arg[60] [235, 236]. Although the discovery is rather recent, more is known about the physiological implications of this modification than of the ADPribosylation. The activity of the kinase is reduced after treatment of cells with nerve growth factor or forskolin, suggesting that cAMP may regulate the phosphorylation of eEF-2 [235, 236, 238]. This is also supported by the observed reduction in kinase activity *in vitro* in the presence of cAMP [239]. The phosphorylation of eEF-2 is increased by mitogenic stimulation of fibroblasts and by estradiol in rat corpus luteum [240, 241]. Phosphorylated eEF-2 has also been found in nerve cells, where the phosphorylation was increased by treatment with dimethylphenylpiperazinium or by preganglionic stimulation [242]. Dephosphorylation of eEF-2 is stimulated by tumour-promoting agents like phorbol esters [243] and it seems likely that the enzyme responsible for the dephosphorylation is a type 2A phosphatase [295].

Phosphorylation of eEF-2 results in reduced protein synthesis and the modified factor is unable to restore protein synthesis in an eEF-2-dependent reticulocyte lysate [303]. The inhibition of translocation is caused by an inability of the phosphorylated factor to associate with pre-translocational ribosomes. Thus, the effect of phosphorylation seems to be partly analogous to that obtained following ADPribosylation.

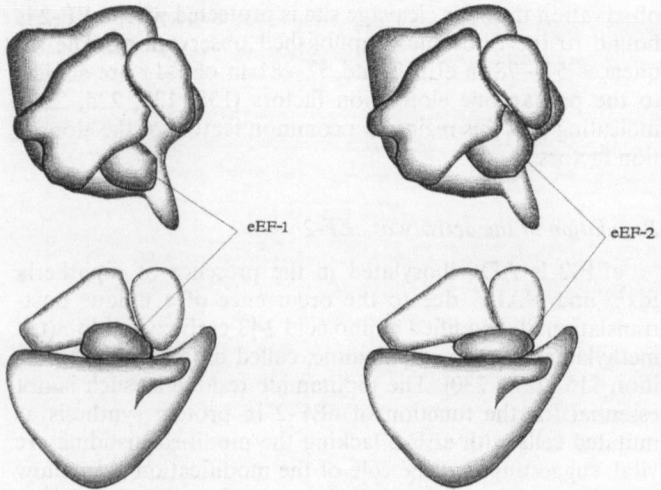

eEF-1

eEF-2

Fig. 8. *Localisation of the ribosomal binding sites for eEF-1α and eEF-2.* See text for details

Hence, it seems reasonable to assume that the two regulatory systems are activated by different signals. Phosphorylation is usually a reversible phenomenon due to the presence of phosphatases. In the case of ADPribosylation, there is no indication of a reactivation of the modified factor by removal of the ADPribose moiety. This could indicate that reduction of the levels of ADPribosylated factor in the cell requires degradation of the modified protein and *de novo* synthesis of functionally active eEF-2.

Ribosomal binding domain for eEF-1α and eEF-2

eEF-1α and eEF-2 requires both ribosomal subunits for association and consequently, the binding domain for both factors consists of proteins from both the large and the small subunit (Table 3). Simultaneous association of eEF-1α and eEF-2 to the ribosome is not possible, but this does not automatically indicate overlapping binding sites since the interaction with the elongation factors depends on the translocational status of the ribosome. However, the finding that some of the proteins belonging to the factor binding sites are common to both sites suggests overlapping binding sites. It can be concluded from the known localization of some of these proteins that both elongation factors associate with the ribosome at or close to the ribosomal interface (Fig. 8). One particularly interesting observation is that the binding site for eEF-2 contains the acidic proteins A33 and P2. In reconstitution experiments, a split protein fraction consisting of the phosphoproteins P1 and P2 was able to restore almost completely the eEF-2-dependent GTPase activity in the presence of core particles derived from 60 S subunits [244, 296]. The same split protein fraction also restored approximately 50% of the eEF-1α-mediated binding of Phe-tRNA to the core particles. Obviously, the two phosphoproteins are of vital importance for the function of the ribosome in the elongation cycle. In *E. coli*, the acidic proteins L7/L12 and L10 are known to be required for the function of the elongation factors and immunoelectron microscopic studies have shown that the two factors EF-Tu and EF-G associate with the ribosome close to the L7/L12 stalk. EF-Tu is positioned at the exterior surface of the 30 S particle, while EF-G seems to be located at a less exposed site near the subunit interface area [245 – 247]. Like L7/L12, P1 and P2 have a high tendency to dimerise and it

has been proposed that the complex between the three acidic proteins P1, P2 and A33 from the eukaryotic large subunit is functionally similar to the protein complex L7/L12 · L10 in the prokaryotic ribosomes [248, 249].

The binding site for eEF-2 on the ribosome is in close proximity to 5 S rRNA [61]. In the bacterial ribosome the 5 S rRNA is located at the central protuberance of the 50 S particle. If we assume that the position of 5 S rRNA is evolutionary conserved, the position of eEF-2 would be close to the central protuberance. It has also been suggested that a complex, consisting of protein L5 and 5 S rRNA, is involved in the eEF-2-dependent GTPase activity [250, 251]. This view is further supported by the observation that both components are within the binding domain for eEF-2 on the ribosome [61, 62]. No indication that ribosomal RNA is involved in the binding of eEF-1α has so far been reported, but recent experiments in prokaryotes have shown that the factor protects nucleotides in loop V of the 23 S rRNA from chemical modification, indicating that 23 S rRNA may be directly involved in binding of the ternary complex [49]. The protection pattern observed with EF-G partly overlaps with that observed for EF-Tu [48]. In addition, EF-G also protects nucleotides 1067 – 1069 and the factor can be cross-linked to A^{1067} [252]. The protected loop V in 23 S rRNA is evolutionary conserved and contains the α-sarcin and ricin modification sites [253]. It is therefore reasonable to assume that 28 S rRNA may also play a role in the eEF-1α – ribosome interaction in eukaryotes.

Two elongation factors or more?

The mechanism for the elongation cycle seems to be highly similar in the eukaryotic and the prokaryotic system. The finding that yeast need a third elongation factor, eEF-3, came therefore as a surprise [254]. The factor is needed only for protein synthesis catalysed by yeast ribosomes while eEF-1α and eEF-2 from yeast are active with rat liver ribosomes even in the absence of eEF-3, suggesting that the activity of eEF-3 is provided by the ribosome itself in higher eukaryotes [255]. Attempts to demonstrate eEF-3 dependence on the elongation cycle with hybrid ribosomes, containing one subunit from yeast and the other from wheat germ, has failed [293]. The gene for eEF-3 has been cloned [256] but the sequence has not yet been published. It is therefore at present not possible to elucidate by sequence comparison if eEF-3 is similar to any other component in the systems from higher eukaryotes. eEF-3s, as are the other elongation factors, a ribosome-dependent GTPase but it also functions as an ATPase [254, 257]. The function of eEF-3 is still unclear, although the factor has been shown to increase the eEF-1α-dependent binding of aminoacyl-tRNA by promoting the catalytic action of eEF-1α [297]. It is remarkable that the function of eEF-3 should still be so completely unknown 14 years after it was first detected. Hopefully, careful examination of the effect of eEF-3 on all the partial reactions in the elongation cycle will clarify its physiological role for protein synthesis in yeast. One approach is the use of genetic tools and a thermosensitive mutant of eEF-3 has been identified [258]. The effect of this mutant on the function of the translational apparatus has, however, not been published.

A further factor that, directly or indirectly, is involved in the yeast protein synthesis elongation cycle is the SUF12 suppressor protein (Table 5) [259]. It was identified as a suppressor of +1 frameshift mutations in glycine and proline codons. SUF12 has a sequence similar to the GDP binding region of EF-Tu but this region is in the centre of the molecule.

The protein is not similar to other known proteins involved in protein synthesis and the function of the wild type is unknown. Identification of the function of SUF12 could shed new light on the mechanism of the elongation cycle as well as answer the question of its ability to suppress frameshifts.

Termination

Termination factors

Elongation of the nascent polypeptide chain is stopped as one of the three termination codons, UAA, UAG and UGA, is translocated into the ribosomal A-site. The termination codons have no corresponding tRNA and instead of supporting ribosomal association of the eEF-1α · GTP · aminoacyl-tRNA complex, these codons stimulate binding of a release factor. The eukaryotic release factor responds to tetranucleotides such as UAG(A) and UAA(A) rather than to the normal codons [260]. However, as the fourth nucleotide can vary without altering the interaction with the release factor, it seems likely that the fourth nucleotide is only necessary for formation of a stable ribosomal complex. The recognition of the termination codon has been postulated to involve an interaction of the termination codon with the ribosomal RNA of the small ribosomal subunit, although the critical nucleotides do not seem to be conserved in all organisms [121, 261]. Recent experiments support this hypothesis. The termination codon UGA was found to be cross-linked to 16 S rRNA but ultraviolet irradiation and the complex formation was stabilised by the release factor (RF) [262]. RF possesses a ribosome-dependent GTPase activity that is markedly stimulated by the termination codon UAA(A) [260]. The ribosomal binding of RF is stimulated by GTP and its non-hydrolysable analogue GuoPP[CH$_2$]P. However, only GTP stimulates the catalytic function of RF suggesting that, in this case also, the hydrolysis of factor-bound GTP is required to destabilise the ribosome − factor interaction. The actual hydrolysis of the ester bond between the nascent peptide chain and the P-site-located tRNA is catalysed by the ribosomal peptidyltransferase center and requires no additional energy in the form of GTP [263, 264]. In reticulocytes, an additional factor that stimulates termination has been identified [265]. This factor has no termination activity per se but allows RF to function at much lower concentrations. Hence, it seems as if the stimulatory factor lowers the K_m for the termination codon in a similar manner to that observed for RF-3 in bacteria.

Ribosomal binding of the release factors in prokaryotes seems to involve the L7/L12 stalk on the large ribosomal subunit [266, 267]. This region overlaps the domain involved in the ribosomal binding of elongation factors and it is therefore not surprising that both A-site-bound tRNA and EF-G bound to the ribosome inhibits association of RF [262, 268, 269]. Ribosomal RNA may also be involved in binding the release factor as ribosomes lacking 5 S rRNA are inactive in termination [270]. However, this effect could be due to the fact that both the binding of RF to the ribosome and the peptidyltransferase activity is lost in these ribosomes.

The peptide chain exit domain

Recent progress in determining the structure of the bacterial ribosome has shown that the synthesised polypeptide chain emerges from the ribosome through a hole in large ribosomal subunit [28, 271]. After the peptide has reached a length of approximately 40 amino acids, the nascent chain protrudes at the exit domain on the ribosome [298]. This domain has been mapped on the E. coli ribosome using antibodies against β-galactosidase. The antibodies were seen associated with the 50 S particle opposite the head of the small ribosomal subunit [272]. By use of the same technique, the exit domain of the eukaryotic ribosome was found to be located at a position equivalent to that in the prokaryotic ribosome [273]. Thus the nascent peptide emerges from the ribosome at a site opposite the suggested location of the peptidyltransferase centre [4].

Ribosome interaction with signal recognition particle and membranes

During translation of secretory proteins, elongation is stopped at an internal position in the amino acid sequence until the translational apparatus has attached to the endoplasmic reticulum [274]. This is apparently achieved by a prolonged pausing of the ribosome at naturally occurring pausing sites [275]. The mechanism behind the elongation arrest is at present unknown, but the signal recognition particle (SRP) obviously interacts directly with the ribosome [276]; it has been suggested that the ribosomal site at which SRP attaches is identical to or partly overlaps the ribosomal A-site [277]. Thereby, the binding of SRP excludes ribosomal interaction with aminoacyl-tRNA and eEF-1α.

The translational arrest caused by the binding of SRP is relieved by association with the membrane-bound docking protein, whereby the translational apparatus becomes attached to the endoplasmic reticulum [278]. In addition to the ribosome − membrane interaction mediated by the nascent polypeptide inserted into the membrane, membrane-bound receptors for the 60 S subunit apparently stabilise the interaction [279]. Two proteins specifically found in the rough endoplasmic reticulum, ribophorin I and II, have been cross-linked to the ribosome and thus suggested to be ribosome receptor candidates [280, 281]. However, recent data suggest that the ribophorins are not directly involved in the binding of ribosomes to membranes [282]. Thus, the proteins responsible for the interaction are not as yet identified. However, from electron microscopic data it can be inferred that membrane binding occurs at a site adjacent to the exit domain of the large ribosomal subunit [4].

We gratefully acknowledge Dr Reinout Amons for valuable suggestions during the preparation of the manuscript and David Herron for improving the language. Our work is supported by a grant (B-Bu-8463-301) from the Swedish Natural Science Research Council.

REFERENCES

1. Wool, I. G. (1979) Annu. Rev. Biochem. 48, 719−754.
2. Noller, H. F. (1984) Annu. Rev. Biochem. 53, 119−162.
3. Gutell, R. R., Weiser, B., Woese, C. R. & Noller, H. F. (1985) Prog. Nucleic Acids Res. 32, 155−216.
4. Lake, J. A. (1985) Annu. Rev. Biochem. 54, 507−530.
5. Frank, J., Verschoor, A., Wagenknecht, T., Radermacher, M. & Carazo, J. M. (1988) Trends Biochem. Sci. 13, 123−127.
6. Tolan, D. R. & Traut, R. R. (1981) J. Biol. Chem. 256, 10129−10136.
7. Gross, B., Westermann, P. & Bielka, H. (1982) EMBO J. 2, 255−260.
8. Meier, N. & Wagner, R. (1984) Nucleic Acids Res. 12, 1473−1487.
8. Uchiumi, T., Kikuchi, M., Terao, K. & Ogata, K. (1985) J. Biol. Chem. 260, 5669−5674.

14

9. Uchiumi, T., Kikuchi, M., Terao, K. & Ogata, K. (1985) *J. Biol. Chem. 260*, 5675–5682.

10. Uchiumi, T., Kikuchi, M. & Ogata, K. (1986) *J. Biol. Chem. 261*, 9663–9667.

11. Hultin, T. (1986) *Biochim. Biophys. Acta 872*, 226–235.

12. Lutsch, G., Noll, F., Theise, H., Enzmann, G. & Bielka, H. (1979) *Mol. Gen. Genet. 176*, 281–291.

13. Lutsch, G., Bielka, H., Enzmann, G. & Noll, F. (1983) *Biomed. Biochim. Acta 42*, 705–723.

14. Stern, S., Powers, T., Changchien, L.-M. & Noller, H. F. (1989) *Science 244*, 783–790.

15. Walleczek, J., Schüler, D., Stöffler-Meilicke, M., Brimacombe, R. & Stöffler, G. (1988) *EMBO J. 7*, 3571–3576.

16. Emmanuilov, I., Sabatini, D. D., Lake, J. A. & Freienstein, C. (1978) *Proc. Natl Acad. Sci. USA 75*, 1389–1393.

17. Bielka, H. (1985) *Prog. Nucleic Acids Res. 32*, 267–284.

18. Spirin, A. S. (1986) in *Ribosome structure and protein biosynthesis*, The Benjamin Cummings Publishing Co., Menlo Park, CA.

19. Kahan, L., Winkelmann, D. A. & Lake, J. A. (1981) *J. Mol. Biol. 145*, 193–214.

20. Lake, J. A. & Strycharz, W. A. (1981) *J. Mol. Biol. 153*, 979–992.

21. Stöffler-Meilicke, M., Epe, B., Steinhäuser, K. G., Wolley, P. & Stöffler, G. (1983) *FEBS Lett. 163*, 94–98.

22. Stöffler-Meilicke, M., Noah, M. & Stöffler, G. (1983) *Proc. Natl Acad. Sci USA 80*, 6780–6784.

23. Serdyuk, I. N., Smirnov, N. I., Ptitsyn, O. B. & Fedorov, B. A. (1970) *FEBS Lett. 9*, 324–326.

24. Serdyuk, I. N. & Grenader, A. K. (1975) *FEBS Lett. 59*, 133–136.

25. Beadry, P., Peterson, H. V., Grunberg-Manago, M. & Jacrot, B. (1976) *Biochem. Biophys. Res. Commun. 72*, 391–397.

26. Struhrman, H. B., Haas, J., Ibel, K., De Wolf, B., Koch, M. H. J., Parfait, R. & Crichton, R. R. (1976) *Proc. Natl Acad. Sci. USA 73*, 2379–2383.

27. Kuhlbrandt, W. & Unwin, P. N. T. (1982) *J. Mol. Biol. 156*, 431–448.

28. Milligan, R. A. & Unwin, P. N. T. (1986) *Nature 319*, 693–695.

29. Yusupov, M. M. & Spirin, A. S. (1986) *FEBS Lett. 197*, 229–233.

30. Burma, D. P., Nag, B. & Tewari, D. S. (1983) *Proc. Natl Acad. Sci. USA 80*, 4875–4878.

31. Herr, W., Chapman, N. M. & Noller, H. F. (1979) *J. Mol. Biol. 130*, 433–439.

32. Meier, N., Goringer, H. U., Kleuvers, B., Scheibe, U., Eberle, J., Szymkowiak, C., Zacharias, M. & Wagner, R. (1986) *FEBS Lett. 204*, 89–95.

33. Tapprich, W. E., Goss, D. J. & Dahlberg, A. E. (1989) *Proc. Natl Acad. Sci USA 86*, 4927–4931.

34. Cover, J. A., Lambert, J. M., Norman, C. M. & Traut, R. R. (1981) *Biochemistry 20*, 2843–2852.

35. Chiam, C. L. & Wagner, R. (1983) *Biochemistry 22*, 1193–1200.

36. Nygård, O. & Nika, H. (1982) *EMBO J. 1*, 357–362.

37. Nygård, O. & Nilsson, L. (1989) *Eur. J. Biochem. 179*, 603–608.

38. Hapke, B. & Noll, H. (1976) *J. Mol. Biol. 105*, 99–109.

39. Conway, T. W. & Lipman F. (1964) *Proc. Natl Acad. Sci. USA 52*, 1462–1469.

40. van Diggelen, O. P., Oostrom, H. & Bosch, L. (1973) *Eur. J. Biochem. 39*, 511–518.

41. Burma, D. P., Srivastava, A. K., Srivastava, S., Tewari, D. S., Dash, D. & Sengupta, S. K. (1984) *Biochem. Biophys. Res. Commun. 124*, 970–978.

42. Spirin, A. S., Baranov, V. I., Polubesov, G. S., Serdyuk, I. N. & May, R. P. (1987) *J. Mol. Biol. 194*, 119–128.

43. Spirin, A. S. (1969) *Cold Spring Harbor Symp. Quant. Biol. 34*, 197–207.

44. Endo, Y. & Wool, I. G. (1982) *J. Biol. Chem. 257*, 9054–9060.

45. Endo, Y., Huber, P. W. & Wool, I. G. (1983) *J. Biol. Chem. 258*, 2662–2667.

46. Endo, Y., Mitsui, K., Matizuki, M. & Tsurugi, K. (1987) *J. Biol. Chem. 262*, 5908–5912.

47. Endo, Y., Tsurugi, K. & Ebert, R. F. (1988) *Biochim. Biophys. Acta 954*, 224–226.

48. Moazed, D., Robertson, J. M. & Noller, H. F. (1988) *Nature 334*, 362–364.

49. Moazed, D. & Noller, H. F. (1989) *Cell 57*, 585–597.

50. Nilsson, L., Asano, K., Svensson, B., Poulsen, F. M. & Nygård, O. (1985) *Biochim. Biophys. Acta 868*, 62–70.

51. Nilsson, L. & Nygård, O. (1986) *Eur. J. Biochem. 161*, 111–117.

52. Terao, K., Uchiumi, T., Endo, Y. & Ogata, K. (1988) *Eur. J. Biochem. 174*, 459–463.

53. Westermann, P., Nygård, O. & Bielka, H. (1980) *Nucleic Acids Res. 8*, 3065–3071.

54. Westermann, P., Nygård, O. & Bielka, H. (1980) *Nucleic Acids Res. 10*, 2387–2396.

55. Bommer, U.-A., Stahl, J., Henske, A., Lutsch, G. & Bielka, H. (1988) *FEBS Lett. 233*, 114–118.

56. Nygård, O. & Westermann, P. (1982) *Nucleic Acids Res. 10*, 1327–1334.

57. Westerman, P. & Nygård, O. (1983) *Biochim. Biophys. Acta 741*, 103–108.

58. Tolan, D. R., Hershey, J. W. B. & Traut, R. R. (1983) *Biochimie 65*, 427–436.

59. Uchiumi, T. & Ogata, K. (1986) *J. Biol. Chem. 261*, 9668–9671.

60. Uchiumi, T., Kikuchi, M., Terao, K., Iwasaki, K. & Ogata, K. (1986) *Eur. J. Biochem. 156*, 37–44.

61. Nygård, O., Nilsson, L. & Westermann, P. (1987) *Biochim. Biophys. Acta 910*, 245–253.

62. Nygård, O. & Nilsson, L. (1987) *Biochim. Biophys. Acta 908*, 46–53.

63. Takahashi, Y. & Ogata, K. (1981) *J. Biochem. (Tokyo) 90*, 1549–1552.

64. Westermann, P. & Nygård, O. (1984) *Nucleic Acids Res. 12*, 8887–8897.

65. Stahl, J. & Kobets, N. D. (1981) *FEBS Lett. 123*, 269–272.

66. Takahashi, Y., Ogata, K. (1985) *Eur. J. Biochem. 152*, 279–286.

67. Nakashima, K. Darzynkiewicz, E., Shatkin, A. J. (1980) *Nature 286*, 226–230.

68. Reboud, A.-M., Dubost, S. & Reboud, J.-P. (1982) *Eur. J. Biochem. 124*, 389–396.

69. Fabijanski, S. & Pelligrini, M. (1983) *Mol. Gen. Genet. 184*, 551–556.

70. Czernilofski, A. P., Collatz, E., Gressner, A. M. & Wool, I. G. (1977) *Mol. Gen. Genet. 153*, 231–235.

71. Vlasov, V. V. & Westermann, P. (1976) *Mol. Biol. Mosc. 10*, 670–674.

72. Metspalu, A., Saarma, M., Villems, R., Ustav, M. & Lind, A. (1978). *Eur. J. Biochem. 91*, 73–81.

73. Ulbrich, N., Lin, A. & Wool, I. G. (1979) *J. Biol. Chem. 254*, 8641–8645.

74. Ulbrich, N., Wool, I. G., Ackermann, E. & Sigler, P. B. (1980) *J. Biol. Chem. 255*, 7010–7016.

75. Villems, R., Saarma, M., Metspalu, A. & Toots, I. (1979) *FEBS Lett. 107*, 66–68.

76. Todokoro, K., Ulbrich, N., Chan, Y.-L. & Wool, I. G. (1981) *J. Biol. Chem. 256*, 7207–7212.

77. Böhm, H., Stahl, J. & Bielka, H. (1979) *Acta Biol. Med. Germ. 38*, 1447–1452.

78. Stahl, J., Dressler, K. & Bielka, H. (1974) *FEBS Lett. 47*, 167–170.

79. Westermann, P., Gross, B. & Heumann, W. (1974) *Acta Biol. Med. Germ. 33*, 699–707.

80. Stahl, J. & Bielka, H. (1983) *Biomed. Biochim. Acta 42*, 47–55.

81. Stahl, J., Böhm, H., Pozdnjakov, V. A. & Girschovich, A. S. (1979) *FEBS Lett. 102*, 273–276.

82. Reyers, R. Vazquez, D. & Bellesta J. P. G. (1976) *Eur. J. Biochem. 67*, 267–274.

83. Reyers, R., Vazquez, D. & Ballesta, J. P. G. (1977) *Eur. J. Biochem. 73*, 25–31.

84. Czernilovsky, A. P., Collatz, E., Gressner, A. M., Wool, I. G. & Küchler, E. (1977) *Mol. Gen. Genet. 153*, 231–235.

85. Westermann, P., Heumann, W., Bommer, U.-A., Bielka, H., Nygård, O. & Hultin, T. (1979) *FEBS Lett. 97*, 101–104.

86. Reboud, A.-M., Dubost, S. & Reboud, J.-P. (1983) *FEBS Lett.* *158*, 285 – 288.
87. Westermann, P., Nygård, O. & Bielka, H. (1981) *Nucleic Acids Res. 9*, 2387 – 2396.
88. Svoboda, A. J. & McConkey, E. H. (1978) *Biochem. Biophys. Res. Commun. 81*, 1145 – 1152.
89. Terao, K., Uchiumi, T. & Ogata, K. (1980) *Biochim. Biophys. Acta 609*, 306 – 312.
90. Reboud, A.-M., Dubost, S. & Reboud, J.-P. (1984) *Eur. J. Biochem. 143*, 303 – 307.
91. Ehresmann, C. & Ofengand, J. (1984) *Biochemistry 23*, 438 – 445.
92. Prince, J. B., Taylor, B. H., Thurlow, D. L., Ofengand, J. & Zimmermann, R. A. (1982) *Proc. Natl Acad. Sci. USA 79*, 5450 – 5454.
93. Ehresmann, C., Ehresmann, B., Millar, R., Ebel, J.-P., Nurse, K. & Ofengand, J. (1984) *Biochemistry 23*, 429 – 437.
94. Denman, R., Nègre, D., Cunningham, P. R., Nurse, K., Colgan, J., Weitzmann, C. & Ofengand, J. (1989) *Biochemistry 28*, 1012 – 1019.
95. Wagenknecht, T., Frank, J., Boublik, M., Nurse, K. & Ofengand, J. (1988) *J. Mol. Biol. 203*, 753 – 760.
96. Collins, J. F., Moon, H. M. & Maxwell, E. S. (1972) *Biochemistry 11*, 4187 – 4195.
97. Brimacombe, R. (1988) *Biochemistry 27*, 4207 – 4214.
98. Wower, J., Hixson, S. S. & Zimmermann, R. A. (1989) *Proc. Natl Acad. Sci. USA 86*, 5232 – 5236.
99. Hardesty, B., Culp, W. & McKeehan, W. (1969) *Cold Spring Harbor Symp. Quant. Biol. 34*, 331 – 345.
100. Wettstein, F. O. & Noll, H. (1965) *J. Mol. Biol. 11*, 35 – 53.
101. Gnirke, A., Geigenmüller, U., Rheinberger, H. J. & Nierhaus, K. H. (1989) *J. Biol. Chem. 264*, 7291 – 7301.
102. Lill, R., Robertson, J. M. & Wintermeyer, W. (1984) *Biochemistry 23*, 6710 – 6717.
103. Moazed, D. & Noller, H. F. (1986) *Cell 47*, 985 – 994.
104. Dahlberg, A. E. (1989) *Cell 57*, 525 – 529.
105. Pain, V. M. (1986) *Biochem. J. 235*, 625 – 637.
106. Proud, C. G. (1986) *Trends Biochem. Sci. 11*, 73 – 77.
107. Roades, R. E. (1988) *Trends Biochem. Sci. 13*, 52 – 56.
108. Sohnenberg, N. (1988) *Prog. Nucleic Acids Res. 35*, 174 – 207.
109. Trachsel, H. & Staehelin, T. (1979) *Biochim. Biophys. Acta 565*, 305 – 314.
110. Raychaudhuri, P., Stringer, E. A., Valenzuela, D. M. & Maitra, U. (1984) *J. Biol. Chem. 259*, 11930 – 11935.
111. Trachsel, H., Erni, B. & Staehelin, T. (1978) *J. Mol. Biol. 116*, 755 – 767.
112. Benne, R. & Hershey, J. W. B. (1978) *J. Biol. Chem. 253*, 3078 – 3087.
113. Peterson, D. T., Merrick, W. C. & Safer, B. (1979) *J. Biol. Chem. 254*, 2509 – 2516.
114. Kozak, M. (1989) *J. Cell Biol. 108*, 229 – 241.
115. Ochoa, S. (1983) *Arch. Biochem. Biophys. 223*, 325 – 349.
116. Sundkvist, I. C. & Staehelin, T. (1975) *J. Mol. Biol. 99*, 401 – 418.
117. Boublik, M., Hellmann, W., Staehelin, T. & Trachsel, H. (1983) *Eur. J. Cell Biol. 32*, 136 – 142.
118. Lutsch, G., Benndorf, R., Westermann, P., Bommer, U.-A. & Bielka, H. (1986) *Eur. J. Cell Biol. 40*, 257 – 265.
119. Bommer, U.-A., Lutsch, G., Behlke, J., Stahl, J., Nesytova, N., Henske, A. & Bielka, H. (1988) *Eur. J. Biochem. 172*, 653 – 662.
120. Behlke, J., Bommer, U.-A, Lutsch, G., Henske, A.-M. & Bielka, H. (1986) *Eur. J. Biochem. 157*, 523 – 570.
121. Shine, J. & Dalgarno, L. (1974) *Proc. Natl Acad. Sci. USA 71*, 1342 – 1346.
122. Stiege, W., Stade, K., Schüler, D. & Brimacombe, R. (1988) *Nucleic Acids Res. 16*, 2369 – 2388.
123. Nombela, C., Redfield, B., Ochoa, S. & Weissbach, H. (1976) *Eur. J. Biochem. 65*, 395 – 402.
124. Slobin, L. I. & Möller, W. (1975) *Nature 258*, 452 – 454.
125. Slobin, L. I. & Möller, W. (1976) *Eur. J. Biochem. 69*, 351 – 366.
126. Maessen, G. D. F., Amons, R., Maassen, J. A. & Möller, W. (1986) *FEBS Lett. 208*, 77 – 83.
127. Maessen, G. D. F., Amons, R., Zeelen, J. P. & Möller, W. (1987) *FEBS Lett. 223*, 181 – 186.
128. Carvalho, J. F., Carvalho, M. & Merrick, W. C. (1984) *Arch. Biochem. Biophys. 234*, 591 – 602.
129. Sacchi, G. A., Aocchi, G. & Gocucci, S. (1984) *Eur. J. Biochem. 139*, 1 – 4.
130. Kemper, W. M. & Merrick, W. C. (1976) *Arch. Biochem. Biophys. 174*, 603 – 612.
131. Tuháckhová, A. & Hradec, J. (1985) *Eur. J. Biochem. 146*, 365 – 370.
132. Rosenberry, T. L., Krall, J. A., Dever, T. E., Haas, R., Louvard, D. & Merrick, W. C. (1989) *J. Biol. Chem. 264*, 7096 – 7099.
133. Whiteheart, S. W., Shenbagamurthi, P., Chen, L., Cotter, R. J. & Hart, G. W. (1989) *J. Biol. Chem. 264*, 14334 – 14341.
134. Janssen, G. M. C. & Möller, W. (1988) *Eur. J. Biochem. 171*, 119 – 129.
135. de Vos, A. M., Tong, L., Milburn, M. V., Matias, P. M., Jancarik, J., Noguchi, S., Nishimura, S., Miura, K., Ohtsuka, E. & Kim, S.-H. (1988) *Science 239*, 888 – 893.
136. Jurnak, F. (1985) *Science 230*, 32 – 36.
137. LaCour, T. F. M., Nyborg, J., Thirup, S. & Clark, B. F. C. (1985) *EMBO J. 4*, 2385 – 2388.
138. van Hemert, F. J., Amons, R., Pluijms, W. J. M., van Ormondt, H. & Möller, W. (1984) *EMBO J. 3*, 1109 – 1113.
139. Kohno, K., Uchida, T., Ohkubo, H., Nakanishi, S., Nakanishi, T., Fukui, T., Ohtsuka, E., Ikehara, M. & Okada, Y. (1986) *Proc. Natl Acad. Sci. USA 83*, 4978 – 4982.
140. Amons, R., Pluijms, W., Roobol, K. & Möller, W. (1983) *FEBS Lett. 153*, 37 – 42.
141. Dever, T. E., Glynias, M. J. & Merrick, W. C. (1987) *Proc. Natl Acad. Sci. USA 84*, 1814 – 1818.
142. Miller, D. L. & Weissbach, H. (1977) in *Nucleic acids – protein recognition* (Vogel, H. J., ed.) pp. 409 – 430, Academic Press, New York.
143. Nolan, R. D., Grasmuk, H., Högenauer, G. & Drews, J. (1974) *Eur. J. Biochem. 45*, 601 – 609.
144. Nagata, S., Iwasaki, K. & Kaziro, Y. (1976) *Arch. Biochem. Biophys. 172*, 168 – 177.
145. Slobin, L. I. & Möller, W. (1978) *Eur. J. Biochem. 84*, 69 – 77.
146. Moon, H.-M., Redfield, B. & Weissbach, H. (1972) *Proc. Natl Acad. Sci. USA 69*, 1249 – 1252.
147. Duffy, L. K., Gerber, L., Johnson, A. E. & Miller, D. L. (1981) *Biochemistry 20*, 4663 – 4666.
148. van Noort, J. M., Kraal, B., Bosch, L., laCour, T. F. M., Nyborg, J. & Clark, B. F. C. (1984) *Proc. Natl Acad. Sci. USA 81*, 3969 – 3972.
149. van Noort, J. M., Kraal, B. & Bosch, L. (1985) *Proc. Natl Acad. Sci. USA 82*, 3212 – 3216.
150. Metz-Boutigue, M.-H., Reinbolt, J., Ebel, J.-P., Ehresmann, C. & Ehresmann, B. (1989) *FEBS Lett. 245*, 194 – 200.
151. Beres, L. & Lucas-Lenard (1973) *Biochemistry 12*, 3998 – 4004.
152. Johnson, A. E. & Slobin, L. I. (1980) *Nucleic Acid. Res. 8*, 4185 – 4200.
153. Furano, A. V. (1975) *Proc. Natl Acad. Sci. USA 72*, 4780 – 4784.
154. Slobin, L. I. (1980) *Eur. J. Biochem. 110*, 555 – 563.
155. Nygård, O. & Nilsson, L. (1984) *Eur. J. Biochem. 145*, 345 – 350.
156. Möller, W., Schipper, A. & Amons, R. (1987) *Biochimie 69*, 983 – 989.
157. Slobin, L. I., Clark, R. V. & Olson, M. O. J. (1981) *Biochemistry 20*, 5761 – 5762.
158. Wittinghofer, A., Frank, R. & Leberman, R. (1980) *Eur. J. Biochem. 108*, 423 – 431.
159. Nilsson, L. & Nygård, O. (1988) *Eur. J. Biochem. 171*, 293 – 299.
160. Grasmuk, H., Nolan, R. D. & Drews, J. (1974) *Eur. J. Biochem. 48*, 485 – 493.
161. Nombela, C. & Ochoa, S. (1973) *Proc. Natl Acad. Sci. USA 70*, 3556 – 3560.
162. Nolan, R. D., Grasmuk, H. & Drews, J. (1975) *Eur. J. Biochem. 50*, 391 – 402.
163. Weissbach, H., Redfield, B. & Moon, H.-M. (1973) *Arch. Biochem. Biophys. 156*, 267 – 275.

16

164. Crechet, J.-B. & Parmeggiani, A. (1986) *Eur. J. Biochem. 161*, 655–660.
165. Thompson, R. C., Dix, D. B., Gerson, R. B. & Karim, A. M. (1981) *J. Biol. Chem. 256*, 81–86.
166. Luce, M. C., Tschanz, K. D., Gotto, D. A. & Bunn C. L. (1985) *Biochim. Biophys. Acta 825*, 280–288.
167. Ellis, N. & Gallant, J. (1982) *Mol. Gen. Genet. 188*, 169–172.
168. Loftfield, R., van der Jagt, D. (1972) *Biochem. J. 128*, 1353–1356.
169. Ninio, J. (1975) *Biochimie 57*, 587–595.
170. Kurland, C. G. (1987) *Trends Biochem. Sci. 12*, 126–128, 169–171, 210–212.
171. Thompson, R. C. (1988) *Trends Biochem. Sci. 13*, 91–93.
172. Vladimirov, S. N., Graifer, D. M., Karpova, G. G., Semenkov, Y. P., Makhno, V. I. & Kirillov, S. V. (1985) *FEBS Lett. 181*, 367–372.
173. Ruusala, T., Andersson, D. T., Egrenberg, M. & Kurland, C. G. (1984) *EMBO J. 3*, 2575–2580.
174. Kirsebom, L. A. & Isaksson, L. A. (1986) *Mol. Gen. Genet. 205*, 240–247.
175. Vijgenboom, E., Vink, T., Kraal, B. & Bosch, L. (1985) *EMBO J. 4*, 1049–1052.
176. Murgola, E. J. (1985) *Annu. Rev. Genet. 19*, 57–80.
177. Sandbaken, M. G. & Culbertson, M. R. (1988) *Genetics 120*, 923–934.
178. Mendenhall, M. D., Leeds, P., Fen, H., Mathison, L., Zwick, M., Sleiziz, C. & Culbertson, M. R. (1987) *J. Mol. Biol. 194*, 41–58.
179. Cummins, C. M., Donahue, T. F., Culbertson, M. R. (1982) *Proc. Natl Acad. Sci. USA 79*, 3565–3569.
180. Winey, M., Mendenhall, M. D., Cummins, C. M. & Culbertson, M. R. (1986) *J. Mol. Biol. 192*, 49–63.
181. Ball, C. B., Mendenhall, M. D., Sandbaken, M. G. & Culbertson, M. R. (1988) *Nucleic Acids Res. 16*, 8712–8713.
182. Blazejovski, C. A. & Webster, C. G. (1983) *Mech. Aging Dev. 22*, 345–356.
183. Richardson, A., Birchenall-Sparks, M. C. & Plesko, M. M. (1984) *Topics in Aging Research in Europe 1*, 3–12.
184. Webster, G. C. & Webster, S. L. (1983) *Mech. Ageing Dev. 22*, 121–128.
185. Moldave, K., Harris, J., Sabo, W. & Sadnik, I. (1979) *Fed. Proc. 38*, 1979–1983.
186. Engelhardt, D. L. & Sarnoski, J. (1975) *J. Cell. Physiol. 86*, 15–30.
187. Fisher, I. & Moldave, K. (1980) *Biochemistry 19*, 1417–1425.
188. Coppard, N. J., Clark, B. F. C. & Cramer, F. (1983) *FEBS Lett. 164*, 330–334.
189. Fonzi, W. A., Katayama, C., Lethers, T. & Sypherd, P. S. (1985) *Mol. Cell. Biol. 5*, 1100–1103.
190. Hiatt, W. R., Garcia, R., Merrick, W. C. & Sypherd, P. S. (1982) *Proc. Natl Acad. Sci. USA 79*, 3433–3437.
191. Davydova, E. K., Sitikov, A. S. & Ovchinnikov, L. P. (1984) *FEBS Lett. 176*, 401–405.
192. Janssen, G. M. C., Maessen, G. D. F., Amons, R. & Möller, W. (1988) *J. Biol. Chem. 263*, 11063–11066.
193. Lührmann, R., Eckardt, H. & Stöffler, G. (1979) *Nature 280*, 423–425.
194. Smith, D. & Yarus, M. (1989) *Proc. Natl Acad. Sci. USA 86*, 4397–4401.
195. Johnson, A. E., Adkins, H. J., Matthews, E. A. & Cantor, C. R. (1982) *J. Mol. Biol. 156*, 113–140.
196. Barta, A., Steiner, G., Brosius, J., Noller, H. F. & Kuechler, E. (1984) *Proc. Natl Acad. Sci. USA 81*, 3607–3611.
197. Noller, H. F., Kap, J. A., Wheaton, V., Brosius, J., Gutell, R. R., Kopylov, A. M., Dohme, F., Herr, W., Stahl, D. A., Gupta, R. & Woese, C. R. (1981) *Nucleic Acids Res. 9*, 6167–6189.
198. Henriksen, O., Robinson, E. A. & Maxwell, E. S. (1975) *J. Biol. Chem. 250*, 720–724.
199. Nurten, R. & Bermek, E. (1980) *Eur. J. Biochem. 103*, 551–555.
200. Nilsson, L. & Nygård, O. (1984) *Biochim. Biophys. Acta 782*, 49–54.
201. Chuang, D.-M. & Weissbach, H. (1972) *Arch. Biochem. Biophys. 152*, 114–124.
202. Nilsson, L. & Nygård, O. (1985) *Eur. J. Biochem. 148*, 299–304.
203. Pestka, S. (1969) *J. Biol. Chem. 244*, 1533–1539.
204. Bergemann, K. & Nierhaus, K. H. (1983) *J. Biol. Chem. 258*, 15105–15113.
205. Tanaka, M., Iwasaki, K. & Kaziro, Y. (1977) *J. Biochem. (Tokyo) 82*, 1035–1043.
206. Robertson, J. M. & Wintermeyer, W. (1987) *J. Mol. Biol. 196*, 525–540.
207. Potapov, A. P. (1982) *Mol. Biol. (Mosc.) 16*, 21–25 (Engl. Transl.).
208. Robertson, J. M. & Wintermeyer, W. (1989) *American Society for Microbiology Conference on Ribosomes*, Abstr. II–12, Montana.
209. Jörgensen, T., Siborska, G. E., Wikman, F. P. & Clark, B. F. C. (1985) *Eur. J. Biochem. 153*, 203–209.
210. Spirin, A. S. & Lim, V. I. (1985) in *Structure, function and genetics of ribosomes* (Hardesty, B. & Kramer, G., eds) pp. 556–572, Springer-Verlag, Berlin.
211. Crothers, D. M. & Cole, P. E. (1978) *Transfer RNA* (Altmann, S., ed.) pp. 196–247, MIT Press, Cambridge, MA.
212. Holschuh, K. & Gassen, H. G. (1982) *J. Biol. Chem. 257*, 1987–1992.
213. Beletsina, N. V. & Spirin, A. S. (1979) *Eur. J. Biochem. 94*, 315–320.
214. Nierhaus, K. H. (1984) *Mol. Cell. Biochem. 61*, 63–81.
215. Matzke, A. J. M., Barta, A. & Kuechler, E. (1980) *Proc. Natl Acad. Sci. USA 77*, 5110–5114.
216. Nygård, O. & Nilsson, L. (1984) *Eur. J. Biochem. 145*, 345–350.
217. Girschovich, A. S., Pozdnyakov, V. A. & Ovchinnikov, T. A. (1976) *Eur. J. Biochem. 69*, 321–328.
218. DeVendettis, E., Masullo, M. & Bocchini, V. (1986) *J. Biol. Chem. 261*, 4445–4450.
219. Corquet, F., Lavergne, J.-P., Paleologue, A., Reboud, J.-P. & Reboud, A.-M. (1987) *Eur. J. Biochem. 163*, 15–20.
220. Mosteller, R., Ravel, J. & Hardesty, B. (1966) *J. Biol. Chem. 241*, 1698–1704.
221. Nurten, R., Aktar, N. B. & Bermek, E. (1983) *FEBS Lett. 154*, 391–394.
222. Nygård, O. & Nilsson, L. (1985) *Biochim. Biophys. Acta 824*, 152–162.
223. Kohno, K. & Uchida, T. (1987) *J. Biol. Chem. 262*, 12298–12305.
224. Omura, F., Kohno, K. & Uchida, T. (1989) *Eur. J. Biochem. 180*, 1–8.
225. Zengel, J. A., Archer, R. H. & Lindahl, L. (1984) *Nucleic Acids Res. 12*, 2181–2192.
226. Arai, K., Clark, B. F. C., Dyffy, L., Jones, M. D., Kaziro, Y., Laursen, R. A., L'Italien, J., Miller, D. L., Nagarkatti, S., Nakamura, S., Nielsen, K. M., Petersen, T. E., Takahashi, K. & Wade, M. (1980) *Proc. Natl Acad. Sci. USA 77*, 1326–1330.
227. van Ness, B. G., Howard, J. B. & Bodley, J. W. (1980) *J. Biol. Chem. 255*, 10710–10716.
228. Dunlop, P. C. & Bodley, J. W. (1983) *J. Biol. Chem. 258*, 4754–4758.
229. Moehring, J. M., Moehring, T. J. & Danley, D. E. (1980) *Proc. Natl Acad. Sci. USA 77*, 1010–1014.
230. Moehring, J. M. & Moehring, T. J. (1988) *J. Biol. Chem. 263*, 3840–3844.
231. Fenderick, J. L. & Iglewski, W. J. (1989) *Proc. Natl Acad. Sci. USA 86*, 554–557.
232. Burns, G., Abraham, K. A. & Vedeler, A. (1986) *FEBS Lett. 208*, 217–220.
233. Sitikov, A. S., Davydova, E. K., Bezlepkina, T. A., Ovchinnikov, L. P. & Spirin, A. S. (1984) *FEBS Lett. 176*, 406–410.
234. Palfrey, H. C. (1983) *FEBS Lett. 157*, 183–190.
235. Nairn, A. C. & Palfrey, H. C. (1987) *J. Biol. Chem. 262*, 17229–17303.

236. Nairn, A. C., Nichols, R. A., Brady, M. J. & Palfrey, H. C. (1987) *J. Biol. Chem. 262*, 14265 – 14272.

237. Ryazanov, A. G., Shestakova, E. A. & Natapov, P. G. (1988) *Nature 334*, 170 – 173.

238. Koizumi, S., Ryazanov, A., Hama, T., Chen, H.-C. & Guroff, G. (1989) *FEBS Lett. 253*, 55 – 58.

239. Sitikov, A. S., Simonenko, P. N., Shestakova, E. A., Ryazanov, A. G. & Ovchinnikov, L. P. (1988) *FEBS Lett. 228*, 327 – 331.

240. Palfrey, H. C., Nairn, A. C., Muldoon, L. L. & Villereal, M. L. (1987) *J. Biol. Chem. 262*, 9785 – 9792.

241. Rao, M. C., Palfrey, H. C., Nash, N. T., Greisman, A., Yayatilak, P. G. & Gibori, G. (1987) *Endocrinology 120*, 1010 – 1018.

242. Cahill, A. L., Applebaum, R. & Perlman, R. L. (1988) *Neurosci Lett. 84*, 345 – 350.

243. Gschwendt, M., Kittstein, W. & Marks, F. (1988) *Biochem. Biophys. Res. Commun. 153*, 1129 – 1135.

244. MacConell, W. P. & Kaplan, N. O. (1982) *J. Biol. Chem. 257*, 5359 – 5366.

245. Langer, J. A. & Lake, J. A. (1986) *J. Mol. Biol. 187*, 617 – 621.

246. Girshovich, A. S., Kurtskhalia, T. V., Ovchinnikov, Y. A. & Vasiliev, V. D. (1981) *FEBS Lett. 130*, 54 – 59.

247. Girshovich, A. S., Bochkareva, E. S. & Vasiliev, V. D. (1986) *FEBS Lett. 197*, 192 – 198.

248. Uchiumi, T., Whaba, A. J. & Traut, R. R. (1987) *Proc. Natl Acad. Sci. USA 84*, 5580 – 5584.

249. Rich, B. E. & Seitz, J. A. (1987) *Mol. Cell Biol. 7*, 4065 – 4074.

250. Grummt, F., Grummt, I. & Erdmann, V. A. (1974) *Eur. J. Biochem. 43*, 343 – 349.

251. Terao, K., Takahashi, T. & Ogata, K. (1975) *Biochim. Biophys. Acta 402*, 230 – 237.

252. Thompson, J., Cundliffe, E. & Dahlberg, A. E., (1988) *J. Mol. Biol. 203*, 457 – 465.

253. Wool, I. (1984) *Trends Biochem. Sci. 9*, 14 – 17.

254. Skogerson, L. & Wakatama, E. (1976) *Proc. Natl Acad. Sci. USA 73*, 73 – 76.

255. Skogerson, L. & Engelhardt, D. (1977) *J. Biol. Chem. 252*, 1471 – 1475.

256. Qin, S., Moldave, K. & McLaughlin, C. S. (1987) *J. Biol. Chem. 262*, 7802 – 7807.

257. Samahapatra, B. & Chakraburrty, K. (1981) *J. Biol. Chem. 256*, 9999 – 10004.

258. Kamath, A. & Chakraburrty, K. (1986) *J. Biol. Chem. 261*, 12593 – 12595.

259. Wilson, P. G. & Culbertson, M. R. (1988) *J. Mol. Biol. 199*, 559 – 573.

260. Beaudet, A. L. & Caskey, C. T. (1971) *Proc. Natl Acad. Sci. USA 68*, 619 – 624.

261. Dalgarno, L. & Shine, J. (1973) *Nature 245*, 261 – 262.

262. Lang, A., Friemert, C. & Gassen, H. G. (1989) *Eur. J. Biochem. 180*, 547 – 554.

263. Tate, W. P. (1984) in *Peptide and protein reviews*, vol. 2, (Hearn, M. T. W., ed.) pp. 173 – 208, Marcel Dekker, New York.

264. Caskey, C. T. (1980) *Trends Biochem. Sci. 5*, 234 – 237.

265. Konecki, D., Aure, K. C., Tate, W. & Caskey, C. T. (1977) *J. Biol. Chem. 252*, 4414 – 4425.

266. Tate, W. P., Caskey, C. T. & Stöffler, G. (1975) *J. Mol. Biol. 93*, 375 – 389.

267. Brot, N., Tate, W. P., Caskey, C. T. & Weissbach, H. (1974) *Proc. Natl Acad. Sci. USA 71*, 89 – 92.

268. Tate, W. P., Beaudet, A. L. & Caskey, C. T. (1973) *Proc. Natl Acad. Sci. USA 70*, 2350 – 2355.

269. Tate, W. P., Horning, H. & Lührmann, R. (1983) *J. Biol. Chem. 258*, 10360 – 10365.

270. Erdmann, V., Fahnstock, S., Hiyo, K. & Nomura, M. (1971) *Proc. Natl Acad. Sci. USA 68*, 2932 – 2936.

271. Yonath, A. & Wittmann, H. G. (1989) *Trends Biochem. Sci. 15*, 329 – 335.

272. Bernabeu, C. & Lake, J. A. (1982) *Proc. Natl Acad. Sci. USA 79*, 3111 – 3115.

273. Bernabeu, C., Tobin, E. M., Fowler, A., Zabin, I. & Lake, J. A. (1983) *J. Cell. Biol. 96*, 1471 – 1474.

274. Walter, P. & Blobel, G. (1981) *J. Cell Biol. 91*, 557 – 561.

275. Wolin, S. L. & Walter, P. (1988) *EMBO J. 7*, 3559 – 3569.

276. Walter, P., Ibrahimi, I. & Blobel, G. (1981) *J. Cell Biol. 91*, 545 – 550.

277. Lipp, J., Dobberstein, B. & Heuptle, M. T. (1987) *J. Biol. Chem. 262*, 1680 – 1684.

278. Meyer, D. I., Krause, E. & Dobberstein, B. (1982) *Nature 297*, 647 – 650.

279. Borgese, N., Mok, W., Kreibich, G. & Sabatini, D. D. (1974) *J. Mol. Biol. 88*, 559 – 580.

280. Kreibich, G., Ulrich, B. L. & Sabatini, D. D. (1978) *J. Cell Biol. 77*, 464 – 487.

281. Kreibich, G., Freienstein, C. M., Pereyra, B. N., Ulrich, B. L. & Sabatini, D. D. (1978) *J. Cell Biol. 77*, 488 – 506.

282. Hortsch, M., Avossa, D. & Meyer, D. I. (1986) *J. Cell Biol. 103*, 241 – 253.

283. Brands, J. H. G. M., Maassen, J. A. van Hemert, F. J., Amons, R. & Möller, W. (1986) *Eur. J. Biochem. 155*, 167 – 171.

284. Linz, J. E., Lira, L. M. & Sypherd, P. S. (1986) *J. Biol. Chem. 261*, 150022 – 150029.

285. Nagata, S., Nagashima, K., Tsunetsugu-Yokota, Y., Fujimura, K., Miyazaki, M. & Kaziro, Y. (1984) *EMBO J. 3*, 1825 – 1830.

286. Pokalsky, A. R., Hiatt, W. R., Ridge, N., Rasmussen, R., Houck, C. M. & Shewmaker, C. K. (1989) *Nucleic Acids Res. 17*, 4661 – 4673.

287. Raeburn, S., Collins, J. F., Moon, H. M. & Maxwell, E. S. (1971) *J. Biol. Chem. 246*, 1041 – 1048.

288. Oleinikov, A. V., Jokhadze, G. G. & Alakhov, Yu. B. (1989) *FEBS Lett. 248*, 131 – 136.

289. Toda, K., Tosaka, M., Mashima, K., Kohno, K., Uchida, T. & Takeuchi, I. (1989) *J. Biol. Chem. 264*, 15489 – 15493.

290. Dasmahapatra, B. & Chakraburrty, K. (1981) *J. Biol. Chem. 256*, 9999 – 10004.

291. Reddington, M. A. & Tate, W. P. (1979) *FEBS Lett. 97*, 335 – 338.

292. Bellé, R., Derancourt, J., Poulhe, R., Capony, J.-P., Ozon, R. & Mulner-Lorillon, O. (1989) *FEBS Lett. 255*, 101 – 104.

293. Moazed, D. & Noller, H. F. (1989) *Nature 342*, 142 – 148.

294. Lill, R., Robertson, J. M. & Wintermeyer, W. (1989) *EMBO J. 8*, 3933 – 3938.

295. Gschwendt, M., Kittstein, W., Mieskes, G. & Marks, F. (1989) *FEBS Lett. 257*, 357 – 360.

296. Lavergne, J.-P., Marzouki, A., Reboud, J.-P. & Reboud, A.-M. (1988) *FEBS Lett. 236*, 345 – 351.

297. Kamath, A. & Chakraburrty, K. (1989) *J. Biol. Chem. 264*, 15423 – 15428.

298. Blobel, G. & Sabatini, D. D. (1970) *J. Cell Biol. 45*, 130 – 145.

299. Lu, X. & Werener, D. (1989) *Nucleic Acids Res. 17*, 442.

300. Rapp, G., Klaudiny, J., Hagendorff, G., Luck, M. R. & Scheit, K. H. (1989) *Biol. Chem. Hoppe-Seyler 370*, 1071 – 1075.

301. Hiremath, L. S., Hiremath, S. T., Rychlik, W., Joshi, S., Domier, L. L. & Rhoades, R. E. (1989) *J. Biol. Chem. 264*, 1132 – 1138.

302. Nygård, O. & Nilsson, L. (1990) *J. Biol. Chem. 265*, 6030 – 6034.

303. Carlberg, U., Nilsson, A: & Nygård, O. (1990) *Eur. J. Biochem.*, in the press.

Eur. J. Biochem. *191*, 257–261 (1990)
© FEBS 1990

Review

Translation and regulation of translation in the yeast *Saccharomyces cerevisiae*

Peter P. MÜLLER and Hans TRACHSEL

Institut für Biochemie und Molekularbiologie der Universität Bern, Switzerland

(Received January 15, 1990) – EJB 90 0039

In recent years the yeast *Saccharomyces cerevisiae* has become a model system for studies of eukaryotic translation and translation regulation. Analysis of mRNA structure, translation initiation factor sequences and the translation initiation pathway indicate, that translation in *S. cerevisiae* is very similar to translation in higher eukaryotes. The availability of powerful genetic techniques lead to the dissection in yeast of individual steps in the translation pathway, the detection of biochemical interactions between components involved in translation and the unravelling of complex regulation phenomena.

The yeast *Saccharomyces cerevisiae* is extensively used as a model system to study various aspects of eukaryotic cellular biology. Among many reasons for working with this experimental system the two most important ones are: (a) basic biological processes are very similar in the lower eukaryote *S. cerevisiae* and higher eukaryotes and (b) the availability of powerful genetic techniques in this system offers advantages for many studies. During the last few years the *S. cerevisiae* system has also evolved into a model system for studies of eukaryotic translation. This review describes the major recent developments in this field. Since it is not possible to cover it extensively in a short review, the reader should consult other reviews [1–8] for additional or more complete information.

MECHANISM OF TRANSLATION

Among the important achievements and fundamental for further progress was the development of a cell-free translation system from *S. cerevisiae* by Gasior et al. [9]. Later, alternative procedures were introduced [10–12] and a detailed discussion of every step of the protocol of Gasior et al. was published by Tuite and Plesset [13]. In short, to prepare this system, cells are converted to spheroplasts, homogenized and the homogenate subjected to centrifugation and gel filtration. Gel-filtered extracts are treated with micrococcal nuclease to obtain cell-free systems dependent on exogenous mRNA for translation. In our hands, the procedure of Gasior et al. gives very active translation systems with good reproducibility.

Correspondence to P. Müller, Institut für Biochemie und Molekularbiologie, Universität Bern, Bühlstrasse 28, CH-3012 Bern, Switzerland

Abbreviations. eIF, eukaryotic initiation factor; EF, elongation factor; GCD (general amino acid control constitutively derepressed) and GCN (general amino acid control nonderepressible), yeast factors required for wild-type regulation of amino acid biosynthesis gene expression; cap, m^7GTP residue at the 5' end of eukaryotic mRNA.

Initiation

Two pathways for translation initiation in eukaryotic cells have been described: ribosomes (with their associated initiation factors) either recognize and bind to the mRNA at (or near) the 5' cap structure and scan the leader region to reach the initiator AUG codon (cap-dependent initiation, reviewed in [14–16]) or ribosomes bind internally to mRNA (internal initiation [17–19]). At present, it is generally believed that cap-dependent translation initiation represents the major pathway. This pathway is schematically shown in Fig. 1. 80S ribosomes dissociate into 40S and 60S subunits and the 40S subunits associate with the eukaryotic initiation factors 3 and 4 (eIF-3 and eIF-4C). Initiation factor eIF-2 then carries the initiator methionyl-tRNA (Met-tRNA$_i^{Met}$) as part of the ternary complex eIF-2–GTP–Met-tRNA$_i^{Met}$ to the 40S subunit. The resulting complex binds in a manner which is dependent on ATP, ATP hydrolysis and eIF-4A, B, E, F at or near the mRNA cap structure, scans the mRNA in the 5'→3' direction and positions itself at the AUG initiator codon. Finally, the 60S subunit joins the 40S subunit in an eIF-5-catalyzed reaction whereby initiation factors are released and GTP is hydrolyzed.

There is quite a large body of evidence suggesting that translation initiation in *S. cerevisiae* closely resembles the cap-dependent initiation pathway in higher eukaryotes: *S. cerevisiae* mRNAs are monocistronic [20, 21], capped [22] and polyadenylated [23]. Ribosomes initiate translation almost exclusively at AUG codons [24, 25] and usually choose the 5' most proximal AUG (reviewed in [4–6]). There is also good evidence to support the model whereby ribosomes reach the first AUG by a scanning mechanism. Sequences rich in G or sequences which can form double-stranded structures inhibit translation initiation probably by interfering with the scanning process [5, 6, 25, 26]. *S. cerevisiae* differs from higher eukaryotes in that the sequences flanking the initiator AUG are not very important for recognition by the *S. cerevisiae* ribosome [4–6, 24]. An A in the −3 position (relative to the A of the AUG codon) stimulates initiation only 2–3-fold compared to an initiator codon with U in the −3 position.

258

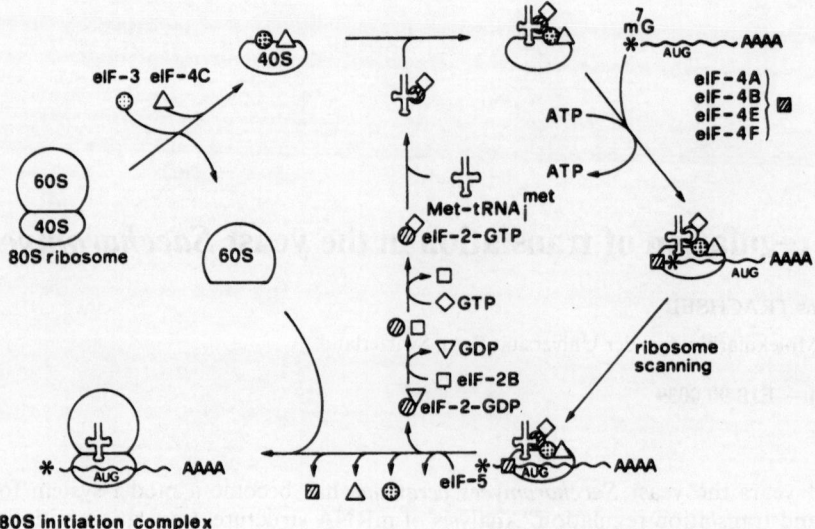

Fig. 1. *Scheme of eukaryotic translation initiation*. The cycle begins with the 80S ribosome and ends with the 80 initiation complex. 40S, small ribosomal subunit; 60S, large ribosomal subunit; eIF, eukaryotic initiation factor; m^{7}G, mRNA cap structure; Met-tRNA$_i^{Met}$, initiator methionyl transfer RNA

Finally, as described for mammalian cells [27], upstream AUG codons in *S. cerevisiae* inhibit initiation from downstream AUG codons [6, 28]. In the context of the scanning model [14], this is thought to be due to sequestering of ribosomes and initiation factors by upstream AUG codons. Leaky scanning ('missing' of AUG codons) or reinitiation (new initiation after translation of upstream reading frames) would allow translation of downstream open reading frames to occur to some extent. At present, however, it is not clear how well *S. cerevisiae* ribosomes can reinitiate translation at a downstream AUG and what the specific requirements for reinitiation are.

Additional similarities between *S. cerevisiae* and higher eukaryotes in the translation initiation pathway include the translation initiation factors. To date, five translation initiation factor peptides from *S. cerevisiae* have been isolated and their genes cloned. These are eIF-2α [29], eIF-2β [30], eIF-4A [31 – 33], eIF-4D [34, 35], eIF-4E [36, 37] and the large subunit of eIF-4F [38], (and C. Goyer and N. Sonenberg, personal communication). The genes encoding eIF-2α and eIF-2β (*SUI2* and *SUI3*) were isolated in Donahue's laboratory [29, 30] as suppressors (*sui2* and *sui3*) of *HIS4* gene initiator codon mutations. Both are single-copy essential genes and the encoded proteins are 58% (eIF-2α) and 42% (eIF-2β) identical at the amino acid sequence level with the corresponding human initiation factor polypeptides [39, 40]. Like the human factor [40], eIF-2β from *S. cerevisiae* contains a Zn(II) finger motif believed to be involved in RNA binding. The initiation factor polypeptides eIF-2α and eIF-2β are subunits of eIF-2 and involved in initiator Met-tRNA$_i^{Met}$ binding to the small (40S) ribosomal subunit (Fig. 1) [15]. The initiation factor eIF-4A of *S. cerevisiae* was found to be encoded by two genes (*TIF1* and *TIF2*) [31]. The nucleotide sequences of the two genes differ but they code for identical proteins. Either one of the two genes is sufficient (and essential) for growth. *S. cerevisiae* eIF-4A is 65% identical with mouse eIF-4A [41]. It is involved, together with eIF-4B and eIF-4F, in mRNA binding to the 40S – Met-tRNA$_i^{Met}$ complex (Fig. 1) [15] and believed to act as an RNA helicase (ATP hydrolysis-dependent unwinding of RNA secondary structure) [42]. *S. cerevisiae* initiation factor eIF-4E, a 24-kDa protein that binds to the

mRNA 5′ cap structure during translation initiation (Fig. 1) [15, 16, 36] is encoded by a single-copy essential gene [37]. The protein is 30% identical with human eIF-4E [43] and contains eight conserved tryptophan residues and flanking sequences thought to be involved in cap recognition [44]. The gene encoding eIF-4E was recently found to correspond to the gene *CDC33* [45]. Certain mutations in this gene lead to cell cycle arrest under non-permissive conditions. The function of eIF-4E in cell cycle control is, however, not known. Very recently, the gene encoding the putative homolog of mammalian p220, the largest subunit of initiation factor eIF-4F (Fig. 1) [15, 16] was cloned and sequenced (C. Goyer and N. Sonenberg, personal communication). The function of the encoded protein in *S. cerevisiae* is not known at present. In mammalian cells, the integrity of this polypeptide is essential for cap recognition during initiation, since its proteolytic cleavage leads to inhibition of capped mRNA translation. Proteolytic cleavage of p220 was shown to correlate with inhibition of host protein synthesis in poliovirus-infected cells [46]. In higher eukaryotes, the gene encoding p220 has not been isolated so far. Finally, a 20-kDa polypeptide (Co-eIF-2A) involved in Met-tRNA$_i^{Met}$ binding to the 40S ribosome has been shown to occur also in *S. cerevisiae* [47].

How similar *S. cerevisiae* and higher eukaryotic translation initiation pathways are, is most dramatically illustrated by the fact that mouse eIF-4E can substitute for yeast eIF-4E in *S. cerevisiae* in vivo [48]. However, interchangeability of translation initiation factors between yeast and mammals is not a general phenomenon: despite even greater similarity, the mammalian eIF-4A cannot replace its yeast counterpart *in vitro* or *in vivo* (S. Blum, P. Linder and H. Trachsel, unpublished results).

Apart from these striking similarities in the cap-dependent translation initiation pathway between *S. cerevisiae* and higher eukaryotes there is evidence suggesting that also the internal initiation pathway exists in *S. cerevisiae* (unpublished results). A dicistronic mRNA with the 5′ proximal open reading frame encoding thymidine kinase and the 3′ proximal open reading frame encoding chloramphenicol acetyltransferase can only program synthesis of the latter enzyme if the 5′ untranslated region of poliovirus RNA is inserted between the two open reading frames. This sequence was previously shown by Pelletier and Sonenberg [18] to allow internal initiation. Like the mammalian system [18] the *S. cerevisiae* translation system is also able to synthesize chloramphenicol acetyltransferase in the absence of thymidine kinase synthesis from such mRNA constructs (unpublished results) indicative of internal initiation of translation.

Exploitation of the *S. cerevisiae* system has already advanced our knowledge about eukaryotic translation initiation beyond that known from experiments with reconstituted mammalian systems (Fig. 1) [15]. For example, elegant genetic experiments performed in Donahue's laboratory established that the initiator Met-tRNA$_i^{Met}$ base-pairs with the initiator AUG codon. Translation initiation on *HIS4* mRNA with initiator tRNA mutated in the anticodon [3'-UCC-5') could be restored by initiation at a complementary AGG codon [49]. Furthermore, mutations in eIF-2α and eIF-2β allow initiation to occur at an UUG codon in the absence of an AUG codon [29, 30], suggesting that initiation factor eIF-2 mediates recognition of the AUG codon during initiation. Rather surprisingly, Sachs and Davis [50] have shown the poly(A) binding protein to be involved in translation initiation *in vivo*. Its precise role in initiation has, however, not yet been elucidated. Finally, recent studies by Altmann et al. [51] using an eIF-4E-dependent *S. cerevisiae* translation system have revealed that certain mRNAs can be translated efficiently *in vitro* in the absence of functional eIF-4E.

Further insight into the mechanism and regulation of translation should be expected from the closer analysis of *S. cerevisiae* mutants originally isolated and partly characterized by Hartwell and McLaughlin [52, 53] (recently reviewed in [1, 3]). For one of these mutants, ts 187, the gene responsible for the defect in translation initiation has been cloned [54, 55] and sequenced [55].

Elongation

During the elongation phase of translation the ribosome moves relative to the mRNA in the 5' → 3' direction incorporating amino acids into the growing polypeptide chain. This process is very similar in prokaryotes, lower and higher eukaryotes. Elongation in *S. cerevisiae* has been reviewed extensively [1, 2] and since no dramatic new discoveries were made in this field very recently, we give here only a short summary.

Aminoacyl-tRNA (aa-tRNA) is carried to the ribosome as a complex with elongation factor 1α (EF-1α) and GTP; the amino acid is incorporated into the polypeptide chain by the ribosome-associated peptidyltransferase. Upon GTP hydrolysis, EF-1α − GDP is released from the ribosome and GDP exchanged with GTP probably through the action of EF-1$\beta\gamma$. Thereafter, an EF-2 − GTP complex induces the relative movement of mRNA and ribosome. EF-2 is released from the ribosome after GTP hydrolysis. The elongation factors 1α, β, γ and 2 have all been purified from *S. cerevisiae*. Furthermore, two genes encoding EF-1α have been cloned and sequenced (for references, see [2]).

Yeast translation elongation is so far unique among eukaryotes in that it requires a third elongation factor, EF-3 [56]. EF-3 has ATPase and GTPase activity and stimulates the binding of the EF-1α − GTP − aa-tRNA complex to the ribosome. It requires ATP or GTP and hydrolysis of the nucleotide to function [57].

Termination

Release of the polypeptide chain from the tRNA and of the ribosome from mRNA occurrs at the termination codons UAA, UAG and UGA and is catalyzed by release factor(s). In contrast to *Escherichia coli* which contains two release factors, mammalian cells appear to have only one release factor [58]. Details of the reaction(s) leading to termination of translation in eukaryotes are not known. No release factors have so far been isolated from *S. cerevisiae*. Analysis of mutations affecting translational suppression of termination codons such as the sal mutations [59] may lead to the identification of genes and gene products involved in translation termination in *S. cerevisiae*.

REGULATION OF TRANSLATION

A number of features are known to affect translational efficiency of mRNA. These include secondary structure, upstream AUG codons in the 5' leader region, RNA sequences flanking the AUG initiation codon and the posttranscriptionally added poly(A) tract at the 3' end. Here, translational regulation is viewed in a narrow sense: we only consider naturally occurring species of cytoplasmic mRNAs which are translated with different efficiencies under appropriate inducing or repressing conditions relative to the translational efficiency of bulk mRNA.

The *CPA1* gene codes for the small subunit of one of two carbamoyl-phosphate synthetases involved in the biosyntheses of pyrimidines and arginine. Expression of *CPA1* has been proposed to be translationally regulated in response to arginine availability (for reviews, see [60, 61]). The first clues for post-transcriptional control came from the comparison of *CPA1* mRNA levels with the amount of *CPA1* protein when cells were grown in the presence or absence of exogenous arginine. A fivefold decrease in protein levels in response to arginine addition was observed under conditions where mRNA levels remained essentially constant [62]. Analysis of the *CPA1* transcript revealed two open reading frames, one in the 5' upstream region encoding a 25-amino-acid peptide and another one further downstream encoding the small subunit of carbamoyl-phosphate synthetase [63, 64]. Mutations in the upstream reading frame abolished the decrease of *CPA1* protein relative to *CPA1* mRNA levels in response to arginine addition, suggesting a role for the peptide in translational repression [64]. The precise mechanism of regulation remains to be elucidated.

Another gene reported to be regulated at the translational level is catalase T. Expression of catalase T has been reported to be repressed post-transcriptionally in the absence of heme [65]. Subsequently, *in vitro* translation products of total yeast mRNA in lysates from heme-deficient yeast cells were analyzed for the presence of catalase T by immunoadsorption [66]. Whereas little catalase was detected in the absence of heme, the addition of hemin to the lysate resulted in a severalfold increase in the amounts of catalase T protein. No stimulation by hemin of catalase A or overall translation products produced in this system were detected, indicating that the hemin effect is specific. This interesting system awaits further characterization.

Probably, the most prominent known example of translational regulation in *S. cerevisiae* concerns *GCN4* mRNA which has been investigated in some detail (reviewed in [61]). *GCN4* mRNA translation is inhibited under normal growth conditions by four small open reading frames in the 5' region upstream of the *GCN4* initiation codon. Amino acid imbalances or uncharged tRNA cause an increase in the translation efficiency and as a consequence GCN4 protein accumulates. GCN4 protein is a transcriptional activator which binds to a short DNA sequence which occurs upstream of amino acid biosynthetic genes. As a result of the stimulation of transcription of these genes, the capacity of amino acid biosynthetic pathways increases. This regulatory system allows yeast cells to grow in the presence of drastic amino acid imbalances in the media. In the absence of upstream codons *GCN4* mRNA translation becomes constitutive under all conditions.

Whereas translation of the first short reading frame is efficient and not regulated, it nevertheless allows efficient initiation at downstream AUG codons. In fact, it is required to suppress the strong inhibition exerted by the remaining three small open reading frames [28]. The precise sequence of the upstream reading frames is not essential for translational regulation, arguing against a role for the encoded peptides. No significant changes are observed in the sole absence of the second and third AUG codons or if the fourth reading frame is substantially elongated. In contrast, changes at or near the first reading frame can drastically reduce the efficiency of expression of downstream reading frames. Despite the overall decrease in translational efficiency in the absence of the first AUG or even in the absence of the first two AUG codons the derepression response can still be observed. This suggests that the function of the first reading frame is to increase the translational efficiency of *GCN4* mRNA whereas other upstream reading frames can substitute for the first in the regulation. Regulation of *GCN4* mRNA translation is presently believed to be due to modulation of reinitiation frequency of ribosomes after translation of one or more of the upstream reading frames. An alternative model for initiation at internal AUG codons has been proposed for translation of viral mRNA in mammalian cells whereby ribosomes bind directly to an internal segment in the mRNA 5′ leader and bypass inhibitory regions further upstream [18, 19]. So far, no experiments have been performed to address directly the possibility of 5′ independent internal initiation of *GCN4* mRNA translation.

Several mutants in *GCN* and *GCD* genes which are deficient in the translational regulation of *GCN4* mRNA have been isolated and analyzed. Interestingly, one of the factors required for derepression, GCN2 has amino acid sequence similarity to protein kinases and histidyl-tRNA synthase, provoking the idea that the GCNz protein may be kinase regulated by the degree of tRNA charging [67]. Various models have been proposed in which GCN and GCD factors modulate translation [68, 69] or act indirectly [70]. Recently, it was found that mutations in the protein synthesis initiation factor subunits 2α and 2β also cause derepression of *GCN4* mRNA translation independent of GCN2 [71]. This finding, and the fact that eIF-2 in mammalian cells is sequestered in an inactive complex when it is phosphorylated (for a review, see [15]), has prompted the idea that GCN2 kinase may phosphorylate initiation factor eIF-2 and modify its activity to increase *GCN4* mRNA translation.

PERSPECTIVE

Recent discoveries in the field of eukaryotic translation including (a) the importance of mRNA 5′ leader secondary structure in initiation, (b) the existence of RNA secondary structure unwinding factors, (c) reinitiation and internal initiation and (d) the existence of specific mRNA binding factors like the protein binding to ferritin and transferrin mRNA [72] make studies of the mechanism and regulation of translation important and attractive. Very likely, the *S. cerevisiae* system will contribute significantly to the elucidation of the regulation of gene expression at the translational level in the future.

We would like to thank M. Altmann for helpful discussions, K. J. Clemetson for critical reading of the manuscript and G. Mengod for the preparation of the figure.

REFERENCES

1. Tuite, M. F. (1989) *The yeasts*, vol. 3, 2nd edn, pp. 161−204, Academic Press, New York.
2. Chakraburtty, K. & Kamath, A. (1988) *Int. J. Biochem. 20*, 581−590.
3. Moldave, K. & McLaughlin, C. S. (1988) *NATO ASI Ser. A Life Sci. 14*, 271−281.
4. Donahue, T. F., Cigan, A. M., de Castilho, B. A. & Yoon, H. (1988) *NATO ASI Ser. A Life Sci. 14*, 361−372.
5. Laz, T., Clements, J. & Sherman, F. (1988) *NATO ASI Ser. A Life Sci. 14*, 353−360.
6. Laz, T., Clements, J. & Sherman, F. (1987) *Translational regulation of gene expression* (J. Ilan, edn) pp. 413−429, Plenum Press, New York.
7. Hinnebusch, A. G., Müller, P. P. & Harashima, S. (1988) *NATO ASI Ser. A Life Sci. 14*, 499−512.
8. Hinnebusch, A. G. & Müller, P. P. (1987) *Translational regulation of gene expression* (J. Ilan, edn) pp. 397−412, Plenum Press, New York.
9. Gasior, E., Herrera, F., Sadnik, I., McLaughlin, C. S. & Moldave, K. (1979) *J. Biol. Chem. 254*, 3965−3969.
10. Szczesna, E. & Filipowicz, W. (1980) *Biochem. Biophys. Res. Commun. 92*, 563−569.
11. Hofbauer, R., Fessl, F., Hamilton, B. & Ruis, H. (1982) *Eur. J. Biochem. 122*, 199−203.
12. Kreutzfeldt, C. & Lochmann, E.-R. (1983) *FEMS Microbiol. Lett. 16*, 179−182.
13. Tuite, M. F. & Plesset, J. (1986) *Yeast 2*, 35−52.
14. Kozak, M. (1983) *Microbiol. Rev. 47*, 1−45.
15. Pain, V. M. (1986) *Biochem. J. 235*, 625−637.
16. Edery, I., Pelletier, J. & Sonenberg, N. (1987) *Translational regulation of gene expression* (J. Ilan, ed.) pp. 413−429, Plenum Press, New York.
17. Pelletier, J., Kaplan, G., Racaniello, V. R. & Sonenberg, N. (1988) *Mol. Cell. Biol. 8*, 1103−1112.
18. Pelletier, J. & Sonenberg, N. (1988) *Nature 334*, 320−325.
19. Bienkowska-Szewczyk, K. & Ehrenfeld, E. (1988) *J. Virol. 62*, 3068−3072.
20. Petersen, N. S. & McLaughlin, C. S. (1973) *J. Mol. Biol. 81*, 33−45.
21. Sherman, F. & Stewart, J. W. (1975) *Proc. FEBS Meet. 38*, 175−191.
22. Sripati, C. E., Groner, Y. & Warner, J. R. (1976) *J. Biol. Chem. 251*, 2898−2904.
23. McLaughlin, C. S., Warner, J. R., Edmonds, M., Nakazato, H. & Vaughan, M. H. (1973) *J. Biol. Chem. 248*, 1466−1471.
24. Clements, J. M., Laz, T. M. & Sherman, F. (1988) *Mol. Cell. Biol. 8*, 4533−4536.
25. Cigan, A. M. & Donahue, T. F. (1987) *Gene 59*, 1−18.
26. Baim, S. B. & Sherman, F. (1988) *Mol. Cell. Biol. 8*, 1591−1601.
27. Kozak, M. (1984) *Nucleic Acids Res. 12*, 3873−3893.
28. Müller, P. P. & Hinnebusch, A. G. (1986) *Cell 45*, 201−207.
29. Cigan, A. M., Pabich, E. K., Feng, L. & Donahue, T. F. (1989) *Proc. Natl Acad. Sci. USA 86*, 2784−2788.
30. Donahue, T. F., Cigan, A. M., Pabich, E. K. & Valavicius, B. C. (1988) *Cell 54*, 621−632.
31. Linder, P. & Slonimski, P. P. (1988) *Nucleic Acids Res. 16*, 10359.
32. Linder, P. & Slonimski, P. P. (1989) *Proc. Natl Acad. Sci. USA 86*, 2286−2290.

33. Blum, S., Mueller, M., Schmid, S. R., Linder, P. & Trachsel, H. (1989) *Proc. Natl Acad. Sci. USA 86*, 6043 – 6046.
34. Schnier, J., Smit-McBride, Z. & Hershey, J. W. B. (1989) *Translational control*, p. 156, Cold Spring Harbor Laboratory, Cold Spring Harbor NY.
35. Sandholzer, U., Duchêne, M. & Lottspeich, F. (1989) *Translational control*, p. 154, Cold Spring Harbor Laboratory, Cold Spring Harbor NY.
36. Altmann, M., Edery, I., Sonenberg, N. & Trachsel, H. (1985) *Biochemistry 24*, 6085 – 6089.
37. Altmann, M., Handschin, C. & Trachsel, H. (1987) *Mol. Cell. Biol. 7*, 998 – 1003.
38. Goyer, C., Altmann, M., Trachsel, H. & Sonenberg, N. (1989) *J. Biol. Chem. 264*, 7603 – 7610.
39. Ernst, H., Duncan, R. F. & Hershey, J. W. B. (1987) *J. Biol. Chem. 262*, 1206 – 1212.
40. Pathak, V. K., Nielsen, P. J., Trachsel, H. & Hershey, J. W. B. (1988) *Cell 54*, 633 – 639.
41. Nielsen, P. J. & Trachsel, H. (1988) *EMBO J. 7*, 2097 – 2105.
42. Rozen, F., Edery, I., Dever, T. E., Merrick, W. C. & Sonenberg, N. (1990) *Mol. Cell. Biol. 10*, 1134 – 1144.
43. Rychlik, W., Domier, L. L., Gardner, P. R., Hellmann, G. M. & Rhoads, R. E. (1987) *Proc. Natl Acad. Sci. USA 84*, 945 – 949.
44. Altmann, M., Edery, I., Trachsel, H. & Sonenberg, N. (1988) *J. Biol. Chem. 263*, 17229 – 17232.
45. Brenner, C., Nakayama, N., Goebl, M., Tanaka, K., Toh-E, A. & Matsumoto, K. (1988) *Mol. Cell. Biol. 8*, 3556 – 3559.
46. Sonenberg, N. (1987) *Adv. Virus Res. 33*, 175 – 204.
47. Ahmad, M. F., Nasrin, N., Banerjee, A. C. & Gupta, N. K. (1985) *J. Biol. Chem. 260*, 6955 – 6959.
48. Altmann, M., Müller, P. P., Pelletier, J., Sonenberg, N. & Trachsel, H. (1989) *J. Biol. Chem. 264*, 12145 – 12147.
49. Cigan, A. M., Feng, L. & Donahue, T. F. (1988) *Science 242*, 93 – 96.
50. Sachs, A. B. & Davis, R. W. (1989) *Cell 58*, 857 – 867.
51. Altmann, M., Sonenberg, H. & Trachsel, H. (1989) *Mol. Cell. Biol. 9*, 4467 – 4472.
52. Hartwell, L. H. & McLaughlin, C. S. (1968) *J. Bacteriol. 96*, 1664 – 1671.
53. Hartwell, L. H. & McLaughlin, C. S. (1969) *Proc. Natl Acad. Sci. USA 62*, 468 – 474.
54. Keierleber, C., Wittekind, M., Qin, S. & McLaughlin, C. S. (1986) *Mol. Cell. Biol. 6*, 4419 – 4424.
55. Hanic-Joyce, P. J., Singer, R. A. & Johnston, G. C. (1987) *J. Biol. Chem. 262*, 2845 – 2851.
56. Skogerson, L. (1979) *Methods Enzymol. 60*, 676 – 685.
57. Kamath, A. & Chakraburtty, K. (1989) *J. Biol. Chem. 264*, 15423 – 15428.
58. Cashey, C. T. (1980) *Trends Biochem. Sci. 5*, 234 – 237.
59. Tuite, M. F., Izgu, F., Grant, C. M. & Crouzet, M. (1988) *NATO ASI Ser. A Life Sci. 14*, 393 – 402.
60. Davis, R. H. (1986) *Microbiol. Rev. 50*, 280 – 313.
61. Hinnebusch, A. G. (1988) *Microbiol. Rev. 52*, 248 – 273.
62. Messenguy, F., Feller, A., Crabeel, M. & Pierard A. (1983) *EMBO J. 2*, 1249 – 1254.
63. Werner, M., Feller, A. & Pierard, A. (1985) *Eur. J. Biochem. 146*, 371 – 381.
64. Werner, M., Feller, A., Messenguy, F. & Pierard, A. (1987) *Cell 49*, 805 – 813.
65. Sledziewski, A., Rytka, J., Bilinski, T., Hörtner, H. & Ruis, H. (1981) *Curr. Genet. 4*, 19 – 23.
66. Hamilton, B., Hofbauer, R. & Ruis, H. (1982) *Proc. Natl Acad. Sci. USA 79*, 7609 – 7613.
67. Wek, R. C., Jackson, B. M. & Hinnebusch, A. G. (1989) *Proc. Natl Acad. Sci. USA 86*, 4579 – 4583.
68. Harashima, S., Hannig, E. M. & Hinnebusch, A. G. (1987) *Genetics 117*, 409 – 419.
69. Tzamarias, D., Roussou I. & Thireos, G. (1989) *Cell 57*, 947 – 954.
70. Niederberger, P., Aeby M. & Hütter, R. (1986) *Curr. Genet. 10*, 657 – 664.
71. Williams, N. P., Hinnebusch, A. G. & Donahue, T. F. (1989) *Proc. Natl Acad. Sci. USA 86*, 7517 – 7519.
72. Klausner, R. D. & Harford, J. B. (1989) *Science 246*, 870 – 872.

Eur. J. Biochem. *192*, 1−15 (1990)
© FEBS 1990

Review

Nonribosomal biosynthesis of peptide antibiotics

Horst KLEINKAUF and Hans von DÖHREN

Institute of Biochemistry and Molecular Biology, Technical University Berlin, Federal Republic of Germany

(Received December 4, 1989) − EJB 89 1449

Peptide antibiotics are known to contain non-protein amino acids, D-amino acids, hydroxy acids, and other unusual constituents. In addition they may be modified by *N*-methylation and cyclization reactions. Their biosynthetic origin has been connected in many cases to an enzymatic system referred to as the 'thiotemplate multienzymic mechanism'. This mechanism includes the activation of the constituent residues as adenylates on the enzymic template, the acylation of specific template thiol groups, epimerization or *N*-methylation at this thioester stage, and polymerization in the sequence directed by the multienzymic structure with the aid of 4'-phosphopantetheine as a cofactor, including possible cyclization or terminal modification reactions. The reaction sequences leading to gramicidin S, tyrocidine, cyclosporine, bacitracin, polymyxin, actinomycin, enniatin, beauvericin, δ-(L-α-aminoadipyl)-L-cysteinyl-D-valine and linear gramicidin are discussed. The structures of the multienzymes, their genetic organization, the biological functions of these peptides and results on related systems are discussed.

Although the enzymatic formation of essential peptides such as glutathione and pantetheine was already known in the 'preribosomal era', the elucidation of the biosynthesis of more complex peptides followed the unravelling of the genetic code in the sixties [1−3]. A prediction of a poly- or multienzymatic pathway to peptides had been presented by Fritz Lipmann as early as 1954 [4] and a similar biosynthetic scheme of multienzymes as templates for polypeptides has now been verified for various types of peptides. Some of these studies have been reviewed elsewhere [5−12], but recent results have advanced the field considerably. We therefore attempt to present here a more general view of this biosynthetic mechanism, and discuss the current areas of research.

Biosynthetic scheme of peptide formation

The scheme presented in Fig. 1 summarizes our present views on peptide biosynthesis. Both ribosomal and multienzymatic processes operate by 'head growth', a principle set forward by Lipmann in 1969 [13]:

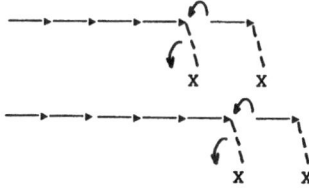

Correspondence to H. von Döhren, Institute of Biochemistry and Molecular Biology, Technical University Berlin, Franklinstrasse 29, D-1000 Berlin 10

Abbreviations. ACV, δ-(L-α-aminoadipyl)-L-cysteinyl-D-valine; GS, gramicidin S; TY, tyrocidine. Unusual amino and hydroxy acids: Aad, 2-aminoadipic acid; Abu, 2-aminobutyric acid; Aib, 2-amino-isobutyric acid; Bmt, (4*R*)-4[(*E*)-2-butenyl]-4-methyl-L-threonine; Da4hPro, D-allo-4-hydroxy-proline; Dbu, 2,4-diaminobutyric acid; Dhb, 2,4-dihydroxybutyric acid; Dpr, 2,3-diaminopropionic acid; Hiv, 2-hydroxyisovaleric acid; HyPic, 3-hydroxypicolinic acid; Ise, isoserine; MePhg, N-methylphenylglycine; Mha, 3-hydroxy-4-methylanthralinic acid; N2mLeu, N,3-dimethyl-L-leucine; Qxa, quin-oxaline-2-carboxylic acid; Sar, sarcosine. MeXaa represents an *N*-methylated amino acid.

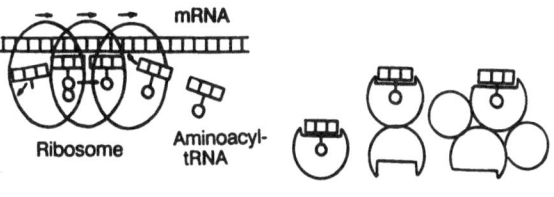

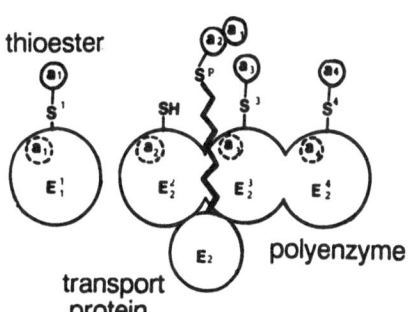

Fig. 1. *Scheme of ribosomal and enzymatic polypeptide biosynthesis.* In the ribosomal system amino acids are activated by aminoacyl-tRNA synthetases. These are symbolized here by their different types of subunit structures. The charged tRNA is then lined up at the ribosome according to the mRNA sequence. In the enzymatic system polyenzymes activate amino and hydroxy acids, and the template itself is aminoacylated, generally as a thioester. A transport system then assembles the lined-up residues, presumably by a sequential mechanism

Carboxyl-activated (X) amino acids (⟶) or peptides (⟶ ⟶) combine to form an activated elongated peptide. Thus growth occurs at the carboxyl terminal, leaving the amino terminal in either a free, formylated, or acylated state. Growing chains are attached to the ribosome as peptidyl-tRNA, or to an enzyme as activated peptide. Initiation and termination reactions in the enzymatic systems show a variety of reaction types. If the mRNA-directed ribosomal system is compared with a multienzymatic organization, one can observe the delicately controlled usage of the amino acids in the protein, manifested by the substrate recognition and correction events of tRNA aminoacylation by the aminoacyl-tRNA synthetases. This is in contrast to the liberal use by the peptide synthetases of not only amino but also hydroxy acids as substrates, the known number of carboxy-activated compounds at the moment being well over 300 [14]. Both systems lead to linear peptide structures, but whereas mRNAs coding for 2000 and more residues have been found, the longest peptide known to be formed by an enzyme system is still alamethicin (I), a 20-residue peptaibol [15].

Ac-Aib-Pro-Aib-Ala-Aib-Ala-Gln-Aib-Val-^{10}-Aib-

-Gly-Leu-Aib-Pro-Val-Aib-Aib-Glu-Gln-^{20}Pheol

(I)

where Aib is α-aminoisobutyric acid.

In the enzymatic system linear precursor peptides may be modified in a variety of ways. Intrinsic enzymatic functions contained in the corresponding multienzymes include (a) precursor modification preceding incorporation into the peptide, or (b) peptide modifications. The precursor amino acids may be epimerized or methylated in their activated state, and several types of cyclization reactions, including piperazinedione formation, direct linkage of α-carboxyl-α-amino groups, α-carboxyls with β, γ, or δ-amino groups, or lactonization, are known to occur.

A general scheme for enzymatic peptide formation is as follows:

aminoacyladenylate   thioester   transfer

$$E^1 + a_1 + MATP^{2-} \rightarrow E(a_1AMP) \rightarrow E^1 - S^1 - a_1 \rightarrow E - S^p - a_1$$
$$\downarrow$$
$$E^2 + {}_2 + MATP^{2-} \rightarrow E(a_2AMP) \rightarrow E^2 - S^2 - a_2 \quad \rightarrow E^2 - S - a_2 a_1$$
$$\rightarrow E - S^p - a_2 a_1$$
$$\downarrow$$
$$E^3 + a_3 + MATP^{2-} \rightarrow E(a_3AMP) \rightarrow E^3 - S^3 - a_3 \quad \rightarrow E - S^3 - a_3 a_2 a_1$$
$$\rightarrow E - S^p - a_3 a_2 a_1$$
$$\downarrow$$
$$\vdots$$
$$\downarrow$$
$$E^n + a_n + MATP^{2-} \rightarrow E(a_nAMP) \rightarrow E^n - S_n - a_n \rightarrow E^n - S_n - a_n a_{n-1} \ldots a_1.$$

In this scheme E_n represents an enzyme site for the activation of the nth amino acid a_n of the peptide, the site being either a specific enzyme *per se* or a specific enzyme site within a multienzyme. The amino acid is activated at the expense of ATP (here indicated as $MATP^{2-}$, with M standing for Mg^{2+} or Mn^{2+}), via the formation of an enzyme-stabilized aminoacyl adenylate (a_nAMP). This then is attached to a specific enzyme thiol group S^nH, located in a specific multienzyme site E^n. The transfer of intermediates is mediated by a pantetheine thiol S^pH, located in a specific multienzyme site E. The growing peptides remain bound to the enzyme at their specific multienzyme sites or to the cofactor.

This scheme has in principle been proposed on the basis of the identification of 4'-phosphopantetheine as a covalently attached cofactor in gramicidin S [17] and tyrocidine synthetases [18], as well as the isolation of peptide intermediates attached to enzyme fragments containing this cofactor [19]. The scheme has been shown to operate in a variety of peptide-forming enzyme systems, as summarized in Table 1, but although it is a general mechanism, it is not the only biosynthetic scheme which is operating. As Bose et al. have

shown for the biosynthesis of mycobacillin (II) in *Bacillus subtilis* B3 [20 – 25], there is another mechanism which operates via phosphate-mediated activation of peptide intermediates, presumably noncovalently enzyme-stabilized, but nonetheless organized in multienzyme structures. Vater et al. recently obtained some evidence for a similar system in surfactin (III) [26, 27] (i3hC$_{14}$ is iso-3-hydroxytetradecanoic acid).

$$\begin{array}{c} \rightarrow Pro \rightarrow DAsp \rightarrow DGlu - (\rightarrow Tyr \rightarrow Asp \rightarrow Tyr \rightarrow Ser) \\ \\ \downarrow(DAsp \leftarrow Ala \leftarrow DAsp) \leftarrow DGlu \leftarrow Leu \leftarrow DAsp \end{array}$$

(II)

$$\begin{array}{c} \rightarrow i3hC_{14} \rightarrow Glu \rightarrow Leu \rightarrow DLeu \\ \\ \leftarrow Leu - DLeu - Asp - Val \leftarrow \end{array}$$

(III)

Enzyme sub-structures have been found to be organized into multienzymes, and the protein template structure functions either as a set of interacting multienzymes or a single multienzymic structure. Which one of the two organizational principles is used in each individual case cannot be predicted at present. So far no differently organized enzyme systems forming identical products have been characterized, in contrast to the case of fatty acid synthases.

Structures of non-ribosomally synthesized peptides

In a recent compilation of peptides with unusual features pointing to their non-ribosomal origin or to extensive enzymatic modification of ribosomally synthesized precursors, all the structures can be described by the above schemes of elongation, with few exceptions [14]. Thus for each peptide structure determined, a linear peptide precursor can be predicted, and such a precursor has been identified in the cases listed in Table 1. The substitution of amino acids by hydroxy acids seems to make no difference with regard to activation and formation of thioester intermediates, as can be concluded from the studies with enniatin, beauvericin, and destruxin.

Peptide structures can be systematically arranged according to secondary reactions related to initiation, modification, or termination procedures. With a linear precursor, the amino terminal may be free, formylated, or acylated with a variety of carboxylic acids. Cysteine and serine residues within a chain may undergo cyclization to thiazole or oxazole derivatives. Termination requires the release of the nascent peptide chain from the enzyme and can occur in a variety of ways: (a) hydrolysis of a thioester (leading to free peptide); (b) aminolysis by a free amine component (resulting in terminal modification); (c) cyclization with the amino terminus (giving a cyclic peptide); (d) cyclization with a side chain amino group (giving a branched cyclopeptide); (e) lactonization with a terminal or side-chain hydroxyl group (giving a lactone or branched lactone, cyclodepsipeptide).

With the exception of the first case (a), which is an analogous reaction to the termination in animal fatty acid synthesis, relevant examples for all these processes have been described, although the reactions have not been studied in detail. It is to be expected that δ-(L-α-aminoadipyl)-L-cysteinyl-D-valine (ACV), the precursor of the β-lactam antibiotics penicillin and cephalosporin, is released by a thioesterase termination step [28, 29].

Table 1. *Current state of the enzymology of peptide synthetases*

The abbreviations used are: P peptide, C cyclopeptide, L lactone, D depsipeptide, R acyl, M modified. The number of residues activated by each (multi)enzyme is indicated in parentheses after the molecular mass; 3×2 indicates the trimerization of two residues. The structural types are defined as follows: P-3 = Aad-Cys-DVal; P-5-M = βTyr-Ise-Dpr-Dha-Gly-spermidine; P-15-M = fVal-Gly-Ala-DLeu-Ala-DVal-Val-DVal-Trp-(DLeu-Trp)₃-ethanolamine; P-19-M = AcAib-Pro-Ala-Aib-Ala-Gln-Aib-Val-Aib-Gly-Leu-Aib-Pro-Val-Aib-Aib-Glu-Gln-Pheol; C-2 = c(anthranilate-Phe); P-C-E-3 = c(Dhb-Ser)₃; C-6 = c[(Gly)₃(OrnOH)₃]; C-(P-5)₂ = c(DPhe-Pro-Val-Orn-Leu)₂; C-10 = c(DPhe-Pro-Phe-DPhe-Asn-Gln-Val-Orn-Leu); C-11-M = c(DAla-MeLeu-MeLeu-MeVal-MeBmt-Abu-Sar-MeLeu-Val-MeLeu-Ala); C-13 = c{Pro-DAsp-DGlu-[Tyr-Asp-Tyr-Ser-DAsp-Leu-DGlu-(DAsp-Ala-DAsp)]}; L-6-M = c(DHiv-Pro-Ile-MeVal-MeAla-βAla); R-(L-5)₂ = Mha-c(Thr-DVal-Pro-Sar-Val); R-L-7 = HyPic-c(Thr-DLeu-Da4hPro-Sar-N2mLeu-Ala-MePhg-); (R-P-4)₂-L = c(Qxa-DSer-Ala-Cys-MeVal)₂; R-P-10-C-7 = Oct-Dab-Thr-Dab-c(Dab-Dab-Leu-Leu-Thr-Dab-Dab); P-12-C-7 = (Ile-Cys)-Leu-DGlu-Ile-c(Lys-DOrn-Ile-Phe-Asn-DAsp-His); D-6a = c(MeVal-DHiv)₃; D-6b = c(MePhe-DHiv)₃. The abbreviations for unusual amino and hydroxy acids are listed in the Abbreviations footnote on the first page. The enzymes are as follows: ACVS, δ-(α-L-aminoadipyl)-L-cysteinyl-D-valine synthetase; BA, bacitracine synthetase; BES, beauvericin synthetase; CPS, cyclopeptine synthetase; CYS, cyclosporine synthetase; DXS, destruxin synthetase; ECS, echinomycin synthetase; ED, edeine synthetase; ENS, enniatin synthetase; FES, ferrichrome synthetase; LG, linear gramicidin synthetase; MYS, myocobacillin synthetase; POS, polymyxin synthetase unusual

Peptide	Organism	Structural type	Enzyme	Size (amino acids activated)	Reference
				kDa	
Linear					
ACV	*Aspergillus nidulans*	P-3	ACVS	230(3)	[29]
Edeine	*Bacillus brevis*	P-5-M	ED1		[5]
			ED2		
Gramicidin	*Bacillus brevis*	P-15-M	LG1	160(2)	[103]
			LG2	350(5)	
			LGx	? (8)	
Alamethicine	*Trichoderma viride*	P-19-M	Alx	? (19)	[16]
Cyclic					
Cyclopeptin	*Penicillium cyclopium*	C-2	CPS	?	[173, 174]
Enterochelin	*Escherichia coli*	P-C-E-3			[151]
Ferrichrome	*Aspergillus quadricinctus*	C-6	FES	?	[175]
Gramicidin S	*Bacillus brevis*	C-(P-5)₂	GS1	120(1)	[7]
			GS2	200(4)	
Tyrocidine	*Bacillus brevis*	C-10	TY1	115(1)	[8]
			TY2	230(3)	
			TY3	450(6)	
Cyclosporine	*Beauveria nivea*	C-11-M	CYS	650?(11)	[85]
Mycobacillin	*Bacillus subtilis*	C-13	MY1		[24]
			MY2		
			MY3		
Lactones					
Destruxin	*Metarhizium anisopliae*	L-6-M	DXS	? (6)	[122]
Actinomycin	*Streptomyces clavuligerus*	R-(L-5)₂	AC1	45(1)	[113]
			AC2	220(2)	
			AC3	280(3)	
Etamycin	*Streptomyces griseus*	R-L-7	ET1	57(1)	[115]
			ETx		
Echinomycin	*Streptomyces echinatus*	(R-P-4)₂-L	EC1	52(1)	[115]
			EC2		
			ECx		
Branched cyclopeptides					
Polymyxin	*Bacillus polymyxa*	R-P-10-C-7	POS	? (10)	[104]
Bacitracin	*Bacillus licheniformis*	P-12-C-7	BA1	330(5)	[64]
			BA2	210(2)	
			BA3	380(5)	
Depsipeptides					
Enniatin	*Fusarium oxysporum*	D-6a	EN	250(3 × 2)	[88]
Beauvericin	*Beauveria bassiana*	D-6b	BE	250(3 × 2)	[127]

It is of interest to see whether certain properties of the enzyme-synthesized peptides are similar, e.g. length or ring sizes of cyclopeptide and lactones. An evaluation of most of the structural data now available [14] is compiled in Table 2. There are four structural types: linear peptides, P-*n*, with *n* indicating the chain length; cyclic peptides, C-*n*, and in the case of a branched structure, P-*m*-C-*n*, where *n* is the number of residues in the ring; lactones, L-*n* or P-*m*-L-*n*, and depsipeptides, D-*n*.

These represent a very limited set of structures, compared to the many more that could be created by synthetic organic chemistry. We do indeed observe a preference for certain ring

Table 2. *Structural types of enzymatically formed peptides [14]*
P represents a peptide chain, R is an acyl residue at the amino terminal, C a cyclopeptide, L is a lactone, D is a depsipeptide (alternating peptide and ester bonds, or more than one ester bond in a peptide chain), and E is structure of hydroxy or amino acids linked by ester bonds. *n* represents the number of residues (amino acid or hydroxy acid) in a chain not including the acyl component

Type	Number for $n =$												
	2	3	4	5	6	7	8	9	10	11	12	13	>13
Linear													
P-n	13	20	7	7	2	22[a]	3	1	—	—	—	—	—
P-$n(\beta,\gamma)$	8	14	—	—	—	—	4	1	—	—	—	—	—
R-p-n	5	—	1	—	—	—	—	1	2	—	—	—	—
P n-C/L-x[b]	—	1	7	—	2	4	3	3	7	1	1	1	1
Cyclic													
C-n	20	—	6	1	11	6	4	—	3	2	1	—	—
C-$n(\beta,\gamma)$	—	—	—	1	—	—	3	—	5	—	—	—	—
R-C-n	—	1	—	—	4	2	—	—	—	—	—	—	—
L-n	—	—	3	—	8	1	—	4	—	—	—	—	—
R-L-n	—	—	—	3	3	2	4	—	—	1	—	—	—
D-n	—	1	2	—	4	—	1	—	—	—	1	—	—
E-n	—	3	2	—	—	—	—	—	—	—	—	—	—
P-x-C/L-n[c]	2	—	2	4	3	7	8	6	—	1	1	—	—
others			5[d]										

[a] The high number of peptides with $n = 7$ can be retraced to extensive screenings in the vancomycin/ristocetin group; these are very closely related glycopeptides

[b] Indicates the length of the peptide chain produced, before cyclization produced a branched cyclic structure; the size of the ring is not considered here

[c] Indicates the ring sizes of branched chain cyclic structures, both cyclopeptides and lactones; the length of the amino-terminal branch is not considered

[d] These are peptides containing a urea-type structure NH-CO-NH, thus introducing a change of direction of the chain; this has not yet been evaluated biosynthetically

sizes, such as cyclohexapeptides (C-6), hexapeptide lactones (L-6), or branched cycloheptapeptides (P-n-C-7), but for many of the structural types no examples have yet been identified. One might also expect to detect certain sequence similarities within these groups of enzymatically formed peptides. Such sequence similarities are surprisingly low (Table 3), but there are peptide analogs containing one or several exchanged positions. The substitutions are often closely related amino acids, such as leucine for phenylalanine, or threonine for serine. Many analogs originate from the limited specificity of the activation reactions, which allow the substitution of leucine by isoleucine or valine, or of phenylalanine by tyrosine or tryptophan. Cell-free enzyme systems permit an even larger frame-work of substitutions (see below).

THE MULTIENZYMIC CYCLE
Amino (hydroxy) acid activation

The substrates are activated directly at their respective enzyme sites. There is no CoA-like activated form of the amino acids, as in polyketide formation. In the majority of cases studied so far, activation proceeds by cleavage of the α,β-phosphate bond of ATP, leading to the formation of an enzyme-stabilized, non-covalently bound amino or hydroxyl acyladenylate. Upon addition of pyrophosphate, the reaction is reversed and ATP is formed. Similarly, under certain conditions, adenosine (5')tetraphospho(5')adenosine may be formed with ATP, a reaction well known in the case of aminoacyl-tRNA synthetases [30]. This amino- or hydroxy-acid-dependent cleavage of ATP is still the most widely used detection procedure for peptide synthetases, particularly since many of the unusual substrates involved indicate the presence of the respective synthetase with a high certainty.

Early ideas that the sites of peptide synthetases resemble aminoacyl-tRNA synthetases, and that enzymatic amino-acylation by peptide synthetases would find some analogy in the reaction sequence leading to aminoacyl-tRNA, have not been verified. The aminoacyl-tRNA synthetases which have so far been studied in detail show no essential thiol functions involved in aminoacyl transfer [31]. The study of the respective ATP binding sites by substrate analog usage implies that there are significant differences between the two types of synthetases, although they catalyze similar reactions [32, 33].

Studies on the amino acid activation reactions have been carried out for gramicidin S synthetases; these have been reviewed elsewhere [7, 34, 35]. A basic feature of these systems is the simultaneous binding of an adenylate and the respective thioester [36, 37]. Mutants of gramicidin S synthetase 1 characterized by Saito et al. [38, 39] are able to catalyze adenylate formation, but have lost the aminoacylation function. Modification studies have shown that aminoacylation can be inhibited by alkylation of a single thiol [40—42], whereas the ability to form adenylate is lost upon modification of two or three additional thiol groups [40—42]. These findings correlate well with the comparison of the amino acid sequences of gramicidin S synthetase 1 and tyrocidine synthetase 1: these multienzymes contain one and three conserved thiols, respectively, in distinct distant regions [43, 44].

Initiation

Initiation of multienzymic peptide biosynthesis is a variable process, and details of the systems investigated so far are listed in Table 4. The N-terminal starter amino acid may retain a free amino group, or it may be formylated or acylated. The

Table 3. *Examples of similarities within enzymatically formed peptides*
The first three cyclopeptides in the tyrocidine/gramicidin S group produced by strains of *Bacillus brevis*, have a conserved pentapeptide sequence Val-Orn-Leu-DPhe-Pro. The apparent similarity of the protein templates was found to be complex. The five residues are located on three different enzymes in tyrocidine, on two enzymes in gramicidin S. Gratisin has been obtained from a mutant of the gramicidin S producer. The stereochemistry of the aromatic residues indicated with an asterisk has not been determined. A similar tripeptide sequence found on one multienzyme in both tyrocidine and gramicidin S has also been detected in the cortinarins, toxins isolated from species of the basidiomycete *Cortinarius*. The branched cyclopeptides in the polypeptin/octapeptin group are formed by strains of *Bacillus circulans*, *B. polymyxa*, and *B. colistinus*, as well as *Aerobacter aerogenes*. They contain a highly similar heptapeptide ring. Residues identical in all of the analogs so far isolated are indicated in bold-face lettering. In the last group, several fungi form highly related 20-peptides, such as *Trichoderma viride* (alamethicin, suzukacillin), *Hypocrea peltata* (hypelcin), *Gliocladium delinquescens* (gliodeliquescin), *Trichoderma resei* (paracelsin), and a 19-peptide, trichorzianine, is formed by *Trichoderma harzianum*. There is a consensus sequence of 13 of the 20 (19) residues, while there are some conservative replacements and one deletion (del). If some of the various analogs of these peptides are compared, sequence deviations are found only in one or two positions

Type of similarity	Group	Peptide	Sequence
Sequence	tyrocidine/ gramicidin S	gramicidin S	c(DPhe→Pro→Val→Orn→Leu→)
		tyrocidine A	c(DPhe→Pro→Phe→DPhe→Asn→Gln→Val→Orn→Leu→)
		gratisin	c(*Phe→Pro→*Tyr→Val→Orn→Leu→)
		cortinarin C	c(Phe→OmTrp→Val→Orn→Leu→Ile→Ala→DThr→Gly→Lys→)
Cycloheptapeptide element/amino acid replacements	polypeptin/ octapeptin	polymyxins	FA→Dbu→Thr→Dbu→Dbu→Dbu→DLeu→Leu→Dbu→Dbu→Leu→
			DSer ↑ DPhe Phe Thr
			Phe Thr
		octapeptins	FA→DDbu→Dbu→Dbu→DLeu→Leu→Dbu→Dbu→Leu→
			↑ DPhe Phe
Amino acid replacements	alamethicin group of peptaibols		AcAibProAibAlaAibAlaGlnAibValAibGlyLeuAibProValAibAibGluGlnPheol
			Ala Aib Leu SerAib DIva Leuol
			Ile Ile Trpol
			Aib
			del

first possibility is of particular importance when subsequent cyclization has to occur, while the latter possibilities usually apply in the case of linear peptides or branched cyclic structures. Clearly no rules can be laid down, either for the state of the terminal amino group or for the use of D-epimers in this reaction. The initiation in some cases requires two enzymes, whereas in others the catalytic sites are assembled on a single multienzyme.

The gramicidin S cycle

This reaction has been studied in most detail by Saito et al. in case of gramicidin S synthetases. The sequence of reactions leading to gramicidin S (I) is [7] given by:

In this scheme, E^1 represents synthetase 1 with a size of 126 kDa, and E^2 synthetase 2 with an estimated size of 280 kDa. The enzymes accept the amino acids (Xaa) phenylalanine, proline, valine, ornithine and leucine to form with ATP (as an Mg^{2+}, Mn^{2+}, or Ca^{2+} complex [34]) the mixed anhydride aminoacyl adenylate as intermediate (XaaAMP). Random substrate binding has been detected [34] and the adenylates formed cause aminoacylation of specific active thiol groups E^i-S^nH, releasing AMP.

Reactions 1 – 3 (adenylate formation, thioester formation, and epimerization) initially led to the term ATP-dependent phenylalanine racemase for gramicidin S synthetase 1 (GS1, EC 5.1.1.11) [45, 46]. These reactions proceed in complete analogy to the formation of tyrocidine (V), a nonsymmetrical

(1, 2) $E^1 + Phe + MATP^{2-} \rightarrow E(PheAMP)$ $\rightarrow E^1$-S^1-Phe
(3) $\rightarrow E^1$-S^1-DL-Phe
(4?) $\rightarrow (E^2$-S^P-DPhe)?
                                            ↓

(5 – 7) $E^2 + Pro + MATP^{2-} \rightarrow E(ProAMP)$ $\rightarrow E^2$-S^2-Pro $\rightarrow E^2$-S-Pro-DPhe
(8) $\rightarrow E^2$-S^P-Pro-DPhe
                                            ↓

(9 – 11) $E^2 + Val + MATP^{2-} \rightarrow E(ValAMP)$ $\rightarrow E^2S^3$-Val $\rightarrow E^2S^3$-Val-Pro-DPhe
(12) $\rightarrow E^2S^P$-Val-Pro-DPhe
                                            ↓

(13, 14) $E^2 + Orn + MATP^{2-} \rightarrow E(OrnAMP)$ $\rightarrow E^2$-S^4-Orn
(15) $\rightarrow E^2$-S^4-Orn-Val-Pro-DPhe
(16) $\rightarrow E^2$-S^P-Orn-Val-Pro-DPhe
                                            ↓

(17, 18) $E^2 + Leu + MATP^{2-} \rightarrow E(LeuAMP)$ $\rightarrow E^2$-S^5-Leu $\rightarrow E^2$-S^5-Leu-Orn-Val-Pro-DPhe
(19) $\rightarrow E^2$-S^P-Leu-Orn-Val-Pro-DPhe
(20?) $\rightarrow$ cyclo(DPhe-Pro-Val-Orn-Leu-)$_2$.
(21?)

Table 4. *Types of initiation reactions in multienzymic peptide biosynthesis*

Donor (thioester)	Acceptor (thioester)	Multi-enzyme(s)	System
Carboxyl group Aad(δ)	amino acid Cys	ACVS	penicillin
Amino acid Ile	amino acid Cys	BA1	bacitracin
Amino acid DPhe DPhe	imino acid Pro Pro	GS1/GS2 TY1/TY2	gramicidin S tyrocidine
Methyl-amino acid MeVal MePhe	hydroxy acid Hiv Hiv	EN BE	enniatin beauvericin
fXaa fVal	amino acid Gly	LG1	linear gramicidin
AcXaa AcAib	imino acid Pro	Alx	alamethicin
acyl-Xaa OctDbu	amino acid Thr	POx	polymyxin

cyclodecapeptide formed by other strains of *Bacillus brevis* [47, 48]. However, the respective tyrocidine synthetase 1 (TY1)

$$\begin{array}{c} \rightarrow DPhe^1 \rightarrow Pro^2 \rightarrow Val^2 \rightarrow Orn^2 \rightarrow Leu^2 \longrightarrow \\ \llcorner Pro \leftarrow DPhe \leftarrow Leu \leftarrow Orn \leftarrow Val \longleftarrow \end{array}$$

(IV)

$$\begin{array}{c} \rightarrow DPhe^1 \rightarrow Pro^2 \rightarrow Phe^2 \rightarrow DPhe^2 \rightarrow Asn^3 \longrightarrow \\ \llcorner Leu^3 \leftarrow Orn^3 \leftarrow Val^3 \leftarrow Tyr^3 \leftarrow Gln^3 \longleftarrow \end{array}$$

(V)

(with a size of 125 kDa) cannot substitute for GS1 in gramicidin S formation [47, 49, 50]. The genes of both multienzymes have recently been cloned [43, 44, 51 – 53]. Both enzymes show extensive similarities, including a preserved His-Cys sequence suspected to be an active thiol [42, 44]. Since the enzymes cannot be interchanged, it follows that specific interaction of GS1 and TY1 with their respective subsequent multienzymes GS2 and TY2 is an essential requirement. The numbers *n* in Xaa" in the formulae above represent the corresponding multienzymes GS1 and GS2 in IV and TY1, TY2 and TY3 in V respectively, which activate the residues concerned.

A set of mutationally altered GS2 multienzymes has been isolated from nonproducer mutants by Saito et al. [54 – 56]. Four of these lack gramicidin S formation when complemented with GS1 and have been shown not to contain the cofactor pantetheine [54, 55]. Although these mutant enzymes activate the constituent amino acids, they fail to initiate peptide formation, and a possible involvement of the cofactor in the transfer of D-Phe from GS1 to GS2 has been postulated. The reaction (4?) in the above scheme has been supported by the isolation of GS2 fragments containing pantothenic acid, together with transfer of labelled phenylalanine to GS2.

Surprisingly, Laland et al. found that the multienzyme GS2 activates both L- and D-phenylalanine, although it was expected to activate only proline, valine, ornithine and leucine [57]. They obtained some evidence that this unexpected acti-

vation takes place at the leucine activation site, and isolated an analog of gramicidin S [57, 58], thought to have the structure VI but this has not yet been proved.

$$\begin{array}{c} \rightarrow DPhe \rightarrow Pro \rightarrow Val \rightarrow Orn \rightarrow Phe \longrightarrow \\ \llcorner Phe \leftarrow Orn \leftarrow Val \leftarrow Pro \leftarrow DPhe \longleftarrow \end{array}$$

(IV)

The activation site had been shown to bind both epimers of leucine, but D-leucine proved inhibitory to product formation [59]. In order to exclude a binding of phenylalanine at this 'wrong' activation site, an excess of leucine has to be added in order to prevent misaminoacylation.

Phenylalanine attached as a thioester to GS1 is apparently only reactive in the D configuration. Kurahashi et al. have shown that in the presence of NH_4^+ ions, D-phenylalanine amide is formed [33]. Thus, in the interaction with GS2 the exclusive transfer of the D-isomer can be assumed, and only the D-isomer has been found in gramicidin S and GS-like peptides.

Isolated GS2 can be assumed to be charged with amino acid and peptide intermediates. This can be concluded from the stability of aminoacyl thioesters [7, 60] and indeed, in studies on the formation of product from a labelled precursor, an initial lag phase has been found [61]. Some evidence for an inhibition of the enzyme upon mis-acylation has come from substrate protection studies performed to aid isolation of the active multienzyme [62]. This suggests a role for a gene like the thioesterase gene, which was identified recently as the first reading frame (*grsT*) in the gramicidin S cluster [44]. This gene, which shows similarities to the medium-chain fatty acyl thioesterases of rat and duck, is followed by the structural genes for GS1 (*grsA*) and GS2 (*grsB*). The cluster is apparently transcribed as a polycistronic message, thus assuring the equimolar stoichiometry of the multienzymes. Evidence for a similar arrangement of the tyrocidine synthetase genes has been obtained recently [51, 52].

Elongation

The elongation cycle, as illustrated schematically in Fig. 2, is not a repeated cycle of reactions as in polyketide formation. Rather it is a single cycle of sequential and similar reactions. The intermediates remain in an active state as thioesters, and their transport to the cofactor phosphopantetheine by successive transthiolations has been postulated [17]. The general principle of the successive transport, and the selection of the sequential reaction sites, is at present not understood. A scheme proposed by Lipmann et al. [2, 19] suggests that intermediate peptides remain attached to their specific elongation sites. Thus, in case of gramicidin S formation, the synthetase 2 remains charged with di-, tri-, and tetrapeptide if the last amino acid is omitted. The classical experiment supporting this model was the direct analysis of the enzyme-bound growing peptide chains by amino-terminal labelling [63]. Similar experiments have been carried out in the studies on tyrocidine [47], bacitracin [64] and alamethicin [15], with varying degrees of success.

The basic problem, namely the stability of these intermediates, has been studied in detail by Vater et al. [7, 41, 65 – 68]. It is known that, depending on the structure of the specific peptide, loss by side reactions may occur. This is especially evident (by analogy to chemical peptide synthesis) in the case of dipeptides. Cyclization of dipeptides has frequently been

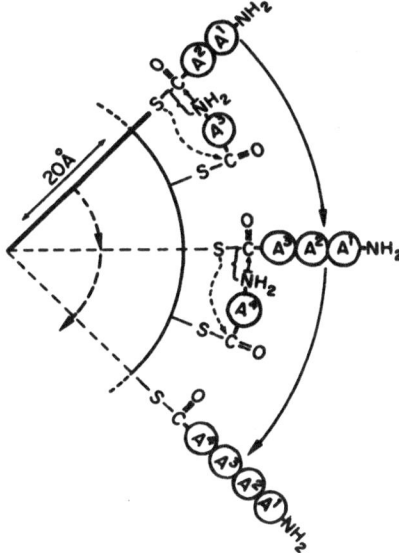

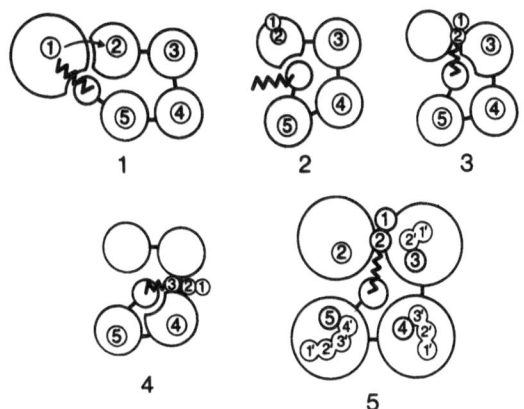

Fig. 2. *Scheme of the postulated transpeptidation/transthiolation mechanisms of the thiotemplate type of polypeptide formation.* Intermediate peptidyl-thioesters are transported by the covalently attached cofactor 4'-phosphopantetheine to the following aminoacyl-residue, where the peptide bond is formed in a transpeptidation reaction. 20 Å = 2 nm

Fig. 3. *Models explaining the sequence of events in polyenzymatic peptide formation.* The model of sequential conformational changes assumes a step by step movement of the cofactor (1 – 4). This movement could be controlled by an 'opening' of a binding site following transthiolation. An alternative model (5) suggests the existence of a peptide recognition site in analogy to substrate recognition sites in enzymatic reactions. The latter model implies a highly specialized multienzyme system, whereas the first model is of a more general type

observed in biosynthetic studies, e.g. in gramicidin S, tyrocidine, bacitracin [64], cyclosporin [69] or actinomycin [70], especially when proline or *N*-methyl amino acids are involved. This side reaction has been used for the preparation of cyclodipeptides of the type D-aromatic-amino-acid − L-imino-acid in the gramicidin S system [62, 71]. Similarly, the cyclization of ornithyl thioesters is sterically favoured. The formation of both cyclo-ornithine and cyclo-ornithyl peptides have been studied as side reactions in gramicidin S biosynthesis [41, 68]. Furthermore, a significant enhancement of hydrolytic cleavage of thioesters by the thiol reducing agents, which are generally included in reaction mixtures, has been noted [41, 67]. It has long been known that certain intermediates can barely be detected, since they undergo cyclization reactions. Besides dipeptides, these are the pentapeptides in gramicidin S formation, which rapidly form head-to-tail dimers, and an intermediate in the trimerization of MeVal-DHiv during enniatin formation [72]. However, the available evidence implies a possible accumulation of such intermediates, indicating the repeated functioning of the pantetheine transport system within one biosynthetic cycle.

Two models for interpreting the regulation of the reaction sequence have been proposed. The first one treats each elongation step as an independent enzymic reaction, by analogy with the known single-step enzymic peptide synthetases involved in the formation of glutathion or (4'-phospho-)pantetheine [9]. This model, as illustrated in Fig. 3, suggests the presence of specific recognition sites for the intermediates. It could be compared to the partial reactions of fatty acid biosynthesis with thioester intermediates, which should undergo a specific modification reaction. A second more general model suggests successive conformational changes of adjacent enzymic sites. This model resulted from the elucidation of the sequence of amino acid activation sites on the GS2 multifunctional chain, in accordance with the sequence of gramicidin S [73]. A similar model has been proposed in the case of mycobacillin, which is not formed by a thiotemplate mechanism, but which nevertheless requires a multienzyme type of system [24]. Here, two molecules of ATP have been estimated to be consumed per peptide bond formed, in contrast to the single molecule/bond

in gramicidin S [74], and it has been proposed that this requirement is at least partially necessary to promote successive conformational changes, thus assuring the correct elongation steps.

Termination

Fragment condensation reactions, such as the head-to-tail cyclization of two pentapeptides in gramicidin S formation, are attractive candidates for intermolecular termination reactions involving a complex of multienzymes. To decide whether a single GS2 enzyme could form two pentapeptides, Laland et al. designed an experiment which involved charging two sets of multienzymes with peptide intermediates, one of these being an imino acid analog (azetidine-2-carboxylate) instead of proline. Upon mixing, the possible formation of hybrid peptides were examined. Since no hybrids were detected, dimerisation of the pentapeptides must have occurred on a single multienzyme, and a 'waiting position' involving an extra thiol group has been postulated [75]. Contrary to this finding is the experiment of Lipmann et al. in which GS2 was charged with the set of four substrate amino(imino) acids, and the multienzyme was precipitated with ammonium sulfate. It had been shown earlier that salting out removes the non-covalently bound aminoacyl adenylates, leaving the acyl thioesters. Thus if synthesis was now initiated with GS1, cyclopeptide formation could only occur if two GS2 enzymes were able to interact. Gramicidin S was indeed formed in the experiment [76]. The two experiments do not necessarily contradict each other, since the kinetic aspects were not considered. Both mechanisms, intermolecular and intramolecular, could operate, but with different efficiencies.

Similarly, in the trimerization reaction leading to the formation of enniatins, no evidence for a complex formation has so far been obtained. However, the transfer of intermediates also implies contact between multienzymes; no stable complex has yet been demonstrated.

The possibility of protein templates being used to direct the cyclization of a peptide chain remains an attractive goal. Each protein template would efficiently direct a specific amino group towards the terminal thioester to ensure correct cyclization.

Enzyme kinetic treatment of the interaction of multienzymes

During studies on the effect of protein concentration on gramicidin S formation in a crude preparation, non-linear behaviour was observed [77]. This was interpreted by postulating a required statistical collision between the multienzymes GS1 and GS2, so that both of their (equimolar) concentrations present enter the rate equation. This interpretation was based on the failure to detect a complex between the two enzymes in solution; it would indeed lead to a significant increase in the rate of peptide formation at high multienzyme levels. One approach considered in some detail the sequence of events during initiation of peptide formation and predicted deviations from the second-order rate function, depending on the rates of certain partial reactions [78]. At present, no detailed measurements of all the elementary reaction steps are available and the rate-limiting step remains to be demonstrated. However, the concept of interacting multienzymes remains an attractive explanation for the difficulties often encountered in obtaining an effective cell-free system for biosynthesis. Consider for example a system of three interacting multienzymes, as in the case of tyrocidine or bacitracin. If the concentration of each of the multienzymes involved were to enter the rate equation, then we would obtain a third-order dependence on protein concentration and a reasonable reaction velocity could be achieved in a 'compartmentalized' situation which is probably encountered *in vivo*.

That the situation could turn out to be even more complicated became apparent in an unpublished study by von Döhren and Froyshov (cited in [5]) on the bacitracin multienzymes. Apparently equimolar amounts of the three multienzymes were not present and the intermediate enzyme, interacting with both synthetases 1 and 3, seemed to dimerize in gradient centrifugation. Similar dimerization has been observed frequently under non-denaturing conditions with other multienzymes. However, no evidence for a catalytic function of this type of association has so far been obtained, either in the case of dimerization of an intermediate, as in gramicidin S formation, or in case of a trimerization, as in enniatin biosynthesis. This is especially evident, since the full catalytic cycle of enniatin synthetase is functioning in the immobilized state [79–82].

In a critical evaluation of the *in vitro* rates of gramicidin S formation, we have pointed out that the performance of an enzyme actually corresponds well with the *in vivo* performance if low productivity conditions in the milligram/litre range are used. It cannot account, however, for high productivity conditions where gram/litre concentrations are formed [78]. Actual enzyme concentrations used *in vitro* have so far been below 1 mg/ml. In addition, a recent study on the multienzyme tyrocidine synthetase 1 suggests a localization at the cytoplasmic membrane (E. Pfeifer and von Döhren, unpublished results). Also the enzyme system forming penicillin G is localized in Golgi-like vesicles in *Penicillium chrysogenum* [83]. Thus, we are far from reproducing the actual situation in the cell and significant improvements in the system's performance can be expected in the future.

OTHER MULTIENZYMIC CYCLES

Cyclopeptides tyrocidine and cyclosporin

These two systems of quite different cyclopeptides have been studied in considerable detail. The linear precursor is formed according to the general scheme presented above, and the substrates are modified by two epimerizations in the case of tyrocidine (V) or by seven *N*-methylations in the case of cyclosporin (VII):

$$
\begin{array}{c}
\rightarrow \text{DAla} \rightarrow \text{MeLeu} \rightarrow \text{MeLeu} \rightarrow \text{MeVal} \rightarrow \text{MeBmt} \\
\text{Ala} \leftarrow \text{MeLeu} \leftarrow \text{Val} \leftarrow \text{MeLeu} \leftarrow \text{Sar} \leftarrow \text{Abu}
\end{array}
$$

(VII)

where Bmt = (4R)-4[(E)-2-butenyl]-4-methyl-L-threonine and Abu = 2-aminobutyric acid.

In order to describe the sequence of reactions schematically, a short notation is introduced below. In tyrocidine biosynthesis there is a multienzyme system catalyzing the following sequence of reactions:

(1)   Phe $\rightarrow$ PheAMP $\rightarrow$ E^1-S-DPhe $\rightarrow$

(2)   Pro $\rightarrow$ ProAMP $\rightarrow$ E^2-S-Pro $\rightarrow$

(3)   Phe $\rightarrow$ PheAMP $\rightarrow$ E^2-S-Phe $\rightarrow$

(4)   Phe $\rightarrow$ PheAMP $\rightarrow$ E^2-S-DPhe $\rightarrow$

(5)   Asn $\rightarrow$ AsnAMP $\rightarrow$ E^3-S-Asn $\rightarrow$

(6)   Gln $\rightarrow$ GlnAMP $\rightarrow$ E^3-S-Gln $\rightarrow$

(7)   Tyr $\rightarrow$ TyrAMP $\rightarrow$ E^3-S-Tyr $\rightarrow$

(8)   Val $\rightarrow$ ValAMP $\rightarrow$ E^3-S-Val $\rightarrow$

(9)   Orn $\rightarrow$ OrnAMP $\rightarrow$ E^3-S-Orn $\rightarrow$

(10)   Leu $\rightarrow$ LeuAMP $\rightarrow$ E^3-S-Leu $\rightarrow$
           cyclization

C-10

in which E^1 is tyrocidine synthetase 1 (the product of the gene *tycA*) with a molecular mass of 125 kDa [33, 37, 38], E^2 is the synthetase 2 (product of *tycB*) of 230 kDa [34, 36, 38] and E^3 is the synthetase 3 (product of *tycC*) of 450 kDa [33, 35, 36, 38]. Upon breaking the producer cells of *Bacillus brevis* ATCC 8185, these multienzymes are soluble, and cannot be isolated as a stable complex [48, 84]. On the other hand, in the case of the cyclosporin synthetase obtained from the eucaryotic organism *Beauveria nivea* (formerly *Tolypocladium inflatum*), all the catalytic functions are found to be integrated in a single multifunctional enzyme E of approximately 800 kDa [85–88], with the following postulated sequence of reactions:

(1)   DAla $\rightarrow$ DAlaAMP $\rightarrow$ E-S-DAla $\rightarrow$

(2)   Leu $\rightarrow$ LeuAMP $\rightarrow$ E-S-Leu $\rightarrow$ E-S-MeLeu $\rightarrow$

(3)   Leu $\rightarrow$ LeuAMP $\rightarrow$ E-S-Leu $\rightarrow$ E-S-MeLeu $\rightarrow$

(4)   Val $\rightarrow$ ValAMP $\rightarrow$ E-S-Val $\rightarrow$ E-S-MeVal $\rightarrow$

(5)   Bmt $\rightarrow$ BmtAMP $\rightarrow$ E-S-Bmt $\rightarrow$ E-S-MeBmt $\rightarrow$

(6)   Abu $\rightarrow$ AbuAMP $\rightarrow$ E-S-Abu $\rightarrow$

(7)   Gly $\rightarrow$ GlyAMP $\rightarrow$ E-S-Gly $\rightarrow$ E-S-Sar $\rightarrow$

(8)   Leu $\rightarrow$ LeuAMP $\rightarrow$ E-S-Leu $\rightarrow$ E-S-MeLeu $\rightarrow$

(9)   Val $\rightarrow$ ValAMP $\rightarrow$ E-S-Val $\rightarrow$

(10) Leu $\rightarrow$ LeuAMP $\rightarrow$ E-S-Leu $\rightarrow$ E-S-MeLeu $\rightarrow$

(11) Ala $\rightarrow$ AlaAMP $\rightarrow$ E-S-Ala $\rightarrow$
         cyclization $\rightarrow$

C-11.

Whereas in the tyrocidine system epimerization is an integrated function, in the cyclosporin system D-alanine is required for initiation. However, this giant multienzyme contains one or several integrated methyl transferase function(s) for the N-methylation of seven intermediate thioesters via S-adenosyl-methionine, prior to the elongation cycle. Pioneering studies on the N-methylation of peptides carried out by Zocher et al. in the enniatin system have been reviewed recently [88].

Branched cyclopeptides bacitracin and polymyxin

Bacitracin (VIII) was the first branched-chain system to be studied, and a nearly complete pathway has been described [64, 89 – 99].

—Ile→Cys→Leu→DGlu→Ile→Lys→DOrn→Ile→DPhe—
 L_____J   ↑ |
                          L—— Asn←DAsp←His ——↓

(VIII)

Polymyxin is an acylpeptide (IX), and although the detailed molecular

iOct→Dbu→Thr→Dbu→Dbu→Dbu→Leu→Leu—
                          ↑ |
                          L—Thr←Dbu←Dbu —↓

(IX)

(Dbu = 2,4-diaminobutyric acid)

properties of the multienzyme(s) involved in its formation are not yet known [100 – 107], it has been demonstrated that an additional protein fraction, an acyl transferase, is required to initiate the cycle, starting from isooctanoyl-CoA [101, 102]. The schematic biosynthetic sequence for bacitracin (VIII) as follows:

(1) Ile → IleAMP → E^1-S-Ile   →
(2) Cys → CysAMP → E^1-S-Cys   →
(3) Leu → LeuAMP → E^1-S-Leu   →
(4) Glu → GluAMP → E^1-S-DGlu →
(5) Ile → IleAMP → E^1-S-Ile   →
(6) Lys → LysAMP → E^2-S-Lys   →
(7) Orn → OrnAMP → E^2-S-DOrn →
(8) Ile → IleAMP → E^3-S-Ile   →
(9) Phe → PheAMP → E^3-S-DPhe →
(10) His → HisAMP → E^3-S-His   →
(11) Asp → AspAMP → E^3-S-DAsp →
(12) Asn → AsnAMP → E^3-S-Asn   →

cyclization(Lys)

P-12-C-7.

The synthetases involved in bacitracin formation have been named as A (BA1, 330 kDa), B (BA2, 210 kDa) and C (BA3, 380 kDa) and studied in *Bacillus licheniformis*. A soluble system has been isolated from growing cells, whereas in the transition and stationary growth phase the enzymatic activity is localized in the particulate fraction [89]. The system has been fractionated into three multienzymes of 330, 210, and 380 kDa, respectively. There are contradictory results concerning the pantetheine content of the intermediate enzyme,

one report indicating a soluble form of the cofactor [93]. The structural gene for this multienzyme has recently been cloned and expressed in *E. coli* [97]. Although the reaction sequence has been demonstrated via the isolation of thioester-bound intermediate peptides, some uncertainty exists concerning the formation of the thiazolidine ring between the amino terminal Ile and Cys. This certainly appears to be an enzyme-catalyzed reaction, since the isolation of intermediates led to the finding of three distinct compounds containing Ile and Cys [64], suggesting that the modification takes place at the dipeptide stage. One report indicated the covalent attachment of a bacitracin precursor, which was subsequently liberated by a proteolytic reaction [108]. It has also been proposed in this connection that the formation of the thiazolidine ring involves a proteolytic enzyme function [98]. The polymyxin synthetase fraction from *Aerobacter aerogenes* has an estimated size of 640 kDa [104]. The involvement of one or several multienzymes in this organism or in *Bacillus polymyxa* has not yet been evaluated; fractions analyzed by denaturating polyacrylamide gel electrophoresis contained several proteins, which have not yet been identified [106, 107].

Lactones actinomycin, quinomycin, and destruxin

The lactone structures currently under biosynthetic investigation are representative members of different structural classes. Actinomycin (X) and quinomycin (XI), both well-known highly toxic DNA-binding peptides formed by several strains of *Streptomyces*, are chromo- or acylpeptides. While actinomycin contains two simple lactone structures on a uniquely dimerizing chromophore, quinomycins represent a head-to-tail lactonization of two tetrapeptide units. Both these

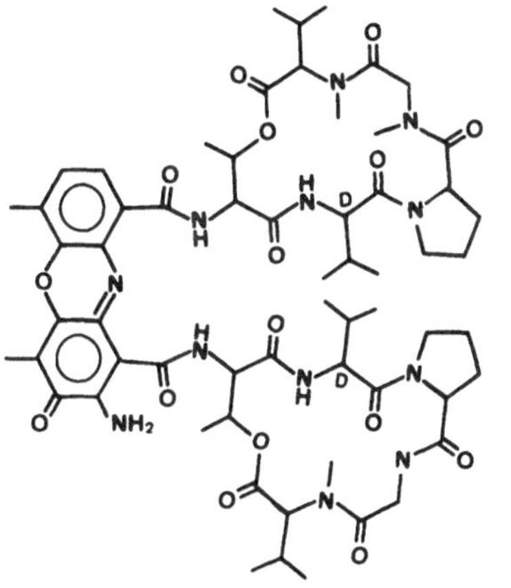

(X)

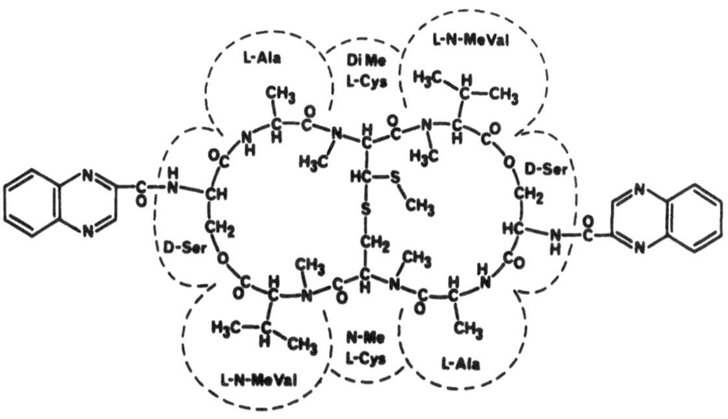

(XI)

systems occur in procaryotic organisms, whereas in contrast destruxin is formed by eucaryotes such as *Metarhizium anisopliae* [109] or *Alternaria brassicae* [110], and a simple lactone (containing however β-alanine) which supposedly should involve a more integrated enzyme system.

Difficulties have been encountered for a long time in attempts to characterize the biosynthetic system leading to formation of actinomycins [111], and a complete *in vitro* synthesis has still not been achieved. Keller has succeeded in elucidating the steps in the reaction, and has also made a partial characterization of the (multi)enzymes involved [112–115]. Considerable similarities with actinomycin have aided the study of the quinomycin system [114, 115]. A short scheme describing the actinomycin system is as follows (Mha = 3-hydroxy-4-methylanthranilic acid):

(A)   Mha → E^1(MhaAMP) →

(1)   Thr → ThrAMP → E^2-S-Thr →

(2)   Val → ValAMP → E^2-S-DVal →

(3)   Pro → ProAMP → E^3-S-Pro →

(4)   Gly → GlyAMP → E^3-S-Gly → E^3-S-Sar →

(5)   Val → ValAMP → E^3-S-Val → E^3-S-MeVal →

lactonization

R-L-5

phenoxazinone –

formation   R*(L-5)$_2$

The acylating enzyme, or actinomycin synthetase 1 (45 kDa), activates 4-methyl-hydroxyanthranilate as adenylate. No thioester or CoA intermediate has been detected [112]. Apparently the threonyl-thioester on synthetase 2 is acylated in the initiation step [113]. Similar acyl (chromophore) activating enzymes have been isolated which act in the synthetic pathways of quinomycins (52 kDa) and etamycin (57 kDa [115], XII). The enzymic organization of these *Actinomycetes* systems has turned out to be surprisingly complex. Two, and in case of the quinomycin type at least two, multienzymes are required to form an acyl-penta or tetrapeptide, respectively. The actinomycin synthetases 2 and 3 (225 and 280 kDa) have been partially characterized [70, 113] and fragments have been expressed as lac fusion proteins in *E. coli* [114]. The final step, namely the oxidation of two acylpentapeptides to the chromophore, can proceed either enzymatically (e. g. in *Streptomyces antibioticus*) or nonenzymatically (e. g. in *Streptomyces clavuligerus*). The corresponding phenoxazinone synthase from *S. antibioticus* has been characterized [116] and cloned [117]. Expression studies in *Streptomyces lividans* revealed the presence of a silent structural gene within this organism, which could be activated by transformed DNA [118]. Such silent genes appear to be common in non-producer strains [117].

A genetic analysis of the actinomycin pathway revealed the chromosomal location of at least some of the biosynthetic enzymes in *Streptomyces chrysomallus* [120, 121]. A fungal lactone system is destruxin synthetase. Evidence for the *in vitro* formation of destruxin B by a high-molecular-mass enzyme fraction has been obtained [122], but the system has yet to be characterized.

Depsipeptides enniatin and beauvericin

The hexadepsipeptide enniatin, produced by strains of *Fusarium*, known as a plant pathogen, was the first fungal

system to be characterized in detail. Enniatins, the analog beauvericin, and the higher analog bassianolide have the general structure XIII: c(Hiv-MeXaa)$_n$ with MeXaa = *N*-methyl amino acid; in enniatin Xaa = Val, Leu, Ile, $n = 3$; in beauvericin Xaa = Phe, $n = 3$; in bassianolide Xaa = Leu, $n = 4$.

Multienzymes of 250 kDa have been characterized by Zocher et al. [79–81, 88, 116, 123–127]. These basic studies have shown that β-hydroxy acids are equivalent to amino acids in enzymic peptide formation. D-Hydroxyisovalerate (DHiv) is activated as adenylate, and acylates a specific thiol group on the multienzyme. Furthermore, it has been shown that the frequent *N*-methylation of peptide antibiotics takes place at the thioester stage, before condensation by a methylase function integrated in the multifunctional peptide synthetase [124]:

(1)   Hiv → HivAMP → E-S-Hiv   →

(2)   Xaa → XaaAMP → E-S-Xaa → E-S-MeXaa →

trimerization

D-6

A methylation pathway completely analogous to that which takes place in enniatin has been verified for beauvericin [127], actinomycin [113, 114], cyclosporin [87], and destruxin [122]. Monoclonal antibodies of the IgM type, affecting the *N*-methylation reaction in enniatin synthetase, showed cross reactions between these peptide synthetases, regardless as to their procaryotic or eucaryotic origin [88, 125]. Photoaffinity labeling of the methyltransferase site, that was also inhibited by these antibodies, led to a 25-kDa fragment by limited proteolytic digestion [88, 124].

The remarkable similarity between the multienzymes with a single altered amino acid recognition sites implies the possible exchange of such multienzyme sub-sites. This accounts for the frequent alterations in peptide sequences found in families of natural peptides. The specificities of the recognition sites vary considerably and can be used for *in vitro* biosynthesis of an even wider variety of analogs as discussed below.

Linear peptides ACV, gramicidin, alamethicin

None of the pathways leading to formation of linear peptides has been characterized completely. With regard to the most simple case of the penicillin precursor peptide, ACV, it has long been assumed, by analogy with glutathione, to be formed by a two-enzyme system [128]. A multienzyme, ACV synthetase, has now been identified from *Aspergillus nidulans* [28, 29], *Cephalosporium acremonium* and *Streptomyces clavuligerus* (von Döhren et al., unpublished results) which catalyzes ACV formation from α-aminoadipate (Aad), cysteine and valine. By analogy with the thiotemplate mechanism, valine is epimerized at the thioester stage. The proposed biosynthetic scheme is:

(1)   Aad → AadAMP → E-S-Aad   →

(2)   Cys → CysAMP → E-S-Cys   →

(3)   Val → ValAMP → E-S-DVal →

hydrolysis(?)

ACV.

Here we have an integrated system in bacterial as well as in fungal organisms. From extensive studies on the isopenicillin N synthase, the enzyme which converts ACV into the β-lactam structure, it has been proposed that the genes encoding the penicillin biosynthetic enzymes are arranged in a cluster. This is in agreement with the mapping data obtained from mutant work and transformation studies [129]. The unusually high G + C content of the fungal genes for isopenicillin N synthase has led to the speculation that this β-lactam cluster has been transferred from procaryotes to the eucaryotic fungi [130, 131]. It remains to be seen whether such an assumption is in agreement with the structure of ACV synthase.

The linear gramicidin is a terminally blocked pentadeca-peptide with an alternating L/D structure: fVal-Gly-Ala-DLeu-Ala-DVal-Val-DVal-Trp-(DLeu-Trp)₃-Etn (f = formyl, Etn = ethanolamine). It is synthesized, together with tyrocidine, by strains of *Bacillus brevis* and presumably has some function in shift-down events such as sporulation [132, 133]. The biosynthetic events have only been partially evaluated, but nevertheless some very important basic discoveries have been made. By analogy with ribosomal peptide synthesis, gramicidin starts with a formylated amino terminal. Although Bauer et al. achieved the *in vitro* synthesis of gramicidin in 1972 [134], it took several years to verify that the initial formylation of the thioester-attached valine takes place on the first multienzyme [135–137]. The synthesis can also proceed at a lower efficiency with a free amino-terminal group. A second important achievement was the elucidation of the carboxy-terminal modification by ethanolamine (which represents the actual termination reaction) by Kubota [138–141]. It was possible to release the peptide chain by addition of ethanolamine [134], to obtain gramicidin. The actual mechanism *in vivo* remained to be solved. By analogy with the biosynthesis of 4'-phosphopantetheine, where the terminal cysteamine originates by decarboxylation of a cysteine residue, a search was conducted for a similar reaction involving a terminal serine. Finally Kubota identified phosphatidylethanolamine as the terminating donor molecule. Since formation of phosphatidylethanolamine via serine is catalysed by membrane-bound enzymes, the termination reaction appears to direct the final product into the membrane. A preliminary scheme for the formation of gramicidin is as follows:

formylation by tetrahydrofolate →

(1) Val → ValAMP → E¹-S-Val →
(2) Gly → GlyAMP → E¹-S-Gly →
(3) Ala → AlaAMP → E²-S-Ala →
(4) Leu → LeuAMP → E²-S-DLeu →
(5) Ala → AlaAMP → E²-S-Ala →
(6) Val → ValAMP → E²-S-DVal →
(7) Val → ValAMP → E²-S-Val →
(8) Val → ValAMP → E³-S-DVal →
(9) Trp → TrpAMP → E*-S-Trp →
(10) Leu → LeuAMP → E*-S-DLeu →
(11) Trp → TrpAMP → E*-S-Trp →
(12) Leu → LeuAMP → E*-S-DLeu →
(13) Trp → TrpAMP → E*-S-Trp →
(14) Leu → LeuAMP → E*-S-DLeu →
(15) Trp → TrpAMP → E*-S-Trp →

phosphatidyl-ethanolamine →

P-15-M.

The first two gramicidin synthetases have been isolated and characterized [135, 142]. Linear-gramicidin synthetase 1 (LG1), with a size estimated as between 140 and 180 kDa (K. Bauer, personal communication; von Döhren, unpublished results), catalyzes both the formylation of the starting thioester and dipeptide formation. The dipeptide is then transferred to LG2 (350 kDa) where it becomes elongated to the heptapeptide. An enzyme fraction shown by gel filtration to have a size above 100 kDa catalyses the formation of the pentadecapeptide, which can be released by the addition of ethanolamine. Some evidence has been collected for the organization of the additional enzymes [103]. Protein fractions have been partially purified, which catalyze the activation of L- and D-valine, leucine and tryptophan. This implies the repeated use of synthetases for the addition of the terminally repeated dipeptide sequence DLeu → Trp. It must be remembered, however, that Trp has been found to be exchanged exclusively by Tyr or Phe *in vivo* [14]. In a re-evaluating study of the activating enzymes, we have attempted to purify these fractions further. Preliminary data indicate the existence of a D-Val/Trp-activating enzyme, rather than the addition of a single residue by one enzyme. The actual number of enzymes involved remains to be established.

The peptaibol alamethicin (I) with its 20 residues is still the longest peptide that has been shown to be formed enzymatically. Peptaibols are linear peptides containing α-amino-butyrate (Abu) and terminated by an aminoalcohol. Biosynthetic studies *in vitro* [15, 16] have shown that the amino-terminal acetyl is introduced as acetyl-CoA in the initiation reaction, whereas the C-terminal alcohol results from the direct addition of phenylalaninol to the nascent nonadeca-peptide chain, thus terminating the reaction sequence. The aminoalcohol has been shown not to inhibit the respective aminoacyl-tRNA synthetase in extracts of *Trichoderma viride* [15]. Structures of other peptaibols have been revised with the result that 20 residues now are still the longest reported. The multienzyme fraction contained two high-molecular-mass proteins, but stability problems have so far hampered a more detailed characterization of this enzyme system.

REGULATION OF PEPTIDE ANTIBIOTIC BIOSYNTHESIS

Secondary and special metabolism

Many of the known peptide antibiotics have been described as secondary metabolites, i.e. dispensable for growth. This view is partly based on the observation that strains kept in laboratory cultures apparently lost their antibiotic-producing abilities. Many investigations with producer and non-producer strains, however, rely on the chromatographic detection of the respective metabolite and cannot really differentiate a non-producer from a low-producer strain. Highly sensitive detection methods in combination with molecular genetics seem indispensable for this task.

Some metabolites have been assigned specific functions, such as the complexing of Fe³⁺ to serve the respective transport system. On the assumption that indeed every secondary metabolite could fulfil such a definite function, the introduction of the term 'special metabolite' has been recommended [143, 144]. Despite the availability of non-producers, the tasks and targets of these metabolites remain unknown; a solution of this problem is currently being sought by molecular genetic techniques.

Organization of biosynthetic genes

From the systems that have been studied so far in sufficient detail, i.e. gramicidin S and tyrocidine in *Bacillus brevis* ATCC 9999 and 8185 [44, 51, 52, 145], penicillin and cephalosporin in *Streptomyces clavuligerus* and several fungi [129, 146, 147], or enterochelin in *Escherichia coli* [148−151], and other non-peptide antibiotics [119, 152−155], biosynthetic genes are arranged in clusters, with either chromosomal or plasmid origin. This reflects the distribution of biosynthetic abilities in a wide variety of organisms.

The few studies available so far on the regulation of expression of such gene clusters indicate that specific regulatory events are embedded in the specific life cycle of each organism. Thus tyrocidine, known to be synthesized in the transition from the vegetative to the stationary phase, has been proposed to function specifically in the early sporulation process [131−133]. The promoter region of the tyrocidine biosynthetic cluster has been inserted with the aid of the transducing phage Tn917 into the chromosome of *Bacillus subtilis* 168, linked to a reporter *lacZ* function. The fusion was repressed in *spoOA*, *spoOB*, and *spoOE* mutants, but not in *spoOE*, *spoOF*, and *spoOH* [156]. The dependence on *spoOA* was bypassed in an *abrB* suppressor mutant, which constitutively expressed the fusion with its sigma-43 promoter. The *tyc* promoter thus proved to be under developmental control in both its source and host organism. The repressor protein, identified as AbrB, also operates in the control of the *B. subtilis* genes *abrB*, *aprE*, *spoOE* and *spoVG* [157−159]. A similar control mechanism is now being searched for in *Bacillus brevis* ATCC 8185.

Gramicidin S has not been linked exclusively to sporulation. It is also produced under conditions of limitation leading to a non-sporulating stationary state, or in continuous culture under carbon source limitation [143]. The promoter of the gramicidin S biosynthetic cluster, *grs*, resembles sequences under the control of specific transcription factors, such as sigma-30 or sigma-37 in *B. subtilis* [44].

In contrast to these developmentally expressed pathways, enniatin synthetase has been shown to be transcribed constitutively [160]. This possibly reflects an essential or advantageous function in growth.

Functions of peptide antibiotics within their producers

With the exception of siderophores functioning in the transport of Fe^{3+} into the cell, no definite functions have been assigned convincingly to peptide antibiotics.

Destruxins are well known insecticidal hexapeptidolactones which play an essential role in the infection of various insects by *Metarhizium anisopliae* [122, 161]. Recently it has been demonstrated that destruxin B is also a key compound in the infection of rape by *Alternaria brassicea* [110]. Strain improvement against certain target insects has been shown to involve a significant increase in the production of destruxin E, which contains an epoxide function in the side chain of the hydroxy acid [122]. Thus in some cases a particular set of circumstance contributes to the selection of a particular product, and a specific function can be assigned.

Other attempts to define the endogenous functions of peptide antibiotics, e.g. as effectors in developmental events (tyrocidine as a transcriptional regulator of early sporulation events [132, 133], bacitracin functioning in Mn^{2+} acquisition [162] or gramicidin S as spore constituent and germination effector [163−170]) have not yet been convincingly proven by the analysis of non-producer mutants. Genetic engineering should help to construct defined mutants, thus overcoming the problems of conventional non-producer identification.

Applications of cell-free multienzymic systems

Multienzymes have found applications as biocatalysts, since they permit the construction of peptide analogs from the respective amino acid precursors. The synthesis proceeds stereospecifically and purification in easily accomplished. The potential of the method has not been fully exploited, but main pathways as well as side reactions have been used in microscale preparative work [57, 58, 62, 71, 80, 81, 85, 86, 113, 171, 172]. A more general approach for the design of enzyme systems as catalysts for specific peptides is now under consideration.

CONCLUSIONS

From the biosynthetic investigations on all types of peptides and depsipeptides which are encountered especially in microorganisms, a general mechanism for non-ribosomal multienzymatic peptide synthesis has been observed. This mechanism operates with more or less integrated enzyme systems of so far unique complexity. Single-chain multienzymes, such as cyclosporin synthetase, which catalyzes the formation of a methylated cyclo-undecapeptide, can carry out as many as 40 sequential reactions. The functional and structural relatedness of these multienzymes indicates that they should be considered as a non-ribosomal system, although the organizational principles are not yet understood.

Work in the authors' laboratory was supported by grants from the *Deutsche Forschungsgemeinschaft* (*Sonderforschungsbereich* 9) and the *Bundesministerium für Forschung und Technologie*.

REFERENCES

1. von Döhren, H. & Kleinkauf, H. (1988) in *The roots of modern biochemistry* (Kleinkauf, H., von Döhren, H. & Jaenicke, R., eds) pp. 355−368, de Gruyter, Berlin, New York.
2. Lipmann, F. (1973) *Science 173*, 875−884.
3. Kleinkauf, H. & von Döhren, H. (1982) in *Trends in antibiotic research*, pp. 220−232, Japan Antibiotics Research Association, Tokyo.
4. Lipmann, F. (1954) in *The mechanism of enzyme action* (McElroy, W. D. & Glass, B., eds) p. 599, Johns Hopkins University Press, Baltimore.
5. Kleinkauf, H. & von Döhren, H. (1987) *Annu. Rev. Microbiol. 41*, 259−289.
6. Kleinkauf, H. & von Döhren, H. (1983) in *Biochemistry and genetic regulation of commercially important antibiotics* (Vining, L. C., ed.) pp. 95−145, Addison-Wesley, London.
7. Vater, J. (1990) in *Biochemistry of peptide antibiotics* (Kleinkauf, H. & von Döhren, H., eds) pp. 33−55, de Gruyter, Berlin.
8. Kleinkauf, H. & von Döhren, H. (eds) (1982) *Peptide antibiotics − biosynthesis and functions*, de Gruyter, Berlin.
9. Kleinkauf, H. & Koischwitz, H. (1978) *Progr. Mol. Subcell. Biol. 6*, 59−112.
10. Kurahashi, K. (1981) in *Antibiotics*, vol. 4, *Biosynthesis* (Corcoran, J. W., ed.) pp. 325−352, Springer Verlag, Berlin, New York.
11. Kleinkauf, H. & von Döhren, H. (1981) *Curr. Top. Microbiol. Immunol. 91*, 129−177.
12. Lipmann, F. (1982) in *Peptide antibiotics − biosynthesis and functions* (Kleinkauf, H. & von Döhren, H., eds) pp. 23−45, de Gruyter, Berlin.
13. Lipmann, F. (1968) *Essays Biochem. 4*, 1−23.
14. von Döhren, H. (1990) in *Biochemistry of peptide antibiotics* (Kleinkauf, H. & von Döhren, H., eds) pp. 411−507, de Gruyter, Berlin.

15. Mohr, H. (1977) Thesis, TU Berlin.
16. Mohr, H. & Kleinkauf, H. (1978) *Biochim. Biophys. Acta 526*, 375−386.
17. Gilhuus-Moe, C. C., Kristensen, T., Bredesen, J. E., Zimmer, T.-L. & Laland, S. G. (1970) *FEBS Lett. 7*, 287−290.
18. Kleinkauf, H., Gevers, W., Roskoski, R., Jr & Lipmann, F. (1970) *Biochem. Biophys. Res. Commun. 41*, 1218−1222.
19. Kleinkauf, H., Roskoski, R., Jr & Lipmann, F. (1971) *Proc. Natl Acad. Sci. USA 68*, 2069−2071.
20. Ghosh, S. K., Mukhopadyay, N. K., Majumder, S. & Bose, S. K. (1983) *Biochem. J. 215*, 539−543.
21. Ghosh, S. K., Mukhopadyay, N. K., Majumder, S. & Bose, S. K. (1985) *Biochem. J. 230*, 785−789.
22. Majumder, S., Ghosh, S. K., Mukhopadyay, N. K. & Bose, S. K. (1985) *J. Gen. Microbiol. 131*, 119−127.
23. Mukhopadyay, N. K., Ghosh, S. K., Majumder, S. & Bose, S. K. (1985) *Biochem. J. 225*, 639−643.
24. Mukhopadyay, N. K., Majumder, S., Ghsoh, S. K. & Bose, S. K. (1986) *Biochem. J. 240*, 265−268.
25. Majumder, S., Mukhopadyay, N. K., Ghosh, S. K. & Bose, S. K. (1988) *J. Gen. Microbiol. 134*, 1147−1153.
26. Vater, J. (1989) in *Biologically active molecules* (Schlunegger, U. P., ed.) pp. 27−38, Springer-Verlag, Berlin.
27. Kluge, B., Vater, J., Salnikow, J. & Eckart (1988) *FEBS Lett. 231*, 107−110.
28. van Liempt, H. (1988) Thesis, TU Berlin.
29. van Liempt, H., von Döhren, H. & Kleinkauf, H. (1989) *J. Biol. Chem. 264*, 3680−3684.
30. Rapaport, E., Remy, P., Kleinkauf, H., Vater, J. & Zamecnik, P. C. (1987) *Proc. Natl Acad. Sci. USA 84*, 7891−7895.
31. Schimmel, P. (1987) *Annu. Rev. Biochem. 56*, 125−158.
32. Kleinkauf, H. & von Döhren, H. (1985) in *Regulation of secondary metabolite formation* (Kleinkauf, H., von Döhren, H., Dornauer, H. & Nesemann, G., eds) pp. 173−207, Verlag Chemie, Weinheim.
33. Freist, W., Sternbach, H. & Cramer, F. (1981) *Hoppe-Seyler's Z. Physiol. Chem. 362*, 1247−1252.
34. Kittelberger, R., Altmann, M. & von Döhren, H. (1982) in *Peptide antibiotics − biosynthesis and functions* (Kleinkauf, H. & von Döhren, H., eds) pp. 209−218, de Gruyter, Berlin.
35. Bothe, D., von Döhren, H., Zschiedrich, H., El-Samaraie, A., Krause, M. & Kleinkauf, H. (1982) in *Peptide antibiotics − biosynthesis and functions* (Kleinkauf, H. & von Döhren, H., eds) pp. 233−241, de Gruyter, Berlin.
36. Gevers, W., Kleinkauf, H. & Lipmann, F. (1969) *Proc. Natl Acad. Sci. USA 63*, 1335−1342.
37. Lipmann, F., Gevers, W., Kleinkauf, H. & Roskoski, R., Jr (1971) *Adv. Enzymol. 35*, 1−34.
38. Kanda, M., Hori, K., Kurotsu, T., Miura, S. & Saito, Y. (1984) *J. Biochem. (Tokyo) 96*, 701−711.
39. Kanda, M., Hori, K., Kurotsu, T., Miura, S., Yamada, Y. & Saito, Y. (1981) *J. Biochem. (Tokyo) 90*, 765−771.
40. Kanda, M., Hori, K., Kurotsu, T., Miura, S., Nozoe, A. & Saito, Y. (1978) *J. Biochem. (Tokyo) 84*, 435−441.
41. Schlumbohm, W., Vater, J. & Kleinkauf, H. (1985) *Biol. Chem. Hoppe-Seyler 366*, 925−930.
42. Schlumbohm, W. (1987) Thesis, TU Berlin.
43. Hori, K., Yamamoto, Y., Minetoki, T., Kurotsu, T., Kanda, M., Miura, S., Okamura, K., Furuyama, J. & Saito, Y. (1989) *J. Biochem. (Tokyo) 106*, 639−645.
44. Krätzschmar, J., Krause, M. & Marahiel, M. A. (1989) *J. Bacteriol. 171*, 5422−5429.
45. Yamada, M. & Kurahashi, K. (1968) *J. Biochem. (Tokyo) 63*, 59−69.
46. Takahashi, H., Sato, E. & Kurahashi, K. (1971) *J. Biochem. (Tokyo) 69*, 973−976.
47. Roskoski, R., Jr, Gevers, W., Kleinkauf, H. & Lipmann, F. (1970) *Biochemistry 9*, 4839−4845.
48. Lee, S. G. & Lipmann, F. (1975) *Methods Enzymol. 43*, 585−607.
49. Fujikawa, K., Sakamoto, Y. & Kurahashi, K. (1971) *J. Biochem. (Tokyo) 69*, 869−879.
50. Kambe, M., Sakamoto, Y. & Kurahashi, K. (1971) *J. Biochem. (Tokyo) 69*, 1131−1133.
51. Mittenhuber, G., Weckermann, R. & Marahiel, M. A. (1989) *J. Bacteriol. 171*, 4881.
52. Mittenhuber, G. (1988) Thesis, TU Berlin.
53. Weckermann, R., Fürbass, R. & Marahiel, M. A. (1988) *Nucleic Acids Res. 16*, 11 841.
54. Hori, K., Kanda, M., Kurotsu, T., Miura, S., Yamada, Y. & Saito, Y. (1981) *J. Biochem. (Tokyo) 90*, 439−447.
55. Hori, K., Kanda, M., Miura, S., Yamada, Y. & Saito, Y. (1981) *J. Biochem. (Tokyo) 93*, 177−188.
56. Hori, K., Kurotsu, T., Kanda, M., Miura, S., Yamada, Y. & Saito, Y. (1981) *J. Biochem. (Tokyo) 91*, 369−379.
57. Laland, S. G., Aarstadt, K. & Zimmer, T.-L. (1982) in *Peptide antibiotics − biosynthesis and functions* (Kleinkauf, H. & von Döhren, H., eds) pp. 185−194, de Gruyter, Berlin.
58. Aarstadt, K., Zimmer, T.-L. & Laland, S. G. (1980) *Eur. J. Biochem. 112*, 335−338.
59. Saxholm, H., Zimmer, T.-L. & Laland, S. G. (1972) *Eur. J. Biochem. 30*, 138−144.
60. Gadow, A., Vater, J., Schlumbohm, W., Palacz, Z., Salnokow, J. & Kleinkauf, H. (1983) *Eur. J. Biochem. 132*, 229−234.
61. Tomino, S., Yamada, M., Itoh, H. & Kurahashi, K. (1967) *Biochemistry 6*, 2552−2560.
62. Kleinkauf, H. & von Döhren, H. (1981) *Adv. Biotechnol. 3*, 83−88.
63. Gevers, W., Kleinkauf, H. & Lipmann, F. (1969) *Proc. Natl Acad. Sci. USA 63*, 1335−1342.
64. Froyshov, O. (1975) *Eur. J. Biochem. 59*, 201−206.
65. Vater, J., Schlumbohm, W., Palacz, Z. & Salnikow, J. (1987) *Eur. J. Biochem. 163*, 297−302.
66. Schlumbohm, W. (1987) Thesis, TU Berlin.
67. Vater, J., Schlumbohm, W., Salnikow, J., Irrgang, K.-D., Miklus, M., Choli, T. & Kleinkauf, H. (1989) *Biol. Chem. Hoppe-Seyler 370*, 1013−1018.
68. Vater, J., Mallow, N., Gerhardt, S. & Kleinkauf, H. (1985) *Biochemistry 24*, 2022−2027.
69. Zocher, R., Nihira, T., Paul, E., Madry, N., Peeters, H., Kleinkauf, H. & Keller, U. (1986) *Biochemistry 25*, 550−553.
70. Stindl, A. & Keller, U. (1989) *Abstracts Workshop Biochemie, Biologie and Molekularbiologie von Streptomyceten*, p. 65, Technical University, Berlin.
71. von Döhren, H. (1982) in *Peptide antibiotics − biosynthesis and functions* (Kleinkauf, H. & von Döhren, H., eds) pp. 169−182, de Gruyter, Berlin.
72. Zocher, R., Keller, U. & Kleinkauf, H. (1983) *Biochem. Biophys. Res. Commun. 110*, 292−299.
73. Koischwitz, H. (1979) *Hoppe-Seyler's Z. Physiol. Chem. 360*, 307.
74. Fujikawa, K., Suzuki, J. & Kurahashi, K. (1966) *Biochim. Biophys. Acta 161*, 232−246.
75. Stoll, E., Froyshov, O., Holm, H., Zimmer, T.-L. & Laland, S. G. (1970) *FEBS Lett. 11*, 348−352.
76. Roskoski, R., Jr, Ryan, G., Kleinkauf, H., Gevers, W. & Lipmann, F. (1971) *Arch. Biochem. Biophys. 143*, 485−492.
77. Koischwitz, H. & Kleinkauf, H. (1976) *Biochim. Biophys. Acta 429*, 1041−1051.
78. Wang, D. I. C. & Hamilton, B. K. (1977) *Biotechnol. Bioeng. 19*, 1225−1232.
79. Zocher, R., Keller, U. & Kleinkauf, H. (1982) *Biochemistry 21*, 43−48.
80. Siegbahn, N., Mosbach, K., Grodzki, K., Zocher, R. & Kleinkauf, H. (1985) *Biotechnol. Bioeng. 7*, 297−302.
81. Madry, N., Zocher, R., Grodzki, K. & Kleinkauf, H. (1984) *Appl. Microbiol. Biotechnol. 20*, 83−86.
82. Kleinkauf, H. & von Döhren, H. (1983) in *Industrial aspects of biochemistry and genetics* (Alaeddinoglu, N. G., Demain, A. L. & Lancini, G., eds) pp. 107−128, Plenum Press, New York, London.
83. Kurylowicz, W., Kurzatkowski, W. & Kurzatkowski, J. (1985) *Arch. Immunol. Ther. Exp. 35*, 699−724.

14

84. Lee, S. G., Roskoski, R., Jr, Bauer, K. & Lipmann, F. (1973) *Biochemistry 12*, 398–405.
85. Billich, A. & Zocher, R. (1987) *J. Biol. Chem. 262*, 17258–17261.
86. Lawen, A., Traber, R., Geyl, D., Zocher, R. & Kleinkauf, H. (1989) *J. Antibiot. 42*, 1283–1289.
87. Lawen, A. & Zocher, R. (1990) *J. Biol. Chem.*, in the press.
88. Billich, A. & Zocher, R. (1990) in *Biochemistry of peptide antibiotics* (Kleinkauf, H. & von Döhren, H., eds) pp. 57–79, de Gruyter, Berlin.
89. Froyshov, O. (1977) *FEBS Lett. 81*, 315–318.
90. Froyshov, O. (1974) *FEBS Lett. 44*, 75–78.
91. Froyshov, O. & Mathiesen, A. (1979) *FEBS Lett. 106*, 275–278.
92. Froyshov, O. & Mathiesen, A. (1982) in *Peptide antibiotics — biosynthesis and functions* (Kleinkauf, H. & von Döhren, H., eds) pp. 307–314, de Gruyter, Berlin.
93. Roland, I., Froyshov, O. & Laland, S. G. (1977) *FEBS Lett. 84*, 22–24.
94. Ishihara, H., Endo, Y., Abe, S. & Shimura, K. (1975) *FEBS Lett. 50*, 43–46.
95. Ishihara, H., Ogawa, I. & Shimura, K. (1979) *FEBS Lett. 99*, 109–112.
96. Ishihara, H., Ogawa, I. & Shimura, K. (1982) in *Peptide antibiotics — biosynthesis and functions* (Kleinkauf, H. & von Döhren, H., eds) pp. 289–296, de Gruyter, Berlin.
97. Ishihara, H., Hara, N. & Iwabuchi, T. (1989) *J. Bacteriol. 171*, 1705–1711.
98. Vitkovíc, L. & Pfaender, P. (1982) in *Peptide antibiotics — biosynthesis and functions* (Kleinkauf, H. & von Döhren, H., eds) pp. 297–306, de Gruyter, Berlin.
99. Froyshov, O. (1984) in *Biotechnology of industrial antibiotics* (Vandamme, E. J., ed.) pp. 665–694, Dekker, New York.
100. Komura, S. & Kurahashi, K. (1979) *J. Biochem. (Tokyo) 86*, 1013–1021.
101. Komura, S. & Kurahashi, K. (1980) *J. Biochem. (Tokyo) 88*, 285–288.
102. Komura, S. & Kurahashi, K. (1980) *Biochem. Biophys. Res. Commun. 95*, 1145–1151.
103. Kurahashi, K., Komura, S., Akashi, K. & Nishio, C. (1982) in *Peptide antibiotics — biosynthesis and functions* (Kleinkauf, H. & von Döhren, H., eds) pp. 275–288, de Gruyter, Berlin.
104. Komura, S. & Kurahashi, K. (1985) *J. Biochem. (Tokyo) 97*, 1409–1417.
105. Vasantha, N., Balakrishnan, R., Kaur, S. & Jayaraman, K. (1980) *Arch. Biochem. Biophys. 200*, 40–44.
106. Balakrishnan, R., Kaur, N. S., Goel, A. K., Padmavathy, S. & Jayaraman, K. (1980) *Arch. Biochem. Biophys. 200*, 45–54.
107. Kaur, N. S., Chandrasekaran, G. N., Vasantha, N. & Jayaraman, K. (1982) in Peptide antibiotics — biosynthesis and functions (Kleinkauf, H. & von Döhren, H., eds) pp. 453–465, de Gruyter, Berlin.
108. Vitkovíc, L. & Sadoff, H. L. (1977) *J. Bacteriol. 131*, 897–905.
109. Pais, M., Das, B. C. & Ferron, P. (1981) *Phytochemistry 20*, 715–723.
110. Bains, P. S. & Tewari, J. P. (1987) *Physiol. Mol. Plant Pathol. 30*, 259–271.
111. Keller, U. (1977) Thesis, TU Berlin.
112. Keller, U., Kleinkauf, H. & Zocher, R. (1984) *Biochemistry 23*, 1479–1484.
113. Keller, U. (1987) *J. Biol. Chem. 262*, 5852–5856.
114. Pahl, A. & Keller, U. (1989) *Abstracts Workshop Biochemie, Biologie and Molekularbiologie von Streptomyceten*, p. 35, Technical University, Berlin.
115. Schlumbohm, W. & Keller, U. (1989) *Abstracts Workshop Biochemie, Biologie und Molekularbiologie von Streptomyceten*, p. 36, Technical University, Berlin.
116. Barry, C. E., III, Nayar, P. G. & Begley, T. P. (1989) *Biochemistry 28*, 6323–6333.
117. Jones, G. A. & Hopwood, D. (1984) *J. Biol. Chem. 259*, 14151–14157.
118. Jones, G. A. & Hopwood, D. (1984) *J. Biol. Chem. 259*, 14158–14164.
119. Martin, J. F. & Liras, P. (1989) *Annu. Rev. Microbiol. 43*, 173–206.
120. Haese, A. (1987) Thesis, TU Berlin.
121. Haese, A. & Keller, U. (1988) *J. Bacteriol. 170*, 1360–1368.
122. Rabie, M. (1990) Thesis, TU Berlin.
123. Zocher, R. & Kleinkauf, H. (1978) *Biochem. Biophys. Res. Commun. 81*, 1161–1167.
124. Billich, A. & Zocher, R. (1987) *Biochemistry 26*, 8417–8423.
125. Billich, A., Zocher, R., Kleinkauf, H., Braun, D. G., Lavanchy, D. & Hochkeppel, H. (1987) *Biol. Chem. Hoppe-Seyler 368*, 521–529.
126. Peeters, H., Zocher, R. & Kleinkauf, H. (1988) *J. Antibiot. 41*, 352–359.
127. Peeters, H. (1988) Thesis, TU Berlin.
128. Nüesch, J., Heim, J. & Treichler, H.-J. (1987) *Annu. Rev. Microbiol. 41*, 51–75.
129. Burnham, M. K. A., Earl, A. J., Bull, J. H., Smith, D. J. & Turner, G. (1989) *Eur. Patent* 0 320 272 A1.
130. Weigel, B. J., Burgett, S. G., Chen, V. J., Skatrud, P. L., Frolik, C. A., Queener, S. W. & Ingolia, T. D. (1988) *J. Bacteriol. 170*, 3817–3826.
131. Miller, J. R. & Ingolia, T. D. (1989) *Mol. Microbiol. 3*, 689–695.
132. Ristow, H. & Paulus, H. (1982) *Eur. J. Biochem. 129*, 395–401.
133. Pschorn, W., Paulus, H., Hansen, J. & Ristow, H. (1982) *Eur. J. Biochem. 129*, 403–407.
134. Bauer, K., Roskoski, R., Jr, Kleinkauf, H. & Lipmann, F. (1972) *Biochemistry 11*, 3266–3271.
135. Akashi, K. & Kurahashi, K. (1978) *J. Biochem. (Tokyo) 83*, 1219–1229.
136. Akashi, K. & Kurahashi, K. (1978) *Biochem. Biophys. Res. Commun. 77*, 259–267.
137. Akashi, K., Kubota, K. & Kurahashi, K. (1977) *J. Biochem. (Tokyo) 81*, 269–272.
138. Kubota, K. (1982) *Biochem. Biophys. Res. Commun. 105*, 688–697.
139. Kubota, K. (1985) in *Cellular recognition and malignant growth* (Ebashi, S., ed.) pp. 187–191, Japan Scientific Society Press, Springer, Berlin.
140. Kubota, K. (1987) *Biochem. Biophys. Res. Commun. 144*, 203–209.
141. Kubota, K. (1988) in *The roots of modern biochemistry* (Kleinkauf, H., von Döhren, H. & Jaenicke, R., eds) pp. 331–337, de Gruyter, Berlin, New York.
142. Akers, H. A., Lee, S. G. & Lipmann, F. (1977) *Biochemistry 16*, 5722–5729.
143. Kleinkauf, H., von Döhren, H., Dornauer, H. & Nesemann, G. (eds) (1985) *Regulation of secondary metabolite formation*, Verlag Chemie, Weinheim.
144. Campbell, I. (1984) *Adv. Microb. Physiol. 25*, 1–60.
145. Krause, M. & Marahiel, M. A. (1988) *J. Bacteriol. 170*, 4669–4674.
146. Smith, D. J., Burnham, K. R., Edwards, J., Earl, A. J. & Turner, G. (1990) *Bio/technology 8*, 39–41.
147. Diez, B., Barredo, J. L., Alvarez, E., Cantoral, J. M., van Solingen, P., Groenen, M. A. M., Veenstra, A. E. & Martin, J. F. (1989) *Mol. Gen. Genet. 218*, 572–576.
148. Braun, V. (1990) in *Biochemistry of peptide antibiotics* (Kleinkauf, H. & von Döhren, H., eds) pp. 103–129, de Gruyter, Berlin.
149. Nahlik, M. S., Brickman, T. J., Ozenberger, B. A. & McIntosh, M. A. (1989) *J. Bacteriol. 171*, 784–790.
150. Liu, J., Duncan, K. & Walsh, C. T. (1989) *J. Bacteriol. 171*, 791–798.
151. Rusnak, F., Faraci, N. S. & Walsh, C. T. (1989) *Biochemistry 28*, 6827–6835.
152. Hopwood, D. A. (1988) in *Biology of actinomycetes '88* (Okami, Y., Beppu, T. & Ogawara, H., eds) pp. 3–10, Japan Scientific Society Press, Tokyo.
153. Epp, J. K., Huber, M. L. B., Turnrt, J. R. & Schoner, B. (1988) in *Biology of actinomycetes '88* (Okami, Y., Beppu, T. & Ogawara, H., eds) pp. 82–85, Japan Scientific Society Press, Toyko.

154. Piepersberg, W., Distler, J., Ebert, A., Heinzel, P., Mansouri, K., Mayer, G. & Pissowotzki, K. (1988) in *Biology of actinomycetes '88* (Okami, Y., Beppu, T. & Ogawara, H., eds) pp. 86–91, Japan Scientific Society Press, Tokyo.
155. Anzai, H., Murakami, T., Hara, O., Kumada, Y., Imai, S., Satoh, A., Nagaoka, K., Holt, T., Raibaud, A. & Thompson, C. J. (1988) in *Biology of actinomycetes '88* (Okami, Y., Beppu, T. & Ogawara, H., eds) pp. 92–96, Japan Scientific Society Press, Tokyo.
156. Marahiel, M. A., Zuber, P., Czekay, G. & Losick, R. (1987) *J. Bacteriol. 169*, 2215–2222.
157. Robertson, J. B., Gocht, M., Marahiel, M. A. & Zuber, P. (1989) *Proc. Natl Acad. Sci. USA 86*, 8457–8461.
158. Strauch, M. A., Spiegelman, G. B., Perego, M., Johnson, W. C., Burbulys, D. & Hoch, J. A. (1989) *EMBO J. 8*, 1615–1621.
159. Ferrari, E., Henner, D. S., Perego, M. & Hoch, J. A. (1988) *J. Bacteriol. 170*, 289–295.
160. Billich, A. & Zocher, R. (1988) *Appl. Environm. Microbiol. 54*, 2504–2509.
161. Tamura, S. & Suzuki, A. (1978) in *Bioactive peptides produced by microorganisms* (Umezawa, H., Takita, T. & Shiba, T., eds) pp. 105–127, Wiley, New York.
162. Haavik, H. I. & Froyshov, O. (1975) *Nature 254*, 79–82.
163. Ristow, H., Russo, J., Stochaj, E. & Paulus, H. (1982) in *Peptide antibiotics — biosynthesis and functions* (Kleinkauf, H. & von Döhren, H., eds) pp. 381–388, de Gruyter, Berlin.
164. Marahiel, M. A., Danders, W., Kraepelin, G. & Kleinkauf, H. (1982) in *Peptide antibiotics — biosynthesis and functions* (Kleinkauf, H. & von Döhren, H., eds) pp. 389–397, de Gruyter, Berlin.
165. Marahiel, M. A. & von Döhren, H. (1982) in *Peptide antibiotics — biosynthesis and functions* (Kleinkauf, H. & von Döhren, H., eds) pp. 375–380, de Gruyter, Berlin.
166. Danders, W. (1983) Thesis, TU Berlin.
167. Daher, E., Rosenberg, E. & Demain, A. L. (1985) *J. Bacteriol. 161*, 47–50.
168. Frangou-Lazaridis, M. & Seddon, B. (1985) *J. Gen. Microbiol. 131*, 437–449.
169. Piret, J. M. & Demain, A. L. (1985) *Arch. Microbiol. 133*, 38–43.
170. Murray, T., Lazaridis, I. & Seddon, B. (1985) *Lett. Appl. Microbiol. 1*, 63–65.
171. Kleinkauf, H. & von Döhren, H. (1988) *Crit. Rev. Biotechnol. 8*, 1–32.
172. Kleinkauf, H. & von Döhren, H. (1990) *Progr. Drug Res. 34*, 287–317.
173. Lerbs, W. & Luckner, M. (1985) *J. Basic Microbiol. 25*, 387–391.
174. Gerlach, M., Schwelle, N., Lerbs, W. & Luckner, M. (1985) *Phytochemistry 24*, 1935–1939.
175. Hummel, W. & Diekmann, H. (1981) *Biochim. Biophys. Acta 617*, 313–320.

Eur. J. Biochem. *192*, 245–261 (1990)
© FEBS 1990

Review

Biologically active products of stimulated liver macrophages (Kupffer cells)

Karl DECKER

Biochemisches Institut der Albert-Ludwigs-Universität, Freiburg i.Br., Federal Republic of Germany

(Received February 14, 1990) — EJB 90 0155

CONTENTS

Correspondence to K. Decker, Biochemisches Institut der Albert-Ludwigs-Universität, Hermann-Herder-Straße 7, D-7800 Freiburg i.Br., Federal Republic of Germany

Abbreviations. Acyl$_2$Gro, diacylglycerol; G protein, guanylate-binding protein; HETE, hydroxyeicosatetraenoic acid; HPETE, hydroperoxyeicosatetraenoic acid; IL, interleukin; LPS, lipopolysaccharide (endotoxin); LT, leukotriene; PAF, platelet-activating factor; PG, prostaglandin; PMA, phorbol 12-myristate 13-acetate; TNF, tumor necrosis factor α (cachectin); TX, thromboxane.

Enzymes. Acyl-CoA synthetase (EC 6.2.1.3); adenylate cyclase (EC 4.6.1.1); angiotensin-converting enzyme (dipeptidyl carboxypeptidase I, EC 3.4.15.1); collagenase (EC 3.4.24.7); prostaglandin H synthase (cyclo-oxygenase; EC 1.13.11.–); diacylglycerol lipase (EC 3.1.1.34); elastase (EC 3.4.21.37); epoxide hydrolase (EC 3.3.2.3); glucose-6-phosphate dehydrogenase (EC 1.1.1.49); glutathione S-transferase (EC 2.5.1.18); γ-glutamyl transferase (EC 2.3.2.2); glycerol-3-phosphate dehydrogenase (EC 1.1.1.8); lipoprotein lipase (EC 3.1.1.34); 5-lipoxygenase (EC 1.13.11.34); lysolecithin acyltransferase (EC 2.3.1.23); lyso-PAF acetyltransferase (EC 2.3.1.–); lysozyme (EC 3.2.1.17); myeloperoxidase (EC 1.11.1.7); NADPH oxidase (EC 1.6.99.–); 6-phosphogluconate dehydrogenase (EC 1.1.1.44); phospholipase A$_2$ (EC 3.1.1.4); phospholipase C (EC 3.1.4.3); pronase (EC 3.4.24.–); prostaglandin-D$_2$ synthase (EC 5.3.99.2); prostaglandin-E$_2$ synthase (EC 5.3.99.3); protein kinase C (EC 2.7.1.37); superoxide dismutase (EC 1.15.1.1); thromboxane synthase (EC 5.3.99.5).

Macrophages belong to the mononuclear phagocyte system that includes the blood monocytes and the various kinds of mobile and sessile macrophages. They are recruited from the stem cells of the bone marrow and differentiate under the influence of specific signals [1] [e.g. the macrophage colony-stimulating factor, the granulocyte/macrophage colony-stimulating factor and interleukin-3 (IL-3)] via several intermediary stages to mature macrophages [2] (Fig. 1). They belong, together with the highly specific immune system, to the body's defensive machinery. Although all macrophages are ultimately derived from the same source (bone marrow) they may in some instances, e.g. in the liver, also propagate at the site of their final destination [3]. Topological factors seem to influence their final differentiation and to endow each type with particular metabolic and structural features. One can clearly recognize specific capabilities in the mobile blood-borne macrophages (monocytes), in the individually moving but sequestered (peritoneal and alveolar) macrophages, and in the various sessile tissue macrophages such as those residing in bone marrow, spleen, brain, skin and liver [4,5].

Despite their functional specialization, the individual types of macrophages have at least four major functions in common: presentation of antigens, phagocytosis, unspecific immune response (immunomodulation) and biochemical attack. Antigen

246

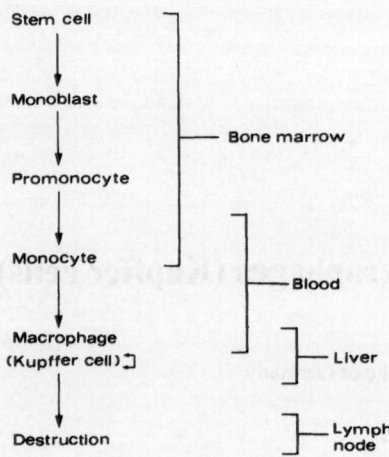

Fig. 1. *Maturation of Kupffer cells from stem cells*. The maturation of Kupffer cells includes various compartments beginning in the bone marrow and ending in the liver sinusoid. Proliferation of Kupffer cells can occur in the liver, particularly after an inflammatory challenge. Migration to adjacent lymph nodes is thought to terminate the life of the Kupffer cells

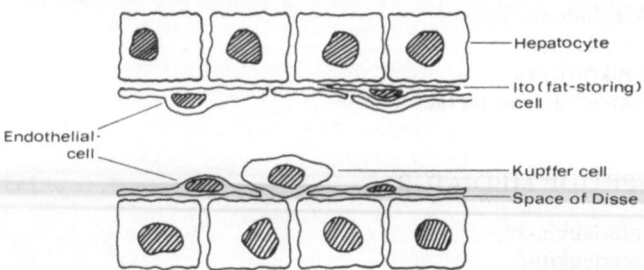

Fig. 2. *Schematic representation of the liver sinusoid*. The sinusoid connects the portal vein (and the hepatic artery) with the central vein and facilitates the exchange of nutrients and metabolites between the blood and the liver parenchyma. The sinusoidal endothelial cells form a fenestrated sheath around the sinusoid that regulates the exchange between the lumen and the space of Disse (the space between the endothelial layer and the hepatocytes). The Ito cells are located in this space; they embrace the endothelial layer with their extended structures. The Kupffer cells reside on the luminal side on top of the endothelium with occasional extensions into the space of Disse. In the rat liver Kupffer cells represent about 15% by cell number and 3% by mass

presentation and the process of phagocytosis *per se* cannot be covered in this contribution, but a survey of substances with known biological roles secreted by stimulated Kupffer cells will be attempted; furthermore, the activating agents, the function of the products and the regulation of their synthesis and action will be discussed.

The Kupffer cells are macrophages residing in the sinusoids of the liver (Fig. 2). These liver macrophages are named after the pathologist C. von Kupffer who appears to have been the first to recognize this non-parenchymal cell type [6, 7]. Kupffer cells from various mammalian livers can be obtained after perfusion with proteolytic enzymes (e.g. collagenase, pronase), density-gradient centrifugation and centrifugal elutriation of the resulting liver cell suspension [8]. They can be kept in primary culture for many days. A number of functional assays as well as histochemical and immunological methods have been developed for the identification of these liver macrophages [9]. With the aid of these techniques

it was possible to identify the bone marrow origin as well as the *in situ* propagation of Kupffer cells in the liver sinusoid [10]. It is likely that monocytes are intermediary stages in the differentiation process [2]; the large granular lymphocytes (Pit cells) [11] found in the liver sinusoid are possible precursors of the Kupffer cells [12].

The liver macrophages are in close contact with the sinusoidal endothelium and may reach into the space of Disse and the neighbourhood of the parenchymal cells, the hepatocytes. Kupffer cells are usually observed in greater number in the periportal area than near the central vein. The topological position within the liver sinusoid makes the Kupffer cells the first macrophages to come into contact with foreign or noxious material including viruses, bacteria, unicellular eukaryotes and their products that enter the circulation by the way of the portal vein. The strategic location (Fig. 2), together with the fact that the Kupffer cells constitute the largest single pool of macrophages in the body [6], attribute to them an important function in clearing the inflow from the splanchnic area of particulate and immunoreactive material, as well as in the removal of migrating tumor cells [13].

STIMULATION OF SIGNAL PRODUCTION AND RELEASE

Receptor-mediated phagocytosis

It has been known for some time that phagocytosis, i.e. the uptake of particles, by macrophages can be accompanied by the release of specific molecules. The contact between a suitable particle and its receptor on the surface of a macrophage elicits the production of compounds that include reactive oxygen species and arachidonate metabolites (eicosanoids). Depending on the kind of receptor engaged, different transduction paths may be activated in a given cell so that not all the possible signals must be produced at the same time. In recent years it has become increasingly clear that particle-stimulated synthesis and release of signal substances does not require the complete internalization of the phagocytosable material. The close contact of the particle to be phagocytosed with a specific receptor or binding site seems to be sufficient to trigger responses [14]. However, lateral contacts of several of these interacting molecules are necessary for signal transduction [15, 16].

The presence of F_c receptors on Kupffer cells (Table 1) is instrumental in the binding and uptake of immunocomplexes. Besides these F_c receptors, specific binding sites have been identified for glycoconjugates and also for colloidal silver or carbon, and for heat-aggregated bovine serum albumin [33]. Some of these may also be instrumental in the binding and internalization of zymosan (glucan- and mannan-containing particles from yeast cell walls), particulate glucan and also of bacteria and unicellular eucaryotes (e.g. yeasts). Whether or not Kupffer cells also possess receptors identified on other macrophages, such as those binding unopsonized erythrocytes, glycosylated albumin, the complement component C3, Mac-1 (cell-adhesion molecules) interleukin-6 or tumor necrosis factor α (TNF), remains to be seen; the ligands listed in Table 1 B were found to elicit responses in Kupffer cells quite similar to those seen in receptor-carrying cells. The receptor for interferon-γ deserves particular attention as it conveys the message of the cytokine, considered as one of the most potent activators of cells of the monocyte/macrophage lineage [34].

Table 1. *Receptors and binding sites on Kupffer cells*
Receptors listed under A have been established for Kupffer cells; those under B are on other macrophages, but are very likely to be also present on Kupffer cells. HDL, high-density lipoproteins

Section	Recognized ligand	Reference
A	F_c part of immunoglobulins	17
	mannose/N-acetylglucosamine	18
	fucose	19
	galactose/N-acetylgalactosamine	20
	platelet-activating factor	21
	apolipoprotein C (HDL)	22
	scavenger receptor (for modified lipoproteins)	23, 24
	complement	25
B	insulin	26
	glucagon	26
	interferon-γ	27
	tumor necrosis factor	28
	interleukin-6	29
	PGE_2	30
	nucleotide triphosphates	31, 32

Table 3. *Effects of interferon-γ*
(Observations compiled from various sources [34, 47, 48]). MHC, major histocompatibility factor; eIF, eukaryotic initiation factor

Level	Effect
Cellular	antiviral activity (first detected effect)
	antiproliferative and antitumor activity against some but not all cells
	increasing the natural killer cell activity (together with IL-2)
	increasing the number of F_c receptors
Molecular	induction of oligoadenylate synthesis and P1/eIF-2α protein kinase (activated by double-stranded RNA)
	induction of G proteins
	induction of MHC class II antigens
	enhancement of lymphotoxin production and its effects
	activation of the Na^+/H^+ exchanger, Ca^{2+} efflux
	enhanced production of superoxide (in some macrophages, but not in Kupffer cells)
	induction of IL-1 and TNF production (in some macrophages)

Table 2. *Activators and immunomodulators of Kupffer cells*
Substances listed under A were shown to stimulate Kupffer cells; those listed under B are known activators of other macrophages only. GM-CSF, granulocyte/macrophage colony-stimulating factor; M-CSF, macrophage colony-stimulating factor

Section	Activator/immunomodulator
A	interferon-γ (macrophage-activating factor)
	lipopolysaccharide (endotoxin)
	tumor necrosis factor α (cachectin)
	virus (Sendai, Newcastle disease)
	platelet-activating factor
	muramyl dipeptide
	nucleotide triphosphates and diphosphates
	PMA
	ionophores (H^+ and Ca^{2+})
B	colony-stimulating factors (GM-CSF, M-CSF)
	macrophage migration inhibitory factor
	macrophage deactivation factor
	transforming growth factor β (?)
	complement component C5a
	chemotactic peptide (fMet-Leu-Phe)

Activators and immunomodulators of Kupffer cells

Besides phagocytosable particles, a great number of soluble substances is able to activate or stimulate macrophages (Table 2). Some of them are also known as biological response modifiers [35]. They may bind to specific receptors on the plasma membrane such as the complement factor C5a, fMet-Leu-Phe, interferon-γ or nucleotide triphosphates, or to binding sites in the cytosol like lipopolysaccharides (LPS) or phorbol esters (e.g. phorbol 12-myristate 13-acetate, PMA).

Their binding leads to the release of mediators or signal molecules. Some activators, e.g. interferon-γ, also confer the ability to kill foreign cells, e.g. tumor cells. This kind of cytotoxicity of macrophages including Kupffer cells is brought about by a mechanism different from that of phagocytosis [36]. It is shared by the natural killer cells; Pit cells (large granular lymphocytes) found in the liver sinusoid also belong to this category. The participation of a galactose/N-

acetylgalactosamine-specific lectin [37] and the dissolution of microfilaments preceding cytolysis [38] were reported in macrophage-dependent cytotoxicity. Priming by interferon-γ includes an activation of protein kinase C, the Na^+/H^+ exchanger and Ca^{2+} efflux from the cells [39]. Full activation of macrophages including the cytotoxic capability is a two-stage process and appears to require a cytokine, e.g. interferon-γ, and an elicitor like LPS. LPS appears to act through its ability to stimulate Kupffer cells to produce cytokines like TNF [40], interferon α/β [42, 43] or IL-1 [44]. It should be noted that apparently closely related macrophages such as mouse and rat Kupffer cells may behave differently towards activators. While the former require the two-stage activation in order to become cytotoxic and to release eicosanoids, the latter are in the primed state already (even when obtained from germ-free rats) and need an elicitor such as LPS [40] or muramyl dipeptide [41] only; interferon-γ has an additional but small stimulatory effect on rat Kupffer cells [45]. In general, however, it is safe to consider interferon-γ as the macrophage-activating factor (Table 3). The advent of improved immunological techniques and of the many possibilities offered by molecular biology (molecular cloning, genetic engineering) have renewed the interest in the therapeutic use of biological response modifiers [46].

Lipopolysaccharides, the endotoxins of Gram-negative intestinal bacteria, play a prominent role as immunomodulators. They may be absorbed in significant amounts from the gut in cases of altered permeability of intestinal cells or be derived directly from Gram-negative microorganisms during episodes of bacterial sepsis. Their interaction with macrophages is probably one of the most important features of inflammation or shock following bacterial invasion. Kupffer cells have been identified as the first and most avid scavenger cells for LPS passing into the liver through the portal vein [49]. Detoxification by partial degradation and clearance [50, 51] of LPS occurs mainly in the liver macrophages, but the spleen has also been confirmed as a site of LPS uptake and modification. Bile salts, in the millimolar range, strongly reduce the uptake of LPS by rat Kupffer cells and enhance its deactivation [52]; the pathophysiological relevance of this observation remains to be established. Labeled LPS administered via the portal vein is found very early in phagosomes, later in lysosomes

Table 4. *Regulation of signalling in Kupffer cells*

Regulation of synthesis, secretion and transport of the signal

Regulation of signal inactivation

Regulation by self-activation or self-destruction of enzymes

Regulation of receptor availability (up- or down-regulation) involving synthesis, internalisation, degradation and recycling of receptor structures

Regulation of signal transduction including G proteins, second messengers, and activation/deactivation of target molecules

Regulation of cell-cell interactions by signal transmission or autostimulation

[53]. Uptake is assumed to be by receptor-mediated endocytosis, but up to now neither a specific receptor nor the mechanism of signal transduction have been identified. Myristoylation of certain proteins in LPS-exposed cells has been associated with the intracellular action of LPS [54].

The role of LPS as the major inducer of peptide mediators such as TNF, IL-6 or IL-1 will be discussed later. LPS also depresses the complement receptor function of Kupffer cells [55] and triggers the production of reactive oxygen species in guinea pig Kupffer cells [56] (not in blood-derived macrophages [57] and rat Kupffer cells [58]), while in others it acts as an enhancer of the stimuli that lead to the release of arachidonic acid and its metabolites [58, 59]. The production of prostaglandin E_2 (PGE_2) by LPS-exposed rat Kupffer cells [45] seems to have great impact on the response to LPS (see below). A very interesting observation connects LPS with the modulation of *c-fos* and *c-myc* expression in macrophages [59]: LPS increases temporarily the mRNA levels of these proto-oncogenes up to 20-fold.

Of particular interest is the recently described [60] ability of extracellular nucleotide triphosphates or diphosphates (mainly ATP and UTP) to stimulate liver macrophages to release eicosanoids. The source of the extracllular nucleotides is not yet sufficiently established; they may originate from activated platelets or from synaptic transmission. Several effects of extracellularly admitted nucleotides on various cells have been reported in recent years [61]. It is assumed that purinergic and perhaps pyrimidinergic receptors are involved in these stimulations [31]. In human neutrophils, G proteins are involved in the activation of NADPH oxidase by purine and pyrimidine nucleotides [62].

SIGNAL PATHS

Activators or immunomodulators require pericellular or intracellular binding sites in order to elicit specific actions in target cells (Table 4). The responses which may be of a rather complex nature, such as modulation of growth or differentiation, can be traced to activation/inactivation and induction/repression, respectively, of enzymes or functional proteins. Fast and slow signal realization is quite often connected with these different mechanisms.

The binding site of the effector and the target protein or structure to be modified are usually topologically separated; a possible exception may be PMA, which is thought to interact directly with protein kinase C leading to its activation [63]. In most cases, however, the signal received at the receptor has to be transduced to the locus of realization.

G proteins

A dominant role in signal transduction from plasma-membrane-associated ligand-receptor interactions is attributed to guanine-nucleotide-binding proteins, G proteins [64]. Enzymes known to be regulated by G proteins include adenylate cyclase, stimulated by G_s and inhibited by G_i, and phospholipase(s) C hydrolyzing phosphatidylinositol phosphates to *myo*-inositol phosphates ($InsP_n$) and diacylglycerol ($acyl_2Gro$) (Fig. 3). All the G-protein-mediated signal-transduction pathways are of relevance to the stimulation of macrophages by activators or immunomodulators.

Adenylate cyclase

Adenylate cyclase activation followed by a rise of intracellular cAMP levels has been observed after PGE_2 and LPS treatment of Kupffer cells [30]. A very short spike of cAMP was also seen after contact of these macrophages with zymosan particles [65]; this agrees with reports of a cAMP rise subsequent to ligation of the F_{c2a} receptors of macrophage-like P388D1 cells [66]. cAMP-mediated processes lead to collagenase synthesis and release by rat Kupffer cells [64]. The extracellular availability of this protease is important for the reorganization of damaged tissues.

Phospholipase C

Regulation of phospholipase C appears to be a major process in signal transduction in macrophages including Kupffer cells. The enhanced activity of this enzyme may lead to activation of protein kinase C (through the $acyl_2Gro$ formed), to Ca^{2+} mobilization from the endoplasmic reticulum (through $InsP_3$) and possibly to Ca^{2+} influx from outside (through $InsP_4$). Thus, a key protein kinase and a second messenger (Ca^{2+}) are directly mobilized while cleavage of arachidonate from $acyl_2Gro$ and activation of phospholipase A_2 by Ca^{2+}/calmodulin are secondary effects leading to the release of important signal molecules, the eicosanoids.

Phospholipase C is coupled to a G protein, designated G_p, but this has not yet been fully identified. It appears to be closely related or even identical to the product of the *ras* oncogene [67]. Phosphatidylinositol 4,5-bisphosphate interacts with components of the cytoskeleton, e.g. actin [68]; it may be instrumental in the changes of macrophage morphology such as occur during phagocytosis or movement.

The two-pronged signal-transduction through $acyl_2Gro$ and $InsP_3$ opens particularly intriguing aspects of regulation as the targets of both second messengers interact [69]: Ca^{2+} is required for binding to the plasma membrane and activation of protein kinase C, while the proteins phosphorylated by protein kinase C may influence Ca^{2+} movement and binding within the cell. Experiments with rat Kupffer cells indicate that protein kinase C is involved in the activation of the NADPH (respiratory burst) oxidase and the Na^+/H^+ antiporter while Ca^{2+} influx is necessary for phospholipase A_2 activation and eicosanoid synthesis [70]. Phorbol esters are thought to mimic $acyl_2Gro$ in creating the lipid environment for membrane anchorage of protein kinase C.

Calcium ions

Ca^{2+} is a well-established second messenger of cellular regulation [71]. It binds to specific modulator molecules, e.g. calmodulin, or directly to effector proteins such as phospho-

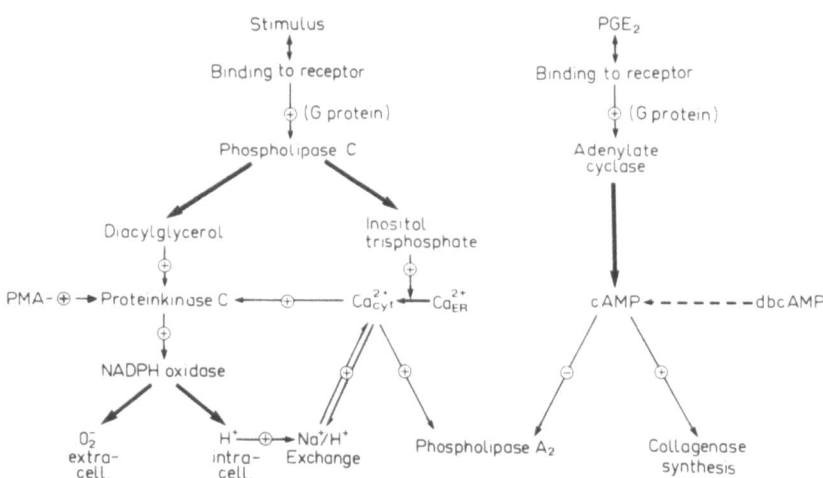

Fig. 3. *Signal paths in rat Kupffer cells.* cyt, cytoplasmic; ER, endoplasmic reticulum; dbcAMP, N^6, 2-*O*-dibutyryl-adenosine 3′,5′-(cyclic)phosphate

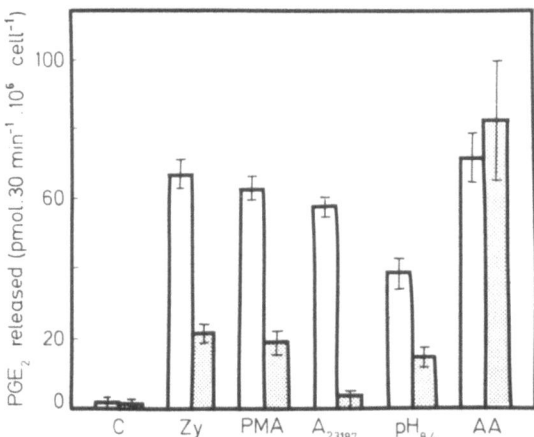

Fig. 4. *Stimulated PGE_2 synthesis in rat Kupffer cells: dependence on extracellular Ca^{2+}.* Rat Kupfer cells in primary culture were incubated in the presence of 1.3 mM Ca^{2+} (open bars) or 1 mM EGTA (dotted bars) and exposed to various stimulating conditions. The intracellular pH was adjusted using 1 μM nigericin and 122 mM K^+. C, control (no stimulation); Zy, zymosan; A 23187, Ca^{2+} ionophore; AA, free arachidonic acid. From Dieter et al. [70]

lipase A_2. The very low intracellular basal level of the free ion (< 0.2 μM) can be changed by the $InsP_3$-stimulated influx from the storage sites (mainly endoplasmatic reticulum, but to some extent also mitochondria), the net uptake from the outside by activation of Ca^{2+} channels, antiporters (2 Na^+/Ca^{2+}), or by modulation of the activity of the ATP-driven Ca^{2+} pump. The intracellular Ca^{2+} level is also related to the internal pH; activation of the Na^+/H^+ antiporter [72] and eicosanoid synthesis [73], e.g. after zymosan addition, in rat Kupffer cells is connected with a transcellular Ca^{2+} influx [70] (Fig. 4).

SUPEROXIDE AND NITRIC OXIDE

Superoxide

Macrophages are able to respond to certain stimuli with the secretion of inorganic compounds (simple ion fluxes not considered). Among these, the reactive oxygen intermediates are the most prominent. They are reported to activate granulocytes but signalling does not appear to be their main function in macrophages. The available evidence suggests that they act as chemical weapons for immediate use. The reactive oxygen intermediates usually do not act in specific ways (through receptors or binding sites) but exploit their chemical reactivity on any suitable compounds with which they collide.

Kupffer cells brought into contact with phagocytosable material (e.g. zymosan particles, opsonized erythrocytes, immunocomplexes) show the well-known phenomenon of oxidative burst, i.e. the rapid uptake of molecular oxygen. The major product of this process in rat Kupffer cells is the superoxide anion radical, $^•O_2^-$ [74]; other phagocytosing cells and polymorphonuclear neutrophils (granulocytes) are able to further convert $^•O_2^-$ to hydrogen peroxide, singlet oxygen and hydroxyl radicals, depending on their possession of superoxide dismutase and myeloperoxidase. The superoxide produced by Kupffer cells helps the macrophage to inactivate and destoy phagocytosed organisms or particles.

Since superoxide is not as strongly reacting as the hydroxyl radical, it appears that a larger concentration and time of exposure is required for effective action. Such a condition exists after the bound particle has been engulfed by coated pits of the plasma membrane and finally sequestered in phagosomes. In line with this view it was found [14] that rat Kupffer cells exposed to immunocomplex-coated latex particles produce superoxide even when the cells were treated with cytochalasin B that prevents the formation of coated pits and phagosomes [75]. In fact, substantial $^•O_2^-$ release into the medium occurred only in the treated cells indicating that the majority of the superoxide formed remained within the boundaries of the coated vesicles.

Superoxide formation is catalyzed by NADPH oxidase residing in the plasma membrane. $^•O_2^-$ is released on the outside while the reductant, NADPH, is supplied from the cytosolic compartment mainly by an increased flux of glucose through the pentose phosphate pathway [74] (Fig. 5). Superoxide production by rat Kupffer cells does not require Ca^{2+}, seems to be regulated by protein kinase C and is not suppressible by glucocorticoids [70, 76]; the latter observation agrees with the lack of superoxide production by LPS-exposed rat Kupffer cells [58].

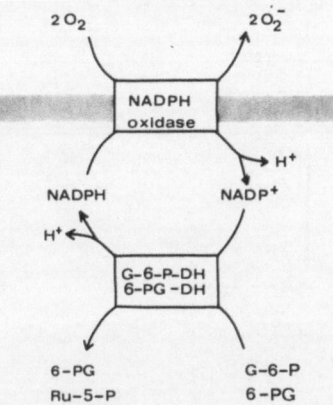

Fig. 5. *Scheme of superoxide synthesis in Kupffer cells.* Note that for every molecule of superoxide formed in the pericellular space an H^+ is generated in the cytoplasm. G-6-P, glucose 6-phosphate; G-6-PDH, glucose-6-phosphate dehydrogenase; 6-PG, 6-phosphogluconate; 6-PG-DH, 6-phosphogluconate dehydrogenase; Ru-5-P, ribulose 5-phosphate

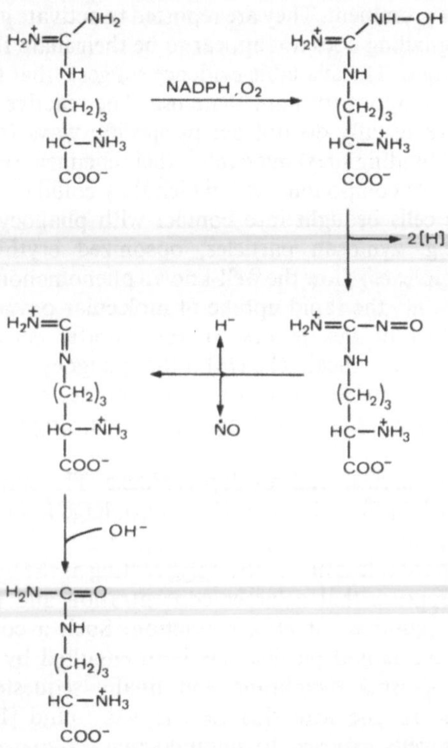

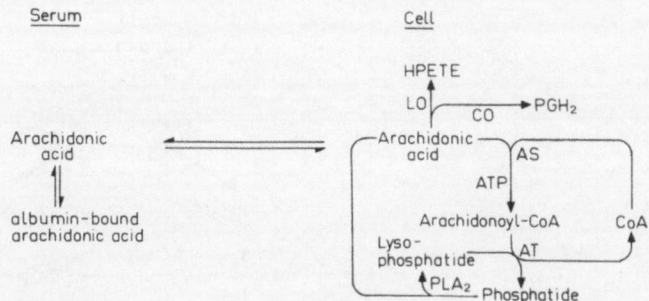

Fig. 7. *Scheme of the arachidonic acid metabolism in Kupffer cells.* AS, acyl-CoA synthetase; AT, lysolecithin acyltransferase; CO, cyclo-oxgenase (PGH synthase); LO, 5-lipoxygenase; PLA_2, phospholipase A_2

protein synthesis in cocultured hepatocytes [79]. The mediator of this inhibition is most likely NO. This compound binds easily to haemoglobin and may thus have a rather narrow range of action. It can be assumed, however, that nitric oxide originating within the liver sinusoid plays a role in the regulation of hepatic blood flow.

EICOSANOIDS
AND PLATELET-ACTIVATING FACTOR

Two major groups of lipid signal substances are presently known to be produced by Kupffer cells: The derivatives of arachidonic acid collectively called eicosanoids and the 1-*O*-alkyl-2-acetyl-*sn*-glycero-3-phosphocholines known as platelet-activating factors (PAF). Lipoproteins and lipoprotein lipase excreted by macrophages including Kupffer cells will not be discussed here.

Arachidonic acid

Arachidonic acid, (6 *Z*, 9 *Z*, 12 *Z*, 15 *Z*)eicosa-6,9,12,15-tetraenoic acid (20:4), is formed from the essential fatty acid, linoleic acid, by chain elongation and desaturation. Deficiency of essential fatty acids or feeding with unsaturated fatty acids like 20-carbon trienoic or pentaenoic acids, interferes with the conversion to biologically active products and affects processes thought to involve the arachidonic acid cascade [80, 81].

The intracellular concentration of free arachidonic acid is very low [82]; almost all of the cellular content is esterified to the *sn*-2 position of diacylphosphoglycerols (Fig. 7). In the extracellular space, arachidonic acid is non-covalently attached to serum proteins (albumin) or diacylphosphoglycerol bound in lipoproteins. Its concentration in serum may reach substantial levels that are sufficient to elicit the synthesis of eicosanoids by potential producer cells. Whether a membrane-associated fatty-acid-binding protein [83] mediates the uptake of arachidonic acid by non-parenchymal liver cells is still unresolved. In many cells, especially in macrophages, arachidonic acid is subject to rapid turnover employing phospholipase A_2 for removal of the acid from the diacylphosphoglycerol and long-chain acyl-CoA synthetase for activation and acyltransferase for reintegration into the phospholipids. Exchange with the albumin-associated arachidonate seems to be rapid and efficient [84]. The combined actions of phospholipase C and diacylglycerol lipase are also able to provide free arachidonic acid to the cells; the relevance of this pathway in Kupffer cells is not well estab-

![Fig. 6 diagram of nitric oxide synthesis]

Fig. 6. *Proposed scheme of nitric oxide synthesis in Kupffer cells.* NO formation from L-arginine. Adapted from Marletta et al. [78]

Nitric oxide

A specific inorganic signal molecule that many cells synthesize and excrete is nitric oxide. It has been identified as endothelium-derived relaxing factor, a powerful agent in adjusting the blood pressure in small vessels [77]. Exposure to NO leads to an enhanced activity of the soluble guanylate cyclase in target cells. Nitric oxide is rapidly converted to nitrite and finally to nitrate. LPS-activated rat Kupffer cells are able to synthesize NO from L-arginine (Gaillard et al., unpublished results) evidently by the mechanism proposed from work with other cell types [78] (Fig. 6). An L-arginine-dependent process in Kupffer cells leads to the inhibition of

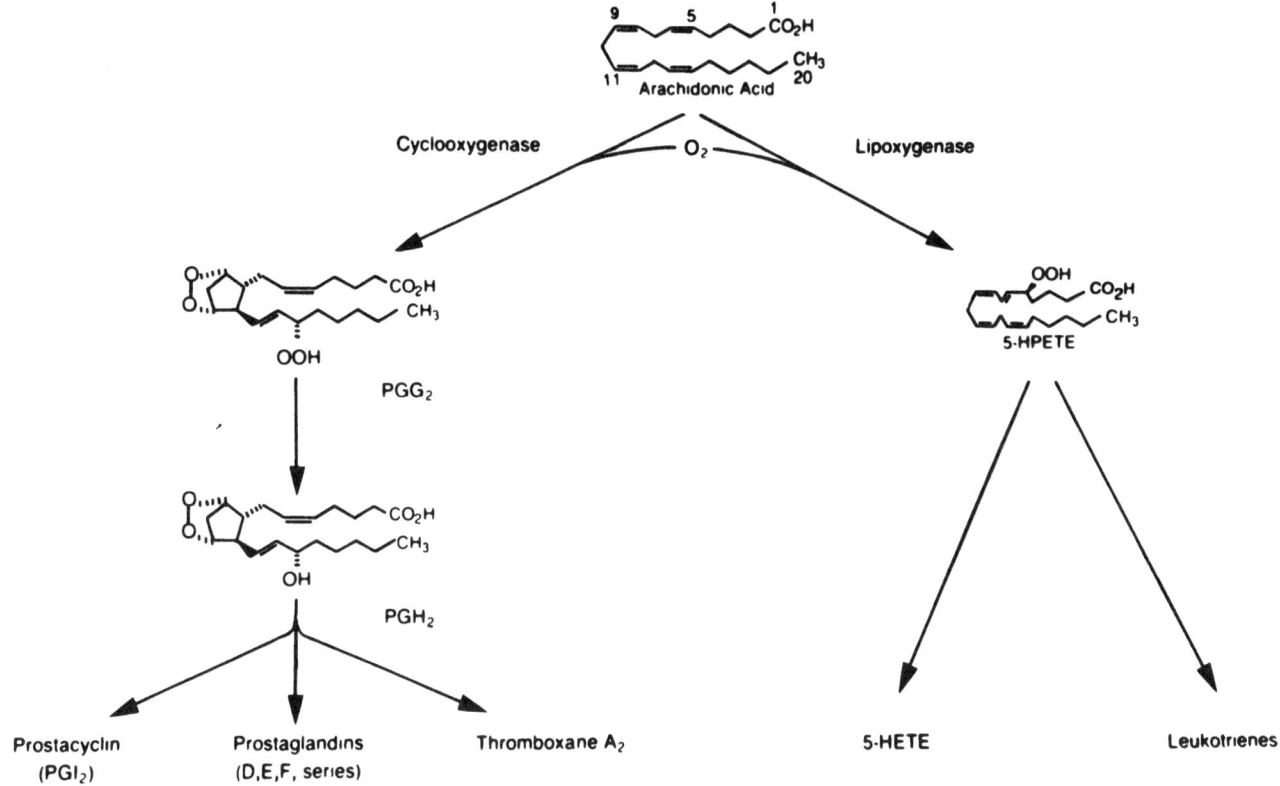

Fig. 8. *The main routes of eicosanoid synthesis*

lished and very likely depends on the nature of the stimulator and the respective signal-transduction pathway. The immediate influence of these enzymes on the availability of free arachidonic acid makes them participants in the regulatory network of eicosanoid metabolism and likely targets for pharmacological interference. Regulation of phospholipase A_2 activity has been reported in Kupffer cells [85] and other macrophages; protein-kinase C-directed phosphorylation of the enzyme [86] or of regulatory proteins (lipocortins) [87] has been invoked; in Kupffer cells a Ca^{2+}- and Mg^{2+}-dependent reversible attachment to cell membranes has also been observed [88]. The physiological role of these modifications of phospholipase A_2 remains to be established.

Prostanoids

The most important signal substances are made from arachidonate by two pathways, one involving prostaglandin H synthase the other 5-lipoxygenase (Fig. 8). The main products of the former are the prostaglandins, prostacyclin and thromboxane. The 5-lipoxygenase pathway produces 5-hydroperoxy-6,8,11,14-eicosatetraenoic acid (5-HPETE), 5-hydroxyeicosatetraenoic acid (5-HETE), LTB$_4$, the lipoxins and the peptidoleukotrienes. Both mechanisms have been found to operate in macrophages though the extent and the resulting pattern differ, not only due to the kind of stimulation of the cells but also to the type of macrophage and the species from which it is derived. In the liver the major producers of eicosanoids are the Kupffer cells [89]. A small contribution, mainly of thromboxane, prostacyclin and PGE$_2$, comes from the sinusoidal endothelial cells [90]; rat hepatocytes do not produce substantial amounts of eicosanoids [91, 92]. It may be expected (but has so far not been proved) that cells within the liver other than hepatocytes and the sinus-lining cells, e.g.

mast cells, epithelial cells of the bile duct or cells lining the walls of the small blood vessels, are able to contribute to a minor extent to the total output of eicosanoids by the stimulated organ. None of the eicosanoids has been found to accumulate or to be stored within liver cells. Furthermore, their half-life in the circulation is short, ranging from seconds for thromboxane A_2 (TXA$_2$), LTC$_4$ and PGI$_2$ to minutes for LTD$_4$ [93, 94]. Thus, the availability of these signal substances depends on their synthesis on demand.

PGH synthase contains two separate catalytic activities: that of a prostaglandin bis-oxygenase (cyclo-oxygenase) and of a glutathione-requiring hydroperoxidase converting arachidonic acid via the endoperoxide, PGG$_2$, to PGH$_2$ [95]. The haem-containing enzyme is membrane-associated, activated by hydroperoxides and inhibited by some non-steroidal anti-inflammatory drugs such as aspirin or indomethacin [96]. Its presence in cell-free extracts of rat Kupffer cells is documented [97]. PGH$_2$ can be converted to PGD$_2$, PGE$_2$ and PGF$_{2\alpha}$, to prostacyclin and to TXA$_2$ by specific enzymes (Fig. 9); in rat Kupffer cells the glutathione-requiring PGD$_2$ synthase resides in the cytoplasm [98]. A side product of PGH synthase activity is hydroxyheptatrienoic acid whose formation is greatly augmented in the presence of haem compounds.

Synthesis of eicosanoids in Kupffer cells requires stimulation. In most instances, e.g. phagocytosis, phorbol ester or Ca^{2+}-ionophore treatment, the regulated step seems to be the provision rather than the conversion of arachidonate. This stems from the fact that a sufficient supply of free arachidonate leads to an amount and profile of prostaglandins released nearly identical to that provided by synthesis from endogeneous sources (phospholipids) after zymosan stimulation (Table 5), with one exception: in rat liver macrophages stimulated by phagocytosis, intracellular Ca^{2+} elevation or

252

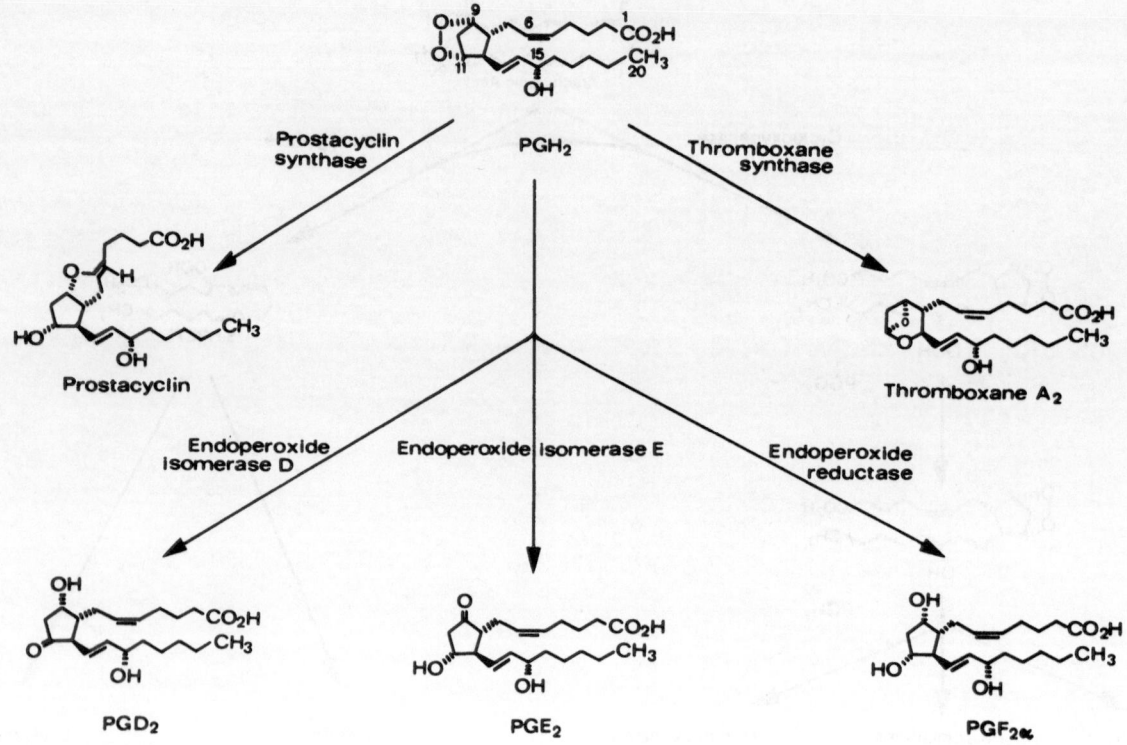

Fig. 9. *PGH₂, the precursor of the prostanoids*

Table 5. *The major eicosanoids produced by rat Kupffer cells*
The data represent the production of an average batch of rat Kupffer cells after 72 h in primary culture. The yield of total eicosanoids may vary between cell batches, but the relative contribution of the different compounds is rather constant. Small amounts of compounds co-eluting with HETE are also formed, probably as side reactions of the PGH synthase. 5-HETE could not be identified as product of rat Kupffer cells. The values for TNF and LPS are averages of the 24 h values. The rates are not linear over the 24 h period. PGD₂ is much less stable in aqueous media than PGE₂. n.d., not determined

Stimulus	Eicosanoid found in the medium			
	PGD_2	PGE_2	$PGF_{2\alpha}$	TX
	$pmol \cdot h^{-1} \cdot 10^6 \ cell^{-1}$			
None	260	9	< 5	< 5
Zymosan	2100	145	65	170
PMA	2350	134	70	165
PAF	n.d.	45	n.d.	n.d.
TNF	12	14	< 10	15
LPS	13	16	< 10	15
A 23187	1780	132	60	145
Arachidonate	2420	152	75	< 20

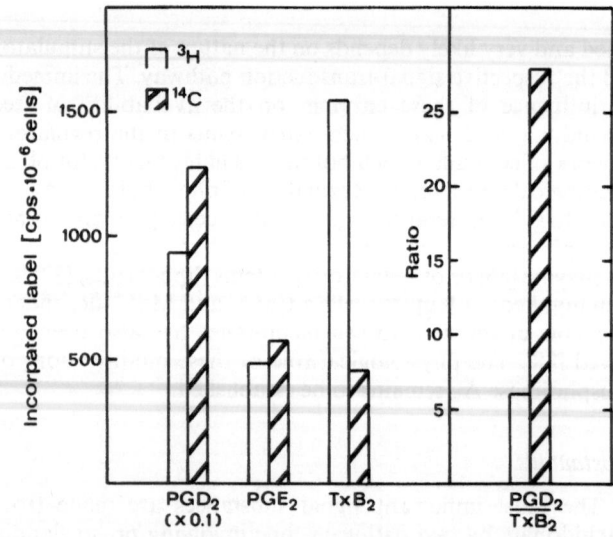

Fig. 10. *Prostanoid synthesis from endogeneous and added arachidonic acid.* Rat Kupffer cells were incubated with [³H]arachidonic acid for 24 h. The media were replaced and free [¹⁴C]arachidonic acid was added together with zymosan. 60 min later, the media were removed and processed for HPLC and isotope analysis. Open bars, ³H; hatched bars, ¹⁴C content. Data are taken from Dieter et al. [97]

phorbol ester the major eicosanoid formed is PGD_2, but thromboxane and PGE_2 are also seen [92]. In the presence of free arachidonate, the prostaglandins are still formed but thromboxane is absent from the released products [99] (Fig. 10). Since thromboxane shows some unique features in its hepatic metabolism, one wonders if particular topological factors in producer cells separate this compound from the other eicosanoids.

PGE_2 production in Kupffer cells is also stimulated by lipopolysaccharides [64], tumor necrosis factor-α [28] and

some viruses (Newcastle disease or Sendai virus) [100]. Exposure of liver macrophages to one of these agents increases dramatically the rate of PGE_2 synthesis (relative to the stimulation by zymosan or phorbol ester) without affecting PGD_2 and thromboxane production [28]; this indicates a regulated event at the level of the PGH_2 metabolizing enzymes. In cell-free extracts of rat Kupffer cells the activity of the isomerase converting PGH_2 to PGE_2 (PGE synthase) is increased under these conditions [97].

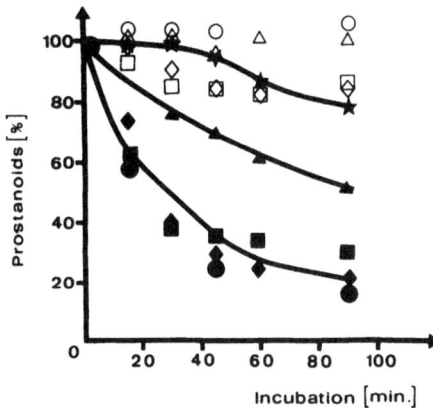

Fig. 11. *Degradation of added prostanoids by rat hepatocytes.* Primary cultures of rat hepatocytes were exposed to 156 pmol PGD_2 (■), 145 pmol PGE_2 (●), 231 pmol $PGF_{2\alpha}$ (◆), 318 pmol 6-oxo-$PGF_{1\alpha}$ (★) and 124 pmol TXB_2 (▲), respectively. Open symbols represent prostanoid content in the absence of hepatocytes. The remaining prostanoids were determined in aliquots by radioimmuno assay. Data are taken from Tran-Thi et al. [19]

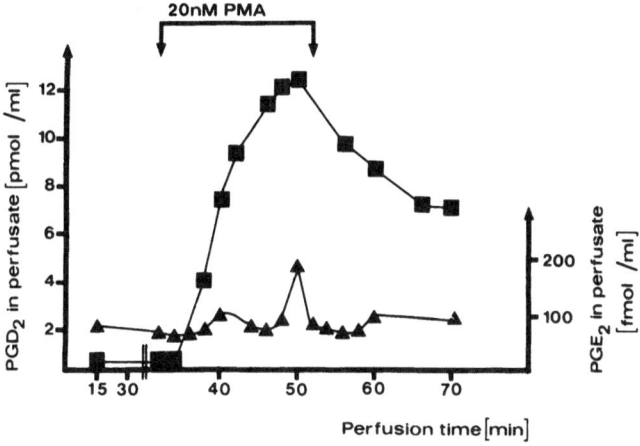

Fig. 12. *Time course of PGD_2 and PGE_2 production by the* in situ *stimulated perfused rat liver.* PGD_2 (■) and PGE_2 (▲) were measured by radioimmunoassay. Data are taken from Tran-Thi et al. [110]

PGE_2 plays an autoregulatory role in Kupffer cells. It inhibits its own production in LPS-elicited, but not in TNF-, zymosan- or phorbol-ester-elicited cell responses [100a]. It also inhibits the LPS-induced synthesis of TNF [45, 101] (see below). On the other hand, PGE_2 quite specifically activates functions of Kupffer cells such as the stimulated production of cAMP followed, e.g. by an increased collagenase synthesis [64].

The prostanoids are seen as local hormones with a short range of action [102]. Thus, inactivation of the signals is a necessary feature of their actions. Rapid uptake and degradation of Kupffer-cell-derived prostaglandins is accomplished by the hepatocytes [91]. The main products found both in cultured hepatocytes [103] and in the isolated perfused rat liver (Tran-Thi et al., unpublished results) result from one and two cycles, respectively, of β-oxidation starting at the carboxyl group; dinor- and tetranor-prostaglandins leave the parenchymal cells readily and may be further metabolized by other cells of the body. Hepatocytes also degrade thromboxane TXB_2 and 6-oxo-$PGF_{2\alpha}$ which are the (spontaneously formed and biologically inactive) primary degradation products of the very labile TXA_2 and prostacyclin (PGI_2), respectively.

The parenchymal cells of the liver were reported to carry receptors for prostaglandins [104, 105]. The PGE_2 receptor is coupled to adenylate cyclase by a G protein [106]. However, it is not clear whether these receptors are instrumental in signal transduction only or also responsible for the uptake of excess prostanoids. The rate of degradation by cultured rat hepatocytes is almost the same for all prostanoids tested [91] (Fig. 11) while their metabolic effects on parenchymal cells differ widely [60, 107, 108]. A substantial translocation into bile of both intact and degraded prostaglandins during liver perfusion [60] also points to a high-capacity, low-specificity transport mechanism involved in the hepatic degradation of prostanoids.

The synthetic capacity of the non-parenchymal cells, the extent of degradation by the liver parenchyma (Kupffer cells and sinusoidal endothelial cells do not contribute significantly to prostanoid degradation) [92] and the importance of the topology of the producing liver macrophages is best demon-

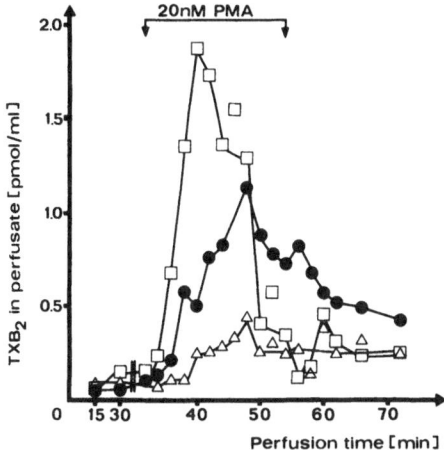

Fig. 13. *Time course of TXB_2 formation by the* in situ *stimulated perfused rat liver.* The stimulation was accomplished by infusion of PMA (●) into the non-recirculating perfusion system. The addition of the thromboxane synthase inhibitor, CGS 13080 (△) led to an almost complete suppression of thromboxane release while the thromboxane receptor antagonist BM 13.177 (□) allowed an enhanced outflow of TXB_2 from the liver

strated in experiments with the isolated perfused liver applying *in loco* stimulation of eicosanoid production [31, 60, 107, 109–113]. The application of 20 nM PMA to the isolated perfused rat liver results in a strongly stimulated output of PGD_2 (Fig. 12) and thromboxane (Fig. 13) from the Kupffer cells [60, 109]. Interestingly, the effects differ depending on the direction of perfusion: in the (unphysiological, central → portal vein) retrograde mode, more eicosanoids are found unchanged in the effluent, and much smaller effects on glucose output and perfusion pressure occur than in the orthograde (portal → central vein) direction [110]. This behaviour is consistent with the preferred location of Kupffer cells in the portal part of the liver sinusoid. Obviously, the products of the *in situ* stimulation of Kupffer cells come more frequently into contact with hepatocytes in the orthograde perfusion mode. The studies employing the isolated perfused liver also showed the relative effectiveness of the various compounds in the intact organ.

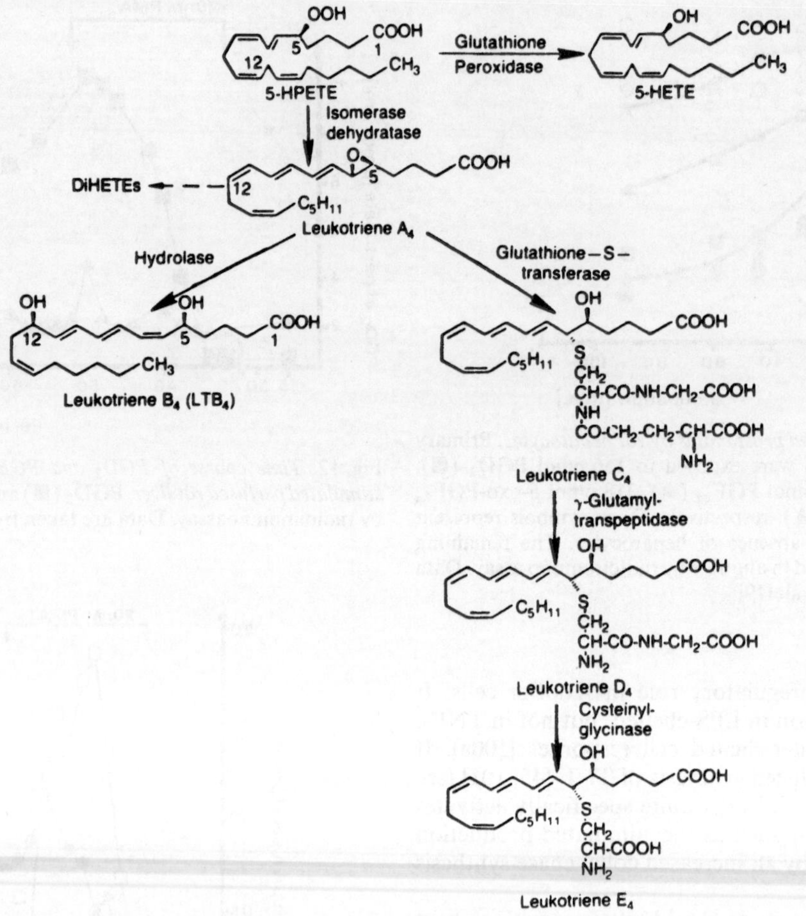

Fig. 14. *Scheme of leukotriene formation from arachidonic acid*

The effects of the prostanoids in the mammalian organism are manifold. They are mediated by specific receptors on target cells and reflect the signal-transducing paths and the responsive enzyme outfits of the recipient cell as much as the respective prostanoid itself. In hepatocytes, $PGF_{2\alpha}$ and PGE_2 increase the glycogenolytic activity [107, 114] though not as much as TXA_2 which was shown to be responsible for more than 60 % of the phorbol-ester-triggered effects in the perfused liver [109]. On the other hand, PGE_2 was found to depress the glucagon-stimulated cAMP formation and glycogenolysis in hepatocytes [115].

Leukotrienes

The second route of arachidonate metabolism is initiated by peroxide formation catalyzed by lipoxygenases (for reviews see Needleman et al. [80] and Samuelsson et al. [116]). Various C atoms of the unsaturated fatty acid can be targets of the position-specific enzymes, e.g. positions 5, 11, 12, 13 or 15. The most important of these reaction products are those synthesized by the ATP- and Ca^{2+}-dependent 5-lipoxygenase (Fig. 14). This enzyme is most abundant in neutrophilic polymorphonuclear leukocytes, eosinophils, basophils and mast cells, but it is also present in monocytes and macrophages. It may be isolated from the cytosolic compartment of the cells but gains its full activity only after attachment to membrane components [117]. The first product of the enzyme, 5-HPETE, can be converted by three different routes: (a) reduction by glutathione leads to 5-HETE which is a

powerful chemoattractant and a typical product of neutrophilic granulocytes; (b) using 15-HPETE as a substrate 5-lipoxygenase can introduce another hydroperoxy group that finally rearranges to 5,6,15- or 5,14,15-trihydroxy-eicosatetra-enoic acids, called lipoxins (A_4 and B_4, respectively) [118]. Alveolar macrophages [119] have been shown as source of these compounds which are able to trigger superoxide formation and degranulation in human neutrophils; (c) 5-lipoxygenase may convert 5-HPETE to the 5,6-epoxy derivative, LTA_4. This compound is either converted by a specific epoxide hydrolase [120] to LTB_4, a strong chemotactic and adhesion-promoting factor of granulocytes and macrophages, or to leukotriene C_4 (LTC_4) by the specific microsomal glutathione-S-transferase-catalyzed reaction [121]. The addition of glutathione occurs readily in stimulated mast cells, granulocytes and monocytes. Peritoneal and some other macrophages are also able to synthesize variable amounts of peptidoleukotrienes [122−124]. γ-Glutamyl transpeptidase [125] rapidly converts LTC_4 to LTD_4 by removal of the glutamyl residue; LTD_4 has about the same biological potency as the parent compound.

Inactivation of the leukotrienes follows several routes: peptidase(s) may further split off the glycine residue to form LTE_4. Leukotrienes are taken up actively by hepatocytes [126] where they may be converted to N-acetyl-LTE_4; a substantial part of these mediators is eliminated through the bile [127]. The uptake of leukotrienes by the liver is inhibited by the immunosuppressive drug, cyclosporin [128]. An oxidative pathway of degradation of leukotrienes also occurring in liver

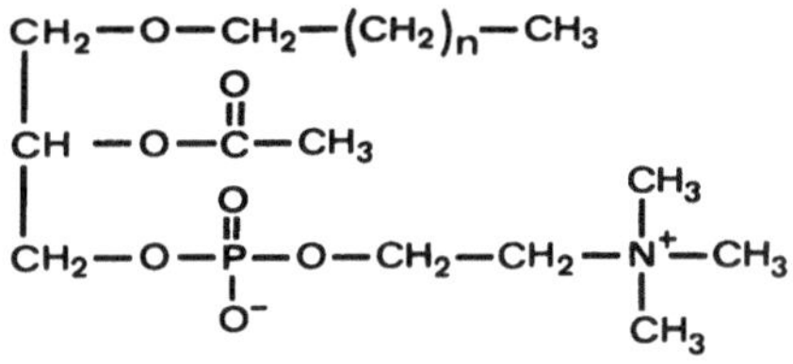

CH₂—O—CH₂—(CH₂)ₙ—CH₃ ...

$$CH_2-O-CH_2-(CH_2)_n-CH_3$$
$$CH-O-\overset{O}{\underset{\|}{C}}-CH_3$$
$$CH_2-O-\overset{O}{\underset{\underset{O^-}{\|}}{P}}-O-CH_2-CH_2-\overset{CH_3}{\underset{CH_3}{N^+}}-CH_3$$

Fig. 15. *Platelet-activating factor*

is initiated by a cytochrome-P_{450}-dependent hydroxylation of the C20 methyl group followed by ω-oxidation [129].

The key enzymes of the arachidonate cascade, PGH synthase and lipoxygenase, may serve as examples of regulation by self-activation and self-destruction, respectively (for a review see Needleman et al. [80]). Hydroperoxides including intermediates of the cyclo-oxygenase reaction act as activators of the reaction. The phenomenon can be seen in the acceleration of the initial sluggish reaction to its maximal velocity; the latter can be obtained immediately if hydroperoxy fatty acids are added to the reaction mixture. Inversely, the highly reactive intermediates of the lipoxygenase reaction, the same is true for thromboxane synthase, inactivate the enzyme by oxidation of some of its amino acid side chains (such as methionine) [130].

The peptidoleukotrienes were first recognized as the 'slow-reacting substance of anaphylaxis' [131]. More recently, they were shown to be important secondary mediators of LPS-triggered shock [132, 133]. Vasoconstriction and broncho-constriction, enhanced vascular permeability and the activation of leukocytes, granulocytes and macrophages are prominent features of their action spectrum. Receptors for LTC_4 and LTD_4 have been identified on many supposed target cells; receptor antagonists such as FPL 55712 or MK 88675, an inhibitor of 5-lipoxygenase activation [134], proved very helpful in elucidating the role of these signal molecules.

Platelet-activating factor

Activation and aggregation of thrombocytes under the influence of products from activated leukocytes was first reported in 1971 [135]. It turned out that the active agent was a lipid of the phosphoglyceride class (for review see Hanahan [136]); its characteristics are the ether bond between the *sn*-1 position and a long-chain (C_{13-18}) alkyl residue, the short-chain (usually acetyl) fatty acid esterified to the *sn*-2 position and a phosphodiester bond with phosphorylcholine at *sn*-3 of the glycerol backbone (Fig. 15).

Neutrophils stimulated sequentially with cytochalasin B and the Ca^{2+} ionophore A 23187 or fMet-Leu-Phe seem to be the best source of PAF [137]. But basophils, endothelial cells, platelets and macrophages were also shown to produce this lipid mediator upon stimulation. Macrophages are known to contain relatively high amounts of long-chain 1-*O*-alkyl-2-acyl-*sn*-glycero-3-phosphocholine [136]. PAF synthesis appears to follow mainly the route from these precursors through phospholipase A_2-catalyzed lyso-PAF formation and reacylation of the *sn*-2 position with acetyl-CoA by a specific Ca^{2+}-dependent lyso-PAF acetyltransferase. Cultured rat Kupffer cells synthesize PAF from lyso-PAF within 10 min after stimulation with the Ca^{2+} ionophore A 23187 [138].

The biological activity of PAF is most likely mediated by cell-surface-integrated receptors; however, isolation and physicochemical characterization has not yet been reported.

PAF elicits the release of vasoactive amines from thrombocytes and the aggregation of platelets with themselves and also with endothelial cells. The strong lytic activity of lyso-PAF may take part in these events. PAF was also shown to induce the degranulation of neutrophils and the activation of macrophages. Rat Kupffer cells exposed to PAF produce superoxide and eicosanoids [76] and show an increase of inositol phosphate production and cytosolic free Ca^{2+} concentrations [139].

Livers perfused with PAF respond with increased release of glucose [140] and perfusion pressure [141]. Hepatocytes, however, do not respond with increased glucose production to PAF; it is likely that the hepatic action of PAF is mediated by the Kupffer cells that have been shown to produce the signal substances (TXA_2, PGE_2) required to elicit the observed effects on hepatocytes.

PEPTIDE MEDIATORS

Work in many laboratories has identified more than 50 compounds of peptidic nature that are synthesized and secreted by stimulated macrophages. They include complement components, fibronectin, apolipoprotein E, transcobalamine, α_2-macroglobulin, plasminogen activator inhibitor, angiogenesis factor and several coagulation factors (for a review see [142]); furthermore, numerous enzymes, mainly hydrolases of various specificities, including collagenase, elastase, angiotensin-converting enzyme, lysozyme and lipoprotein lipase. A functional significance, however, can be attributed to enzyme release only if leakiness of the activated (and sometimes damaged) cells is excluded [143]. It is questionable, whether macrophages, in contrast to polymorphonuclear neutrophils, secrete lysosomal enzymes actively. Kupffer cells produce and secrete collagenase [64], fibronectin [144] and apolipoprotein E [145].

Of particular interest are the typical signal polypeptides, the cytokines, of which interleukins IL-1 and IL-6, TNF and interferon-α/β are well-defined products of Kupffer cells. Research on these mediators has been greatly enhanced by the availability of the recombinant molecules from several species as well as of monoclonal antibodies against them. Of course, one should bear in mind that recombinant cytokines lack the carbohydrate side chains present on most native molecules and may possess somewhat different affinities to enzymes and receptors.

Interleukin-1

IL-1, formerly also known as lymphocyte-activating factor or as endogenous pyrogen, is mainly produced by blood monocytes [146], but sessile macrophages were also shown to synthesize and secrete this mediator of inflammation and pain. It is best elicited by LPS. Kupffer cells produce significant amounts of IL-1 after stimulation with LPS [147]. The molecular mass of the mature compound is 14 kDa [148] by removal of the four N-terminal amino acid residues catalyzed by a peptidyl aminopeptidase [149]. An inhibitor of IL-1 is reported to be coproduced with IL-1 after LPS treatment of rat Kupffer cells [147]. IL-1 is well known to stimulate IL-2 production by T cells; it also affects and finally destroys the insulin-producing B cells of the pancreas [150].

Interleukin-6

IL-6 was formerly known as interferon-$\beta2$, B cell stimulatory factor 2 or hepatocyte-stimulatory factor; this points

256

Table 6. *Effects of interleukin-6*

Induction of acute phase protein synthesis in hepatocytes

Induction of terminal B cell maturation

Stimulation of activated B cells to secrete IgM, and IgA

Enhancement of proliferation of hemopoetic progenitor cells

Induction of cellular differentation in PC12 nerve cells

Release of adrenocorticotropin from cultured pituitary cells

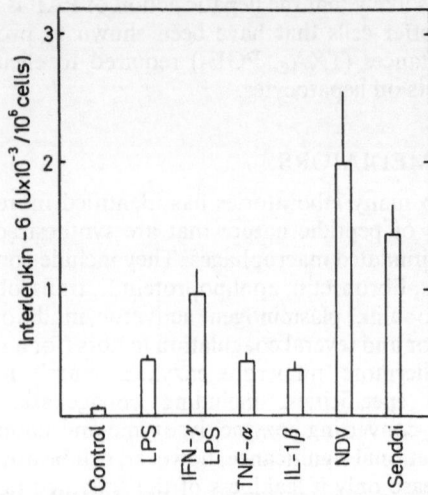

Fig. 16. *Release of interleukin-6 from stimulated rat Kupffer cells.* Rat Kupffer cells in primary culture were exposed to the indicated stimuli. IL-6 in the medium was determined by bioassay [100]. NDV, Newcastle disease virus. IFN, interferon

to the numerous physiological functions which must be attributed to this cytokine (Table 6). For recent reviews see Bauer [29] and Kishimoto [151].

The mature, predominantly N-glycosylated IL-6 has a molecular mass of about 26 kDa [152]; it arises from a precursor with a 26 amino acid extention (signal sequence). IL-6 is synthesized upon inflammatory stimulation (e.g. LPS, TNF, IL-1) or viruses by a variety of cells including fibroblasts, endothelial cells, monocytes, macrophages and also some tumors. The hardly detectable level of IL-6 in normal human serum rises up to 100-fold during inflammation, dependent on the severity of the condition. IL-6 is considered a sensitive indicator of meningitis, graft rejection and rheumatoid arthritis.

As Kupffer cells are located in close proximity to hepatocytes their capacity to release IL-6 is of particular interest. Exposure to 1 ng LPS/ml stimulates the liver macrophages to an IL-6 secretion of $300-400$ pg/10^6 cells [153] as measured by the B cell hybridoma growth test [154]. Dexamethasone at 1 μM completely suppresses LPS-stimulated IL-6 synthesis. PGE_2 that also inhibits TNF release is without effect on the IL-6 production. Besides LPS and TNF some viruses (Sendai, Newcastle disease) are able to elicit rat Kupffer cells to produce IL-6 (Fig. 16).

IL-6 is a useful example for the regulation at the receptor level [155]. IL-6 is the most potent stimulator known for the elicitation of acute-phase protein synthesis in hepatocytes [156]. Little mRNA for this receptor is found in quiescent

parenchymal cells; however, after exposure to glucocorticoid hormones, its intracellular level is greatly enhanced. This observation seems to explain the permissive effect of these hormones in the acute phase response. IL-6 receptors are also found on blood monocytes; their synthesis is repressed by LPS that induces IL-6 formation and is also down-regulated by IL-6 and IL-1; thus, self-stimulation of the monocytes by their own products is prevented. In the normal situation, IL-6 receptors are present on monocytes and on hepatocytes, but in the inflammatory state they are low in monocytes and high in hepatocytes (and also in B cells). Positive autocrine cooperation was observed in multiple myeloma cells: here, IL-6 is a growth factor for the cells which produce at the same time the IL-6 receptor.

Interferon-α/β

The interferons comprise a group of molecules that have initially been defined by their antiviral activity (for review see Pestka and Langer [27]). Presently, two major families of interferons with overlapping activities but unrelated structures are known: the interferon-γ mentioned already as stimulator of macrophages and the closely related interferons-α and -β. The former has first been described as product of peripheral leukocytes, the latter as products of fibroblasts. Due to their structural and functional similarities, interferons-α and -β are usually considered as an operational entity.

Both interferons appear to be produced in several variants their molecular masses ranging over $16-27$ kDa for human interferon-α and around 20 kDa for the β form. Murine interferon-β was obtained as a 26-kDa and a 35-kDa species. Both interferons-α and -β are predominantly found in glycosylated forms.

Activated macrophages secrete interferon-α/β [157]. The cytokine released by murine Kupffer cells [42] augments the liver-associated natural killer activity [43]. In general, viruses and double-stranded RNA have proven to be the most efficient inducers of interferon-α/β expression. Besides their antiviral property, this cytokine inhibits tumor growth, modulates cell differentiation and induces the expression of class I major histocompatibility complex antigens including β_2-microglobulin [158, 159] and of F_c receptors [160, 161]. In some cases they antagonize interferon-γ, e.g. in the down-regulation of the mannosyl fucosyl receptor [162]. Specific receptors for interferon-α/β are present on susceptible cells which are down-regulated by an excess of the ligand [163, 164].

Tumor necrosis factor α (cachectin)

TNF is a very important component of the group of signal molecules produced by macrophages after inflammatory stimulation; it is also known as cachectin (for a review see Beutler and Cerami [165]). It was described independently as a mediator of tumor necrosis seen after inflammation and as the agent responsible for the anorexia in chronic neoplastic or inflammatory diseases. Molecular cloning of the respective compounds brought definite proof of their identity [166]. Major producers are the blood monocytes and the macrophages. In the rat, the Kupffer cells seem to be the most abundant source of this cytokine [45]. TNF synthesis and release is strongly stimulated after exposure of producer cells to LPS [167] or to some viruses, e.g. Sendai or Newcastle disease virus [100]. The presence of interferon-γ is mandatory

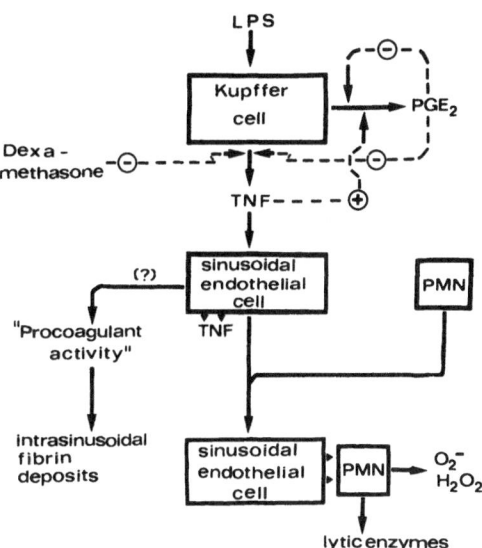

Fig. 17. *Self-limiting regulatory circuit of LPS-stimulated rat Kupffer cells.* Endotoxin action on the liver. In addition to the regulatory network displayed in the upper half of the scheme, the firm attachment of TNF to the sinusoidal endothelial cells is indicated. The bound TNF triggers sticking of polymorphonuclear neutrophilic granulocytes (PMN), and their activation is followed by release of reactive oxygen intermediates and lytic enzymes, e.g. elastase

Table 7. *Specific suppressive effect of PGE$_2$ on cytokine production by rat Kupffer cells*
Rat Kupffer cells in primary culture were exposed to LPS (100 ng/ml) and 100 nM PGE$_2$, respectively, and TNF and IL-6 measured. With regard to IL-6, the same results were obtained using 1 ng LPS/ml. For IL-1 synthesis, 10 µg LPS and 1 µM LPS and 1 µM PGE$_2$ were used. The data were taken from Busam et al. [153] and Shirahama et al. [147]. n.d., not determined

Addition	Cytokine formed		
	TNF	IL-6	IL-1
—	U $\cdot$ 4 h^{-1} $\cdot$ 10^6 cell^{-1}	U $\cdot$ 24 h^{-1} $\cdot$ 10^6 cell^{-1}	% control
None	< 20	44 ± 12	n.d.
LPS	680 ± 256	364 ± 44	100
PGE$_2$	< 20	52 ± 16	n.d.
LPS/PGE$_2$	64 ± 16	332 ± 38	15

Table 8. *Tumor necrosis factor α of rat Kupffer cells*

Synthesized by Kupffer cells, not by hepatocytes or sinusoidal endothelial cells

Synthesis is elicited by LPS and viruses, not by phagocytosis, PMA or Ca^{2-} ionophore

Synthesis is instantly suppressed by dexamethasone or PGE$_2$

Rate of PGE$_2$ synthesis is increased preferentially (like LPS)

Mediates most (if not all) effects of LPS

Replaces LPS in the galactosamine-enhanced cytotoxicity

Elicits sticking of granulocytes to sinusoidal cells

Fails to trigger superoxide production in Kupffer cells

Fails to induce acute-phase reaction in hepatocytes

in some producer cells, and in others, such as rat Kupffer cells [45], it enhances TNF synthesis.

The cytokine is synthesized as a precursor having an N-terminal extension of 76 (man) and 79 (mouse) amino acids [165]. The glycosylated polypeptide appears to be the normal form although minor unglycosylated species have been observed. Membrane-associated TNF is present in producer cells; it is not yet clear, however, whether the membrane-integrated form (is it the same as in cytotoxic cells [168] or different?) is an obligatory intermediate in the release of the soluble polypeptide. The TNF found in the extracellular space consists of 157 (man) or 156 (mouse) amino acids.

TNF released from the producing macrophage binds to a plasma membrane receptor of target cells. This receptor whose mass is about 300 kDa [169] appears to be widely distributed in the animal body. The mechanisms of signal transduction from the liganded receptor to the various intracellular effector systems are not yet understood.

TNF is one of the signal molecules that is able to trigger the same cells that produced it (autostimulation, autacoid). It activates macrophages including Kupffer cells to become cytotoxic and to produce PGE$_2$, IL-1 and IL-6 [28, 153, 170]. The intricate interaction between TNF and PGE$_2$ forms a self-limiting regulatory cycle (Fig. 17) that may explain the kinetics of both the TNF and PGE$_2$ production by rat Kupffer cells [45].

The cytokine/prostaglandin couples are valuable examples for the understanding of the autoregulation of signal systems in Kupffer cells: TNF and other cytokines are produced by these macrophages after exposure to lipopolysaccharides (or viruses) and interferon-γ. Although gene expression following stimulation increases only about threefold, mRNA levels may reach up to 1 % of all messenger RNAs and the extracellular cytokine will become the major secreted protein [165]. On the other hand, macrophages from LPS-resistant mice can form high levels of mRNA after LPS, but hardly secrete any TNF. Their synthesis is controlled partly at the post-transcriptional

level involving a untranslated U,A-rich extension of the mRNA at the 3' end [171]. This region is thought to shorten the half-life of the mRNA; as a corollary to this instability, one would expect that activation of the cells that leads to a rapid onset of cytokine production should remove or suppress this element. The finding [45, 172] that inhibitors of the elicited TNF synthesis in rat Kupffer cells such as PGE$_2$ or dexamethasone prevent cytokine formation, even when given simultaneously with LPS, point in this direction. The suppressive effect of PGE$_2$ on TNF production is also remarkable for its specificity: other prostaglandins, e.g. PGD$_2$ or PGF$_{2\alpha}$, are ineffective [45]; on the other hand, PGE$_2$ suppresses the synthesis of IL-1 [147] and TNF [45], but not that of IL-6 [153] (Table 7).

The overall effects of TNF are manifold; those seen in rat liver are listed in Table 8. This cytokine is now thought to mediate most if not all the consequences of LPS exposure in sensitive animals [173]. It was shown to be the endogenous pyrogen of rabbits [174]; the action of LPS on animals can be suppressed by antibodies raised against TNF [175]. Anaemia is produced by this cytokine through reduction of erythrocyte synthesis and lifespan [165]. Fever is the result of the TNF-mediated stimulation of PGE$_2$ synthesis in the hypothalamus, as well as of interleukin-1 induction in peripheral cells [165].

Vascular endothelial cells respond to TNF by up-regulation of class I major histocompatibility complex antigens

258

and of intercellular adhesion molecules [175]. These, as well as the sinusoidal endothelial cells of rat liver [176], bind TNF and elicit the adherence of granulocytes concomitant with their activation (release of hydrolytic enzymes and reactive oxygen intermediates). Furthermore, the cytokine induces co-agulant activity [177] and IL-1 production [178] in endothelial cells. It is obvious that these effects are able to make a major contribution to the pathological changes seen after LPS administration or in endotoxin shock.

Most of the regulatory phenomena in the world of signal transmission involve more than one cell type although autocrine stimulations and inhibitions are also known. In a compact organ such as the liver, the interplay of parenchymal and non-parenchymal cells can be expected to be an important event in the functional regulation of the whole organ. As an example the various influences between liver macrophages and hepatocytes may be mentioned. IL-6 that is produced upon exposure to inflammatory substances (LPS, viruses) by the Kupffer cells in the neighbourhood of the parenchymal cells elicits the production of the anti-inflammatory acute-phase proteins [156]. At the same time, thromboxane and PGE_2 triggered by the same stimuli increase the rate of glyco-genolysis thereby supplying the hepatocytes and other cells with the energy-providing substrate, glucose.

On the other hand, hepatocytes do not produce hormones and mediators but respond to many of them; they also inacti-vate most of these substances and in this way serve the signal character of the mediators. For example, eicosanoids are synthesized by the non-parenchymal cells, mainly by Kupffer cells, while the inactivation, degradation and elimination (into bile) is almost exclusively done by the hepatocytes [89]. There are also strong indications for an important interaction be-tween Kupffer cells and fat-storing (Ito) cells in their trans-formation into collagen-producing fibroblast-like cells in the course of liver fibrosis [179].

A number of malignantly transformed cells are lyzed by TNF-producing macrophages. The mechanism of this cyto-toxic action of the cytokine is not fully understood, but the involvement of a cytolytic proteinase activated in stimulated macrophages is widely assumed [36]. A cell-associated form of TNF serves in the killing of tumor cells by activated cytotoxic murine macrophages [168]. In fibroblasts, TNF profoundly affects gene regulation stimulating the *fos* and *myc* genes [174].

The anorexia often observed in malignant or inflammatory states appears to result from biochemical events in adipocytes elicited by TNF leading to a preferentially lipolytic, catabolic state of this tissue [165]: the expression of lipoprotein lipase, the fatty-acid-binding protein, glycerol-3-phosphate dehydro-genase and probably also of some enzymes of triglyceride synthesis from acetate is suppressed while the hormone-sensi-tive lipase of the fat cells is activated.

Another cytokine has been identified in the circulation of man and other mammalian species whose physiological activities are very much like those of TNF. It is called lymphotoxin or TNF-β. It is structurally related to TNF (28% sequence similarity) and encoded on the same gene [180]. It differs, however, with regard to the source and the elicitors. While TNF is synthesized mainly by macrophages in response to LPS or viruses [181], lymphotoxin is produced by chal-lenged T cells or lymphoblastoid cell lines [182].

The work carried out in the author's laboratory was supported by grants from the *Deutsche Forschungsgemeinschaft*, Bonn, through SFB 154, and from *Fonds der Chemischen Industrie*, Frankfurt (FRG).

REFERENCES

1. Golde, D. W. & Gasson, J. C. (1988) in *Inflammation. Basic principles and clinical correlates* (Gallin, J. I., Goldstein, I. M. & Snyderman, R., eds) pp. 253—264, Raven Press, New York.
2. van Furth, R. (1988) in *Inflammation. Basic principles and clinical correlates* (Gallin, J. I., Goldstein, I. M. & Snyderman, R., eds) pp. 281—295, Raven Press, New York.
3. Bouwens, L., Baekeland, M. & Wisse, E. (1984) *Hepatology 4*, 213—219.
4. Walker, W. S. (1976) in *Immunobiology of the macrophage* (Nelson, S. W., ed.) pp. 91—111, Academic Press, New York, San Francisco, London.
5. Gordon, S., Crocker, P. R., Morris, L., Lee, S. H., Perry, V. H. & Hume, D. A. (1986) *Ciba Found. Symp. 118*, 54—67.
6. Aterman, K. (1977) in *Kupffer cells and other sinusoidal cells* (Wisse, E. & Knook, D. L., eds) pp. 3—20, Elsevier/North-Holland Biomedical Press, Amsterdam, New York, Oxford.
7. Wake, K. (1980) *Int. Rev. Cytol. 66*, 303—353.
8. Brouwer, A., Barelds, R. J. & Knook, D. L. (1984) in *Centrifuga-tion, a practical approach* (Rickwood, D., ed.) pp. 183—218, IRL Press, Oxford.
9. Wake, K., Decker, K., Kirn, A., Knook, D. L., McCuskey, R. S., Bouwens, L. & Wisse, E. (1989) *Int. Rev. Cytol. 190*, 173—229.
10. Diesselhoff-den Dulk, M. M. C., Crofton, R. W. & van Furth, R. (1979) *Immunology 37*, 7—14.
11. Bouwens, L., Remels, L., Baekeland, H., Van Bossuyt, E. & Wisse, E. (1987) *Eur. J. Immunol. 17*, 7—42.
12. Decker, T., Baccarini, M. & Lohmann-Matthes, M.-L. (1986) *Eur. J. Immunol. 16*, 693—699.
13. Malter, M., Süss, R. & Friedrich, E. (1986) in *Cells of the hepatic sinusoid* (Kirn, A., Knook, D. L. & Wisse, E., eds) vol. 1, pp. 439—444, Kupffer Cells Foundation, Rijswijk.
14. Latocha, G., Dieter, P., Schulze-Specking, A. & Decker, K. (1989) *Biol. Chem. Hoppe-Seyler 370*, 1055—1061.
15. Ezekowitz, R. A. B. & Gordon, S. (1986) in *Ciba Found. Symp. 118*, pp. 127—136.
16. Dini, L. & Kolb-Bachofen, V. (1989) *Exp. Cell Res. 184*, 235—240.
17. Munthe-Kaas, A. C., Kaplan, G. & Seljelid, R. (1976) *Exp. Cell Res. 103*, 201—212.
18. Praaning-van Dalen, D. P., De Leeuw, A. M., Brouwer, A. & Knook, D. L. (1987) *Hepatology 7*, 672—679.
19. Haltiwanger, R. S., Lehrman, M. A., Eckhart, A. E. & Hill, R. L. (1986) *J. Biol. Chem. 261*, 7433—7439.
20. Kolb-Bachofen, V., Schlepper-Schäfer, J., Vogell, W. & Kolb, H. (1982) *Cell 29*, 859—866.
21. Chao, W., Liu, H., DeBuysere, M., Hanahan, D. J. & Olson, M. S. (1989) *J. Biol. Chem. 264*, 13591—13598.
22. Kleinherenbrink-Stins, M. F., Schouten, D., Brouwer, A., Knook, D. L. & van Berkel, T. J. C. (1989) in *Cells of the hepatic sinusoid* (Wisse, E., Knook, D. L. & Decker, K., eds) vol. 2, pp. 105—108, Kupffer Cell Foundation, Rijswijk.
23. Sparrow, C. P., Parthasarathy, S. & Steinberg, D. (1989) *J. Biol. Chem. 264*, 2599—2604.
24. Nagelkerke, J. F., Barto, K. P. & van Berkel, T. J. C. (1983) *J. Biol. Chem. 258*, 12221—12227.
25. Loegering, D. J. (1986) *Circ. Shock 20*, 321—333.
26. Kreusch, J. (1984) *Thesis*, Faculty of Medicine, The University of Freiburg, FRG.
27. Pestka, S. & Langer, J. A. (1987) *Annu. Rev. Biochem. 56*, 727—777.
28. Peters, T. & Decker, K. (1989) in *Cells of the hepatic sinusoid* (Wisse, E., Knook, D. L. & Decker, K., eds) vol. 2, pp. 182—185, Kupffer Cell Foundation, Rijswijk.
29. Bauer, J. (1989) *Klin. Wochenschr. 67*, 697—706.

30. Bhatnagar, R., Schade, U., Rietschel, T. E. & Decker, K. (1982) *Eur. J. Biochem. 125*, 125 – 130.
31. Häussinger, D. (1989) *J. Hepatol. 8*, 259 – 266.
32. Kuroki, M. & Minakami, S. (1989) *Biochem. Biophys. Res. Commun. 162*, 377 – 380.
33. De Leeuw, A. M., Praaning-van Dalen, D. P., Brouwer, A. & Knook, D. L. (1989) in *Cells of the hepatic sinusoid* (Wisse, E., Knook, D. L. & Decker, K., eds) vol. 2, pp. 94 – 98, Kupffer Cell Foundation, Rijswijk.
34. Russell, S. W. & Pace, J. L. (1987) *Vet. Immunol. Imunopath. 15*, 129 – 165.
35. Oldham, R. K. (1983) *J. Natl Cancer Inst. 70*, 789 – 796.
36. Adams, D. O. & Hamilton, T. A. (1984) *Annu. Rev. Immunol. 2*, 283 – 318.
37. Oda, S., Sato, M., Toyoshima, S. & Osawa, T. (1989) *J. Biochem. (Tokyo) 105*, 1040 – 1043.
38. Scanlon, M., Laster, S. M., Wood, J. G. & Gooding, L. R. (1989) *Proc. Natl Acad. Sci. USA 86*, 182 – 186.
39. Uhing, R. J. & Adams, D. O. (1989) *Agents Actions 26*, 1 – 2.
40. Decker, T., Lohmann-Matthes, M.-L., Karck, U., Peters, T. & Decker, K. (1989) *J. Leukocyte Biol. 45*, 139 – 147.
41. Daemen, T., Veninga, A., Roerdink, F. H. & Scherphof, G. L. (1989) *Biochim. Biophys. Acta 991*, 145 – 151.
42. Kirn, A., Koehren, F. & Steffan, M. A. (1982) *Hepatology 2*, 670 – 679.
43. Werner-Wasik, M., von Muenchhausen, W., Nolan, J. P. & Cohen, S. A. (1989) *Cancer Immunol. Immunother. 28*, 107 – 115.
44. Hori, K., Mihich, E. & Ehrke, M. J. (1989) *Cancer Res. 49*, 2606 – 2614.
45. Karck, U., Peters, T. & Decker, K. (1988) *J. Hepatol. 7*, 352 – 361.
46. Foon, K. A. (1989) *Cancer Res. 49*, 1621 – 1639.
47. Celada, A., Gray, P. W., Rinderknecht, E. & Schreiber, R. D. (1984) *J. Exp. Med. 160*, 55 – 74.
48. Kishimoto, T. (1987) *J. Clin. Immunol. 7*, 343 – 355.
49. Freudenberg, M. A., Freudenberg, N. & Galanos, C. (1982) *Br. J. Exp. Pathol. 63*, 56 – 65.
50. Freudenberg, M. A. & Galanos, C. (1985) *Eur. J. Biochem. 152*, 353 – 359.
51. Fox, E. S., Thomas, P. & Broitman, S. A. (1989) *Gastroenterology 96*, 456 – 461.
52. Van Bossuyt, H., Zanger, R. B. & Wisse, E. (1988) *J. Hepatol. 7*, 325 – 337.
53. Van Bossuyt, H., Zanger, R. B. & Wisse, E. (1988) *J. Hepatol. 7*, 325 – 337.
54. Aderem, A. A. (1988) *J. Cell Sci. Suppl. 9*, 151 – 167.
55. Commins, L. M., Loegering, D. J. & Minnear, F. L. (1989) *Circ. Shock 27*, 237 – 244.
56. Rieder, H., Ramadori, G. & Meyer zum Büschenfelde, K.-H. (1988) *J. Hepatol. 7*, 338 – 344.
57. Rellstab, P. & Schaffner, A. (1989) *J. Immunol. 142*, 2913 – 2820.
58. Birmelin, M., Karck, U., Dieter, P., Freudenberg, N. & Decker, K. (1986) in *Cells of the hepatic sinusoid* (Kirn, A., Knook, D. L. & Wisse, E., eds) vol. 1, pp. 295 – 300, Kupffer Cell Foundation, Rijswijk.
59. Introna, M., Hamilton, T. A., Kaufman, R. E., Adams, D. O. & Bast, R. C. (1986) *J. Immunol. 137*, 2711 – 2715.
60. Tran-Thi, T.-A., Häussinger, D., Gyufko, K. & Decker, K. (1988) *Biol. Chem. Hoppe-Seyler 369*, 65 – 68.
61. Gordon, J. L. (1986) *Biochem. J. 233*, 309 – 319.
62. Seifert, R., Burde, R. & Schultz, G. (1989) *Biochem. J. 259*, 813 – 819.
63. Nishizuka, Y. (1986) *Science 233*, 305 – 312.
64. Gilman, A. G. (1987) *Annu. Rev. Biochem. 56*, 615 – 649.
65. Birmelin, M. & Decker, K. (1984) *Eur. J. Biochem. 142*, 219 – 225.
66. Nitta, T., Saito-Taki, T. & Suzuki, T. (1984) *J. Leukocyte Biol. 36*, 493 – 504.
67. Wakelam, M. J. O., Davies, S. A., Houslay, M. D., McKay, I., Marshall, C. J. & Hall, A. (1986) *Nature 323*, 173 – 176.

68. Lassing, I. & Lindberg, U. (1985) *Nature 314*, 472 – 474.
69. Berridge, M. J. (1987) *Annu. Rev. Biochem. 56*, 159 – 193.
70. Dieter, P., Schulze-Specking, A. & Decker, K. (1988) *Eur. J. Biochem. 177*, 61 – 67.
71. Carafoli, E. (1987) *Annu. Rev. Biochem. 56*, 395 – 433.
72. Dieter, P., Schulze-Specking, A., Karck, U. & Decker, K. (1987) *Eur. J. Biochem. 170*, 201 – 206.
73. Birmelin, M. & Decker, K. (1983) *Eur. J. Biochem. 131*, 539 – 543.
74. Bhatnagar, R., Schirmer, R., Ernst, M. & Decker, K. (1981) *Eur. J. Biochem. 119*, 171 – 175.
75. Davis, P., Fox, R. I., Polyzomis, M., Allison, A. C., Phil, D. & Haswell, A. D. (1973) *Lab. Invest. 28*, 16 – 22.
76. Dieter, P., Schulze-Specking, A. & Decker, K. (1986) *Eur. J. Biochem. 159*, 451 – 457.
77. Palmer, R. M., Ferrige, A. G. & Moncada, S. (1987) *Nature 327*, 524 – 526.
78. Marletta, M. A., Yoon, P. S., Iyengar, R., Leaf, C. D. & Wishnock, J. S. (1988) *Biochemistry 27*, 8706 – 8711.
79. Billiar, T. R., Curran, R. D., Stuehr, D. J., West, M. A., Bentz, B. G. & Simmons, R. L. (1989) *J. Exp. Med. 169*, 1467 – 1472.
80. Needleman, P., Turk, J., Jakschik, B. A., Morrison, A. R. & Lefkowith, J. B. (1986) *Annu. Rev. Biochem. 55*, 69 – 102.
81. Croft, K. D., Beilin, L. J., Vandongen, R., & Mathews, E. (1984) *Biochim. Biophys. Acta 795*, 196 – 207.
82. Irvine, R. F. (1982) *Biochem. J. 204*, 3 – 16.
83. Stremmel, W., Strohmeyer, G., Borchard, F., Kochwa, S. & Berk, P. D. (1985) *Proc. Natl Acad. Sci. USA 82*, 4 – 8.
84. Dieter, P., Krause, H. & Schulze-Specking, A. (1990) *Eicosanoids 3*, 45 – 51.
85. Birmelin, M., Marmé, D., Ferber, E. & Decker, K. (1984) *Eur. J. Biochem. 140*, 55 – 61.
86. Kaever, V., Pfannkuche, H.-J., Wessel, K. & Resch, K. (1989) *Agents Actions 26*, 175 – 177.
87. Wijkander, J. & Sundler, R. (1989) *Biochim. Biophys. Acta 1010*, 78 – 87.
88. Krause, H., Dieter, P., Schulze-Specking, A. & Decker, K. (1989) *J. Hepatol. 9*, Suppl. 1, S 178.
89. Decker, K. (1985) *Semin. Liver Dis. 5*, 175 – 190.
90. Eyhorn, S., Schlayer, H.-J., Henninger, H. P., Dieter, P., Hermann, R., Woort-Menker, M., Becker, H., Schaefer, H. E. & Decker, K. (1988) *J. Hepatol. 6*, 23 – 35.
91. Tran-Thi, T.-A., Gyufko, K., Henninger, H., Busse, R. & Decker, K. (1987) *J. Hepatol. 5*, 322 – 331.
92. Decker, K. (1987) in *Modulation of liver cell expression* (Reutter, W., Popper, H., Arias, I. M., Heinrich, P. C., Keppler, D. & Landmann, L., eds) pp. 397 – 409, MTP Press, Lancaster.
93. Hamberg, M., Svensson, J. & Samuelsson, B. (1975) *Proc. Natl Acad. Sci. USA 72*, 2994 – 2998.
94. Smith, D. L., Stone, K. J. & Willis, A. L. (1987) in *Handbook of eeicosanoids: prostaglandins and related compounds* (Willis, A.L., ed.) vol. 1, pp. 245 – 301, CRC Press, Boca Raton.
95. Ohki, S., Ogino, N., Yamamoto, S. & Hayaishi, O. (1979) *J. Biol. Chem. 254*, 829 – 836.
96. Flower, R. J. & Vane, J. R. (1973) in *Prostaglandin inhibitors* (Robinson, H. J. & Vane, J. R., eds) pp. 9 – 18, Raven Press, New York.
97. Grewe, M., Schulze-Specking, A. & Decker, K. (1989) in *Cells of the hepatic sinusoid* (Wisse, E., Knook, D. L. & Decker, K., eds) vol. 2, pp. 206 – 207, Kupffer Cell Foundation, Rijswijk.
98. Urade, Y., Ujihara, M., Horiguchi, Y., Ikai, K. & Hayaishi, O. (1989) *J. Immunol. 143*, 2982 – 2989.
99. Dieter, P., Peters, T., Schulze-Specking, A. & Decker, K. (1989) *Biochem. Pharmacol. 38*, 1577 – 1581.
100. Busam, K., Hoss, A., Zawatzki, R., Bauer, J., Gerok, W. & Decker, K. (1989) *Eur. J. Biochem. 191*, 577 – 582.
100a. Peters, R., Karck, U. & Decker K. (1990) *Eur. J. Biochem. 191*, 583 – 589
101. Kunkel, S. L., Spengler, M., May, M. A., Spengler, R., Larrick, J. & Remick, D. (1988) *J. Biol. Chem. 263*, 5380 – 5384.
102. Ferreira, S. H. & Vane, J. R. (1967) *Nature 216*, 868 – 873.

260

103. Okumura, T., Nakayama, R., Sago, T. & Saito, K. (1985) *Biochim. Biophys. Acta 837*, 197−207.
104. Garrity, M. J. & Robertson, R. P. (1983) *Adv. Prostaglandin Thromboxane Leukotriene Res. 12*, 279−282.
105. Kuiper, J., Zijlstra, F. J., Kamps, J. A. A. M. & van Berkel, T. J. C. (1989) *Biochem. J. 262*, 195−201.
106. Garrity, M. J., Reed, M. E. & Brass, E. (1989) *J. Pharmacol. Exp. Ther. 248*, 979-983.
107. Häussinger, D., Stehle, T., Tran-Thi, T.-A., Decker, K. & Gerok, W. (1987) *Biol. Chem. Hoppe-Seyler 368*, 1509−1513.
108. Athari, A. & Jungermann, K. (1989) *Biochem. Biophys. Res. Commun. 163*, 1235−1242.
109. Tran-Thi, T.-A., Gyufko, K., Reinke, M. & Decker, K. (1988) *Biol. Chem. Hoppe-Seyler 369*, 1179−1184.
110. Tran-Thi, T.-A., Gyufko, K., Häussinger, D. & Decker, K. (1988) *J. Hepatol. 6*, 151−157.
111. Gardemann, A., Strulik, H. & Jungermann, K. (1987) *Am. J. Physiol. 253*, E 238−245.
112. Häussinger, D., Stehle, T. & Gerok, W. (1988) *Biol. Chem. Hoppe-Seyler 369*, 97−107.
113. Kuiper, J., Casteleyn, E. & van Berkel, T. J. C. (1989) *Agents Actions 26*, 201−202.
114. Gómez-Foix, A. M., Rodriguez-Gil, J. E., Guinovart, J. J. & Bosch, F. (1989) *Biochem. J. 261*, 93−97.
115. Brass, E. P. & Garrity, M. J. (1984) *FEBS Lett. 169*, 293−296.
116. Samuelsson, B., Dahlén, S.-E., Lindgren, J.-A., Rouzer, C. A. & Serhan, C. N. (1987) *Science 237*, 1171−1176.
117. Rouzer, C. A. & Samuelsson, B. (1985) *Proc. Natl Acad. Sci. USA 82*, 6040−6044.
118. Fitzsimmons, B. J., Adams, J., Evans, J. F., Leblanc, Y. & Rokach, J. (1985) *J. Biol. Chem. 260*, 13008−13012.
119. Kim, S. J. (1988) *Biochem. Biophys. Res. Commun. 150*, 870−876.
120. Radmark, O., Shimizu, T., Jörnvall, M. & Samuelsson, B. (1984) *J. Biol. Chem. 259*, 12339−12345.
121. Yoshimoto, T., Soberman, R. J., Lewis, R. A. & Austen, K. F. (1985) *Proc. Natl Acad. Sci. USA 82*, 8399−8403.
122. Lüderitz, T., Schade, U. & Rietschel, E. Th. (1986) *Eur. J. Biochem. 155*, 377−382.
123. Sakagami, Y., Mizoguchi, Y., Seki, S., Kobayashi, K., Morisawa, S. & Yamamoto, S. (1989) *Prostaglandins leukotrienes Essent. fatty acids 36*, 125−128.
124. Sakagami, Y., Mizoguchi, Y., Seki, S., Kobayashi, K., Morisawa, S. & Yamamoto, S. (1988) *Biochem. Biophys. Res. Commun. 156*, 217−221.
125. Tate, S. S. & Meister, A. (1981) *Monogr. Cell Biochem. 39*, 357−368.
126. Uehara, N., Ormstad, K., Orrenius, S., Örning, L. & Hammarström, S. (1983) *Adv. Prostaglandin Thromboxane Leukotriene Res. 11*, 147−150.
127. Hagmann, W., Denzlinger, C., Rapp, S., Weckbecker, G. & Keppler, D. (1986) *Prostaglandins 31*, 239−251.
128. Hagmann, W., Parthé, S. & Kaiser, I. (1989) *Biochem. J. 261*, 611−616.
129. Orning, L. (1987) *Eur. J. Biochem. 170*, 77−85.
130. Rapoport, S., Härtel, B. & Hausdorf, G. (1984) *Eur. J. Biochem. 139*, 573−576.
131. Murphy, R. C., Hammarström, S. & Samuelsson, B. (1979) *Proc. Natl Acad. Sci. USA 76*, 4275−4279.
132. Leffer, A. M. (1986) *Biochem. Pharmacol. 33*, 123−127.
133. Keppler, D., Hagmann, W. & Rapp, S. (1987) *Rev. Infect. Dis. 9*, S 580−584.
134. Gillard, J., Ford-Hutchinson, A. W., Chan, C., Charleson, S., Denis, D., Foster, A., Fortin, R., Leger, S., McFarlane, C. S., Morton, H., Piechuta, H., Riendeau, D., Rouzer, C., Rokach, J., Young, R., MacIntyre, D. E., Peterson, L., Bach, T., Eiermann, G., Hopple, S., Humes, J., Hupe, L., Luell, S., Metzger, J., Meurer, R., Miller, D. K., Opas, E. & Pacholok, S. (1989) *Can. J. Physiol. Pharmacol. 67*, 456−464.
135. Benveniste, J., Henson, P. M. & Cochrane, C. G. (1972) *J. Exp. Med. 136*, 1356−1377.
136. Hanahan, D. J. (1986) *Annu. Rev. Biochem. 55*, 483−509.
137. Ludwig, J. C., McManus, L. M., Clark, P. O., Hanahan, D. J. & Pinckard, R. N. (1984) *Arch. Biochem. Biophys. 232*, 102−110.
138. Chao, W., Siafaka-Kapadai, A., Olson, M. S. & Hanahan, D. J. (1989) *Biochem. J. 257*, 823−829.
139. Fisher, R. A., Sharma, R. V. & Bhalla, R. C. (1989) *FEBS Lett. 251*, 22−26.
140. Lapointe, D. S. & Olson, M. S. (1989) *J. Biol. Chem. 264*, 12130−12133.
141. Buxton, D. B., Fisher, R. A., Hanahan, D. J. & Olson, M. S. (1986) *J. Biol. Chem. 261*, 644−649.
142. Takemura, R. & Werb, Z. (1984) *Am. J. Physiol. 246*, C1−C9.
143. Dieter, P., Schulze-Specking, A., Karck, U. & Decker, K. (1988) *J. Hepatol. 6*, 167−174.
144. Rieder, H., Birmelin, M. & Decker, K. (1982) in *Sinusoidal liver cells* (Knook, D. L. & Wisse, E., eds) pp. 193−200, Elsevier Biomedical Press, Amsterdam, New York, Oxford.
145. Dawson, P. A., Lukaszewski, L. M., Ells, P. F., Malbon, C. C. & Williams, D. L. (1989) *J. Lipid Res. 30*, 403−413.
146. Dinarello, C. A. (1984) *N. Engl. J. Med. 311*, 1413−1418.
147. Shirahama, M., Ishibashi, H., Tsuchiya, Y., Kurokawa, S., Hayashida, K., Okumura, Y. & Niho, Y. (1988) *Scand. J. Immunol. 28*, 719−725.
148. Mizel, S. B. & Mizel, D. (1981) *J. Immunol. 126*, 834−837.
149. Dalboge, H., Bayne, S., Christensen, T. & Hejnaes, K. R. (1989) *FEBS Lett. 246*, 89−93.
150. Spinas, G. A., Hansen, B. S., Linde, S., Kastern, W., Molvig, J., Mandrup-Poulsen, T., Dinarello, C. A., Nielsen, J. H. & Nerup, J. (1987) *Diabetologia 30*, 474−480.
151. Kishimoto, T. (1989) *Blood 74*, 1−10.
152. Bauer, J., Ganter, U., Geiger, T., Jacobshagen, U., Hirano, T., Matsuda, T., Kishimoto, T., Andus, T., Acs, G., Gerok, W. & Ciliberto, G. (1988) *Blood 72*, 1134−1140.
153. Busam, K., Bauer, T., Bauer, J., Gerok, W. & Decker, K. (1989) *J. Hepatol.*, in the press.
154. Aarden, L., Lansdorp, P. & De Groot, E. (1985) *Lymphokines 10*, 175−185.
155. Bauer, J., Bauer, T. M., Kalb, T., Taga, T., Lengyel, G., Hirano, T., Kishimoto, T., Acs, G., Mayer, L. & Gerok, W. (1989) *J. Exp. Med. 170*, 1537−1549.
156. Andus, T., Geiger, T., Hirano, T., Kishimoto, T., Tran-Thi, T.-A., Decker, K. & Heinrich, P. C. (1988) *Eur. J. Biochem. 173*, 287−293.
157. Havell, E. A. & Spitalny, G. L. (1983) *J. Reticuloendothel. Soc. 33*, 369−376.
158. Heron, I., Hokland, M. & Berg, K. (1978) *Proc. Natl Acad. Sci. USA 75*, 6215−6219.
159. Imai, K., Ng, A. K., Glassy, M.C. & Ferrone, S. (1981) *J. Immunol. 127*, 505−509.
160. Yoshie, O., Aso, H., Sakakibara, A. & Ishida, N. (1985) *J. Interferon Res. 5*, 531−540.
161. Fridman, W. H., Gresser, I., Bandu, M. T., Aguet, M. & Neauport-Sautes, C. (1980) *J. Immunol. 124*, 2436−2441.
162. Alan, R., Ezekowitz, B., Hill, M. & Gordon, S. (1986) *Biochem. Biophys. Res. Commun. 136*, 737−744.
163. Branca, A. A. & Baglioni, C. (1982) *J. Biol. Chem. 257*, 13197−13200.
164. Sarkar, F. H. & Gupta, S. L. (1984) *Eur. J. Biochem. 140*, 461−470.
165. Beutler, B. & Cerami, A. (1988) *Endocrine Rev. 9*, 57−66.
166. Beutler, B., Greenwald, D., Hulmes, J. D., Chang, M., Pan, Y.-C. E., Mathison, J., Ulevich, R. & Cerami, A. (1985) *Nature 316*, 552−554.
167. Carswell, E. A., Old, L. J., Kassel, R. L., Green, S., Fiore, N. & Williamson, B. (1975) *Proc. Natl Acad. Sci. USA 72*, 3666−3670.
168. Decker, T., Lohmann-Matthes, M.-L. & Gifford, G. E. (1987) *J. Immunol. 138*, 957−962.
169. Smith, R. A., Kirstein, M. & Baglioni, C. (1986) *J. Biol. Chem. 261*, 14871−14874.
170. Bachwich, P. R., Chensue, S. W., Larrick, J. W. & Kunkel, S. L. (1986) *Biochem. Biophys. Res. Commun. 136*, 94−101.

171. Caput, D., Beutler, B., Hartog, K., Brown-Shimer, S. & Cerami, A. (1986) *Proc. Natl Acad. Sci. USA 83*, 1670 – 1674.
172. Decker, K. (1989) in *Cells of the hepatic sinusoid* (Wisse, E., Knook, D. L. & Decker, K., eds) vol. 2, pp. 171 – 175, Kupffer Cell Foundation, Rijswijk.
173. Lehmann, V., Freudenberg, M. A., & Galanos, C. (1987) *J. Exp. Med. 165*, 657 – 663.
174. Dinarello, C. A., Cannon, J. G., Wolff, S. M., Bernheim, H. A., Beutler, B., Cerami, A., Palladino, M. A. & O'Connor, J. V. (1986) *J. Exp. Med. 163*, 1433 – 1450.
175. Pober, J. S., Gimbrone, Jr., M. A., Lapierre, L. A., Mendrick, D. L., Fiers, W., Rothlein, R. & Springer, T. A. (1986) J. . *Immunol. 137*, 1893 – 1896.
176. Schlayer, H. J., Laaf, H., Peters, T., Woort-Menker, M., Estler, C., Karck, U., Schaefer, H. E. & Decker, K. (1988) *J. Hepatol. 7*, 239 – 249.
177. Stern, D. M. & Nawroth, P. P. (1986) *J. Exp. Med. 163*, 740 – 745.
178. Nawroth, P. P., Bank, I., Handley, D., Cassimeris, J., Chess, L. & Stern, D. (1986) *J. Exp. Med. 163*, 1363 – 1375.
179. Tanaka, M. & Ishikawa, E. (1986) in *Cells of the hepatic sinusoid* (Kirn, A., Knook, D. E. & Wisse, E., eds) vol. 1, pp. 259 – 262, Kupffer Cell Foundation, Rijswijk.
180. Nedwin, G. E., Naylor, S. L., Sakaguchi, A. Y., Smith, D., Jarrett-Nedwin, J., Pennica, D., Goeddel, D. V. & Gray, P. W. (1985) *Nucleic Acids Res. 13*, 6361 – 6373.
181. Männel, D. N., Moore, R. N. & Mergenhagen, S. E. (1980) *Infect. Immun. 30*, 523 – 530.
182. Ruddle, N. H., Powell, M. B. & Conta, B. S. (1983) *Lymphokine Res. 2*, 23 – 29.

Eur. J. Biochem. *192*, 563—576 (1990)
© FEBS 1990

Review

Techniques in plant molecular biology — progress and problems

Richard WALDEN and Jeff SCHELL

Max-Planck-Institut für Züchtungsforschung, Köln, Federal Republic of Germany

(Received March 7, 1990) — EJB 90 0258

Progress in plant molecular biology has been dependent on efficient methods of introducing foreign DNA into plant cells. Gene transfer into plant cells can be achieved by either direct uptake of DNA or the natural process of gene transfer carried out by the soil bacterium *Agrobacterium*. Versatile gene-transfer vectors have been developed for use with *Agrobacterium* and more recently vectors based on the genomes of plant viruses have become available. Using this technology the expression of foreign DNA, the functional analysis of plant DNA sequences, the investigation of the mechanism of viral DNA replication and cell to cell spread, as well as the study of transposition, can be carried out. In addition, the versatility of the gene-transfer vectors is such that they may be used to isolate genes not amenable to isolation using conventional protocols. This review concentrates on these aspects of plant molecular biology and discusses the limitations of the experimental systems that are currently available.

Many of the advances in plant molecular biology have been promoted by the development of experimental systems for the insertion of foreign DNA into plant cells. While developments in plant transformation have often been hailed as being of potential benefit in agriculture, experimental systems utilising transgenic plants also allows the functional analysis of plant DNA. The ease of plant transformation, allied with a short life cycle, extensive genetics and simplicity in producing large numbers of transgenic individuals, may prove that in the future plants will provide an important contribution in attempts to understand the molecular basis of the control of gene expression. The aim of this review is to describe critically some of the recent progress in plant molecular biology with particular emphasis on gene transfer and discuss some of the limitations of the techniques that are currently available.

Transfer of DNA to plant cells

Transformation is now a routine technique for many plant species including some important crop plants [1]. One of two

---

Correspondence to R. Walden, Max-Planck-Institut für Züchtungsforschung, Carl von Linné Weg 10, D-5000 Köln 30, Federal Republic of Germany

Abbreviations. DMGT, DNA-mediated gene transfer; NPT(II), neomycin phosphotransferase; CAT, chloramphenicol acetyl transferase; WDV, wheat dwarf virus; BMV, brome mosaic virus; TMV, tobacco mosaic virus; CLV, cassava latent virus; CaMV, cauliflower mosaic virus; MSV, maize streak virus; BCTV, beet curly top virus; TGMV, tomato golden mosaic virus; CHS, chalcone synthase.

Note. While this paper was in the press, several significant relevant papers have appeared and include further detailed analysis of the 35S RNA promoter [Benfey, P. N., Ren, L. & Chua, N.-H. (1990) *EMBO J. 9*, 1677—1684; Benfey, P. N., Ren, L. & Chua, N.-H. (1990) *EMBO J. 9*, 1685—1696], factors that bind to the promoters of plant genes [Gilmatin, P. M., Sarokin, L., Memelink, J. & Chua, N.-H. (1990) *The Plant Cell 2*, 369—378] and the tagging of the arabidopsis agamous gene by T-DNA [Yanofsky, M. F., Ma, H., Bowman, J. L., Drews, G., Feldmann, K. A. & Meyerowitz, E. M. (1990) *Nature 346*, 35—39].

general types of transformation technique may be adopted, one utilizing the soil bacterium *Agrobacterium tumefaciens*, the other involving the application of naked DNA (DNA-mediated gene transfer, DMGT; for a review of the techniques of transformation see [2, 3] (Fig. 1). Frequent multimerisation and rearrangement of foreign DNA sequences prior to integration observed following DGMT make techniques involving *Agrobacterium* the most favoured for producing stable transgenic tissue. On the other hand DMGT has been used extensively in studying transient gene expression.

Genetic markers for use in plant cells

Regardless of transformation strategy, genetic markers set the criteria by which a plant cell is judged to be transformed by providing novel nucleic acid sequences that can be detected by Southern analysis and unique assayable enzymatic activities. An increasing number of genetic markers are becoming available and include dominant selectable markers for the direct selection of transgenic tissue, screenable markers which allow the detailed analysis of gene expression and markers which allow negative selection (see Table 1). Generally, the genetic markers that have been developed for use in plant cells were derived from either bacterial or plant sources although their expression is directed by plant-specific promoters.

Markers used in plants vary in their characteristics and do not function uniformly in all species. Hence markers required for dominant selection may differ between not only different plant species but also different cell types within an individual. One of the most extensively used selectable markers, neomycin phosphotransferase II (NPTII), conferring kanamycin resistance [4] can be used as a dominant marker with tobacco protoplasts, callus, tissue explants and whole plants [21]. In contrast, whereas chloramphenicol acetyltransferase (CAT) can be used to confer resistance to chloramphenicol in germinating seedlings it is not effective at other developmental

564

Table 1. *Genetic markers for use in plant cells*
EPSP, *enol*pyruvylshikimate-3-phosphate

Enzyme activity	Dominant selection	Assay	Reference
Neomycin phosphotransferase	yes	yes	4
Hygromycin phosphotransferase	yes	yes	5
Dihydrofolate reductase	yes	yes	6
Chloramphenicol acetyltransferase	yes	yes	6
Gentamycin acetyltransferase	yes	yes	7
Nopaline synthase	no	yes	8
Octopine synthase	no	yes	8
β-Galactosidase	no	yes	9
β-Glucuronidase	no	yes	10
Streptomycin phosphotransferase	yes	yes	11
Bleomycin resistance	yes	no	12
Firefly luciferase	no	yes	13
Bacterial luciferase	no	yes	14
Threonine dehydratase	yes	yes	15
Metallothionein II	yes	yes	16
EPSP synthase	yes	no	17
Phosphinothricin acetyltransferase	yes	yes	18
Acetolactate synthase	yes	no	19
Bromoxynil nitrilase	yes	no	20

stages. Hence in attempting to use a dominant marker it is important to devise a selection protocol for the plant species and the developmental stage at which selection is to be carried out.

Routinely assayable markers, such as CAT, have been used effectively for measuring gene activity in whole plant extracts or in transient assays in protoplasts in a manner analogous to animal systems (for an example see [22]). However, significant advances have been made recently in the development of marker genes which allow histochemical detection of enzymatic activity in plant tissue. These include the luciferase genes from either firefly or bacteria [23, 24], β-glucuronidase [10] and most recently β-galactosidase [25]. These have allowed the detailed analysis of the cell-specific expression directed by individual plant promoters. However, it remains difficult to quantify the levels of expression of such markers due to differences in cell size, metabolic activity and penetration of substrate into the cell (for discussion see [26]).

Such markers also begin to allow us to construct fate maps of cells. In this case an individual cell might have a chimeric gene introduced into it and the clonal tissue arising from it can be assayed for activity [27]. Currently the only limitation to this approach is that where β-glucuronidase is used, the cell is killed during the process of visualizing the enzymatic activity. This problem might be overcome by using luciferase which can be visualized in living tissue. However, in this case difficulties may be encountered in impregnating the tissue with the enzyme substrates and visualising the enzyme activity.

Negative selection markers have been developed which are lethal or interfere with normal development. These markers could be used to isolate regulatory mutants or study gene inactivation. An example of this type of marker has been developed taking advantage of the genes involved in auxin

biosynthesis contained on the T-DNA of the Ti plasmid of *Agrobacterium* [28, 29]. The product of gene 1 converts tryptophan to indole-3-acetamide and this or α-naphthalene acetamide can be converted to indole-3-acetic acid or α-naphthalene acetic acid by an aminohydrolase encoded by gene 2. High levels of α-naphthalene acetic acid are toxic to plant cells, hence plant cells with an active gene 2 will be unable to grow on a media containing high concentrations of α-naphthalene acetamide. This type of selection can be applied at the protoplast or plant level.

Transformation using Agrobacterium

Strategies for the production of transgenic plants using the soil bacteria *Agrobacterium tumefaciens* and *Agrobacterium rhizogenes*, rely on their natural ability to form tumours and hairy root disease, respectively, on dicotyledonous plants. The process of tumour formation has been studied in most detail and it is generally considered that the establishment of the hairy root phenotype is analogous.

The presence of wounded plant cells triggers a cascade of molecular events mediated by the virulence (*vir*) region of the tumour inducing, or Ti plasmid, of *Agrobacterium* [30]. The *vir* region encodes proteins which sense and respond to the presence of phenolic compounds released by wounded plant cells [31]. Moreover, *vir* proteins are responsible for the appearance of a single-stranded DNA, the T-strand, and the transfer of this to the plant cell [32]. The T-strand is a copy of the T-DNA of the Ti plasmid which is delimited by a 25-bp direct repeat sequence. The T-strand is transferred from the bacterium to the plant cell by a process thought to be analogous to conjugational transfer of bacterial plasmids [33, 34]. The genes that are encoded by the T-DNA are functional in plant cells and their products disrupt the normal hormonal balance of the cell resulting in the tumourous phenotype [35].

The mechanism by which the T-DNA integrates into the plant genome is not known and is currently one of the most intriguing questions in plant molecular biology. Usually single or low copy inserts occur although with one type of *Agrobacterium* strain, C58, tandem copies of inserted DNA can be preferentially obtained [36]. Multiple insertions of the T-DNA can take place and may result from the *Agrobacterium* transferring more than one copy of the T-DNA to the plant cell or replication of the T-DNA prior to its integration [37]. Until recently, it was assumed that the T-DNA inserted at random sites within the genome, however, data obtained using T-DNA to tag promoter sequences (see later) indicates that the T-DNA preferentially inserts into potentially transcribed DNA [38]. Although chloroplast transformation with T-DNA has been reported [39], organelle transformation has not been routinely observed.

The transformation vectors used with *Agrobacterium* are divided into two types, cointegrative or binary vectors. Both rely on the observations that during tumour formation, the T-DNA genes play no part in transfer, the *vir* region acts in *trans* and DNA that is located between the 25-bp T-DNA border repeats is transferred from the Ti plasmid to the plant genome (for reviews see [2, 3, 40–42].

Cointegrative vectors are based on the Ti plasmids which have had the majority of the T-DNA between the border repeats replaced by a specific sequence of DNA [43, 44]. DNA which is to be inserted into the plant genome is cloned into an intermediate cloning vector, based on a bacterial plasmid, which contains a sequence homologous to that present between the border repeats of the cointegrative vector. The inter-

Agrobacterium
Mediated Transformation

DNA Mediated
Gene Transfer

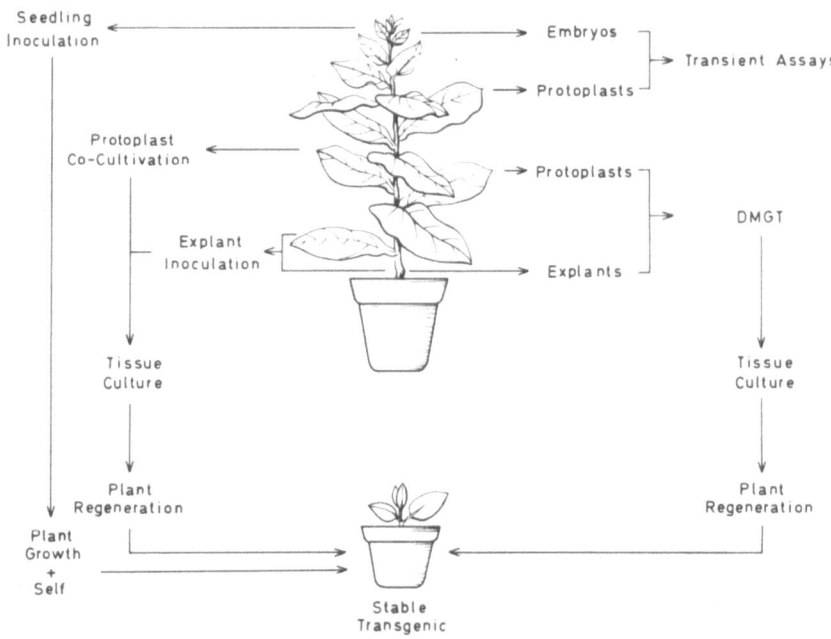

Fig. 1. *Production of transgenic plants.* Schema for producing transgenic plants by either *Agrobacterium*-mediated transformation or by DNA-mediated transformation

mediate vector is introduced into the *Agrobacterium* harboring the cointegrative vector by conjugation [45, 46]. As the intermediate vector lacks an origin of replication functional in *Agrobacterium* it is lost. However, homologous recombination between the intermediate and cointegrative vectors results in the foreign DNA being integrated between the border repeats of the T-DNA and can be selected for by the appropriate genetic markers. Cointegrative vectors have the advantage that they are generally stable in *Agrobacterium*.

Binary vectors contain origins of replication and selectable markers functional in *Escherichia coli* and *A. tumefaciens* and hence they can be maintained in both types of bacteria. These vectors can be transferred between bacteria by either conjugation or introduced directly by either transformation [47] or electroporate [48]. Moreover, binary vectors contain the 25-bp border repeats between which are located appropriate cloning sites for the insertion of foreign DNA and selectable markers functional in plant cells [49, 50]. The *Agrobacterium* host for plant transformation with a binary vector harbors a Ti plasmid from which the T-DNA has been deleted. While binary vectors have proven to be versatile they tend to be unstable in *Agrobacterium* and require selection to confirm their maintenance.

The production of transgenic plants using *Agrobacterium* relies on providing the bacteria with wounded cells which can be transformed and subsequently cultured to divide and eventually to regenerate into plants. With *A. tumefaciens* explant inoculation involves the incubation of sectioned tissue with the bacteria followed by culturing on media which allows the induction of the growth of callus and the subsequent formation of shoots. The shoots are subcultured to form roots. Transformant selection can be applied at the initiation of callus growth as well as at the stage of transfer of the shoots to rooting media. This approach has proved successful with

leaf [51], stem [52], roots [53], tubers [54] and epidermal strips derived from flowering branches [55]. Plant regeneration can be rapid with plants being regenerated in as short a period as 6–8 weeks see Fig. 1. When using *A. rhizogenes*, explants inoculated with the bacteria form hairy roots at the site of inoculation and these can be induced to form shoots [56–59].

Large populations of isolated cells can be transformed by protoplast co-cultivation [37, 60]. In this technique isolated protoplasts which have been allowed to partially regenerate their cell walls are incubated with *Agrobacteria*. Following the co-cultivation period the protoplasts are cultured further to form callus in the presence of antibiotics to prevent the growth of remaining bacteria and allow selection of transgenic tissue. Shoot and root growth is initiated by subculturing the callus on appropriate media.

A third transformation strategy is arabidopsis seedling inoculation [61]. Here imbibed seeds are inoculated with *Agrobacterium* and the resulting plants are selfed and progeny are germinated in the presence of selection to obtain transgenic tissue. Although the mechanism by which transformation takes place remains unknown, a selectable marker can be inherited in a Mendelian fashion following selfing.

Explant inoculation has proven to be the simplest method to produce plants containing a specific insert of DNA. However, it ceases to be a feasible approach when the target of the experiment is to generate a large number of individual transformants, for example in insertional mutagenesis. In this case protoplast co-cultivation is more appropriate although resulting plant populations may be subject to somaclonal variation. Moreover, co-cultivation is currently limited to the solanaceous species where high frequencies of cell division and subsequent growth can be obtained. Seedling inoculation overcomes the potential difficulties of somaclonal variation and allows the generation of a large number of transgenic individuals however it is thusfar limited to arabidopsis.

Although it is now becoming increasingly clear that *Agrobacterium* can transfer DNA to the cells of monocotyledonous plants [62—65] the limitation in producing transgenic individuals from these species remains with regenerating plants from isolated cells or explants. This is especially true with the graminaceous monocotyledonous species as well as the Leguminoseae. However, with the increasing number of reports of plant regeneration from suspension cultures of cereal cells [66] it is likely to be only a matter of time until the production of a wide range of transgenic cereals and other species is achieved.

Transformation by naked DNA uptake

Many strategies involving DMGT in plant transformation have been shown to be successful and include the fusion of protoplasts with bacterial spheroplasts [67], or liposomes containing foreign DNA [68], microinjection [69], macroinjection [70], treatment of protoplasts with DNA in the presence of polyvalent cations [71], electroporation [72] and microbombardment [73]. However, the latter three techniques have proved to be the most generally adopted because of their relative ease. The high frequencies of insertion of DNA into protoplasts are such that these systems can be used to carry out transient expression assays.

Transient expression assays in protoplasts or wounded embryos have been used extensively in promoter analysis (for example see [22, 72, 74]). However, there are limitations to this approach, not least because the cells used are derived from developed tissue and might not express gene constructs normally active at another developmental stage. Moreover, protoplasts may respond differently to an external stimulus when compared to cells within the whole plant. Nevertheless it is significant that protoplasts do respond to some stimuli in a manner similar to cells in the intact plant which has allowed the study, for example, of the expression of the alcohol dehydrogenase promoter in protoplasts under low oxygen tension [22] and the chalcone synthase (CHS) promoter in response to fungal elicitors and ultraviolet light [75].

Transfection of protoplasts with cloned DNAs representing viral genomes or RNA transcribed *in vitro* from DNA representing viral genomes has also been used to successfully analyse viral replication and expression. This approach was used to demonstrate that DNA A but not DNA B of the cassava latent virus (CLV) genome contained all of the viral sequences required for replication [76]. Similary, transient assays have been used to indicate the regions of the brome mosaic virus (BMV) and tobacco mosaic virus (TMV) [77, 78] and wheat dwarf virus (WDV) [79] required for replication as well as being able to direct the expression of marker genes engineered into the viral genome (see later).

One of the limitations that remains with transient expression assays is that there is no simple method of assaying the percentage of protoplasts that have taken up the applied DNA. Hence, in expression assays it is difficult to distinguish whether a few protoplasts are expressing the foreign DNA at a high level or a larger proportion of protoplasts have taken up the DNA and are expressing it at a lower level. Similarly, with plant cells there is no equivalent of the plaque assay system used to study viral replication in animal cells. Currently the only method to quantitatively assess expression in protoplasts *in situ* is to measure the fluorescence obtained in protoplasts expressing a luciferase construct by either autoradio-graphy, where dishes containing transfected protoplasts are placed next to X-ray film [13] or use a video-intensified microscope camera and image-intensifying system [80].

Stable transgenic plants can be obtained following appropriate culture of cells that have been subjected to DMGT [71, 81]. This approach can be applied to transfer DNA for which no direct selection exists [82] as well as gene targeting [83]. In the latter case the initial frequencies obtained with targeting a specific fragment of DNA into the genome were low but comparable with some mammalian cell systems suggesting that the specific modification of a plant gene either by mutagenesis, modification or replacement may be feasible.

Following DMGT the mechanism by which DNA integrates into the plant genome is unknown. However, the integrated DNA displays a complex organisational array suggesting that replication, ligation and recombination of the DNA takes place prior to the integration into one or two sites in the genome [84]. There is evidence suggesting that the highest rates of transformation can be obtained when DNA uptake is carried out in cells synchronised prior to nuclear division [85]. Hence, one might speculate that stable integration takes place as the chromatin is relaxed, the DNA undergoing replication (or transcription) and that the foreign DNA can serve as a substrate for the enzymes involved in DNA replication and repair. Organelle transformation has been achieved in *Chlamydomonas* where microbombardment has been used to introduce DNA into the chloroplast [86]. In this case integration took place as a result of homologous recombination.

Analysis of transgenic plants

Whether *Agrobacterium*-mediated transformation or DMGT is adopted for producing transgenic tissue, methods relying on tissue culture will produce plant populations which are prone to somaclonal variation. Somaclonal variation is apparently induced by the passage of plant material through tissue culture and appears to be the result of chromosomal modifications and rearrangements [87]. The variation is more pronounced in plant populations derived from protoplasts than from leaf discs. Nevertheless, regardless of approach adopted, caution is required when attempting to attribute a specific phenotype to a specific insert of foreign DNA. Hence it is important to demonstrate that the phenotype genetically co-segregates with the insert of foreign DNA.

Many workers have observed variation in the levels of expression of a particular gene construct between different individual transformants. This phenomenon has come to be known as 'position effect'. Position effect generally appears to influence the quantitative rather than the qualitative pattern of expression of the introduced gene and may act differentially on different gene constructs introduced into the plant on the same DNA (for example see [88]). The reasons for position effect remain unclear and might result from a number of different factors including insertion of foreign DNA into regions of the genome not normally transcribed or near sequences which control expression in a positive or negative manner. Moreover, methylation of the foreign DNA following stable integration into the genome can effect its expression [89, 90]. Copy number of the insert does not generally appear to account for the variation in levels of expression observed [91], nor can it necessarily be overcome by nesting the gene construct of interest between flanking sequences of up to 10 kb in size [92]. Position effects will cause difficulties when attempting to correlate differing levels of gene expression with different gene constructs, for example in promoter analysis. In

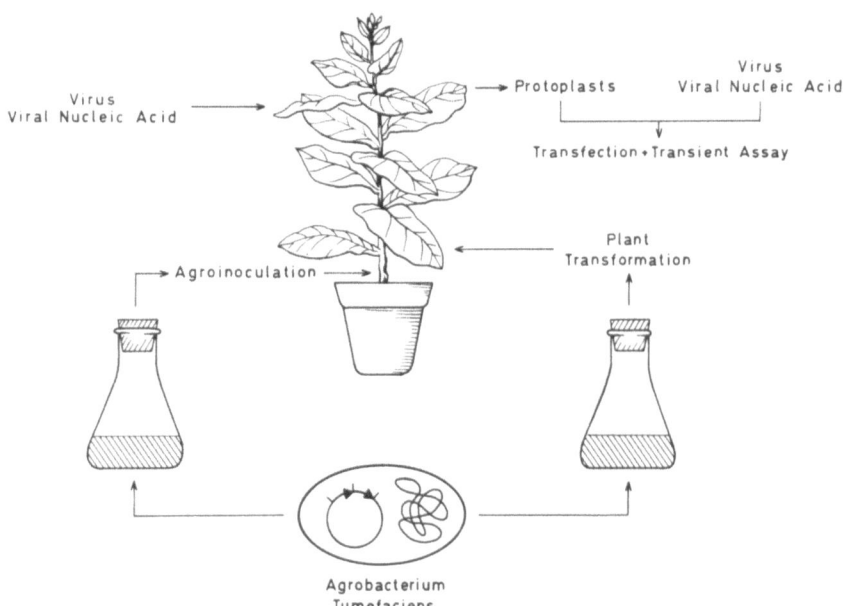

Fig. 2. *Methods of introducing viral nucleic acids into plant cells.* For agroinoculation and the production of transgenic tissue in which viral nucleic acids can escape from the nuclear genome, dimers of the viral genome are cloned between the T-DNA borders at the Ti plasmid

this case it is necessary to screen a large number of individual transformants in order to obtain accurate quantitative data or link the gene construct of interest to a reference gene in tandem [93] or have the genes transcribed divergently from a closely linked dual promoter [94, 95]. At present it is not clear whether, in an analogous manner to animal cells, the relative strength of the promoter construct is important in the variation of levels of expression with the weaker promoter elements being more susceptible to position effects [96]. In addition, it remains to be seen whether position effects can be removed by the addition of a DNase-I-hypersensitive site to the foreign DNA insert as has been found in transgenic mice [97].

Viral genomes as potential transformation vectors

Viral genomes have the potential of being engineered into vectors which, once inside the plant cell, can replicate to high copy number and provide the possibility of expressing foreign DNA at high levels. Such replicating vectors can be used in protoplasts and if able to pass from cell to cell could be used to infect host plants by mechanical inoculation, thus overcoming the drawbacks of plant tissue culture in producing transgenic tissue. There are three general ways of experimentally infecting plants with nucleic acids representing cloned viral genomes (Fig. 2). Mechanical inoculation is the simplest and involves applying the nucleic acid to wounded plant tissue. The second approach, agroinfection (or agroinoculation), involves cloning DNA representing dimers of the viral genome within the T-DNA and inoculating host plants with the *Agrobacteria* containing the construct [98, 99]. Finally, DNA representing dimers of the viral genome can be stably inserted into the plant genome by *Agrobacterium*-mediated transfer. In the latter two cases escape of the viral genome from the T-DNA can result from either intramolecular recombination or transcription.

Viral vectors for use in protoplasts

Although the replication of a variety of viruses has been studied in protoplasts following the inoculation of viral nucleic

acid or virions [100], there are fewer reports of the replication of viral nucleic acids following the inoculation of isolated cells with either cloned viral genomes or RNAs derived *in vitro* from them. This is due, in part, to being able to engineer viral escape from cloned DNA. This can be circumvented with RNA viruses where the genome, represented by a cDNA, may be transcribed *in vitro* to produce an infectious RNA which can be used to inoculate protoplasts. This approach has proved successful with, for example, tobacco mosaic virus [101], turnip crinkle virus [102], cowpea mosaic virus [103] and beet yellow vein virus [104]. Apart from allowing the detailed functional analysis of the viral genome this ultimately can lead to the replacement of regions of the viral genome nonessential in replication with sequences of foreign DNA. An example of this is provided by BMV, which can infect protoplasts derived from a barley suspension cell line [77]. In this example, three cDNAs representing the three RNAs which comprise the BMV genome were cloned and the coat protein gene on one of the components was replaced with a CAT reporter gene. Protoplasts were inoculated with the three RNAs representing the components of the BMV genome transcribed *in vitro*. The modified RNA component replicated in the presence of the normal viral RNAs and the protoplasts were found to express high levels of CAT activity. Analogous work has been reported with TMV which has a single RNA component [78].

Infection of protoplasts with cloned derivatives of gemini virus genomes such as CLV and WDV has been obtained with either linearised cloned viral DNA or a dimer of the viral genome contained in the cloning vector [76, 79]. In the latter case viral escape resulted from intramolecular recombination. Gemini viruses comprise a single-stranded DNA genome and replication of viral DNA in the protoplasts appear to be linked to cell division suggesting that host factors are required for the replication. In the case of WDV it has been found that when the coat protein is replaced with CAT, NPTII or β-galactosidase the viral DNA is able to replicate and that the highest levels of enzyme activity appear when maximal amounts of the viral DNA are present [79]. Similar results were obtained in cells derived from wheat, maize and rice suggesting that the WDV genome may prove to be able to

568

function as a replicating vector in a variety of monocotyledonous cells. The accumulation of the viral DNA was followed for 20 days and no rearrangement of the cloned sequences could be detected. However, the fate of the viral DNA in suspension cell cultures over a longer period of time, to our knowledge, has not been investigated.

Viral vectors for intact plants

The potential of vectors based on viral genomes to spread throughout a host plant to produce systemic infection has provided the spur for work with virus vectors which are likely to replicate in plant species such as the cereals. However, although a great deal of interest has been paid in this approach, problems have been encountered because of a lack of understanding of the molecular basis of viral replication and passage throughout the infected plant, as well as the difficulty of infecting host plants by mechanical inoculation of some viruses.

Much initial interest in producing vectors based on viral genomes was paid to cauliflower mosaic virus (CaMV) [105]. The potential advantages of using CaMV as a vector were that it is well characterised, comprises a single double-stranded DNA genome which, once cloned, remains infectious by mechanical inoculation. *In vitro* mutagenesis revealed that the majority of the coding regions are required for normal infection to take place although it is not clear whether loss of infectivity results from a block in the replication of the virus, virion assembly or the inability of the virus to passage throughout the plant. Nevertheless, two regions of the genome can be modified without destroying viral infectivity. However, only small inserts of DNA can be inserted into these sites without interfering with viral function probably because of packaging constraints. Moreover, extensive recombination between inoculated viral DNAs preclude the use of a helper virus system [106]. Use of CaMV as a vector is complicated further by the finding that the virus replicates via an RNA intermediate and that the mode of expression of viral genes is complex, possibly involving a polycistronic mRNA [106]. Nevertheless, bearing these difficulties in mind, CaMV has proved successful as a vector for transferring small genes into plants and has shown that high levels of expression of foreign DNA can be obtained [16, 107, 108]. In these cases open reading frame II, whose product is required for insect transmission but not replication and systemic infection of plants, was replaced by the dihydrofolate reductase and metallothionine genes in such a way that they would not interfere with the translation of viral polypeptides. The cloned DNA was inoculated directly onto a host of CaMV, turnip. In both cases the viral-vector-initiated infection and was able to systemically spread throughout the whole plant. The foreign genes were expressed and resulted in a relatively high level of enzyme activity [11, 32, 99].

In order to increase the versatility of CaMV as a vector a more detailed understanding of its replication is required. Once the regions of the CaMV genome required in both *cis* and *trans* for replication, expression and virion assembly are defined one can envisage that the capacity of the genome to tolerate larger insertions of foreign DNA might be increased. In order to carry this out an efficient protocol for studying the replication of cloned CaMV derivatives in protoplasts is required. Regions of the genome important in cell-to-cell spread of the virus may be defined by inserting derivatives of the viral genome into plant genomes using T-DNA. These could be engineered so that viral deletion derivatives containing foreign DNA might be subsequently inoculated into transgenic plants and replicate under the control of factors produced by the viral DNA integrated into the genome.

The gemini viruses have received much attention as potential vectors because as a group they infect a wide range of host plants including monocotyledonous species [79, 109]. Generally, they can be classified according to their insect vector. Leaf-hopper-transmitted gemini viruses comprise a single genome and infect monocotyledonous plants (e. g. maize streak virus, MSV, WDV) and dicotyledonous plants (e. g. beet curly top virus, BCTV). Gemini viruses transmitted by whitefly have a bipartite genome and infect dicotyledonous plants (e. g. CLV, tomato golden mosaic virus, TGMV). The whitefly-transmitted viruses can be mechanically inoculated onto test plants and infection can also be initiated when test plants are inoculated with cloned DNAs representing the two genomes. In contrast, experimental infection of leaf-hopper-transmitted viruses can only be achieved by agroinoculation.

There is a high degree of sequence similarity between different viral genomes (Fig. 3) with transcription appearing to take place divergently from the common region which is conserved in both genomes of the bipartite genome viruses. With CLV three conserved open reading frames are arranged in DNA A in the complementary sense (AC1 – AC3), one in the virion sense (AV1) and one in both the complementary (BC1) and in the virion sense (BV1) in DNA B. Similarities in the organisation of the genomes are observed between DNA A and the single-component gemini viruses. Although both DNAs are required to produce systemic infection DNA A alone is able to replicate and form virions indicating that DNA B is only required for cell-to-cell spread. AC3 and AC2 can be mutated without disrupting replication but AC1 and AC2 are essential for systemic infection. In addition, both BC1 and BV1 are required for systemic infection. The observation that coat protein mutants retain their infectivity has allowed its replacement either with viral genes [110] or a CAT gene [111]. In the later example the CAT gene was fused in frame with the amino-terminal of the coat protein and following inoculation, produced systemic infection and CAT expression throughout the plant. Similar work has been carried out with the MSV genome [112]. Here coat protein mutants are unable to systemically spread throughout the plant but following inoculation of coat protein replacement vectors containing the CAT gene, enzymatic activity could be detected in inoculated tissue.

The gemini virus genome can be exploited further as a vector by utilising *Agrobacterium* in both establishing transgenic plants containing viral genomes and in agroinoculation. Both DNA A and B of TGMV have been inserted as tandem repeats into plants [113]. In transgenic plants containing tandem DNA A inserts freely replicating and encapsidated DNA A accumulates demonstrating that this component contains all of the requirements for replication and virion assembly [114]. This work raised the possibility of establishing replicating viral systems in plants where one component, a 'master copy', is inserted into the genome and a second is introduced into the plant by agroinoculation. This has been carried out by replacing the coat protein of TGMV with an NPTII gene and constructing partial dimers of both DNA A and B and using these to produce stable transgenics and in agroinoculation [115]. The NPTII gene replacement in a partial dimer of DNA A was either inserted into the T-DNA alone (A1.6*neo*) or with dimers of DNA B (A1.6*neo*B2). In both cases a freely replicating DNA A containing the NPTII gene was found in transgenic tissue. Likewise, following agroinoculation of

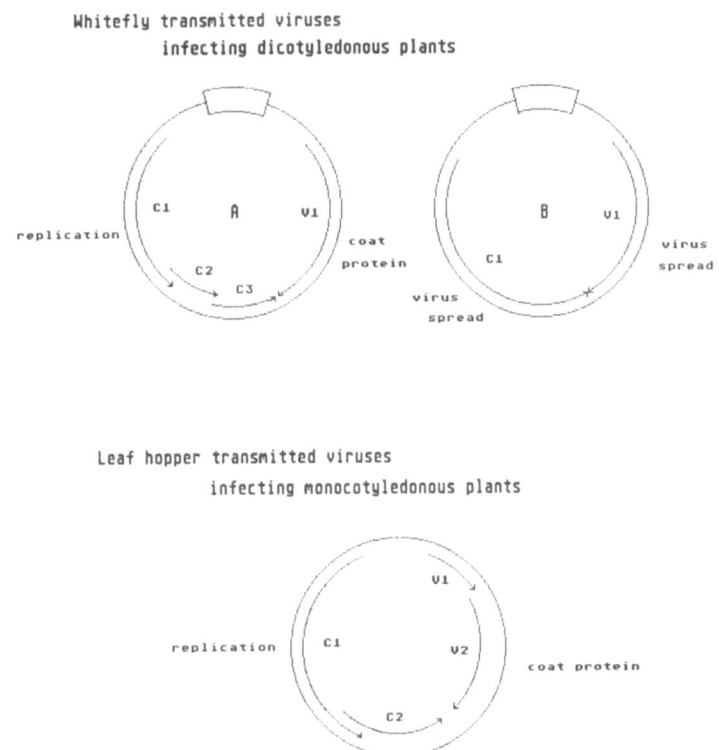

Fig. 3. *Functional orgainisation of the gemini virus genome*. Arrows correspond to the open reading frames and direction of transcription; boxes refer to regions of sequence homology common to both whitefly-transmitted virus genomes

plants with A1.6*neo*B2, replicating DNA A containing the NPTII gene were detected. A similar result was found when plants engineered to contain dimers of DNA B were agroinoculated with A1.6*neo*. The levels of NPTII activity that were observed in the tissue correlated with the copy number of the replicating DNA A containing the NPTII gene [115]. This type of approach greatly enhances the versatility of the virus vector so that a two-component replicating system might be established. One might envisage that the genes important in encoding replicative functions might be engineered into the plant genome under the control of a constitutive promoter and elements of the viral DNA able to replicate and passage through the plant, yet not contain their own replication functions, might be introduced by agroinoculation.

Dissection of promoter function and factors interacting with promoter sequences

An important control point of differential gene expression is transcription and a large number of plant promoters which direct transcription in either an inducible or tissue-specific manner have been studied in detail [26]. Plant transformation has allowed the functional analysis of promoter sequences in both transient assays or in stable transformants [116]. Generally it has been found that sequences 5′ to the coding regions of the gene in question contain many, if not all, of the signals necessary for correct developmental expression. By studying the pattern of expression directed by deletions it appears that a promoter can comprise a combination of several sequence motifs which are responsible for the quantitative and qualitative (i.e. tissue-specific and/or developmentally regulated) control of gene expression. These sequence elements are likely to control RNA polymerase activity by interacting directly or indirectly with regulatory proteins. Although each promoter may have a unique array of *cis* positive or negative

regulatory elements which can overlap, in some cases they can be exchanged so that novel patterns of gene expression can be conferred on a marker gene [26]. Of those plant promoters which have been studied in detail the general architecture of the promoter is such that sequence motifs important in qualitative expression (i.e. developmental or in response to environmental stimulii) are located close to the CAAT and TATA boxes (or their equivalents) and strong positive elements directing high levels of activity are located 5′ to these regions [117–121].

Apart from those of the T-DNA, promoters from non-plant sources work poorly, if at all, in plants [122]. Plant specific promoters derived from pathogens, i.e. T-DNA or CaMV, have been used most extensively to direct the expression of foreign genes in plants [116]. Initially it was considered that these promoters were constitutive in their mode of expression. However, with the use of sensitive markers such as β-glucuronidase, luciferase or β-galactosidase able to detect cell-specific expression it has become clear that these promoters can be affected by subtle environmental and developmental changes in the levels of growth substances within the plant. For example, the nopaline synthase promoter of T-DNA has been found to be most active in the basal region of plants with its activity decreasing in the upper parts of the plant suggesting that it may respond to differing levels of auxin [123–125]. Following flowering, the activity of the promoter in vegetative organs declines whereas in flower tissue there is a general increase in activity [125]. Other T-DNA promoters also appear to respond to endogenous plant growth substances. The gene 5 promoter is most active in callus, stems, and roots while activity in leaf tissue is barely detectable [126] whereas the 1′/2′ dual promoter is most active in the basal region of the plant suggesting that auxins control its expression [127].

570

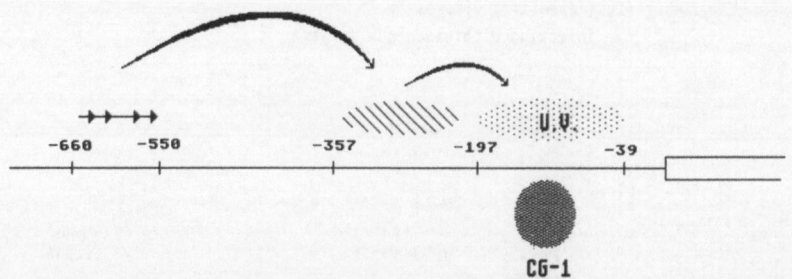

Fig. 4. *Functional organisation of the* A. majus *CHS promoter*. The region −357 to −197 acts to maximise the effect of the ultraviolet (U. V.) responsive element (−197 to −39). The upstream enhancer (−660 to −550) acts to boost expression levels further. CG-1 represents a factor identified which binds to the ultraviolet-responsive sequence of the promoter

The promoter of the 35S RNA of CaMV has been found to direct high levels of gene expression in a wide variety of plant species. Although, once again, the levels of its activity can differ in different tissue and it appears to be most active in leaf tissue [10]. Expression directed by the region +8 to −90 is strongest in embryo tissue, the radicle pole of the endosperm and in root, whereas the region −90 to −343 directed strongest expression in cotyledons, leaves and stem [128]. This suggests that the promoter contains multiple *cis* elements able to individually confer tissue specific expression which in the intact 35S promoter work in concert in order to confer constitutive expression.

The 35S RNA promoter also contains a strong transcriptional enhancer [129]. This sequence can increase transcription from a heterologous promoter [129−132] and appears to be orientation independent [132]. Moreover, the multiplication of the region can result in a 10-fold increase in transcription [132−134], with the enhancement of expression able to act over distances of up to 2 kb [134].

The plant promoters that have been studied in detail are those that respond to environmental stimulii such as light, anaerobic or heat stress and wounding or developmental cues such as leaf, seed, flower and fruit development. Moreover, tissue specific expression has been investigated in green tissue, seeds, nodules, tubers and flowers [26].

An example of such a promoter is provided by that of CHS from *Antirrhinum majus* [135]. CHS catalyses the first step in the flavonoid-specific branch of the phenylproponoid pathway, can be induced by ultraviolet light, infection of pathogens and wounding as well as by normal developmental cues triggering leaf and pigment formation. Hence, it is likely that the CHS promoter responds to a variety of signals in directing gene expression. A 1.1-kb *A. majus* promoter fragment has been found to be sufficient for ultraviolet induction in dark-adapted tobacco callus cells [136]. Internal deletions demonstrated the presence of at least two regulatory regions, one located between −1200 and −357 was essential for optimal expression following ultraviolet induction. Transient assays in parsley cells have found that the region −197 to −39 is sufficient for ultraviolet inducibility irrespective of orientation. Moreover, this region is able to render a heterologous minimal promoter ultraviolet responsive. Sequences between −197 and −357, although insufficient for ultraviolet induction, are important in directing high levels of expression following ultraviolet induction and an enhancer is located between −564 and −661 [137] which, although orientation independent, requires other flanking sequences for maximal activity (Fig. 4; Fritze et al., unpublished results).

With a preliminary understanding of 5′ sequence elements important in the *cis* regulation of promoters work has moved towards identifying the factors, and the genes that encode them, which act in *trans* by binding to these regions. The −62 to −83 region of the 35S RNA promoter has been found to bind a nuclear factor [138] and specifically a factor that binds to a TGACG motif [139]. This motif is also present in the promoter region of the wheat histone H3 and nopaline synthase genes and the nuclear factor, called activation sequence factor 1, also binds to these. cDNAs representing the mRNAs encoding factors binding to the TGACG motif have been isolated and sequence analysis of two independent clones show that the protein they encode contains a conserved basic domain. This domain is similar to the DNA binding protein products of yeast GCN4 and human cAMP-response-element-binding proteins. Similarly, factors have been identified which bind to the ultraviolet-responsive sequence [140] of the *A. majus* CHS promoter. The exact role that these factors play in directing gene expression remains to be seen. On the other hand, an example of a *trans*-acting factor which may play a direct role in controlling gene expression in plants is provided by the protein encoded by the *c* 1 locus of maize. This locus regulates the tissue-specific expression of the anthocyanin biosynthetic enzymes. The protein encoded by the *c* 1 locus contains an acidic and basic domain with the latter bearing homology to the DNA binding domain of the *myb* proto-oncogene [141]. The overall structure of the protein is reminiscent of the products of the transcriptional activator genes, *GAL 4* and *GCN 4* of yeast and it is tempting to speculate that *c* 1 acts in a similar way.

To date, the expression of most plant genes that have been studied at the molecular level are induced by specific environmental or developmental stimuli. Nevertheless, the pattern of expression generally observed is somewhat leaky and it is difficult to find tissue in which a promoter is entirely inactive. Such a promoter would be invaluable, for example, in studying the effects of expression of genes encoding plant growth substances. Attempts in carrying this out with either the auxin [142] or cytokinin [143] biosynthetic genes from *Agrobacterium* have utilised plant, or *Drosophila* heat-shock promoters. The idea here is that plant morphology can be compared under normal, non-inducing conditions or following growth under inducing conditions at elevated temperatures. The limitations of this approach however, are that the heat-shock promoters are not entirely inactive under noninducing conditions and it is difficult to be able to discount any subtle side effects that the heat shock might have on the hormone balance within the plant. The question remains whether there may be an inducible promoter, possibly from a non-plant source, which can be induced uniquely in plant cells. Of course in this case it will be important for the plant cell to possess the appropriate signal-transduction pathway to perceive and transmit the inducing signal from the cell surface to the promoter. Steps in this direction have been taken with the observation that the *Tn10* encoded *tet* repressor can regulate the expression of a chimeric gene containing the *tet* operator in transient assays in protoplasts [144]. Similarly, *GAL4*, a transcriptional activator from yeast, can serve to stimulate transcription of a marker gene linked to the *GAL4* DNA-binding domain [145].

Novel approaches to the isolation of plant genes

Generally plant genes have been isolated via the construction of cDNA libraries followed by screening of clones with an appropriate hybridisation probe derived from purified RNA, a predetermined sequence or differential hybridisation. These types of approach, while versatile, are unlikely to be able to identify genes whose products are extremely rare or for which no selection protocol exists, for example, those genes involved in controlling development.

The alternative approach in isolating plant genes involves the production and analysis of mutants. Mutants displaying novel phenotypes have been proven to be powerful tools in molecular genetics and in addition to producing mutations in structural components of the plant are also likely to allow the isolation of genes involved in the control of gene expression and pattern formation. However, although a variety of procedures such as ultraviolet light, ethyl methane sulphonate and X-rays have been used to produce mutant plants and despite a long history of mutant selection, there are relatively few examples where there is an understanding of the primary molecular defect from which a mutant phenotype arises [146].

Where the molecular basis of a mutation is understood there is the potential of rescuing genes by complementation. However, in plants, this powerful technique has had only limited success and the examples of complementation that have been reported have been achieved by using genes that had previously been characterised. For example, an auxotrophic mutant of *Nicotiana plumbaginifolia* requiring isoleucine has been complemented by a yeast gene encoding threonine dehydratase which had been cloned between the *nos* promoter and the octopine synthase poly(A) addition site [15]. Another example has been provided by the complementation of a flower pigment mutation of *Petunia hybrida* with a maize gene previously cloned between the 35S RNA promoter and the octopine synthase poly(A) addition site. In this case the maize gene was able to catalyse the conversion of an intermediate metabolite in anthocyanin biosynthesis which accumulated in the mutant plant thus changing the flower colour of the plant from white to brick red [147]. Both of these examples demonstrate the power of complementation but do not address the question of isolating genes by this method. Moreover, the techniques employed did not attempt to mimic those that would be required to rescue a particular gene from a complex mixture of DNAs, such as those representing a complete plant genome.

To date, to our knowledge, there have been no reports of the rescue of a plant gene from a shotgun library of plant genomic DNA by complementation, although several studies concerning the feasibility of this approach have been reported. These have included rescue of a gene encoding resistance to kanamycin [148] or chlorosulfuron [149] from the genome of arabidopsis. In these cases the genomic DNA was cloned into a plant transformation vector, introduced into *Agrobacterium* and used in leaf disc inoculation of either petunia or tobacco, respectively. Transformants were selected for resistance to

either kanamycin or the herbicide. While these experiments establish the feasibility of such an approach it needs to be borne in mind that powerful genetic markers were used and some plant genes might not be so easily selected for, particularly if the selection is carried out at the stage of callus growth.

In order to successfully carry out rescue by complementation several criteria need to be met. (a) A plant mutant that is well defined at the molecular level and has an easily definable phenotype is required. (b) A genomic library of the wild-type plant DNA is needed as a source of complementing DNA. Obviously, the genome size is important as this will determine the number of clones required represent it. (c) It is necessary to have a plant transformation vector in which genomic DNA can be cloned and remains stable, prior to and during transfer to the plant. (d) A means of selection of the complemented phenotype is required.

There have been several reports concerning the development of vectors suitable for the construction of plant genomic libraries for use in complementation [149, 150]. These binary vectors contain between the T-DNA borders sequences which allow cloning of DNA, selection of transgenic plants, act as *E. coli* plasmid origins of replication, contain bacterial selectable markers and have a *cos* site which allow rescue of large fragments of plant DNA in *E. coli*. The source of the plasmid cloning vector is important in determining the stability of the foreign DNA within the bacteria. One report describes that 60% of the plant clones appeared to be unstable [150], another suggests that 20 out of 21 random clones appeared to be stable [149]. In complementation experiments it is obviously important that the library is efficiently transferred from the *Agrobacterium* to the plant cell and correctly integrated into the genome. The feasibility of this has been studied in detail by investigating the frequency of transfer of a small library containing 10 fragments of DNA derived from bacteriophage T7 into the tobacco genome [151]. Analysis of 66 transgenic plants obtained showed that each of the 10 library members were present in at least two plants. The majority of the plants contained a single insert of DNA although some contained two inserts and one contained four. However, 20% of all of the transferred DNA appeared to be rearranged when compared with the original clone. In another study where a sample of seven random clones from an arabidopsis library were transferred to the plant genome [149] 10 out of 18 plants contained unrearranged DNA, six contained deletions and the remaining two could not be investigated because the original clone contained repetitive DNA sequences. Hence it is clear that gene rescue by complementation in plants is by no means straightforward. Certainly choice of vector appears to be important. The development of protocols for electroporation of *Agrobacteria* [48] can overcome any instability in the clones which arises during conjugational transfer into *Agrobacteria* from *E. coli* although at the moment it is difficult to perceive of improvements which can be incorporated into the process of plant transformation whereby the efficiency of transfer can be improved and the stability of the transferred DNA can be assured.

Gene isolation by tagging

Because of the difficulties discussed in the preceding section attention has been focused on the molecular analysis of mutant phenotypes which have been induced by the introduction of exogenous DNA into the germ line. The insertion of DNA into, or near a gene might cause a mutation by disruption of either its structure or expression. This approach has the additional advantage that an altered phenotype might provide clues to the function of the product of the gene in question as well as facilitating its isolation. Using exogenous DNA as a tag has the potential that any gene sequence might be isolated from the plant genome providing that the conditions of the experiment are such that random insertion of the exogenous DNA takes place and that the production of sufficient numbers of individual insertion events ensures the possibility that the majority of the plant genome is mutated. Gene tagging in plants has been proposed with two different systems, one using transposable elements, the other utilising T-DNA insertion.

Transposon tagging

Plant transposons have proven to be a powerful tool to produce mutations in the plant genome [152, 153]. Although there have been a large number of transposable element families described in a variety of plant species the best characterised system remains the *Ac/Ds* system in maize [154]. *Ac* (activator) is an autonomous element able to transpose by itself and *Ds* (dissociation) is non-autonomous and can only transpose when *Ac* is present in the plant genome. The *Ac/Ds* system has been used extensively to tag genes in maize [153, 155] and the isolation and characterisation of individual *Ac* and *Ds* elements has allowed their reintroduction into heterologous host plants via T-DNA-mediated transformation. The *Ac/Ds* system is functional in tobacco, arabidopsis, carrot, potato and tomato [156 – 160]. This greatly extends the potential of transposon tagging in gene isolation because the *Ac/Ds* system may be used to induce mutations in species where information concerning endogenous transposons may be lacking. The mechanism by which *Ac/Ds* elements transpose is becoming increasingly clear at the molecular level [161] and this has allowed their modification so that assays can be devised which detect not only excision but also reinsertion of the elements in transgenic tissue. The excision of a *Ds* element can be assayed by inserting it into the 5′ untranslated region of a chimeric NPTII gene in such a manner that upon transposition cells containing the construction become resistant to kanamycin (Fig. 5) [162]. This type of approach has been extended using a marker gene encoding resistance to streptomycin or the *GUS* reporter gene [27, 163].

Ac and *Ds* alone confer no selectable or screenable phenotypes. Hence gene tagging would be greatly facilitated by an element which itself confers a phenotype. This would allow the direct measurement of frequency of loss of the element during transposition as well as the ability to simply monitor co-segregation of the transposon with a putative transposon associated allele. Such a marked transposon has been developed in which a *Ds* element which itself was inserted into the 5′ untranslated region of an NPTII selectable marker, was engineered to contain a second selectable marker encoding resistance to methotrexate (Fig. 5) [158]. When this construct was transferred into the plant genome it conferred resistance to methotrexate. If *Ac* was present in the host genome transposition of the *Ds* element could be monitored by selection on kanamycin. In the cases where kanamycin resistance was obtained it was found that in approximately 70% of the cases the *Ds* element containing the methotrexate resistance gene had reinserted into another position of the plant genome [158].

With these types of genetic markers gene tagging becomes more feasible simply by crossing plants containing *Ac* elements which by themselves are unable to transpose as a result of mutagenesis [164] with plants containing marked *Ds* elements

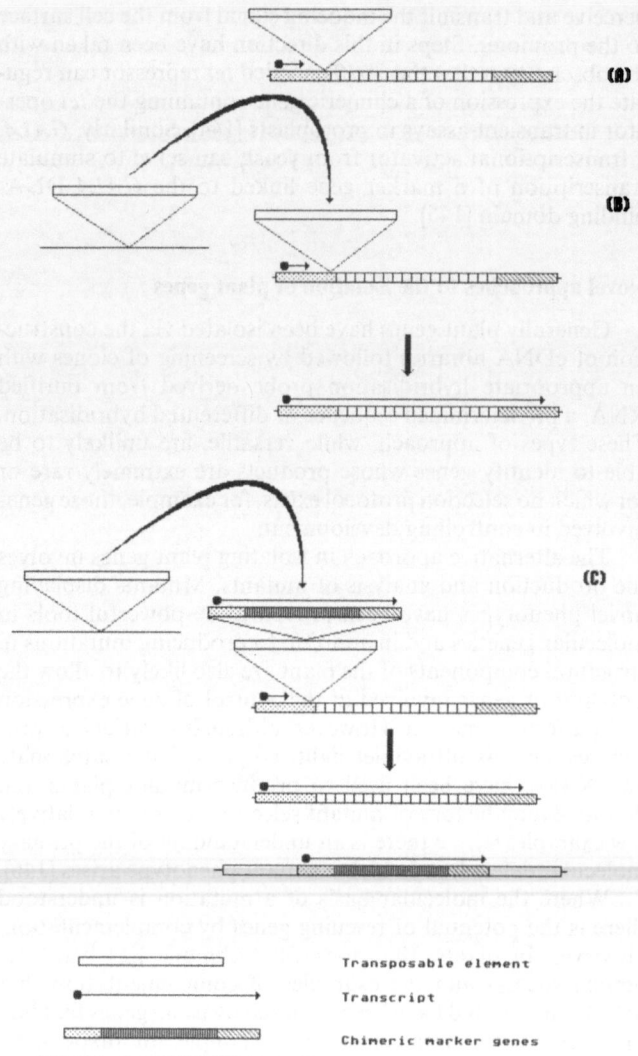

Fig. 5. *Engineering transposable elements in order to study transposition and gene tagging.* In order to assay transposition, a non-autonomous transposable element can be inserted into the 5′ untranslated region of a chimeric dominant or screenable marker gene (A). In the presence of an autonomous transposon acting in *trans* the element can excise from the marker gene restoring its activity (B). Should the non-autonomous transposon be engineered to contain a second marker this can be used to monitor reinsertion of the element into the genome (C)

[165]. These then will transpose so that the genome will be peppered with insertion events. Further crossing is then carried out to remove the *Ac* element from the genome and so stabilise the marked *Ds* elements. The advantage of this system is that multiple, independent insertion events will reduce the number of plants required to isolate a particular gene. Moreover, the potential difficulties of novel phenotypes arising from somaclonal variation can be overcome because transposon tagging takes place after crossing, i.e. in the absence of tissue culture. However, it has the problem that mutant phenotypes can only be selected for following removal of the *Ac* due to the instability of the *Ds* element. While it can be argued that the instability of the element might aid the linking of a particular element to a specific phenotype the possibility that imprecise excision of both elements resulting in a maintained mutant phenotype remains. Moreover, the detection of the

individual insertion responsible for an altered phenotype requires detailed genetic analysis.

To date most mutable alleles which have been studied are genes whose products are involved in the synthesis of scorable features such as flower, seed or leaf pigmentation [161]. Scorable phenotypes which can be screened following transposon tagging include resistance to pathogens or modified growth characteristics. In such cases once the transposable element system has been constructed, crosses and screening can be carried out by conventional field tests.

T-DNA as an insertional mutagen

An alternative approach to transposon tagging is to use T-DNA as an insertional mutagen. The advantage of this approach is that single, stable insertions can be obtained. This simplifies mutant analysis although it increases the number of individual transformants that need to be analysed in order to tag a particular gene. In addition, if the T-DNA itself also contains an origin of replication and a selectable marker functional in *E. coli*, any tagged plant DNA flanking the T-DNA can be isolated by cutting the DNA at a restriction site that is not present in the T-DNA, religating the DNA and transforming directly into *E. coli*. On the other hand, T-DNA tagging involves the passage of plant tissue through culture and tagged plant populations may be prone to somaclonal variation. Hence caution is required in analysing the resultant plants and to be successful it must be proven that the T-DNA insertion is responsible for any novel phenotype.

T-DNA insertional mutagenesis has been used successfully to isolate plant promoters in both tobacco and arabidopsis [38, 166 – 168]. In this case the T-DNA was engineered so that the right border sequence was fused to a marker gene lacking a promoter region so that if the T-DNA inserted into the plant genome in a region that was able to act as a promoter the marker gene would be expressed. Using constructs which contained a marker gene at the right border sequence and a separate selectable marker gene within the T-DNA, extensive analysis of the transformants obtained with both tobacco and arabidopsis indicated that in both cases a remarkably high portion of the transformants (approximately 8%) expressed the marker gene [38]. This result is quite unexpected when it is considered that the transgenics contained, one, two or more inserts in a 1:1:1 ratio and that the relative sizes of tobacco and arabidopsis genomes differ greatly (1.6×10^6 and 7×10^4 kb/haploid genome, respectively) along with the relative amount of repeated sequences in the two genomes (60% in tobacco and low in arabidopsis). This suggests that the T-DNA preferentially inserts into potentially transcribed regions of the plant genome which is reminiscent of observations concerning the insertion of transposable elements in yeast [169], P elements in *Drosophila* [170] and retroviruses in animals [171].

This finding increases the versatility of T-DNA as a tag to isolate regions of the genome important in directing gene expression and reduces the number of individual transformants which are required to isolate a gene of interest.

Insertional mutagenesis has been used successfully to produce phenotypic mutants in arabidopsis. In one example a variety of different phenotypes including floral variants, embryo lethals, stems lacking tricomes and dwarfs were obtained [172]. The dwarf plant contained a single insert of 4 – 6 T-DNAs in a tandem arrangement which was genetically linked to the dwarf phenotype. In another example, a T-DNA induced mutation, *pale*, was found to result from insertion into a gene encoding a novel chloroplast polypeptide [168]. In this case, the T-DNA was used as a tag to isolate the gene responsible for the chloroplast polypeptide and final proof that this gene had been mutated was provided by complementation.

The versatility of T-DNA mutagenesis may be exploited further in order to isolate regions of the genome that act in *cis* to control gene expression. An example of how this can be achieved is provided by engineering the T-DNA to contain markers allowing the selection of either positive or negative regulatory elements. This could be carried out by replacing the marker gene used in promoter sniffing with a reporter gene containing a minimal promoter able only to direct low levels of gene expression. Following insertion into the plant genome, T-DNA elements directing high levels of gene expression can be screened for. Conversely, a reporter gene conferring negative selection could be used to isolate elements which can decrease gene expression.

Insertional mutagenesis, by its very nature, is likely to disrupt gene function and a phenotype, which might prove to be lethal, will only be seen when the mutant is in a homozygous state. An alternative is to produce dominant mutations in which expression of a particular gene is not disrupted and this can be achieved by producing over-expression mutants. In order to do this the T-DNA can be engineered so that it contains enhancer sequences which serve to activate transcription of flanking sequences following insertion into the plant genome. In this way mutants can be obtained which express genes in a developmentally independent manner.

Like transposon tagging, T-DNA tagging is limited by the ability to be able to select for a required phenotype. This difficulty is compounded by the size of the plant genome which demands a large number of individual transformants to be screened to be certain of recovering a desired mutant. In addition, insertions might be made into individual members of a gene family and may not produce a scorable phenotype. Insertion mutations in diploid plants are likely to be recessive and will only become obvious in an homozygous progeny. This may be overcome by using haploid plants however, these are prone to somaclonal variation. Alternatively selection *in vitro* of transformed protoplasts following co-cultivation can be carried out. However, in this case, although biochemical selection of protoplasts can be attempted, this severely limits the phenotypes which can be selected for and relies on the expectation that the phenotype of the regenerated plant will be the same. Moreover, protoplast selection effectively limits the plant in which gene tagging can be carried out members of the solanacea. Although a protocol for the transformation of arabidopsis with naked DNA is available [173] successful transformation by protoplast co-cultivation has yet to be reported.

Conclusions

The successful application of the techniques of gene transfer to plant cells has had a significant impact on plant molecular biology. Transformation in plants can be used in an analogous manner to other eucaryotic and procaryotic systems to study the structure and function of DNA sequences. Nevertheless, as illustrated by the example of gene isolation by complementation, there are still limitations in the experimental systems and the techniques that are available.

Further understanding of the mechanism of integration of DNA, particularly T-DNA, into the plant genome may allow further improvement of the methods of transformation. It is

574

already clear that the versatility of T-DNA is such that it can be used not only as a vector but also as a gene tag. With the observation that T-DNA preferentially inserts into potentially transcribed regions of the genome it may be used in the future to probe chromatin structure.

Tobacco remains the plant system of choice in gene-transfer experiments simply because it is so amenable to tissue culture. In time, arabidopsis may take its place because of the benefits it offers by its small genome and stature, rapid life cycle and extensive genetics [174]. Nevertheless, arabidopsis can be difficult to manipulate in tissue culture and a protoplast co-cultivation protocol has yet to be developed.

Totipotency, the unique ability of the isolated plant cell to regenerate into a plant has facilitated the advances in plant transformation. Moreover, the developmental flexibility of the plant cell, under the control of different plant growth substances, is likely to provide special opportunities to study the molecular basis of development. Genes which are either directly involved in the synthesis of auxins or cytokinins, or affect the response of the cell to them have been cloned and reintroduced into plants [142, 143, 175]. This along with the potential of being able to modify the biochemistry of the plant cell by either antisense [176] or ribozyme [177] technology raises the possibility of significantly changing the form and function of plant cells. Allied to this, significant advances have been made recently in isolating genes encoding auxin-binding proteins (i.e. putative receptors) [178], G proteins [179], calmodulin [180] and protein kinase homologues [181] all of which are thought to play key roles in signal transduction, the process by which the cell responds to external changes in growth substances. This may provide an alternative approach to *cis* and *trans* analysis to understanding the molecular basis of gene expression as well as provide unique opportunities to study plant development.

Our thanks to our co-workers for valuable discussions and particularly to Desy Rosahl, Bob Masterson, Ruth Wingender and Ellen Buschfeld for their comments on the manuscript. Thanks also to Frau Furkert for her help with the figures.

REFERENCES

1. Gasser, C. S. & Fraley, R. T. (1989) *Science 244*, 1293−1299.
2. Draper, J., Scott, R., Armitage, P. & Walden, R. (1988) *Plant genetic transformation and gene expression. A laboratory manual*, Blackwell Scientific.
3. Walden, R. (1988) *Genetic transformation in plants*, Open University Press, Milton Keynes.
4. Herrera-Estrella, L., DeBlock, M., Messens, E., Hernalsteens, J.-P., van Montagu, M. & Schell, J. (1983) *EMBO J. 4*, 2987−2995.
5. Waldron, C., Murphy, E. B., Roberts, J. L., Gustafson, G. D., Armour, S. L. & Malcolm, S. K. (1985) *Plant Mol. Biol. 5*, 103−108.
6. Herrera-Estrella, L., DeBlock, M., Messens, E., Hernalsteens, J.-P., van Montagu, M. & Schell, J. (1983) *EMBO J. 4*, 2987−2995.
7. Hayford, M. B., Medford, J. I., Hoffman, N. L., Rogers, S. G. & Klee, H. (1988) *Plant Physiol. 86*, 1216−1222.
8. Otten, L. A. B. & Schilperoot, R. A. (1978) *Biochem. Biophys. Acta 527*, 497−500.
9. Helmer, G., Casadaban, M., Bevan, M., Kayes, L. & Chilton, M.-D. (1984) *Bio-Technology 2*, 520−527.
10. Jefferson, R. A., Kavanagh, T. A. & Bevan, M. W. (1987) *EMBO J. 6*, 3901−3907.
11. Jones, J. D. G., Svab, Z., Harper, E. C., Horwitz, C. D. & Maliga, P. (1987) *Mol. Gen. Genet. 210*, 86−91.
12. Hille, J., Roelvink, A., Franssen, H., Van Kammen, A. & Zabel, P. (1986) *Plant Mol. Biol. 7*, 171−176.
13. Ow, D., Jacobs, J. D. & Howell, S. H. (1987) *Proc. Natl Acad. Sci. USA 84*, 4870−4874.
14. Koncz, C., Olsson, O., Langridge, W. H. R., Schell, J. & Szalay, A. (1987) *Proc. Natl Acad. Sci. USA 84*, 131−135.
15. Colau, D., Negrutiu, I., van Montagu, M. & Hernalsteens, J.-P. (1987) *Mol. Cell. Biol. 7*, 2552−2557.
16. Lefebvre, D. D., Miki, B. L. & Laliberte, J.-F. (1987) *Bio-Technology 5*, 1053−1056.
17. Shah, D., Horsch, R. B., Klee, H. J., Kisherer, G. M., Winter, J. A., Tumer, N., Hironaka, C. M., Sanders, P. R., Gasser, C. S., Aykent, S., Siegel, N. R., Rogers, S. G. & Fraley, R. T. (1986) *Science 233*, 478−481.
18. DeBlock, M., Batterman, J., Vandewiele, M., Dockx, M., Theon, C., Gassele, V., Rao Movva, N., Thompson, C., van Montagu, M. & Leemans, J. (1987) *EMBO J. 6*, 2513−2518.
19. Haughn, G. W., Smith, J., Mazur, B. & Somerville, C. (1988) *Mol. Gen. Genet. 211*, 266−271.
20. Stalker, D., McBride, K. E. & Malyj, L. D. (1988) *Science 242*, 419−423.
21. DeBlock, M., Herrera-Estrella, L., van Montagu, M., Schell, J. & Zambryski, P. (1984) *EMBO J. 3*, 1681−1689.
22. Howard, E. A., Walker, J. C., Dennis, E. S. & Peacock, W. J. (1987) *Planta (Berl.) 170*, 535−540.
23. Olsson, O., Koncz, C. & Szalay, A. (1988) *Mol. Gen. Genet. 215*, 1−9.
24. Ow, D., Wood, K. V., DeLuca, L., DeWet, J. R., Helsinki, D. & Howell, S. H. (1986) *Science 234*, 856−859.
25. Teeri, T. H., Lehväslaiho, H., Franck, M., Uotila, J., Heino, P., Palva, E. T., van Montagu, M. & Herrera-Estrella, L. (1989) *EMBO J. 8*, 343−350.
26. Benfey, P. N. & Chua, N.-H. (1989) *Science 244*, 174−181.
27. Finnegan, E. J., Taylor, B. H., Craig, S. & Dennis, E. S. (1989) *The Plant Cell 1*, 757−764.
28. Budar, F., Deboek, F., van Montagu, M. & Hernalsteens, J.-P. (1986) *Plant Sci. 46*, 195−206.
29. Depicker, A. G., Jacobs, A. M. & van Montagu, M. (1988) *Plant Cell Rep. 7*, 63−66.
30. Zambryski, P. (1988) *Annu. Rev. Genet. 22*, 1−30.
31. Stachel, S. E. & Zambryski, P. (1986) *Cell 46*, 325−333.
32. Stachel, S. E., Timmermann, B. & Zambryski, P. (1987) *EMBO J. 6*, 3889−3891.
33. Zambryski, P. (1989) in *Mobile DNA* (Berg, D. B. & Howe, M. M., eds) pp. 309−333, American Society for Microbiology, Washington, D.C.
34. Zambryski, P., Tempe, J. & Schell, J. (1989) *Cell 56*, 193−201.
35. Kahl, G. & Schell, J. (1982) *Molecular biology of plant tumors*, Academic Press, London.
36. Jorgensen, R., Snyder, C. & Jones, J. D. G. (1987) *Mol. Gen. Genet. 207*, 471−477.
37. Depicker, A. G., Herman, L., Jacobs, A., Schell, J. & van Montagu, M. (1987) *Mol. Gen. Genet. 201*, 477−484.
38. Koncz, C., Martini, N., Mayerhofer, R., Koncz-Kalman, Z., Körber, H., Redei, G. P. & Schell, J. (1989) *Proc. Natl Acad. Sci. USA 86*, 8467−8471.
39. DeBlock, M., Schell, J. & van Montagu, M. (1985) *EMBO J. 4*, 1367−1372.
40. Klee, H. J., Horsch, R. & Rogers, S. (1987) *Annu. Rev. Genet. 20*, 467−486.
41. Schell, J. (1987) *Science 237*, 1176−1183.
42. Walden, R., Koncz, C. & Schell, J. (1990) *Methods Mol. Biol. 5*, 175−194.
43. Fraley, R. T., Rogers, S. G., Horsch, R. B., Eichholtz, D. A., Flick, C. L., Hoffman, N. L. & Saunders, P. R. (1985) *Bio-Technology 3*, 629−635.
44. Zambryski, P., Joos, H., Genetello, C., Leemans, J., van Montagu, M. & Schell, J. (1983) *EMBO J. 2*, 2143−2154.
45. Ditta, G., Stanfield, S., Corbin, D. & Helsinki, D. R. (1980) *Proc. Natl Acad. Sci. USA 77*, 7347−7351.
46. van Haute, E., Joos, H., Maes, S., Warren, G., van Montagu, M. & Schell, J. (1983) *EMBO J. 2*, 411−418.

47. Ebert, P. R., Ha, S. B. & An, G. (1987) *Proc. Natl Acad. Sci. USA 84*, 5745–5749.
48. Mattanovich, D., Rüker, F., da Camara Machado, A., Laimer, M., Regner, F., Steinkeliner, H., Himmler, G. & Katinger, H. (1989) *Nucleic Acids Res. 17*, 6747.
49. Bevan, M. (1984) *Nucleic Acids Res. 12*, 8711–8721.
50. Hoekema, A., Hirsch, P. R., Hooykaas, P. J. & Schilperoot, R. A. (1983) *Nature 303*, 179–181.
51. Rogers, S. G., Horsch, R. B. & Fraley, R. T. (1986) in *Methods Enzymol. 118*, 627–640.
52. An, G., Watson, B. D. & Chiang, C. C. (1986) *Plant Physiol. 81*, 301–305.
53. Valvekans, D., van Montagu, M. & Lijbettens, M. (1988) *Proc. Natl Acad. Sci. USA 85*, 5536–5540.
54. Ishida, B. K., Snyder, G. W. & Belknop, W. R. (1989) *Plant Cell Rep. 8*, 325–328.
55. Trinh, T. H., Manto, S., Pua, E.-C. & Chua, N.-H. (1987) *Bio-Technology 5*, 1081–1084.
56. Simpson, R. B., Spielmann, A., Margossian, L. & McKnight, T. D. (1986) *Plant Mol. Biol. 6*, 403–415.
57. Stougaard, J., Abildstein, D. & Marcker, K. A. (1987) *Mol. Gen. Genet. 207*, 251–255.
58. Stougaard Jensen, J., Marcker, K. A., Otten, L. & Schell, J. (1986) *Nature 321*, 669–674.
59. Tepfer, D. (1989) in *Plant microbe interactions: molecular and genetic perspectives* (Kosuge, T. & Nester, E. W., eds) vol. 3, pp. 294–342, McGraw Hill, New York.
60. Marton, L., Wullems, G. J., Molendijk, L. & Schilperoort, R. A. (1979) *Nature 227*, 129–131.
61. Feldmann, K. A. & Marks, M. D. (1981) *Mol. Gen. Genet. 208*, 1–9.
62. Grimsley, N., Hohn, B., Ramos, G., Kado, C. & Radowky, P. (1989) *Mol. Gen. Genet. 216*, 309–316.
63. Hernalsteens, J.-P., Thia-Thoong, L., Schell, J. & van Montagu, M. (1984) *EMBO J. 3*, 3039–3041.
64. Hooykaas-van Slogteren, G. M. S., Hooykaas, P. J. J. & Schilperoort, R. A. (1984) *Nature 311*, 763–764.
65. Schäfer, W., Görz, A. & Kahl, G. (1987) *Nature 327*, 529–532.
66. Lörz, H., Göbel, E. & Brown, P. (1988) *Plant Breeding 100*, 1–25.
67. Hain, R., Steinbiss, H.-H. & Schell, J. (1985) *Plant Cell Rep. 3*, 60–64.
68. Deshayes, A., Herrera-Estrella, L. & Caboche, M. (1985) *EMBO J. 4*, 2731–2737.
69. Crossway, A., Oakes, J. V., Irvine, J. M., Ward, B., Knauf, V. C. & Shewmaker, C. K. (1986) *Mol. Gen. Genet. 203*, 179–185.
70. DeLaPena, A., Lörz, H. & Schell, J. (1987) *Nature 325*, 274–276.
71. Paskowski, J., Shillito, R. D., Saul, M., Mandak, V., Hohn, T., Hohn, B. & Potrykus, I. (1984) *EMBO J. 3*, 2717–2722.
72. Fromm, M., Taylor, L. P. & Walbot, V. (1985) *Proc. Natl Acad. Sci. USA 82*, 5824–5828.
73. Klein, T. M., Wolf, E. D., Wu, R. & Stanford, J. C. (1987) *Nature 327*, 70–73.
74. Töpfer, R., Pröls, M., Schell, J. & Steinbiss, H.-H. (1988) *Plant Cell Rep. 7*, 225–228.
75. Dangl, J., Hauffe, K. D., Lipphardt, S., Hahlbrock, K. & Scheel, D. (1987) *EMBO J. 6*, 2551–2556.
76. Townsend, R., Stanley, J., Curson, S. J. & Short, M. N. (1985) *EMBO J. 4*, 33–37.
77. French, R., Janda, M. & Ahlquist, P. (1986) *Science 231*, 1294–1296.
78. Takamatsu, N., Ishikawa, M., Meshi, T. & Okada, Y. (1987) *EMBO J. 5*, 307–311.
79. Gronenborn, B. & Matzeit, V. (1989) in *Cell culture and somatic genetics of plants: molecular biology of plant nuclear genes* (Schell, J. & Vasil, I. K., eds) pp. 69–101, Academic Press, New York.
80. Gallie, D. R., Lucas, W. J. & Walbot, V. (1989) *The Plant Cell 1*, 301–311.
81. Shillito, R. D., Paszkowski, J. & Potrykus, I. (1985) *Bio-Technology 3*, 1099–1103.
82. Schöchter, R. J., Shillito, R. D., Saul, M., Paszkowski, J. & Potrykus, I. (1986) *Bio-Technology 4*, 1093–1096.
83. Paszkowski, J., Baur, M., Bogucki, A. & Potrykus, I. (1988) *EMBO J. 7*, 4021–4026.
84. Potrykus, I., Paszkowski, J., Saul, M., Petruska, J. & Shillito, R. D. (1985) *Mol. Gen. Genet. 199*, 169–177.
85. Meyer, P., Walgenbach, E., Bussmann, K., Hombrecher, G. & Saedler, H. (1985) *Mol. Gen. Genet. 198*, 513–518.
86. Boynton, J. E., Gillham, N. W., Harris, E. H., Hosler, J. P., Johnson, A. M., Jones, A. R., Randolph-Anderson, B. L., Robertson, D., Klein, T. M., Shark, K. B. & Sanford, J. C. (1988) *Science 240*, 1534–1558.
87. Scowcroft, W. R. & Larkin, P. J. (1988) in *Applications in plant cell and tissue culture* (Bock, G. & Marsh, J., eds) pp. 21–35, John Wiley, New York.
88. Nagy, F., Morelli, G., Fraley, R. T., Rogers, S. G. & Chua, N.-H. (1985) *EMBO J. 4*, 3063–3068.
89. Hepburn, A. G., Clarke, L. E., Pearson, L. & White, J. (1984) *J. Mol. Appl. Genet. 2*, 315–329.
90. Van Slogteren, G. M. S., Hooykaas, P. J. J. & Schilperoort, R. A. (1984) *Plant Mol. Biol. 3*, 333–336.
91. Jones, J., Gilbert, D., Grady, K. & Jorgenson, R. (1987) *Mol. Gen. Genet. 207*, 478–485.
92. Dean, C., Favreau, M., Tamaki, S., Bond-Nutter, D., Dunsmiur, P. & Bedbrook, J. (1988) *Nucleic Acids Res. 16*, 7601–7617.
93. Dean, C., Jones, J., Favreau, M., Dunsmiur, P. & Bedbrook, J. (1988) *Nucleic Acids Res. 16*, 9267–9283.
94. Velten, J. & Schell, J. (1985) *Nucleic Acids Res. 13*, 6981–6998.
95. Vaeck, M., Reynaerts, A., Hofte, H., Jansens, S., DeBeukeleer, M., Dean, C., Zabeau, M., van Montagu, M. & Leemans, J. (1987) *Nature 328*, 33–37.
96. Gordon, J. W. (1989) *Int. Rev. Cytol. 115*, 171–229.
97. Grosveld, F., van Assendelft, G. B., Greaves, D. R. & Kollias, G. (1987) *Cell 51*, 975–985.
98. Grimsley, N., Hohn, T., Davies, J. W. & Hohn, B. (1987) *Nature 325*, 177–179.
99. Grimsley, N., Hohn, B., Hohn, T. & Walden, R. (1986) *Proc. Natl Acad. Sci. USA 83*, 3282–3286.
100. Sanders, E. & Mertes, G. (1985) *Adv. Virus Res. 29*, 215–262.
101. Dawson, W. O., Beck, D. L., Knorr, D. A. & Grantham, G. L. (1986) *Proc. Natl Acad. Sci. USA 83*, 1832–1836.
102. Simon, A. E. & Howell, S. H. (1987) *Virology 156*, 146–152.
103. Vos, P., Jeagle, M., Wellink, J., Verver, J., Eggen, R., van Kammen, A. & Goldbach, R. (1988) *Virology 166*, 33–41.
104. Ziegler-Graff, V., Bouzoubaa, S., Jupin, I., Guilley, H., Jonard, G. & Richards, K. (1988) *J. Gen. Virol. 69*, 2347–2357.
105. Covey, S. & Hull, R. (1985) *Oxf. Surv. Plant Mol. Cell. Biol. 2*, 339–346.
106. Mason, W. S., Taylor, J. M. & Hull, R. (1987) *Adv. Virus Res. 32*, 35–96.
107. Brisson, N., Paszkowski, J., Penswick, J., Gronenborn, B., Potrykus, I. & Hohn, T. (1985) *Nature 310*, 511–514.
108. DeZoeten, G. A., Penswick, J. R., Horisberger, M. A., Ahl, P., Schultze, M. & Hohn, T. (1989) *Virology 172*, 213–222.
109. Davies, J. S. & Stanley, J. (1989) *Trends Genet. 5*, 77–81.
110. Etessami, P., Callis, R., Ellwood, S. & Stanley, J. (1988) *Nucleic Acids Res. 16*, 4811–4829.
111. Ward, A., Etassami, P. & Stanley, J. (1988) *EMBO J. 7*, 1583–1587.
112. Lazarowitz, S. G., Pinder, A. J., Damsteegt, V. D. & Rogers, S. G. (1989) *EMBO J. 8*, 1023–1032.
113. Rogers, S. G., Bisaro, D. M., Horsch, R. B., Fraley, R. T., Hoffman, N. L., Brand, L., Elmer, J. S. & Lloyd, A. M. (1986) *Cell 45*, 593–600.
114. Sunter, G., Gardiner, W. E., Rushing, A. E., Rogers, S. G. & Bisaro, D. (1988) *Plant Mol. Biol. 8*, 477–484.
115. Hayes, R. J., Petty, I. T. D., Coutts, R. H. A. & Buck, K. W. (1988) *Nature 334*, 179–181.
116. Weising, K., Schell, J. & Kahl, G. (1988) *Annu. Rev. Genet. 22*, 421–477.
117. de Bruijn, F. J., Felix, G., Grunnenberg, H. J., Metz, B., Ratet, P., Simons-Schreier, A., Szabados, L., Welters, P. & Schell, J. (1989) *Plant Mol. Biol. 13*, 319–325.

576

118. Chen, Z.-L., Pan, N.-S. & Beachy, R. N. (1988) *EMBO J. 7*, 297—302.
119. Poulson, C. & Chua, N.-H. (1988) *Mol. Gen. Genet. 214*, 16—23.
120. Simpson, J., Schell, J., van Montagu, M. & Herrera-Estrella, L. (1986) *Nature 323*, 551—552.
121. Timko, M. P., Kausch, A. P., Castrescena, C., Fassler, J., Herrera-Estrella, L., Van den Broek, G., van Montagu, M., Schell, J. & Cashmore, A. R. (1985) *Nature 318*, 579—582.
122. Koncz, C., Kreuzaler, F., Kalman, Z. & Schell, J. (1984) *EMBO J. 3*, 1929—1937.
123. An, G., Costa, M. A., Mitra, A., Ha, S.-B. & Marton, L. (1988) *Plant Physiol. 88*, 547—552.
124. An, G., Ebert, P. R., Yi, B.-Y. & Choi, C.-H. (1986) *Mol. Gen. Genet. 203*, 245—250.
125. Mitra, A. & An, G. A. (1989) *Mol. Gen. Genet. 215*, 294—299.
126. Koncz, C. & Schell, J. (1986) *Mol. Gen. Genet. 204*, 383—396.
127. Langridge, W. H. R., Fitzgerald, K. J., Koncz, C., Schell, J. & Szalay, A. A. (1989) *Proc. Natl Acad. Sci. USA 86*, 3219—3223.
128. Benfey, P. N., Ren, L. & Chua, N.-H. (1989) *EMBO J. 8*, 2195—2202.
129. Odell, J. T., Nagy, F. & Chua, N.-H. (1985) *Nature 313*, 810—812.
130. Fluhr, R., Kuhlemeyer, C., Nagy, F. & Chua, N.-H. (1986) *Science 232*, 1106—1112.
131. Odell, J. T., Knowlton, S., Lin, W. & Mauvais, C. J. (1988) *Plant Mol. Biol. 10*, 263—272.
132. Ow, D., Jacobs, J. D. & Howell, S. H. (1987) *Proc. Natl Acad. Sci. USA 84*, 4870—4874.
133. Fang, R.-X., Nagy, F., Sivasubramaniam, S. & Chua, N.-H. (1989) *The Plant Cell 1*, 141—150.
134. Kay, R., Chan, A., Daly, M. & McPherson, J. (1987) *Science 236*, 1299—1302.
135. Dangl, J., Hahlbrock, K. & Schell, J. (1988) in *Cell culture and somatic cell genetics of plants* (Vasil, I. K. & Schell, J., eds) pp. 155—173, Academic Press, New York.
136. Kaulen, H., Schell, J. & Kreuzaler, F. (1986) *EMBO J. 5*, 1—8.
137. Lipphardt, S., Brettschneider, R., Kreuzaler, F., Schell, J. & Dangl, J. (1988) *EMBO J. 7*, 4027—4033.
138. Prat, S., Willmitzer, L. & Sanchez-Serrano, J. J. (1989) *Mol. Gen. Genet. 217*, 209—214.
139. Katagiri, F., Lam, F. & Chua, N.-H. (1989) *Nature 340*, 727—730.
140. Staiger, D., Kaulen, H. & Schell, J. (1989) *Proc. Natl Acad. Sci. USA 86*, 6930—6934.
141. Paz-Ares, J., Ghosal, D., Wienand, U., Peterson, P. A. & Saedler, H. (1988) *EMBO J. 6*, 3553—3558.
142. Medford, J. & Klee, H. (1989) in *The molecular basis of plant development* (Goldberg, R., ed.) pp. 211—220, Alan R. Liss, New York.
143. Schmülling, T., Beinsberger, S., De Greef, J., Schell, J., Van Onckelen, H. & Spena, A. (1989) *FEBS Lett. 249*, 401—406.
144. Gatz, C. & Quail, P. H. (1988) *Proc. Natl Acad. Sci. USA 85*, 1394—1397.
145. Ma, J., Przibilla, E., Hu, J., Bogorad, L. & Ptashne, M. (1988) *Nature 334*, 631—633.
146. Maliga, P. (1984) *Annu. Rev. Plant Physiol. 35*, 519—542.
147. Meyer, P., Heidmann, I., Forkmann, G. & Saedler, H. (1987) *Nature 330*, 677—678.
148. Klee, H. J., Hayford, M. B. & Rogers, S. (1987) *Mol. Gen. Genet. 210*, 282—287.
149. Olszewski, N. E., Martin, F. B. & Ausubel, F. (1988) *Nucleic Acids Res. 16*, 10765—10782.
150. Simeons, C., Alliate, Th., Mendel, R., Muller, A., Sciemann, J., Van Lijsebettens, M., Schell, J., van Montagu, M. & Inze, D. (1986) *Nucleic Acids Res. 14*, 8073—8090.
151. Prosen, D. E. & Simpson, R. B. (1987) *Bio-Technology 5*, 966—971.
152. Federoff, N. V. (1983) in *Mobile genetic elements* (Shapiro, J. A., ed.) pp. 1—63, Academic Press, New York.
153. Shepherd, N. S. (1988) in *Plant molecular biology; a practical approach* (Shaw, C. H., ed.) pp. 187—220, IRL Press, Oxford.
154. McClintock, B. (1951) *Cold Spring Harbor Symp. Quant. Biol. 16*, 13—47.
155. Sutton, W. D., Gerlach, W., Schwartz, D. & Peacock, W. J. (1984) *Science 223*, 1265—1268.
156. Baker, B., Schell, J., Lörz, H. & Federoff, N. (1986) *Proc. Natl Acad. Sci. USA 83*, 4844—4848.
157. Knapp, A., Coupland, G., Uhrig, U., Starlinger, P. & Salamini, F. (1988) *Mol. Gen. Genet. 213*, 285—290.
158. Masterson, R. V., Furtek, D., Grevelding, C. & Schell, J. (1989) *Mol. Gen. Genet. 219*, 461—466.
159. Van Sluys, M. A., Tempe, J. & Federoff, N. V. (1987) *EMBO J. 6*, 3881—3889.
160. Yoder, J. I., Palys, J., Alpert, K. & Lossner, M. (1988) *Mol. Gen. Genet. 213*, 291—296.
161. Döring, H.-P. & Starlinger, P. (1989) *Annu. Rev. Genet. 20*, 175—200.
162. Baker, B., Coupland, G., Federoff, N., Starlinger, P. & Schell, J. (1987) *EMBO J. 6*, 1547—1554.
163. Jones, J. D. G., Garland, F. M., Maliga, P. & Dooner, H. K. (1989) *Science 244*, 204—207.
164. Coupland, G., Baker, B., Schell, J. & Starlinger, P. (1988) *EMBO J. 7*, 3653—3659.
165. Hehl, R. & Baker, B. (1989) *Mol. Gen. Genet. 217*, 53—59.
166. Andre, D., Colau, D., Schell, J., van Montagu, M. & Hernalsteens, J.-P. (1988) *Mol. Gen. Genet. 204*, 512—518.
167. Teeri, T. H., Herrera-Estrella, L., Depicker, A., van Montagu, M. & Palva, E. T. (1986) *EMBO J. 5*, 1755—1760.
168. Koncz, C., Meyerhofer, R., Koncz-Kalman, Z., Nawrath, C., Reiss, B., Redei, G. P. & Schell, J. (1990) *EMBO J. 9*, 1337—1346.
169. Goebl, M. G. & Pedes, T. D. (1986) *Cell 46*, 983—992.
170. Simmons, M. J. (1986) *Genetics 111*, 869—884.
171. Gridley, T., Soriano, P. & Jaenisch, R. (1987) *Trends Biochem. Sci. 3*, 162—166.
172. Feldmann, K. A., Marks, M. D., Christianson, M. L. & Quatrano, R. S. (1989) *Science 243*, 1351—1354.
173. Damm, B. & Willmitzer, L. (1988) *Mol. Gen. Genet. 213*, 15—20.
174. Meyerowitz, E. (1987) *Annu. Rev. Genetics 21*, 93—111.
175. Spena, A., Schmülling, T., Koncz, C. & Schell, J. (1987) *EMBO J. 6*, 3891—3889.
176. Sandler, S. J., Slayton, M., Townsend, J. A., Ralston, M., Bedbrook, J. R. & Dunsmuir, P. (1988) *Plant Mol. Biol. 11*, 301—310.
177. Haseloff, J. & Gerlach, W. L. (1988) *Nature 334*, 585—591.
178. Hesse, T., Feldwisch, J., Balshüsemann, D., Bauvo, G., Puype, D., Vandekerkhove, J., Löbler, M., Klämpt, D., Schell, J. & Palme, K. (1989) *EMBO J. 8*, 2453—2461.
179. Palme, K., Diefenthal, T., Sander, C., Vingron, M. & Schell, J. (1989) in *The guanine nucleotide binding proteins* (Bosch, L., Kraal, B. & Parmeggiani, A., eds) pp. 273—284, Plenum Publishing, New York.
180. Jena, P. K., Reddy, A. S. N. & Pooviah, B. W. (1989) *Proc. Natl Acad. Sci. USA 86*, 3644—3648.
181. Lawton, M. A., Yamamoto, R. T., Hanks, S. K. & Lamb, C. J. (1989) *Proc. Natl Acad. Sci. USA 86*, 3140—3144.

Eur. J. Biochem. *193*, 1–18 (1990)
© FEBS 1990

Review

myo-Inositol metabolites as cellular signals

C. Peter DOWNES and Colin H. MACPHEE

Department of Biochemistry, Medical Sciences Institute, University of Dundee, Dundee, Scotland
and Smith Kline & Beecham Ltd, The Frythe, Welwyn, England

(Received May 14, 1990) — EJB 90 0545

The discovery of the second-messenger functions of inositol 1,4,5-trisphosphate and diacylglycerol, the products of hormone-stimulated inositol phospholipid hydrolysis, marked a turning point in studies of hormone function. This review focusses on the *myo*-inositol moiety which is involved in an increasingly complex network of metabolic interconversions. *myo*-Inositol metabolites identified in eukaryotic cells include at least six glycerophospholipid isomers and some 25 distinct inositol phosphates which differ in the number and distribution of phosphate groups around the inositol ring. This apparent complexity can be simplified by assigning groups of *myo*-inositol metabolites to distinct functional compartments. For example, the phosphatidylinositol 4-kinase pathway functions to generate inositol phospholipids that are substrates for hormone-sensitive forms of inositol-phospholipid phospholipase C, whilst the newly discovered phosphatidylinositol 3-kinase pathway generates lipids that are resistant to such enzymes and may function directly as novel mitogenic signals. Inositol phosphate metabolism functions to terminate the second-messenger activity of inositol 1,4,5-trisphosphate, to recycle the latter's *myo*-inositol moiety and, perhaps, to generate additional signal molecules such as inositol 1,3,4,5-tetrakisphosphate, inositol pentakisphosphate and inositol hexakisphosphate. In addition to providing a more complete picture of the pathways of *myo*-inositol metabolism, recent studies have made rapid progress in understanding the molecular basis underlying hormonal stimulation of inositol-phospholipid-specific phospholipase C and inositol 1,4,5-trisphosphate-mediated Ca^{2+} mobilisation.

myo-Inositol is the predominant isomer of cyclohexane hexol that occurs in eukaryotic cells. Perhaps its best known function is as a precursor of the inositol phospholipids which, when cleaved by a hormone-stimulated inositol-phospholipid-specific phospholipase C (PIC), generate the ubiquitous second messengers, inositol 1,4,5-trisphosphate [$Ins(1,4,5)P_3$] and diacylglycerol. The principal structural feature of Ins is that the hydroxyl group located at the 2-position is out of step with the others: at neutral pH, the preferred conformation has an axial 2-hydroxyl with all the remaining hydroxyl groups being equatorial to the plane of the six-membered ring. A consequence of this structure is that derivatisation of Ins, for example by the insertion of a monoester phosphate group, can generate a large number of structurally related molecules, including up to 63 distinct inositol phosphate species [1]. In functionally important molecules, such as $Ins(1,4,5)P_3$, the precise location of the phosphate groups is fundamental; the

Correspondence to C. P. Downes, Department of Biochemistry, Medical Sciences Institute, The University, Dundee DD1 4HN, Scotland

Abbreviations. Ins*P*, inositol phosphate; PtdIns*P*, phosphatidylinositol phosphate; PIC, phosphatidylinositol-specific phospholipase C; G-protein, guanine-nucleotide-dependent regulatory protein; GTP[S], guanosine 5'-[γ-thio]triphosphate; GDP[S], guanosine 5'-[β-thio]triphosphate; PDGF, platelet-derived growth factor; EGF, epidermal growth factor; CSF, colony-stimulating factor; cIns*P*$_3$, inositol 1,2-(cyclic)phosphate; cIns*P*$_3$, inositol 1,2-(cyclic)phosphate 4,5-bisphosphate; ADP[S], adenosine 5'-[β-thio]triphosphate.

structurally similar metabolite, inositol 1,3,4-trisphosphate [$Ins(1,3,4)P_3$], is at least 100-fold weaker than $Ins(1,4,5)P_3$ itself in assays of their binding to the $Ins(1,4,5)P_3$ receptor [2] which functions as a ligand-gated Ca^{2+} channel in the endoplasmic reticulum [3].

A burgeoning complexity of Ins metabolites has been uncovered during the last five years and it seems likely that the pathways involved subserve a number of cellular functions in addition to the generation of diacylglycerol and $Ins(1,4,5)P_3$ second-messengers. These additional pathways include the extraordinarily complex metabolism of $Ins(1,4,5)P_3$ [4]; the metabolism of inositol polyphosphates such as inositol pentakisphosphates and hexakisphosphates [5]; the emergence of a novel inositol phospholipid pathway initiated by phosphatidylinositol 3-kinase which appears to be regulated by tyrosine kinases [6], and the synthesis of PtdIns-glycans which provide the membrane-anchoring domains of a variety of cell-surface proteins and may be precursors of water-soluble inositol glycans of ill-defined structure, which have been proposed as putative second messengers of insulin action. The latter is a highly specialised and complex field in its own right and the reader is referred to the excellent review by Low [7] for a comprehensive discussion of the topic, which will not be considered further in this article. In this review we shall describe the scope of the metabolic and functional possibilities which result from the structural features of Ins depicted in Fig. 1.

2

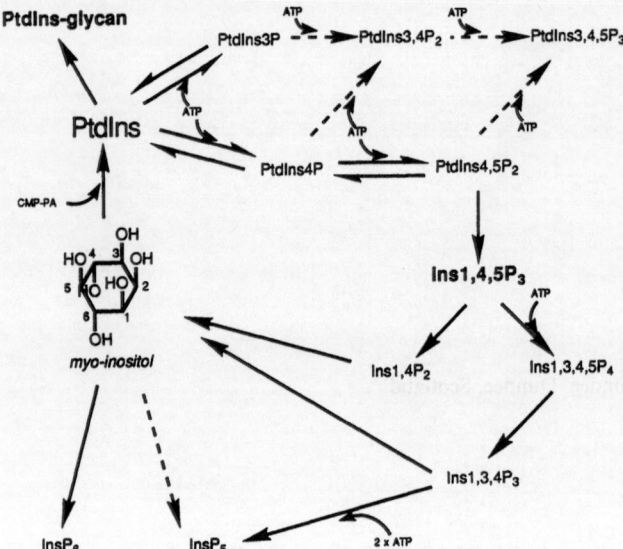

Fig. 1. *myo-Inositol metabolism. myo*-Inositol is depicted in the Fisher projection with the numbering positions indicated. Lipid metabolites are grouped at the top of the figure with water-soluble compounds at the bottom. Parentheses have been omitted from the standard abbreviations to simplify the diagram. Dotted arrows indicate reactions or pathways that are suspected, but not yet confirmed. Further details of the intermediates in inositol phosphate metabolism and the enzymes catalysing these reactions are given in the text and Fig. 2. CMP-PA, cytidine monophosphoryl phosphatidate

CELLULAR INOSITOL AND ITS METABOLIC FATE

All cells appear to maintain substantial levels of Ins in their cytosol, often above that found in the surrounding medium, but precisely how this is achieved has been a neglected area of investigation. Essentially, there are three potential sources of cellular Ins; uptake from the medium, *de novo* synthesis and recycling of Ins-containing compounds such as inositol phosphates [8, 9]. Na^+-dependent, active Ins-transport mechanisms have been described in several types of cell [10, 11], but most cells appear to transport Ins via a very-low-affinity, passive mechanism, whose precise characteristics and specificity have not been fully defined. The low-affinity transporters in liver and parotid gland, for example, are not saturated by Ins concentrations as high as 10 mM, some 200 times the expected serum concentration [12, 13].

Certain tissues, notably brain and testis, synthesise large amounts of Ins from glucose. The rate-limiting step in this process is the synthesis of inositol 3-phosphate (Ins3*P*) from glucose 6-phosphate. Ins3*P* is then dephosphorylated by inositol monophosphate phosphomonoesterase (see the section on inositol phosphate metabolism) to give free Ins [14]. The activity of Ins3*P* synthase is ordinarily low in most tissues, but may be under hormonal control in the reproductive organs and liver of male rats [15]. In fact, since several cultured cell lines appear able to grow in customised media lacking Ins, it appears that regulation of Ins synthesis may be a common occurrence. In AR4-2J pancreatoma cells, *de novo* synthesis of Ins occurred in media that contained less than 400 µM Ins which implies that this pathway is under feedback regulation by the cellular Ins content [16].

Inositol phosphates formed by the hormonal stimulation of inositol phospholipid breakdown are efficiently recycled to the free Ins pool by a series of inositol-phosphate phosphomonoesterases (see the section on inositol phosphate metabolism). This pathway appears to be an important source of Ins during hormonal stimulation and it is efficiently blocked by therapeutically relevant concentrations of Li^+, which is used in the treatment of manic depressive psychoses, but it clearly cannot contribute to net accumulation of Ins by cells [9, 17].

Synthesis of phosphatidylinositol

The majority of the compounds to be discussed in this article are ultimately derived from an inositol phospholipid species and their formation requires the incorporation of Ins into PtdIns. This is achieved by PtdIns synthase (CDP-diacylglycerol: *myo*-inositol 3-phosphatidyltransferase, Fig. 1). The subcellular distribution of this key enzyme is currently of interest because of the evidence that many cells contain metabolically segregated pools of inositol phospholipids. One pool may provide the substrates for hormone-stimulated inositol phospholipid breakdown and is characterised by a high metabolic turnover rate, especially in the presence of hormone, whilst the bulk of cellular PtdIns turns over more slowly and is not immediately affected by hormonal stimulation (for a critical assessment of this field see the review by Michell et al. [18]). The maintenance of such segregated pools would require the occurrence of PtdIns synthase activity in each compartment, and there is at least preliminary evidence for distinct forms of this enzyme located in the endoplasmic reticulum and plasma membrane of GH_3 pituitary cells [19, 20]. Confirmation of this important observation in a variety of cells, and a description of the characteristics and control of distinct forms of PtdIns synthase, are now required in order to understand how the synthesis of PtdIns that is destined for specific cellular functions is regulated.

Alternative routes of Ins metabolism

A large number of inositol phosphate isomers have recently been identified in cell extracts from a variety of sources. Whilst many of these compounds are likely to be direct or indirect products of inositol phospholipid hydrolysis, the metabolic origins of some of them remain to be established. For example, although inositol 1,3,4,5,6-pentakisphosphate [Ins(1,3,4,5,6)P_5] and hexakisphosphate (InsP_6) are two of the oldest known Ins metabolites, their widespread occurrence in mammalian tissues has only recently been appreciated and relatively little is known about how they are made. One possibility is that at least some cellular inositol phosphates may be synthesised through direct phosphorylation of Ins itself. An Ins kinase activity which converts Ins to DIns1*P* or LIns1*P* was reported as long ago as 1965 in mung bean seeds (which make large amounts of InsP_6) and in bovine liver [21]. A similar activity has now been demonstrated in the slime-mould *Dictyostelium*, and a putative pathway to InsP_6 constructed [21 a]. These results suggest that direct, inositol-phospholipid-independent, phosphorylation of Ins may provide additional molecules (e.g. Ins$P_{n \geqslant 1}$) of regulatory significance, although they require confirmation in a variety of organisms and tissues to establish the extent of their involvement in Ins$P_{n \geqslant 1}$ metabolism. Further consideration of these aspects will be deferred to the section on inositol phosphate metabolism.

THE PTDINS 4-KINASE PATHWAY:
INOSITOL PHOSPHOLIPIDS THAT ARE SUBSTRATES
FOR HORMONE-STIMULATED PHOSPHOLIPASE C

The only established function of a cellular Ins metabolite is the role of Ins(1,4,5)P_3 as a second messenger for Ca^{2+} mobilisation from intracellular stores [22] in response to stimulation by a wide variety of hormones, neurotransmitters and growth factors (for a complete and up-to-date list of the cell-surface receptors which utilise this signal-transduction pathway [23 – 25]). Ins(1,4,5)P_3 is cleaved from its phospholipid precursor, PtdIns(4,5)P_2, by PIC. This process simultaneously generates 1,2-diacylglycerol which is the second messenger for activation of protein kinase C [26], emphasising the central importance of PIC activity in coordinating cellular responses to hormonal stimulation. As cells contain only a small pool of PtdIns(4,5)P_2 that needs to be constantly replenished, and as PIC activities assayed *in vitro* are generally capable of hydrolyzing PtdIns and PtdIns4P as well as PtdIns(4,5)P_2, it is also important to consider how these phospholipids are interconverted in cell membranes in order to appreciate the stoichiometry of formation of each of these second messengers [27].

Interconversion of PtdIns, PtdIns4P and PtdIns(4,5)P_2

PtdIns(4,5)P_2 is synthesised from PtdIns by sequential phosphorylation catalysed by PtdIns 4-kinase and PtdIns4P 5-kinase. Specific phosphomonoesterases also exist which promote the reversal of these reactions (see [25] and references therein). There has been considerable recent progress in the isolation and characterisation of these inositol phospholipid kinases and some limited evidence for their regulation as discussed below.

PtdIns 4-kinase

Several isozymes of PtdIns 4-kinase are now known to exist although the possible role(s) of each in receptor-stimulated inositol phospholipid metabolism remain to be elucidated. Endemann and colleagues [28] were the first to provide definitive evidence for the existence of two quite distinct PtdIns 4-kinases from bovine brain. The most extensively studied enzyme appears to be an integral membrane protein and has been purified to apparent homogeneity by several groups [29 – 31]. Its main distinguishing characteristics are as follows: it is activated by detergents, has a reported K_m for ATP of 20 – 70 µM, is sensitive to inhibition by adenosine ($K_i \approx$ 20 µM), and has an estimated native molecular mass of approximately 55 kDa. The kinetic parameters for the other substrate of this enzyme, PtdIns, are more difficult to summarize due to substantial variability in the assay conditions employed by different laboratories. Detergent-free assays using sonicated substrate have been reported, but the most commonly used assay procedure employs detergent/PtdIns mixed micelles but with varying type and molar ratio detergent. This is a recurring problem in lipid biochemistry and stems from the obvious difficulty in attempting to reconstruct relevant features of the phospholipid environment which exists in cellular membranes.

The other major PtdIns 4-kinase activity reported, although not yet purified to homogeneity, can readily be distinguished by a number of features. Firstly, substantial activity can be detected, along with PtdIns 3-kinase (see later), in the soluble fraction of human placental extracts (Macphee and

Downes, unpublished observation). This activity is stimulated by detergents, has a high K_m for ATP of between 250 – 750 µM, is relatively insensitive to adenosine inhibition (K_i approximately 1 mM), and has an apparent molecular mass of around 200 kDa ([28, 32], Macphee and Downes, unpublished data). Additional reports describe the substantial purification and characterisation of what would appear to be two further mammalian isozymes of PtdIns 4-kinase. Saltiel and coworkers [33] have reported that a 45-kDa membrane-bound PtdIns kinase purified from bovine brain myelin also has the capacitiy to phosphorylate PtdIns$P$, although the enzyme has a much higher affinity for PtdIns than previously characterised activities. Finally, also from brain, an 80-kDa membrane-bound PtdIns kinase has been characterised which exclusively phosphorylates PtdIns, is stimulated by detergents, but only weakly inhibited by adenosine (K_i = 200 µM) [34]. Isolation of the genes coding for each of the prescribed PtdIns 4-kinases together with amino acid sequence data should provide insights into the structural relationships between different subtypes of PtdIns 4-kinase.

A more complete understanding of the roles, regulation, and true cellular location of each enzyme will result from the use of specific antisera and molecular biology technologies. Although such molecular tools are only now forthcoming, reports do exist describing subcellular location and possible modes of regulation of PtdIns 4-kinase. As previously mentioned, PtdIns 4-kinase activities can be found both in soluble and membrane fractions of cells. Recent data suggest that the majority of membrane-associated activity is not located in the plasma membrane but instead in low-density membranes possibly involved in membrane trafficking [35]. This is in agreement with other findings demonstrating PtdIns kinase concentrated in chromaffin granules [36] and endocytic coated vesicles [37], both of which are transport vesicles. One theory is that, following the synthesis of PtdIns in the endoplasmic reticulum, the phosphorylation to PtdIns4P occurs intracellularly during translocation to the plasma membrane. This cannot be the whole story, however, because mammalian erythrocytes, which lack intracellular membranes, have substantial PtdIns 4-kinase activity. In addition, highly purified nuclei from liver, Friend erythroleukemia cells and Swiss 3T3 cells have also been shown to possess PtdIns 4-kinase activity [38 – 40], although the function of nuclear PtdIns$P_{n \geqslant 1}$ is not currently known.

There is little conclusive evidence for the activation of PtdIns 4-kinase in hormone-stimulated cells. An apparent PtdIns 4-kinase has been shown to be rapidly activated in membranes derived from epidermal-growth-factor (EGF)-treated A431 cells [41]. Although the actual mechanism(s) for this activation has still to be elucidated it appeared to involve a decrease in the K_m for PtdIns. Phorbol esters, activators of protein kinase C, have been demonstrated to markedly increase the levels of PtdIns$P$ in some cells [42, 43]. However, in each case, it was not clear whether the response was due to inositol phospholipid kinase activation or to inhibition of the phosphomonoesterase. Other reports have suggested a role for cAMP in regulating PtdIns kinase activity. Suga and colleagues [44] concluded that PtdIns kinase was activated following an elevation of cAMP in intact platelets. Similar reports of cAMP-dependent activation of PtdIns4-kinase in lymphocyte membranes [45], however, were not confirmed in a more recent study [46]. An intriguing new proposal has arisen from data suggesting that cGMP-dependent protein kinase enhances the activity of the plasmalemmal Ca^{2+} pump of smooth muscle cells indirectly, apparently via the phosphorylation/activation of an associated, adenosine-sensitive PtdIns kinase [47]. Finally, the polyamines, spermidine and spermine, have been shown to enhance PtdIns kinase acitivities in isolated membranes [48, 49]. The potential physiological relevance of this mode of activation is discussed in the following section.

*PtdIns4*P *5-kinase*

PtdIns*P* kinase activity can be found in both the cytosolic and membrane fractions of tissues [50, 51]. Unlike membrane-bound PtdIns kinases which require detergents for solubilisation, PtdInsP kinases can be released from membranes by salt extraction [52, 53]. Although the majority of investigators have simply assumed that the phosphorylation of PtdIns*P* occurs at the 5-position of the inositol ring, Ling and co-workers [52] have demonstrated such a specificity for their purified enzyme. In view of the recent discovery of novel inositol phospholipids phosphorylated at the 3-position of the inositol ring (see later), definitive structural characterisation of the products of lipid kinases will be a vital feature of future work in this area. The human-erythrocyte-membrane-associated PtdIns4*P* 5-kinase appeared to have a subunit molecular mass of 53 kDa by SDS/PAGE and an apparent native size of approximately 150 kDa. It also had a K_m for ATP of 2 μM and was inhibited by its product, PtdIns(4,5)P_2. The enzyme showed a number of similarities to a soluble PtdIns*P* kinase from rat brain [50, 51] which was identified as a 45-kDa protein and which migrated as a 110-kDa complex in the non-denatured form. Another highly purified PtdIns*P* kinase from bovine brain membranes has recently been characterised and showed high specificity for PtdIns*P* since phosphorylation of PtdIns was not observed [53]. This enzyme had an estimated molecular mass of 110 kDa but the position on which it phosphorylates PtdIns*P* was not defined. As yet, there is no definitive proof of the existence of different isozymes of PtdIns*P* kinase and further studies are necessary to establish, for example, the true relationships between soluble and particulate activities.

Although both soluble and particulate PtdIns*P* kinase activities exist, the consensus of opinion suggests that much of the membrane-associated activity is located in the plasma membrane [54−56], where it needs to be in order to contribute to receptor-mediated PtdIns metabolism. However, there is also unequivocal evidence for the occurrence of PtdIns*P* kinase in nuclei [38−41]. In addition, a number of laboratories have reported data which supports the notion that PtdIns4*P* kinase becomes activated during receptor stimulation [42, 57, 58]. The proposed mechanism(s) underlying such an activation remains undefined. One possibility is the removal of product feedback inhibition, through loss of PtdIns(4,5)P_2, but this is unlikely to be the sole contributor. Recent work demonstrating phorbol-ester-mediated increases in PtdInsP_2 levels in intact cells suggest that protein kinase C may be involved [42, 43, 59]. In contrast, Gispen and colleagues have suggested that protein kinase C could indirectly inhibit PtdIns*P* kinase via the phosphorylation of a brain-specific protein termed B-50 [60−62]. It was proposed that the B-50 phosphorylation could modulate PtdIns*P* kinase by alternating between a phosphorylated (inhibitory) and dephosphorylated form.

An interesting recent observation has been the finding that the activity of membrane-bound, but not soluble, PtdIns*P* kinase can be enhanced upon the addition of guanosine 5′-[γ-thio]triphosphate (GTP[S]) [63−65]. The effect apparently was not mediated by inhibition of phosphomonoesterase ac-

tivity nor activation of PIC, but whether this indicates the involvement of a specific, receptor-coupled guanine-nucleotide-dependent regulatory protein (G-protein) is not known at present.

As previously mentioned, polyamines have been observed to stimulate both PtdIns and PtdIns*P* kinase activities with the effect being much more pronounced on the latter [49]. A similar report by Lundberg et al. [66] noted an approximate threefold stimulation of plasma membrane PtdIns*P* kinase upon spermine addition (median effector concentration ≈ 0.2 mM) with only a minimal effect on PtdIns*P* kinase activity. The data argued against a role for either intermembrane effects or inhibitory action on PtdIns(4,5)P_2 metabolism and extended a previous report by this group which demonstrated polyamine-mediated activation of a soluble PtdIns*P* kinase [67]. From these data, they speculated that polyamines might be physiological regulators of the PtdIns4*P* kinase reaction. In keeping with this notion, Sherdorf and co-workers have suggested that PtdIns*P* kinase activity could be regulated during the cell cycle of the sea urchin egg by the known fluctuations in polyamine levels [68].

Classification of PICs

A PIC of 68 kDa was originally purified from rat liver [69], and since that time several distinct isozymes have been identified which differ from one another with respect to their molecular mass, charge, divalent cation requirements, substrate specificity and the form of the released products (either cyclic or non-cyclic phosphates) [70]. Amongst the important landmarks in this field have been the isolation of two immunologically distinct isozymes of PIC from sheep seminal vesicles [71]; purification to homogeneity of three isozymes from bovine brain and the preparation of isozyme-specific antisera (this enabled the first systematic attempts to classify PICs based upon their immunological characteristics) [70, 72−74] and the isolation of cDNA clones encoding the structural genes of four distinct forms of PIC with predicted molecular masses of 56 kDa, 86 kDa, 138−139 kDa and 148 kDa [75−79]. There is surprisingly little sequence similarity between these four enzymes but, with the exception of the 56-kDa protein, they contain two conserved sequences that are likely to represent a common catalytic domain.

PIC activities have now been purified from a wide variety of sources (reviewed in [70, 80−83]) and for the most part they can be grouped into one of five immunologically distinguishable isoforms. The substrate specificities of these enzymes are highly dependent upon the conditions of assay. This was evident from early experiments utilising crude brain cytosol fractions which presumably contained several PIC isozymes. Such crude preparations were able to hydrolyse PtdIns, PtdIns4*P* and PtdIns(4,5)P_2 equally efficiently when assayed under a variety of non-physiological conditions. However, when assayed with inositol phospholipid substrates in a lipid mixture designed to mimic that in the inner leaflet of the plasma membrane and using an assay medium comparable to the intracellular ionic environment (pH 7, high K^+, pCa^{2+} < 6) it was found that PtdIns(4,5)P_2 hydrolysis exceeded that of PtdIns by a factor of about 100 [84]. This observation was later confirmed using homogeneous PIC preparations from ram seminal vesicles and subsequently other sources [85]. Rhee et al. compared the substrate specificities of PIC β, γ and δ purified from bovine brain and concluded that, at neutral pH, PICβ had the greatest specificity for PtdIns(4,5)P_2 [70]. As these studies utilised pure phospholipid substrates presented as small unilamellar vescles it is not clear what they tell us about the preference of these enzymes when faced with their substrates embedded in a real membrane.

Several attempts have been made to determine the substrate specificity of hormone-stimulated PIC expressed in intact cells, by monitoring alterations in the levels of inositol phospholipids and specific inositol phosphate isomers [27]. The rates of turnover of individual inositol phosphate pools has been determined using parotid acinar cells that were first stimulated with a cholinergic agonist and the stimulus then terminated abruptly by the addition of atropine. The results of this study suggest that the rate of turnover of $Ins(1,4,5)P_3$ is sufficient to account for the accumulation of inositol monophosphates and biphosphates which implies that the receptor-stimulated PIC is acting specifically upon $PtdIns(4,5)P_2$ [86]. Analysis of the accumulation of inositol cyclic phosphates, which are minor, slowly metabolised products of PIC, in pancreatic acinar cells, led to the opposite conclusion, namely that as much as 50% of the accumulated products arose from direct hydrolysis of PtdIns [87]. Similarly conflicting data from earlier studies has been summarised in past reviews [27] and it seems that this debate will continue, at least for the time being.

In the past, the much larger amount of diaglyglycerol than $Ins(1,4,5)P_3$ that accumulates in cells after stimulation has been used to argue that direct hydrolysis of PtdIns must occur, especially during prolonged exposure to hormones. However, it is now clear that much of the diaclyglycerol accumulating in stimulated cells arises from phosphatidylcholine hydrolysis by the combined activities of a phospholipase D and phosphatidate phosphohydrolase [88–90]. Phospholipase D activity appears to be regulated by activators of protein kinase C, although there is limited support for a more direct activation by receptors and a G-protein [91]. At present there is no information to indicate the subcellular location of diacylglycerol formation from this source, nor to indicate whether it functions to prolong the activation of protein kinase C.

Regulation of PIC by G-proteins

The distinctive tissue distributions of the different PIC isozymes [92] and the occurrence of putative regulatory domains showing sequence similarity with regions of the *src* encoded familiy of tyrosine kinases in PICγ, but not in other forms, suggests the possibility of multiple regulatory mechanisms for these enzymes [70]. However, few clues have emerged from studies of the purified enzymes to indicate how they may be regulated. A clearer picture of the mechanism underlying hormonal stimulation of PIC activity has come instead from studies of permeabilised cells and membrane preparations. This calls attention to the problem of defining the relationship between membrane-associated and soluble forms of PIC.

All of the purified PICs noted above are found predominantly in cytosol fractions after tissue homogenisation, but the first membrane-associated PIC to be purified (from bovine brain) [76] was found to be essentially identical to cytosolic PICβ [93]. Significant amounts of membrane-associated PICα and γ were also detected using radioimmunoassay procedures which suggests that each of these enzymes may have a dual distribution and, more speculatively, that translocation to the membrane may play an important part in their activation. Alternatively, PIC may be only weakly membrane associated

so that dissociation could occur following cellular disruption during homogenisation.

G-proteins play an essential role in the hormonal regulation of adenylyl cyclase and rhodopsin-mediated activation of cGMP phosphodiesterase [94] and it is now clear that functionally similar G-proteins are responsible for hormonal regulation of PIC. Functionally assigned G-proteins are heterotrimers with α-subunits of molecular mass 39–52 kDa, and are substrates for ADP-ribosylation by cholera and/or pertussis toxins. In the inactive state the GDP-liganded α-subunit is associated with β- and γ-subunits of molecular mass 35–36 kDa and 6–8 kDa, respectively. Activation by receptors stimulates exchange of GDP for GTP and dissociation of the α-subunit from the $\beta\gamma$-subunit complex. It is this GTP-liganded α-subunit that is thought to be responsible for effector activation, but its lifetime is limited by an intrinsic GTPase activity which converts bound GTP to GDP. Reassociation with the $\beta\gamma$-subunit complex completes the activation/inactivation cycle so that continued receptor stimulation is required to maintain regulatory influence upon the effector [94].

The circumstantial evidence suggesting the involvement of a G-protein in receptor-mediated regulation of PIC has been reviewed elsewhere [25, 80]. The most convincing early studies were reported by Cockroft and Gomperts [95] and Litosh et al. [96], who demonstrated that stable analogues of GTP could stimulate PIC activity present in cell-free membrane preparations, and that guanine nucleotides were required for receptor-mediated activation of the enzyme.

The apparent similarities between hormonal regulation of adenylyl cyclase and PIC activation have been further strengthened by experiments utilising turkey erythrocyte membranes. These membanes have a PIC activity that is regulated by P_{2y} purinergic receptors in an exquisitely GTP-dependent manner. Activation by GTP[S] obeys first-order kinetics and rate of activation is enhanced by purinergic agonists, as expected if the receptor functions to accelerate a nucleotide-exchange reaction on the putative G-protein [97]. In the turkey erythrocyte system, in common with many others, stable analogues of GTP achieve a much greater degree of activation of PIC than does GTP itself, providing indirect support for a mechanism of inactivation involving GTP hydrolysis. Confirmation of this idea was provided following kinetic experiments in which inactivation of pre-stimulated PIC by GDP[S] was rapid if GTP had been used to activate, but much slower if a GTPase-resistant analogue had been used.

The identities of G-proteins responsible for activating PIC are currently unknown, but the occurrence of both pertussis-toxin-sensitive and pertussis-toxin-insensitive responses suggests there are at least two. Moreover, they are likely to be heterotrimeric G-proteins because they are capable of sensitive functional interactions with $\beta\gamma$-subunits purified from a variety of known G-proteins [98, 99]. Indirect evidence suggests that the toxin-sensitive G-protein, G_{i2}, may be responsible for coupling chemotactic peptide receptors to PIC in neutrophils [100] suggesting that this protein may mediate pertussis-toxin-sensitive PIC activation. However, another toxin-sensitive G-protein, termed G_0, can activate PIC when GTP[S]-liganded α-subunits are injected into *Xenopus* oocytes [101]. There are currently no clues to the identity of pertussis-toxin-insensitive G-proteins mediating hormonal activation of PIC. Hormonal inhibition of PIC activity has also been reported, but it is not yet clear whether such responses reflect the activity of a distinct inhibitory G-protein (reviewed in [102]).

Most studies of hormone- and guanine-nucleotide-stimulated PIC have utilised membranes or permeabilised cells containing endogenous radiolabelled substrates, but some reports have appeared in which activity was expressed against exogenous PtdInsP_2 [103–105]. Potentially, such a responsive system would overcome many of the kinetic disadvantages of utilising membranes which contain a limited amount of substrate. However, Litosch has reported that guanine-nucleotide-dependent hydrolysis of PtdIns(4,5)P_2 by brain membranes requires prior intercalation of some of the substrate into the membranes [106]. Studies of this type, therefore, are likely to identify factors that influence the distribution of substrate between micelles and receptor-bearing membranes, which could be confused with those factors required for regulation of PIC itself.

There are several reports of guanine nucleotides stimulating the activity of soluble PIC in the absence of membranes. These results appear mainly to reflect non-specific activation by guanine nucleotides. In the case of a soluble PIC from platelets, however, activation by guanine nucleotides depended upon the presence of GTP-binding activity which co-purified with PIC [107]. The relationship between this soluble guanine-nucleotide-sensitive PIC/G-protein complex and receptor-stimulated events in platelets, however, is not clear at present.

The PIC activity present in turkey erythrocyte cytosol has now been purified to homogeneity. It is relatively specific for PtdInsP_2 over PtdIns and has a moleclear mass on SDS gels of 150 kDa [108]. This purified enzyme has also been successfully reconstituted with PIC-depleted turkey erythrocyte membranes containing endogenous labelled substrates and the P_{2y} receptor/G-protein systems that controls PIC activity in the intact cells and erythrocyte ghosts [109]. The reconstituted enzyme had little effect on inositol phospholipid breakdown in the absence of guanine nucleotides, but in the presence of GTP or its non-hydrolysable analogues, a dramatic activation was observed that was also potentiated by purinergic agonists. The reconstituted system showed the same kinetics of activation, and the same guanine nucleotide and purinergic agonist sensitivity that had previously been determined for PIC-containing erythrocyte ghosts. Although the turkey erythrocyte enzyme is immunologically distinct from previously characterised mammalian enzymes, polyclonal antibodies raised against the avian erythrocyte PIC preferentially recognise mammalian PICβ (T. K. Harden, personal communication). The results demonstrate that a PIC activity isolated from cell cytosol, possibly the avian form of PICβ, is capable of functional interaction with a membrane-bound receptor-regulated G-protein, suggesting that this form of PIC functions to couple such receptors to inositol phospholipid breakdown in intact cells.

Regulation of PIC by phosphorylation

The ability of platelet derived growth factor (PDGF) and EGF to stimulate inositol phospholipid breakdown in some, but apparently not in all cells, has been well documented [110]. The receptors for these growth factors have an intracellular tyrosine kinase domain and are structurally quite different from the classical G-protein linked receptors. A functional tyrosine kinase domain is required for this response because kinase-deficient EGF or PDGF receptor mutants were unable to initiate inositol phospholipid breakdown [111, 112]. Circumstantial evidence exists to suggest that PDGF-induced inositol phospholipid breakdown in Swiss 3T3 cells occurs by a different mechanism to that employed by bombesin, whose receptor is linked to a G-protein [113]. It has now been demonstrated that PICγ, but apparently not other PIC isozymes, is a receptor tyrosine kinase substrate both *in vitro* and in intact cells [114–117]. For example, PICγ is phosphorylated on tyrosine residues to high stoichiometry within seconds of stimulating A431 cells with EGF and in fibroblasts responding to PDGF. Moreover, the accumulation of inositol phosphates following EGF stimulation of A431 cells was efficiently blocked by a cell-penetrating inhibitor of the EGF receptor tyrosine kinase [118].

It is particularly interesting that c-*fms*, which encodes the colony-stimulating factor-1 (CSF-1) receptor tyrosine kinase, shows sufficient similarity with the gene for the PDGF receptor to suggest that they both evolved from the same ancestral gene. However, when CSF-1 receptors were expressed in fibroblasts, addition of CSF-1 did not stimulate inositol phospholipid hydrolysis [119]. These results suggest that PICγ can act as a substrate for the PDGF receptor tyrosine kinase, but not for that of the closely related receptor for CSF-1.

PICγ can also be phosphorylated on serine/threonine residues by protein kinase C or cAMP-dependent protein kinase *in vitro*. Phosphorylation of PICγ occurs in C_6Bu_1 cells stimulated by agents which raise the intracellular cAMP concentration and appears to be associated with inhibition of agonist-induced inositol phosphate accumulation in these cells [120].

Despite the circumstantial evidence noted above, implicating reversible phosphorylation in the regulation of PIC, it has not proved possible, using available assay procedures, to demonstrate a change in the activity state of PICγ after phosphorylation *in vitro* by either cAMP-dependent protein kinase or growth factor receptors. This is particularly intriguing in the case of EGF-receptor-mediated tyrosine phosphorylation. The growth factor phosphorylates four tyrosine residues *in vitro*, at least two of which are phosphorylated *in vivo* [121]. Failure to detect altered catalytic activity of the phosphoprotein led to the suggestion that phosphorylation may affect the interaction of PICγ with putative modulatory proteins. An alternative suggestion, based on early observations by Irvine and his colleagues, that soluble PICs utilise true membrane bilayer substrates very poorly [122], is that phosphorylation, or interaction with appropriate G-proteins, modulates the capacity of specific PICs to degrade inositol phospholipids embedded in the inner leaflet of the plasma membrane. Which of these two hypotheses is closest to the truth remains to be established.

INOSITOL PHOSPHATE METABOLISM

The current picture of inositol phosphate metabolism is one of extreme complexity. As we have pointed out before, however, when viewed in a functional context, this complexity is more apparent than real [9]. The metabolism of inositol phosphates thus appears to fulfill three basic functions: the rapid metabolism, and hence control of the cellular concentration, of Ins(1,4,5)P_3; recycling of inositol phosphates to regenerate the intracellular pools of Ins, and the synthesis of specific Ins$P_{n>1}$ which may have distinctive functions, for example Ins(1,3,4,5)P_4, InsP_5 and InsP_6. Known and suspected routes of inositol phosphate metabolism are illustrated in Fig. 2, but it should be realised that this represents an assemblage of information from a variety of tissues and there is no homogeneous cell type in which all of the reactions have been demonstraed to occur and all of the proposed intermediates have been formally identified. The latter point is critical because many of the proposed intermediates in Fig. 2 have been identified as products of the metabolism of a particular inositol phosphate catalysed by enzymes present in cell homogenates, permeabilised cells, or by preparations at various stages of purification. In all but a few cases the identification of cellular inositol phosphates has been by potentially ambiguous methods such as co-chromatography with known standards. Note also that NMR techniques applied to inositol phosphates, whilst providing structurally definitive information, are not comprehensive because the methods are incapable of distinguishing between enantiomeric pairs. Another problem is that the orientation of the arrows denoting the direction of flow of inositol phosphate metabolism is based mainly on *in vitro* observations. The formal identification of all the inositol phosphates that occur in eukaryotic cells and clarification of the metabolic interrelationships betwen them, still represent substantial biochemical challenges.

Excellent and comprehensive reviews have appeared recently on the pathways of inositol phosphate metabolism [123], the separation of inositol phosphates by HPLC techniques [124], and methods for the mass analysis of inositol phosphates [125]. We shall, therefore, focus on the most recent information and illustrate some of the critical points noted above by reference to newly emerging information on the pathways of $InsP_5$ metabolism.

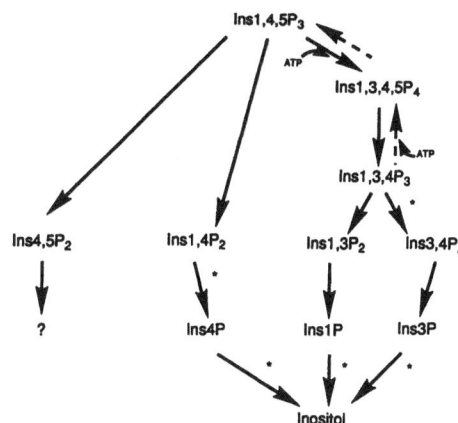

Fig. 2. *Known and suspected routes of $Ins(1,4,5)P_3$ metabolism.* $Ins(1,4,5)P_3$ is metabolised by at least three different enzymes which remove either the 5- or the 1- phosphate or phosphorylate the 3-position. Of these, the pathways initiated by 5-phosphomonoesterase and 3-kinase activities generate products which are inactive in assays of calcium mobilisation and whose further metabolism has been delineated in the most detail. Since it retains the vicinal 4,5-bis-phosphates, $Ins(4,5)P_2$ retains significant calcium-mobilising activity. The function of the 1-phosphomonoesterase, therefore, is not known, nor do we know if it has significant activity *in vivo*. Asterisks denote reactions catalysed by lithium-sensitive enzymes

Inactivation of $Ins(1,4,5)P_3$ and recovery of inositol

Much of the complexity of inositol phosphate metabolism arises from the fact that $Ins(1,4,5)P_3$ is metabolised by two routes, each of which can generate a distinctive fingerprint of intermediates. $Ins(1,4,5)P_3$ 5-phosphomonoesterase and 3-kinase activities, whose products are $Ins(1,4)P_2$ and $Ins(1,3,4,5)P_4$, are critical enzymes because they control the cellular concentration and metabolic half-life of an important second messenger. Moreover, it has been suggested that $Ins(1,3,4,5)P_4$ also functions as a second messenger, in concert with $Ins(1,4,5)P_3$ to regulate Ca^{2+} influx and/or Ca^{2+} translocation between intracellular stores [126] (see the section on functions of inositol phosphates).

A number of tissues contain both soluble and particulate $Ins(1,4,5)P_3$ 5-phosphomonoesterase and, amongst the soluble activities that have been purified to date, there is considerable molecular mass heterogeneity, suggesting the existence of multiple isozymes. This expectation has been confirmed with the isolation from rat brain of two enzymes which differ with respect to substrate specificity and molecular mass [127]. Type I 5-phosphomonoesterase has a molecular mass of 66 kDa and degrades both $Ins(1,4,5)P_3$ and $Ins(1,3,4,5)P_4$, whilst the type II enzyme is apparently specific for $Ins(1,4,5)P_3$ and has a molecular mass of 160 kDa. There is also an abundance of evidence indicating heterogeneous regulation of these enzymes, from a variety of tissues, by Ca^{2+}/calmoldulin- or protein-kinase-C-dependent phosphorylations [127–129] (reviewed by Shears [123]). Membrane-bound forms of $Ins(1,4,5)P_3$ 5-phosphomonoesterase have not yet been purified to homogeneity, so their relationship to the soluble isozymes described above has not been established.

$Ins(1,4,5)P_3$ 3-kinase is also subject to multiple regulatory mechanisms. When isolated from a variety of tissues, 3-kinase activity is stimulated by Ca^{2+}/calmodulin and can be purified using calmodulin affinity columns. Recent reports indicate that 3-kinase activity is highly sensitive to low levels of the

Ca^{2+}/calmodulin complex [130, 131] suggesting that, in cells with a large excess of calmodulin, small increases in cytosol Ca^{2+} concentration will be sufficient to accelerate the formation of $Ins(1,3,4,5)P_4$. In addition, $Ins(1,4,5)P_3$ 3-kinase may be subject to acute regulation by protein-kinase-C-dependent phosphorylation and chronic regulation during development, the latter involving enhanced synthesis of the enzyme [132, 133].

3-Kinase activities isolated from a variety of sources also show considerable molecular mass heterogeneity, but it now seems likely that many low-molecular-mass forms result from proteolytic cleavage by the Ca^{2+}-sensitive protease, calpain [134]. The rat brain enzyme has recently been purified to homogeneity and its cDNA cloned and sequenced revealing a gene encoding a protein of 449 amino acids with a predicted molecular mass of 50 kDa [134]. The sequence contains six regions that are likely to provide cleavage sites for calpain and the occurrence of low molecular mass protein bands in 3-kinase preparations from rat brain could be inhibited by a specific calpain inhibitor peptide. As a panel of monoclonal antibodies raised against the 53-kDa rat brain 3-kinase recognised the low-molecular-mass protein bands noted above, it is most probable that the peptides are degradation products of the 53-kDa protein. Whether the 3-kinase from pig aortic muscle, with a denatured molecular mass of 93 kDa [135], represents a distinct isoenzyme is not known at present.

The $Ins(1,4)P_2$ and $Ins(1,3,4,5)P_4$ formed by the above reactions are metabolised to replenish the cellular inositol pool by a complex series of reactions that have been discussed in previous reviews [4, 123] and are depicted in Fig. 2. As noted earlier, it is still not clear whether any of the $Ins(1,4)P_2$ and Ins1P that accumulate in intact cells are formed by direct PIC action upon PtdIns4P and PtdIns, respectively. Two of the enzymes which facilitate these pathways are potently inhibited by lithium, which is an important drug used in the treatment of manic depression. One of these enzymes is inositol monophosphate phosphomonoesterase which occupies a pivotal position in inositol phosphate metabolism because it lies at the convergence of the pathways for *de novo* synthesis and recycling of Ins. Lithium inhibits this enzyme in an uncompetitive manner which means that the drug binds preferentially to the enzyme/substrate complex [4, 14, 136]. Consequently, the intensity of inhibition increases as the substrate concentration rises. This forms the basis of the inositol-depletion hypothesis of lithium action, in which the drug is postulated to prevent the recycling of Ins in cells undergoing active stimulation of PIC-coupled receptors where the concentration of inositol phosphates is higher than in unstimulated cells [9].

Inositol 1,2-(cyclic)phosphates (cInsPs) are minor products of PIC attack upon inositol phospholipids, resulting from nucleophilic attack by the 2-OH of the inositol ring, instead of water, upon the phosphodiester bond [137]. Since they are metabolised extremely slowly, they can accumulate during prolonged stimulation of cells [138, 139], but they have no functions that are known at present. The little that is known of the metabolism of inositol 1,2-(cyclic)phosphate-4,5-biphosphate suggests it is dephosphorylated by sequential removal of the 5- and then the 4-phosphates to give cIns$P$ [4]. The latter compound is then converted to Ins1P by a widespread, soluble cyclic phosphodiesterase [140]. cInsP_3 is not a substrate for Ins(1,4,5)P_3 3-kinase, so its metabolism appears less complex than that of the non-cyclic compounds [4].

In addition to the established pathways of inositol phosphate metabolism described above, recent studies indicate that Ins(1,4,5)P_3 can also be metabolised by removal of the 1-phosphate [123], a reaction that appears to be particularly prominent in the slime mould, *Dictyostelium discoideum* [141]. How the Ins(4,5)P_2 formed by this reaction is further metabolised is not known at present. More puzzling are the observations that Ins(1,3,4,5)P_4 can be converted back to Ins(1,4,5)P_3 by a 3-phosphomonoesterase [142, 143] and that Ins(1,3,4)P_3 can be rephosphorylated by a 5-kinase to give Ins(1,3,4,5)P_4 [144], apparently establishing two kinase/phosphomonoesterase futile cycles. Under certain incubation conditions, the above reactions achieve substantial rates relative to the previously characterised metabolic pathways, but a crucial question is whether they are significant *in vivo*. Limited evidence from experiments with parotid glands suggest they might not be important in that tissue. If the futile cycles are sufficiently active then the half lives of Ins(1,4,5)P_3, Ins(1,3,4,5)P_4 and Ins(1,3,4)P_3 would be expected to be very similar; in fact they appear to be very different (7.6 s, 43.1 s and 8.6 min for Ins(1,4,5)P_3, Ins(1,3,4,5)P_4 and Ins(1,3,4)P_3, respectively) [84]. More recent experiments have further indicated that InsP_5 may be the true substrate for the 3-phosphomonoesterase as the former is a potent competitive inhibitor *in vitro* [145].

Synthesis of InsP_5 and InsP_6

Ins(1,3,4,5,6)P_5 and InsP_6 were two of the earliest inositol phosphates to be identified, primarily because they are highly concentrated in avian erythrocytes and plant seeds, respectively. These compounds have now been detected essentially in all cells where they have been sought and their concentrations in animal tissues have been shown, by ^{31}P NMR, to range over $5 - 15$ µM [1, 146]. A characteristic feature of these compounds appears to be that they incorporate [3H]inositol very much more slowly than does Ins(1,4,5)P_3 and its immediate metabolites. As the compounds must be synthesised and de-

graded without inadvertently generating significant quantities of potent signals, such as Ins(1,4,5)P_3 and perhaps Ins(1,3,4,5)P_4, the question of how this is achieved is of paramount importance. Possible functions of Ins$P_{n > 1}$ are discussed in the following section.

Ins(1,3,4,5,6)P_5 is the major InsP_5 isomer present in avian erythrocytes and probably also in other cells. As there is no evidence for the occurrence in cells of a PtdInsP_4 it seems likely that precursors will be found in the cellular InsP_4 pool. Studies using a variety of cell types have established the common occurrence of at least three InsP_4 isomers; Ins(1,3,4,5)P_4, Ins(1,3,4,6)P_4 and Ins(3,4,5,6)P_4. The latter compound, whose metabolic origins are currently unknown, was unequivically identified originally in extracts of avian erythroytes and murine macrophages [144], and serves as a substrate for a widespread kinase which phosphorylates the 1-position to give the anticipated species of InsP_5 [148]. Ins(1,3,4,5)P_4 is not readily phosphorylated by cellular homogenates in the presence of ATP, but its hydrolysis product, Ins(1,3,4)P_3, can be phosphorylated to Ins(1,3,4,6)P_4 by a kinase originally described in rat liver and adrenal glomerulosa cells [149, 150]. Cellular Ins(1,3,4,6)P_4 was subsequently identified unequivocally in extracts of avian erythrocytes [151]. The latter InsP_4 isomer potentially links PtdInsP_2 hydrolysis to InsP_5 synthesis because it also serves as a substrate for a 5-kinase which converts it to InsP_5 [151, 152].

Initial attempts to determine whether the above pathways contribute to InsP_5 synthesis in intact cells have focussed upon the effects of agonist stimulation on the levels of InsP_5 and each of the InsP_4 isomers. Very similar results have been obtained in several cell systems: adrenal glomerulosa cells stimulated with angiotensin [152]; vasopressin stimulated WRK-1 cells [153]; fMet-Leu-Phe-stimulated HL-60 promyelocytic leukaemia cells [154]; turkey erythrocytes stimulated with adenosine 5′-[β-thio]diphosphate (ADP[S]) [155]; bombesin-stimulated AR4-2J cells (J. W. Putney, personal communication). The level of Ins(1,3,4,5)P_4 is acutely, and often dramatically, sensitive to agonist stimulation of these cells, whilst the amount of radioactively labelled Ins(1,3,4,6)P_4 increases to a lesser extent and with a distinctly lagged timecourse. The amount of Ins(3,4,5,6)P_4 also appears to increase at even longer times, but variable results have been reported for Ins(1,3,4,5,6)P_5, perhaps because this compound turns over its Ins moiety very slowly, so that, even after prolonged periods of labelling, it may still not have reached isotopic equilibrium. Consequently, some of the reported changes may be due to an increase in its specific radioactivity rather than in mass [154, 155]. In situations where equilibrium labelling was probably achieved, or where mass analysis was carried out, then a small decrease in the cellular content of Ins(1,3,4,5,6)P_4 has been detected [154, 155a].

One interpretation of these results is that the synthesis of Ins(1,3,4,5,6)P_5 during stimulation may involve the pathway from Ins(1,4,5)P_3 that was first delineated in full using rat brain homogenates [151] (see Fig. 2). Two groups have also suggested that agonists stimulate the breakdown of Ins(1,3,4,5,6)P_5, most likely to Ins(3,4,5,6)P_5, which would explain the decline in the mass of the former [154, 155a]. In this scheme, therefore, Ins(3,4,5,6)P_4 is seen as a metabolic product, rather than a precursor of Ins(1,3,4,5,6)P_5. Evidence in favour of a common (presumably phospholipid) precursor of Ins(1,3,4,5)P_4 and Ins(1,3,4,6)P_4, which might support the above scheme, was obtained using WRK-1 cells dual labelled, to isotopic equilibrium with [14C]inositol, then relatively briefly with [3H]inositol [156]. The [3H] appeared relatively

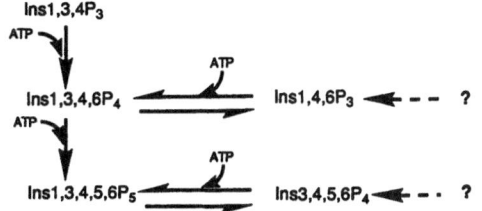

Fig. 3. *Possible reactions contributing to* InsP_5 *biosynthesis.* Ins$(1,3,4,5,6)P_5$ can be synthesised from Ins$(1,4,5)P_3$ via the intermediates shown in the figure. ^{32}P-labelling experiments using chick erythrocytes, however, indicate that Ins$(1,4,6)P_3$ and Ins$(3,4,5,6)P_4$ are the major precursors of Ins$(1,3,4,6)P_4$ and Ins$(1,3,4,5,6)P_5$, respectively. Such a labelling pattern could occur if the kinase/phosphomonoesterase cycles indicated in the figure are active in intact cells. Evidence which may support such a proposal is discussed in the text

quickly in Ins$(1,3,4,5)P_4$ and Ins$(1,3,4,6)P_4$, but not in Ins$(3,4,5,6)P_4$, suggesting that synthesis of the latter compound involves a different or, at the very least, a more complex pathway.

The above ideas were tested directly using chick erythrocytes that were labelled for several hours with [^{3}H]inositol, then briefly with [^{32}P]orthophosphate. By using a variety of characterized dephosphorylation reactions it was possbile to determine the specific radioactivity of each of the InsP_4 isomers and InsP_5, as well as the relative and absolute specific radioactivities of the individual phosphate groups for all of these compounds [157]. This form of analysis revealed that Ins$(3,4,5,6)P_4$ is the major precursor of InsP_5 in chick erythrocytes (in keeping with the fact that in the tissues examined, the activity of Ins$(3,4,5,6)P_4$ 1-kinase was very much greater than that of Ins$(1,3,4,6)P_4$ 5-kinase). More surprising was the conclusion that Ins$(1,3,4)P_3$ is apparently not the major precursor of cellular Ins$(1,3,4,6)P_4$; this function would appear to be fulfilled by Ins$(1,4,6)P_3$. More recent experiments have confirmed these basic conclusions using the inositol phosphates extracted from chick erythrocytes stimulated with ADP[S] rather than from resting cells [21a].

One way in which the above observations could be reconciled is illustrated in Fig. 3. This assumes that InsP_5 is ultimately synthesised from PtdIns$(4,5)P_2$ via Ins$(1,4,5)P_3$, but that at least two futile cycles of phosphorylation and dephosphorylation exist which distort the ^{32}P-labelling patterns of the inositol phosphates. For example, if the activity of the proposed cycle Ins$(1,4,6)P_3 \rightleftharpoons$ Ins$(1,3,4,6)P_4$, is very much greater than the reaction catalysed by Ins$(1,3,4)P_3$ 6-kinase, then the 3-phosphate of Ins$(1,3,4,6)P_4$ would label more quickly than the others, as is observed. Evidence for such a mechanism is provided by the observation that Ins$(1,4,6)P_3$ is present in chick erythrocytes and is a major product of Ins$(1,3,4,6)P_4$ metabolism catalysed by an enzyme activity present in rat brain homogenates [158].

Whilst it is evident that InsP_6 is likely to be synthesised from an InsP_5, the question of which of the six possible isomers is involved is obviously an important question. A kinase which phosphorylates Ins$(1,2,4,5,6)P_5$, but apparently not Ins$(1,3,4,5,6)P_5$ has been observed in rat brain homogenates, and the former InsP_5 has also been identified in extracts of mammalian cells, indicating that it may be a physiologically relevant precursor of InsP_6 (P. T. Hawkins and M. R. Hanley, personal communication). The metabolic origins of Ins$(1,2,4,5,6)P_5$ have not yet been defined.

As noted earlier, there is now convincing evidence for direct phosphorylation of Ins to InsP_6 in the slime-mould, *D. discoideum* [21a]. The first intermediate in this pathway appears to be Ins3P, which can also arise in cells through dephosphorylation of Ins$(1,3,4,5,)P_4$ or as an intermediate in the synthesis of Ins from glucose. Ins3P may, therefore, function as an intermediate which coordinates the synthesis of InsP_6 from each of these sources.

As InsP_5 and InsP_6 must be synthesised and degraded without inadvertently generating Ins$(1,4,5)P_3$, it seems likely that cells may contain several InsP_3 species that have hitherto escaped detection, or separation from previously described compounds. Indeed, six InsP_3 isomers were identified in chick erythrocytes [158]. Two of these were Ins$(1,4,5)P_3$ itself and its metabolite, Ins$(1,3,4)P_3$; the remaining four [Ins$(3,4,5)P_3$, Ins$(3,4,6)P_3$, Ins$(1,4,6)P_3$ and Ins$(1,3,6)P_3$] are probably metabolites and/or precursors of Ins$(1,3,4,5,6)P_5$. Radenberg et al. [159] have also reported two further InsP_3 isomers in turkey erythrocytes [Ins$(1,5,6)P_3$ and Ins$(2,4,5)P_3$], whose metabolic origins and enantiomeric configurations are not known.

FUNCTIONS OF INOSITOL PHOSPHATES

The complex metabolism of inositol phosphates, which results in the accumulation in intact cells of the inositol phosphate species described in the preceding section, has prompted speculation that many of these compounds may function as intracellular signals. Experimental evidence, ranging from the first observations of Ins$(1,4,5)P_3$-stimulated Ca^{2+} mobilisation in permeabilised cell preparations to the recent purification, cDNA sequencing and functional reconstitution of the Ins$(1,4,5)P_3$ receptor, provide unequivocal evidence that Ins$(1,4,5)P_3$ is the second messenger responsible for hormone-stimulated intracellular Ca^{2+} release. Although Ins$(1,3,4,5)P_4$ has been implicated in Ca^{2+} signalling (see below) and Ins$(1,4)P_2$ apparently can activate a low-affinity form of DNA polymerase α [160], it seems unlikely that many of the rapidly formed metabolites of Ins$(1,4,5)P_3$ serve distinctive functions other than to act as intermediates in the recovery of Ins. The ubiquitous occurrence of Ins$P_{n>1}$, such as InsP_5 and InsP_6, in eukaryotic cells, and the complex pathways concerned with their synthesis and degradation, suggests that these compounds are functionally significant, but their cellular roles remain obscure at present.

Ins$(1,4,5)P_3$-stimulated Ca^{2+} release

Early studies, reviewed extensively elsewhere, established that Ins$(1,4,5)P_3$, injected into cells or applied to a variety of permeabilised cell preparations, stimulates Ca^{2+} release from non-mitochondrial stores presumed to be in a part of the endoplasmic reticulum [22, 161]. Since this response is relatively insensitive to reduced temperature (i. e. down to 4 °C) it is evident that Ins$(1,4,5)P_3$ somehow functions by opening a Ca^{2+} channel in the appropriate Ca^{2+}-storage organelle without necessarily requiring the involvement of an enzyme such as a protein kinase. The Ca^{2+}-release process has distinctive pharmacological characteristics being unaffected by a variety of calcium-channel blockers, but sensitive to agents such as cinnarizine and flunarizine and the potassium channel inhibitor, tetraethylammonium. The latter observation, together with the fact that Ins$(1,4,5)P_3$ cannot release Ca^{2+} in a K$^+$-free medium, suggest that K$^+$ uptake into the

Ins(1,4,5)P_3-sensitive Ca^{2+} store is necessary to maintain electrical neutrality during Ca^{2+} release ([22, 161, 162] and references therein). Another important pharmacological agent which stimulates Ca^{2+} release from non-mitochondrial stores is the sesquiterpene lactone tumour promoter, thapsigargin. This compound appears to act by inhibiting ATP-dependent Ca^{2+} uptake although it does not distinguish between Ins(1,4,5)P_3-sensitive and insensitive Ca^{2+} stores [163]. A synthetic agent, 2,5-di-butyl-1,4-benzohydroquinone, which is a potent inhibitor of liver microsomal ATP-dependent Ca^{2+} uptake and which evokes Ca^{2+} release from the Ins(1,4,5)P_3 sensitive pool in hepatocytes, may act in much the same way [164].

Structure activity relationships for Ca^{2+} release were initially restricted to several naturally occurring inositol phosphates or simple analogues thereof [e.g. glycerophosphoinositol 4,5-bisphosphate obtained by deacylation of PtdIns(4,5)P_2]. Nevertheless, such studies indicated that Ins(1,4,5)P_3 binds to a specific receptor in order to elicit Ca^{2+} release. Remarkable progress has been made during the last three years in defining the pharmacological characteristics, molecular identity and subcellular localisation of specific Ins(1,4,5)P_3 receptors. This has been facilitated by the availability of high-specific-activity radiolabelled Ins(1,4,5)P_3 and the development of a radioligand binding assay; stereospecific synthesis of DIns(1,4,5)P_3 and LIns(1,4,5)P_3 as well as other inositol phosphates and analogues such as photolabile 'caged' Ins(1,4,5)P_3, and the synthesis of the metabolically stable phosphorothioate analogues, inositol 1,4,5-triphosphorothioate and inositol 1,4-bisphosphate-5-phosphorothioate [165, 166].

Spät et al. were the first to identify high-affinity binding sites for radiolabelled Ins(1,4,5)P_3 (in liver and neutrophils) and their original observations [167, 168] were quickly followed by similar reports for rat brain, adrenal cortex and anterior pituitary gland. As the optimum conditions for each assay are quite different, it is perhaps not surprising that the binding affinities of a variety of inositol phosphates are generally much higher than their apparent affinities determined by functional assays of Ca^{2+} release. Measurements of Ins(1,4,5)P_3 binding to liver membranes and Ca^{2+} release from permeabilised hepatocytes carried out under identical conditions, however, apparently yielded the same affinities [169]. Moreover, the rank orders of potency for displacement of bound Ins(1,4,5)P_3 by a variety of inositol phosphate analogues were similar amongst a number of tissues and identical to that observed for stimulation of Ca^{2+} release. Thus the high-affinity binding site displays strict stereospecificity with DIns(1,4,5)P_3 being > 3000-fold more potent than the L-isomer in rat cerebellum, adrenal cortex and liver [165]. These characteristics, together with estimated molecular masses for the high-affinity Ins(1,4,5)P_3-binding sites in liver and cerebellum of 258 kDa determined by radiation inactivation analysis, also distinguish such binding sites from enyzmes involved in Ins(1,4,5)P_3 metabolism [169].

The above results indicated that the high-affinity binding sites for Ins(1,4,5)P_3 that are present in a variety of tissues represent physiological Ins(1,4,5)P_3 receptors [170] and, therefore, that the binding assay could be used in the purification of the receptor. The Purkinje cells in the rat cerebellum are a particularly rich source of Ins(1,4,5)P_3 receptors which, by virtue of their affinity for heparin and concanavalin A, have been purified to near homogeneity [171]. The receptor is a membrane glycoprotein with a relative molecular mass of approximately 250 kDa that is a major substrate for cAMP-dependent protein kinase. Phosphorylation appears to both inhibit agonist binding and block Ins(1,4,5)P_3-stimulated Ca^{2+} release.

The extraordinary abundance of this protein led to its isolation as long ago as 1976 [172], but its true function obviously was not appreciated at that time. Nevertheless, the availability of monoclonal antibodies to the originally purified protein (termed P400) enabled Furuichi et al. [173] to isolate its complementary DNA and deduce its sequence. At the same time, Mignery et al. [174] identified a major Purkinje cell transcript as part of a heparin-binding protein containing several putative membrane spanning domains which they deduced must be an integral membrane channel protein. This protein is essentially identical to that described above as it binds Ins(1,4,5)P_3 and contains a 500-amino-acid sequence which differs from the deduced C terminal sequence of P400 by only one amino acid. The above studies reveal that the Ins(1,4,5)P_3 receptor consits of 2749 amino acids with a deduced molecular mass of 313 kDa. Reconstitution using lipid vesicles demonstrated that the purified protein possesses both the ligand recognition site and Ca^{2+}-channel activity [3]. The native molecular mass of the Ins(1,4,5)P_3 receptor is estimated at > 1000 suggesting that it occurs as a tetramer in vivo. This may explain the cooperativitiy of Ins(1,4,5)P_3-stimulated Ca^{2+} release [175] assuming that at least three subunits must be occupied by the second messenger in order to generate the open state of the channel. A particularly striking feature of the receptor's amino acid sequence is that it closely resembles that of the ryanodine receptor which is the Ca^{2+} channel found in the sarcoplasmic reticulum of skeletal muscle. Both proteins contain several putative transmembrane domains; the ryanodine receptor has four, placing both its N and C termini on the cytosolic side of the membrane, but the precise number of transmembrane domains within the Ins(1,4,5)P_3 receptor, and hence its likely orientation, is not clear at present.

A number of reports have indicated that several peripheral tissues possess Ins(1,4,5)P_3 receptors with relatively high affinity (K_d 1 − 5 nm) whilst the binding site in rat cerebellum has an affinity about 10-fold lower [165], raising the possibility of receptor heterogeneity (but see reference [169]) which reports almost identical affinities of 6.5 nm and 7.8 nm for liver and cerebellum, respectively). If the cerebellum protein is the first of a family of Ins(1,4,5)P_3 receptors to be identified then its sensitivity to phosphorylation by cAMP-dependent protein kinase, inhbition by heparin, and its interaction with a Ca^{2+}-sensitive inhibitor protein [171] are likely to be valuable points of comparison between the family members.

It has been known for some time that Ins(1,4,5)P_3 releases Ca^{2+} from only a portion of the non-mitochondrial ATP-dependent intracellular stores, for example in liver the Ins(1,4,5)P_3-sensitive compartment is about one third of the total [176]. Based upon their observations of the intracellular distribution of a calsequestrin-like protein in non-muscle cells, Volpe et al. [177] have proposed the existence of a discrete Ca^{2+}-storage organelle, distributed throughout the cytosol, which they have termed the calciosome. The calsequestrin-like protein was reported to copurify with markers of the Ins(1,4,5)P_3-sensitive Ca^{2+} store, suggesting that calciosomes may be the site of action of Ins(1,4,5)P_3. However, immunohistochemistry, utilising antibodies raised against the purified Ins(1,4,5)P_3 receptor, suggests that the receptor is located in the rough endoplasmic reticulum and related membranes such as the Golgi and nuclear envelope [174, 178]. Further data appear to identify some receptors in the plasma membrane [173] which may explain results implicating $Ins(1,4,5)P_3$ as a signal for receptor-mediated Ca^{2+} entry (but see below).

Subcellular fractionation is an alternative approach towards identifying the $Ins(1,4,5)P_3$-sensitive compartment. Such studies have been frought with difficulty due to the relative lack of responsiveness to $Ins(1,4,5)P_3$ of isolated microsomal preparations compared with permeabilised cells. These difficulties were partly resolved when it was realised that $Ins(1,4,5)P_3$ stimulated Ca^{2+} release could be greatly augmented by GTP using permeabilised smooth muscle cells [179] or rat liver microsomes [180] in which close apposition and/or fusion of membranes was encouraged by including poly(ethylene glycol) in the incubations. When smooth muscle microsomes were fractionated on sucrose density gradients, Gill's group isolated two types of Ca^{2+}-pumping vesicles: relatively dense vesicles, corresponding to rough endoplasmic reticulum markers, which are sensitive to both $Ins(1,4,5)P_3$ and GTP and are oxalate permeable, and a lighter vesicle fraction that is impermeable to oxalate and sensitive only to GTP [181]. GTP stimulates Ca^{2+} release in the vesicle preparations whether oxalate is present or not, but in permeabilised cells, GTP stimulates Ca^{2+} accumulation in the oxalate-loaded $Ins(1,4,5)P_3$-sensitive compartment. These studies suggest that GTP may function to translocate Ca^{2+} between intracellular compartments, presumably via some sort of junctional complex, and this may represent a fundamental control mechanism influencing the size of the Ca^{2+} pool accessed by $Ins(1,4,5)P_3$.

Gill and his colleagues have further pointed out that the sensitivity and specificity of GTP-stimulated Ca^{2+} release, which is blocked both by GDP and non-hydrolysable analogues of GTP, are similar to the binding characteristics of several small molecular mass GTP-binding proteins present in GTP-responsive vesicle preparations. Such proteins include the Ras family, sec4p and Ypt1p, the latter being implicated in Ca^{2+} homeostatic mechanisms in yeast and in mediating vesicular transport between endoplasmic reticulum and Golgi in mammalian cells, a process that is also dependent upon GTP hydrolysis [181]. Whether conformational switches induced in such a protein by guanine nucleotide exchange and GTP hydrolysis serve to translocate Ca^{2+} between intracellular organelles is currently a matter of speculation.

Functions of $Ins(1,3,4,5)P_4$

$Ins(1,3,4,5)P_4$ does not bind significantly to the $Ins(1,4,5)P_3$ receptor and is inactive in experiments designed to demonstrate the latter's Ca^{2+}-mobilising activity [165]. Primarily, therefore, $Ins(1,4,5)P_3$ 3-kinase serves to inactivate a potent second messenger. The co-existence of a 5-phosphatase pathway which also fulfills the function of inactivating $Ins(1,4,5)P_3$, the apparent energy cost of the 3-kinase pathway and the short half-life of cellular $Ins(1,3,4,5)P_4$, have all contributed to speculation that the latter has some specific function as an intracellular signal molecule. Several reports have documented a variety of $Ins(1,3,4,5)P_4$-mediated responses including a relatively weak effect on the release of Ca^{2+} from internal stores, potentiation of $Ins(1,4,5)P_3$ evoked Ca^{2+} release, enhancement of Ca^{2+} sequestration by Ca^{2+} storage organelles and the activation of a heparin-sensitive preparation containing protein phosphatase-1 (see [162] and references therein). The notion that has dominated this area, however, is that $Ins(1,3,4,5)P_4$ somehow acts in concert with

$Ins(1,4,5)P_3$ to prolong agonist-stimulated Ca^{2+} signals [126]. In order to understand how this idea has developed and the evidence which supports it, it is necessary to understand some of the characteristics of Ca^{2+} signals initiated by activation of PIC-coupled receptors.

Many agonists which function through inositol phospholipid hydrolysis stimulate Ca^{2+} influx as well as Ca^{2+} release from intracellular stores. These two components can be resolved by comparing the Ca^{2+} signals obtained when cells are stimulated in Ca^{2+}-free or Ca^{2+}-containing media. In the former case responses are transient as the intracellular, $Ins(1,4,5)P_3$-sensitive pool is emptied and can only refill if Ca^{2+} is added to the extracellular medium. In the presence of Ca^{2+}, the initial Ca^{2+} transient is inevitably followed by a sustained Ca^{2+} signal which varies in magnitude depending on the receptor being stimulated [182]. Attempts to distinguish the onset times for Ca^{2+} influx and release have produced mixed results, but at least in parotid acinar cells, where the response to cholinergic agonists is probably mediated through a single class of receptors coupled to PIC activation, the lag times for influx and release are indistinguishable [183].

A possible role for $Ins(1,3,4,5)P_4$ in regulating the Ca^{2+}-influx phase of the response was first suggested following experiments in which the putative signal molecule was injected into certain species of sea urchin oocytes [126, 184]. Egg activation, initiated by the rasing of a fertilisation envelope, occured in response to co-injection of $Ins(1,3,4,5)P_4$ and $Ins(2,4,5)P_3$, the latter being able to mimic the Ca^{2+}-releasing activity of $Ins(1,4,5)P_3$ without being an effective substrate for the 3-kinase. Injection of either substance alone was ineffective. As egg activation in these particular oocytes appeared to depend upon extracellular Ca^{2+} (but see [185] for a critical discussion) it was initially suggested that $Ins(1,4,5)P_3$ and $Ins(1,3,4,5)P_4$ act together, the former initiating intracellular Ca^{2+} mobilisation and the latter stimulating Ca^{2+} influx.

One problem with the idea of independent actions of these two inositol phosphates was that neither appeared to be effective on its own. This led to the proposal of a coupled mechanism in which the $Ins(1,4,5)P_3$-sensitive Ca^{2+} pool is constantly replenished by some process that is controlled by $Ins(1,3,4,5)P_4$ [126].

Convincing support for such a coupled mechanism of action of inositol phosphates has come from detailed studies of the responses of lachrymal acinar cells to acetylcholine or to internal perfusion of inositol phosphates [186]. Cholinergic agonists cause a prolonged stimulation of a Ca^{2+}-dependent outward K^+ current, but intracellular application of $Ins(1,4,5)P_3$ results only in transient activation of the channel. The combined presence of $Ins(1,4,5)P_3$ and $Ins(1,3,4,5)P_4$ in the intracellualr perfusion buffer leads to prolonged activation that is typical of the response to receptor stimulation. This effect of $Ins(1,3,4,5)P_4$ is not due to inhibition of $Ins(1,4,5)P_3$ metabolism and shows an appropriately high degree of structural specificity, but it is still not clear why $Ins(1,4,5)P_3$ is ineffective on its own because it might be expected to generate $Ins(1,3,4,5)P_4$ if there is sufficient 3-kinase activity in the perfused cells (presumably, this is how a cholinergic stimulus is able to cause a prolonged response).

One model which seeks to explain the relationship between the Ca^{2+} release and Ca^{2+} influx phases of the responses of non-excitable cells to agonist stimulation is based mainly on experimental evidence from a closely related cell to that used above, the parotid acinar cell. The capacitance model argues that the empty state of the $Ins(1,4,5)P_3$-sensitive Ca^{2+} store activates an influx pathway that refills the store [187, 188]. In the original version of this model, external Ca^{2+} was thought somehow to gain direct access to the internal store without traversing the cytosol, presumably via some sort of junctional complex between the membrane surrounding the store and the plasma membrane. This would be analogous, if not identical, to the proposed GTP-stimulated channels linking $Ins(1,4,5)P_3$-dependent and $Ins(1,4,5)P_3$-independent Ca^{2+} stores described above. Irvine and his colleagues have, therefore, proposed that $Ins(1,3,4,5)P_4$ might function as the physiological regulator of these putative channels, thereby controlling Ca^{2+} influx, directly or indirectly, to the $Ins(1,4,5)P_3$-sensitive store [126].

Recent experiments have provided evidence both in favour of and against this attractive hypothesis of interdependent second messenger functions. It now appears that refilling of the $Ins(1,4,5)P_3$-sensitive Ca^{2+} pool, after stimulation of parotid acinar cells in Ca^{2+}-free medium followed by addition of Ca^{2+}, is accompanied by a modest rise in cytosol $[Ca^{2+}]$ detected using direct and more sensitive assays than those available previously [189]. This argues against the existence of a 'secret channel' linking extracellular Ca^{2+} directly to the internal stores. Moreover, the influx of Ca^{2+} upon its addition to the medium, required $Ins(1,4,5)P_3$-sensitive pools to be empty, as predicted by the 'capacitance' model, yet still occurred many minutes after termination of the initial stimulus by receptor blockade, when the levels of both $Ins(1,4,5)P_3$ and $Ins(1,3,4,5)P_4$ would have returned to basal. This suggests that the pool refilling is not regulated by either of these inositol phosphates.

Evidence from the same cell system which might support the hypothesis of $Ins(1,3,4,5)P_4$ function arises from experiments designed to determine the half-lives of signals in parotid acinar cells stimulated with a cholinergic agonist and then abruptly blocked with atropine. When atropine was added after a brief period of stimulation and during the internal release phase of the response, the Ca^{2+} signal declined with a half-life mirroring that of $Ins(1,4,5)P_3$. If atropine was added much later, however, the Ca^{2+} signal declined significantly more slowly than did the level of $Ins(1,4,5)P_3$, more nearly mirroring the half-life of $Ins(1,3,4,5)P_4$ [86]. One interpretation of these results is that the latter inositol phosphate is more important during prolonged stimulation when maintenance of a Ca^{2+} signal depends upon extracellular Ca^{2+}.

In conclusion, a variety of experimental observations suggest that $Ins(1,3,4,5)P_4$ might serve some specific cellular function, the most concerted argument being that it acts in concert with $Ins(1,4,5)P_3$ to sustain Ca^{2+} signals, particularly during prolonged stimulation. The identification of specific, high-affinity binding sites for $Ins(1,3,4,5)P_4$ in several tissues [165] may presage the isolation of a receptor whose properties may provide a molecular basis for many of the current observations. Until then, or until sophisticated approaches such as whole-cell perfusion can be applied to a variety of cell types, any role for $Ins(1,3,4,5)P_4$ is likely to remain ill-defined.

Inositol phosphates and calcium signals in single cells

The almost routine use of Ca^{2+}-sensitive fluorescent indicators and the availability of image analysers have greatly improved knowledge of how inositol-lipid-dependent Ca^{2+}-signalling operates in individual cells. The Ca^{2+} signal observed in a population of identical cells is obviously an average of all of the individual responses. If the latter are out of phase then considerable detail and potential heterogeneity will be obscured. This is indeed found to be the case and the experiments using single cells clearly demonstrate that Ca^{2+} signals may be initiated from discrete sites within the cell, they can propagate in waves and they are frequently found to oscillate (reviewed in [161, 190–192]).

Theories attempting to explain the phenomenon of Ca^{2+} oscillations are numerous, but fall into two distinct categories according to whether it is believed that the intracellular concentration of $Ins(1,4,5)P_3$ oscillates in time with the Ca^{2+} signal (receptor-based models) or whether a constant level of $Ins(1,4,5)P_3$ is believed to drive an intracellular Ca^{2+} oscillator [190]. Unfortunately, it is not yet possible to monitor the concentration of $Ins(1,4,5)P_3$ in a single cell and so direct evaluation of the above hypothesis has not been achieved. Cobbold and his colleagues, however, have argued in favour of a receptor-based model because the shapes of Ca^{2+} transients obtained in individual rat hepatocytes are highly dependent upon which of several PIC-coupled receptors is being stimulated [193]. There is also evidence in favour of an intracellular, receptor-independent oscillatory mechanism; for example, intracellular injection of a non-metabolisable analogue of $Ins(1,4,5)P_3$ induced Ca^{2+} oscillations in pancreatic acinar cells [194], so it seems unlikely that a single model explains the oscillatory behaviour of all cells. The reader is referred to a series of recent reviews for a thorough discussion of this topic [161, 190–192].

Inositol cyclic phosphates

As noted earlier, inositol 1,2-(cyclic)phosphates are minor products of PIC activity (under physiological conditions) which nevertheless accumulate in cells due to their relatively slow rate of metabolism. Initial experiments suggested that $cInsP_3$ was equipotent with its non-cyclic counterpart as a Ca^{2+}-mobilising agonist. Since these compounds reach similar levels following prolonged stimulation of some cells, the formation of the cyclic compounds seemed likely to be functionally significant [195]. More recent studies, however, have consistently found that $cInsP_3$ has much lower affinity for the $Ins(1,4,5)P_3$ receptor and correspondingly lower Ca^{2+}-mobilising potency than $Ins(1,4,5)P_3$ itself [196]. Either the cyclic compounds have some other, unsuspected function, or their formation is a metabolic accident which has been limited to only about 1% of the PIC-mediated cleavage of $PtdIns(4,5)P_2$, and perhaps has no cellular role at all.

Inositol 1,3,4,5,6-pentakisphosphate and inositol hexakisphosphate

The widespread occurrence of these compounds has only recently been fully appreciated and most of the earlier proposals concerned putative functions within tissues in which they were found at particularly high concentrations. For example, millimolar levels of $Ins(1,3,4,5,6)P_5$ are present in avian erythrocytes which replaces 2,3-bisphosphoglycerate as the major non-nucleotide phosphorus-containing compound a few days after hatching in the chick [5]. As $Ins(1,3,4,5,6,)P_5$ binds tightly to haemoglobin, reducing its affinity for oxygen, it is possible that this compound acts as a physiological regulator of oxygen delivery to avian tissues. If so, then it might be anticipated that the $InsP_5$ content of avian blood may be regulated by stress factors which influence the oxygen tension experienced in the lungs and other tissues.

One of the more interesting recent contributions is the observation that $InsP_6$, which is particularly abundant in some plant seeds, forms an iron chelate that augments Fe^{2+} mediated oxygen reduction and blocks the formation of hydroxyl radicals, thus supressing lipid peroxidation [197]. Suppression of ·OH radical formation appears to underlie the ability of $InsP_6$ to reduce the incidence of large intestinal cancer induced in rats by the chemical carcinogen, azoxymethane [198], so the antioxidant properties noted above may have far-reaching implications.

$InsP_6$ has also been proposed to function as a phosphorus and/or inositol reserve in plant seeds, but this would not explain its widespread distribution in animal cells. Another specialised function for $InsP_{n>1}$ was proposed by Vallejo et al. [199], based upon their observation of differential distribution of $InsP_5$ and $InsP_6$ in rat brain. They suggested these compounds might have extracellular functions as neurotransmitters and supported this idea by showing that injection of $InsP_5$ or $InsP_6$ into the nucleus tractus solitarius (a brain stem nucleus implicated in cardiovascular regulation) caused potent and profound reductions in blood pressure and heart rate of rats. Whilst other anticipated properties of neurotransmitters have not yet been demonstrated for these compounds (e.g. specific release mechanisms), they have been shown to bind to high-affinity sites in discrete regions of mammalian brain [200] and to be degraded by extracellular enzyme(s) of NIH 3T3 cells, which show specificity towards $InsP_{n>1}$ [201]. These activites may be related to phytases which have been extensively studied in plants where they are responsible for degradation of $InsP_6$ in seeds. Phytases are also found in the intestinal epithelium and presumably function to convert dietary inositol phosphates to Ins for absorption (of Ins and of chelated metal ions).

Additional neuronal effects of $Ins(1,3,4,5,6)P_5$ and $InsP_6$ have recently been described following their microinjection into identified neurons in the abdominal ganglia of the marine mollusc, Aplysia [202, 203]. The injection of $InsP_6$, achieving a calculated intracellular concentration of 0.2 μM, evoked a biphasic response consisting of an initial inward current carried mainly by Na^+ and Ca^{2+}, followed by a prolonged hyperpolarisation due to stimulation of an outward K^+ channel. The latter effect was attributed to the initial rise in Ca^{2+} caused by the inward current, rather than to a direct effect of $InsP_6$. Injection of a similar amount of $Ins(1,3,4,5,6)P_5$ into the same neurones activated only a slow, non-selective inward current. Neither compound induced any response when applied to the cell exterior, but the specificity of the observed effects was not established using any other inositol phosphate species. Although the concentrations of $InsP_{n>1}$ in Aplysia neurones are not known, 10–100-μM levels are common in many eukaryotic cells, so the physiological relevance of the low amounts used in the above studies is questionable.

THE PTDINS 3-KINASE PATHWAY: NOVEL INOSITOL PHOSPHOLIPIDS THAT ARE RESISTANT TO PIC-MEDIATED CLEAVAGE

An important principle, which is evident from information given in the preceding sections, is that the precise distribution of phosphate groups around the inositol ring is an essential feature which dictates the biological activity of inositol phosphates. Since the elucidation of the head-group structures of PtdIns4P and $PtdIns(4,5)P_2$, a number of unsuccessful attempts have been made to detect additional $PtdInsP_{n>1}$

species. More recent studies, however, have identified PtdIns3P as a product of a novel PtdIns kinase and as a minor constituent of several cell types [6, 204–212]. It now seems likely that distinct pathways, initiated by PtdIns 4-kinase and PtdIns 3-kinase, respectively, may serve to generate inositol phospholipid isomers with indepedent cellular functions. As inositol phospholipids with a monoester phosphate in the 3-position have recently been shown to be resistant to phosphodiesteratic cleavage by a variety of purified PICs [213–215], it is probable that the phospholipid products of the PtdIns 3-kinase pathway themselves, rather than inositol phosphates derived from them, are functionally significant.

Distinct forms of PtdIns kinase were initially reported in fibroblasts by Whitman et al. [216]. The most prominent activity detected in such cells is potently inhibited by adenosine (see previous section) and phosphorylates the 4-position on the inositol ring to give PtdIns4P. The other activity found in fibroblasts is strongly inhibited by non-ionic detergents such as Triton X-100, is relatively insensitive to inhibition by adenosine and phosphorylates the 3-position to give PtdIns3P [6]. Cantley and his colleagues originally termed these enzymes types II and I PtdIns kinases, respectively, before their distinct products had been definitively identified. A more informative nomenclature, therefore, employs the terms, PtdIns 4-kinase (equivalent to type II) and PtdIns 3-kinase (equivalent to type I). To confirm that PtdIns 3-kinase activity is expressed in intact cells, it was necessary to demonstrate the occurrence of PtdIns3P in cell membranes. Stereochemically definitive identification of cellular PtdIns3P has thus far only been achieved for the phospholipid extracted from 1321 N1 astrocytoma cells where it comprised about 5% of the PtdInsP pool [204]. Preliminary evidence has been obtained, however, for the occurrence of PtdIns3P as a normal constituent of all cells in which it has been sought (see Table 1). In all such studies the analysis of inositol phospholipid isomers has been by deacylation using methylamine and separation of the glycerol derivatives by anion-exchange HPLC. One of the more remarkable findings is that, whilst PtdIns3P comprises 1–5% of the PtdInsP pool in most of the mammalian cells examined, it is a major inositol phospholipid in the yeast, Saccharomyces cerevisiae, where it comprises 50–60% of the total PtdInsP [212]. Whether this indicates an important function of PtdIns 3-kinase in yeast is not known at present.

Recent studies have indicated that PtdIns3P is one component in a novel phospholipid pathway in which a $PtdInsP_2$ and a $PtdInsP_3$ also participate [205, 209, 217]. Upon deacylation, the glycerol derivatives of the latter two lipids have chromatographic properties consistent with the structures $PtdIns(3,4)P_2$ and $PtdIns(3,4,5)P_3$ for their parent molecules. Interest in this pathway has intensified because it is now clear that it is subject to acute regulation by intracellular tyrosine kinases, growth factor receptor tyrosine kinases and certain G-protein-coupled receptors, leading to speculation that the novel lipids may have important regulatory functions.

Regulation by protein tyrosine kinases was first indicated by the tendency of PtdIns 3-kinase to physically associate with complexes of the polyoma virus middle T antigen and pp60[c-src] (the cellular c-src gene product), with pp60[v-src] (the v-src gene product of avian sarcoma virus, a constitutively active tyrosine kinase) and with the agonist-occupied PDGF receptor [218]. Stimulation of PDGF receptors in smooth muscle cells did not appear to greatly affect the cellular level of PtdIns3P, but instead caused rapid increases in the levels of putative $PtdIns(3,4)P_2$ and $PtdIns(3,4,5)P_3$ [205]. PtdIns 3-kinase activity is readily detected in anti-receptor or anti-

Table 1. *Regulation of the PtdIns 3-kinase pathway*

Cell type	Stimulus	Response detected	References
Fibroblast	viral transformation	a	6, 216, 226
Fibroblast	PDGF	a	6, 216
Fibroblast	CSF-1 (in *c-fms* transfected cells)	a	208
Vascular smooth muscle	PDGF	a/b	205
Transfected CHO cells	insulin	a/b	206/206
MA-10 Leydig tumour cells	EGF	b	207
P388 D1 macrophage	CSF-1	b	208
Human neutrophil	fMet-Leu-Phe	b	209/217
Murine bone-marrow***d macrophage	fMet-Leu-Phe	b	d
Platelet	thrombin/GTP[S]	b/c	210/211
Yeast	glucose addition	c	e

Activation of the PtdIns 3-kinase pathway was indicated as follows: (a) detection of PtdIns 3-kinase activity in anti-receptor or anti-phosphotyrosine immunoprecipitates; (b) a rapid increase in the cellular content of PtdIns(3,4)P_2 and/or PtdIns-(3,4,5)P_3 following application of the stimulus indicated; (c) enhanced activity of PtdIns 3-kinase activity in permeabilised cells or membranes; (d) S. Vallance, C. Macphee, C. P. Downes and A. D. Whetton, unpublished results; (e) C. Macphee, J. Piggott and C. P. Downes, unpublished results. CHO, Chinese hamster ovary

phosphotyrosine immunoprecipitates immediately after stimulation of fibroblasts or smooth muscle cells with PDGF, and co-purifies with an 85-kDa protein that is phosphorylated on tyrosine *in vivo*. These immunoprecipitates also contain activities that phosphorylate PtdIns4P and PtdIns(4,5)P_2 to give the novel lipids noted above, but is not yet clear whether the 85-kDa phosphoprotein is PtdIns 3-kinase or whether a single enzyme is responsible for phosphorylating all three substrates; nor have the structures of the novel PtdInsP_2 and PtdInP_3 that appear in the stimulated cells been unambiguously worked out.

Stimulation of human neutrophils with the chemotactic peptide, fMet-Leu-Phe, both stimulates PIC and causes the rapid accumulation of the putative PtdIns 3-kinase pathway lipids, PtdIns(3,4)P_2 and PtdIns(3,4,5)P_3 [209]. The response is blocked by pertussis toxin and does not appear to be a consequence of the Ca^{2+} signal generated by fMet-Leu-Phe or of the simultaneous activation of protein kinase C [217]. These results suggest that chemotactic peptide receptors may couple, relatively directly, to a novel lipid kinase via a pertussis-toxin-sensitive G-protein. Similar results have been obtained using fMet-Leu-Phe-stimulated murine macrophages (S. Vallance, C. Macphee, C. P. Downes and A. D. Whetton, unpublished results) and thrombin-stimulated platelets [210, 211]. A particularly important result is that GTP[S] directly stimulates the synthesis of putative PtdIns(3,4)P_2 and PtdIns(3,4,5)P_3 in saponin-permeabilised platelets [210] thus strengthening the argument in favour of G-protein-mediated regulation.

A spate of recent publications has extended the initial observations implicating PDGF receptors in the regulation of PtdIns 3-kinase to include additional growth factor tyrosine kinases, notably the insulin [206] and CSF-1 [208] receptors. Table 1 gives a contemporary listing of receptors and other mechanisms which appear to activate the PtdIns 3-kinase pathway. A crucial role in mitogenesis is suggested by several studies utilising mutants of the genes for polyoma middle T, v-src and the PDGF receptor [219]. For example, the PDGF receptor contains a 100-amino-acid region known as the tyrosine kinase insert domain which appears to direct the ligand-stimulated association of the receptor with target proteins including PtdIns 3-kinase. Deletion of an 83-amino-acid segment of this domain, or replacement of Tyr751 which lies

within it (one of two autophosphorylation sites) with phenylalanine, blocks PDGF-induced mitogenesis and prevents association of the recepetor with PtdIns 3-kinase [220, 221]. As noted earlier, the CSF-1 recpetor, encoded by the *c-fms* protooncogene, is homologous with the PDGF receptor, but has a significantly shorter kinase insert domain, which is not required for mitogenesis and which has no tyrosine residue homologous with Tyr751 [222]. Addition of CSF-1 to fibroblasts that had been transfected with *c-fms* and which expressed large numbers of CSF-1 receptors activated the PtdIns 3-kinase pathway [208], but in a similar study did not activate PIC [119]. This contrasts with the activity of endogenous PDGF receptors which are capable of activating both pathways in such cells. These transfected cell lines may, therefore, provide valuable tools to evaluate the structural features of growth factor receptors that enable functional interactions with putative effector proteins such as PIC and PtdIns 3-kinase.

The occurrence of the PtdIns 3-kinase pathway and its regulation by G-protein-coupled receptors in platelets, neutrophils and macrophages raises some intriguing questions. Such cells are largely post-mitotic (platelets having no nucleus), suggesting that the PtdIns 3-kinase pathway, in addition to its putative role in mitogenesis, may have some specific functions in differentiated cells. It may be especially relevant that platelets are an abundant source of a pp60[c-src]-related tyrosine kinase which is activated by thrombin treatment [223] and that peripheral blood granulocytes, monocytes and tissue macrophages express the *c-fgr* protooncogene which encodes another protein tyrosine kinase (pp55[c-fgr]) of the *src* gene family. The latter undergoes a translocation from secretory vesicles to the plasma membrane upon stimulation of human neutrophils by chemotactic peptides and may be involved in the process of exocytosis [224].

As noted earlier, PtdIns 3-kinase pathway phospholipids are poor substrates for known PICs and appear to be metabolised solely by the action of specific phosphomonoesterases [215], suggesting a functional role for the lipids themselves, rather than a metabolite such as an inositol phosphate. One possible function that can be immediately considered is in the regulation of actin polymerisation by gelsolin. The conventional PtdIns$P_{n \geqslant 1}$ have been shown to be capable of inhibiting gelsolin's role in the mechanical severing of actin filaments

[225]. When sufficient of the novel lipids become available it will be important to assess whether they exert even more potent effects upon actin polymerisation. It is perhaps significant that gelsolin is known to associate with the plasma membranes of macrophages and platelets and that stimuli which cause PtdIns$P_{n \geqslant 1}$ breakdown inhibit this association. As noted above, these same stimuli also regulate the PtdIns 3-kinase pathway. It is, therefore, tempting to propose a role for these novel lipids in controlling the cytoskeletal rearrangements that must occur during the mitotic cell cycle and in post-mitotic cells during exocytosis or the shape changes which characterise the responses of platelets and neutrophils to external stimuli. Whatever the answer, there will be intense effort to understand the signalling roles of these lipids during the next few years.

CONCLUDING REMARKS

In 1980 most studies of *myo*-inositol metabolism were concerned with understanding the role of the enhanced metabolic turnover of PtdIns that accompanied hormonal stimulation of a variety of cells. At that time there were just a handful of studies concerning more exotic metabolites such as the PtdIns$P_{n \geqslant 1}$. The past decade has seen not only the resolution of a 30-year-old mystery, but also, with the introduction of sophisticated chromatographic procedures, a new era has emerged in which the true complexity of *myo*-inositol metabolism is becoming apparent. Within the last few years the natural occurrence of some 25 different inositol phosphate species has been reported and it seems likely that many of these compounds will be present in the majority of eukaryotic cells. For a time the discoveries of so many novel inositol phosphates diverted attention from the inositol phospholipids, but the wheel has come full circle so that now, at the start of a new decade, a further three PtdIns$P_{n \geqslant 1}$ have been described. Uncovering the functions of each of these compounds will surely occupy the minds and laboratories of many scientists during the next ten years!

I am grateful to the following colleagues for discussion of the contents of this review and for sending reprints of papers in press: M. Hanley, S. G. Rhee, T. K. Harden, R. H. Michell, L. Stepens, L. Cantly, C. Taylor, J. Putney, P. Cobbold and S. Shears. Work in the author's laboratory during the time of writing was supported by a Royal Society Research Grant and by a grant from Smith Kline & French Research Ltd.

REFERENCES

1. Downes, C. P. (1989) *Biochem. Soc. Trans. 17*, 259 – 268.
2. Nahorski, S. R. (1988) *Trends Neurosci. 11*, 444 – 448.
3. Ferris, C. D., Huganir, R. L., Supattapone, S. & Snyder, S. H. (1989) *Nature 342*, 87 – 89.
4. Majerus, P. W., Connolly, T. M., Bansal, V. S., Inhorn, R. C., Ross, T. S. & Lips, D. L. (1988) *J. Biol. Chem. 263*, 3051 – 3053.
5. Cosgrove, D. J. (1980) *Inositol phosphates: their chemistry, biochemistry and physiology*, Elsevier, Netherlands.
6. Whitman, M., Downes, C. P., Keeler, T. K. & Cantley, L. (1988) *Nature 332*, 644 – 646.
7. Low, M. G. (1989) *Biochim. Biophys. Acta 988*, 427 – 454.
8. Downes, C. P. (1986) *Neurochem. Int. 9*, 211 – 230.
9. Berridge, M. J., Downes, C. P. & Hanley, M. R. (1989) *Cell 59*, 411 – 419.
10. Greene, D. A., Lewis, R. A., Lattimer, S. A. & Brown, M. J. (1982) *Diabetes 31*, 573 – 578.
11. Gillon, K. R. W. & Hawthorne, J. N. (1983) *Life Sci. 32*, 1943 – 1947.
12. Prpic, B., Blackmore, P. F. & Exton, J. H. (1982) *J. Biol. Chem. 257*, 11315 – 11322.
13. Downes, C. P. & Stone, M. A. (1986) *Biochem. J. 234*, 199 – 204.
14. Sherman, W. R., Leavitt, A. L., Honchar, M. P., Hallcher, L. M. & Phillips, B. E. (1981) *J. Neurochem. 36*, 1947 – 1951.
15. Hasegawa, K. & Eisenberg, F. Jr. (1981) *Proc. Natl Acad. Sci. USA 78*, 4863 – 4866.
16. Horstman, D., Takemura, H. & Putney, J. W. Jr. (1988) *J. Biol. Chem. 263*, 15297 – 15303.
17. Hallcher, L. M. & Sherman, W. R. (1980) *J. Biol. Chem. 255*, 10896 – 10091.
18. Michell, R. H., Kirk, C. J., Maccallum, S. H. & Hunt, P. A. (1988) *Phil. Trans. R. Soc. Lond. B 320*, 239 – 246.
19. Imai, A. & Gershengorn, M. C. (1987) *Nature 325*, 726 – 728.
20. Cubitt, A. B. & Gershengorn, M. C. (1989) *Biochem. J. 257*, 639 – 644.
21. Dietz, M. & Albersheim, P. (1965) *Biochem. Biophys. Res. Commun. 19*, 598 – 603.
21a. Stephens, L. R. & Irvine, R. F. (1990) *Nature 346*, 580 – 583.
22. Berridge, M. J. & Irvine, R. F. (1989) *Nature 341*, 197 – 205.
23. Michell, R. H. (1989) *Current Opinion in Cell Biology 1*, 201 – 205.
24. Downes, C. P. & Carter, A. N. (1990) *Current Opinion in Cell Biology 2*, 185 – 191.
25. Downes, C. P. & Michell, R. H. (1985) in *Molecular mechanisms of transmembrane signalling* (Cohen, P. & Houslay, M. D., eds) pp. 3 – 56, Elsevier, Netherlands.
26. Nishizuka, Y. (1988) *Nature, 334*, 661 – 665.
27. Downes, C. P., Hawkins, P. T. & Stephens, L. R. (1989) in *Inositol lipids in cell signalling* (Michell, R. H., Drummond, A. H. & Downes, C. P., eds), pp. 3 – 37. Academic Press, London.
28. Endemann, G., Dunn, S. N. & Cantley, L. C. (1987) *Biochemistry 26*, 6845 – 6852.
29. Hou, W.-M., Zhang, Z.-L. & Tai, H.-H. (1988) *Biochim. Biophys. Acta 959*, 67 – 75.
30. Porter, F. D., Li, Y.-S. & Denel, T. F. (1988) *J. Biol. Chem. 263*, 8989 – 8995.
31. Walker, B. H., Dougherty, N. & Pike, L. J. (1988) *Biochemistry 27*, 6504 – 6511.
32. Li, Y.-S., Porter, F. D., Hoffman, R. M. & Denel, T. F. (1989) *Biochem. Biophys. Res. Commun. 160*, 202 – 209.
33. Saltiel, A. R., Fox, J. A., Sherline, P., Sahyoun, N. & Cuatrecasas, P. (1987) *Biochem. J. 241*, 759 – 763.
34. Yamakawa, A. & Takenawa, T. (1988) *J. Biol. Chem. 263*, 17555 – 17560.
35. Lunberg, G. A. & Jergil, B. (1988) *FEBS Lett. 240*, 171 – 176.
36. Husebye, E. S. & Flatmark, T. (1988) *Biochim. Biophys. Acta 968*, 261 – 265.
37. Campbell, C. R., Fishman, J. B. & Fine, R. E. (1985) *J. Biol. Chem. 260*, 10948 – 10951.
38. Smith, C. D. & Wells, W. W. (1983) *J. Biol. Chem. 258*, 9368 – 9373.
39. Cocco, L., Gilmour, R. S., Ognibene, A., Letcher, A. J., Manzoli, F. A. & Irvine, R. F. (1987) *Biochem. J. 248*, 765 – 770.
40. Cocco, L., Martelli, A. M., Gilmour, R. S., Ognibene, A., Manzoli, F. A. & Irvine, R. F. (1988) *Biochem. Biophys. Res. Commun. 154*, 1266 – 1272.
41. Walker, D. H. & Pike, L. T. (1987) *Proc. Natl Acad. Sci. USA 84*, 7513 – 7517.
42. Taylor, M. V., Metcalfe, J. C., Hesketh, T. R., Smith, G. A. & Moore, J. P. (1984) *Nature 312*, 462 – 465.
43. Halenda, S. P. & Feinstein, M. B. (1984) *Biochem. Biophys. Res. Commun. 124*, 507 – 513.
44. Suga, F., Kambayashi, J., Kawasaki, T., Sahon, M. & Mori, T. (1986) *Thromb. Res. 44*, 155 – 163.
45. Sarkadi, B., Enyedi, A., Farago, A., Meszaros, G., Kremmer, T. & Gardos, G. (1983) *FEBS Lett. 152*, 195 – 198.
46. Coco, A. & Macara, I. G. (1986) *Am. J. Physiol. 251*, C883 – C886.

16

47. Vrolix, M., Raeymaekers, L., Wuytock, F., Hofmann, F. & Casteels, R. (1988) *Biochem. J. 255*, 855—863.
48. Vogel, S. & Hoppe, J. (1986) *Eur. J. Biochem. 154*, 253—257.
49. Smith, C. D. & Snyderman, R. (1988) *Biochem. J. 256*, 125—130.
50. Van Dongen, C. J., Zwiers, H. & Gispen, W. H. (1984) *Biochem. J. 223*, 197—203.
51. Cochet, C. & Chambaz, E. M. (1986) *Biochem. J. 237*, 25—31.
52. Ling, L. E., Schulz, J. T. & Cantley, L. C. (1989) *J. Biol. Chem. 264*, 5080—5088.
53. Moritz, A., DeGraan, P. N. E., Ekhart, P. F., Gispen, W. H. & Wirtz, K. W. A. (1990) *J. Neurochem. 54*, 351—354.
54. Jergil, B. & Sundler, R. (1983) *J. Biol. Chem. 258*, 7968—7973.
55. Seyfred, M. A. & Wells, W. W. (1984) *J. Biol. Chem. 259*, 7659—7665.
56. Cockroft, S.,Taylor, J. A. & Judah, J. D. (1985) *Biochim. Biophys. Acta 845*, 163—170.
57. Imai, A. & Gershengorn, M. E. (1986) *Proc. Natl Acad. Sci. USA 83*, 8540—8544.
58. Renard, D., Poggioli, J., Berthon, B. & Claret, M. (1987) *Biochem. J. 243*, 391—398.
59. Chaffoy de Courcelles, D., Roevers, P. & Van Belle, H. (1985) *J. Biol. Chem. 260*, 15762—15770.
60. Jolles, J., Zwiers, H., Van Dongen, C. J., Schotman, P., Wirtz, K. W. A. & Gispen, W. H. (1980) *Nature 286*, 623—625.
61. Aloyo, V. J., Swiers, H. & Gispen, W. H. (1983) *J. Neurochem. 41*, 649—653.
62. Van Dongen, C. J., Zwiers, H., De Graan, P. N. E. & Gispen, W. H. (1985) *Biochem. Biophys. Res. Commun. 128*, 1219—1227.
63. Urumow, T. & Wieland, O. H. (1986) *FEBS Lett. 207*, 253—257.
64. Smith, C. D. & Chang, K.-J. (1989) *J. Biol. Chem. 264*, 3206—3210.
65. Strosznajder, J. & Strosznajder, R. P. (1989) *Neurochem. Res. 14*, 717—723.
66. Lundberg, G. A., Sundler, R. & Jergil, B. (1987) *Biochim. Biophys. Acta 992*, 1—7.
67. Lundberg, G. A., Gergil, B. & Sundler, R. (1986) *Eur. J. Biochem. 161*, 257—262.
68. Oberdorf, J., Vilar-Rojas, C. & Epel, D. (1989) *Dev. Biol. 131*, 236—242.
69. Takenawa, T. & Nagai, Y. (1981) *J. Biol. Chem. 256*, 6769—6775.
70. Rhee, S. G., Suh, P. G., Ryu, S. H. & Lee, S. Y. (1989) *Science 244*, 546—550.
71. Hoffman, S. L. & Majerus, P. W. (1982) *J. Biol. Chem. 257*, 6461—6469.
72. Ryu, S. H., Cho, K. S., Lee, K. Y., Suh, P. G. & Rhee, S. G. (1986) *Biochem. Biophys. Res. Commun. 141*, 137—144.
73. Ryu, S. H., Cho, K. S., Lee, K. Y., Suh, P. G. & Rhee, S. G. (1987) *J. Biol. Chem. 262*, 12511—12518.
74. Ryu, S. H., Suh, P. G., Cho, K. S., Lee, K. Y. & Rhee, S. G. (1987) *Proc. Natl Acad. Sci. USA 84*, 6649—6653.
75. Suh, P. G., Ryu, S. H., Moon, K. H., Suh, H. W. & Rhee, S. G. (1988) *Cell 54*, 161—169.
76. Katan, M., Kriz, R. W., Totty, N., Philp, R., Meldrum, E., Aldape, R. A., Knopf, J. L. & Parker, P.J. (1988) *Cell 54*, 171—177.
77. Bennett, C. F., Balcarek, J. M., Varrichio, A. & Crooke, S. T. (1988) Nature 334, 268—270.
78. Stahl, M. L., Ferenz, C. R., Kelleher, K. L., Kriz, R. W. & Knopf, J. (1988) *Nature 332*, 269—272.
79. Suh, P. G., Ryu, S. H., Moon, K. H., Suh, H. W. & Rhee, S. G. (1988) *Proc. Natl Acad. Sci. USA 85*, 5419—5423.
80. Boyer, J. L., Hepler, J. R. & Harden, T. K. (1989) *Trends Pharmacol. Sci. 10*, 360—364.
81. Meldrum, E., Katan, M. & Parker, P. (1989) *Eur. J. Biochem. 182*, 673—677.
82. Baldassare, J. J., Henderson, P. A. & Fisher, G. J. (1989) *Biochemistry 28*, 6010—6016.
83. Ohta, S., Matsui, A., Nazawa, Y. & Kagawa, Y. (1989) *FEBS Letts. 242*, 31—35.
84. Irvine, R. F., Letcher, A. J. & Dawson, R. M. C. (1984) *Biochem. J. 218*, 177—185.
85. Wilson, D. B., Bross, T. E., Hoffman, S. L. & Majerus, P. W. (1984) *J. Biol. Chem. 259*, 11718—11724.
86. Hughes, A. R. & Putney, J. W. Jr. (1989) *J. Biol. Chem. 264*, 9400—9407.
87. Dixon, J. F. & Hokin, C. E. (1989) *J. Biol. Chem. 264*, 11721—11724.
88. Cockroft, S. & Allan, D. (1984) *Biochem. J. 222*, 557—559.
89. Billah, M. M., Eckel, S., Mullmann, T. J., Egan, R. W. & Siegel, M. I. (1989) *J. Biol. Chem. 264*, 17069—17077.
90. Billah, M. M., Pai, J.-K., Mullman, T. J., Egan, R. W. & Siegel, M. I. (1989) *J. Biol. Chem. 264*, 9069—9076.
91. Bocckino, S. B., Blackmore, P. F., Wilson, P. B. & Exton, J. H. (1987) *J. Biol. Chem. 262*, 15309—15315.
92. Choi, W. C., Gerfen, C. R., Suh, P. G. & Rhee, S. G. (1989) *Brain Res. 499*, 193—197.
93. Lee, K. Y., Ryu, S. H., Suh, P. G., Choi, W. C. & Rhee, S. G. (1987) *Proc. Natl Acad. Sci. USA 84*, 5540—5544.
94. Gilman, A. G. (1987) *Annu. Rev. Biochem. 56*, 615—649.
95. Cockroft, S. & Gomperts, B. D. (1985) *Nautre 314*, 534—536.
96. Litosch, I., Wallis, C. & Fain, J. N. (1985) *J. Biol. Chem. 260*, 5464—5471.
97. Boyer, J. L., Downes, C, P. & Harden, T. K. (1989) *J. Biol. Chem. 264*, 884—890.
98. Boyer, J. L., Waldo, G. L., Evans, T., Northup, J. K., Downes, C. P. & Harden, T. K. (1989) *J. Biol. Chem. 264*, 13917—13992.
99. Moriarty, T. M., Gillo, B., Carty, D. J., Premont, R. T., London, E. M. & Iyengar, R. (1988) *Proc. Natl Acad. Sci. USA 85*, 8865—8869.
100. Polakis, P. G., Uhing, R. J. & Snyderman, R. (1988) *J. Biol. Chem. 263*, 4969—4976.
101. Moriarty, T. M., Padrell, E., Carty, D. J., Omri, G., Landau, E. M. & Iyengar, R. (1990) *Nature 343*, 79—82.
102. Linden, J. L. & Delahunty, T. M. (1989) *Trends Pharmacol. Sci. 10*, 114—120.
103. Litosch, I. & Fain, J. N. (1985) *J. Biol. Chem. 260*, 16052—16055.
104. Taylor, S. J. & Exton, J. H. (1987) *Biochem. J. 248*, 791—799.
105. Claro, E., Wallace, M. A., Lee, H. M. & Fain, J. N. (1989) *J. Biol. Chem. 264*, 18288—18295.
106. Litosch, I. (1989) *Biochem. J. 261*, 325—331.
107. Baldassare, J. J., Knipp, M., Henderson, P. A. & Fisher, G. J. (1988) *Biochem. Biophys. Res. Commun. 154*, 351—357.
108. Morris, A. J., Waldo, G. L., Downes, C. P. & Harden, T. K. (1990) *J. Biol. Chem.*, in the press.
109. Morris, A. J., Waldo, G. L., Downes, C. P. & Harden, T. K. (1990) *J. Biol. Chem.*, in the press.
110. Pouysségur, J., Chambard, J. C., L'Allermain, G., Magnaldo, I. & Seuwen, K. (1988) *Philos. Trans. R. Soc. Lond. B. 320*, 191—200.
111. Moolenaar, W. H., Bierman, A. J., Tilly, B. C., Verlaan, I., Defize, L. H., Honegger, A. M., Ullrich, A. & Schlessinger, J. (1988) *EMBO J. 7*, 707—710.
112. Keating, M. T., Escobedo, J. A. & Williams, L. T. (1988) *J. Biol. Chem. 263*, 12805—12808.
113. Cattaneo, M. B. & Vicentini, M. (1989) *Biochem. J. 262*, 665—668.
114. Meisenhelder, J., Sho, P. G., Rhee, S. G. & Hunter, T. (1989) *Cell 57*, 1109—1122.
115. Nishibe, S., Wahl, M. I., Rhee, S. G. & Carpenter, G. (1989) *J. Biol. Chem. 264*, 10335—10338.
116. Wahl, M. I., Olashaw, N. E., Nishibe, S., Rhee, S. G., Pledger, W. J. & Carpenter, G. (1989) *Mol. Cell Biol. 9*, 2934—2943.
117. Wahl, M. I., Nishibe, S., Suh, P. G., Rhee, S. G. & Carpenter, G. (1989) *Proc. Natl Acad. Sci. USA 86*, 1568—1572.
118. Margolis, B., Rhee, S. G., Felder, S., Mernic, M., Lyall, R., Levitzki, A., Ullrich, A., Zilberstein, A. & Schlessinger, J. (1989) *Cell 57*, 1101—1107.
119. Downing, J. R., Margolis, B. J., Zilberstein, A., Ashmun, R. A., Ullrich, A., Sherr, C. J. & Schlessinger, J. (1989) *EMBO J. 8*, 3345—3350.

120. Kim, U. H., Kim, J. W. & Rhee, S. G. (1989) *J. Biol. Chem. 264*, 20167–20170.
121. Kim, J. W., Sim, S. S., Kim, U. H., Nishibe, S., Wahl, M. I., Carpenter, G. & Rhee, S. G. (1990) *J. Biol. Chem. 265*, 3940–3943.
122. Irvine, R. F., Hemington, N. & Dawson, R. M. C. (1979) *Eur. J. Biochem. 99*, 525–530.
123. Shears, S. B. (1989) *Biochem. J. 260*, 313–324.
124. Dean, N. M. & Beavan, M. A. (1989) *Anal. Biochem. 183*, 199–209.
125. Palmer, S. & Wakelam, M. J. O. (1989) *Biochim. Biophys. Acta. 1014*, 239–246.
126. Irvine, R. F., Moor, R. M., Pollock, W. K., Smith, P. M. & Wreggett, K. A. (1988) *Philos. Trans. R. Soc. Lond. B. 320*, 281–298.
127. Hansen, C. A., Johanson, R. A., Williamson, M. T. & Williamson, J. R. (1987) *J. Biol. Chem. 262*, 17319–17326.
128. Connolly, T. M., Lawing, W. J. Jr. & Majerus, P. W. (1986) *Cell 46*, 951–958.
129. Imboden, J. B. & Pattison, G. (1987) *J. Clin. Invest. 79*, 1538–1541.
130. Li, G., Comte, M., Wollheim, C. B. & Cox, J. A. (1989) *Biochem. J. 260*, 771–775.
131. Takazawa, K., Passaveiro, H., Dumont, J. E. & Erneux, C. (1989) *Biochem. J. 261*, 483–488.
132. King, W. & Rittenhouse, S. E. (1989) *J. Biol. Chem. 264*, 6070–6074.
133. Moon, K. H., Lee, S. Y. & Rhee, S. G. (1989) *Biochem. Biophys. Res. Commun. 164*, 370–374.
134. Choi, K. Y., Kim, H. K., Lee, S. Y., Moon, K. H., Sim, S. S., Kim, J. W. & Rhee, S. G. (1990) *Science 248*, 64–66.
135. Yamaguchi, K., Hirata, M. & Kuriyama, H. (1988) *Biochem. J. 251*, 129–134.
136. Gee, N. S., Ragan, C. I., Watling, K. J., Aspley, S., Jackson, R. G., Reid, G. R., Gani, D. & Shute, J. K. (1988) *Biochem. J. 249*, 883–889.
137. Wilson, D. B., Connolly, T. M., Bross, T. E., Majerus, P. W., Sherman, W. R., Tyler, A. N., Rubin, L. J. & Brown, J. E. (1985) *J. Biol. Chem. 260*, 13496–13501.
138. Sekar, M. C., Dixon, J. F. & Hokin, L. E. (1987) *J. Biol. Chem. 262*, 340–344.
139. Hughes, A. R., Takemura, H. & Putney, J. W. Jr. (1988) *J. Biol. Chem. 263*, 10314–10319.
140. Dawson, R. M. C. & Clarke, R. (1972) *Biochem. J. 127*, 113–118.
141. Van Lookeren Campagne, M. M., Erneux, C., Van Eijk, R. & Van Haastert, P. J. M. (1988) *Biochem. J. 254*, 343–350.
142. Cunha-Melo, J. R., Dean, N. M., Ali, H. & Beavan, M. A. (1988) *J. Biol. Chem. 263*, 14245–14250.
143. Doughney, C., McPherson, M. A. & Dormer, R. L. (1988) *Biochem. J. 251*, 927–929.
144. Shears, S. B. (1989) *J. Biol. Chem. 264*, 19879–19886.
145. Hughes, P. J. & Shears, S. J. (1990) *J. Biol. Chem. 265*, 9869–9875.
146. Szwergold, B. S., Graham, R. A. & Brown, T. R. (1987) *Biochem. Biophys. Res. Commun. 149*, 874–881.
147. Stephens, L. R., Hawkins, P. T., Carter, N., Chahwala, S. B., Morris, A. J., Whetton, A. D. & Downes, C. P. (1988) *Biochem. J. 249*, 271–282.
148. Stephens, L. R., Hawkins, P. T., Morris, A. J. & Downes, C. P. (1988) *Biochem. J. 249*, 283–292.
149. Shears, S. B., Parry, J. B., Tang, E. K. Y., Irvine, R. F., Michell, R. H. & Kirk, C. J. (1987) *Biochem. J. 246*, 139–147.
150. Balla, T., Guillemette, G., Baukal, A. J. & Catt, K. J. (1987) *J. Biol. Chem. 262*, 9952–9955.
151. Stephens, L. R., Hawkins, P. T., Barker, C. J. & Downes, C. P. (1988) *Biochem. J. 253*, 721–733.
152. Balla, T., Baukal, A. J., Hunyady, L. & Catt, K. J. (1989) *J. Biol. Chem. 264*, 13605–13611.
153. Barker, C. J., Morris, A. J., Kirk, C. J. & Michell, R. H. (1988) *Biochem. Soc. Trans. 16*, 984–985.
154. Pittet, D., Schlegel, W., Lew, D. P., Monod, A. & Mayr, G. (1989) *J. Biol. Chem. 264*, 18489–18493.
155. Berrie, C. P., Hawkins, P. T., Stephens, L. R., Harden, T. K. & Downes, C. P. (1989) *Mol. Pharmacol. 35*, 526–532.
155a. Menniti, F. S., Oliver, K. G., Nogimori, K., Obie, J. F., Shears, S. B. & Putney, J. W. Jr (1990) *J. Biol. Chem. 265*, 11167–11176.
156. Maccallum, S. H., Hunt, P. A., Michell, R. H. & Kirk, C. J. (1989) *Biochem. Soc. Trans. 17*, 88–89.
157. Stephens, L. R. & Downes, C. P. (1990) *Biochem. J. 265*, 435–452.
158. Stephens, L. R., Hawkins, P. T. & Downes, C. P. (1989) *Biochem. J. 262*, 727–737.
159. Radenberg, T., Scholz, P., Bergmann, G. & Mayr, G. W. (1989) *Biochem. J. 264*, 323–333.
160. Sylvia, V., Curtin, G., Norman, J., Stec, J. & Busbee, D. (1988) *Cell 54*, 651–658.
161. Berridge, M. J. & Irvine, R. F. (1984) *Nature 312*, 315–321.
162. Joseph, S. K. & Williamson, J. R. (1989) *Arch. Biochem. Biophys. 273*, 1–15.
163. Hanley, M. R., Jackson, T. R., Vallejo, M., Patterson, S. I., Thastrup, O., Lightman, S., Rogers, J., Henderson, G. & Pini, A. (1988) *Philos. Trans. R. Soc. Lond. B. 320*, 381–398.
164. Kass, G. E. N., Duddy, S. K., Moore, G. A. & Orrenius, S. (1989) *J. Biol. Chem. 264*, 15192–15198.
165. Nahorski, S. R. & Potter, B. V. L. (1989) *Trends Pharmacol. Sci. 10*, 139–144.
166. Walker, J. W., Somlyo, A. V., Goldman, Y., Somlyo, A P. & Trentham, D. R. (1987) *Nature 327*, 249–252.
167. Spät, A., Bradford, P. G., McKinney, J. S., Rubin, R. P. & Putney, J. W. Jr. (1986) *Nature 319*, 514–516.
168. Spät, A., Fabiato, A. & Rubin, R. P. (1986) *Biochem. J. 233*, 929–932.
169. Nunn, D. C., Potter, B. V. L. & Taylor, C. W. (1990) *Biochem. J. 265*, 393–398.
170. Worley, P. F., Baraban, J. M., Supattapone, S., Wilson, V. S. & Snyder, S. H. (1987) *J. Biol. Chem. 262*, 12132–12136.
171. Supattapone, S., Worley, P. F., Baraban, J. M. & Snyder, S. H. (1988) *J. Biol. Chem. 263*, 1530–1534.
172. Mallet, J., Huchet, M., Pougelois, R. & Changeux, J. P. (1976) *Brain Res. 103*, 291–312.
173. Furuichi, T., Yoshikawa, S., Miyawaki, A., Wada, K., Maeda, N. & Mikoshiba, K. (1989) *Nature 342*, 32–38.
174. Mignery, G. A., Südhof, T. C., Takei, K. & De Camilli, P. (1989) *Nature 342*, 192–195.
175. Meyer, T., Holowka, D. & Stryer, L. (1988) *Science 240*, 653–656.
176. Taylor, C. W. & Putney, J. W. Jr. (1985) *Biochem. J. 232*, 435–438.
177. Volpe, P., Krause, K. H., Sadamitsu, H., Zorsato, F., Pozzan, T., Meldolesi, J. & Low, P. B. (1988) *Proc. Natl Acad. Sci. USA 85*, 1091–1095.
178. Ross, C. A., Meldolesi, J., Milner, T. A., Satoh, T., Supattapone, S. & Snyder, S. H. (1989) *Nature 339*, 468–470.
179. Ghosh, T. K., Eis, P. S., Mullaney, J. M., Ebert, C. L. & Gill, D. L. (1988) *J. Biol. Chem. 263*, 11075–11079.
180. Dawson, A. P. (1985) *FEBS Lett. 185*, 147–150.
181. Ghosh, T. K., Mullaney, J. M., Tarazi, F. I. & Gill, D. L. (1989) *Nature 340*, 236–239.
182. Putney, J. W. Jr. (1986) *Annu. Rev. Physiol. 48*, 75–88.
183. Merritt, J. E. & Rink, T. J. (1987) *J. Biol. Chem. 262*, 4958–4960.
184. Irvine, R. F. & Moor, R. M. (1987) *Biochem. Biophys. Res. Commun. 146*, 284–290.
185. Crossley, I., Swann, K., Chambers, E. & Whitaker, M. (1988) *Biochem. J. 252*, 257–262.
186. Morris, A. P., Gallacher, D. V., Irvine, R. F. & Peterson, O. H. (1987) *Nature 330*, 653–655.
187. Casteels, R. & Droogmans, G. (1981) *J. Physiol. 317*, 263–279.
188. Putney, J. W. Jr. (1986) *Cell Calcium 7*, 1–12.
189. Takemura, H. & Putney, J. W. Jr. (1989) *Biochem. J. 258*, 409–412.
189. Berridge, M. J., Cobbold, P. H. & Cuthbertson, K. S. R. (1988) *Philos. Trans. R. Soc. Lond. B. 320*, 325–343.

18

191. Cobbold, P., Dixon, J., Sanchez-Bueno, A., Woods, N., Daly, M. & Cuthbertson, R. (1990) *Transmembrane signalling, intracellular messengers and implications for drug development* (S. Nahorski, ed.) pp. 185–206, John Wiley.
192. Jacob, R. (1990) *Biochem. Biophys. Acta 1052*, 427–438.
193. Woods, N. M., Cuthbertson, K. S. R. & Cobbold, P. H. (1986) *Nature 319*, 600–602.
194. Wakui, M., Potter, B. V. L. & Peterson, O. H. (1989) *Nature 339*, 317–320.
195. Wilson, D. B., Connolly, T. M., Bross, T. E., Majerus, P. W., Sherman, W. R., Tyler, A. N., Rubin, L. J. & Brown, J. E. (1985) *J. Biol. Chem. 260*, 13496–13501.
196. Willcocks, A. L., Strupish, J., Irvine, R. F. & Nahorski, S. R. (1989) *Biochem. J. 257*, 297–300.
197. Graf, E., Empson, K. L. & Eaton, J. W. (1987) *J. Biol. Chem. 262*, 11647–11650.
198. Shamsuddin, A. M., Elsayed, A. M. & Ullah, A. (1988) *Carcinogenesis 9*, 577–580.
199. Vallejo, M., Jackson, T. R., Lightman, S. L. & Hanley, M. R. (1987) *Nature 330*, 656–658.
200. Hawkins, P. T., Reynolds, D. J. N., Poyner, D. R. & Hanley, M. R. (1990) *Biochem. Biophys. Res. Commun. 167*, 819–827.
201. Carpenter, D., Hanley, M. R., Hawkins, P. T., Jackson, T. R., Stephens, L. R. & Vallejo, M. (1989) *Biochem. Soc. Trans. 17*, 3–5.
202. Sawada, M., Ichinose, M. & Maeno, T. (1989) *Neurosci. Lett. 106*, 328–333.
203. Sawada, M., Ichinose, M. & Maeno, T. (1989) *Brain Res. 503*, 167–169.
204. Stephens, L., Hawkins, P. T. & Downes, C. P. (1988) *Biochem. J. 259*, 267–276.
205. Augur, K. R., Serunian, L. A., Soltoff, S. P., Libby, P. & Cantley, L. C. (1989) *Cell 57*, 167–175.
206. Endemann, G., Yonezawa, K. & Roth, R. A. (1990) *J. Biol. Chem. 265*, 396–400.
206a. Ruderman, N. B., Kapeller, R., White, M. F. & Cantley, L. C. (1990) *Proc. Natl Acad. Sci. USA 87*, 1411–1415.
207. Pignataro, O. P. & Ascoli, M. (1990) *J. Biol. Chem. 265*, 1718–1723.
208. Varticovski, L., Druker, B., Morrison, D., Cantley, L. & Roberts, T. (1989) *Nature 324*, 699–702.
209. Traynor-Kaplan, A. E., Harris, A. L., Thompson, B. L., Taylor, P. & Sklar, L. A. (1988) *Nature 334*, 353–356.
210. Kucera, G. L. & Rittenhouse, S. E. (1990) *J. Biol. Chem.*, in the press.
211. Nolan, R. D. & Lapetina, E. G. (1990) *J. Biol. Chem. 265*, 2441–2445.
212. Augur, K. R., Carpenter, C. L., Cantley, L. C. & Varticovski, L. (1989) *J. Biol. Chem. 264*, 20181–20184.
213. Lips, D. L., Majerus, P. W., Gorga, F. R., Young, A. T. & Benjamin, T. L. (1989) *J. Biol. Chem. 264*, 8759–8763.
214. Serunian, L. A., Haber, M. T., Fukui, T., Kim, J. W., Rhee, S. G., Lowensten, J. M. & Cantley, L. C. (1989) *J. Biol. Chem. 264*, 17809–17815.
215. Lips, D. L. & Majerus, P. W. (1989) *J. Biol. Chem. 264*, 19911–19915.
216. Whitman, M., Kaplan, D. R., Roberts, T. M. & Cantley, L. C. (1987) *Biochem. J. 247*, 165–174.
217. Traynor-Kaplan, A. E., Thompson, B. L., Harris, A. L., Taylor, P., Omann, G. M. & Sklar, L. A. (1989) *J. Biol. Chem. 264*, 15668–15673.
218. Kaplan, D. R., Whitman, M., Schaffhausen, B., Pallas, D. C., White, M., Cantley, L. & Roberts, T. M. (1987) *Cell 50*, 1021–1029.
219. Whitman, M., Kaplan, D. R., Schaffhausen, B., Cantley, L. & Roberts, T. (1985) *Nature 315*, 239–242.
220. Coughlin, S. R., Escobedo, J. A. & Williams, L. T. (1989) *Science 243*, 1191–1193.
221. Kazlauskas, A. & Cooper, J. A. (1989) *Cell 58*, 1121–1133.
222. Taylor, G. R. Reedijk, M., Rothwell, V., Rohrschneider, L. & Pawson, T. (1989) *EMBO J. 8*, 2029–2037.
223. Farrell, J. E. Jr. & Martin, S. G. (1988) *Mol. Cell Biol. 8*, 3603–3610.
224. Gutkind, J. S. & Robbins, K. C. (1989) *Proc. Natl Acad. Sci. USA 86*, 8783–8787.
225. Yin, H. L. (1988) *Bioessays 7*, 176–179.
226. Fukui, Y., Kornbluth, S., Jong, S. M., Wang, L. H. & Hanafusa, J. (1989) *Oncogene Res. 4*, 283–292.

Eur. J. Biochem. *193*, 599 – 622 (1990)
© FEBS 1990

Review

Structural and functional aspects of calcium homeostasis in eukaryotic cells

Daniela PIETROBON[1], Francesco Di VIRGILIO[1] and Tullio POZZAN[2]

[1] Consiglio Nazionale delle Ricerche, Unit for the Study of the Physiology of Mitochondria and Institute of General Pathology,
University of Padova, Padova, Italy
[2] Institute of General Pathology, University of Ferrara, Ferrara, Italy

(Received February 6, 1990) — EJB 90 0120

The maintenance of a low cytosolic free-Ca^{2+} concentration, ($[Ca^{2+}]_i$) is a common feature of all eukaryotic cells. For this purpose a variety of mechanisms have developed during evolution to ensure the buffering of Ca^{2+} in the cytoplasm, its extrusion from the cell and/or its accumulation within organelles. Opening of plasma membrane channels or release of Ca^{2+} from intracellular pools leads to elevation of $[Ca^{2+}]_i$; as a result, Ca^{2+} binds to cytosolic proteins which translate the changes in $[Ca^{2+}]_i$ into activation of a number of key cellular functions.

The purpose of this review is to provide a comprehensive description of the structural and functional characteristics of the various components of $[Ca^{2+}]_i$ homeostasis in eukaryotes.

It is well known how Sidney Ringer recognized the role of Ca^{2+} in muscle contraction [1] and how fortunate it was that he was working with frog heart muscle, rather than with skeletal muscle, given that contractility in the latter muscle is basically independent of extracellular Ca^{2+}. Less appreciated is the fact that it took a century before a wide-range test of the 'Ca^{2+} hypothesis' in the mediation of cellular responses became possible. So many years in fact elapsed between the publication of the seminal paper by Ringer in 1883 [1] and the report in 1982 by Tsien and co-workers [2] describing the application of quin 2 to the measurement of intracellular free Ca^{2+} ($[Ca^{2+}]_i$) in small mammalian cells.

The exploitation of fluorescent tetracarboxylate/penta-carboxylate dyes for measuring Ca^{2+} has enabled a thorough testing of Ca^{2+} involvement in a host of cellular responses spanning fertilization to muscle contraction, proliferation to neurotransmitter release and cell motility to cell death. The fluorescent dye technique has also complemented biochemistry, electrophysiology and molecular biology, and allowed the description in real time of the actual changes in $[Ca^{2+}]_i$ following the activation of membrane Ca^{2+} influx/efflux pathways. Additional levels of complexity have emerged from the application of these techniques to the study of Ca^{2+} homeostasis and new problems have arisen, yet for the first time we are beginning to understand Ca^{2+} signaling as the integration of multiple pathways.

Ca^{2+} is maintained in the cytoplasm of mammalian cells at a concentration that is 10000-fold lower than that of the extracellular milieu. This is achieved by means of sophisticated homeostatic mechanisms given that with a negative interior-plasma membrane potential of -60 mV, a $[Ca^{2+}]_i$ of $0.1 - 0.2$ M should be expected, if Ca^{2+} were distributed at electrochemical equilibrium [3]. The first barrier to Ca^{2+} overflow into the cytoplasm is the natural impermeability of phospholipid bilayers to charged species. However, permeability to Ca^{2+} is significant even across a pure lipid bilayer and even more so across the plasma membrane of living cells that also possess specific pathways (channels) for Ca^{2+} influx. These pathways can be opened by ligands or by depolarization and allow passive diffusion of Ca^{2+}. Ca^{2+} can also be released from intracellular depots via a complex chain of intracellular messengers, in part still unknown. Once Ca^{2+} has entered the cytoplasm, it is bound by a number of proteins that transduce the Ca^{2+} signal or simply buffer this ion. The long-term maintainance of steady and low $[Ca^{2+}]_i$, even in the absence of a massive Ca^{2+} influx through specific channels, depends on the operation of Ca^{2+}-extrusion mechanisms. Excess cytoplasmic Ca^{2+} can also be taken up by mitochondria via an electrophoretic uniport and/or accumulated into non-mitochondrial intracellular stores. The homeostasis of $[Ca^{2+}]_i$ will be the result of all these interacting mechanisms.

THE PLASMA MEMBRANE

In all eukaryotes, the primary barrier to Ca^{2+} inflow into the cytoplasm is represented by the plasma membrane, where the ultimate long-term regulators of $[Ca^{2+}]_i$ homeostasis, i.e. the Ca^{2+}-efflux mechanisms, are located. It is evident that intracellular Ca^{2+}-binding proteins or organelles capable of

Correspondence to T. Pozzan, Institute of General Pathology, University of Ferrara, Via Borsari 46, I-35121 Ferrara, Italy

Abbreviations. $[Ca^{2+}]_i$, cytosolic free-Ca^{2+} concentration; VOC, voltage-operated channel; ROC, receptor-operated channel; SMOC, second messenger-operated channel; GOC, G-protein-operated channel; EC coupling, excitation-contraction coupling; MeDAsp, *N*-methyl D-aspartate; Ins(1,4,5)P_3, inositol 1,4,5-trisphosphate; Ins(1,3,4,5)P_4, inositol 1,3,4,5-tetrakisphosphate; PtdIns(4,5)P_2, phosphatidylinositol (4,5) bisphosphate; SR, sarcoplasmic reticulum; ER, endoplasmic reticulum; TC, terminal cisternae; E1, enzyme form 1 of ATPase; E2, enzyme form 2 of ATPase; G-protein, guanyl-nucleotide-binding protein.

accumulating Ca^{2+} can only transiently buffer $[Ca^{2+}]_i$ increases. The final resetting of $[Ca^{2+}]_i$ must involve the transfer of Ca^{2+} out of the cell. Ca^{2+} extrusion is accomplished by means of membranous Ca^{2+}-binding proteins that reversibly complex and translocate Ca^{2+} outside the cytosol. Two main types of Ca^{2+}-extruding systems are known to exist in the plasma membrane involving (a) Ca^{2+}-ATPases and (b) a Na^+/Ca^{2+} exchanger. These two transport systems differ in their Ca^{2+} affinity and V_{max} of Ca^{2+} transport.

Although in the steady state the net flux of Ca^{2+} across the plasma membrane is zero, there are transient situations in which large increases of Ca^{2+} influx from the extracellular milieu occur. This rapid change in the permeability of the plasma membrane to Ca^{2+} depends on the opening of the plasma membrane Ca^{2+} channels. Several types of Ca^{2+} channels have been described; they differ in ionic selectivity, conductance, tissue distribution, subcellular localization and mechanism of activation. Therefore two main systems for the regulation of $[Ca^{2+}]_i$ homeostasis are present on the plasma membrane: Ca^{2+}-extrusion mechanisms and Ca^{2+}-influx pathways.

Ca^{2+}-EXTRUSION MECHANISMS

The plasma membrane Ca^{2+}-ATPase

Ca^{2+}-pumping activity was first described in the plasma membrane of erythrocytes by Schatzmann in 1966 (see [4] for a recent review). This enzyme was subsequently found in all eukaryotic cells investigated [5] where it accounts for only 0.1% of the total membrane protein [6]. The plasma membrane Ca^{2+}-ATPases belong to the family of so called E1/E2 transport ATPases which exist in two conformational states (E1 and E2) during the reaction cycle. Energy liberated by ATP hydrolysis is stored intramolecularly as an aspartyl residue [4].

The discovery that calmodulin regulates the enzyme by directly binding to it suggested a purification procedure using a Sepharose column coupled to calmodulin. This allowed the fast and reproducible isolation of large amounts of protein [7]. Initial purification from erythrocyte plasma membrane and later isolation of complementary DNA from rat brain, gave a monomeric enzyme having a molecular mass of approximately 130 kDa (138 kDa for erythrocytes; 129.5 and 132.6 kDa for the two rat-brain isoforms) [8, 9]. Amino acid similarity and hydropathy profile comparison indicated a common transmembrane organization for Na^+/K^+-ATPase, sarcoplasmic reticulum Ca^{2+}-ATPase and the isoforms of plasma membrane Ca^{2+}-ATPase isolated from rat brain [8, 9]. Since both the N and C termini of the plasma membrane Ca^{2+}-ATPase have been shown to be on the cytoplasmic side of the membrane, it is obvious that there must be an even number of transmembrane domains. The actual number of transmembrane domains is uncertain,. but there may be as many as ten; four transmembrane domains have been identified in the N-terminal half of the molecule and four to six are likely to be present in the C-terminal half [9]. Six to ten transmembrane domains have also been suggested for the membrane-spanning region of the sarcoplasmic reticulum (SR) Ca^{2+}-ATPase [10]. The transmembrane domains of the plasma membrane Ca^{2+}-ATPase also contain seven carboxylate groups that may have a role in Ca^{2+} translocation. Both isoforms cloned from rat brain possess, near the C termini, a sequence typical of calmodulin-binding domains [9]. Numerous studies suggest the presence of two ATP-binding

sites with different affinities for ATP [11 – 13], although it is not entirely clear whether the two sites are separate or the same site is undergoing cyclic changes in affinity. The high-affinity ATP-binding site ($K_m \approx 1\ \mu M$) is the catalytic site, while the low-affinity site (K_m 0.1 – 0.3 mM) has been suggested to have a regulatory role [4]. Nucleotide analysis of two cDNAs recently isolated from rat brain gave further support to this suggestion [9]. Most ATPases so far characterized possess two binding sites for ATP.

As noted above, the discovery of modulation of the plasma membrane Ca^{2+}-ATPase by calmodulin was instrumental for its purification [5, 7, 14, 15]. Calmodulin modulates this enzyme by direct interaction, not by a calmodulin-dependent phosphorylation, and shifts the enzyme to the high Ca^{2+}-affinity state [5]. Following the interaction with calmodulin, the affinity for Ca^{2+} of the transport site increases 30-fold (giving an apparent K_d for Ca^{2+} of 1 μM and 0.2 μM at pH 7.0 and 7.8, respectively) and the affinity for ATP (of the low-affinity ATP-binding site) 100-fold [16 – 18]. The state of phosphorylation of the enzyme remains, however, the same. At resting physiological $[Ca^{2+}]_i$ and cytosolic free-Mg^{2+} concentrations, calmodulin is dissociated from the ATPase; as $[Ca^{2+}]_i$ increases, Ca^{2+}-calmodulin–ATPase association proceeds slowly; consequently pump activity also increases slowly in response to elevations in $[Ca^{2+}]_i$ [18]. This mechanism imparts hysteretic behavior to the control of $[Ca^{2+}]_i$ homeostasis and allows overshoot of $[Ca^{2+}]_i$ before a new steady state is achieved [19]. The activity of the enzyme may also be increased, in the absence of calmodulin, by stimulation with acidic phospholipids and long-chain polyunsaturated fatty acids, or by controlled trypsin digestion. In view of the recent interest for the second-messenger functions of the metabolites of the phosphatidylinositol cycle, it is of interest that the plasma membrane Ca^{2+}-ATPase can be activated by phosphorylated metabolites of phosphatidylinositol. Phosphorylation by the cAMP-dependent kinase also activates the enzyme isolated from heart sarcolemma [20], but not all isoforms of plasma membrane Ca^{2+}-ATPases appear to contain a consensus sequence for cAMP-dependent phosphorylation. This is consistent with the observation that in numerous cell types, elevation of cAMP does not result in an increased Ca^{2+} extrusion [21]. In a reconstituted liposomal system the enzyme functions as a $Ca^{2+}/2\,H^+$ exchanger having a Ca^{2+} stoichiometry with respect to hydrolyzed ATP of 1:1 [8, 22].

Protein kinase C activation by either phorbol esters or endogenously produced diacylglycerol has been shown to activate Ca^{2+} extrusion from intact cells [23 – 25] (T. Pozzan, unpublished results) and also to increase Ca^{2+} accumulation by inside-out plasma membrane vesicles [23]. It is not clear yet, however, whether this positive modulation is a consequence of direct phosphorylation of the plasma membrane Ca^{2+}-ATPase, or whether it results from indirect effects.

The Na^+/Ca^{2+} exchanger

The Na^+/Ca^{2+} exchanger was discovered in the late 1960s in cardiac muscle and invertebrate nerve [26, 27]. Its presence has been well documented in the plasma membrane of a variety of excitable cells, while its presence (and importance) in non-excitable cells is debated. The coupling ratio of Na^+/Ca^{2+} is 3:1, therefore the exchanger is electrogenic and voltage sensitive. Na^+/Ca^{2+} exchanger is not energized by ATP, nonetheless it appears that the exchanger is phosphorylated under normal intracellular concentrations of ATP [28 – 30].

In the phosphorylated form the exchanger has high affinity for intracellular Ca^{2+} and extracellular Na^+. Furthermore, internal Ca^{2+} in the physiological range $0.1-1.0$ µM, activates the exchanger, while internal Na^+ is ineffective. Were the Na^+/Ca^{2+} exchanger the only regulator of $[Ca^{2+}]_i$, this would drive $[Ca^{2+}]_i$ at equilibrium to a concentration given by the following equation:

$$[Ca^{2+}]_i = [Ca^{2+}]_o \frac{[Na]_i^3}{[Na]_o^3} \exp \frac{(n-2)FV_m}{RT}$$

where V_m is the transmembrane potential, F is the Faraday number, R is the gas constant, T is absolute temperature and n is the coupling factor. The electrogenicity of the system carries with it the consequence that the direction in which Ca^{2+} (and Na^+) is transported depends on the plasma membrane potential. If the transmembrane potential is positive to the reverse potential of the exchanger, Na^+ is driven out and Ca^{2+} in; the reverse happens for transmembrane potentials negative with respect to the reverse potential. The exchanger will cause Ca^{2+} influx when the plasma membrane is depolarized and Ca^{2+} efflux when it repolarizes. It will therefore work in parallel with the voltage-gated Ca^{2+} channels in increasing $[Ca^{2+}]_i$ during the action potential and with the membrane Ca^{2+}-ATPase in extruding Ca^{2+} from the cytosol during the repolarization phase. The K_m of the exchanger is ≈ 2 µM, while the V_{max} is much larger than that of the plasma membrane Ca^{2+}-ATPase [31]. This has led to the suggestion that the main role of the exchanger is in the gross resetting of $[Ca^{2+}]_i$ and not in the fine tuning; however, it cannot be excluded that in cells expressing a high level of the exchanger, and under favourable thermodynamic conditions, the antiporter could also contribute to the regulation of resting $[Ca^{2+}]_i$.

The Na^+/Ca^{2+} exchanger is a crucial point of intersection between Na^+ and Ca^{2+} homeostasis, explaining how any alteration of the Na^+/K^+ pump has immediate consequences on $[Ca^{2+}]_i$ homeostasis. For example, the positive inotropic action of cardiac glycosides can be explained in terms of an increase in $[Ca^{2+}]_i$ as follows: cardiac glycosides bind to and inhibit the Na^+/K^+ pump; the cytosolic free-Na^+ concentration is elevated; elevation of the cytosolic free-Na^+ concentration drives Ca^{2+} in via the Na^+/Ca^{2+} exchanger; $[Ca^{2+}]_i$ is increased. An elevated $[Ca^{2+}]_i$ can either directly increase cardiac muscle tension, or cause initially a larger accumulation of Ca^{2+} into the SR. The increased release of Ca^{2+} from the SR, when the muscle is activated, increases muscle tension [32].

PLASMA MEMBRANE Ca²⁺ CHANNELS

Calcium channels are membrane proteins which, in the open conformation, allow the passive flux of calcium ions across the membrane down the electrochemical gradient. Plasma membrane calcium channels can be divided into two major groups depending on the mechanism controlling the transition(s) from closed to open conformation: A) Channel in which the gating depends on voltage; B) Channels in which the gating depends on ligand binding. The calcium channels of intracellular organelles can be grouped in a different category, the so called Ca^{2+}-release channels.

Voltage-operated Ca²⁺ channels, Ca²⁺ VOCs

Ca^{2+} VOCs are found not only in excitable cells such as neurons, muscle and heart, but also in some non-excitable cells such as fibroblasts (for review see [33–35]). In recent years, it has become clear that several types of Ca^{2+} VOCs exist, which differ in their biophysical properties, pharmacological sensitivity and tissue distribution. Three main classes of Ca^{2+} VOCs have been clearly defined, and will be discussed separately, but there is evidence for the existence of additional types of Ca^{2+} VOCs [36, 37]. Besides the steep voltage dependence of activation, the different classes of Ca^{2+} VOCs all have in common a very high selectivity for Ca^{2+} over Na^+ in physiological solutions.

Ca²⁺ VOCs

The nomenclature adopted for Ca^{2+} VOCs in that proposed by Nowycky et al. [122].

A convenient defining property of L-type Ca^{2+} VOCs is the presence of allosterically linked, high-affinity binding sites for dihydropyridines, phenylalkylamines and benzothiazepines [38–40]. The functional properties of L-type Ca^{2+} VOCs in cardiac, secretory and neuronal cells are similar, but differ markedly from those of the dihydropyridine-sensitive Ca^{2+} VOCs from skeletal muscle, as will be described at the end of the next section.

The single-channel conductance of L-type Ca^{2+} VOCs is $20-25$ pS and $7-8$ pS with Ba^{2+} (0.1 M) and Ca^{2+} (0.1 M) serving as charge carriers, respectively [41, 42]. Activity is characterized by clusters of bursts of openings of short duration (mean open time 1 ms), separated by long periods (tens of milliseconds to several seconds) of inactivity [43, 44]. Dihydropyridine agonists greatly prolong the average opening time of the channel [43].

In the absence of divalent ions in the extracellular medium, L-type channels become permeable to monovalent ions [45, 46]; however, in the presence of micromolar concentrations of Ca^{2+} [45, 46] on the outer side of the plasma membrane, the monovalent current is blocked. The high rate of influx, despite the high affinity (micromolar) of Ca^{2+} for the channel, has been explained either with more than one intrachannel Ca^{2+}-binding site and mutual repulsion of Ca^{2+} ions [45, 46], or with a high-affinity Ca^{2+}-binding site, externally located, that controls selectivity and a single low-affinity intrachannel Ca^{2+}-binding site [47].

L-type calcium channels are potently inhibited by dihydropyridine blockers, such as nifedipine and nitrendipine. Both the binding of dihydropyridine blockers and the inhibition of Ca^{2+} current are voltage dependent, increasing with increasing (less negative) holding potentials [39, 48–51]. The voltage and time dependence associated with dihydropyridine blockers can be understood assuming that dihydropyridine antagonists bind more tightly to the inactivated state of the channel (which predominates at relatively depolarized holding potentials) than to the resting state (which predominates at more negative potentials) [48, 49]. The difference in binding affinities between the two states accounts for the discrepancy between binding data (obtained at 0 mV) and inhibition of biological response (obtained at negative potentials) [48]. Because of the strong voltage and time dependence of dihydropyridine blockers, these drugs are not straightforward pharmacological tools; in fact in neurons and neurosecretory cells both the relatively negative resting potential and the short duration of the action potential make dihydropyridines poor inhibitors of L-type Ca^{2+} VOCs under physiological conditions [50, 51].

Under situations close to physiological conditions, whole cell L-type calcium currents start to activate at membrane

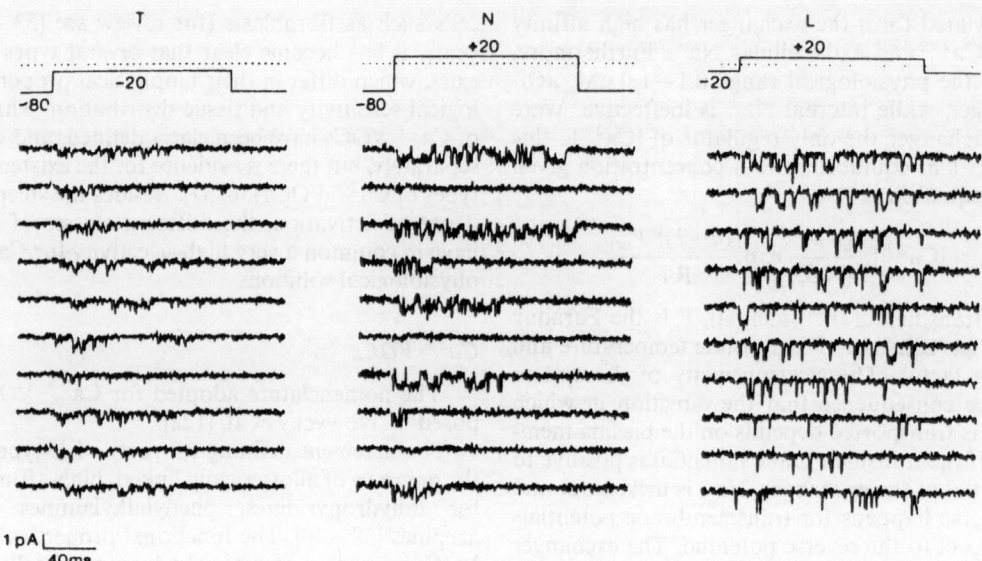

Fig. 1. *The three types of Ca^{2+} channels in chick dorsal-root ganglion neurons.* Cell-attached recordings with 110 mM Ba^{2+} in the pipette. The voltage clamp protocol is shown above the current traces. Reproduced with permission from Nowicky et al. [122]

potentials more positive than -30 mV to -20 mV, with time constants in the millisecond time range [41, 52]. At holding potentials more positive than -30 mV to -40 mV, steady-state inactivation starts to appear [49, 51]. The time course of voltage-dependent inactivation appears to be one of the most variable parameters between individual L-type Ca^{2+} VOCs. The most rapidly inactivating L-type Ca^{2+} currents are found in cardiac and vascular muscle (time constants of hundreds of milliseconds, with Ba^{2+} as the charge carrier) [53]. In neuronal and secretory cells, voltage-dependent inactivation is much slower (time constants of seconds or longer) if not totally absent [51, 52, 54]. In addition to voltage-dependent inactivation, L-type currents also show Ca^{2+}-dependent inactivation [54, 57 – 61]. In cardiac and smooth muscle, with Ca^{2+} as a charge carrier, L-type current results in quite rapid inactivation with time constants of 20 – 100 ms [53]. In secretory cells and neurones, a steady-state non-inactivatable component of Ca^{2+} current still remains when Ca^{2+} is the charge carrier [56, 57, 61].

Hormones and drugs which activate adenylate cyclase increase cardiac L-type Ca^{2+} current as a consequence of cAMP-dependent phosphorylation either of the channel or of a protein closely associated with it [62 – 66]. The cAMP-dependent increase in current is due to a higher opening probability of single L-type channels, primarily as a result of shorter intervals between bursts of openings and shorter periods of inactivity between clusters of bursts of openings [67, 68]. In addition to the cAMP-dependent stimulation there is evidence for a parallel direct stimulatory effect of the GTP-binding proteins on cardiac L-type Ca^{2+} VOCs [69, 70]. There are some indirect indications that protein kinase C might also have some stimulatory effects on cardiac and smooth muscle L-type Ca^{2+} channels [71 – 73]. cAMP-dependent phosphorylation does not seem to increase the opening probability of all L-type Ca^{2+} VOCs, since Ca^{2+} currents of some neuronal, secretory and smooth muscle cells are not increased by the elevation of intracellular cAMP [74 – 79].

L-type Ca^{2+} VOCs of skeletal muscle differ significantly from those in other cells in two major functional aspects; (a) they are activated more than one order of magnitude more

slowly [80 – 82] and (b) have a smaller unitary conductance for both monovalent and divalent ions [83, 84] and a somewhat different ionic selectivity. The activity of skeletal L-type channels, as seen in their counterparts in cardiac cells, is stimulated by β-adrenergic agonists and internal cAMP [85] and also directly by the GTP-binding proteins [86]. Inositol 1,4,5-trisphosphate [Ins(1,4,5)P_3] has also been reported to increase the opening probability of single L-type Ca^{2+} VOCs of T-tubules [87], and stimulation of protein kinase C by phorbol esters to increase the number of dihydropyridine-binding sites and the ^{45}Ca influx in skeletal muscle cells [88].

Since the highest concentration by far of dihydropyridine receptors is found in the T-tubules of skeletal muscle, most of the biochemical characterization of the L-type Ca^{2+} VOCs has been performed on this tissue. The skeletal muscle dihydropyridine-sensitive Ca^{2+} VOCs have been purified by several laboratories [89 – 93]. The purified dihydropyridine-sensitive Ca^{2+} VOCs have then been reconstituted into phospholipid vesicles and lipid bilayers, and shown to be functional by assays for Ca^{2+} and Ba^{2+} uptake into vesicles [94, 95] and by single-channel measurements in artificial bilayers [96 – 99].

The purified channel from skeletal muscle is a 1:1:1:1 complex of four polypeptide chains (see [100 – 102] and references therein) which have been referred to as the $\alpha1$ subunit (apparent molecular mass 155 – 200 kDa as analyzed by SDS/polyacrylamide gel electrophoresis), the $\alpha2$ subunit, formerly called α (165 – 220 kDa in non-reducing conditions; 130 – 150 kDa in reducing conditions), the β subunit (52 – 65 kDa) and the γ subunit (30 – 35 kDa). Upon reduction of the $\alpha2$ subunit, a fifth subunit appears, the δ subunit (24 – 33 kDa). The $\alpha1$ and β subunits are substrates for cAMP-dependent protein kinase [103 – 105], for a Ca^{2+}/calmodulin-dependent protein kinase [104, 106], for protein kinase C [107], for the cGMP-dependent protein kinase [106] and for a protein kinase intrinsic to skeletal muscle triads [105].

The primary amino acid sequences of the $\alpha1$, $\alpha2$ and β subunits of the skeletal muscle dihydropyridine-sensitive Ca^{2+} VOCs have been deduced from their cloned cDNAs [108 – 110]. The $\alpha1$ subunit (212 kDa), which bears a high degree of similarity with the Na^+ channel [111] has four repeating domains of high homology and each domain contains five, presumably α-helical membrane-spanning hydrophobic segments and one positively charged segment, named S4. S4 contains five or six Arg or Lys residues at every third position, with mostly non-polar residues intervening between the basic residues and has approximately the right length to cross the membrane. The S4 sequence is highly conserved in the corresponding segment of the Na^+ [111] and K^+ channels [112, 113]. Recently, site-directed mutagenesis of the Na^+ channel expressed into oocytes has provided evidence that the S4 segment is involved in the voltage-sensing mechanism for activation of the channel [114]. The sequence of the $\alpha 1$ subunit contains seven potential phosphorylation sites for cAMP-dependent protein kinase and no obvious Ca^{2+}-binding sites corresponding to EF hands or similar structures present in intracellular Ca^{2+}-binding proteins.

The L-type Ca^{2+} current is practically absent in mice with muscular dysgenesis. In these animals excitation/contraction (EC) coupling fails [115] and the level of mRNA encoding the $\alpha 1$ subunit is greatly reduced [116]. Microinjection of an expression plasmid carrying the $\alpha 1$ subunit cDNA into nuclei of cultured skeletal muscle cells from dysgenic mice is able to restore functional dihydropyridine-sensitive Ca^{2+} VOCs, as well as normal EC coupling [116]. The observation that EC coupling in this system is preserved also in the absence of extracellular Ca^{2+} or in the presence of Cd^{2+}, a potent blocker of Ca^{2+} permeation through the channels, indicates that EC coupling depends on functional dihydropyridine receptors, but not on their ability to conduct an inward Ca^{2+} current [116]. This evidence strongly supports the view that the dihydropyridine receptor in the T-tubule membrane has a dual function as a Ca^{2+} channel and as the voltage sensor for EC coupling [117].

Several lines of evidence suggest that the dihydropyridine-sensitive L-type Ca^{2+} VOCs of non-skeletal muscle cells are different structural entities from those of skeletal muscle. Dysgenic mice lack dihydropyridine receptors in muscle, but have a normal density of dihydropyridine receptors and of dihydropyridine-sensitive Ca^{2+} VOCs in heart and sensory neurons [115, 116]. RNA blot analyses with probes for the $\alpha 1$ [109] and β [110] subunits of skeletal muscle, show very weak ($\alpha 1$) or no hybridization (β) to RNA from heart or to RNA from smooth muscle or brain. Specific antibodies raised against the $\alpha 1$ subunit of skeletal muscle are not able to recognize the analogous subunit in heart and other tissues [118, 119]. However, both RNA blot analysis and antibody crossreactivity indicate that the $\alpha 2$ subunits of different tissues are homologous [109, 119]. The purified $\alpha 1$ subunit of cardiac muscle is significantly larger (185 – 195 kDa) than its counterpart in skeletal muscle [118, 120] and surprisingly it could not be phosphorylated in detergent solution by cAMP-dependent protein kinase [118]. Finally, the primary amino acid sequence of the $\alpha 1$ subunit of the cardiac L-type channel has been recently deduced from the cloned cDNA and shown to be 66% homologous to that of skeletal muscle [121].

N-type Ca^{2+} VOCs

N-type (N-type) Ca^{2+} VOCs have been described first in chick dorsal-root ganglion neurons [122] and are probably neuron specific since they have been found only in cells of neuronal origin and their expression in PC12 cells is enhanced by nerve growth factor [35, 37, 123, 124]. N-type Ca^{2+} VOCs

probably play a prevalent, if not exclusive role, in controlling transmitter release in certain neurons [123].

According to the original work of Tsien and collaborators in sensory and sympathetic neurons, N-type channels activate over a similar voltage range as L-type channels, but can be distinguished at the single channel level from L-type channels by their lower single-channel conductance (11 – 15 pS with 0.1 M Ba^{2+} as a charge carrier), by their insensitivity to dihydropyridine drugs and by their almost complete inactivation at holding potentials more positive than -20 mV to -30 mV [55, 122, 125, 126]. In whole cell recordings, N-type channels were previously considered to give rise to a component of high-threshold current, inactivating completely, with time constants (near 0 mV) of tens of milliseconds, in dorsal-root ganglion neurons, and hundreds of milliseconds in sympathetic neurons [55, 122, 126]. Recently, however, the view that N-type channels contribute only to the decaying component of the whole cell Ca^{2+} current, and that L-type channels underlie all of the late sustained current, has been shown to be an over simplification. In both sensory and sympathetic neurons in fact, the average current from single N-type channels can have a sustained component which is not inactivated at the end of test depolarizations of several hundred milliseconds [37, 127, 128]. Moreover, in some dorsal-root ganglions and sympathetic neurons, dihydropyridine blockers, even when applied at depolarized potentials, have almost no effect on currents that decay very little during long (6 s) depolarizations [129]. In general, despite large variability (that probably reflects the different sub-populations of neurons of spinal and cervical ganglia) the pharmacology of calcium currents in sensory and sympathetic neurons (in particular their irreversible block by ω-conotoxin, which has been shown to have no irreversible effect on L-type channels [37, 127]) is consistent with the view that N-type channels contribute to most (or in some cells all) of the high-voltage-activated calcium current [37, 126 – 131].

Several lines of evidence suggest that N-type Ca^{2+} VOCs are probably a heterogeneous group of channels [36, 37, 127, 128]. A Ca^{2+} VOC from cerebellar Purkinje cells, which is neither dihydropyridine sensitive nor ω-conotoxin sensitive, has been recently described and shown to be inhibited by a toxin from a spider venom [36]. Furthermore, another level of Ca^{2+} VOC heterogeneity could arise when comparing the channels expressed in the cell body and in the nerve terminal.

T-type Ca^{2+} VOCs

T-type Ca^{2+} VOCs with similar functional properties, originally described in vertebrate cells by Carbone and Lux [132], are found in a wide variety of excitable and non-excitable cells [34, 35]. T-type channels are also called low-threshold-activated channels since they start to activate at potentials 30 – 40 mV more negative than L-type and N-type Ca^{2+} VOCs. The low threshold of T-type channels renders them well suited for participating in pacemaking in many cells. During test-pulse depolarization they are inactivated quite rapidly ($t_{1/2}$ 10 – 40 ms at -20 mV) and completely [133 – 137] in a purely voltage-dependent manner (no Ca^{2+}-dependent inactivation) [138]. Steady-state inactivation is substantial at quite negative holding potentials [54, 55, 137, 139].

T-type Ca^{2+} VOCs have a single-channel conductance of 5 – 10 pS with either Ba^{2+} or Ca^{2+} as charge carriers (0.1 M) [122, 125, 137, 140]. They are blocked more effectively than L-type or N-type channels by Ni^{2+} [55, 135, 137, 141], they are insensitive to micromolar concentrations of dihydropyridine antagonists [55, 122, 133, 137] and only weakly and reversibly blocked by ω-conotoxin [130, 142]. A few drugs, such as tetramethrin [137], amiloride [143], diphenylhydantoin [144] and octanol [145] have been shown to inhibit T-type channels, but none of them very specifically [34, 35].

In Fig. 1 different electrophysiological characteristics of the three types of Ca^{2+} VOCs in dorsal-root ganglion neurons are summarized.

Agonist-dependent Ca^{2+} VOC inhibition

Ca^{2+} VOCs play an essential role in a number of cellular functions, from muscle contraction to neurotransmitter and hormone secretion. Thus the discovery that Ca^{2+} VOCs can be modulated by a number of physiological agonists has been of the utmost importance. We have mentioned above the stimulatory effects of cAMP and the GTP-binding proteins on L-type Ca^{2+} VOCs. The complex kinetics and heterogeneity of N-type channels, on the other hand, makes it difficult to establish which types of Ca^{2+} VOCs are the targets for the widespread inhibitory modulation of neuronal-Ca^{2+} currents by neurotransmitters such as noradrenaline (via $\alpha2$ adrenoreceptors) [146–151]; 4-aminobutyric acid (via B receptors) [146, 147]; dopamine (via D2 receptors) [149]; acetylcholine (via M1 receptors) [37, 74, 152]; adenosine (via A1 receptors) [153]; opioids (via k and δ receptors) [154–156]; somatostatin (via unknown receptors) [150, 154]. In a few cases one can safely conclude that the targets for inhibitory modulation are N-type Ca^{2+} VOCs [37, 150, 151, 153].

Most, if not all, of the reported inhibitory modulation of neuronal Ca^{2+} current seems to be independent of intracellular cAMP but dependent on the activation of a pertussis-toxin-sensitive, GTP-binding protein [37, 74, 147–156]. As previously suggested by reconstitution studies [156–158], the G-protein mediating the effect of neurotransmitters has been shown recently, using specific antibodies, to be the most abundant G-protein in neurons, denoted as the G_o-protein [159]. In sensory neurons, part of the inhibition of the Ca^{2+} current, mediated by the G_o-protein, requires protein kinase C activation [158, 160]. However, the participation of protein kinase C in inhibitory modulation of neuronal Ca^{2+} VOCs is not a general feature. Depending on the type of neuron, there is evidence for [146, 160–163] and against [74, 151] an involvement of protein kinase C. There are some indications that the targets for protein kinase C modulation in sensory neurons are L-type channels [163]. We tentatively propose, as an interesting working hypothesis, that the same neurotransmitter can modulate through a different molecular mechanism both N-type (possibly through direct G-protein/channel interaction) and L-type (possibly through protein kinase C) Ca^{2+} VOCs.

The modulation of Ca^{2+} currents in neurosecretory cells shares many features with that in neuronal cells. Thus, in some pituitary cell lines the inhibitory modulation by somatostatin of a slowly inactivating, high voltage-activated Ca^{2+} current was shown to be cAMP independent and pertussis toxin sensitive [76, 164]. Activators of protein kinase C inhibit voltage-dependent Ca^{2+} currents in many secretory cells [76, 165, 166] and in some cases they were shown to affect dihydropyridine-sensitive L-type Ca^{2+} VOCs [166].

Very few data are available on the inhibition of T-type Ca^{2+} VOCs by neurotransmitters. In sensory neurons, norepinephrine and dopamine inhibit T-type Ca^{2+} currents by an unknown mechanism [129, 149]. In GH3 cells, activation of protein kinase C has been reported to inhibit the T-type

Ca^{2+} current [165] while cAMP displays no effect in several cell types [133, 141, 167]. In fibroblasts, T-type Ca^{2+} VOCs can be selectively suppressed by transforming oncogenes [136].

Receptor-activated and second-messenger-activated Ca^{2+} channels

Until a few years ago the Ca^{2+} channels with opening mechanisms not dependent on membrane-potential depolarization, but rather on the activation of specific plasma membrane receptors, were termed receptor-operated Ca^{2+} channels or Ca^{2+} ROCs. Recently [33, 168] it has been proposed to distinguish this type of Ca^{2+} channel into three major groups.

(a) ROCs are those channels in which the ligand-binding site and the channel are either on the same polypeptide or in the same molecular complex [33]. Within the group of agonist-activated Ca^{2+} channels, Ca^{2+} ROCs are by far the best characterized in molecular and electrophysiological terms and incude the nicotinic channel, the N-methyl D-aspartate (MeDAsp) channel and perhaps, the ATP (external) gated channel(s).

(b) The channels that are linked to the receptor via a soluble second messenger belong to a second group, and are termed SMOCs [33]. The Ca^{2+}-activated Ca^{2+} channels, found in different cell types, are the prototype of the Ca^{2+} SMOCs. Evidence has been also provided for the existence on the plasma membrane of $Ins(1,4,5)P_3$-regulated Ca^{2+} channels, and/or inositol 1,3,4,5-tetrakisphosphate $[Ins(1,3,4,5)P_4]$-regulated Ca^{2+} channels, but in this case a general consensus has not yet been reached.

(c) The third group may be represented by Ca^{2+} channels coupled to the receptor via a G-protein. This latter type of Ca^{2+} channel is termed a G-protein-operated channel (GOC) [168].

Common characteristics of receptor and second-messenger-activated Ca^{2+} channels are the much lower selectivity for Ca^{2+} over monovalent cations, compared to Ca^{2+} VOCs.

Ca^{2+} ROCs

ROCs are epitomized by the nicotinic receptor/channel. However, this channel, although permeable also to Ca^{2+} under physiological conditions, transports mainly Na^+ and K^+ and therefore the nicotinic channel will not be discussed here (see [169]). The only Ca^{2+} ROC extensively characterized is the glutamate-activated channel, usually indicated as the MeDAsp receptor [170], to distinguish it from the other receptors ligated by glutamate, i.e. those preferring (S)-α-amino-3-hydroxy-5-methyl-4-isoxazolepropionic acid and kainic acid [171]. The MeDAsp channel has been extensively investigated by electrophysiological techniques, but no information is yet available about its molecular structure. The MeDAsp-channel receptor has been found so far only in vertebrate neurons mostly in the central nervous system (but see [172] for other locations). This channel is permeable to both Na^+ and Ca^{2+} and at resting (negative) plasma membrane potential, is powerfully inhibited by even micromolar Mg^{2+} concentrations [173]. Mg^{2+} blockade can be reversed by depolarization [173]. In Mg^{2+}-free solution, the MeDAsp channel exhibits a linear current-voltage relationship and a reverse potential near 0 mV. The channel conductance is 50 pS (with 140 mM CsCl as a charge carrier). When Mg^{2+} is present, MeDAsp causes fast current bursts suggesting that the channel undergoes cycles of blocking and unblocking by Mg^{2+}. The

MeDAsp channel is positively modulated by glycine [174]. The MeDAsp channel is currently the object of great interest as it is increasingly appreciated that it is involved in the long-lasting modification of neuronal excitability, known as long-term potentiation [175] and in neuronal cell death following ischemia [176].

Another member of the ROC family might be the channel(s) activated by extracellular ATP [177]. In recent years several ATP-activated Ca^{2+} fluxes have been described and in a few instances, detailed electrophysiological characterization has been carried out. In smooth muscle, exogenous ATP opens a channel with a Ca^{2+}/Na^+ selectivity ratio of 3:1 at physiological concentrations and with a conductance of 5 pS [177]. This channel resists inhibition by classical inhibitors of voltage-gated Ca^{2+} channels and by Mg^{2+}. Besides Ca^{2+} and Na^+, the channel also admits glucosamine, albeit with a permeability sixfold lower in comparison to that for Na^+ [177]. The channel is activated by ATP concentrations as low as 10 nM and without the involvement of soluble second messengers [177]. Recently, ATP-activated channels with similar features have also been described in heart and neuronal cells [178, 179] and we believe that more examples will be found as their presence will be investigated more thoroughly. Besides opening this low-conductance channel, in many cell types (i.e. macrophages, mast cells, mouse neuroblastoma cells) exogenous ATP also causes large increases in whole-cell conductance depending on the opening of large non selective channel(s) that admit aqueous solutes of molecular mass $\leqslant 0.9$ kDa [180, 181].

Ca^{2+} SMOCs

In a number, and possibly all cell types, the activation of receptors coupled to the hydrolysis of phosphatidylinositol 4,5-bisphosphate [PtdIns(4,5)P_2] causes both release of Ca^{2+} from intracellular stores and increased Ca^{2+} influx through the plasma membrane [33]. This latter event is generally thought to involve Ca^{2+} channels, mostly because the influx of Ca^{2+} depends on the transmembrane potential, i.e. influx is increased by hyperpolarization and depressed by depolarization [182, 183]. More direct evidence for the involvement of Ca^{2+} channels in the influx, activated by receptors coupled to PtdIns(4,5)P_2 hydrolysis, is however still lacking in most systems.

Ca^{2+} ions (presumably released from intracellular stores) might serve as the second messenger that regulates the opening of plasma membrane cation channels [184]. Such channels have been described in a variety of cells, both excitable and non-excitable, but either the [Ca^{2+}]$_i$ necessary to activate them is outside the physiological range, or their permeability to Ca^{2+} is so low that their relevance for maintaining Ca^{2+} influx is doubtful. This is not the case for the cation channel activated by micromolar [Ca^{2+}]$_i$ in neutrophils [185], whose permeability to Na^+ is approximately similar to that for Ca^{2+}. The importance of this Ca^{2+}-activated Ca^{2+} channel in the sustained increase of [Ca^{2+}]$_i$ caused by chemotactic peptides is however challenged by the observation that the Ca^{2+} (or Mn^{2+}) influx, activated by receptor stimulation in neutrophils, can persist even when [Ca^{2+}]$_i$ increases (due to Ca^{2+} redistribution) are blunted [186−188]. Thus, at best, this type of channel could contribute to the initial phase of a [Ca^{2+}]$_i$ increase after receptor triggering, but probably has no role in the sustained phase of Ca^{2+} influx.

The other major candidate for activation of Ca^{2+} SMOCs is Ins(1,4,5)P_3, alone or in combination with Ins(1,3,4,5)P_4.

Kuno and Gardner [189] reported that Ins(1,4,5)P_3 can increase the opening probability of a cation channel in excised patches from lymphocyte plasma membrane; more recently, Penner et al. [183] showed that intracellular Ins(1,4,5)P_3 can induce sustained increases in the [Ca^{2+}]$_i$ in mast cells although they were unable to demonstrate the activation of an Ins(1,4,5)P_3-dependent inward current. Snyder et al. [190] showed that in sea urchin eggs intracellularly injected Ins(1,4,5)P_3 could induce a sustained increase of Ca^{2+}-dependent Cl^- conductance (tens of minutes), that was considered to reflect sustained Ca^{2+} influx. In this experiment, however, the effect of intracellularly injected Ins(1,4,5)P_3 was observed even several minutes after microinjection of the stimulus, when presumably all the Ins(1,4,5)P_3 had been metabolized. In the experiments reported by Snyder et al., Ins(1,3,4,5)P_4 could not mimic the Ins(1,4,5)P_3 effects on Cl^- current [190]. In contrast to Snyder et al., Morris et al. [191] failed to observe sustained effects of intracellular Ins(1,4,5)P_3 on Ca^{2+}-activated K^+ permeability in rat lacrimal acinar cells, thus arguing against a role for Ins(1,4,5)P_3 as the sole mediator of the persistent Ca^{2+} influx triggered by muscarinic receptors in these cells. Rather, they showed [191] that the inclusion of Ins(1,4,5)P_3 and Ins(1,3,4,5)P_4 in the intracellular perifusion medium could mimic the effect of extracellularly applied acetylcholine. In a more recent paper however, Petersen and coworkers [192] showed that Ins(1,3,4,5)P_4 could affect K^+ currents also in the absence of extracellular Ca^{2+}, thus raising doubts about the interpretation of their previous experiments. Thus at present one has to conclude that clear evidence for the existence of Ins(1,4,5)P_3-dependent Ca^{2+} SMOCs is still lacking in most systems and alternative possibilities must be taken into consideration. An attractive possibility would be a coupling of the receptor to the channel by direct interaction mediated by a G-protein, similar to the muscarinic K^+ channel of heart myocytes [193].

Do Ca^{2+} GOCs exist? Admittedly, with one relevant exception, the demonstration that Ca^{2+} channels can be gated by G-proteins is still lacking in most systems. However, several reports in the last two years have provided indirect evidence that G-proteins may indeed modulate the activity of Ca^{2+} channels linked to receptors coupled to PtdIns(4,5)P_2 hydrolysis. Penner et al. [183] demonstrated that intracellularly injected guanosine 5'-[γ-thio]triphosphate could cause the opening of an unselective cation channel, but they were unable to correlate the opening of this channel with any sustained rise in [Ca^{2+}]$_i$, nor could they exclude the fact that guanosine 5'-[γ-thio]triphosphate was causing channel opening indirectly via generation of a second messenger. Fasolato et al. [194] reported evidence, in PC12 cells, for the opening, by the nonapeptide bradikinin, of a Ca^{2+} channel which was not regulated by any known second messenger; the authors suggested that G-protein-dependent Ca^{2+} channels could explain their data.

The only Ca^{2+} channel for which convincing evidence in favour of G-protein-mediated control has been provided, is that triggered by insulin-like growth factor II described in BALB/c 3T3 fibroblasts by Ogata and coworkers [195, 196]. The following evidence indicating that insulin-like-growth-factor-I-activated and insulin-like-growth-factor-II-activated channels are GOCs and not SMOCs is convincing [195, 196] (E. Ogata, personal communication): (a) GTP or non-hydrolyzable GTP analogues activate channel opening in excized patches; (b) the activation of the channel is inhibited by treatment with pertussis toxin; (c) in the cell-attached mode, channel opening can be triggered by insulin-like growth factor

II in the patch pipette, but not by bath application of the stimulus.

Comparable to other Ca^{2+} SMOCs or ROCs and to the influx activated by bradykinin in PC12 cells [194] the insulin-like-growth-factor-activated GOC is also an unspecific cation channel, permeable to Ca^{2+}. It remains to be demonstrated whether the Ca^{2+} GOC described in BALB/c 3T3 fibroblasts [195, 196] is specific for the receptors activated by insulin-like growth factors I and II, or whether this channel is the first example of a much larger family.

The capacitative model for Ca^{2+} influx

An alternative model for explaining receptor-activated Ca^{2+} influx has been proposed in 1986 by Putney [197] and further elaborated by Merritt and Rink [198]. In this model the signal that activates Ca^{2+} influx, after triggering receptors coupled to PtdIns(4,5)P_2 hydrolysis, depends somehow on the level of Ca^{2+} within the intracellular Ins(1,4,5)P_3-sensitive stores. According to the modification of Merritt and Rink [198] of Putney's original model [197], Ins(1,4,5)P_3 first empties the intracellular stores and, as a consequence of the low Ca^{2+} concentration in the lumen of the store, a gap-junction-like connection with the plasma membrane is formed; Ca^{2+} thus flows from the extracellular medium directly into the store through this channel and, provided the Ins(1,4,5)P_3-sensitive channel is open, into the cytoplasm. According to this model, Ca^{2+} influx into the cytoplasm will occur through the store and not through typical plasma membrane Ca^{2+} channels. The strongest argument in favour of the capacitative model is the observation that refilling of empty stores often occurs with minimal or no increase in $[Ca^{2+}]_i$ upon addition of Ca^{2+} to the medium. Recently, Putney and coworkers [199] have reformulated their hypothesis suggesting that once the Ins(1,4,5)P_3-sensitive stores are empty, a soluble messenger is released into the cytoplasm that in turn activates the opening of a traditional plasma-membrane-Ca^{2+} channel. This second model of Putney's is thus only a variation on the theme of Ca^{2+} SMOCs. The capacitative model is at present an intellectually challenging problem, though the evidence supporting it is still rather indirect.

The pharmacology of Ca^{2+} SMOCs or GOCs, unlike that of Ca^{2+} VOCs, is to date rather primitive, since the only investigative methods available are unspecific and/or not applicable to live animals. A new family of drugs, one prototype being now available from Smith, Kline and French [200], although not very specific, appears rather promising.

In conclusion, unlike in the case of Ca^{2+} ROCs and VOCs, which have become routinely amenable to investigation by molecular biology and electrophysiological techniques, Ca^{2+} SMOCs or GOCs are still in their infancy.

Ca^{2+}-SEQUESTERING ORGANELLES

All mammalian cells, except erythrocytes, are endowed with organelles capable of accumulating Ca^{2+} against the electrochemical gradient in an energy-dependent way. Under appropriate conditions, Ca^{2+} can be released from these organelles either slowly (e.g. mitochondria or secretory granules) or rapidly [e.g. sarcoplasmic reticulum (SR) or Ca^{2+} stores of non-muscle cells]. Many similarities exist between muscle SR and the rapidly exchangeable pools of non muscle cells, yet some of their characteristics are unique and these two types of organelles will therefore be discussed separately.

Mitochondrial Ca^{2+} accumulation appears to be quite different from that of other organelles and they seem incapable of regulated fast Ca^{2+} release (see below). A third type of Ca^{2+}-sequestering structure are the secretory granules; they too, however, are not involved in the regulation of the metabolically responsive pool of Ca^{2+}. Scattered reports also exist in the literature about the Ca^{2+}-accumulating capacity of the Golgi network or of lysosomes [5]. Last, but not least, it has been suggested that Ca^{2+} gradients exist across the nuclear membrane, although this observation has not been confirmed by all investigators.

As will be discussed below, any Ca^{2+}-storage organelle in rapid equilibrium with the cytoplasm must be endowed with three basic characteristics: (a) a Ca^{2+}-accumulation mechanism; (b) an intravesicular Ca^{2+}-buffering system; (c) a Ca^{2+}-release channel. Each Ca^{2+}-storage organelle will be discussed according to this scheme.

SARCOPLASMIC RETICULUM

Striated muscle SR is a complex network of tubules and cisternae wrapped around the myofibrils [201] and specific for Ca^{2+} accumulation and release [202]. Anatomically, SR is distinguished in longitudinal SR, composed mainly of tubules, and small cisternae, and terminal cisternae (TC). TC comprise the large expansions of the SR facing the T-tubules, the invaginations of the plasma membrane where coupling between depolarization and the signal causing Ca^{2+} release from TC occurs [202, 203]. Longitudinal SR and TC differ not only in their morphological features, but also in their biochemical properties.

SR Ca^{2+}-ATPase

The SR Ca^{2+}-ATPase may represent up to 90% of total SR protein. Thanks to its abundance it has been purified rather early and has been the object of fruitful biochemical, physiological and molecular studies. The SR Ca^{2+}-ATPase, a member or the E1/E2 family, is a single protein of molecular mass 110 kDa that interacts with Ca^{2+} with high affinity (K_D 0.5 μM at approximately pH 7.0) and transports it with a stoichiometric ratio of 2:1 with respect to ATP [204]. Genes encoding Ca^{2+}-ATPases of fast-twitch and cardiac SR have been isolated and the amino acid sequence determined. The SR ATPase, according to most recent reconstructions, is organized with a Ca^{2+}-binding domain, a transduction domain, a phosphorylation domain, a nucleotide-binding domain, a hinge domain and six or ten membrane-spanning α-helices [10, 205]. Both the C and N termini face the cytoplasm [10, 205] and the membrane-spanning α-helices constitute the ion channel, although it is not clear whether all the helices contribute to its formation. Determination of the amino acid sequence has allowed a mechanistic hypotheses to be proposed for the transmembrane Ca^{2+} translocation, that fit previous interpretations based on biochemical and biophysical data. The Ca^{2+} ATPase of fast-twitch skeletal muscle differs significantly with regard to its primary sequence and immunological reactivity compared to that of the slow skeletal/cardiac isoforms [10, 205 – 207]. The Ca^{2+}-ATPase of smooth muscle SR resembles closely the cardiac isoform, but the mRNAs of cardiac and smooth muscle SR Ca^{2+}-ATPase differ in their 3′-end sequence [208].

The Ca^{2+}-ATPase from heart, slow skeletal muscle and smooth muscle SR is regulated by an acidic proteolipid called

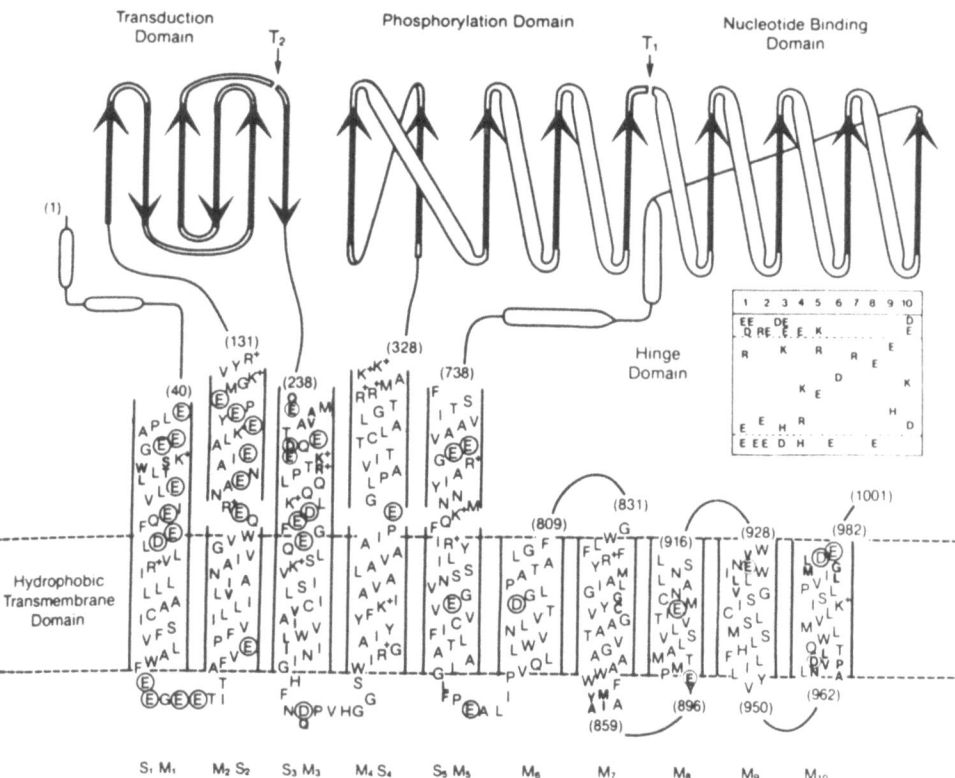

Fig. 2. *Schematic diagram of the sarcoplasmic reticulum* Ca^{2+}-*ATPase*. Amino acid differences between the fast-twitch and slow-twitch Ca^{2+}-ATPases are indicated by two residues, the slow above the fast. Acidic residues are encircled and basic residues are highlighted by a plus sign. In the inset the location of negative (D, E) and positive (K, R, H) charges close to or within the transmembrane helices are shown. 1 – 10 refer to trans-membrane helices. Reproduced with permission from Brandl et al. [10]

phospholamban [209]. This proteolipid binds ATP with a stoichiometry of 1:1 and can be phosphorylated by both cAMP and Ca^{2+} −calmodulin-dependent kinases. Recent experiments [210] have confirmed the previous suggestion [211] that phospholamban, in its dephosphorylated form, is bound to the ATPase and inhibits its activity; phosphorylation of phospholamban causes detachment from and activation of the ATPase.

The Ca^{2+}-ATPase of SR has been the object of intense investigation during the last decade and several excellent reviews have appeared dealing with most of the basic molecular and functional properties of this enzyme (for review see [20, 212 – 214]). Fig. 2 shows the predicted structure of SR Ca^{2+}-ATPase.

Ca²⁺ buffering within SR, calsequestrin

Among the several Ca^{2+}-binding proteins isolated from the SR, calsequestrin is the most abundant and because of its location and number of Ca^{2+}-binding sites, it is thought to represent the major Ca^{2+} buffer of the SR. Calsequestrin was initially described by MacLennan and Wong in 1971 [215] and defined as a high capacity medium/low-affinity Ca^{2+}-binding protein, specific of striated muscle. At least two isoforms of calsequestrin exist, one specific for fast-twitch skeletal muscle and the other for cardiac muscle [216]. Calsequestrin has been isolated from mammals, amphibians, birds, etc. Despite some variability in the molecular mass among the different animals and muscles, all calsequestrins share a number of properties, the most relevant being the presence of a large number of low-affinity Ca^{2+}-binding sites (K_d in the millimolar range with the number of sites from 30 – 50 mol/mol protein) [215 – 217].

Calsequestrin isolated from rabbit fast-twitch skeletal muscle is an acidic glycoprotein (P_i 4.2) with anomalous electrophoretic behaviour on SDS gels, i.e. slower mobility at alkaline pH than at neutral pH [217]. According to the predicted amino acid sequence [218], the real molecular mass of calsequestrin is 42 kDa. Calsequestrin has been very recently crystallized [219] but no structural data based on the X-ray analysis is yet available. Thus information on the structure of the protein is so far based on the hydropathy plot of the primary structure, circular dichroism measurements, fluorescence properties of the endogenous tyrosine and tryptophan, and chromatographic techniques [216 – 221]. Calsequestrin in striated muscles is localized exclusively in the lumen of the TC [222], in close proximity to the ryanodine-sensitive Ca^{2+} channels. This localization of calsequestrin is strategic for its function; the protein is not only essential to buffer intravesicular [Ca^{2+}] (the total Ca^{2+} content of the terminal cisternae may well be > 30 mM) but also to concentrate Ca^{2+} near the release sites, thus increasing the speed of Ca^{2+} release.

Although it is undisputed that calsequestrin contains many low-affinity Ca^{2+}-binding sites/mol protein [215 – 223], the conclusion that the protein contains only one class of site with the same affinity for the cation [223] may be an oversimplification. In fact there is no repeating distribution of acidic residues along the primary sequence [218] and the putative Ca^{2+}-binding sites are often clustered, thus making it unlikely that they would not influence each other, at least electrostatically.

The sorting of calsequestrin after synthesis in the rough endoplasmic reticulum (ER) is still unresolved. Calsequestrin does not contain a KDEL C-terminus region, as do proteins retained in the ER [224]. Evidence has been provided indicating that calsequestrin is transported via coated vesicles to the Golgi [225, 226] where further glycosylation occurs. From the Golgi (cis or intermediate portion), calsequestrin is then directed to its final destination and actually calsequestrin-containing structures may form nucleation centers for SR biogenesis [225, 226].

Calcium-release channels of the SR

Two channels for Ca^{2+} release from intracellular organelles have so far been described: the SR Ca^{2+}-release channel and the $Ins(1,4,5)P_3$-sensitive channel (see below).

The SR Ca^{2+}-release channel has been the object of intense investigation since its electrophysiological, biochemical and molecular features are of the utmost relevance for understanding the mechanism of EC coupling. Two alternative hypotheses have been proposed to explain EC coupling. (a) The chemical hypothesis, whereby a soluble messenger, released from the T-tubule membrane upon depolarization, opens the ryanodine sensitive Ca^{2+}-release channel of the TC. The nature of the second messenger is also a matter of discussion. Ca^{2+} itself (Ca^{2+}-induced Ca^{2+} release) or $Ins(1,4,5)P_3$ are the most popular candidates. (b) The electromechanical hypothesis which proposes direct physical contact between the voltage sensor on the T-tubule and the Ca^{2+}-release channel on the TC membrane. Neither of these mechanisms have, as yet, been proved conclusively by experimental evidence.

The Ca^{2+}-release mechanism from the SR has been studied using permeabilized muscle fibers (skinned fibers), isolated SR vesicles, SR membrane vesicles fused with planar lipid bilayers and purified channels inserted into black lipid films [227–229]. In both skeletal and cardiac SR vesicles, Ca^{2+} release is stimulated by extravesicular Ca^{2+} (half-maximal concentrations 0.5–1.5 μM in the absence of Mg^{2+}), by ATP (1–5 mM), caffeine, $Ins(1,4,5)P_3$ and by a variety of pharmacological agents [230–235], whilst Ca^{2+} release is inhibited by Ca^{2+} itself (at concentrations 2–3 orders of magnitude higher than those stimulating Ca^{2+} release), Mg^{2+} and ruthenium red [230–233, 236–239].

There are important differences between skeletal and cardiac SR vesicles in their sensitivity to activation or inhibition of Ca^{2+} release by the stimuli above mentioned. Ca^{2+} and caffeine are more effective in stimulating Ca^{2+} release from cardiac than from skeletal SR vesicles, while ATP is more effective in skeletal than in cardiac SR vesicles [231, 232, 238]. The presence of Ca^{2+} on the trans (cytoplasmic) side of the channel seems to be an essential requirement for opening the cardiac Ca^{2+}-release channel, whereas the skeletal channel can be partially activated by ATP even at nanomolar trans Ca^{2+} concentrations [238]. The strict dependence of the cardiac SR Ca^{2+}-release channel on $[Ca^{2+}]_i$ is consistent with the results obtained from skinned fibres suggesting an essential role for Ca^{2+}-induced Ca^{2+} release in cardiac muscle [240].

The other potential physiological activator of the SR Ca^{2+}-release channel is $Ins(1,4,5)P_3$. The first reports of $Ins(1,4,5)P_3$-induced Ca^{2+} release from skeletal muscle SR appeared in 1985 [234, 235]. The evidence in favour of $Ins(1,4,5)P_3$ may be summarized as follows: (a) $Ins(1,4,5)P_3$ releases Ca^{2+} both from isolated SR terminal cisternae [234] and from skinned fibre preparations [234, 235, 241–244]; (b) the entire machinery for the synthesis of the $Ins(1,4,5)P_3$ precursor $PtdIns(4,5)P_2$, $Ins(1,4,5)P_3$ formation (G-protein-dependent phospholipase C) and $Ins(1,4,5)P_3$ degradation [$Ins(1,4,5)P_3$ 5-phosphatase] is present in the plasma membrane of striated muscles [245–248] and in a few cases, a

selective enrichment of these synthetic and metabolic pathways has been found in T-tubules [245, 248]; (c) $Ins(1,4,5)P_3$ levels increase dramatically upon electrical stimulation of skeletal muscles [235]; (d) $Ins(1,4,5)P_3$, at micromolar concentrations, increases the opening probability of the Ca^{2+}-release channel of the SR after vesicle fusion with lipid bilayers [249]. A number of objections have been raised against the physiological role of $Ins(1,4,5)P_3$ as a mediator in EC coupling in striated muscle, the most important being: (a) the release of Ca^{2+} from skinned skeletal muscle fibers activated by photohydrolyzed $Ins(1,4,5)P_3$ is orders of magnitude slower than that observed under physiological conditions [250]; (b) the formation of $Ins(1,4,5)P_3$ in electrically stimulated cells is observed only after tetanus [235]; (c) fiber contraction is not blocked by heparin, a known inhibitor preventing $Ins(1,4,5)P_3$ binding to its receptor in smooth muscle and in non-muscle cells [251]. However, a high rate of contractions can be obtained by pressure injection of $Ins(1,4,5)P_3$ [252] and heparin has no effect on the $Ins(1,4,5)P_3$-induced contraction and Ca^{2+} release in skinned skeletal muscle fibers [253]. In turn, this result suggests that $Ins(1,4,5)P_3$-binding in skeletal muscle occurs on a receptor pharmacologically distinct from that of smooth muscle fibers and other non muscle cells. Very recently, the ryanodine receptor has been expressed in Chinese-hamster-ovary cells and shown to confer caffeine- and ryanodine-sensitive Ca^{2+} release [254]. On the other hand, the transfected channel appears to be insensitive to $Ins(1,4,5)P_3$. A possible interpretation of this result is that the ryanodine receptor is itself insensitive to $Ins(1,4,5)P_3$ and that an additional component, present in the TC membrane, is required for the expression of $Ins(1,4,5)P_3$ sensitivity. This hypothesis is supported by recent date by Valdivia et al. [400].

Taken together, these observations support the physiological relevance of $Ins(1,4,5)P_3$-induced Ca^{2+} release in skeletal muscle, yet they do not allow any conclusion as to whether $Ins(1,4,5)P_3$ functions as the main chemical messenger in stimulus-contraction coupling or whether it has only an ancillary role [255].

Single-channel recordings after fusion of SR vesicles with planar lipid bilayers have shown that Ca^{2+} release is mediated by a high-conductance channel ($\approx$ 100 pS with 50 mM Ca^{2+} in the lumenal side), with low selectivity for divalent over monovalent ions ($pCa^{2+}/pK^+ \approx 5$) [238, 256, 257]. Ca^{2+}, ATP [256, 257] and $Ins(1,4,5)P_3$ [249] on the cytoplasmatic side affect the opening probability of the channel. In agreement with Ca^{2+}-efflux experiments from SR vesicles, the cardiac Ca^{2+}-release channel is more sensitive to activation by Ca^{2+} and less sensitive to activation by ATP than the skeletal channel [238]. No data are yet available on the effect of $Ins(1,4,5)P_3$ on the cardiac SR channel reconstituted in lipid bilayers.

There are conflicting reports on the effect of Mg^{2+} on single Ca^{2+}-release channels [238, 257]. There is also evidence of voltage dependence on the opening probability of the Ca^{2+}-release channel in the lipid bilayers [238, 257]. Ryanodine at concentrations close to the K_d of high-affinity binding (5–50 nM) increases the opening probability of the channel during its bursting periods [258], while at micromolar concentrations ryanodine locks the channel in a low-conductance state with an opening probability near unity [258].

Using ryanodine, the Ca^{2+}-release channel has been purified from both skeletal and cardiac muscle [259–262]. The purified ryanodine receptor has been shown to function as a Ca^{2+}-release channel when reconstituted into planar lipid bilayers with biophysical properties similar, but not identical,

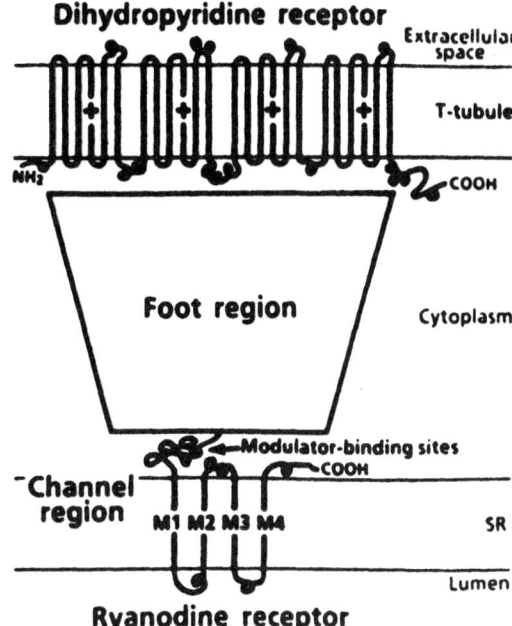

Fig. 3. *Structural model of the junctional region of the sarcoplasmic reticulum.* Plus signs within the transmembrane segments of the T-tubular dihydropyridine receptor indicate the putative voltage sensor. Reproduced with permission from Takeshima et al. [270]

to those of the channel from the SR membrane vesicles [259, 261–266]. Electron microscopy and ryanodine binding suggest that the purified ryanodine receptor forms a homotetrameric complex [261, 267, 268]. Reconstructed three-dimensional images of the purified receptor from negative-stain electron microscopy reveal a four-leaved clover (quatrefoil) structure $(27 \times 27 \times 14 \text{ nm})$ similar to that previously described for the 'feet' structures that in intact muscle span the junctional gap between the T-tubule and the junctional SR [269].

Recently, the sequence of the 5037 amino acids composing the ryanodine receptor from the rabbit skeletal muscle SR has been deduced by cloning and sequencing of the complementary DNA [270]. The primary sequence of the human ryanodine receptor has also been established [271]. Only the 500 or so amino acids at the C terminus of the deduced sequence are sufficiently hydrophobic to span the membrane. The four potential transmembrane sequences identified in this region resemble the membrane-spanning segments of the subunits of the nicotinic and related receptors, thus supporting the suggestion that in the ryanodine receptor there may be a single aqueous pore formed by amino acid residues from the C-terminal domain of each of the four monomeric units. 90% of the N-terminal domain of the protein is hydrophilic and most probably forms the enormous quatrefoil foot structure shown by electron microscopy [270]. The demonstration that the ryanodine receptor, the foot structure and the SR-release channel are the same protein and that skeletal muscle dihydropyridine-receptors are voltage sensors for EC coupling, independent of their function as Ca^{2+} channels, may suggest that the large cytoplasmic region of the ryanodine receptor directly interacts with the cytoplasmic region of the dihydropyridine receptor to effect EC coupling [272] (cf. electromechanical coupling hypothesis of Schneider and Chandler) [273]. A cartoon describing the molecular organiza-

tion of the triad junction of the skeletal muscle is shown in Fig. 3.

Ironically, despite the fact that the ryanodine receptor of striated muscle SR is by far the best characterized Ca^{2+}-release channel, the key question of the coupling between T-tubule depolarization and the opening of this channel still remains unanswered [255].

Ca^{2+} STORES IN NON-MUSCLE CELLS

The notion that microsomal fractions isolated from a variety of non-muscle cells accumulate Ca^{2+} in an ATP-dependent way is nearly 20 years old [274], but only in the last decade has the physiological importance of this Ca^{2+} store become obvious [275–277]. After the initial demonstration that agonist stimulation rapidly mobilizes a non-mitochondrial Ca^{2+} pool [278] (and the unraveling of the role played by $Ins(1,4,5)P_3$ in this phenomenon [275, 279]) the identification of an $Ins(1,4,5)P_3$-sensitive store with the entire endoplasmic reticulum was generally accepted. Only in the last two years has the complexity of the Ca^{2+} pools of non-muscle cells become apparent. On the one hand it has been proposed that non-muscle cells possess a specialized structure, apparently distinct from ER and related to muscle SR, named 'calciosome' [280–282]; on the other hand, several reports suggest the existence of Ca^{2+} pools insensitive to $Ins(1,4,5)P_3$, but sensitive to other agents (Ca^{2+} itself, caffeine etc., for review see [277, 283, 284]). Two detailed reviews on the nature, morphology, biochemistry and dynamics of the Ca^{2+} pools in non-muscle cells have been published recently and the reader should refer to these [283, 284] for a more comprehensive appraisal.

Ca^{2+} accumulation

Unlike the SR, where it is firmly established that Ca^{2+} accumulation depends on the activity of the Ca^{2+}-ATPase, in non-muscle cells the mechanism of Ca^{2+} accumulation is still a matter of discussion. Without doubt the Ca^{2+} stores in non-muscle cells possess a Ca^{2+}-ATPase similar, but not identical, to that in the SR [280–282, 285, 286]; in addition, it has also been recently suggested that the $Ins(1,4,5)P_3$-sensitive store may accumulate Ca^{2+} via the coordinated action of a H^+-ATPase and a $2 H^+/Ca^{2+}$ exchanger [287, 288].

Ca^{2+}-ATPase(s)

The activity of Ca^{2+}-ATPase in microsomal fractions enriched in ER markers is about two orders of magnitude lower than in muscle SR, suggesting that either this enzyme is scattered at low density throughout the ER or it is selectively enriched in a specific subcompartment(s). In fact, it has been demonstrated that, in a number of non-muscle cells, the SR-type Ca^{2+}-ATPase is concentrated in specialized structures, the calciosomes, distinct from the bulk of the ER, and also containing a calsequestrin-like Ca^{2+}-binding protein [280–282]. Later studies have reported the molecular cloning of cDNAs encoding ATPases that closely resemble cardiac (human) or fast- and slow-twitch (rat) Ca^{2+}-ATPases [289]. These proteins have a molecular mass of approximately 110 kDa and contain the conserved domains of E1/E2 ATPases. Like other members of this family of enzymes, these ATPases form phosphorylated intermediates [290] and are inhibited by millimolar concentrations of vanadate [291]. In one study [292]

610

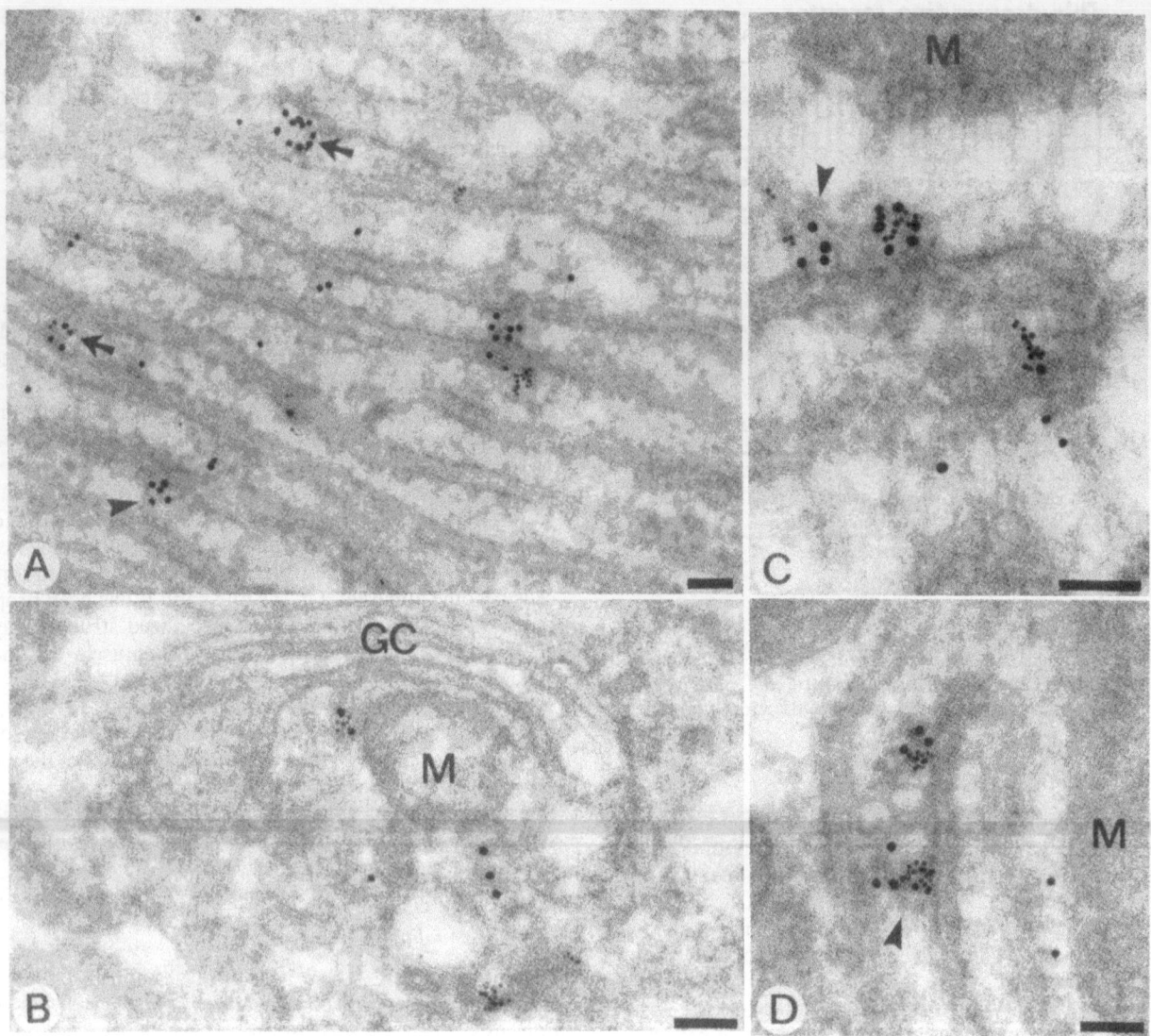

Fig. 4. *Dual labelling of ultrathin liver cryosections with anti-(ATPase) antibodies (large dots) and anti-(calsequestrin) antibodies (small dots).* The two antigens appear often, but not always, localized in the calciosomes. M, mitochondrion; GC, Golgi complex. Note that the bulk of ER is unlabelled. Bar, 0.1 μm. Reproduced with permission from Hashimoto et al. [281]

using monoclonal antibodies raised against cardiac Ca^{2+}-ATPase, a high concentration of cross reactive antigen was found in the Purkinje cells of the chicken cerebellum, where the highest concentration of $Ins(1,4,5)P_3$-binding sites is also found [293]. The predicted amino acid sequence from cDNA of the non-muscle Ca^{2+}-ATPase isoforms has a characteristic extended C-terminus that may either provide the binding site for cellular regulatory factors or contain a signal sequence that targets the protein to it's final subcellular location [285]. Using two distinct monoclonal antibodies raised against the Ca^{2+}-ATPase of skeletal muscle SR, it has been shown that they stain different compartments in adrenal chromaffin cells [286]. It has been suggested that the two bands recognized by these two antibodies (apparent molecular mass 110 kDa and 130 kDa) represent two different types of Ca^{2+}-ATPases that may belong to two intracellular Ca^{2+} stores with different functional properties [286]. Admittedly no evidence has been provided so far indicating that the two proteins recognized by the two antibodies are indeed Ca^{2+}-ATPases.

$Ca^{2+}/2 H^+$ exchanger

Schultz and coworkers [287, 288] have recently suggested that Ca^{2+} accumulation by the $Ins(1,4,5)P_3$-sensitive store does not depend on the activity of a Ca^{2+}-ATPase, but rather on a more complicated mechanism involving first the formation of a pH gradient (acidic interior), due to a H^+-ATPase and subsequently an electro-neutral exchange of 2 H^+/Ca^{2+}. In this model two Ca^{2+} pools are supposed to exist; one is insensitive to $Ins(1,4,5)P_3$ but with high vanadate sensitivity and fills via a classical Ca^{2+}-ATPase of the SR type; the other is sensitive to $Ins(1,4,5)P_3$ but with low vanadate sensitivity and utilizes the 2 H^+/Ca^{2+} exchanger. However, observations made in our laboratory do not support the generality of the model suggested by Shultz and coworkers. In fact, in intact cells treatment with a number of drugs capable of collapsing internal pH gradients, such as nigericin, monensin or NH_4Cl, results neither in significant depletion of $Ins(1,4,5)P_3$-sensitive Ca^{2+} pools, nor in inhibition of Ca^{2+} refilling into this store

[294] (T. Pozzan, unpublished results). Furthermore it can be argued that if Schultz's model were true, Ca^{2+} ionophores, such as ionomycin and A23187, should facilitate Ca^{2+} accumulation into the intracellular $Ins(1,4,5)P_3$-sensitive store and not counteract it. These ionophores, in fact, function essentially as $2 H^+/Ca^{2+}$ exchangers and therefore should act in parallel with the natural exchanger rather than counteracting it (D. G. Nicholls, personal communication). On the contrary, it is a common observation that Ca^{2+} ionophores cause complete depletion of $Ins(1,4,5)P_3$-mobilizable stores, both in intact cells and in isolated microsomes. In our opinion, therefore, the existence and relevance of the electro-neutral exchanger in the $Ins(1,4,5)P_3$-sensitive store is still dubious.

Ca^{2+} buffering within non-muscle stores, calreticulin

In the last two years proteins with properties similar to those of striated muscle calsequestrin have been described by several groups in many non-muscle tissues and they may be present in most eukaryotic cells [280 – 282, 295 – 299]. Quite surprisingly, recent data obtained in our and other laboratories [299 – 301], demonstrate that these calsequestrin-like proteins are in fact proteins which have already been known for some time and which have been given different names: SR high-affinity calcium-binding protein [223]; calregulin [302]; CRP55 [298]. These proteins share the same N-terminal sequence and several other biochemical and immunological characteristics and thus all these different names presumably refer to the same protein.

In two recent papers [301, 303] SR high-affinity calcium-binding protein and CRP55 have been cloned and sequenced and the common name calreticulin proposed. The reason for this confusing nomenclature and for not including calreticulin in the calsequestrin family is probably due to the fact that calreticulin possesses one high-affinity Ca^{2+}-binding site/mol protein [223, 302]. However MacLennan and coworkers, who first described calreticulin [223], had already reported its capacity to bind more than 20 mol Ca^{2+}/mol protein with low affinity. We have recently characterized calreticulin from rat liver and observed that (in addition to the high-affinity site) it also binds up to 50 mol Ca^{2+}/mol protein with low affinity, ($K_d = 1$ mM) [299]. Calreticulin cross-reacts with a few anti-(skeletal muscle calsequestrin) antibodies, particularly in its native form [299]. In ultrathin cryosections, anti-calreticulin antibodies, coated with gold particles, localize in vesicular structures which are morphologically indistinguishable from calciosomes [299]. Calreticulin is a glycoprotein (molecular mass from cDNA 47 kDa); the polysaccharide side chain is of the Golgi type and contains a terminal galactose suggesting that the protein has reached the trans Golgi network [304]. However, the C-terminus of the mRNA codes for KDEL [301, 303], the signal peptide believed to be characteristic of proteins retained in the ER [224]. Thus there is on the one hand, a discrepancy between the morphology and carbohydrate side chain, with both suggesting that calreticulin is not in the ER but rather contained in post Golgi structures [280 – 282] and, on the other hand, the primary sequence [301, 303] which suggests an ER localization of the protein. It must be pointed out however, that recently it has been shown that the C-terminus KDEL only retards, but does not block, the sorting of proteins into the Golgi apparatus [305]. The labelling of ultrathin liver cryosections with anti-calsequestrin antibodies and anti-ATPase antibodies is shown in Fig. 4.

Other proteins with low-affinity Ca^{2+}-binding sites have been described in microsomal fractions [289, 304]. None of them however appear to have the characteristics (K_d, number of binding sites, cellular concentration and ubiquitous distribution in tissues) to function as the primary Ca^{2+} buffer within Ca^{2+}-storage compartments.

The Ins(1,4,5)P$_3$-sensitive channel

Five years have elapsed since it was first demonstrated that $Ins(1,4,5)P_3$ is capable of releasing Ca^{2+} from a non-mitochondrial store [279] and the purification of an $Ins(1,4,5)P_3$ receptor [306]. In these five years, hundreds of papers have documented the existence of receptors coupled to $Ins(1,4,5)P_3$ formation, the correlation between $Ins(1,4,5)P_3$ production and Ca^{2+} mobilization from intracellular stores, and the presence of high-affinity $Ins(1,4,5)P_3$-binding sites in microsomal fractions from several tissues (see [277] for a recent review). This long gap is in large part due to the very low numbers of $Ins(1,4,5)P_3$-binding sites in most cells and the lack of specific, high-affinity ligands for the $Ins(1,4,5)P_3$ receptor. The discovery [293] that the cerebellum, in particular the Purkinje cells, contains an extrordinary high level of $Ins(1,4,5)P_3$-binding sites, was thus a serendipitous bonus. The other side of the coin is that the Purkinje cell $Ins(1,4,5)P_3$ receptor could be a specific isoform of the central nervous system and of a few types of neurons therein [307 – 309]. Furnichi et al. [401] have recently shown that RNA blot analysis with probes for the $Ins(1,4,5)P_3$ receptor of cerebellum show hybridization with in RNA extracted from several tissues. All these mRNA have a molecular mass similar to that of the cerebellum isoform.

The Purkinje cell $Ins(1,4,5)P_3$ receptor is a single polypeptide (molecular mass, based on the cDNA sequence, 313 kDa) [310] that was first described in 1976 [311] as a specific protein of the cerebellum Purkinje cells (named P400 from its apparent molecular mass in SDS gels). This protein is phosphorylated in a cAMP-dependent fashion [312], binds $Ins(1,4,5)P_3$ with high affinity [306, 313] and when reconstituted into liposomes, catalyzes $Ins(1,4,5)P_3$-dependent Ca^{2+} efflux [314]. The $Ins(1,4,5)P_3$ receptor binds strongly to heparin [293], while binding of $Ins(1,4,5)P_3$ to the receptor is highly sensitive to pH and $[Ca^{2+}]_i$ [308]. By scanning the sequence for hydrophobic stretches, up to six membrane-spanning regions could be predicted, all close to the C-terminus [310]. Based on structural constrains and reactivity with different monoclonal antibodies, it has been proposed that the N-terminus is exposed to the cytoplasm, while the C-terminus is localized within the vesicle lumen [310]. The protein has numerous potential glycosylation sites and at least two potential phosphorylation sites for cAMP-dependent kinase [310]. Both in situ hybridization and immunohistochemistry indicate that the protein is expressed at very big levels in the Purkinje cell layer [307, 310, 313]. Alignment of the amino acid sequence of the $Ins(1,4,5)P_3$ receptor and of the ryanodine receptor revealed strong similarities between parts of the sequences, in particular between Asp269 and Pro426 of the $Ins(1,4,5)P_3$ receptor and Asp4867 and Pro5023 of the ryanodine receptor (47% identity over 136 amino acids) [313].

Three different studies have been published recently concerning the subcellular localization of the $Ins(1,4,5)P_3$ receptor in the Purkinje cells [307, 313, 315]. In the first of these studies [307], performed using immunoperoxidase-tagged antibodies, the $Ins(1,4,5)P_3$ receptor has been found on a number of intracellular cysternae (smooth and rough ER, subplasmalemmal cisternae) but particularly striking was its high concentration on the nuclear membrane and its absence from

612

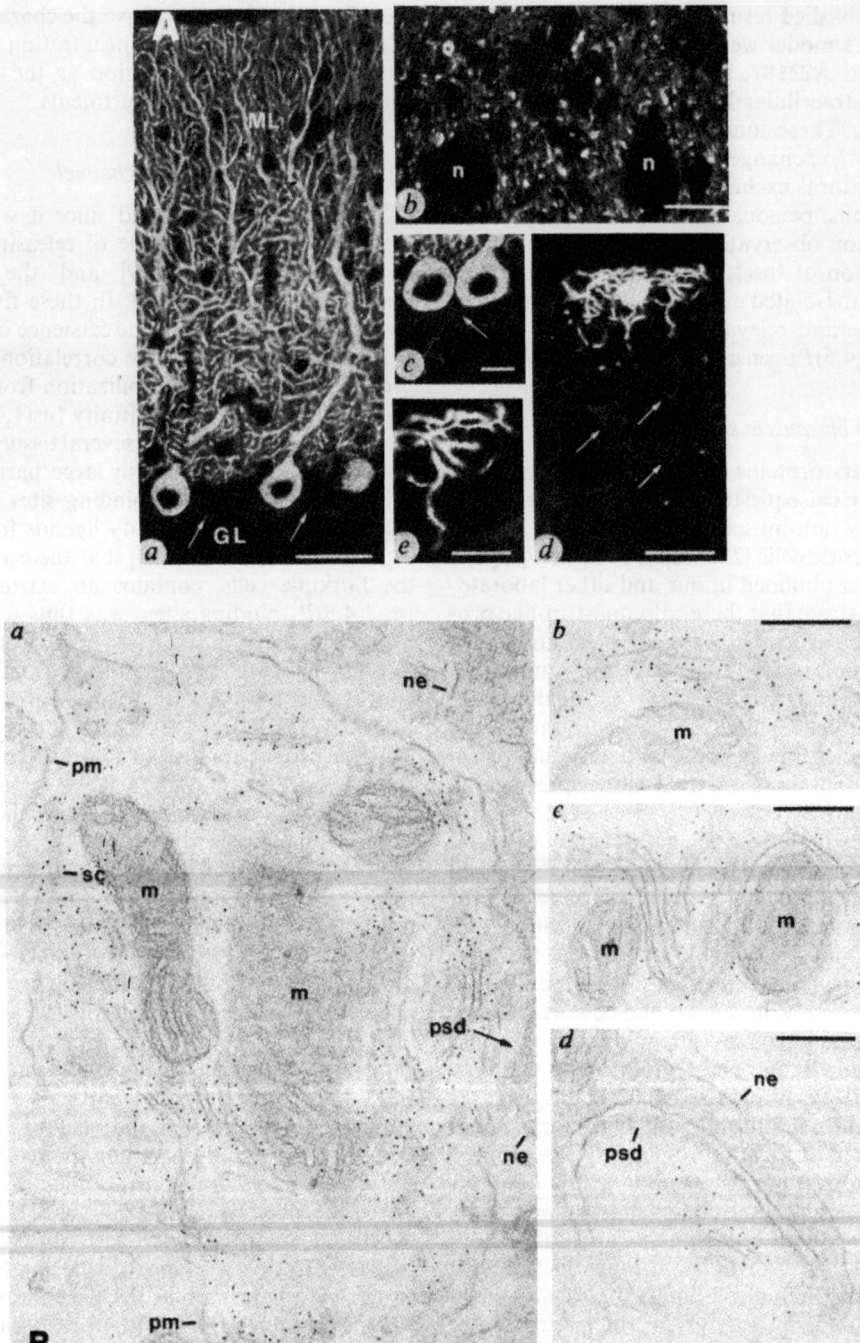

Fig. 5. *Sublocalization of the Purkinje-cell Ins(1,4,5)P₃ receptor visualized by immunofluorescence spectroscopy and immunoelectron microscopy.*
(A) Immunofluorescence spectroscopy: GL, granule cell layer; n, neuronal perikarya (unlabelled). (B) Immunoelectron microscopy: m, mitochondria; pm, plasma membrane; sc, subplasmalemma cisterna; psd, post-synaptic density; ne, nerve ending. Note the presence of a positive cisternum in a dendritic spine (d). Reproduced with permission from Mignery et al. [313]

the plasma membrane. In a second study, using gold-labelled antibodies, Mignery et al. [313] confirmed the widespread distribution of the Ins(1,4,5)P₃ receptor on polymorphic structures described previously [307] but did not confirm the high level of labelling of the nuclear membrane. Fig. 5 shows the immunolabelling of Purkinje cells with an anti-[Ins(1,4,5)P₃ receptor] antibody.

A recent study by Meldolesi's group [316] also performed using gold-labelled antibodies, confirmed that the very high level of labelling of the nuclear membrane probably arises

from an artefact of the peroxidase technique. Rather, it was shown that the highest level of labelling by far (greater than fivefold compared to the nuclear membrane) is observed in special types of smooth stacked cysternae, often continuous with rough ER. Positive cysternae were found in the cell body, in axons and dendrites, and of major physiological relevance, in the dendritic spines [316]. Using peroxidase-coupled anti-[Ins(1,4,5)P₃ receptor] antibodies, Maeda et al. [315] demonstrated the presence of antibodies on the plasma membrane and on the post-synaptic antibody-binding density, though this result may depend again on the diffusion of the immunoprecipitate from closely apposed subplasmalemmal cisternae [316]. A recent study by Otsu et al. [402] confirms the results of Satoh et al. [316].

Since the molecular identity of the peripheral receptor is uncertain, its characterization is still mainly based on functional studies on microsomal preparations. Summarizing a wealth of data on this receptor we can conclude the following: (a) the peripheral $Ins(1,4,5)P_3$ receptor is present at picomolar concentrations within most cells [317]; (b) it has an affinity for $Ins(1,4,5)P_3$ in the nanomolar range [317, 318]; (c) it gates a channel opening within less than 20 ms upon addition of $Ins(1,4,5)P_3$ [319]; (d) opening of the channel by $Ins(1,4,5)P_3$ shows some signs of cooperativity [319]. The characteristics of a channel gated by $Ins(1,4,5)P_3$ have been recently described by Ehrlich and Watras [320] using microsomes from smooth muscle cells fused with black lipid films. The electrophysiological and pharmacological properties of this channel are strikingly different from those of the ryanodine receptor; the conductance is much smaller and caffeine, ryanodine and Ca^{2+} have no effect either on the conductance or on the opening probability.

Are there one or more types of Ca^{2+} stores in non-muscle cells?

This is a much debated issue and considerable evidence has been presented suggesting the existence in non-muscle cells of Ca^{2+} pools insensitive to $Ins(1,4,5)P_3$, but sensitive to caffeine, Ca^{2+} or GTP. Subcellular fractionation studies on a variety of tissues have given conflicting results as far as the correlation between ER markers, Ca^{2+} accumulation, $Ins(1,4,5)P_3$ binding and $Ins(1,4,5)P_3$-sensitive Ca^{2+} release. Concerning the $Ins(1,4,5)P_3$-sensitive store, the correlation with ER marker enzymes was poor or absent in some studies [280, 321, 322], while in others it was excellent, particularly with markers or rough ER [323, 324]. It is not known whether these discrepancies reflect the different tissues analyzed or whether they depend on the different isolation procedures. The picture emerging from these studies is that if any correlation exists between ER and the $Ins(1,4,5)P_3$-sensitive Ca^{2+} store, then this is probably with a subfraction of the ER and not with the whole structure. On the other hand the morphological approach followed by our group [280–282] revealed the existence of structures, named calciosomes, that are enriched with a SR-type Ca^{2+} ATPase and with calreticulin, but are devoid of typical ER markers, i.e. cytochrome $P450$. We initially proposed that the calciosomes were the $Ins(1,4,5)P_3$-sensitive stores; however, this tentative identification never went beyond a plausible working hypothesis, since no marker was available for the peripheral $Ins(1,4,5)P_3$ receptor. As more data accumulates on the newly described $Ins(1,4,5)P_3$-insensitive stores, the possibility that the calciosomes should be identified with this latter store must be considered and is actually suggested by some recent data [286].

With respect to GTP-induced Ca^{2+} release [324, 325] our understanding of the current interpretation of the data is that this nucleotide does not directly gate a Ca^{2+} channel, but rather permits the communication between $Ins(1,4,5)P_3$-sensitive and $Ins(1,4,5)P_3$-insensitive pools [326]. It must also be pointed out that GTP can cause fusion between vesicles, although at present it is unclear how much this phenomenon contributes to overall GTP-induced Ca^{2+} release.

Caffeine-induced Ca^{2+} redistribution in non-muscle cells has been documented in a variety of neurons or neuro-endocrine cells [286, 327, 328]. The sensitivity to caffeine has been interpreted as evidence for the existence of a Ca^{2+}-induced Ca^{2+} release mechanism similar to that of striated muscle SR. It is still unclear however, whether the caffeine-sensitive pool is distinct from the $Ins(1,4,5)P_3$-sensitive pool, or whether the sensitivity to this drug demonstrates the existence of two different channels in the same pool or even of two gating modes of the same channel. Since these issues are discussed in much detail in two recent reviews [283, 284] the reader is referred to these contributions for further details.

MITOCHONDRIA

Had this review been written in the late seventies or early eighties, the section on mitochondrial Ca^{2+} homeostasis would be one of the lengthiest and most complex. Today the interest of the scientific community for mitochondrial Ca^{2+} homeostasis has dropped dramatically, although many aspects of Ca^{2+} handling by mitochondria remain mysterious. Here we will emphasize only the most controversial points, referring the reader to other reviews for a more detailed and exhaustive discussion [329, 330].

Mitochondrial Ca^{2+} uptake depends on the electrical gradient across the inner membrane and does not require ATP hydrolysis. Neither the nature of the transport mechanism (carrier or channel) has been established with certainty, nor the molecular components involved in Ca^{2+} uptake been purified. The only attempt to discriminate between channel- and carrier-mediated transport is, to our knowledge, that of Bragadin et al. [331] who showed that the activation energy of Ca^{2+} uptake in mitochondria is relatively low (33.6–37.8 kJ/mol), with no discontinuity in the Arrhenius plot, as in the case of typical ion carriers in mitochondria. The molecular components of the mitochondrial Ca^{2+} uptake system have been the matter of much investigation until 1982–1983 [332, 333]. Despite some claims in the past that a glycoprotein of molecular mass 30 kDa could be involved [332] the observation has neither been confirmed nor followed up by the authors themselves. A major obstacle in applying molecular biology techniques to this problem is that neither an assay for the reconstituted system is available, nor is a specific blocker of mitochondrial Ca^{2+} uptake on the market. The well-known and widely used lanthanides and ruthenium red are in fact by no means specific.

Our understanding of Ca^{2+} efflux has not improved significantly even after the discovery that because of the existence of electro-neutral $Ca^{2+}/2 H^+$ or electrogenic $Ca^{2+}/3 Na^+$ exchangers, mitochondrial Ca^{2+} accumulation does not reach electrochemical equilibrium (the concept of a mitochondrial set point [334–336]). It was shown that the exchange mechanism can be modulated by the membrane potential [337] and that it is sensitive to a number of drugs known to be fairly unspecific inhibitors of voltage-gated Ca^{2+} channels of the plasma membrane [338, 339]. The median effective concentration for inhibition by these drugs is however rather high (several tens of micromolar) [338, 339] thus affinity purification of the exchanger by this means appears problematic.

The intramitochondrial buffering system is also rather poorly understood. The Ca^{2+} content of mitochondria in intact cells or of carefully isolated mitochondria, [340] is much lower than previously thought, yet a major discrepancy still exists between the total Ca^{2+} content (tens of micromoles/liter of matrix water) and the free mitochondrial Ca^{2+} concentration (<1 µM) calculated indirectly on the basis of the activity of intramitochondrial Ca^{2+}-activated dehydrogenases [341] and more recently, directly with fluorescent probes [342].

The significance in the intact cell of the elaborate mitochondrial Ca^{2+} homeostatic mechanism still remains the most relevant physiological question. The only point that is generally agreed upon is that mitochondrial Ca^{2+} transport plays an essential role in intramitochondrial metabolism [341]. In fact, intramitochondrial-free Ca^{2+} is a regulator of a number of key dehydrogenases and its accurate control is vital for their coordinated function. It is also well established that mitochondria are not the pool responsible for receptor-stimulated Ca^{2+} mobilization, since no physiological agent capable of triggering fast release of mitochondrial Ca^{2+} is known. Recently however, it has been reproposed that mitochondria can serve as intracellular buffering organelles, at least in some cell types. In fact, a significant proportion of the calcium released from the $Ins(1,4,5)P_3$-sensitive pool apparently ends up in the mitochondria [343, 344]. Therefore these organelles may be tentatively included in the elusive $Ins(1,4,5)P_3$-insensitive Ca^{2+} pool.

The involvement of mitochondria as the last high-capacity low-affinity Ca^{2+} store, that will serve to rescue cells when pathological increases of $[Ca^{2+}]_i$ occur, has been largely discussed in the past (see [345] for review).

CYTOPLASM

The highly complex $[Ca^{2+}]_i$ homeostatic mechanisms serve two complementary functions: (a) maintenance of a low $[Ca^{2+}]_i$, essential for cell survival; (b) allowance of controlled changes of the second-messenger level when appropriate stimuli (plasma-membrane-potential depolarization and/or triggering of specific receptors) activate Ca^{2+}-dependent pathways. Thus, the ultimate targets of $[Ca^{2+}]_i$ changes reside in the cytosol and are represented by calcium-binding proteins.

Calcium-binding proteins can be classified according to different criteria. A very comprehensive classification of the EF-hand superfamily of calcium-binding proteins based on their evolutionary relationships has been proposed recently by Kretzinger et al. [346]. In the present contribution we classify cytosolic calcium-binding proteins on the basis of their function.

Given that in all eukaryotic cells studied to date the $[Ca^{2+}]_i$, under physiological conditions, can vary at most from a tenth of a micromole to a few micromoles [347] the obvious prediction is that the useful range of Ca^{2+} affinities for a cytosolic protein must be in the submicromolar to low micromolar range. Many exceptions to this rule however exist and cells have circumvented the problem of the intrinsic low Ca^{2+} affinity of some calcium-binding proteins (i.e. protein kinase C) in various ways (see below). From a functional point of view, cytosolic high-affinity calcium-binding proteins can be divided into two major groups: (a) proteins whose only function is that of buffering Ca^{2+} and (b) proteins with modulatory activity on cellular functions. In the latter family some i.e. calmodulin, troponin C etc., regulate the function of target proteins; others, such as protein kinase C, have catalytic activity of their own.

A detailed discussion of each member of cytosolic calcium-binding proteins would require more room than is available here. Therefore we will briefly describe the major characteristics of a few of the most representative members of each group.

The Ca^{2+} buffers: parvalbumin and related proteins

Parvalbumins are a group of homologous proteins with molecular masses ranging over 9–13 kDa. These proteins were initially described by Deutike in 1934 [348] as soluble 'low-molecular mass albumins' in frog muscles, hence the term parvalbumin (*parva* being latin for small). Later, they were isolated from different fish and amphibian skeletal muscle. For some time parvalbumins were considered to be specific to white muscles of lower vertebrates, but more recently they were shown to exist in almost all mammals and not only in striated muscles (for a review see [349]). Parvalbumins have a special place in the history of calcium-binding proteins not only because they were the first to be crystallized and studied by X-ray diffraction [350], but also because some basic concepts of the structural arrangement of high-affinity Ca^{2+}-binding sites were based on the structure of carp parvalbumin. Three-dimensional structure analysis indicates that carp parvalbumin contains six helical regions (A–F) connected by loops. The A–B domain represents an aborted Ca^{2+}-binding site. The Ca^{2+}-binding sites are located in the loops between the two helices C–D and E–F. The structure of the last two helices (and loops), the EF-hand structure (see Fig. 6), has been synonymous for a long time with the high-affinity Ca^{2+}-binding site. It is now clear that this is an oversimplification and new types of cytosolic calcium-binding proteins, not belonging to the EF superfamily, have been discovered in the last few years, (the annexin family [351], gelsolin [352], protein kinase C [353, 354] etc.). Nonetheless the sequence constraints of the EF-hand structure are so rigid that a protein can be classified in the EF-hand family simply by knowing its amino acid sequence [355]. The Ca^{2+}-binding affinities of the two Ca^{2+}-binding sites of parvalbumin are very similar, i.e. no evidence for positive or negative cooperativity has been provided [356]. The reported K_d values vary over 0.1–4 μM depending on the concentration of Mg^{2+}, pH, ionic strength, source of the protein and methodology employed [356, 359]. The two high-affinity Ca^{2+}-binding sites also bind Mg^{2+} (Ca^{2+}/Mg^{2+} mixed sites) so that under resting conditions, most of the parvalbumin is in the Mg^{2+} bound form.

The only known function of parvalbumin is to bind Ca^{2+}, despite many attempts in the past to look for enzymatic or regulatory functions of the protein. The most widely accepted idea for the role of parvalbumin *in vivo* is that it functions as a soluble Ca^{2+} buffer in the cytoplasm. Consistent with this hypothesis are observations that parvalbumin in skeletal muscle is present almost exclusively in fast twitch fibers [349] and there is a good correlation between the parvalbumin content and the speed of muscle relaxation [349, 356]. According to this model, Ca^{2+} released from the SR will first bind to troponin C, initiating contraction (the dissociation of Mg^{2+} from the Ca^{2+}/Mg^{2+} mixed sites of parvalbumin is relatively slow); the muscle relaxation would be initially due to Ca^{2+} binding to parvalbumin and subsequently to Ca^{2+} uptake by the SR. This model essentially applies to fast-twitch fibers, but it is still largely debated [349, 356, 357].

Parvalbumin or parvalbumin-like proteins have more recently been found in many other tissues, in particular in the central nervous system, in which only a subpopulation of neurons appear to contain the protein and there is some correlation between the parvalbumin and the 4-aminobutyric acid content [349, 356, 358]. Parvalbumins (or related proteins) have also been found in Leydig cells of the testis, in the ovary and in mineralizing tissues [358].

Other proteins with high affinity for Ca^{2+}, whose only known function is Ca^{2+} buffering, have been described in

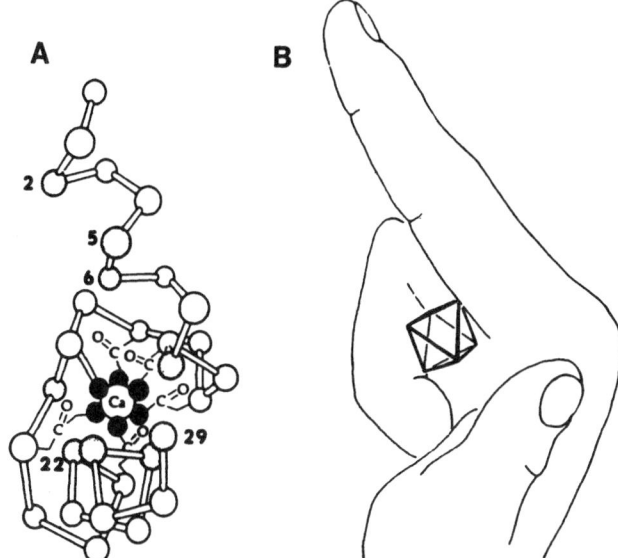

Fig. 6. *The EF-hand structure*. (A) Ball and stick model and (B) symbolic representation of the EF-hand structure. Reproduced with permission from Persechini et al. [399]

recent years. Among them the S100* intestinal Ca^{2+}-binding protein family [359], oncomodulin [360], a parvalbumin-like protein found specifically in tumor tissues, and the calbindin family [361]. All the proteins mentioned above contain two or more EF-hand motifs.

It must be stressed that energy-independent buffering of the $[Ca^{2+}]_i$ depends not only on high-affinity calcium-binding proteins, but also upon low-affinity binding sites, not exclusively provided by negative sites on proteins. Low-affinity Ca^{2+}-binding sites ($K_d > 10$ μM) in fact not only appear to be present in the cytoplasm [362], but their importance as cytoplasmic Ca^{2+} buffers cannot be neglected; it is easy to calculate that if the number of the low affinity sites is large (as it probably is) then even negligible saturation of these sites (< 1%) would provide significant Ca^{2+} buffering of cytoplasmic calcium.

The Ca^{2+} sensors: calmodulin and related proteins

Although discovered relatively late (1970) [363, 364] no other calcium-binding protein has received so much attention by the scientific community as calmodulin [365, 366]. This small acidic protein (molecular mass 16 kDa, isoelectric point 4.5) is present in large quantities in the cytoplasm of all eukaryotes (up to 1% of the total protein mass) [366]. No natural isoform of calmodulin in different tissues has been described so far (in chickens, a calmodulin pseudogene has been found [367]) and the primary structure of calmodulin has been highly conserved during evolution. The crystal structure of calmodulin (at 0.3 nm resolution) was determined in 1985 by Babu et al. [368] and the protein looks like a dumb-bell with the globular N- and C-terminal halves linked by a long α-helical portion (2.0 nm long) which contains the phenothiazine-binding domain. The N- and C-terminal globular domains are remarkably similar and two EF-hand Ca^{2+}-binding sites are localized within each half molecule. It should be emphasized that the three-dimensional structure of calmodulin was determined on crystals grown at pH 5 [368] and thus extrapolation of the data to neutral, physiological conditions may not be completely correct.

Despite the fact that Ca^{2+} binding to calmodulin has been investigated by nearly all available techniques, no consensus has been yet reached on the affinity for Ca^{2+} of the four Ca^{2+}-binding sites [366]. Two alternative models have been proposed. According to one hypothesis, the four Ca^{2+}-binding sites of calmodulin have very similar affinities with little or no cooperativity among the sites [369]; at pH 7.0 and ionic strength 150 mM, the Ca^{2+}-saturation curve fits an ideal Langmuir isotherm with four independent sites of equal affinities (K_d 10 μM). Cox [370] has however demonstrated that the experimental data can also fit a second model [371], according to which the N- and C-terminal Ca^{2+}-binding pairs differ markedly in their affinities for Ca^{2+}. This latter model is supported by kinetic and conformational studies. In fact, based on $^{43}Ca^{2+}$-NMR [372] and stopped-flow fast kinetics [373], it has been shown that the N-terminal half of calmodulin has a k_{off} of $300-500$ s^{-1} while the C-terminus has a k_{off} of $10-40$ s^{-1}. Assuming a diffusion controlled on-rate constant, these data suggest the existence of high-affinity sites in the C-terminus and low-affinity sites in the N-terminus of the protein. Evidence has also been provided suggesting the existence of strong cooperativity between the two sites of each pair, particularly in the C-terminal half [374]. In conclusion, at present it is difficult to solve the controversy among the different models and the introduction of new and more sophisticated techniques should eventually solve the discrepancies.

A number of other cations, notably Mg^{2+} and H^+ can bind to calmodulin on the so called auxiliary sites [375]. Under physiological conditions part of the six auxiliary sites (distinct from the four EF-hand Ca^{2+}-specific sites) are occupied by Mg^{2+}.

Binding of the Ca^{2+}-calmodulin complex to a target enzyme increases the Ca^{2+}-binding constants and leads to the appearance of strong positive cooperativity [376]. In particular a dramatic increase in affinity occurs in the third Ca^{2+}-binding constant. Most studies indicate that the high-affinity complex Ca^{2+}-calmodulin-enzyme is only formed if at least three Ca^{2+} bind to calmodulin [377]. It goes beyond the purpose of this review to discuss in detail the complicated kinetic models for the activation of Ca^{2+}-calmodulin-dependent enzymes. It suffices here to stress three points: (a) the Ca^{2+}-bound form of calmodulin has an affinity four orders of magnitude higher for the target enzyme than the Ca^{2+}-free form [378]; (b) the complex between metal-free calmodulin and the enzyme is inactive [378]; (c) the activation of the target enzyme by Ca^{2+}-calmodulin causes derepression, rather than direct activation [379]. This last concept, initially formulated for the activation of myosin-light-chain kinase, has emerged relatively recently [379] (the pseudosubstrate hypothesis) but it might have major consequences in the design of new pharmacological strategies for specific inhibitors of Ca^{2+}-calmodulin-dependent reactions. According to this model the calmodulin-binding domain, in the absence of calmodulin, prevents access of the natural substrate to the enzyme's active site; binding of calmodulin removes this block. The discovery of this mechanism of activation explains the observation that limited proteolysis of Ca^{2+}-calmodulin-dependent enzymes removes calmodulin activation, while permanently activating the enzymatic activity [380].

The most common trigger for the activation of the Ca^{2+}-calmodulin-dependent reaction in intact cells is an increase in the $[Ca^{2+}]_i$. Thus more Ca^{2+}-calmodulin complex forms and the target enzymes are activated. There is another way to activate Ca^{2+}-calmodulin-dependent reactions, at least in theory, i.e. increasing the calmodulin concentration at fixed

[Ca^{2+}]. This simple observation is well known to biochemists studying calmodulin *in vitro*, but should not be underestimated *in vivo*, since the absolute concentration of calmodulin not only varies in different cells [365, 366, 369], but can also change during the cell cycle. In particular, a number of authors [381 – 383] have shown that the content of calmodulin abruptly increases as cells cross the G1/S boundary. In addition, the absolute level of calmodulin increases in transformed cells [384 – 386]. This has led to the hypothesis that calmodulin might regulate important steps in the cell cycle and in particular some crucial event in the G1 phase. Consistent with this interpretation are a number of results: (a) reversible inhibition of progression through the cell cycle by anti-calmodulin drugs [387]; (b) complete halt of vegetative growth after disruption of the calmodulin gene in the fission yeast *Schizosaccaromyces pombe* [388]; (c) selective reduction of the G1 duration in cells which overexpress calmodulin after transfection with an exogenous calmodulin gene [389]. It is obvious that increasing the calmodulin concentration is a slow and energetically expensive way to activate Ca^{2+} – calmodulin-dependent reactions, but unlike the changes in the [Ca^{2+}]$_i$, which are normally very transient, the increased content of calmodulin might have long lasting effects on cell functions.

The number of enzymatic reactions that are known to be activated by Ca^{2+}-calmodulin is impressive [365, 367, 369]. Two points will be mentioned here that are of general relevance. (a) Despite the fact that so many reactions are known to depend on Ca^{2+}-calmodulin when studied with isolated enzymes, there are much fewer physiological pathways in intact cells that can unambiguously be attributed to Ca^{2+} – calmodulin-dependent reactions. For some time the inhibition by anti-calmodulin drugs has been used by pharmacologists and cell biologists as the main criterion. Although this criterion is still used to assess the involvement of calmodulin in biological reactions, it must be stressed that all calmodulin inhibitors are hydrophobic, perturb membrane functions and are by no means specific *in vivo*. Their effects are so unpredictable, when used in intact cells, that their use should be limited to very few and well-controlled situations. (b) In many instances the amplification of the Ca^{2+} signal is mediated by the Ca^{2+}-calmodulin complex via activation of protein phosphorylation [365, 366, 369]: in this respect it is of major interest that Ca^{2+}-calmodulin kinase II, a multifunctional protein kinase [390] very abundant in the central nervous system (particularly in the hyppocampus), becomes Ca^{2+} – calmodulin-independent after autophosphorylation, i.e. the enzymatic activity becomes autonomons from the second messenger [391]. Deactivation of kinase II can occur either by dephosphorylation or by further autophosphorylation. This phenomenon represents one of the clearest example of biochemical memory whereby once the system has been activated, removal of the stimulator (Ca^{2+}) does not result in deactivation.

Troponin C [392] is closely related structurally to calmodulin, but functionally, the two proteins are quite distinct. In particular troponin C, unlike calmodulin which is a soluble cytosolic protein reversibly bound to a variety of target enzymes, is part of the troponin complex, where the three subunits (C, T and I) behave as a single unit. Furthermore, at least two isoforms (cardiac/slow skeletal and fast skeletal) of troponin C exist with different structural and functional characteristics [393]. Finally, until now troponin C has been found only in striated fibers and has one and only one cellular function, contraction (for review see [393]).

Skeletal muscle troponin C (molecular mass 18 kDa, isoelectric point 4.2) contains four EF-hand Ca^{2+}-binding sites [393]. Of these four sites, the two in the C-terminal domain have a higher affinity for Ca^{2+}, but also bind Mg^{2+}, while the two low-affinity sites of the N-terminus are Ca^{2+} specific [393]. Positive cooperativity in Ca^{2+} binding has been observed in the two high-affinity sites [394].

The basic structure of troponin C closely resembles that of calmodulin [395] in having two globular domains each containing two-EF hand Ca^{2+}-binding sites connected by a long α-helix. The rate of Ca^{2+} binding to troponin C is very fast ($k_{on} > 4 \times 10^7$ s^{-1}) [396] thus explaining why parvalbumin does not compete with troponin C for Ca^{2+} during the onset of contraction. It must be noted that, resembling the effect of target enzymes on calmodulin, the affinity of troponin C for Ca^{2+} in the troponin complex is increased by nearly one order of magnitude [393]. It suffices here to say that troponin C is an integral part of the thin-muscle filament and that its role under resting conditions prevents the interaction of the myosin heads with actin. Upon increase in the [Ca^{2+}]$_i$, Ca^{2+} binds to troponin C and the Ca^{2+}-troponin C complex moves away from the thin filament permitting actin-myosin interaction [392 – 396].

The phylogenetic variability of troponin C is much higher than that of calmodulin and within the same animal species, troponin C isolated from heart is significantly different from that expressed in fast skeletal muscle, but identical to that of slow skeletal fibers. Cardiac/slow skeletal troponin C contains only three EF-hand motifs [397].

Strictly speaking several other proteins should be included in the Ca^{2+}-sensor family: myosin, α-actinin, aequorin and other photoproteins; the Ca^{2+}-ATPases; the Ca^{2+} channels and a variety of Ca^{2+}-activated enzymes such as protein kinase C; neutral proteases; calcineurin (a protein phosphatase); annexins (Ca^{2+}-dependent phospholipid-binding proteins); a number of proteins of lower organisms etc. Some of these were dealt with in the preceeding paragraphs of this review and for the others the more specialized literature should be consulted.

CONCLUSIONS

In this review we have attempted to give a brief but comprehensive description of the basic homeostatic mechanisms of [Ca^{2+}]$_i$ in eukaryotic cells. By necessity we have been selective in quotations and biased in some debated aspects. In particular we have mainly stressed the structural and molecular properties of the many components of the [Ca^{2+}]$_i$ homeostatic machinery. This is however only the starting point for understanding [Ca^{2+}]$_i$ handling in living cells. As pointed out by Berridge [398], future emphasis will also have to be put on the spatial and temporal aspects of [Ca^{2+}]$_i$ homeostasis.

We are indebted to all the colleagues who helped us with suggestions and discussions. In particular we wish to thank Drs P. Hess, J. Cox, C. Heizmann, G. Shull and J. Meldolesi for permitting discussion of unpublished data. We also thank Ms C. Marchioro for skilful secretarial assistance. This work was supported in part by grants from the Italian Ministry of Public Education (40% and 60%), by CNR special project 'Biotecnology' and strategic project 'Transmembrane Signalling' and by a grant from the Italian Association for Cancer Research (AIRC).

REFERENCES

1. Ringer, S. (1883) *J. Physiol.* 4, 29 – 43.
2. Tsien, R. Y., Pozzan, T. & Rink, T. J. (1982) *Nature* 295, 68 – 71.

3. Baker, P. F. (1970) in *Calcium and cellular functions* (Cuthbert, A. W., ed.) pp. 96–107, McMillan, London.
4. Schatzmann, H. J. (1989) *Annu. Rev. Physiol. 51*, 473–485.
5. Carafoli, E. (1987) *Annu. Rev. Biochem. 56*, 395–433.
6. Penniston, J. T., Filoteo, A. G., McDonough, C. S. & Carafoli, E. (1988) *Methods Enzymol. 157*, 340–351.
7. Niggli, V., Penniston, J. T. & Carafoli, E. (1979) *J. Biol. Chem. 254*, 9955–9958.
8. Niggli, V., Adunyah, E. S., Penniston, J. T. & Carafoli, E. (1981) *J. Biol. Chem. 256*, 395–401.
9. Shull, G. E. & Greeb, J. (1988) *J. Biol. Chem. 263*, 8646–8657.
10. Brandl, C. J., Green, N. M., Korkczak, B. & McLennan, D. H. (1986) *Cell 44*, 597–607.
11. Richards, D. E., Rega, A. F. & Garrahan, P. J. (1978) *Biochim. Biophys. Acta 511*, 194–201.
12. Muallen, S. & Karlish, S. J. D. (1980) *Biochim. Biophys. Acta 597*, 631–636.
13. Stieger, J. & Luterbacher, S. (1981) *Biochim. Biophys. Acta 641*, 270–275.
14. Caroni, P. & Carafoli, E. (1983) *Eur. J. Biochem. 132*, 451–460.
15. Hakim, G., Itano, T., Verma, A. K. & Penniston, J. T. (1982) *Biochem. J. 207*, 225–231.
16. Scharf, O. (1978) *Biochim. Biophys. Acta 512*, 309–317.
17. Rossi, J. P. F. C., Rega, A. F. & Garrahan, P. J. (1985) *Biochim. Biophys. Acta 816*, 379–386.
18. Scharf, O. & Foder, B. (1982) *Biochim. Biophys. Acta 691*, 133–143.
19. Scharf, O., Foder, B. & Skibsted, U. (1983) *Biochim. Biophys. Acta 730*, 295–305.
20. Carafoli, E. (1988) *Methods Enzymol. 157*, 3–11.
21. Gatti, G., Madeddu, L., Pandiella, A., Pozzan, T. & Meldolesi, J. (1988) *Biochem. J. 255*, 753–760.
22. Niggli, V., Sigel, E. & Carafoli, E. (1982) *J. Biol. Chem. 257*, 2350–2356.
23. Lagast, J., Pozzan, T., Waldvogel, F. A. & Lew, D. P. (1984) *J. Clin. Invest. 73*, 878–884.
24. Pollock, W. K., Sage, S. O. & Rink, T. J. (1987) *FEBS Lett. 210*, 132–134.
25. Rickard, J. E. & Sheterline, P. (1987) *Biochem. J. 231*, 623–628.
26. Reuter, H. & Seitz, N. (1968) *J. Physiol. 195*, 451–470.
27. Blaunstein, M. P. & Hodgkin, A. L. (1969) *J. Physiol. 200*, 496–527.
28. Caroni, P. & Carafoli, E. (1983) *Eur. J. Biochem. 132*, 451–460.
29. Blaunstein, M. P. (1970) *Biophys. J. 20*, 79–111.
30. Reinlib, L., Caroni, P. & Carafoli, E. (1987) *FEBS Lett. 126*, 74–76.
31. Reeves, J. P. & Sutko, J. L. (1979) *Proc. Natl Acad. Sci. USA 76*, 590–594.
32. Blaunstein, M. P. (1988) *J. Cardiovasc. Pharmacol. 12*, S56–S68.
33. Meldolesi, J. & Pozzan, T. (1987) *Exp. Cell Res. 171*, 271–283.
34. Hess, P. (1990) *Annu. Rev. Neurosci. 13*, 337–356.
35. Bean, B. P. (1989) *Annu. Rev. Physiol. 51*, 367–384.
36. Llinas, R., Sugimori, M., Lin, J.-W. & Cherskey, B. (1989) *Proc. Natl Acad. Sci. USA 86*, 1689–1693.
37. Plummer, M. R., Loghothetis, D. E. & Hess, P. (1989) *Neuron 2*, 1453–1463.
38. Fosset, M., Jaimovich, E., Delpont, E. & Lazdunsky, M. (1983) *J. Biol. Chem. 258*, 6086–6092.
39. Kokubun, S., Prod'hom, B., Becker, C., Porzig, H. & Reuter, H. (1986) *Mol. Pharmacol. 30*, 571–584.
40. Glossmann, H., Ferry, D. R., Striessing, J., Goll, A. & Moosburger, K. (1987) *Trends Pharmacol. Sci. 8*, 95–100.
41. Reuter, H., Stevens, C. F., Tsien, R. W. & Yellen, G. (1982) *Nature 297*, 501–504.
42. Hess, P., Lansman, J. B. & Tsien, R. W. (1986) *J. Gen. Physiol. 88*, 293–319.
43. Hess, P., Lansman, J. B. & Tsien, R. W. (1984) *Nature 311*, 538–544.
44. Cavalié, A., Pelzer, D. & Trautwein, W. (1986) *Pflügers Arch. 406*, 241–258.
45. Hess, P. & Tsien, R. W. (1984) *Nature 309*, 453–456.
46. Almers, W. & McCleskey, E. W. (1984) *J. Physiol. 353*, 585–608.
47. Kostyuk, P. G., Mironov, S. L. & Shuba, Y. M. (1983) *J. Membr. Biol. 76*, 82–93.
48. Bean, B. P. (1984) *Proc. Natl Acad. Sci. USA 81*, 6388–6392.
49. Sanguinetti, M. C. & Kass, R. S. (1984) *Circ. Res. 55*, 336–348.
50. Rane, S. G., Holz, I. V. G. G. & Dunlap, K. (1987) *Pflügers Arch. 409*, 361–366.
51. Cohen, C. J. & McCarthy, R. T. (1987) *J. Physiol. 387*, 195–225.
52. Fenwick, E. M., Marty, A. & Neher, E. (1982) *J. Physiol. 331*, 599–635.
53. Kass, R. S. & Sanguinetti, M. C. (1984) *J. Gen. Physiol. 84*, 705–726.
54. Fedulova, S. A., Kostyuk, P. K. & Veselovski, N. S. (1985) *J. Physiol. 359*, 431–446.
55. Fox, A. P., Nowycky, M. C. & Tsien, R. W. (1987) *J. Physiol. 394*, 149–172.
56. Kay, A. R. & Wong, R. K. (1987) *J. Physiol. 392*, 603–616.
57. Kalman, D., O'Lague, P. H., Erxleben, C. & Armstrong, D. L. (1988) *J. Gen. Physiol. 92*, 531–548.
58. Plant, T. D. (1988) *J. Physiol. 404*, 731–747.
59. Lee, K. S., Marban, E. & Tsien, R. W. (1985) *J. Physiol. 364*, 395–411.
60. Akaike, N., Tsuda, Y. & Oyama, Y. (1988) *Neurosci. Lett. 84*, 46–50.
61. Di Virgilio, F., Milani, D., Leon, A., Meldolesi, J. & Pozzan, T. (1987) *J. Biol. Chem. 262*, 9189–9195.
62. Reuter, H. & Scholz, H. (1977) *J. Physiol. 264*, 49–62.
63. Osterrieder, W., Brum, G., Hescheler, J., Trautwein, W., Flockerzi, V. & Hoffmann, F. (1982) *Nature 298*, 576–578.
64. Bean, B. P., Nowycky, M. C. & Tsien, R. W. (1984) *Nature 307*, 371–375.
65. Kameyama, M., Hofmann, F. & Trautwein, W. (1985) *Pflügers Arch. 405*, 285–293.
66. Hartzell, H. C. & Fischmeister, R. (1986) *Nature 323*, 273–275.
67. Cachelin, A. B., de Peyer, J. E., Kokubun, S. & Reuter, H. (1983) *Nature 304*, 462–464.
68. Brum, G., Osterrieder, W. & Trautwein, W. (1984) *Pflügers Arch. 401*, 111–118.
69. Yatani, A., Codina, J., Imoto, Y., Reeves, J. P., Birnbaumer, L. & Brown, A. M. (1987) *Science 238*, 1288–1292.
70. Yatani, A. & Brown, A. M. (1989) *Science 245*, 71–74.
71. Dosemeci, A., Dhallan, R. S., Cohen, N. M., Lederer, W. J. & Rogers, T. B. (1988) *Circ. Res. 62*, 347–357.
72. Lacerda, A. E., Rampe, D. & Brown, A. M. (1988) *Nature 335*, 249–251.
73. Fish, R. D., Spert, G., Colucci, W. S. & Clapham, D. (1988) *Circ. Res. 62*, 1049–1054.
74. Wanke, E., Ferroni, A., Malgaroli, A., Ambrosini, A., Pozzan, T. & Meldolesi, J. (1987) *Proc. Natl Acad. Sci. USA 84*, 4313–4317.
75. Hescheler, J., Rosenthal, W., Hinsch, K. D., Wulfern, M., Trautwein, W. & Schultz, G. (1988) *EMBO J. 7*, 619–624.
76. Rosenthal, W., Hescheler, J., Hinsch, K. D., Spicher, K., Trautwein, W. & Schultz, G. (1988) *EMBO J. 7*, 1627–1633.
77. Benham, C. D. & Tsien, R. W. (1988) *J. Physiol. 404*, 767–784.
78. Miller, R. (1987) *Science 235*, 46–52.
79. Gray, R. & Johnston, D. (1987) *Nature 327*, 620–622.
80. Donaldson, P. L. & Beam, K. G. (1983) *J. Gen. Physiol. 82*, 449–468.
81. Beam, K. G., Knudson, C. M. & Powell, J. A. (1986) *Nature 320*, 168–170.
82. Cota, G. & Stefani, E. (1986) *J. Physiol. 370*, 151–163.
83. Coronado, R. & Affolter, H. (1986) *J. Gen. Physiol. 87*, 933–953.
84. Rosenberg, R. L., Hess, P., Reeves, J. P., Smilovitz, H. & Tsien, R. W. (1986) *Science 231*, 1564–1566.
85. Cognard, C., Romey, G., Galizzi, J. P., Fosset, M. & Lazdunski, M. (1986) *Proc. Natl Acad. Sci. USA 83*, 1518–1522.
86. Yatani, A., Imoto, Y., Codina, J., Hamilton, S. L., Brown, A. M. & Birnbaumer, L. (1988) *J. Biol. Chem. 263*, 9887–9895.

618

87. Vilven, J. & Coronado, R. (1988) *Nature 336*, 587–589.
88. Navarro, J. (1987) *J. Biol. Chem. 262*, 4649–4652.
89. Curtis, B. M. & Catteral, W. A. (1984) *Biochemistry 23*, 2113–2118.
90. Borsotto, M., Barhanin, J., Fosset, M. & Lazdunski, M. (1985) *J. Biol. Chem. 290*, 14255–14263.
91. Flockerzi, V., Oeken, H. J. & Hofman, F. (1986) *Eur. J. Biochem. 161*, 217–224.
92. Leung, A. T., Imagawa, T. & Campbell, K. P. (1987) *J. Biol. Chem. 262*, 7943–7946.
93. Striessnig, J., Moosburger, K., Goll, A., Ferry, D. R. & Glossmann, H. (1986) *Eur. J. Biochem. 261*, 603–609.
94. Curtis, B. M. & Catterall, W. A. (1986) *Biochemistry 25*, 3077–3083.
95. Horne, W. A., Abdel-Ghani, M., Racker, E., Weiland, G. A., Oswald, R. E. & Cerione, R. A. (1988) *Proc. Natl Acad. Sci. USA 85*, 3718–3722.
96. Flockerzi, V., Oeken, H. J., Hofmann, F., Pelzer, D., Cavalié, A. & Trautwein, W. (1986) *Nature 323*, 22–68.
97. Talvenheimo, J. A., Worley III, J. F. & Nelson, M. T. (1987) *Biophys. J. 52*, 891–899.
98. Smith, J. S., McKenna, E. J., Ma, J., Vilven, J., Vaghy, P. L., Schwartz, A. & Coronado, R. (1987) *Biochemistry 26*, 7182–7188.
99. Hymel, L., Sriessnig, J., Glossmann, H. & Schindler, H. (1988) *Proc. Natl Acad. Sci. USA 85*, 4290–4294.
100. Catteral, W. A., Seagar, M. J. & Takahashi, M. (1988) *J. Biol. Chem. 263*, 3535–3538.
101. Campbell, K. P., Leung, A. T. & Sharp, A. H. (1988) *Trends Neurosci. 11*, 425–430.
102. Froehner, S. C. (1988) *Trends Neurosci. 11*, 90–92.
103. Curtis, B. M. & Catterall, W. A. (1985) *Proc. Natl Acad. Sci. USA 82*, 2528–2532.
104. Hosey, M. M., Borsotto, M. & Lazdunski, M. (1986) *Proc. Natl Acad. Sci. USA 83*, 3733–3737.
105. Imagawa, T., Leung, A. T. & Campbell, K. P. (1987) *J. Biol. Chem. 262*, 8333–8339.
106. Jahn, H., Nastainczyk, W., Rohrkasten, P., Schneider, T. & Hofmann, F. (1988) *Eur. J. Biochem. 178*, 535–542.
107. Nastainczyk, W., Rohrkasten, A., Sieber, M., Rudolph, C., Schachtele, C., Marmé, D. & Hofmann, F. (1987) *Eur. J. Biochem. 169*, 137–142.
108. Tanabe, T., Takeshima, H., Mikami, A., Flockerzi, V., Takahashi, H., Kangawa, K., Kojima, M., Matsuo, H., Hirose, T. & Numa, S. (1987) *Nature 328*, 313–318.
109. Ellis, S. B., Williams, M. E., Ways, N. R., Brenner, R., Sharp, A. H., Leung, A. T., Campbell, K. P., McKenna, E., Koch, W. J., Hui, A., Schwartz, A. & Harpold, M. M. (1988) *Science 241*, 1661–1664.
110. Ruth, P., Rohrkasten, A., Biel, M., Bosse, E., Regulla, S., Meyer, H. E., Flockerzi, V. & Hoffmann, F. (1989) *Science 245*, 1115–1118.
111. Noda, M., Shimuzu, S., Tanabe, T., Takai, T., Kayano, T., Ikeda, T., Takahashi, H., Nakayama, H., Kanaoka, Y., Minamino, N., Kangawa, K., Matsu, H., Raftery, M. A., Hirose, T., Inayama, S., Hayashida, H., Miyata, T. & Numa, S. (1984) *Nature 312*, 121–127.
112. Papazian, D. M., Schwarz, T. L., Tempel, B. L., Jan, Y. N. & Yan, L. J. (1987) *Science 237*, 749–753.
113. Pongs, O., Kecskemethy, N., Mueller, R., Krah-Jentzens, I., Baumann, A., Kiltz, H. H., Canal, T., Llamazares, S. & Ferrus, A. (1988) *EMBO J. 7*, 1087–1096.
114. Stuhmer, W., Conti, F., Suzuki, H., Wang, X., Noda, M., Yahagi, N., Kubo, H. & Numa, S. (1989) *Nature 339*, 597–603.
115. Beam, K. G., Knudson, C. M. & Powell, J. A. (1986) *Nature 320*, 168–170.
116. Tanabe, T., Beam, K. G., Powell, J. A. & Numa, S. (1988) *Nature 336*, 134–139.
117. Rios, E. & Brum, G. (1987) *Nature 325*, 717–720.
118. Chang, F. C. & Hosey, M. M. (1988) *J. Biol. Chem. 263*, 18929–18937.
119. Morton, M. E. & Froehner, S. C. (1989) *Neuron 2*, 1499–1506.
120. Schneider, T. & Hofmann, F. (1988) *Eur. J. Biochem. 174*, 369–375.
121. Mikami, A., Imoto, K., Tanabe, T., Niidome, T., Mori, Y., Takeshima, H., Narumiya, S. & Numa, S. (1989) *Nature 340*, 230–233.
122. Nowycky, M. C., Fox, A. P. & Tsien, R. W. (1985) *Nature 316*, 440–443.
123. Tsien, R. W., Lipscombe, D., Madison, D. V., Bley, K. R. & Fox, A. P. (1988) *Trends Neurosci. 11*, 431–438.
124. Kongsamut, S. & Miller, R. J. (1986) *Proc. Natl Acad. Sci. USA 83*, 2243–2247.
125. Fox, A. P., Nowycky, M. C. & Tsien, R. W. (1987) *J. Physiol. 394*, 173–200.
126. Hirning, L. D., Fox, A. P., McCleskey, E. W., Olivera, B. M., Thayer, S. A., Miller, R. J. & Tsien, R. W. (1988) *Science 239*, 57–61.
127. Aosaki, T. & Kasai, H. (1989) *Pflügers Arch. 414*, 150–156.
128. Kongsamut, S., Lipscombe, D. & Tsien, R. W. (1989) *Ann. N. Y. Acad. Sci. 560*, 312–333.
129. Bean, B. P. (1989) *Ann. N. Y. Acad. Sci. 560*, 334–345.
130. Kasai, H., Aosaki, T. & Fukuda, J. (1987) *Neurosci. Res. 4*, 228–235.
131. Jones, S. W. & Marks, T. N. (1989) *J. Gen. Physiol. 94*, 151–157.
132. Carbone, H. E. & Lux, H. D. (1984) *Nature 310*, 501–511.
133. Bean, B. P. (1985) *J. Gen. Physiol. 86*, 1–30.
134. Bossu, J. L., Feltz, A. & Thomann, J. M. (1985) *Pflügers Arch. 403*, 360–368.
135. Carbone, E. & Lux, H. D. (1987) *J. Physiol. 386*, 547–570.
136. Chen, C., Corbley, M. J., Roberts, T. M. & Hess, P. (1988) *Science 239*, 1024–1026.
137. Hagiwara, N., Irisawa, H. & Kameyama, M. (1988) *J. Physiol. 395*, 233–253.
138. Suarez-Kurtz, G., Katz, G. W. & Reuben, J. P. (1987) *Pflügers Arch. 410*, 345–347.
139. Peres, A., Zippel, R., Sturani, E. & Mostacciuolo, G. (1988) *J. Physiol. 401*, 639–655.
140. Nilius, B., Hess, P., Lansam, J. B. & Tsien, R. W. (1985) *Nature 316*, 443–446.
141. Bonvallet, R. (1987) *Pflügers Arch. 408*, 540–542.
142. McCleskey, E. W., Fox, A. P., Feldman, D. H., Cruz, L. J., Olivera, B. M., Tsien, R. W. & Yoshikami, D. (1987) *Proc. Natl Acad. Sci. USA 84*, 4327–4331.
143. Tang, C.-M., Presser, F. & Morad, M. (1988) *Science 240*, 213–215.
144. Yaary, Y., Hamon, B. & Lux, H. D. (1987) *Science 235*, 680–682.
145. Llinas, R. & Yarom, Y. (1986) *Soc. Neurosci. Abstr. 12*, 174.
146. Holz, G. G., Rane, S. G. & Dunlap, K. (1986) *Nature 319*, 670–672.
147. Dunlap, K. & Fischbach, G. D. (1981) *J. Physiol. 317*, 519–535.
148. Forscher, P., Oxford, G. S. & Schulz, D. (1986) *J. Physiol. 379*, 131–144.
149. Marchetti, C., Carbone, E. & Lux, H. D. (1986) *Pflügers Arch. 406*, 104–111.
150. Bean, B. P. (1989) *Nature 340*, 153–156.
151. Lipscombe, D., Kongsamut, S. & Tsien, R. W. (1989) *Nature 340*, 639–642.
152. Toselli, M. & Lux, H. D. (1989) *Pflügers Arch. 413*, 319–321.
153. Kasai, H. & Aosaki, T. (1989) *Pflügers Arch. 414*, 145–149.
154. Tsunoo, A., Yoshii, M. & Narahashi, T. (1986) *Proc. Natl Acad. Sci. USA 83*, 9832–9836.
155. Gross, R. A. & MacDonald, R. L. (1987) *Proc. Natl Acad. Sci. USA 84*, 5469–5473.
156. Heschler, J., Rosenthal, W., Trautwein, W. & Schultz, G. (1987) *Nature 325*, 445–447.
157. Ewald, D. A., Sternwlis, P. C. & Miller, R. J. (1988) *Proc. Natl Acad. Sci. USA 85*, 3633–3637.
158. Ewald, D. A., Pang, I. H., Sternweis, P. C. & Miller, R. J. (1989) *Neuron 2*, 1185–1193.

159. McFadzean, I., Mullaney, I., Brown, D. A. & Milligan, G. (1989) *Neuron 3*, 177−182.

160. Rane, S. G., Walsh, M. P., McDonald, J. R. & Dunlap, K. (1989) *Neuron 3*, 239−245.

161. Rane, S. G. & Dunlap, K. (1986) *Proc. Natl Acad. Sci. USA 83*, 184−188.

162. Doerner, D., Pitler, T. & Alger, B. E. (1988) *J. Neurosci. 8*, 4069−4078.

163. Ewald, D. A., Matthies, H. J. G., Perney, T. M., Walker, M. W. & Miller, R. J. (1988) *J. Neurosci. 8*, 2447−2451.

164. Lewis, D. L., Weight, F. F. & Luini, A. A. (1986) *Proc. Natl Acad. Sci. USA 83*, 9035−9039.

165. Marchetti, C. & Brown, A. M. (1988) *Am. J. Physiol. 254*, C206−C210.

166. Di Virgilio, F., Pozzan, T., Wollheim, C. B., Vicentini, L. M. & Meloblesi, J. (1989) *J.Biol. Chem. 261*, 32−35.

167. Tytgat, J., Niulius, B., Vereeke, J. & Carmeliet, E. (1988) *Pflügers Arch. 411*, 704−706.

168. Milani, D., Malgaroli, A., Guidolin, D., Fasolato, C., Skaper, S. D., Meldolesi, J. & Pozzan, T. (1990) *Cell Calcium 11*, 191−200.

169. Popot, J. L. & Changeaux, J. P. (1984) *Phyiol. Rev. 64*, 1162−1188.

170. Cotman, C. W. & Iversen, L. L. (1987) *Trends Neurosci. 10*, 263−265.

171. Watkins, J. C. & Olverman, H. J. (1987) *Trends Neurosci. 10*, 265−272.

172. Moroni, F., Luzzi, S., Franchi-Micheli, S. & Ziletti, L. (1986) *Neurosci. Lett. 68*, 57−62.

173. Nowak, L., Bregetowsky, P., Asher, P., Herbet, A. & Prochiantz, A. (1984) *Nature 307*, 462−465.

174. Johnson, J. W. & Asher, P. (1987) *Nature 325*, 529−531.

175. Malenka, R. C., Kauser, J. A., Perkel, D. J. & Nicoll, R. A. (1989) *Trends Neurosci. 12*, 444−450.

176. Choi, D. W. (1988) *Trends Neurosci. 11*, 465−469.

177. Benham, C. D. & Tsien, R. W. (1987) *Nature 328*, 275−278.

178. Friel, D. D. & Bean, B. P. (1987) *J. Gen. Physiol. 91*, 1−27.

179. Bean, B. P. (1990) *J. Neurosci. 10*, 1−10.

180. Gomperts, B. D. (1983) *Nature 306*, 64−66.

181. Steinberg, T. H., Newmann, A. S., Swanson, J. A. & Silverstein, S. C. (1987) *J. Biol. Chem. 262*, 8884−8888.

182. Di Virgilio, F., Lew, D. P., Andersson, T. & Pozzan, T. (1987) *J. Biol. Chem. 262*, 4574−4579.

183. Penner, R., Matthews, G. & Neher, E. (1988) *Nature 334*, 499−504.

184. Partridge, L. D. & Swandulla, D. (1988) *Trends Neurosci. 11*, 69−72.

185. Von Tscharner, V., Prod'hom, B., Baggiolini, M. & Reuter, H. (1986) *Nature 324*, 369−372.

186. Lew, D. P., Wollheim, C. B., Waldvogel, F. A. & Pozzan, T. (1984) *J. Cell Biol. 99*, 1212−1218.

187. Nasmith, P. E. & Grinstein, S. (1987) *FEBS Lett. 221*, 95−100.

188. Merritt, J. E., Jacob, R. & Hallam, T. J. (1989) *J. Biol. Chem. 264*, 1522−1527.

189. Kuno, M. & Gardner, P. (1987) *Nature 326*, 301−304.

190. Snyder, P. M., Krause, K. H. & Welsh, M. J. (1988) *J. Biol. Chem. 263*, 11048−11051.

191. Morris, A. P., Gallacher, D. V., Irvine, R. F. & Petersen, O. H. (1987) *Nature 330*, 653−655.

192. Changya, L., Gallacher, D. V., Irvine, R. F., Potter, B. V. & Petersen, O. H. (1989) *J. Membr. Biol. 109*, 85−93.

193. Codina, J., Yatani, A., Grenet, D., Brown, A. M. & Birnbaumer, L. (1987) *Science 236*, 442−445.

194. Fasolato, C., Pandiella, A., Meldolesi, J. & Pozzan, T. (1988) *J. Biol. Chem. 263*, 17350−17359.

195. Nishimoto, I., Hata, Y., Ogata, E. & Kojima, I. (1987) *J. Biol. Chem. 262*, 12120−12126.

196. Matsumaga, H., Nishimoto, I., Kojima, I., Yamashita, N., Kurokawa, K. & Ogata, E. (1988) *Am. J. Physiol. 255*, C442−C446.

197. Putney, J. W. (1986) *Cell Calcium 7*, 1−12.

198. Merritt, J. E. & Rink, T. J. (1987) *J. Biol. Chem. 262*, 17362−17369.

199. Takemura, H., Hughes, A. R., Thastrup, O. & Putney, J. W. (1989) *J. Biol. Chem. 264*, 12266−12271.

200. Merritt, J. E., Armstrong, W. P., Hallam, T. J., Jaxa-Chamiec, A., Leigh, B. K., Moores, K. E. & Rink, T. J. (1989) *Br. J. Pharmacol. 98*, 674p.

201. Porter, K. R. & Palade, G. E. (1957) *J. Biophys. Biochem. Cytol. 3*, 269−298.

202. Campbell, K. P. (1986) *Sarcoplasmic reticulum in muscle physiology* pp. 65−92, CRC Press, Boca Raton, FL.

203. Costello, B., Chadwick, C., Saito, A., Maurer, A. & Fleisher, S. (1986) *J. Cell Biol. 103*, 741−753.

204. MacLennan, D. H. (1970) *J. Biol. Chem. 245*, 4508−4518.

205. Clarke, D. M., Loo, T. W., Inesi, G. & MacLennan, D. H. (1989) *Nature 339*, 476−478.

206. Damiani, E., Betto, R., Salvatori, S., Volpe, P., Salviati, G. & Margreth, A. (1981) *Biochem. J. 197*, 245−248.

207. Brandl, C. J., de Leon, S., Martin, D. R. & MacLennan, D. H. (1987) *J. Biol. Chem. 262*, 3768−3774.

208. de la Bastie, D., Wisnewski, C., Schwartz, K. & Lompré, A. M. (1988) *FEBS Lett. 229*, 45−48.

209. Tada, M., Kirchberger, M. A., Li, H. C. & Katz, A. M. (1975) *J. Biol. Chem. 250*, 2640−2646.

210. James, P., Inui, M., Chiesi, M. & Carafoli, E. (1989) *Nature 342*, 90−92.

211. Inoui, M., Chamberlain, B. K., Saito, A. & Fleischer, S. (1986) *J. Biol. Chem. 261*, 1794−1800.

212. Fleischer, S. & Tonomura, Y. (1985) *Structure and function of sarcoplasmic reticulum*, Academic Press, New York.

213. MacLennan, D. H., Zubrzycka-Gaarn, E. & Jorgensen, A. O. (1985) *Curr. Top. Membr. Transp. 24*, 338−345.

214. Inesi, G. (1985) *Annu. Rev. Physiol. 47*, 573−601.

215. MacLennan, D. H. & Wong, P. T. S. (1971) *Proc. Natl Acad. Sci. USA 68*, 1231−1235.

216. Slupsky, J. R., Ohnishi, M., Carpenter, M. R. & Reitmeier, R. A. F. (1987) *Biochemistry 26*, 6539−6544.

217. Michalak, M., Campbell, K. P. & MacLennan, D. H. (1980) *J. Biol. Chem. 255*, 1317−1326.

218. Fliegel, L., Ohnishi, M., Carpenter, M. R., Khanna, V. K., Reithmeier, R. A. F. & MacLennan, D. H. (1987) *Proc. Natl Acad. Sci. USA 84*, 1167−1171.

219. Maurer, A., Tanaka, M., Ozawa, T. & Fleisher, S. (1988) *Methods Enzymol. 157*, 321−328.

220. Ikemoto, N., Nagy, B., Bhatnagar, G. M. & Gergely, J. (1974) *J. Biol. Chem. 249*, 2357−2365.

221. Cala, S. E. & Jones, L. R. (1983) *J. Biol. Chem. 258*, 11932−11936.

222. Jorgensen, A. O., Shen, A. C. Y., Campbell, K. P. & MacLennan, D. H. (1983) *J. Cell. Biol. 97*, 1573−1581.

223. Ostwald, T. J. & MacLennan, D. H. (1974) *J. Biol. Chem. 249*, 974−979.

224. Munro, S. & Pelham, H. R. B. (1987) *Cell 48*, 899−907.

225. Jorgensen, A. O., Kolnins, V., Subrzycka, E. & MacLennan, D. H. (1977) *J. Cell Biol. 74*, 287−298.

226. Thomas, K., Navarro, J., Benson, R. J. J., Campbell, K. P., Rotundo, R. L. & Fine, R. E. (1988) *J. Biol. Chem. 264*, 3140−3146.

227. Volpe, P., Zorzato, F., Pozzan, T., Salviati, G. & Di Virgilio, F. (1987) *Methods Enzymol. 141*, 3−18.

228. Fill, M. & Coronado, R. (1988) *Trends Neurosci. 11*, 453−457.

229. Fleischer, S. & Inui, M. (1989) *Ann. Rev. Biophys. Biophys. Chem. 18*, 333−364.

230. Meissner, G. (1984) *J. Biol. Chem. 259*, 2365−2374.

231. Meissner, G., Ruling, E. & Helveth, J. (1986) *Biochemistry 25*, 236−244.

232. Meissner, G. & Henderson, J. S. (1987) *J. Biol. Chem. 262*, 3065−3073.

233. Palade, P. (1987) *J. Biol. Chem. 262*, 6135−6141.

234. Volpe, P., Salviati, C., Di Virgilio, F. & Pozzan, T. (1985) *Nature 316*, 347−349.

235. Vergara, J., Tsien, R. Y. & Delay, M. (1985) *Proc. Natl Acad. Sci. USA 82*, 6352−6356.

236. Chu, A., Volpe, P., Costello, B. & Fleischer, S. (1986) *Biochemistry 25*, 8315−8324.

620

237. Fleischer, S., Ogunbunmi, E. M., Dixon, M. C. & Fleer, E. A. M. (1985) *Proc. Natl Acad. Sci. USA 82*, 7256—7259.
238. Rousseau, E., Smith, J., Henderson, J. S. & Meissner, G. (1986) *Biophys. J. 50*, 1009—1014.
239. Meissner, G. (1986) *J. Biol. Chem. 261*, 6300—6306.
240. Fabiato, A. (1983) *Am. J. Physiol. 245*, C1—C14.
241. Donaldson, S. K., Goldberg, N. D., Walseth, T. F. & Huttermann, D. A. (1987) *Biochim. Biophys. Acta 927*, 92—99.
242. Donaldson, S. K., Goldberg, N. D., Walseth, T. F. & Huttermann, D. A. (1988) *Biophys. J. 53*, 468a.
243. Rojas, E., Nassar-Gentina, V., Luxoro, M., Pollard, M. E. & Carrasco, M. A. (1987) *Can. J. Physiol. Pharmacol. 65*, 672—680.
244. Nosek, T. M., Williams, M. F., Zeigler, J. T. & Godt, R. E. (1986) *Am. J. Physiol. 259*, C807—C810.
245. Hidalgo, C., Carrasco, M. A., Magendzo, K. & Jaimovitch, E. (1986) *FEBS Lett. 202*, 69—73.
246. Di Virgilio, F., Volpe, P., Pozzan, T. & Salviati, G. (1986) *EMBO J. 5*, 259—262.
247. Varsanyi, M., Messner, M. & Brandt, N. R. (1989) *Eur. J. Biochem. 179*, 473—479.
248. Milani, D., Volpe, P. & Pozzan, T. (1988) *Biochem. J. 254*, 525—529.
249. Suarez-Isla, B. A., Irribarra, V., Oberhauser, A., Larralde, L., Bull, R., Hidalgo, C. & Jaimovitch, E. (1988) *Biophys. J. 54*, 737—741.
250. Walker, J. W., Somlyo, A. V., Goldman, Y. E., Somlyo, A. P. & Trentham, D. R. (1987) *Nature 327*, 249—252.
251. Pape, P. C., Konishi, M., Baylor, S. M. & Somlyo, A. P. (1988) *FEBS Lett. 235*, 57—62.
252. Vergara, J., Asotra, K. & Delay, M. (1987) in *Cell calcium and the control of membrane transport* (Mandel, L. J. & Eaton, D. G., eds) pp. 133—151, Rockefeller University Press, New York.
253. Hidalgo, C., Suarez-Isla, B. & Jaimovich, E. (1989) *Int. Congr. Physiol. Sci. Abstr.*, S2044.
254. Penner, R., Neher, E., Takeshima, H., Nishimura, S. & Numa, S. (1989) *FEBS Lett. 259*, 217—221.
255. Volpe, P., Pozzan, T. & Di Virgilio, F. (1988) in *Neuromuscular junction* (Sellin, L. C., Libelius, R. & Thesleff, S. H., eds) pp. 381—393, Elsevier, Amsterdam.
256. Smith, J. S., Coronado, R. & Meissner, G. (1985) *Nature 316*, 446—449.
257. Smith, J. S., Coronado, R. & Meissner, G. (1986) *J. Gen. Physiol. 88*, 573—588.
258. Bull, R., Marengo, J. J., Suarez-Isla, B. A., Donoso, P., Sutko, J. L. & Hidalgo, C. (1989) *Biophys. J. 56*, 749—756.
259. Imagawa, T., Smith, J. S., Coronado, R. & Campbell, K. P. (1987) *J. Biol. Chem. 262*, 1740—1747.
260. Inui, M., Saito, A. & Fleischer, S. (1987) *J. Biol. Chem. 262*, 15637—15643.
261. Lai, F. A., Erickson, H. P., Rousseau, E., Liu, Q. Y. & Meissner, G. (1988) *Nature 331*, 315—319.
262. Lai, F. A., Erickson, H. P., Rousseau, E., Liu, Q. Y. & Meissner, G. (1988) *Biochem. Biophys. Res. Commun. 151*, 441—447.
263. Hymel, L., Inui, M., Fleischer, S. & Schindler, H. (1988) *Proc. Natl Acad. Sci. USA 85*, 441—445.
264. Smith, J. S., Imagawa, T., Ma, J., Fill, M., Campbell, K. P. & Coronado, R. (1988) *J. Gen. Physiol. 92*, 1—26.
265. Ma, J., Fill, M., Knudson, M., Campbell, K. P. & Coronado, R. (1988) *Science 242*, 99—102.
266. Liu, Q.-Y., Lai, F. A., Rousseau, E., Jones, R. V. & Meissner, G. (1989) *Biophys. J. 55*, 415—424.
267. Saito, A., Inui, M., Radermacher, M., Frank, J. & Fleischer, S. (1988) *J. Cell. Biol. 107*, 211—219.
268. Wagenknecht, T., Grassucci, R., Frank, J., Saito, A., Inui, M. & Fleischer, S. (1989) *Nature 338*, 167—170.
269. Franzini-Armstrong, C. (1970) *J. Cell Biol. 47*, 488—499.
270. Takeshima, H., Nishimura, S., Matsumoto, T., Ishida, H., Kagawa, K., Minamino, N., Matsuo, H., Ueda, M., Hanaoka, M., Hirose, T. & Numa, S. (1989) *Nature 339*, 439—445.
271. Zorzato, F., Fuji, J., Otsu, K., Phillips, M., Green, M. N., Lai, F. A., Meissner, G. & MacLennan, D. H. (1989) *J. Biol. Chem. 265*, 2244—2256.
272. Block, B. A., Imagawa, T., Campbell, K. P. & Franzini-Armstrong, C. (1988) *J. Cell Biol. 107*, 2587—2600.
273. Schneider, M. F. & Chandler, W. K. (1973) *Nature 242*, 747—751.
274. Moore, L., Chen, T., Knapp, H. R. Jr & Landon, E. J. (1975) *J. Biol. Chem. 250*, 4562—5468.
275. Berridge, M. J. & Irvine, R. F. (1984) *Nature 312*, 315—321.
276. Berridge, M. J. (1987) *Annu. Rev. Biochem. 56*, 159—193.
277. Berridge, M. J. & Irvine, R. F. (1989) *Nature 341*, 197—205.
278. Pozzan, T., Arslan, P., Tsien, R. Y. & Rink, T. J. (1982) *J. Cell Biol. 94*, 335—345.
279. Streb, H., Irvine, R. F., Berridge, M. J. & Schultz, I. (1983) *Nature 306*, 67—69.
280. Volpe, P., Krause, K. H., Hashimoto, S., Zorzato, F., Pozzan, T., Meldolesi, J. & Lew, D. P. (1988) *Proc. Natl Acad. Sci. USA 85*, 1091—1095.
281. Hashimoto, S., Bruno, B., Lew, D. P., Pozzan, T., Volpe, P. & Meldolesi, J. (1988) *J. Cell Biol. 107*, 2523—2531.
282. Pozzan, T., Volpe, P., Zorzato, F., Bravin, M., Krause, K. H., Lew, D. P., Hashimoto, S., Bruno, B. & Meldolesi, J. (1988) *J. Exp. Biol. 139*, 181—193.
283. Jacob, R. (1990) *Biochim. Biophys. Acta*, in the press.
284. Meldolesi, J., Madeddu, L. & Pozzan, T. (1990) *Biochim. Biophys. Acta*, in the press.
285. Lytton, J. & MacLennan, D. H. (1988) *J. Biol. Chem. 263*, 15024—15031.
286. Burgoyne, R. D., Cheek, T. R., Morgan, A., O'Sullivan, A. J., Moreton, R. B., Berridge, M. J., Mata, M. A., Colyer, J., Lee, A. G. & East, J. M. (1989) *Nature 342*, 72—74.
287. Thevenod, F., Kemmer, T. P., Christian, A. L. & Shultz, I. (1989) *J. Membr. Biol. 107*, 263—275.
288. Thevenod, F., Dehlinger-Kremer, M., Kemmer, T., Christian, A. L., Potter, B. & Schultz, I. (1989) *J. Membr. Biol. 109*, 173—187.
289. Gunteski-Hablin, A. M., Greeb, J. & Shull, G. E. (1988) *J. Biol. Chem. 263*, 15032—15040.
290. Heilmann, C., Spamer, C. & Gerok, W. (1984) *J. Biol. Chem. 259*, 11139—11144.
291. Muallem, S., Becker, T. J. & Fimmel, C. T. (1987) *Biochem. Biophys. Res. Commun. 149*, 213—220.
292. Kaprelian, Z., Campbell, A. N. & Fambrough, D. H. (1989) *Mol. Brain Res. 6*, 55—60.
293. Worley, P. F., Baraban, J. M., Supattapone, S., Wilson, V. S. & Snyder, S. H. (1987) *J. Biol. Chem. 262*, 12132—12136.
294. Pozzan, T., Gatti, G., Dozio, N., Vincentini, L. M. & Meldolesi, J. (1984) *J. Cell Biol. 99*, 628—638.
295. Damiani, E., Spamer, C., Heilmann, C., Salvatori, S. & Margreth, A. (1988) *J. Biol. Chem. 263*, 340—343.
296. Oberdorf, J. A., Lebeche, D., Head, J. F. & Kaminer, B. (1988) *J. Biol. Chem. 263*, 6806—6809.
297. Krause, K. H. & Campbell, K. P. (1988) *FASEB J. 2*, 542—544.
298. Macer, D. R. J. & Koch, G. L. E. (1988) *J. Cell Sci. 91*, 61—70.
299. Treves, S., De Mattei, M., Lanfredi, M., Green, G., MacLennan, D. H., Meldolesi, J. & Pozzan, T. (1989) *Biochem. J.*, in the press.
300. Fliegel, L., Burns, K., Opas, M. & Michalak, M. (1989) *Biochim. Biophys. Acta 982*, 1—8.
301. Smith, M. J. & Koch, G. L. E. (1989) *EMBO J. 8*, 3581—3586.
302. Khanna, N. C., Tokuda, M. & Waisman, D. M. (1987) *Methods Enzymol. 139*, 36—50.
303. Fliegel, L., Burns, K., MacLennan, D. H., Reithmeier, R. A. F. & Michalak, M. (1989) *J. Biol. Chem. 269*, 21522—21528.
304. Van, P. N., Peter, F. & Solig, H. D. (1989) *J. Biol. Chem. 264*, 17494—17501.
305. Zagouras, P. & Rose, J. K. (1989) *J. Cell Biol. 109*, 2633—2640.
306. Supattapone, S., Worley, P. F., Baraban, J. M. & Snyder, S. H. (1988) *J. Biol. Chem. 263*, 1530—1534.
307. Ross, C. A., Meldolesi, J., Milner, T. A., Sato, T., Supattapone, S. & Snyder, S. H. (1989) *Nature 326*, 468—470.

308. Varney, M. A. & Watson, S. P. (1989) *7th International Conference on Cyclic nucleotides, calcium and protein phosphorylation*, abstr. 171.

309. Kamada, H., Yabe, Y., Kagawa, Y., Hirata, H. & Kusaka, I. (1989) *7th International Conference on Cyclic nucleotides, calcium and protein phosphorylation*, abstr. 173.

310. Furuichi, T., Yoshikawa, S., Miyawaki, A., Wada, K., Maeda, N. & Mikoshiba, K. (1989) *Nature 342*, 32 – 38.

311. Mallett, J., Huchet, M., Pougeois, R. & Changeaux, J. P. (1976) *Brain Res. 103*, 291 – 312.

312. Yamamoto, H., Maeda, N., Niinobe, M., Miyamoto, E. & Mikoshiba, K. (1989) *J. Neurochem. 53*, 917 – 923.

313. Mignery, G. A., Sudhof, T. C., Takey, K. & De Camilli, P. (1989) *Nature 342*, 192 – 195.

314. Ferris, C. D., Hunganir, R. L., Supattapone, S. & Snyder, S. H. (1989) *Nature 342*, 87 – 89.

315. Maeda, N., Niinobe, M., Inoue, Y. & Mikoshiba, K. (1989) *Dev. Biol. 133*, 67 – 76.

316. Satoh, T., Ross, C. A., Supattapone, S., Pozzan, T., Snyder, S. H. & Meldolesi, J. (1990) *J. Cell Biol. 111*, 615 – 624.

317. Spat, A., Bradford, P. G., MacKinney, J. S., Rubin, R. P. & Putney, J. W. (1986) *Nature 319*, 514 – 516.

318. Baukal, A. J., Guillemette, G., Rubin, R. P., Spat, A. & Catt, K. J. (1985) *Biochem. Biophys. Res. Commun. 133*, 532 – 538.

319. Champeil, P., Combette, L., Berthon, B., Ducet, E., Orlowski, S. & Claret, M. (1989) *J. Biol. Chem. 264*, 17665 – 17673.

320. Ehrlich, B. E. & Watras, J. (1988) *Nature 336*, 583 – 586.

321. Guillemette, G., Balla, T., Baukal, A. J., Spat, A. & Katt, K. J. (1987) *J. Biol. Chem. 262*, 1010 – 1015.

322. Dunlop, M. E. & Larkins, R. G. (1988) *Biochem. J. 247*, 407 – 415.

323. Prentki, M., Biden, T. J., Janjic, D., Irvine, R. F., Berridge, M. J. & Wollheim, C. B. (1984) *Nature 309*, 562 – 564.

324. Gill, D. L., Ueda, T., Chueh, S. H. & Noel, M. W. (1986) *Nature 320*, 461 – 464.

325. Mullaney, J. M., Chueh, S. H., Ghosh, T. K. & Gill, D. L. (1987) *J. Biol. Chem. 262*, 13865 – 13872.

326. Gosh, T. K., Mullaney, J. M., Tarazi, F. I. & Gill, D. L. (1989) *Nature 340*, 236 – 239.

327. Lipscombe, D., Madison, D. V., Poenie, M., Reuter, H., Tsien, R. W. & Tsien, R. Y. (1988) *Neuron 1*, 355 – 365.

328. Malgaroli, A. & Meldolesi, J. (1990) *J. Biol. Chem.*, in the press.

329. Crompton, M. (1985) *Curr. Top. Membr. Transp. 25*, 231 – 276.

330. Nicholls, D. G. & Akerman, K. (1982) *Biochim. Biophys. Acta 683*, 57 – 88.

331. Bragadin, M., Pozzan, T. & Azzone, G. F. (1979) *FEBS Lett. 104*, 347 – 351.

332. Sottocasa, G. L., Sandri, G., Panfili, E. & De Bernard, B. (1971) *FEBS Lett. 17*, 100 – 105.

333. Jeng, A. Y., Ryan, T. & Shamoo, A. E. (1978) *Proc. Natl Acad. Sci. USA 75*, 2125 – 2129.

334. Pozzan, T., Bragadin, M. & Azzone, G. F. (1977) *Biochemistry 16*, 5618 – 5624.

335. Crompton, M., Capano, M. & Carafoli, E. (1976) *Eur. J. Biochem. 69*, 453 – 462.

336. Nicholls, D. G. (1978) *Biochem. J. 176*, 463 – 474.

337. Bernardi, P. & Azzone, G. F. (1983) *Eur. J. Biochem. 134*, 377 – 383.

338. Vaghy, P. L., Johnson, J. D., Matlib, M. A., Wang, T. & Schwartz, A. (1982) *J. Biol. Chem. 257*, 6000 – 6002.

339. Rizzuto, S., Bernardi, P., Favaron, M. & Azzone, G. F. (1987) *Biochem. J. 246*, 271 – 277.

340. Somlyo, A. P., Bond, M. & Somlyo, A. V. (1985) *Nature 322*, 632 – 635.

341. MacCormak, J. G. & Denton, R. M. (1984) *Biochem. J. 218*, 235 – 247.

342. Lucaks, G. L. & Kapus, A. (1987) *Biochem. J. 248*, 609 – 613.

343. Biden, T. J., Wollheim. C. B. & Schlegel, W. (1986) *J. Biol. Chem. 261*, 7223 – 7229.

344. Eisen, A. & Reynolds, G. T. (1985) *J. Cell Biol. 100*, 1522 – 1527.

345. Carafoli, E. (1982) in *Pathophysiology of shock, anoxia and ischemia* (Cowley, R. A. & Trump, B. P., eds) pp. 95 – 112, Williams & Wilkins, Baltimore, London.

346. Kretzinger, R., Moncrief, N. D., Goodman, M. & Czelusniak, J. (1988) in *The calcium channel: structure function and implication* (Morad, M., Nayler, W., Kazda, S. & Schzamen, M., eds) pp. 16 – 38, Springer-Verlag, Berlin, Heidelberg.

347. Tsien, R. Y. (1988) *Trends Neurosci. 11*, 419 – 424.

348. Deuticke, H. J. (1934) *Hoppe-Seyler's Z. Physiol. Chem. 224*, 216 – 228.

349. Heizmann, C. W., Rohrenbeck, J. & Kamphuis, W. (1989) in *Calcium binding proteins in normal and transformed cells* (Pochet, R., Lawson, D. E. M. & Herzmann, C. W., eds) Plenum Press, New York, pp. 57 – 66.

350. Kretzinger, R. H. & Nockolds, C. E. (1973) *J. Biol. Chem. 248*, 3313 – 3326.

351. Klee, C. B. (1988) *Biochemistry 27*, 6645 – 6653.

352. Kwiatkowski, D., Stossel, T. P., Orkin, S. H., Mole, J. E., Colten, H. R. & Yin, H. L. (1986) *Nature 323*, 455 – 458.

353. Parker, P. J., Coussens, L., Tolty, N., Rhee, L., Young, S., Chen. E., Sabel, S., Waterfield, M. D. & Ullrich, A. (1986) *Science 233*, 853 – 866.

354. Nishizuka, Y. (1988) *Nature 334*, 661 – 665.

355. Weeds, A. G. & MacLachlan, A. D. (1974) *Nature 252*, 646 – 649.

356. Heizmann, C. W. (1984) *Experientia 40*, 910 – 921.

357. Wnuk, W. (1988) in *In Calcium and calcium binding proteins* (Gerday, H. C., Gilles, R. & Bolis, L., eds) pp. 46 – 48, Springer-Verlag, Berlin, Heidelberg.

358. Heizmann, C. W. & Celio, M. R. (1987) *Methods Enzymol. 139*, 552 – 557.

359. Donato, R. (1986) *Cell Calcium 7*, 123 – 145.

360. MacManus, J. P. & Brewer, L. M. (1987) *Methods Enzymol. 139*, 156 – 168.

361. Hunziker, W. (1986) *Proc. Natl Acad. Sci. USA 83*, 7578 – 7582.

362. Baker, P. F. & Sclaepfer, W. W. (1978) *J. Physiol. 276*, 103 – 125.

363. Cheung, W. Y. (1970) *Biochem. Biophys. Res. Commun. 38*, 533 – 538.

364. Kakiuki, S. & Yamazaki, R. (1970) *Biochem. Biophys. Res. Commun. 41*, 1104 – 1110.

365. Cheung, W. Y. (1980) *Science 207*, 19 – 27.

366. Means, A. R., Tash, J. S. & Chafouleas, J. G. (1982) *Physiol. Rev. 62*, 1 – 39.

367. Epstein, P., Simmen, R. C. M., Tanaka, T. & Means, A. R. (1987) *Methods Enzymol. 139*, 217 – 229.

368. Babu, Y. S., Sack, J. S., Greenbough, T. J., Bugg, C. E., Means, A. R. & Cook, W. J. (1985) *Nature 315*, 37 – 40.

369. Burger, D., Cox, J. A., Compte, M. & Stein, E. A. (1984) *Biochemistry 23*, 1966 – 1971.

370. Cox, J. A. (1988) *Biochem. J. 249*, 621 – 629.

371. Vogel, H. J. (1988) in *Handbook of experimental pharmacology. Calcium in drug action* (Baker, P. E., ed.) pp. 57 – 87, Springer-Verlag, Berlin, Heidelberg.

372. Teleman, A., Drakenberg, T. & Forsen, S. (1986) *Biochim. Biophys. Acta 873*, 204 – 213.

373. Martin, S. R., Andersson-Teleman, A., Bayley, M., Drakenberg, T. & Forsen, S. (1985) *Eur. J. Biochem. 151*, 543 – 550.

374. Minowa, O. & Yagi, K. (1984) *J. Biochem. 96*, 1175 – 1182.

375. Milos, M., Schaer, J. J., Compte, M. & Cox, J. A. (1986) *Biochemistry 25*, 6279 – 6287.

376. Burger, D., Stein, E. A. & Cox, J. A. (1983) *J. Biol. Chem. 258*, 14733 – 14739.

377. Malnoe, A., Cox, J. A. & Stein, E. A. (1982) *Biochim. Biophys. Acta 714*, 84 – 92.

378. Huang, C. Y. & King, M. M. (1985) *Curr. Top. Cell Regul. 27*, 437 – 446.

379. Kemp, B. E., Pearson, R. B., Guerriero, V. Jr & Means, A. R. (1987) *J. Biol. Chem. 262*, 2542 – 2548.

380. Klee, C. B. & Vanaman, T. C. (1982) *Adv. Protein Chem. 35*, 213 – 321.

381. Chafouleas, J. G., Bolton, W. E., Hidaka, H., Boyd, A. E. & Means, A. R. (1982) *Cell 28*, 41 – 50.

622

382. Sasaki, Y. & Hidaka, H. (1982) *Biochem. Biophys. Res. Commun.* *104*, 451−456.
383. Chafouleas, J. G., Lagace, L., Bolton, W. E., Boyd, A. E. & Means, A. R. (1984) *Cell 36*, 73−81.
384. Chafouleas, J. C., Pardue, R. L., Brinkley, B. R. & Means, A. R. (1981) *Proc. Natl Acad. Sci. USA 78*, 996−1000.
385. Connor, C. G., Moore, P. B., Brady, R. C., Horn, J. P., Arlinghaus, R. B. & Dedman, J. R. (1983) *Biochem. Biophys. Res. Commun. 112*, 647−654.
386. La Porte, D. C., Gidwitz, S., Weber, M. J. & Storm, D. R. (1980) *Biochem. Biophys. Res. Commun. 86*, 1169−1177.
387. Chafouleas, J. C., Bolton, W. E. & Means, A. R. (1984) *Science 224*, 1346−1348.
388. Takeda, T. & Yamamoto, M. (1987) *Proc. Natl Acad. Sci. USA 84*, 3580−3584.
389. Rasmussen, C. D. & Means, A. R. (1987) *EMBO J. 6*, 3961−3968.
390. Nairn, A., Hemmings, H. & Greengard, P. (1985) *Annu. Rev. Biochem. 54*, 931−976.
391. Schworer, C. M., Colbran, R. J. & Soderling, T. R. (1986) *J. Biol. Chem. 261*, 8581−8584.
392. Ebashi, S. (1963) *Nature 200*, 1010−1011.
393. Kohama, K. (1979) *J. Biochem. 88*, 591−599.
394. Ebashi, S. & Ogawa, Y. (1988) in *Handbook of Experimental Pharmacology. Calcium in drug action* (Baker, P. F., ed.) pp. 31−56, Springer-Verlag, Berlin, Heidelberg.
395. Iida, S. (1988) *J. Biochem. 103*, 482−486.
396. Herzberg, O. & James, M. N. G. (1985) *Nature 313*, 653−689.
397. Van Eerde, J. P. & Takahashi, K. (1976) *Biochemistry 15*, 1171−1180.
398. Berridge, M. J., Cobbold, P. H. & Cuthbertson, K. S. R. (1988) *Philos. Trans. R. Soc. Lond. B Biol. Sci. 320*, 325−343.
399. Persechini, A., Moncrief, N. D. & Kretsinger, R. H. (1989) *Trends Neurosci. 12*, 462−467.
400. Valdivia, C., Valdivia, H. H., Potter, B. V. L. & Cozonado, R. (1990) *Biophys. J. 57*, 1233−1243.
401. Furnichi, T., Shiota, C. & Mikoshiba, K. (1990) *FEBS Lett. 267*, 85−88.
402. Otsu, H., Yemamoto, A., Maede, N., Mikoshibe, K. & Toshiro, Y. (1990) *Cell Struct. Funct. 15*, 163−173.

Eur. J. Biochem. *194*, 1−8 (1990)
© FEBS 1990

Review

The inhibitory glycine receptor:
A ligand-gated chloride channel of the central nervous system

Dieter LANGOSCH, Cord-Michael BECKER and Heinrich BETZ
Zentrum für Molekulare Biologie, Universität Heidelberg, Federal Republic of Germany

(Received May 7, 1990) − EJB 90 0512

The postsynaptic glycine receptor (GlyR) is a major inhibitory chloride channel protein in the central nervous system. The affinity-purified receptor contains polypeptides of 48 kDa, 58 kDa, and 93 kDa. The 48-kDa (α) and 58 kDa (β) subunits span the postsynaptic membrane in a pentameric arrangement to form the anion channel of the receptor. The 93-kDa polypeptide is cytoplasmically localized and may have an anchoring function. Molecular cloning revealed that different structural characteristics are shared by the membrane-spanning subunits of the GlyR and those of other ligand-gated ion channel proteins. Developmental regulation of the GlyR is characterized by alterations in antagonist binding, heterogeneity of α subunits, and increased levels of the 93-kDa polypeptide. Glycine receptor function can be reconstituted by expression of cloned α subunits in heterologous cell systems. Positive charges found at the presumed mouths of the GlyR channel appear to be important determinants of ion selectivity. These data establish the anion-conducting GlyR as a homolog of other ligand-gated ion channel proteins and suggest that the diversity of these channels originates from divergent evolution of a primordial channel protein early in phylogeny.

Rapid communication between excitable cells crucially depends on transmembrane ion fluxes through selective ion channel proteins. Their different gating mechanisms, i.e. activation by changes in transmembrane potential, second messengers, or binding of extracellular ligands correspond to different stages of the intercellular signalling cascade. A wave of depolarization travelling towards the synaptic terminal of the axon results from the temporarily coupled opening and closing of voltage-sensitive sodium and potassium channels [1]. At the nerve ending, depolarization induces calcium influx through voltage-sensitive channels to trigger release of neurotransmitters into the synaptic cleft [2]. Subsequent binding of released neurotransmitters to specific receptors in the postsynaptic membrane induces a variety of physiological effects depending on the effector system activated. Whereas stimulation of second messenger systems by G-protein-coupled receptors is a relatively slow (seconds) process, activation of ligand-gated ion channel proteins changes the membrane potential within milliseconds [3]. Allowing for flux rates of up to 10^7 ions/s, ligand-gated ion channels are among the most efficient proteins known [4].

For a sensible transduction of signals, the neuronal membrane potential has to be altered in a graded way. This is accomplished by the concerted action of excitatory and inhibitory receptors. Excitation, i.e. membrane depolarization, usually results from transient sodium influx through ligand-gated cation channels. Concurrent chloride influx elicited via ligand-gated anion channels antagonizes membrane depolarization and thus suppresses neuronal firing (reviewed in [4]).

In 1965, glycine was first proposed to act as a neurotransmitter in mammalian spinal cord [5]. Subsequently, release of this amino acid upon stimulation [6], its inhibitory action on moto-neurons [7, 8] and its high-affinity uptake by spinal cord neurons and synaptosomes established glycine as a major inhibitory transmitter in the central nervous system (reviewed in [9]). Both glycine and the other major inhibitory amino acid, γ-aminobutyric acid (GABA), produce an increase in chloride permeability of the neuronal membrane [10 − 12].

A receptor specific for glycine which differs from the inhibitory γ-aminobutyric acid receptor both in pharmacology and localization has been identified previously by ligand binding studies [13, 14]. To elucidate the structure of the inhibitory glycine receptor (GlyR), this protein has been isolated from mammalian central nervous system and its primary structure determined by cDNA cloning [15]. Here, we review the data currently available on structure/function relationships pertinent to this ligand-gated anion channel protein and its regulation during mammalian development.

The GlyR is a pentameric channel protein

Ligand binding properties

The GlyR is activated by a number of α- and β-amino acids (Fig. 1). Their potency in eliciting inhibition of spinal cord neurons decreases in the order glycine $>$ β-alanine $>$ taurine $\gg$ L-alanine, L-serine $>$ proline [8, 16, 17]. Concerning

---

Correspondence to H. Betz, ZMBH, Im Neuenheimer Feld 282, W-6900 Heidelberg, Federal Republik of Germany
Abbreviation. GlyR, glycine receptor.

2

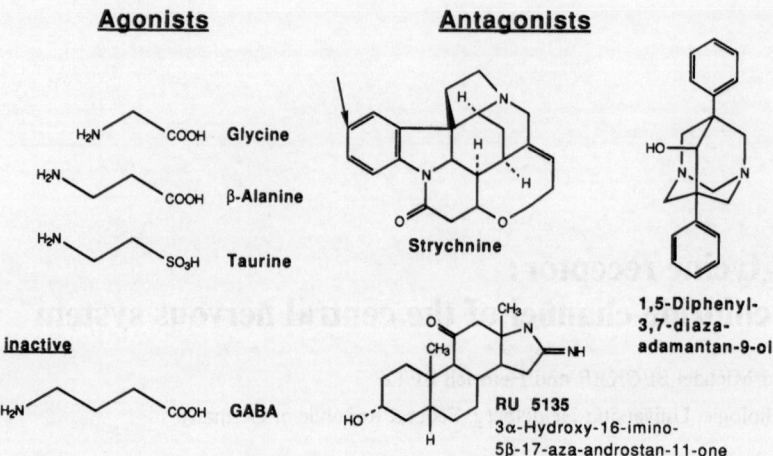

Agonists

H₂N—COOH **Glycine**

H₂N—COOH **β-Alanine**

H₂N—SO₃H **Taurine**

inactive

H₂N—COOH **GABA**

Antagonists

Strychnine

1,5-Diphenyl-3,7-diaza-adamantan-9-ol

RU 5135
3α-Hydroxy-16-imino-5β-17-aza-androstan-11-one

Fig. 1. *Structures of glycinergic ligands.* The order of ω-amino acids reflects decreasing affinity to the GlyR. RU-5135 displays the highest affinity ($K_d \approx 3$ nM) of all glycinergic ligands known. Binding of strychnine to the GlyR is not affected by α-amidation at the 2-position of the benzene ring indicated by the arrow. GABA = γ-aminobutyric acid

antagonists, there are only a few compounds which bind to the GlyR with nanomolar affinities (Fig. 1): the plant alkaloid strychnine and its derivatives, the steroid RU 5135, and 1,5-diphenyl-3,7-diazaadamantan-9-ol (reviewed in [15]). In vertebrates, strychnine poisoning abolishes glycinergic inhibition by blocking GlyR function. The consequence is overexcitation of the motor system resulting in muscular convulsions. Binding of [³H]strychnine to spinal cord membranes is antagonized by the above-mentioned amino acids in their relative order of inhibitory potency. This indicates a largely competitive interaction of these GlyR agonists and antagonist. Further, protein modification and variations in ionic conditions affect the displacement of [³H]strychnine by glycine, but not by the unlabelled antagonist [14, 18]. The binding sites for glycine and strychnine are therefore assumed to be closely related but not identical. This view is compatible with the hypothesis that part of the strychnine molecule resembles glycine in topology and electronic charge distribution [19].

The glycine response produced in *Xenopus* oocytes injected with poly(A)-rich RNA from brain or spinal cord is characterized by Hill coefficients of $n = 2.7-3.3$ [20, 21]. This suggests that at least three glycine molecules are required to activate the GlyR channel. The physiological meaning of cooperativity in agonist binding to channel proteins is not clear. It is commonly argued that the steep dose/response curve characteristics of cooperatively activated receptors illustrates their sensible response to changes in agonist concentration. More recently, theoretical considerations have suggested that binding of more than one agonist molecule to a receptor is necessary to provide sufficient energy for channel activation [22].

Functional characteristics of the chloride channel

Single channel properties of the GlyR have been investigated in detail by Sakmann and coworkers using patch clamp analysis of cultured neurons dissociated from fetal mouse spinal cord [12, 23]. In membrane patches isolated from these cells, glycine elicits bursts of single channel activity displaying a variety of subconductance states with a main conductance of 45 pS in 145 mM chloride solution. This channel is ideally anion-selective; its open diameter has been determined to be 0.52 nm by biionic reversal potential measurements. Com-

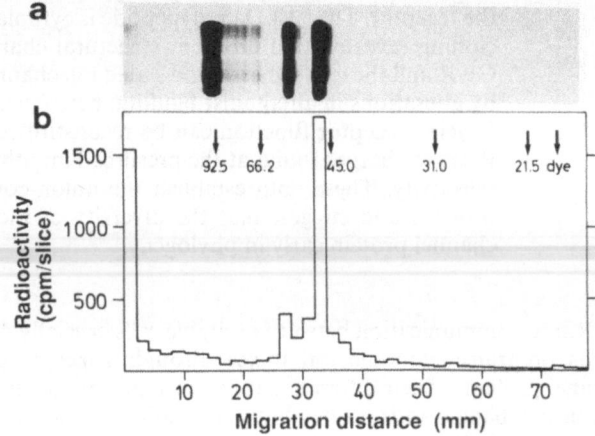

Fig. 2. *Biochemical properties of the GlyR.* (a) Polypeptide pattern of the affinity-purified receptor after SDS/polyacrylamide gel electrophoresis and silver staining. (b) Incorporation of [³H]strychnine into GlyR polypeptides upon ultraviolet-induced crosslinking. Incorporated ³H radioactivity was determined after electrophoresis by counting individual gel slices. Note that the radioactive label exhibits a peak at the position of the α subunit

parison of conductance and permeability sequences showed that ions with a high permeability (e.g. SCN⁻) have a low conductance and vice versa. In mixtures of permeants anomalous mole fraction behaviour of the single channel conductance was observed. The latter findings strongly suggest the existence of at least two sequentially occupied anion binding sites within the channel [12].

Analysis of glycine-induced steady-state currents in the whole cell configuration revealed a pronounced outward rectification of the GlyR channel [12, 24]. As, however, the instantaneous I/V relations in voltage-jump experiments were linear, this voltage dependency is to be attributed to the gating mechanism. In other words, the open probability of the channel is low when the cytoplasmic face of the plasma membrane is negatively charged, but rises with membrane depolarization. This is a meaningful property if one recalls that inhibitory chloride fluxes are to antagonize depolarization of the cell.

Purification of the GlyR

Receptors involved in synaptic signal transmission are usually rare proteins. Frequently, however, they exhibit high affinities to naturally occurring toxins, allowing the development of efficient purification schemes. In the case of the GlyR, its high-affinity binding of strychnine was exploited for affinity purification. Upon solubilization from rat spinal cord membrane fractions by the nonionic detergent Triton X-100, the strychnine binding complex was originally shown to behave as a large glycoprotein [25]. Subsequently, its purification to homogeneity was achieved in a one-step procedure using an affinity matrix consisting of 2-aminostrychnine coupled to agarose beads [26]. SDS/polyacrylamide gel electrophoresis separates the purified GlyR into three polypeptides of 48 kDa, 58 kDa, and 93 kDa (Fig. 2a). The same subunit composition was found for GlyR preparations from pig [27] and mouse [28].

By ultraviolet illumination, [³H]strychnine was covalently incorporated into the 48-kDa polypeptide of affinity-purified GlyR (Fig. 2b, and see [26, 27]). Glycine prohibited this labelling by strychnine, thus defining the 48-kDa polypeptide as the ligand-binding α subunit of the GlyR. Glycine-displaceable photoaffinity labelling of the α subunit with [³H]strychnine was also shown using spinal cord membranes [28−30]. Subsequent proteolytic digestion of the labelled membranes removed an 11-kDa extracellular fragment, but not the radioactive label. Strychnine is thus assumed to bind to a domain close to the transmembrane regions of the α subunit [30].

Proteolytic fingerprinting of purified GlyR polypeptides revealed similar digestion patterns of the 48-kDa (α) and 58-kDa (β) subunits. Also, monoclonal antibodies were raised against purified GlyR. One of them (mAb 4a) recognizes primarily the α polypeptide but also binds to the β chain with lower affinity. These experiments formed the basis for early speculations that the membrane-spanning α and β subunits of the GlyR have a similar primary structure and are evolutionarily related [31].

Upon blotting of purified GlyR onto nitrocellulose, only the α and β subunits, but not the 93-kDa polypeptide, bound the lectins concanavalin A and wheat germ agglutinin [27, 32]. Quantitative precipitation of purified GlyR is obtained with immobilized concanavalin A. Only partial precipitation was, however, accomplished by wheat germ agglutinin Sepharose (D. Langosch and W. Hoch, unpublished), suggesting that the α and β subunits are glycosylated with a high-mannose type of carbohydrate.

The 93-kDa polypeptide is extracted from spinal cord membranes with reagents suitable to remove peripheral membrane proteins, i.e. alkaline pH or dimethyl maleic anhydride [32]. Immunoelectron microscopy demonstrated its cytoplasmic localization at postsynaptic membranes [33, 34]. It thus appears that the 93-kDa polypeptide is a peripheral component associated with cytoplasmic domains of the GlyR subunits. On sucrose gradients, a large complex (> 15 S), predominantly consisting of 93-kDa polypeptide, migrates separately from a structure containing only α and β subunits [32]. The molecular mass of the latter has been determined to about 250 kDa and it is thought to represent the ligand-binding 'core' of the GlyR containing its intrinsic chloride channel [15, 36].

The function of the 93-kDa polypeptide is not clear. Its cytoplasmic localisation as well its propensity to form large aggregates are consistent with an anchoring function to immobilize the GlyR complex in the postsynaptic membrane [32, 37].

Quaternary structure

The subunit arrangement of the GlyR has been investigated by intramolecular crosslinking of its polypeptides using reagents of varying side-chain specificities and lengths [37]. Crosslinking of the purified protein produced a series of adducts up to an apparent molecular mass of 260 kDa which, by compositional analysis, were shown to contain the α and β subunits, but not the 93-kDa polypeptide. Based on these data, the largest adduct was inferred to represent the completely crosslinked core of the GlyR existing as a pentameric structure. The same pattern of adducts was generated upon crosslinking synaptic membrane preparations, indicating preservation of subunit composition during the purification procedure. As the α subunit consistently predominates over the β polypeptide in all species examined [26−28], a model of the GlyR (as shown in Fig. 3a) is proposed. Accordingly, three copies of the α subunit assemble with two copies of the β subunit to form the ion channel of the receptor [37].

A pentameric structure of membrane-spanning subunits is well established for the most thoroughly characterized ligand-gated ion channel, the cation-selective nicotinic acetylcholine receptor of *Torpedo* electric organ and skeletal muscle [38−40]. We have proposed, therefore, that a quasisymmetrical pentameric arrangement of polypeptides around a central ion channel may be a conserved architectural feature of ligand-gated ion channel proteins [37].

Subunit primary structures define the GlyR as a member of the ligand-gated channel protein superfamily

Screening of cDNA libraries with oligonucleotides derived from partial amino acid sequences of the α subunit allowed isolation of corresponding cDNA clones [41]. Analysis of the deduced primary structure revealed several features which are shared with subunits of other ligand-gated ion channel proteins, the nicotinic acetylcholine receptor (e.g. [42−44]) and the γ-aminobutyric acid receptor [45]. A tentative folding model predicted from the sequence data is shown in Fig. 4. It has the following features.

a) Hydropathy analysis of the amino acid sequence revealed four hydrophobic segments (M1−M4) long enough to span the hydrophobic core of a lipid bilayer as α-helices. Their positions are almost identical to those in the subunits of the nicotinic acetylcholine receptor and the γ-aminobutyric acid receptor, suggesting a similar transmembrane topology of the respective polypeptides.

b) Within segment M1, a proline residue is shared by all ligand-gated ion channel subunits. A kink introduced by this residue into the proposed α-helical structure of M1 has been proposed to provide structural flexibility required for ion channel gating [41, 45].

c) In the large extracellular domain, a pair of cysteine residues at positions 138 and 152 is conserved within subunits of ligand-gated ion channel proteins. In the nicotinic acetylcholine receptor, a disulfide bridge between these residues has been demonstrated [46] and, based on site-directed mutagenesis, implied to be essential for stabilization of tertiary structure and/or subunit assembly [47]. Also, a potential *N*-glycosylation site is found within the N-terminal region in accord with the glycoprotein nature of this subunit.

4

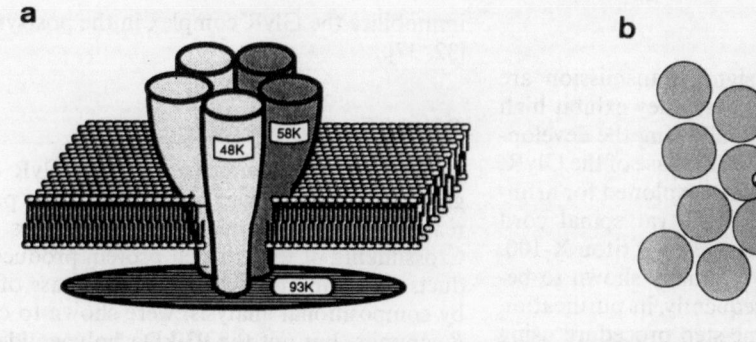

Fig. 3. *A model of the postsynaptic GlyR.* (a) The pentameric receptor complex is composed of multiple copies of the 48-kDa α subunit and 58-kDa β subunit; their nearest-neighbour relationships are tentative. The copurifying 93-kDa polypeptide is localized at the cytoplasmic face of the postsynaptic membrane. (These proteins are indicated by 48 K, 58 K and 93 K in the figure.) (b) Cross-section through the transmembrane core showing a putative arrangement of membrane-spanning segments within the chloride channel region. Positive charges located at the channel mouths are thought to confer anion selectivity. 5.2 Å = 0.52 nm

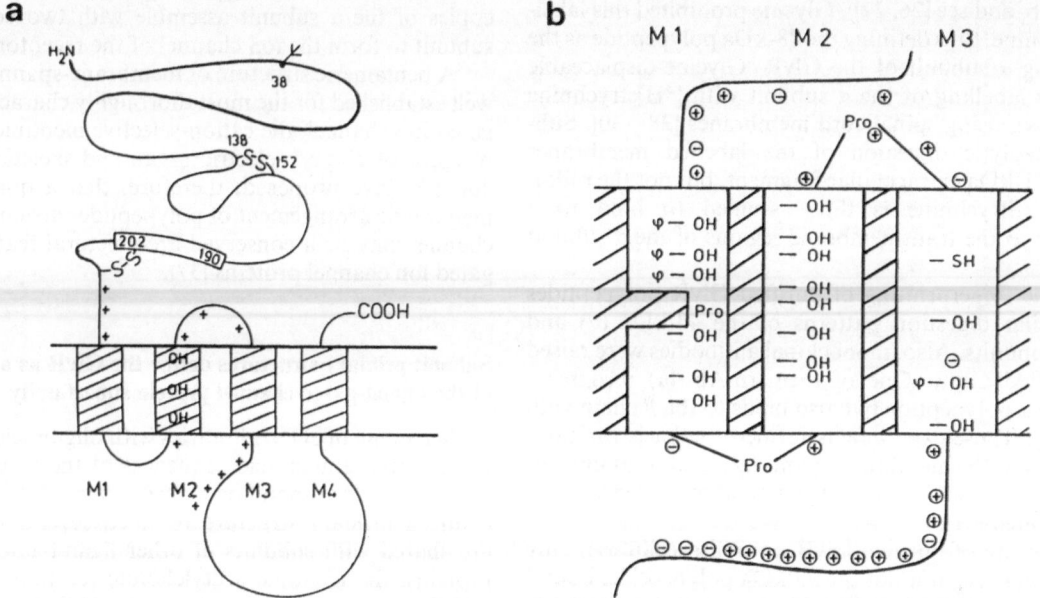

Fig. 4. *Model of the α subunit.* (a) Predicted transmembrane topology. ⌀ denotes a potential glycosylation site. (b) Enlargement of segments M1 − M3 showing the charged and hydroxylated amino acids contained within these domains. -OH, serine or threonine residues; φ-OH, tyrosine; + and −, charged amino acid residues

d) A hydrophilic loop separating transmembrane segments M3 and M4 provides most of the cytoplasmically localized mass of the transmembrane polypeptide.

These features are also conserved in several recently analyzed variants of the α subunit (see below). More recently, partial amino acid sequences were also obtained from the 58-kDa β chain, and corresponding oligo-nucleotides used to identify the gene [48]. The deduced primary sequence exhibits considerable similarity to the α subunit (47% identical amino acids) and displays the same molecular organization as outlined above.

In conclusion, all transmembrane subunits of the GlyR share the same principal organisation of the polypeptide chains including four conserved hydrophobic domains. In contrast, a preliminary analysis of the primary structure of the 93-kDa polypeptide indicates a highly hydrophilic protein without membrane-spanning segments (P. Prior, G. Grenningloh and H. Betz, unpublished data).

Glycine receptor subtypes

Developmental heterogeneity of the GlyR

Recent biochemical and cDNA sequence data have established subtype diversity as a widespread phenomenon for different types of receptors. For the GlyR, subtype heterogeneity was first detected during rodent spinal cord development [35]. There, a neonatal isoform (GlyR$_N$) prevalent at birth differs from the adult receptor (GlyR$_A$) in pharmacological, immunological and biochemical properties. In particular, GlyR$_N$ is characterized by a low strychnine binding affinity, and thus has escaped detection in previous [³H]strychnine binding experiments. Furthermore, its α subunit differs from the adult 48-kDa polypeptide in both molecular mass (49 kDa) and antigenic epitopes [35]. Within 2 − 3 weeks after birth, GlyR$_N$ is replaced in spinal cord by the adult form of the receptor, resulting in a postnatal increase of high-affinity [³H]strychnine

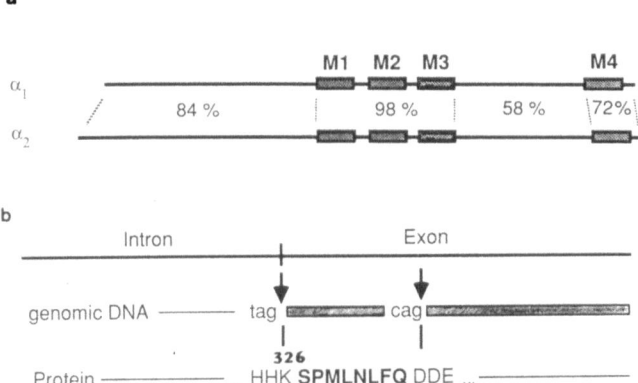

Fig. 5. *Heterogeneity of the α subunit*. (a) α_1- and α_2-subunit isoforms share transmembrane topology and a highly similar primary sequence (amino acid identity is given for different segments of the polypeptide chain). (b) Alternative splicing contributes to α-subunit heterogeneity. By alternate splice acceptor site selection an α-subunit variant (α_{ins}) is created containing an eight-amino-acid insertion between transmembrane segments M3 and M4

binding [35, 49] (and C.-M. Becker and H. Betz, unpublished data).

Cultured neurons from fetal rodent spinal cord are highly sensitive to glycine application [50] and therefore have been widely used for analysing the GlyR channel by the patch clamp technique [12]. Interestingly, the GlyR protein expressed in these cultures is predominantly of the neonatal isotype, a finding which facilitated its biochemical characterization [51]. $GlyR_N$ purified from [^{35}S]methionine-labelled cultures displays only a major polypeptide band of 49 kDa, suggesting a possible homo-oligomeric structure of the neonatal receptor. Furthermore, pulse labelling experiments showed that $GlyR_N$ is a metabolically stable protein (half-life $\approx$ 2 days) which, after polypeptide synthesis, is only slowly assembled into a ligand binding complex (time of assembly $\approx$ 30 min; see [51]).

Evidence for developmental heterogeneity of GlyRs also comes from expression studies. Injection of poly(A)-rich RNA isolated from rat cortex or spinal cord into *Xenopus* oocytes generates glycine-gated chloride channels in the oocyte membrane (see below). By sucrose density centrifugation, two classes of GlyR mRNA could be separated that both gave rise to functional channels upon oocyte expression [52]. A rapidly sedimenting mRNA was abundant in spinal cord of adult animals, whereas preparations from both neonatal spinal cord and adult cortex mainly contained a 'light' GlyR mRNA. The glycine-induced currents produced by these fractions were sensitive to the antagonists strychnine and picrotoxin, but differed in glycine sensitivity and desensitization kinetics [52].

Hybrid arrest experiments employing antisense oligonucleotides [53] complementary to the rat α-subunit cDNA sequence [41] allowed a further differentiation of the high-molecular-mass mRNA species. Whereas expression of this mRNA fraction from adult spinal cord was efficiently suppressed by the antisense oligonucleotides, the same fractions from neonatal poly(A)-rich RNA preparations efficiently produced GlyR channels under the same conditions [53]. Thus, at least three different mRNA species encoding functional GlyR were proposed to occur in the rat CNS.

α-Subunit variants

The heterogeneity of GlyRs observed at the protein and mRNA expression level reflects the existence of several hom-

ologous GlyR α-subunit genes. By screening cDNA and genomic libraries under conditions of low stringency, variants of the originally isolated GlyR α subunit (now termed α_1, see [54]) have been isolated. The novel α_2-subunit sequences predicted from human and rat cDNAs display 79% amino acid identity to their α_1 counterparts (Fig. 5a) and probably correspond to ligand binding subunits of the neonatal GlyR, as α_2 transcripts are abundant in spinal cord of newborn rats [54, 55]. Upon heterologous expression in *Xenopus* oocytes or a human embryonic kidney cell line, all these α subunits assemble into functional glycine-gated chloride channels [21, 54, 56]. With the exception of channels formed by the rat α_2 polypeptide, the expressed α_1 and α_2 receptors are highly sensitive to strychnine. Furthermore, all α-subunit channels respond to the classical glycinergic agonists β-alanine and taurine but with significantly different efficacy. In other words, GlyR α subunit heterogeneity may be important for creating functional diversity of the GlyR. In addition to these subunit variants, an α_3 sequence has been isolated from rat [57], and clones encoding exons of a fourth GlyR polypeptide, α_4, have recently been identified in mouse genomic libraries (J. Kuhse and Y. Maulet, unpublished data). Thus, considerable diversity of ligand binding subunits exists for the GlyR. In conclusion, not only one, but several GlyRs displaying different pharmacological characteristics appear to exist in the vertebrate central nervous system.

Alternative splicing further contributes to GlyR heterogeneity in spinal cord. For the rat α_1 subunit, a variant cDNA has been identified that originates from alternate splice acceptor site selection at an exon encoding the cytoplasmic domain adjacent to transmembrane segment M3 (Fig. 5b). An insertion containing eight additional residues may create novel sites for protein modification and/or association to the cytoskeleton. S1 nuclease mapping and polymerase chain reaction indicate expression of this splice variant at all stages of postnatal development (M. L. Malosio, G. Grenningloh, J. Kuhse, V. Schmieden, B. Schmitt, P. Prior and H. Betz, unpublished results).

Reconstitution of GlyR function

Heterologous expression

Expression in heterologous cell systems followed by electrophysiological analysis is now widely used to investigate functional properties of different receptors and channel proteins. *Xenopus* oocytes are capable of translating and assembling functional glycine receptors, as glycine-gated chloride currents can be recorded after injection of spinal cord and brain poly(A)-rich RNA [20, 21, 52]. Recently, the oocyte expression system has been used to characterize the GlyR channels formed by expression of cloned rat and human α subunits [21, 54]. Upon injection of the corresponding cRNA and 2 – 4 days of expression, glycine application elicited large whole cell currents which were blocked by nanomolar concentrations of strychnine and reversed at the equilibrium potential for chloride. This behaviour parallels results obtained with poly(A)-rich RNA fractions and corresponds to functional characteristics of the GlyR as determined by patch clamp analysis of intact neurons [12, 23]. Obviously, GlyR α subunits are sufficient to form a fully functional receptor.

Transient expression of the rat α_1 subunit has also been accomplished by transfection of cultured mammalian cells [56]. Again, electrophysiology revealed the formation of glycine-gated chloride channels closely resembling their natural

6

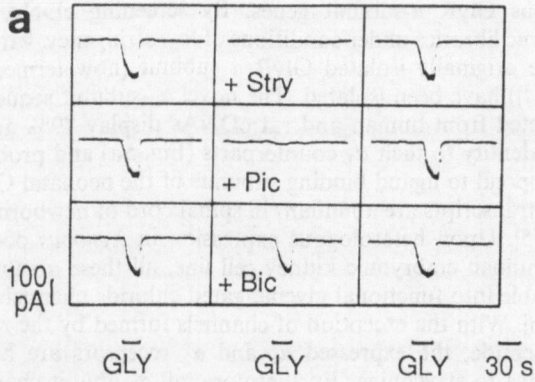

a

+ Stry

+ Pic

+ Bic

100 pA|

GLY   GLY   GLY   30 s

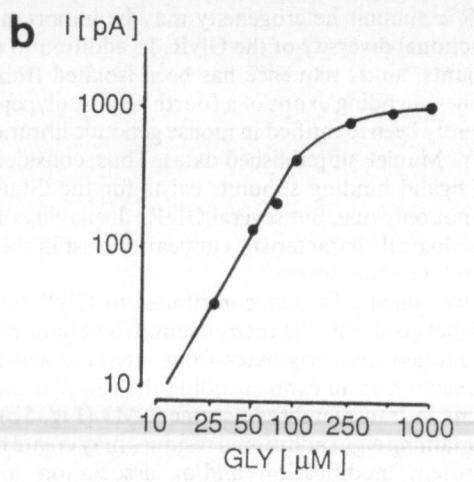

b   I [pA]

1000

100

10

10 25 50 100 250 1000
GLY [µM]

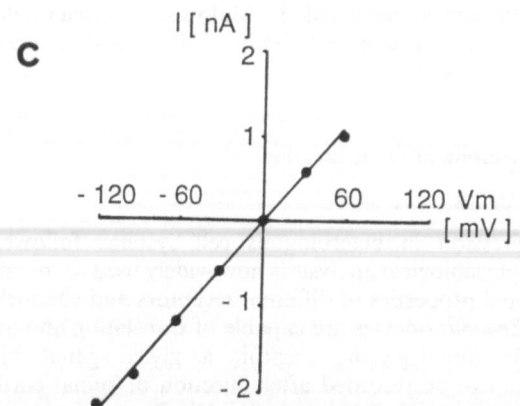

c

I [nA]

2

1

- 120 - 60   60   120 Vm
                          [mV]

- 1

- 2

Fig. 6. *Functional properties of the GlyR* α_1 *subunit expressed in a human kidney cell line.* (a) Glycine-elicited whole cell currents are reversibly blocked by strychnine (Stry, 100 µM) and picrotoxin (Pic, 100 µM), but not by the γ-aminobutyric acid receptor antagonist bicuculline (Bic, 50 µM). (b) Dose/response curve for glycine-evoked peak currents. (c) Current/voltage relationship of responses induced by 100 µM glycine. Current reversal close to the chloride equilibrium potential in these cells (≈ 0 mV) indicates anion selectivity of the channel. The outward rectification displayed by native GlyR (as discussed in text) appears to be only marginal upon expression of single subunits

counterpart (Fig. 6). Expression in cell lines also allows biochemical characterisation of the expressed protein, as large quantities of membranes can be collected. After solubilization

of transfected cells, gradient sedimentation revealed a significant fraction of the expressed α_1 subunit in complexes with hydrodynamic properties indistinguishable from those of native GlyR. It thus appears that GlyR α subunits are capable of assembling into a functional homo-oligomeric channel protein.

Reconstitution of the purified receptor

Functional reconstitution of the GlyR protein was reported by Mayor and coworkers [58]. There, affinity-purified receptor was incorporated into lipid vesicles. The rate of anion influx into these vesicles as determined by dye fluorescence quenching was accellerated by the glycinergic agonists glycine, β-alanine, and taurine; this agonist-induced flux was blocked by different GlyR antagonists. Thus, the basic functional properties of the GlyR are preserved during purification.

Reconstitution of a peptide model

In a different experimental approach, an attempt was made to identify transmembrane domains involved in channel formation by synthesizing individual segments of the membrane-spanning portion of the α_1 subunit. It is generally assumed that the lining of ion channels is provided by an assembly of α-helical transmembrane segments. Within the hydrophobic domain of nicotinic acetylcholine receptor subunits, segment M2 appears to be a major determinant of channel function. It can be photolabelled by non-competitive channel blockers [59, 60] and determines the single channel characteristics of chimaeric nicotinic acetylcholine receptor molecules expressed in *Xenopus* oocytes [61]. Interestingly, segment M2 displays a very high degree of similarity between γ-aminobutyric acid receptor and GlyR α subunits which suggests that this region is also crucial for the function of chloride channels [62]. This view is supported by the presence of eight conserved hydroxylated side chains in M2, which are thought to provide a hydrophilic lining of the channel, as well as by positively charged amino acids bordering this segment (Fig. 4b). As pointed out by Bormann et al. [12], a pentameric arrangement of α-helical peptide segments creates a central bore 0.58 nm in diameter which closely matches the experimentally determined sizes of both GlyR and γ-aminobutyric acid receptor channels. Therefore, each of the five subunits may contribute its M2 region to form the inner wall of the ion channel [37]. Segment M4, on the other hand, is composed of very hydrophobic amino acids and conserved to a much lower degree. It therefore may be localized at the protein/lipid interface of the membrane-spanning part of the GlyR (Fig. 3b).

In an effort to reconstitute GlyR channel function, we synthesized a peptide corresponding to segment M2 of the rat α_1 chain including its N- and C-terminal adjoining arginine residues [63]. After incorporation into planar lipid bilayers, this peptide readily formed randomly gated single channels. Interestingly, the ion selectivity of these channels changed upon inverting the terminal charges of the peptide which suggests ion binding to these residues. Thus the positively charged amino acids at the presumed mouths of the GlyR channel may be the physical correlate of the anion binding sites determined electrophysiologically (D. Langosch, K. Harting, E. Grell, E. Bamberg and H. Betz, unpublished results). Consistent with this view, rings of negatively charged residues bordering the M2 regions of the nicotinic acetylcholine receptor have been shown to determine the conductivity of this cation channel [64].

Implications for the evolution of channel proteins

Comparing primary sequences is a means of establishing phylogenetic relationships between proteins. Within the extracellular domain and the transmembrane regions M1 – M3, an amino acid identity of 18 – 24% is found between GlyR and nicotinic acetylcholine receptor subunits. Total amino acid identities with γ-aminobutyric acid receptor subunits amount to about 35 – 40% [65]. Considering in addition conservative amino acid exchanges, similar residues between GlyR α subunits and nicotinic acetylcholine receptor or γ-aminobutyric acid receptor proteins are around 40% or 55 – 60%, respectively. Conservation is highest within extracellular domains and the putative membrane-spanning regions M1 – M3. Considerable divergence characterizes the intracellular loop separating M3 from M4 and segment M4.

These findings suggest that the different ligand-gated ion channels have evolved from a common primordial channel protein. Certain building principles, such as a pentameric core structure and a common transmembrane topology, apparently persisted in the course of phylogeny. On the other hand, primary sequences grossly diverged during evolution to generate molecules with different ligand specificities and ion selectivities to suit the requirements of complex nervous systems [62]. In addition, convergent evolution may have resulted in common architectural designs of functionally related proteins that otherwise are unrelated in primary sequence. For example, the recently cloned kainate receptor proteins thought to constitute a subtype of excitatory glutamate receptors in the vertebrate central nervous systems do not display recognizable homology to the ligand-gated ion channel family discussed here [66]. Interestingly, however, their predicted transmembrane topology resembles that of the ligand-gated ion channel super-family encompassing the nicotinic acetylcholine receptor, the γ-aminobutyric acid receptor, and the GlyR.

We gratefully acknowledge our colleagues for communication of unpublished data, Ina Baro and Stefan Dübel for help with graphics, and Imke Veit-Schirmer for secretarial assistance. Work in our laboratory is supported by grants from *Bundesministerium für Forschung und Technologie* (BCT 365/1), the *Deutsche Forschungsgemeinschaft* (SFB 317 and Leibniz Program), the German-Israeli Foundation and *Fonds der Chemischen Industrie*. D. L. received a predoctoral fellowship from the *Fonds der Chemischen Industrie*.

REFERENCES

1. Hodgkin, A. L. & Huxley, A. F. (1952) *J. Physiol., 117*, 500 – 544.
2. Katz, B. (1966) *Nerve, muscle and synapse*, McGraw Hill, New York.
3. Strange, D. G. (1988) *Biochem. J. 249*, 309 – 318.
4. Hille, B (1984) *Ionic channels of excitible membranes*, Sinauer, Sunderland.
5. Aprison, M. H. & Wermann, R. (1965) *Life Sci. 4*, 2075 – 2083.
6. Hopkin, J. M. & Neal, M. J. (1970) *Proc. Br. Pharmac. Soc. 4*, 136 – 137P.
7. Werman, R., Davidoff, R. A., & Aprison, M. H. (1966) *Physiologist, 9*, 318.
8. Curtis, D. R., Hösli, L., & Johnston, G. A. R. (1968) *Exp. Brain Res. 6*, 1 – 18.
9. Aprison, M. H. & Daly, E. (1978) *Adv. Neurochem. 3*, 203 – 294.
10. Coombs, J. S., Eccles, J. C. & Fatt, P. (1955) *J. Physiol. (Lond.) 130*, 326 – 373.
11. Barker, J. L. & Ransom, B. R. (1978) *J. Physiol. (Lond.) 280*, 331 – 354.
12. Bormann, J., Hamill, O. P. & Sakmann, B. (1987) *J. Physiol. (Lond.) 385*, 243 – 286.
13. Young, A. B. & Snyder, S. H. (1973) *Proc. Natl Acad. Sci. USA, 70*, 2832 – 2836.
14. Young, A. B. & Snyder, S. H. (1974) *Mol. Pharmacol. 10*, 790 – 809.
15. Betz, H. & Becker, C.-M. (1988) *Neurochem. Int. 13*, 137 – 146.
16. Werman, R., Davidoff, R. A. & Aprison, M. H. (1968) *J. Neurophysiol. 31*, 81 – 95.
17. Davidoff, R. A., Aprison, M. A. & Werman, R. (1969) *Neuropharmacology 8*, 191 – 194.
18. Marvizon, J. C. G., Vazques, J., Calvo, M. G., Mayor, F. Jr., Gomez, A. R., Valdivieso, F. & Benavides, J. (1986) *Mol. Pharmacol. 30*, 590 – 597.
19. Aprison, M. H., Lipkowitz, K. B. & Simon, J. R. (1987) *J. Neurosci. Res. 17*, 209 – 213.
20. Gundersen, C. B., Miledi, R., & Parker, I. (1984) *Proc. R. Soc. Lond. B, 221*, 235 – 244.
21. Schmieden, V., Grenningloh, G., Schofield, P. & Betz, H. (1989) *EMBO J. 8*, 695 – 700.
22. Jackson, M. B. (1988) *Proc. Natl Acad. Sci. USA 86*, 2199 – 2203.
23. Hamill, O. P., Bormann, J. & Sakmann, B. (1983) *Nature 305*, 805 – 808.
24. Faber, D. S. & Korn, H. (1987) *J. Neurosci. 7*, 807 – 811.
25. Pfeiffer, F. & Betz, H. (1981) *Brain. Res. 226*, 273 – 279.
26. Pfeiffer, F., Graham, D. & Betz, H. (1982) *J. Biol. Chem. 257*, 9389 – 9393.
27. Graham, D., Pfeiffer, F., Simler, R. & Betz, H. (1985) *Biochemistry 24*, 990 – 994.
28. Becker, C.-M., Hermans-Borgmeyer, I., Schmitt, B. & Betz, H. (1986) *J. Neurosci. 6*, 1358 – 1364.
29. Graham, D., Pfeiffer, F. & Betz, H. (1981) *Biochem. Biophys. Res. Commun. 102*, 1330 – 1335.
30. Graham, D., Pfeiffer, F., & Betz, H. (1983) *Eur. J. Biochem. 131*, 519 – 525.
31. Pfeiffer, F., Simler, R., Grenningloh, G. & Betz, H. (1984) *Proc. Natl Acad. Sci. USA 81*, 7224 – 7227.
32. Schmitt, B., Knaus, P., Becker, C.-M. & Betz, H. (1987) *Biochemistry 26*, 805 – 811.
33. Triller, A., Cluzeaud, F., Pfeiffer, F., Betz, H. & Korn, H. (1985) *J. Cell Biol. 101*, 683 – 688.
34. Altschuler, R. A., Betz, H., Parakkal, M. H., Reeks, K. A. & Wenthold, R. J. (1986) *Brain Res. 369*, 316 – 320.
35. Becker, C.-M., Hoch, W. & Betz, H. (1988) *EMBO J. 7*, 3717 – 3726.
36. Betz, H. (1987) *Trends Neurosci. 10*, 113 – 117.
37. Langosch, D., Thomas, L. & Betz, H. (1988) *Proc. Natl Acad. Sci. USA 85*, 7394 – 7398.
38. Hucho, F., Bandini, G. & Suarez-Isla, B. A. (1978) *Eur. J. Biochem. 83*, 335 – 340.
39. Brisson, A. & Unwin, P. N. T. (1985) *Nature 315*, 474 – 477.
40. Kubalek, E., Ralston, S., Lindstrom, J. & Unwin, N. (1987) *J. Cell Biol. 105*, 9 – 18.
41. Grenningloh, G., Rienitz, A., Schmitt, B., Methfessel, C., Zensen, M., Beyreuther, K., Gundelfinger, E. D. & Betz, H. (1987) *Nature 328*, 215 – 220.
42. Noda, M., Furutani, Y., Takahashi, H., Toyosato, M., Tanabe, T., Shimizu, S., Kikyotani, S., Kayano, T., Hirose, T., Inayama, S. & Numa, S. (1983) *Nature 305*, 818 – 823.
43. Boulter, J., Evans, K., Goldman, D., Martin, G., Treco, D., Heinemann, S. & Patrick, J. (1986) *Nature 319*, 368 – 374.
44. Hermans-Borgmeyer, I., Zopf, D., Ryseck, R.-P., Hovemann, B., Betz, H. & Gundelfinger, E. D. (1986) *EMBO J. 5*, 1503 – 1508.
45. Schofield, P. R., Darlison, M. G., Fujita, N., Burt, D. R., Stephenson, F. A., Rodriguez, H., Rhee, L. M., Ramachandran, J., Reale, V., Glencorse, T. A. & Seeburg, P. H. (1987) *Nature 328*, 221 – 227.
46. Moscovitz, R. & Gershoni, J. M. (1988) *J. Biol. Chem. 263*, 1017 – 1022.
47. Mishina, M., Tobimatsu, T., Imoto, K., Tanaka, K., Fujita, Y., Fukuda, K., Kurasaki, M., Takahashi, H., Morimoto, Y., Hirose, T., Inayama, S., Takahashi, T., Kuno, M. & Numa, S. (1985) *Nature 313*, 364 – 369.
48. Grenningloh, G., Pribilla, I., Prior, P., Multhaup, G., Beyreuther, K., Taleb, O. & Betz, H. (1990) *Neuron 4*, 963 – 970.

8

49. Benavides, J., Lopez-Lahoya, J., Valdivieso, F. & Ugarte, M. (1981) *J. Neurochem. 7*, 315−320.
50. Nelson, P. G., Ransom, B. R., Henkart, M. & Bullock, P. N. (1977) *J. Neurophysiol. 40*, 1178−1187.
51. Hoch, W., Betz, H. & Becker, C.-M. (1989) *Neuron 3*, 339−348.
52. Akagi, H. & Miledi, R. (1988) *Science 242*, 270−273.
53. Akagi, H., Patton, D.E . & Miledi, R. (1989) *Proc. Natl Acad. Sci. USA 86*, 8103−8107.
54. Grenningloh, G., Schmieden, V., Schofield, P. R., Seeburg, P. H., Siddique, T., Mohandas, T. K. & Betz, H. (1990) *EMBO J. 9*, 771−776.
55. Kuhse, J., Schmieden, V. & Betz, H. (1990) *Neuron*, in the press.
56. Sontheimer, H., Becker, C.-M., Pritchett, D. B., Schofield, P. R., Grenningloh, G., Kettenmann, H., Betz, H. & Seeburg, H. (1989) *Neuron 2*, 1491−1497.
57. Kuhse, J., Schmieden, V. & Betz, H. (1990) *J. Biol. Chem.*, in the press.
58. Garcia-Calvo, M., Ruiz-Gomez, A., Vazques, J., Morato, E., Valdivieso, F. & Mayor, F., Jr. (1989) *Biochemistry 28*, 6405−6409.
59. Hucho, F. L., Oberthur, W. & Lottspeich, F. (1986) *FEBS Lett. 205*, 137−142.
60. Giraudat, J., Dennis, M., Heidmann, T., Chang, J.-Y. & Changeux, J.-P. (1986) *Proc. Natl Acad. Sci. USA 83*, 2719−2723.
61. Imoto, K., Methfessel, C., Sakmann, B., Mishina, M., Mori, Y., Konno, T., Fuduka, K., Kurasaki, M., Bujo, H., Fujita, Y. & Numa, S. (1986) *Nature 324*, 670−674.
62. Grenningloh, G., Gundelfinger, E. D., Schmitt, B., Betz, H., Darlison, M. G., Barnard, E. A., Schofield, P. & Seeburg, P. H. (1987b) *Nature 330*, 25−26.
63. Reference deleted.
64. Imoto, K., Busch, C., Sakmann, B., Mishina, M., Konno, T., Nakai, J., Bujo, H., Mori, Y., Fukuda, K. & Numa, S. (1988) *Nature 335*, 645−648.
65. Barnard, E., Darlison, M.G. & Seeburg, P. (1987) *Trends Neurosci. 10*, 502−509.
66. Hollmann, M., O'Shea-Greenfield, A., Rogers, S. W. & Heinemann, S. (1990) *Nature 342*, 643−648.

Eur. J. Biochem. *194*, 317–321 (1990)
© FEBS 1990

Review

Recent advances in the molecular analysis of inherited disease

Susan MALCOLM

Mothercare Department of Paediatric Genetics, Institute of Child Health, London, England

(Received June 18, 1990) — EJB 90 0702

Many important human genes have been cloned during the last ten years. In some cases, using reverse genetic techniques [Orkin, S. H. (1986) *Cell 47*, 845 – 850], disease-causing genes have been isolated whose product was previously unknown. Important examples include the dystrophin protein which, when mutated, gives rise to either Duchenne or Becker muscular dystrophy [Koenig, M., Hoffman, E. P., Bertelson, C. J., Monaco, A. P., Feener, C. and Kunkel, L. M. (1987) *Cell 50*, 509 – 517; Monaco, A. P., Bertelson, C. J., Liechti-Gallati, S. & Kunkel, L. M. (1988) *Genomics 2*, 90 – 95; Koenig, M., Monaco, A. P. & Kunkel, L. M. (1988) *Cell 53*, 219 – 228] and the cystic fibrosis transmembrane conductance regulator (CFTR) [Riordan, J. R., Rommens, J. M., Kerem, B.-S., Alon, N., Rozmahel, R., Grzelczak, Z., Zielenski, J., Lok, S., Plavsic, N., Chou, J.-L., Drumm, M. L., Ianuzzi, M. C., Collins, F. S. & Tsui, L.-C. (1989) *Science 245*, 1066 – 1073]. Recently the technology for systematically detecting single base-pair changes by chemical methods, enzymatic methods or direct DNA sequencing has greatly expanded and simplified. In addition to providing structural information about these clinically important genes and information on disease-causing mutations, these studies have led to an increased understanding of mechanisms of mutation, to the discovery of novel genetic mechanisms and to important clinical applications of carrier detection and pre-natal diagnosis. The recent rapid progress has been made possible by the development of DNA amplification using the polymerase chain reaction (pcr) invented by Saiki and colleagues [Saiki, R. K., Chang, C-A., Levenson, C. H., Warren, T. C., Boehm, C. D., Kazazian, H. H. & Ehrlich, H.A. (1988) *N. Engl. J. Med. 319*, 537 – 541].

Detection of mutations

Until recently it was time consuming and expensive to analyse directly mutations in a gene leading to disease, as it involved cloning the gene in several fragments from a library made from the patient and sequencing the entire product, often many thousands of base pairs. Even when a base-pair change was found, it had to be established that this was indeed the disease causing mutation as many polymorphisms exist, particularly in the intervening sequence DNA. Several methods have been established recently for simplifying this procedure with the chemical cleavage of mismatched base method (also called HOT after the chemicals used) developed by Cotton [1] being extremely effective. Following pcr amplification of the stretch of sequence under survey, a DNA heteroduplex is formed between mutant and wild-type sequences. Hydroxylamine (H) will react with mismatched C bases and osmium tetroxide (OT) with mismatched T bases. These positions can be subsequently cleaved with piperidine. Mismatched A and G bases can be detected by use of the probe in the opposite strand which converts them to mismatched T and C bases, with the result that single base-pair changes

or small length differences are detectable by this method. Mismatch analysis has been applied to factor-IX mutations in haemophilia B [2], ornithine transcarbamylase deficiency [3] and variations within the dystrophin gene [4]. A method for detecting base changes in homoduplexes involves the use of denaturing gradient gel electrophoresis (DGGE) on an amplified segment. As the duplex migrates into the gel it will eventually reach a concentration of denaturant such that the lowest melting domain will start to melt and this will drastically slow the progress of the molecule through the acrylamide. If the two sequences under comparison differ by point mutations in this lower-melting domain, then they will separate on the gel, either travelling further (A or T changed to C or G) or slower (C or G changed to A or T). Mutations in the catalytic domain of factor IX have been demonstrated by this method [5]. A mutation falling in the higher-melting domain would not normally be detected but the technique can be modified by the addition of a 40-bp G/C-rich sequence (GC clamp) to the end of the amplified products [6]. This can easily be incorporated via the oligonucleotides used to prime amplification (amplimers). At least one clinical application of DGGE has been reported [7] in which a restriction-fragment-length polymorphism close to the adult polycystic kidney disease locus is routinely detected by DGGE following amplification of a 428-bp fragment. A variation is to exploit the abnormal migration of heteroduplexes in acrylamide gels. Amplification of the DNA around the major cystic fibrosis mutation (a 3-bp deletion) in a carrier would be expected to give linear-products

---

Correspondence to S. Malcolm, Mothercare Department of Paediatric Genetics, Institute of Child Health, 30 Guilford Street, London WC1N 1EH, England

Abbreviations. pcr, polymerase chain reaction; CFTR, cystic fibrosis transmembrane regulator; DGGE, denaturing gradient gel electrophoresis; RFLF, altered length fragment; HPRT, hypoxanthine phosphoribosyl transferase.

3 bp different in length, which can be separated by polyacrylamide gel electrophoresis. In addition, however, a much slower migrating band is observed. This is a heteroduplex formed during the denaturing and reannealling of the pcr products [8]. As this pcr product doses not have a perfect duplex structure, migration through polyacrylamide will be retarded. With certain amplimers it could be possible to detect two heteroduplexes arising from both combinations of mutant and wild-type strands. Detection of a heteroduplex has also been used diagnostically to detect carriers of Tay-Sachs disease which was caused by a 4-bp insertion mutation [9]. Kogan and Gitschier [10] have shown that mixing of wild-type and mutant pcr products combined with denaturing gel electrophoresis can reveal extra mutations. They found three mutations in a systematic survey of the factor-8 gene in 228 haemophiliacs, one of which was only found after heteroduplex formation.

The pcr has also revolutionised DNA sequencing by removing the need to subclone first. The amplified product can be sequenced directly either with a separate internal primer or with the original primers. Generation of a single-stranded product especially suitable for sequencing can be produced by using different molar amounts of the two amplification primers [11]. In the first rounds of amplification the normal double-stranded product will be produced, but as the low-concentration primer runs out, only one strand will be further amplified. Oligonucleotide primers can be chosen so that the first stretches to be sequenced are from domains most likely to carry important mutations of regions containing several 5'CpG3' sequences (see below).

Analysis of polymorphism

Traditionally DNA polymorphisms have been detected by analysing point mutations in restriction-enzyme sites which give rise to altered length fragments (RFLP). These are detected using DNA transfer or Southern blotting. These polymorphisms can be detected quicker and more economically by pcr amplification of a stretch of DNA across the site followed by detection of the presence or absence of the restriction-enzyme cutting site on ethidium stained agarose gels [12]. However, pcr amplification also makes it easier to detect point mutations which do not occur within restriction-enzyme sites by allele-specific oligonucleotide (ASO) hybridisation [13]. Without prior amplification the signal/noise ratio is too small for convenient use.

The major disadvantages of using two allele polymorphisms are the expense of restriction enzymes, the labour intensive nature of DNA transfer experiments and the frustration ensuing when a key person in a family is found to be uninformative (i.e. homozygous for a polymorphism). All these problems are overcome by the increasing use of polymorphisms which arise through variation in the number of repeats in a stretch of DNA containing variable numbers of tandem repeats of simple sequences, e.g. $(CA)_n$ [14, 15]. These are detected by using amplimers of single copy sequence from the flanking regions and separating the radioactively labelled products, which differ in length by 2 bp or multiples of 2 bp, on a DNA sequencing gel. Many such simple repetitive motifs exist in the human genome, apparently distributed frequently and at random, based on either 4-bp or 2-bp repeats but a further source of hypervariability is found in the Alu repetitive elements distributed throughout the genome [16]. Alu repetitive elements have a tract of adenine [poly(A)] at the 3'-end

which in fact vary in sequence between different chromosomes. For example the Alu sequence occurring in the 5'-flanking sequence of adenosine deaminase on chromosome 20 has seven alleles which vary due to different numbers of TAAA repeated units. Variable sequences within Alu repeats will be particularly useful as the presence of an Alu repeat is often used to separate human from rodent sequences in libraries made from somatic-cell hybrids, or radiation hybrids containing only a small amount of human DNA.

A modification of pcr in which the primer oligonucleotide is complementary to the site of base change can be used to detect single-base-pair changes. In this technique, called the amplification refractory mutation system (ARMS), the 3'-end of the primary oligonucleotide i.e. that to which the new bases add, corresponds to the base-pair change. The perfectly matched oligonucleotide will prime far better than that with a mismatch and this effect will be increased during the cycles of amplification [17, 18]. The effect is particularly marked when the two oligonucleotides compete [19].

It will be very advantageous from every point of view when radioactive detection can be replaced by colorimetry. Methods of nucleic acid hybrid detection based on biotin/avidin interactions were introduced some years ago [20] but have failed to compete with radioactivity in terms of sensitivity. A radically new and convenient method has been developed by Chehab and Kan [21]. In this method, two primer reactions are carried out with the amplimers labelled with different fluorescent probes yielding differently coloured products or a mixed and distinguishable colour in the case of both products being produced, i.e. a rhodamine-labelled product (red) and a fluorescein-labelled product (green). For example in the detection of the α-globin gene deletion in hydrops fetalis, the α-globin product, if present, is labelled red and the β-globin product used as a control is labelled green. If both are present, i.e. patient non-affected, either red and green products can be seen after electrophoresis of if unincorporated primers are removed a yellow colour observed. If the deletion is present then a green product only will be seen.

Origins of mutations

The CpG effect

It has been known for many years that the dinucleotide, 5'CpG3', is under represented in the human genome and that the C-residue of this dinucleotide is the major site of methylation. The methylation is symmetrical as the opposite strand is of course also 5'CpG3'. It is believed that deamination of mC to form T, followed by repair of the resulting G · T mismatch to A · T, is responsible for this deficiency. Consequently it would be expected that mutations of CpG to TpG and in the other strand, CpG to CpA would contribute considerably to the point mutations causing human genetic disease. In addition, at non-coding sites in the genome, CpG sites would be expected to contribute considerably to polymorphism. Both these expectations have borne out. The restriction enzymes TaqI and MspI, both of which contain CpG in their recognition site are particularly useful in detecting RFLP.

A systematic study of factor-IX mutations in haemophilia B patients revealed that 6 of 11 single-base substitutions were CpG to TpG transitions [22]. Similarly the analysis of exon 7 of the phenylalanine hydroxylase gene, which contains 5 of the total 22 CpG doublets, in phenylketonuria patients, revealed two novel missense mutations (CG to CA and CG to TG). A less systematic but very quick method of screening is to screen *Taq*I and *Msp*I sites within the coding sequence. For example in the gene for ornithine transcarbamylase, an enzyme of the urea cycle encoded on the X chromosome, a *Taq*I-recognition site (TCGA) occurs at codon 109 so that the CGA triplet codes for arginine. In 4 out of 24 families this *Taq*I site is altered [23], in two of the cases because of a G to A mutation on the sense strand which results in glutamine replacing arginine and in one case because a C to T transition converts arginine to a termination codon, resulting in a premature stop [24].

Deletions

Duchenne and Becker muscular dystrophy. Prior to the isolation of the dystrophin gene there were several apparently unusual features about gene mutations at the locus giving rise to muscular dystrophy. Isolation of the gene has gone a long way towards explaining the relationship between Becker and Duchenne muscular dystrophy and the high rate of mutation. It is clear that the outstanding feature of the gene is its enormous length. The messenger RNA is 14-kb long, which although long is not unprecedented, but the coding sequences are spread over at least 65 exons covering 2 Mb of DNA [25, 26]. This enormous target for mutations explains the high mutation rate. Systematic surveys of patients with Duchenne and Becker muscular dystrophy have revealed a high proportion, up to 70% in some surveys, where the disease is caused by a deletion, or less commonly a duplication, of part of the dystrophin gene [25, 27 – 32]. This is much higher than in other genetic disease and is presumably a result of the unique gene structure. As less than 1% of the gene consists of coding sequences, most point mutations will fall in introns and not produce a disease-causing mutation. However, many deletions will start in introns but extend through one or more exons. There is in general a hot spot for mutations towards the centre of the gene which correlates well with a particularly large intron of over 100 kb [29, 30].

The cDNA sequence of dystrophin predicted a rod-shaped cytoskeletal protein [33] and immunochemical studies confirmed a deficiency of dystrophin at the muscle cell surface in Duchenne muscular dystrophy [34]. Becker muscular dystrophy has a more severe clinical course than Duchenne muscular dystrophy. Studies of the deletions have shown that, with a few exceptions, Becker muscular dystrophy arises when the gene deletion results in no frame shift, i.e. the exons removed encode a multiple of 3 bp, whereas Duchenne muscular dystrophy arises when the deletion causes a frame shift [35 – 38].

Cystic fibrosis. The gene giving rise to cystic fibrosis has recently been isolated, sequenced and the structure of the gene product predicted [39 – 41]. Named the cystic fibrosis transmembrane regulator (CFTR) it is predicted to be a transmembrane protein with two nucleotide-binding domains and a further cytoplasmic domain involved in regulatory events such as phosphorylation. In contrast to Duchenne and Becker muscular dystrophies which have many mutations arising independently, there is a major mutation in cystic fibrosis accounting for approximately 70% of all chromosomes in the North American and Northern European population [41, 42]. This is a 3-bp deletion causing the removal of a single amino acid, phenylalanine, at position 508 which falls within one of the nucleotide-binding domains (ΔF508). The presence of such a widespread polymorphism and the relative ease of detection [8] raises the possibility of offering a population screening programme, at least in Caucasian populations where the carrier frequency of cystic fibrosis is around 1:20. However, the absence, so far, of a second major mutation among the other 30% of chromosomes makes this a very complex procedure, because of the high proportion of couples who would be faced with an increased risk of having an affected child, as one partner carried the ΔF508 mutation but without any means of determining whether the other partner carried one of the minor mutations. This is currently the source of much debate [43].

Mechanisms modifying inheritance

Germline and gonadosomal mosaicism

It is commonly assumed that a woman who has had two sons suffering from an X-linked genetic disease such as Duchenne muscular dystrophy must herself be a carrier of the disease. Careful thought analysing the biological processes leading to sperm and egg formation reveals that at least some of these women should exhibit a germ-line or even germ-line and somatic tissue mosaicism for the mutant gene [44]. Direct molecular evidence proving this is available now that mutations of the dystrophin gene can be found directly [45 – 47]. Thus it has been shown that a woman has twice passed on a chromosome carrying a deleted dystrophin gene although she clearly has two complete copies of the gene herself. Direct evidence also exists for gonadal mosaicism in ornithine transcarbamylase deficiency [48], X-linked agammaglobulinaemia [49] and the autosomal dominantly inherited osteogenesis imperfecta [50]. In the last case, a father had two children with lethal osteogenesis imperfecta by separate mothers. The gene must have come from him as he was the common parent but he was clearly not affected himself. The gene mutation in the collagen α-1(I) chain was traced to a point mutation of G to A resulting in substitution of aspartic acid for glycine at position 883 of the triple helix. This resulted in the entire triple-helix domain being overmodified. The point mutation disrupts a *Bgl*I restriction site and could therefore be traced through different tissues in the father. Mosaicism was found in two separate sperm samples (gonadal) and in DNA extracted from white blood cells and hair root bulbs (somal). An appreciation of the possibly wide spread occurrence of mosaicism will make genetic counselling considerably more complex.

Uniparental disomy and genomic imprinting

Autosomal recessive inheritance normally results from a mutant gene being inherited from each parent. The finding of a child suffering from cystic fibrosis who had two identical copies of the maternal sequences for much or all of chromosome 7 [51, 52] showed that uniparental disomy in an individual with a normal chromosome analysis is a novel mechanism for the occurrence of human genetic disease. In the child, the cystic fibrosis had presumably arisen by homozygotization of a mutant gene in the carrier mother and statistically 1 in 20 Caucasians will be carriers.

It has traditionally been assumed that the gene coming from the mother and the father will have the same effect. However, transgenic mouse experiments [53] and the study of the loss of alleles in tumours [54] suggests that this is not always the case and that genomic imprinting can occur. Genomic imprinting has been implicated in two complex childhood neurological disorders, Angelman syndrome and Prader-Willi syndrome [55, 56]. Both are associated with a small chromosomal deletion of chromosome 15. Using both cytogenetic and molecular genetic techniques, these deletions are so far indistinguishable and the only difference found is that in Prader-Willi syndrome, the deletion arises on chromosome 15 inherited from the father whereas in Angelman syndrome the deletion on chromosome 15 is inherited from the mother [57], suggesting that contributions from both mother and father are essential for normal development, at least for gene(s) in this region of chromosome 15. This idea was further strengthened by a report from Nicholls and colleagues [58] who found that Prader-Willi syndrome can also arise when both of the apparently normal chromosome 15s have been inherited from the mother (uniparental maternal heterodisomy) with, in effect, the loss of the paternal contribution. Although uniparental disomy is likely to be a rare event it may explain some previously unexplained phenomena.

X-inactivation

In females, inactivation of one or other of the X chromosomes occurs early in embryogenesis. The progeny of any individual cell expresses the genes carried on the same (active) X chromosome. If the woman is a carrier for an X-linked gene mutation she will, in effect, be a mosaic for that trait. If a gene is necessary for differentiation or survival of a particular cell type then a carrier woman will have a population of those cells using only one (i.e. the non-mutant) active X chromosome. This non-random X-chromosome usage can be detected in a number of ways. The active X chromosome can be captured in somatic cell hybrids which are reliant on the expression of the hypoxanthine phosphoribosyl transferase (HPRT) gene from the active human X chromosome for their survival [59, 60] or methylation-pattern differences specific to active and inactive X chromosomes can be exploited [61]. In the latter method, developed originally by Vogelstein and colleagues to study the clonality of tumours, the two X chromosomes were distinguished using restriction fragment length polymorphisms associated with phosphoglycerol kinase (PGK) or HPRT genes. The methylation patterns were revealed by comparing the pattern of digestion with the isoschizomeric restriction enzymes MspI (methylation insensitive) and HpaII (methylation sensitive). If a cell population exhibits random X-chromosome usage both polymorphic bands will remain after digestion with HpaII. In contrast in a non-random population only one band will remain. The technique has been developed recently by using a highly polymorphic probe, 27 β [62], which is technically simpler to use and more informative. Applications of the study of non-random X-chromosome usage have included: identification of carriers of X-linked immunodeficiencies by studying the lymphocytes of potential women carriers [59, 63, 64], differentiating between X-linked and autosomal forms of immunodeficiencies [63]; investigating the pattern of gene expression [64, 65] and proving that a paternal germ line mosaic was the origin of a familial case of X-linked agammaglobulinaemia [49]. Other applications are likely to be many including the study of females expressing an X-linked disease and the use of X chromosomes with structural variants.

SM has been generously supported by the Muscular Dystrophy Group of Great Britain, the Medical Research Council, Action Research for the Crippled Child and the Child Health Research Appeal Trust.

REFERENCES

1. Cotton, R. G. H., Rodrigues, N. R. & Campbell, R. D. (1988) Proc. Natl Acad. Sci. USA 85, 4397—4401.
2. Montandon, A. J., Green, P. M., Gianelli, F., Bentley, D. R. (1985) Nucleic Acids Res. 17, 3347—3358.
3. Grompe, M., Muzny, D. M. & Caskey, C. T. (1989) Proc. Natl Acad. Sci. USA 86, 5888—5892.
4. Roberts, R. G., Montandon, A. J., Bobrow, M. & Bentley, D. (1989) Nucleic Acids Res. 17, 5961—5971.
5. Attree, O., Vidand, D., Vidand, M., Amselem, S., Lavergne, J.-M. & Goossens, M. (1989) Genomics 4, 266—272.
6. Sheffield, V. C., Cox, D. R., Lerman, L. S. & Myers, R. M. (1989) Proc. Natl Acad. Sci. USA 86, 232—236.
7. Saris, J. J., Breunig, M. H., Dauwerse, H. G., Snijdewint, F. G. M., Tope, B., Fodde, R., van Ommen, G.-J. B. (1990) Lancet I, 1102—1103.
8. Rommens, J., Kerem, B.-S., Greer, W., Chang, P., Tsui, L.-C., Ray, P. Am J. Hum. Genet. (in the press).
9. Trigg-Raine, B. L. & Gravel, R. A. (1990) Am. J. Hum. Genet. 46, 183—184.
10. Kogan, S. & Gitschier, J. (1990) Proc. Natl Acad. Sci. USA 87, 2092—2096.
11. Gyllensten, U. & Ehrlich, H. (1988) Proc. Natl Acad. Sci. USA 85, 7652—7656.
12. Roberts, R. G., Cole, C. G., Hart, K. A., Bobrow, M. & Bentley, D. R. (1989) Nucleic Acids Res. 17, 811.
13. Kerem, B.-S., Rommens, J. R., Buchanan, J. A., Markiewicz, D., Cox, T. K., Chakravati, A., Buchwald, M. & Tsui, L.-C. (1989) Science 245, 1073—1080.
14. Weber, J. L. & May, P. E. (1989) Am. J. Hum. Genet. 44, 388—396.
15. Litt, M. & Luty, J. A. (1989) Am. J. Hum. Genet. 44, 397—401.
16. Economou, E. P., Bergen, A. W., Warren, A. C. & Antonarakis, S. E. (1989) Proc. Natl Acad. Sci. USA 87, 2951—2954.
17. Newton, C. R., Graham, A., Heptinstall, L. E., Powell, S. J., Summers, C., Kalsheker, N., Smith, J. C. & Markham, A. F. (1989) Nucleic Acids Res. 17, 2503—2516.
18. Ballabio, A., Gibbs, R. A. & Caskey, C. T. (1990) Nature 343, 220.
19. Gibbs, R. A., Nguyen, P.-N. & Caskey, C. T. (1989) Nucleic Acids Res. 17, 2437—2448.
20. Leary, J. J., Brigati, D. J. & Ward, D. C. (1983) Proc. Natl Acad. Sci. USA 80, 4045—4049.
21. Chehab, F. F. & Kan, Y. W. (1989) Proc. Natl Acad. Sci. USA 86, 9178—9182.
22. Green, P. M., Bentley, D. R., Mibashan, R. S., Nilsson, I. M. & Gianelli, F. (1989) EMBO J. 8, 1067—1072.
23. Spence, J. E., Maddalena, A., O'Brien, W. E., Fernbach, S. D., Batshaw, M. L., Leonard, C. O. & Beaudet, A. L. (1989) Paediatr. 114, 582—588.
24. Maddalena, A., Spence, J. E., O'Brien, W. E. & Nussbaum, R. L. (1988) J. Clin. Invest. 82, 1353—1357.
25. Koenig, M., Hoffman, E. P., Bertelson, C. J., Monaco, A. P., Feener, C. & Kunkel, L. M. (1987) Cell 50, 509—517.
26. Burmeister, M., Monaco, A. P., Gillard, E. F., van Ommen, G.-J., Affara, N. A., Ferguson-Smith, M. A., Kunkel, L. M. & Lehrach H. (1988) Genomics 2, 189—202.
27. Den Dunnen, J. T., Bakker, E., Klein Breteler, E. G., Pearson, P. L. & van Ommen, G. J. B. (1987) Nature 329, 640—642.
28. Darras, B. T., Koenig, M., Kunkel, L. M. & Francke, U. (1988) Am. J. Med. Genet. 29, 713—726.
29. Forrest, S. M., Cross, G. S., Thomas, N. S. T., Harper, P. S., Smith, T. J., Read, A., Mountford, R. C., Geirsson, R. T. & Davies, K. E. (1987) Lancet II, 1294—1296.
30. Blonden, L. A. J., Den Dunnen, J. T., van Paassen, H. M. B., Wapenaar, M. C., Grootscholten, P. M., Ginjuar, H. B., Bakker, E., Pearson, P. L. & van Ommen, G. J. B. (1989) Nucleic Acids Res. 17, 5611—5621.
31. Hart, K. A., Abbs, S., Wapenaar, M. C., Cole, C. G., Hodgson, S. V. & Bobrow, M. (1989) Clin. Genet. 35, 251—261.
32. Den Dunnen, J. T., Grootscholten, P. M., Bakker, E., Blonden, L. A. J., Ginjaar, H. B., Wapenaar, M. C., van Paasen, H. M.

B., van Broekhoven, C. & Pearsen, P. L. (1989) *Am. J. Hum. Genet. 45*, 835−847.

33. Koenig, M., Monaco, A. P. & Kunkel, L. M. (1988) *Cell 53*, 219−228.

34. Bonilla, E., Saitt, C. E., Miranda, A. F., Hays, A. P., Salviati, G., DiMauro, S., Kunkel, L. M., Hoffman, E. P. & Rowland, L. P. (1988) *Cell 54*, 447−452.

35. Malhotra, S. B., Hart, K., Klamut, H. J., Thomas, N. S. T., Bodrug, S. E., Burhes, A. H. M., Bobrow, M., Harper, P. S., Thompson, M. W., Ray, P. N. & Worton, R.G. (1988) *Science 242*, 755−759.

36. Monaco, A. P., Bertelson, C. J., Liechti-Gallati, S. & Kunkel, L. M. (1988) *Genomics 2*, 90−95.

37. Koenig, M., Beggs, A. H., Moyer, M. & Kunkel, L. M. (1989) *Am. J. Hum. Genet. 45*, 498−506.

38. Gillard, E. F., Chamberlain, J. S., Murphy, E. G., Duff, C. L., Smith, B., Burghes, A. H. M., Thompson, M. W., Sutherland, J., Oss, I., Bodrug, J. S. E., Klamut, H. J., Ray, P. N. & Worton, R. G. (1989) *Am. J. Hum. Genet. 45*, 507−520.

39. Riordan, J. R., Rommens, J. M., Kerem, B.-S., Alon, N., Rozmahel, R., Grzelczak, Z., Zielenski, J., Lok, S., Plavsic, N., Chou, J.-L., Drumm, M. L., Ianuzzi, M. C., Collins, F. S. & Tsui, L.-C. (1989) *Science 245*, 1066−1073.

40. Rommens, J. M., Ianuzzi, M. C., Kerem, B.-S., Drumm, M. L., Melmer, G., Dean, M., Rozmahel, R., Cole, J. L., Kennedy, D., Hidaka, N., Zsiga, M., Buchwald, M., Riordan, J. R., Tsui, L.-C. & Collins, F. S. (1989) *Science 245*, 1059−1065.

41. Kerem, B.-S., Rommens, J. R., Buchanan, J. A., Markiewicz, D., Cox, T. K., Chakravati, A., Buchwald, M. & Tsui, L.-C. (1989) *Science 245*, 1073−1080.

42. McMahon, C. J., Genet, S. A., Middleton-Price, H. R., Rutland, P., Pembrey, M. E., Malcolm, S. *Hum. Genet.* (in the press).

43. Roberts, L. (1990) *Science 247*, 1296−1297.

44. Edwards, J. H. (1989) *Ann. Hum. Genet. 53*, 33−47.

45. Darras, B. T. & Francke, U. (1987) *Nature 329*, 556−558.

46. Bakker, E., Veenema, H., Den Dunnen, J. T., van Broeckhoven, C., Grootscholten, P. M., Bonten, E. J., van Ommen, G. J. B. & Pearson, P. L. (1989) *J. Med. Genet. 26*, 553−559.

47. Boileau, C. & Junien, C. (1989) *J. Med. Genet. 26*, 790−792.

48. Maddalena, A., Sosonoski, D. M., Berry, G. T. & Nussbaum, R. L. (1988) *N. Engl. J. Med. 319*, 999−1003.

49. Hendriks, R. W., Mensink, E. J. B. M., Kraakman, M. E. M., Thompson, A. & Schuurman, R. K. B. (1989) *Hum. Genet. 83*, 267−270.

50. Cohn, D. H., Starman, B. J., Blumberg, B. & Byers, P. H. (1990) *Am. J. Hum. Genet. 46*, 591−601.

51. Spence, J. E., Perciaccante, R. G., Greig, G. M., Willard, H. F., Ledbetter, D. H., Hejtmancik, J. F., Pollack, M. S., O'Brien, W. E. & Beaudet, A. L. (1988) *Am. J. Hum. Genet. 42*, 217−226.

52. Warburton, D. (1988) *Am. J. Hum. Genet. 42*, 215−216.

53. Cattenach, B. M. & Kirk, M. (1985) *Nature 315*, 496−498.

54. Schroeder, W. T., Chao, L., Dao, D. D., Strong, L. C., Pathak, S., Riccardi, V., Lewis, W. H. & Saunders, G. (1987) *Am. J. Hum. Genet. 40*, 413−420.

55. Angelman, H. (1965) *Dev. Med. Child. Neurol. 7*, 7681−7683.

56. Butler, M. G. (1990) *Am. J. Med. Genet. 35*, 319−331.

57. Knoll, J. H. M., Nicholls, R. D., Magenis, R. E., Graham, J. M., Lalande, M. & Latt, S. A. (1989) *Am. J. Med. Genet. 32*, 285−290.

58. Nicholls, R. D., Knoll, J. H. M., Butler, M. G., Karam, S. & Lalande, M. (1989) *Nature 342*, 281−285.

59. Puck, J. M., Nussbaum, R. L., Smead, D. L. & Conley, M. E. (1989) *Am. J. Hum. Genet. 44*, 724−730.

60. Puck, J. M., Krauss, C. M., Puck, S. M., Buckley, R. H. & Conley, M. E. (1990) *N. Engl. J. Med. 322*, 1063−1066.

61. Vogelstein, B., Fearon, E. R., Hamilton, S. R. & Feinberg, A. P. (1985) *Science 227*, 642−645.

62. Abrahamson, G., Fraser, N. J., Boyd, Y., Craig, I. & Wainscoat, J. S. (1990) *Br. J. Haematol. 74*, 371−377.

63. Goodship, J., Malcolm, S., Lau, Y. L., Pembrey, M. E. & Levinsky, R. J. (1988) *Lancet I*, 729−732.

64. Greer, W. L., Kwong, P. C., Peacocke, M., Ip, P., Rubin, L. A. & Siminovitch, K. A. (1989) *Genomics 4* 60−67.

65. Conley, M. E., Brown, P., Pickard, A. R., Buckley, R. H., Miller, D. S., Raskind, W. H., Singer, J. W. & Fialkow, P. J. (1986) *N. Engl. J. Med. 315*, 564−567.

Eur. J. Biochem. *194*, 693–698 (1990)
© FEBS 1990

Review

Self recognition by the immune system

Harald VON BOEHMER

Basel Institute for Immunology, Basel, Switzerland

(Received July 3, 1990) — EJB 90 0776

In each organism, the immune system must acquire the ability to distinguish self from nonself. Experiments in T cell receptor transgenic mice indicate that this process involves the selection of lymphocytes in the thymus.

The immune system is often regarded as a system that protects us from harmful microorganisms such as bacteria and viruses. Apart from this beneficial effect, the immune system can also cause grief because it can reject transplanted organs such as a kidney or a heart. While the immune system can attack the kidney from a foreign donor, or nonself, it usually does not attack organs from the same organism, or self. This means that the immune system can distinguish between self and nonself. The aim of the experiments described here was to elucidate the principles and mechanisms of this distinction. In order to understand these experiments, one needs to know a few facts about the components as well as the development of the immune system.

CELLS AND MOLECULES OF THE IMMUNE SYSTEM

The immune system is, to a great part, represented by a portion of white blood cells called lymphocytes. Morphologically, one lymphocyte looks like another, but in reality each lymphocyte differs from all others because it carries a unique receptor on its surface. These receptors allow lymphocytes to recognize specific antigens. The antigen receptors are proteins that consist of constant and variable parts. The variable part binds antigen and makes each lymphocyte unique [1–4]. Antigen can be anything, such as toxins from bacteria, proteins from viruses or even synthetic chemicals that do not occur in nature.

There are various classes of lymphocytes which have different roles in the immune system. The most 'cruel' class consists of so-called killer lymphocytes [5]. Their major task is to screen other cells in the body, whether they are infected by virus, develop into cancer cells or are abnormal in any other way. The antigen receptors of these cells recognize other cells containing fragments of proteins, called peptides, which should not be there. Proteins are synthesized and degraded in cells, and peptides are part of the proteins. By an astonishing mechanism, the killer lymphocytes can be informed of what is going on inside other cells in the body. For instance, when a cell is infected with a virus, the virus has entered the cell and multi-

plies inside the living cell. Although it is inside the cell, the virus is not hidden from the immune system: peptides from the virus will bind to so-called major histocompatibility complex (MHC) molecules inside the cell, which then transport the peptides to the cell surface and present them to a killer lymphocyte, or better, a pre-killer cell with an appropriate antigen receptor (Fig. 1) [6–9]. This antigen receptor is composed of α and β receptor chains [10] and binds to both the peptide as well as the presenting MHC molecule: by receptor-gene transfection experiments it could be shown that, in fact, one $\alpha\beta$ receptor has specificity for both the peptide and the presenting MHC molecule [11, 12]. The binding of the $\alpha\beta$ receptor to MHC plus peptide is, however, not sufficient to activate the pre-killer cell. The activation requires another nonvariable receptor on the pre-killer cell, the CD8 coreceptor which also binds MHC molecules (Fig. 1) [13, 14]. When the variable $\alpha\beta$ receptor and the invariable CD8 coreceptor are brought together by binding to different sites on the same MHC molecule [15, 16], the pre-killer cell receives a signal to divide and to differentiate, such that the initially inactive pre-killer lymphocyte produces many active killer cells which can lyse the virus-infected target cell. This makes the virus accessible to other components of the immune system, normally the antibodies, that will eliminate it. The role of the killer cell in the elimination of a virus is, in fact, twofold: first, it deprives the virus of its support, the living cell, and second, it makes the virus accessible to antibodies. Without the killer cells the virus could multiply enormously inside the cell and finally destroy it. Thousands of virus particles could then infect other cells, and this cycle could be repeated until the organism is destroyed.

When discussing the activation of a pre-killer cell by a foreign peptide, one should realize that the MHC molecules not only present peptides derived from foreign proteins, but also peptides from proteins that are normal constituents of the cell. This is the reason why, when a transplanted kidney comes from a genetically distinct donor whose cell proteins differ from that of the recipient, pre-killer cells will be activated by peptides from nonself proteins, and destroy the kidney.

Very briefly, two other classes of lymphocytes are considered here, namely the helper cells and the antibody-producing cells. Killer and helper lymphocytes are called T lymphocytes because they are formed in the thymus [17]. The anti-

Correspondence to H. von Boehmer, Basel Institute for Immunology, Grenzacherstrasse 487, CH-4005 Basel, Switzerland

Abbreviations. MHC, major histocompatibility complex; TCR, T cell receptor.

694

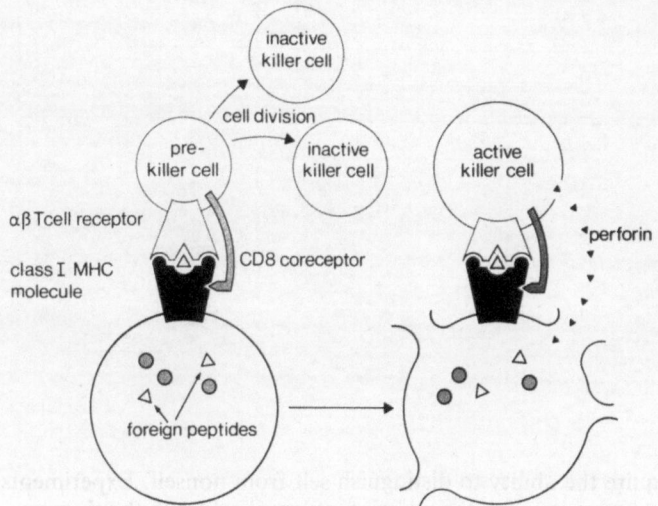

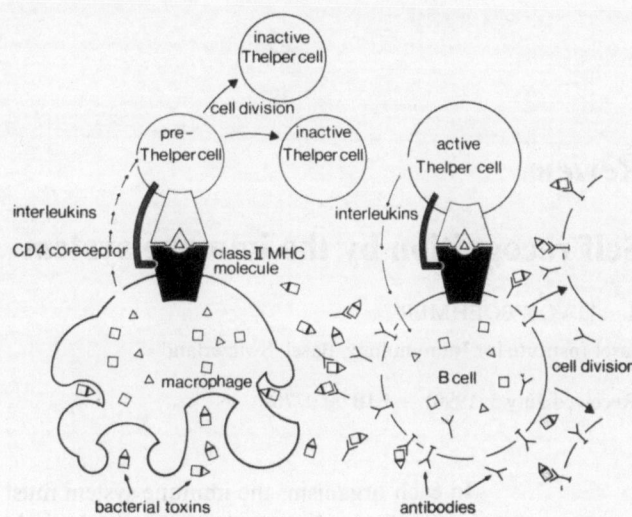

Fig. 1. *The activation of a pre-killer T lymphocyte by peptides and MHC molecules from a virus infected cell.* A pre-killer T lymphocyte with an $\alpha\beta$ TCR (white) and a CD8 coreceptor (grey) binds to both the MHC molecule (black) as well as the peptide (triangle) which is derived from a virus protein made inside the cell. The peptide binds to the MHC molecule inside the cell and the whole complex is transported to the cell surface and 'presented' to the T cell. As a result of the binding of both the $\alpha\beta$ TCR and the CD8 coreceptor to the same MHC molecule, the pre-killer cells receives a signal, starts to divide and differentiate into an active killer cell. When the receptor of an activated killer cell binds to MHC molecules and peptides, the killer cell will produce a substance called perforin which destroys the virus-infected cell. Note that MHC molecules not only present peptides from foreign proteins as shown here, but also peptides from proteins which occur normally in cells

Fig. 2. *The activation of a pre-helper cell and antibody-producing B cells by a bacterial toxin.* A bacterial toxin (square with triangle) is ingested and degraded into peptides by a macrophage. A certain peptide (triangle) binds to a class II MHC molecule (black) and is presented to a pre-helper cell. As a result of the binding of both the $\alpha\beta$ TCR (white) and the CD4 coreceptor (black) to the same class II MHC molecule, the pre-helper cell receives a signal, starts to divide and differentiate into an active helper cell. A B cell can only internalize the toxin when it is bound by an antibody on the surface of a cell. Inside the B cell, the toxin is degraded and peptides are transported to the cell surface by class II MHC molecules. When the $\alpha\beta$ TCR and the CD4 coreceptor of an activated helper cell bind to a class II MHC molecule and peptide, the helper cell will produce interleukins which stimulate the B cell to divide and secrete large amounts of antibodies. These antibodies can then bind to free toxins and neutralize them

body-producing cells are formed in the bone marrow and are called B lymphocytes. T helper cells and B cells cooperate in the antibody response to antigens, like bacterial toxins, which again are proteins [18]. Once a toxin is released into the blood stream, it will be taken up by macrophages; cells which ingest a large variety of substances. Inside the macrophage, the toxin will be degraded, and peptides from the toxin will be presented by MHC molecules. There are two types of MHC molecules: the class I MHC molecules that present peptides from proteins which are made inside the cell (like viral proteins), and the class II MHC molecules that present peptides from proteins which enter the cell (like bacterial toxins) [8]. The $\alpha\beta$ T cell receptor (TCR) of a T helper cell binds to the class-II-MHC-presented peptide, and another invariable CD4 coreceptor [19], characteristic of pre-helper cells, binds to the same class II MHC molecule [20] (like the CD8 coreceptor of a killer cell binds to class I MHC molecules). The pre-helper cell divides and differentiates, and in this way produces many active helper cells (Fig. 2). In the meantime, the toxin has also bound to a specific antibody molecule on a B cell and the whole complex, i.e. the antibody and the toxin, has been ingested by the B cell. The complex is degraded and peptides from the toxin are again transported to the cell surface by class II MHC molecules [21]. Such B cells are now detected by the active T helper cells whose $\alpha\beta$ receptors and CD4 coreceptors bind to the peptide MHC complex at the surface of the B cell. As a result, the T helper cells produce factors, called interleukins, that stimulate the B cells to divide, differentiate and secrete large amounts of antibodies which can neutralize the free toxin in the blood stream (Fig. 2). Thus, both T killer cells as well as

T helper cells have a variable $\alpha\beta$ TCR, while killer cells have an invariable CD8 and helper cells have an invariable CD4 receptor. T cells can only be activated by peptides presented by MHC molecules because T cell activation does not require binding of the $\alpha\beta$ TCR, but also binding of the CD4 or CD8 to the same class II or class I MHC molecule, respectively. T cell activation does not occur when the $\alpha\beta$ TCR binds to one MHC molecule, and the CD4 or CD8 coreceptor to another MHC molecule on the same cell.

Fig. 3 illustrates how an $\alpha\beta$ TCR might bind to a peptide presented by MHC molecules [12]. One important feature of antigen recognition by the immune system that is not illustrated in this figure is the great variability on the interacting molecules. Thus, a very large number of different peptides can be recognized which involve an equally large number of TCR and a diverse set of MHC molecules, each of which can present many, but not all, peptides. The variability of the antigen receptor is somatically generated: the genes encoding these receptors do not exist as a continuous stretch of genetic information but occur in the germline as little pieces or gene segments. These segments recombine randomly in lymphocytes, and by this mechanism form a very large number of different genes encoding in the order of 10^8 different receptors [22] (Fig. 4). The variability of the MHC molecules is largely an allelic variability; this means that one individual contains a limited set of different MHC molecules and that different individuals have different allelic forms of this set [23].

Both the randomly generated receptor diversity and the allelic MHC polymorphism preclude a heritable basis for self/nonself discrimination by the immune system. Thus, in each

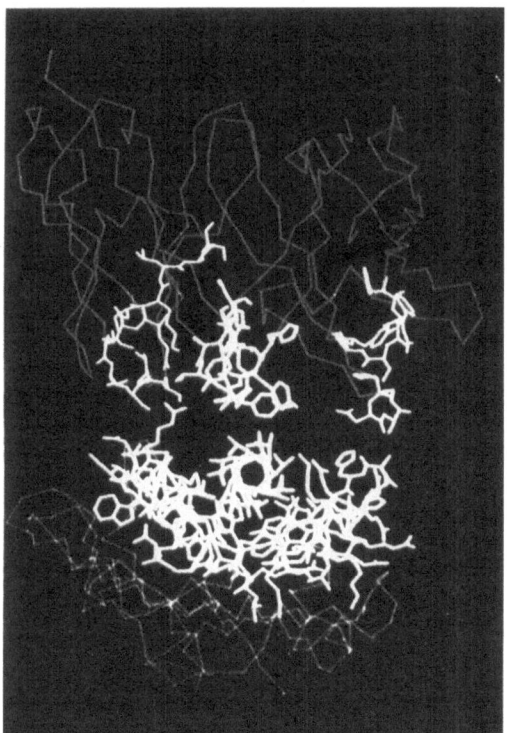

Fig. 3. *A computer model of how an αβ TCR could possibly bind to a class I MHC molecule and peptide.* Taken from [12]. In the lower part, a class I MHC molecule containing a peptide (circular structure in the middle) is shown. In the upper part the TCR is aligned over the MHC molecule. Contacts between the MHC molecule, peptide and T cell receptor are in white

The Molecules of the Immune System

The proteins that recognize foreign invaders are the most diverse proteins known. They are encoded by hundreds of scattered gene fragments, which can be combined in millions or billions of ways

by Susumu Tonegawa

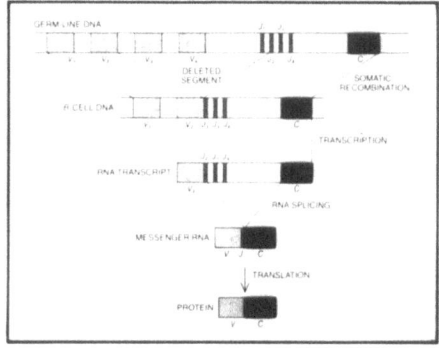

Fig. 4. *The title of and a picture from an article written by Susumu Tonegawa [53] on the origin of antibody diversity.* While the picture describes the gene segments encoding antibodies, very similar events are involved in the generation of TCRs

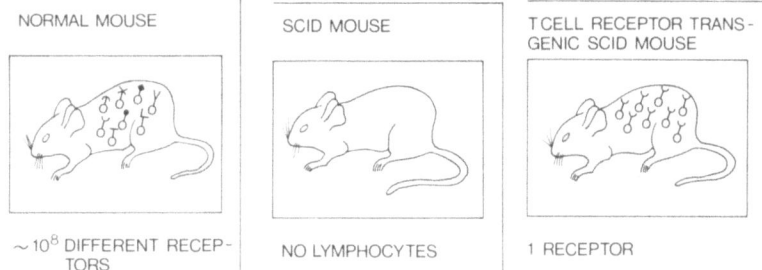

Fig. 5. *The construction of TCR transgenic* scid *mice.* A normal mouse may contain as many as 10^8 different TCRs (left). In a *scid* mouse, the rearrangement mechanism, generating productive TCR-antibody and antibody genes, is defective. As a consequence, this mouse does not produce any antigen receptors nor lymphocytes. The introduction of productively rearranged α and β TCR genes cures the defect, at least partially, and T lymphocytes, all expressing the same receptor, are generated

individual, the immune system must 'learn' to distinguish self from nonself. With regard to T cell receptors, this means that initially, the developing immune system also generates receptors specific for self peptides presented by self MHC molecules. Cells with these receptors are potentially harmful to the organism and thus, the immune system must have ways to preclude the activation of such cells. The developing immune system will also generate receptors which cannot bind to self MHC molecules at all. Cells with these receptors are useless to the organism because they cannot detect foreign peptides presented by self MHC molecules.

This scenario of antigen recognition poses two major questions with regard to self/nonself dsicrimination. First, what happens to potentially harmful pre-killer cells which have receptors specific for self peptides presented by self MHC

molecules? Second, what happens to useless T cells which cannot bind to self MHC molecules at all, and therefore cannot recognize nonself peptides presented by self MHC molecules?

SELF/NONSELF DISCRIMINATION
IN T CELL RECEPTOR TRANSGENIC MICE

We have learned from experiments during the last decade that one cannot study these questions in any conclusive detail in normal animals, because normal animals have about 10^8 different T cell receptors [24–33]. For this reason it is very difficult to follow a single cell with one particular receptor throughout lymphocyte development. To overcome this prob-

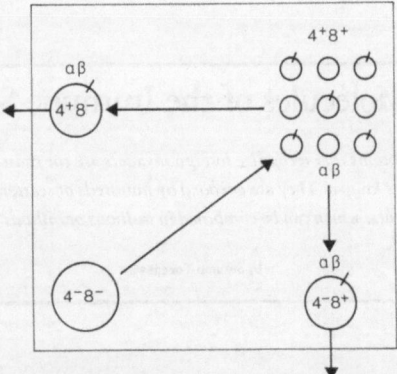

Fig. 6. *T cell development in the thymus*. The thymus is represented by the rectangle. Large dividing cells (lower left) enter the thymus and begin to express TCR (symbolized by /) as well as CD4 and CD8 coreceptors. As a consequence many CD4$^+$8$^+$ immature thymocytes are produced (5×10^7 cells/day). Only a few mature further and become CD4$^+$8$^-$ pre-helper and CD4$^-$8$^+$ pre-killer cells which leave the thymus (1×10^6 cells/day)

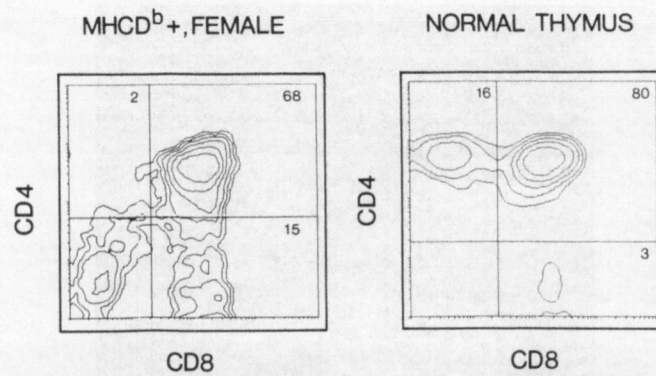

Fig. 8. *The cellular composition of the thymus of a TCR transgenic mouse (a normal thymus for comparison on the right hand side) expressing the MHC molecule (class I MHC D^b) but not the peptide (male antigen) recognized by the transgenic TCR*. Staining was carried out as in Fig. 7. CD4$^-$8$^+$ pre-killer cells are selected (selection of useful cells) from CD4$^+$8$^+$ precursors. Specificity of receptor determines whether cells become CD4$^+$ helpers and CD8$^+$ killers. No CD4$^+$8$^-$ pre-helper cells are detectable

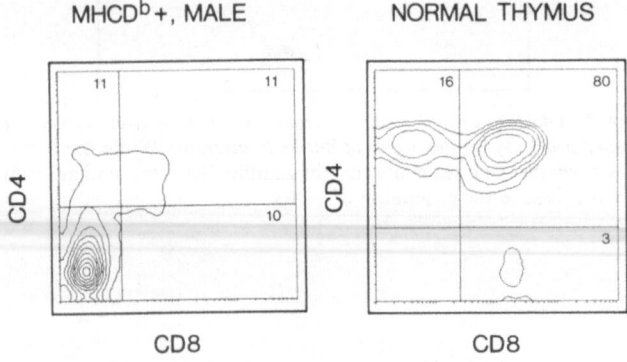

Fig. 7. *The cellular composition of the thymus of a TCR transgenic mouse (a normal thymus for comparison on the right hand side) expressing both the peptide (male antigen) and the MHC molecule (class I MHC D^b) recognized by the transgenic TCR*. The thymocytes are stained by antibodies against CD4 (red fluorescence on *y*-axis) and against CD8 (green fluroescence on *x*-axis) coreceptor. CD4$^+$8$^+$ thymocytes are eliminated (deletion of harmful cells at an immature stage)

lem, we decided to study self/nonself discrimination in TCR transgenic mice which have only one receptor with one known specificity [34–38]. We have constructed such a mouse by introducing productively rearranged α and β TCR genes from one particular killer lymphocyte into a mouse strain which suffers from a genetic defect, the *scid* mutation. *scid* stands for severe combined immune deficiency, which results from the inability of this mouse strain to properly rearrange antigen receptor gene segments. Therefore, mice from this strain cannot produce any antigen receptors [39]. The introduction of one functionally rearranged α and one functionally rearranged β TCR gene into the germ line of this mouse strain should result in a new strain with an immune system expressing only one receptor instead of 10^8 (Fig. 5). With the help of various collaborators we produced such a strain and could address questions concerning the self/nonself discrimination of the immune system in a much more conclusive way.

For our purposes we chose a receptor specific for an intracellular peptide that is present in male but absent in female mice (HY peptide), and which is presented by D^b MHC mol-

ecules. The genes encoding this receptor were isolated from a killer cell with a CD8 coreceptor and the productively rearranged α and β TCR genes were introduced into *scid* mice. In various *scid* mice we could then analyze the fate of T lymphocytes expressing the transgenic receptor: we could study the fate of potentially harmful cells in transgenic mice that have both the HY peptide as well as H-2D^b MHC molecules. We could study the fate of potentially useful cells in female transgenic mice that do not have the HY peptide, but do have D^b MHC molecules. Finally, we could analyze the fate of the useless T cells in female transgenic mice, that lack both the HY peptide as well as D^b MHC molecules. The various transgenic mice were obtained by introducing the T cell receptor transgenes in one particular mouse strain, and by further crossing this transgenic strain with other nontransgenic strains with MHC molecules which were appropriate for our purposes.

T LYMPHOCYTE DEVELOPMENT IN THE THYMUS

For a better understanding of the results obtained with the various transgenic mice, a few facts about T cell development need to be introduced here (Fig. 6). T cell development takes place in the thymus. Throughout life the thymus is colonized by hemopoietic cells that contain the precursors of T cells [40, 41]. As soon as the precursors enter the thymus they start to rearrange their receptor gene segments and express receptors and coreceptors on the cell surface [42–44]. Initially, the immature T cells each express one particular αβ TCR and CD4 as well as CD8 coreceptors. These cells are produced in large quantities (in a mouse 5×10^7 cells/day) and most of them die within a few days [45]. A few of them, however, become mature pre-killer or pre-helper cells with CD8 and CD4 coreceptors, respectively, and leave the thymus (Fig. 6). These mature cells can be stimulated by antigen to become effector cells [46], while their immature CD4$^+$8$^+$ precursors cannot be induced in this way, even though they express αβ TCR, as well as both coreceptors on the cell surface [16, 47]. The overproduction of immature T cells has been suspected to reflect some cellular selection in the thymus [47]. The analysis of the various transgenic mice has confirmed this suspicion

TRANSGENIC RECEPTOR: E^d+ INFLUENZA PEPTIDE

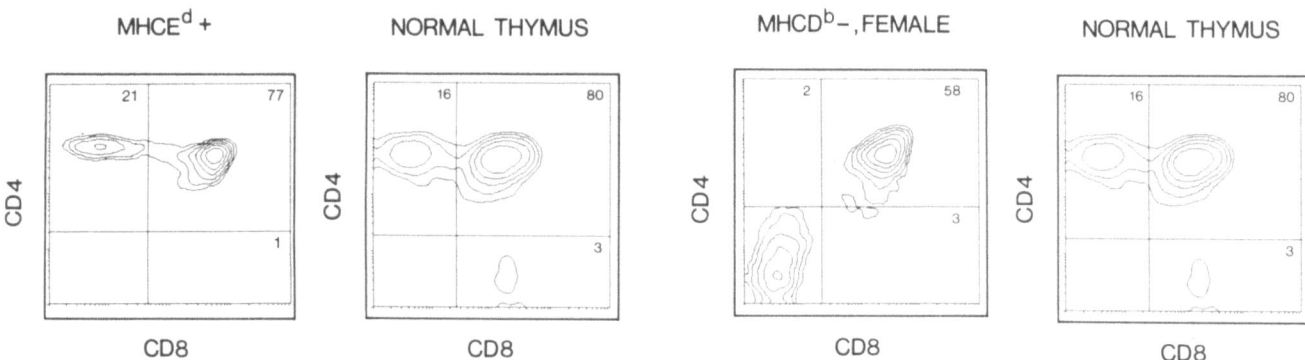

Fig. 9. *The cellular composition of the thymus of a TCR transgenic mouse (a normal thymus for comparison on the right side) expressing the MHC molecule (class II MHC E^d) but not the peptide (influenza hemagglutinin) recognized by the transgenic TCR.* Staining was carried out as in Fig. 7. CD4$^+$8$^-$ pre-helper cells are selected (selection of useful cells) from CD4$^+$8$^+$ precursors. No CD4$^-$8$^+$ pre-killer cells are detectable

Fig. 10. *The cellular composition of the thymus of a TCR transgenic mouse (a normal thymus for comparison on the right hand side) expressing neither the MHC molecule (class I MHC D^b) nor the peptide (male antigen) recognized by the transgenic TCR.* Staining was carried out as in Fig. 7. There is no selection of CD4$^-$8$^-$ or CD4$^-$8$^+$ T cells and development is arrested at the CD4$^+$8$^+$ immature stage (useless cells are neglected)

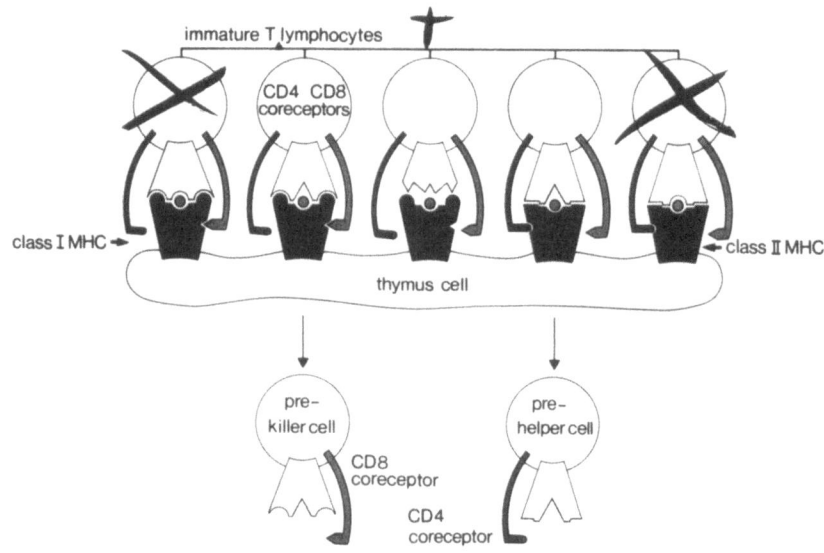

Fig. 11. *Negative and positive selection of cells in the normal thymus.* Cells with various $\alpha\beta$ TCRs (white) and CD8 (black) and CD4 (black) coreceptors are depicted. If the $\alpha\beta$ TCR binds to both the peptide and MHC molecules, the cell is deleted (extreme left and right). If the receptor binds to the MHC molecule but not to the specific peptide (the MHC molecule may contain some other peptide), the cell is selected for further maturation: binding to class I MHC molecules results in mature CD4$^-$8$^+$ pre-killer cells and binding to class II MHC molecules in mature CD4$^+$8$^-$ pre-helper cells. Cells with receptors which do not bind to any ligand die after a short lifespan

and uncovered the principles of these selection processes which take place in the thymus.

SOME EXPERIMENTAL DATA

The experimental data which concern the subsets of thymocytes in the various transgenic mice are shown in Figs 7 – 10. For comparison, the cellular composition of the thymus from a normal nontransgenic mouse is also shown. The thymocytes from the various mice are stained with fluorochrome-marked antibodies against CD4 and CD8 coreceptors. The analysis shows that in the first type of transgenic mouse that has both the HY peptide, as well as the D^b MHC molecule, there are no mature cells, neither CD4$^-$8$^+$ pre-killer

cells nor CD4$^+$8$^-$, pre-helper cells. Even the immature CD4$^+$ 8$^+$ precursors are absent (Fig. 7). Obviously, in this mouse, the potentially harmful T cells are destroyed even before they mature and become dangerous [34, 48]. In the second type of transgenic mouse that lacks the HY peptide but has D^b MHC molecules, potentially useful T cells, which can recognize nonself peptides presented by self MHC molecules, mature and become CD8 positive pre-killer cells (Fig. 8) [35, 36, 49]. Interestingly, there are no mature pre-helper cells with CD4 coreceptors in this mouse. This indicates that the specificity of the $\alpha\beta$ TCR for class I or class II MHC molecules determines whether a cell will develop into a pre-killer or pre-helper cell, respectively [16, 35, 36, 49]. In fact, if one makes a transgenic mouse with an $\alpha\beta$ TCR derived from a CD4$^+$8$^-$ helper cell, one obtains only CD4$^+$8$^-$, but not CD4$^+$8$^-$ mature thymocytes in the trangenic mouse (Fig. 9) [50, 51]. Finally, in the third mouse which lacks both the HY peptide as well as the D^b MHC molecule, cells expressing the transgenic receptor are useless. They die after a short lifespan and mature T cells do not emerge (Fig. 10) [37]. Thus, it is the interaction of the TCR with peptides and MHC molecules in the thymus that determine whether a cell is potentially harmful and, therefore, deleted, potentially useful and selected for maturation, or useless and not selected at all [52].

CONCLUSION

Extrapolating from these data, we arrive at the following picture of self/nonself discrimination in a normal mouse (Fig. 11). Immature CD4$^+$8$^+$ thymocytes express a large variety of receptors that are generated by random recombination of receptor gene segments and α and β TCR chain pairing. If the receptor does not bind to any ligand in the thymus, the cell will die after about three days. If the receptor binds to both the peptide and the presenting MHC molecule, the cell will be deleted. If the receptor binds to the MHC molecule only (which may be occupied by some other peptide), the cell will be selected for further maturation. Depending on whether the receptor binds to a class I or class II MHC molecule, the selected cell will assume the CD4$^-$8$^+$ or CD4$^+$8$^-$ phenotype, respectively. Thus, the initial learning by the immune system does involve selection of lymphocytes: the useful are selected, the useless are neglected and the harmful are deleted.

Prof. Pawel Kisielow's contribution to these studies is especially acknowledged. Other important collaborators include Drs Anton Berns, Horst Blüthmann, Peter Borgulya, Hiroyuki Kishi, Bernadette Scott, Michael Steinmetz, Hung-Sia Teh and Yasushi Uematsu. The expert technical support by Katrin Hafen and Verena Stauffer throughout the conduction of these experiments is gratefully acknowledged. The help of Nicole Schoepflin in the preparation of this manuscript is also gratefully acknowledged.

REFERENCES

1. Jerne, N. K. (1955) *Proc. Natl Acad. Sci. USA 41*, 849–857.
2. Burnet, F. M. (1957) *Aust. J. Sci. 20*, 67.
3. Edelman, G. M. (1970) *Sci. Am. 223*, 34–42.
4. Jerne, N. K. (1973) *Sci. Am. 229*, 52–60.
5. Cerottini, J. C., Nordin, A. A. & Brunner, K. T. (1970) *Nature 227*, 72–73.
6. Babbitt, B. P., Allen, P. M., Matsueda, G., Haber, E. & Unanue, E. R. (1985) *Nature 317*, 359–361.
7. Buus, S., Sette, A., Colon, S. M., Jenis, D. M. & Grey, H. M. (1986) *Cell 47*, 1071–1077.
8. Townsend, A. R., Rothbard, J., Gotch, F. M., Bahadur, G. & Wraith, D. (1986) *Cell 44*, 959–968.
9. Bjorkman, P. J., Saper, M. A., Samraoni, B., Bennett, W. S. & Strominger, J. L. (1987) *Nature 329*, 509–512.
10. Allison, J. P. & Lanier, L. L. (1987) *Annu. Rev. Immunol. 5*, 503–540.
11. Dembic, Z., Haas, W., Weiss, S., McCubrey, J. & Kiefer, H. (1986) *Nature 320*, 232–238.
12. Davis, M. M. & Bjorkman, P. J. (1988) *Nature 334*, 395–402.
13. Dembic, Z., Haas, W., Zamoyska, R., Parnes, J. & Steinmetz, M. (1987) *Nature 326*, 510–511.
14. Connolly, J. M., Potter, T. A., Wormstall, E. & Hansen, T. H. (1988) *J. Exp. Med. 168*, 325–341.
15. Emmerich, F., Strittmatter, V. & Eichmann, K. (1986) *Proc. Natl Acad. Sci. USA 83*, 8298–8302.
16. von Boehmer, H. (1986) *Immunol. Today 7*, 333–336.
17. Miller, J. F. A. (1961) *Lancet 2*, 748.
18. Mitchell, G. F. & Miller, J. F. (1968) *J. Exp. Med. 128*, 821–837.
19. Dialynas, D. P., Quan, Z. S., Wall, K. A., Pierres, A. & Quintans, J. (1983) *J. Immunol. 131*, 2445–2451.
20. Doyle, C. & Strominger, J. L. (1987) *Nature 330*, 256–259.
21. Lanzavecchia, A. (1985) *Nature 314*, 537–539.
22. Tonegawa, S. (1983) *Nature 302*, 575–581.
23. Klein, J. (1979) *Science 203*, 516–521.
24. von Boehmer, H. & Sprent, J. (1976) *Transplant. Rev. 29*, 3–23.
25. Bevan, M. J. (1977) *Nature 269*, 417–418.
26. Zinkernagel, R. M., Callahan, G. N., Althage, A., Cooper, S. & Klein, P. A. (1978) *J. Exp. Med. 147*, 882–896.
27. von Boehmer, H., Haas, W. & Jerne, N. K. (1978) *Proc. Natl Acad. Sci. USA 75*, 2439–2442.
28. Zinkernagel, R. M., Althage, A., Waterfield, E., Kindred, B. & Welsh, R. M. (1980) *J. Exp. Med. 151*, 376–399.
29. Wagner, H., Hardt, C., Stockinger, H., Pfizenmaier, K. & Bartlett, R. (1981) *Immunol. Rev. 58*, 95–129.
30. Matzinger, P. (1981) *Nature 292*, 497–501.
31. von Boehmer, H., Teh, H. S., Bennick, H. R., Haas, W. (1985) in *Recognition and regulation in cell-mediated immunity* (Watson, J. D. & Mabrook, J., eds) pp. 89–101, Dekker, New York and Basel.
32. Ishii, N., Nagy, Z. A. & Klein, J. (1982) *Nature 295*, 531–533.
33. Goverman, J., Hunkapillar, T. & Hood, L. E. (1986) *Cell 45*, 475–484.
34. Kisielow, P., Blüthmann, H., Staerz, U. D., Steinmetz, M. & von Boehmer, H. (1988) *Nature 333*, 742–746.
35. Teh, H. S., Kisielow, P., Scott, B., Kishi, H. & Uematsu, Y. (1988) *Nature 335*, 229–233.
36. Kisielow, P., Teh, H. S., Blüthmann, H. & von Boehmer, H. (1988) *Nature 335*, 730–733.
37. Scott, B., Blüthmann, H., Teh, H. S. & von Boehmer, H. (1989) *Nature 338*, 591–593.
38. von Boehmer, H. (1990) *Annu. Rev. Immunol. 8*, 531–556.
39. Bosma, G. C., Custer, P. R. & Bosma, M. J. (1983) *Nature 301*, 527–529.
40. Moore, M. A. & Owen, J. J. (1967) *J. Exp. Med. 126*, 715–726.
41. Le Douarin, N. M. & Jotereau, F. V. (1975) *J. Exp. Med. 142*, 17–40.
42. Snodgrass, R. H., Dembic, Z., Steinmetz, M. & von Boehmer, H. (1985) *Nature 315*, 232–233.
43. Raulet, D. H., Garman, R. D., Saito, H. & Tonegawa, S. (1985) *Nature 314*, 103–107.
44. Crisanti, A., Colantoni, A., Snodgrass, R. H. & von Boehmer, H. (1986) *EMBO J. 11*, 2837–2843.
45. Shortman, K. & Jackson, H. (1974) *Cell. Immunol. 12*, 230–246.
46. Kisielow, P., von Boehmer, H. & Haas, W. (1982) *Eur. J. Immunol. 12*, 463–467.
47. von Boehmer, H. (1988) *Annu. Rev. Immunol. 6*, 309–326.
48. Sha, W. C., Nelson, C. A., Newberry, R. D., Kranz, D. M. & Russell, J. H. (1988) *Nature 336*, 73–76.
49. Sha, W. C., Nelson, C. A., Newberry, R. D., Kranz, D. M. & Russell, J. H. (1988) *Nature 335*, 271–274.
50. Berg, L. J., Pullem, A. M., Fazekas de St. Groth, B., Mathis, D. & Benoist, C. (1989) *Cell 58*, 1035–1046.
51. Kaye, J., Hsu, M. L., Sauron, M. E., Jameson, J. C. & Gascoigne, R. J. (1989) *Nature 341*, 746–749.
52. von Boehmer, H., Teh, H. S. & Kisielow, P. (1989) *Immunol. Today. 10*, 57–61.
53. Tonegawa, S. (1985) *Sci Am. 253*, 122–131.

Eur. J. Biochem. *194*, 699–712 (1990)
© FEBS 1990

Review

Eukaryotic DNA replication

Enzymes and proteins acting at the fork

Pia THÖMMES and Ulrich HÜBSCHER

Department of Pharmacology and Biochemistry, University Zürich-Irchel, Switzerland

(Received July 3, 1990) – EJB 90 0784

A complex network of interacting proteins and enzymes is required for DNA replication. Much of our present understanding is derived from studies of the bacterium *Escherichia coli* and its bacteriophages T4 and T7. These results served as a guideline for the search and the purification of analogous proteins in eukaryotes. Model systems for replication, such as the simian virus 40 DNA, lead the way.

Generally, DNA replication follows a multistep enzymatic pathway. Separation of the double-helical DNA is performed by DNA helicases. Synthesis of the two daughter strands is conducted by two different DNA polymerases: the leading strand is replicated continuously by DNA polymerase δ and the lagging strand discontinuously in small pieces by DNA polymerase α. The latter is complexed to DNA primase, an enzyme in charge of frequent RNA primer syntheses on the lagging strand. Both DNA polymerases require several auxiliary proteins. They appear to make the DNA polymerases processive and to coordinate their functional tasks at the replication fork. $3' \rightarrow 5'$-exonuclease, mostly part of the DNA polymerase δ polypeptide, can perform proof-reading by excising incorrectly base-paired nucleotides. The short DNA pieces of the lagging strand, called Okazaki fragments, are processed to a long DNA chain by the combined action of RNase H and $5' \rightarrow 3'$-exonuclease, removing the RNA primers, DNA polymerase α or β, filling the gap, and DNA ligase, sealing DNA pieces by phosphodiester bond formation. Torsional stress during DNA replication is released by DNA topoisomerases. In contrast to prokaryotes, DNA replication in eukaryotes not only has to create two identical daughter strands but also must conserve higher-order structures like chromatin.

Introduction and pending problems

DNA replication is the process that guarantees the faithful duplication of the genetic information in advance of division of any living cell. This universal event is crucial for the correct maintenance of the genetic information. Research progress in the last few years suggested that mechanisms of DNA replication might be similar in most organisms investigated so far. DNA synthesis at the advancing replication fork requires the coordinated action of different types of enzymes which appear to be in a higher-order structure to ensure the precise and rapid duplication of the more than 10^9 base pairs in higher eukaryotic cells.

Some basic rules emerging from enzymatic studies of DNA replication proved to be valid for both prokaryotes and eukaryotes (see Kornberg, 1988; Cohen et al., 1988; Kelly and Stillman, 1988; van der Vliet, 1989; Spadari et al., 1989; Baker

and Kornberg, 1990; Richardson and Lehman, 1990; Engler and Wang, 1990 for recent reviews, symposia and monograph; see also Table 1):

a) Sequence-specific DNA binding proteins recognize and possibly open specific DNA sequences called origins of replication.

b) Before the duplication process itself can start, strand separation is achieved by enzymes called DNA helicases.

c) With a few exceptions chain initiation is generally performed through RNA priming.

d) The DNA polymerases involved hydrolyze deoxyribonucleoside 5'-triphosphates adding deoxyribonucleoside monophosphates to the growing chain and releasing pyrophosphate.

e) They are guided by a template via the Watson-Crick base-pairing rule.

f) The direction of DNA polymerization is $5' \rightarrow 3'$.

g) In addition to DNA polymerases, various auxiliary proteins are required for priming, efficient primer recognition, processivity, fidelity, coordination and recycling. This results in different subassemblies for leading and lagging strand synthesis.

h) Release of topological stress is achieved by DNA topoisomerases.

The detailed research in the field of DNA replication has one common goal, which is to understand how the duplication of the genetic material is achieved on the biochemical level and, even more important, how it can be regulated. Several

---

Correspondence to U. Hübscher, Department of Pharmacology and Biochemistry, University Zürich-Irchel, Winterthurerstrasse 190, CH-8057 Zürich, Switzerland

Abbreviations. SV40, simian virus 40; T-ag, the large T antigen; HSV, herpes simplex virus; PCNA, proliferating cell nuclear antigen; SSB, single-stranded DNA binding protein; RF, replication factor; RP, replication protein; HIV, human immunodeficiency virus.

Enzymes. DNA polymerase/primase (EC 2.7.7.7); ribonuclease H (EC 3.1.26.4); DNA ligase (EC 6.5.1.1); DNA topoisomerase I (EC 5.99.1.2); DNA topoisomerase II (EC 5.99.1.3); $3' \rightarrow 5'$ exonuclease (EC 3.1.16.1); $5' \rightarrow 3'$ exonuclease (EC 3.1.4.1).

Table 1. *Enzymes with their likely roles at the eukaryotic replication fork*

DNA primase is always complexed to DNA polymerase α. Details of the five eukaryotic DNA polymerases are given in Table 2. The 3′ — 5′ exonuclease appears either as a separate enzyme or as part of the DNA polymerase δ or ε catalytic polypeptide (Bambara and Jessee, 1990) or, in the case of *D. melanogaster*, of the DNA polymerase α catalytic subunit (Cotterill et al., 1987)

Enzyme	Role
DNA helicase	unwinding of DNA
DNA primase	synthesis of primers
DNA polymerases	synthesis of the DNA strands
Ribonuclease H, 5′ → 3′ exonuclease	removal of primers
DNA ligase	ligation of DNA pieces
3′ → 5′ exonuclease	proof-reading for the DNA polymerases
DNA topoisomerase I	conversion of topological DNA isomers
DNA topoisomerase II	segregation of DNA strands

model systems have helped to come closer to this aim. The bacteriophage T4 provides a model for an asymmetric replication fork in which auxiliary proteins modulate the properties of the replicative enzymes (Alberts, 1987). The *Escherichia coli* DNA polymerase III holoenzyme provides insight into the functional mode of a very complex replication machinery (McHenry, 1988). This enzyme seems to exist as an asymmetric dimer in its functional form which is composed of 22 polypeptides forming nine different subunits (Maki et al., 1988). The eukaryotic virus simian virus 40 (SV40) has been particularly useful for the functional identification of cellular proteins that are needed for its DNA replication (Kelly, 1988; Challberg and Kelly, 1989; Stillman, 1989). It is now possible to replicate the DNA of this virus completely *in vitro*. The replication proteins of the yeast *Saccharomyces cerevisiae* are amazingly similar to the ones of higher eukaryotes, in structure as well as in primary sequence (Burgers, 1989). In addition yeast has the advantage that it can be manipulated genetically, thus providing a functional approach for each of the enzymes involved in DNA replication. Finally, calf thymus tissue is a rich source for purification of mammalian replicative enzymes.

Although these model systems have led to considerable increase of knowledge, many problems still remain unsolved. Questions on how two DNA polymerases coordinate their functional duties at the replication fork should be addressed soon. What are the roles of DNA polymerase auxiliary proteins in such a coordination process? Do DNA polymerases interact with DNA helicases? How are the Okazaki pieces processed? Is there an interaction of the replisomal enzymes with DNA topoisomerases? How can the replication machinery deal with chromatin and other structural elements in the nucleus? To answer these questions the complete isolation of a functionally active replisome is required.

This review focuses exclusively on known mechanistic details occurring at the replication fork itself, while initiation and termination events are neglected. In particular, the following paragraphs address the enzymology of six major processes: (a) the separation of the double-stranded DNA; (b) the synthesis of the two daughter strands; (c) the processing of the Okazaki fragments; (d) the proof-reading of the two DNA strands; (e) the release of torsional stress and (f) the interaction of the replicative complex with chromatin.

Separation of the double helix

Before semiconservative duplication of the genetic material can take place, the double-helical DNA must be opened so that the two single strands can serve as templates for the DNA polymerase. This is performed by enzymes called DNA helicases which break up the hydrogen bonds between the base pairs of the DNA strands (see Matson and Kaiser-Rogers, 1990; Thömmes and Hübscher, 1990a, for recent reviews). Since the helical structure is energetically favored, the unwinding of the helix is an energy-consuming process. Helicases obtain their energy from the hydrolysis of nucleoside 5′-triphosphates so that they can translocate along one strand of the DNA, thus displacing the other one. For example, the opening of one base pair requires hydrolysis of two molecules of ATP as has been exemplified for the *E. coli* rep helicase (Kornberg et al., 1978).

Helicases are needed not only during DNA replication but also in other processes such as recombination and repair. It appears that the helicases are specialized to a great extent for their respective tasks so that the multiplicity of DNA transactions requires even more helicases. For an organism as simple as *E. coli*, ten different helicases have been characterized up to now (Matson and Kaiser-Rogers, 1990). An even higher number of biochemically and functionally different helicases is to be expected in the much more complex eukaryotes. However, only a few eukaryotic helicases have been isolated and characterized so far (Thömmes and Hübscher, 1990a).

Viral systems have been used as models for eukaryotic processes and have been useful to analyse DNA replication. The virus that has been studied most thoroughly is the papova virus, simian virus 40 (SV40; Stillman, 1989; Challberg and Kelly, 1989). It relies completely on the replication machinery of its eukaryotic host with exception of a single viral protein, the large T-antigen (T-ag). This multifunctional protein exhibits, among many other properties, DNA binding and ATP-dependent unwinding activities (Stahl et al., 1986; Stahl and Knippers, 1987). The helicase activity of T-ag is required for local unwinding at the origin of replication as well as during the elongation stages (Stahl et al., 1985; Dean et al., 1987; Wold et al., 1987). Only after T-ag has bound to the origin and has opened up a few nucleotides the DNA polymerase α/primase complex of the cellular replication machinery can enter and start replication.

T-ag is a processive helicase with the ability to unwind DNA stretches as long as several thousand base pairs at a rate of 75 – 100 base pairs/min (Wiekowski et al., 1988). These features make it a likely candidate as the only helicase that is needed for the replication of the whole SV40 genome of 5243 base pairs. Strand displacement requires the hydrolysis of ribo- or deoxyribonucleoside triphosphates, ATP supporting the reaction best. Other nucleotides can be used to a lesser extent. T-ag unwinds DNA 3′ → 5′ with respect to the DNA strand to which it is bound, thus implying sliding along the leading strand of the replication fork. It binds preferentially to 3′ overhangs of single-strand/double-strand transitions from where it starts the unwinding reaction (Goetz et al., 1988; Wiekowski et al., 1988). If no single-stranded stretches are present, unwinding can start internally on fully double-stranded DNA with a minimal length of 60 base pairs (Scheffner et al., 1989). It depends on the reaction conditions whether a T-ag binding site is necessary for internal binding or not (Dodson et al., 1987).

Another very well characterized model system for eukaryotic DNA replication is herpes simplex virus 1 (HSV1). In contrast to SV40, this virus encodes itself for all the proteins that are necessary for its replication (Knipe, 1989). Among the seven viral genes that are essential for replication are UL5, UL8 and UL52 (UL = unique L component reading frame), whose gene products form a helicase/primase complex (Crute et al., 1989). The ratio of proteins within this complex is roughly 1:1:1 (J. J. Crute, personal communication). While the primase activity could be assigned to the 70-kDa UL8 subunit, both the UL5 and the UL52 gene products seem to be necessary for helicase function. When expressed in baculovirus-infected insect cells neither of these proteins is active on its own (J. J. Crute, personal communication). However, upon double or triple infection helicase or helicase/primase activity can be restored (Dodson et al., 1989). HSV1 DNA helicase hydrolyses ATP as well as GTP and to a lesser extent other nucleotides (Crute et al., 1988). This probably reflects the presence of two different catalytic sites, one of which is completely dependent on ATP while the other one is less strict in its nucleotide utilization (J. J. Crute, personal communication). The helicase activity is dependent on the presence of a 3' unpaired tail on the substrate indicating a 5' → 3' direction of movement on the strand to which the enzyme is bound (Crute et al., 1988). This implies binding to the lagging strand of the replication fork, a feature which is stressed by the presence of the primase in the same complex as the helicase. HSV1 encodes for a second DNA helicase, which is the gene product of UL9. This protein binds to the origin of replication and unwinds in 3' → 5' direction, suggesting a function during origin unwinding as can be seen for SV40 T-ag (M. D. Challberg, personal communication).

Since these two well characterized viral systems encode for their own DNA helicase, they could not help in the search of their cellular counterparts. Accordingly little is known about eukaryotic DNA helicases involved in DNA replication or other cellular processes. The usual approach in search of helicases was to follow DNA-dependent ATPases using the energy dependency of the helicase reaction as an assay (Tarawagi et al., 1984). However, in most cases this strategy was unsuccessful. The first assay for the detection of DNA helicases used double-stranded DNA as substrate and followed the creation of single-stranded DNA (Abdel-Monem et al., 1976). Besides being very laborious, this assay proved to be unsuitable to detect helicase activity in eukaryotic cell extracts, since these are rich in nucleases which destroy the substrate before it can be unwound. In recent years some progress has been achieved by using single-stranded DNA as a substrate to which short labelled fragments were hybridized (Matson et al., 1983). This assay allowed quantification of activity, the direct analysis of the reaction products and the determination of the unwinding direction.

From the yeast S. cerevisiae two DNA helicases have been purified so far. ATPase III was first purified as an DNA-dependent ATPase and later identified as a helicase (Sugino et al., 1986). The ATPase activity is completely dependent on single-stranded DNA which can only poorly be substituted by double-stranded DNA or polyribonucleotides. The characteristic feature of this enzyme is its stimulatory effect on yeast DNA polymerase α while DNA polymerase from other organisms are not affected. Its helicase activity is dependent on ATP or dATP. The second yeast helicase, the RAD3 gene product, is probably involved in DNA repair and therefore not discussed here (Sung et al., 1987).

A DNA helicase from Xenopus laevis has been purified extensively but not to homogeneity (Poll and Benbow, 1988). This enzyme is different from other helicases since it is stimu-lated by monovalent cations. The mode of action is distributive since double-stranded stretches of 16 base pairs but not 26 base pairs can be unwound. The direction of unwinding is 3' → 5' and a 3' single-stranded stretch is required for enzyme binding (E. H. A. Poll, personal communication). These properties suggest that the X. laevis enzyme is not a replicative helicase.

By screening for DNA-dependent ATPases in mouse FM3A cells, four different enzymes have been partially characterized (Tarawagi et al., 1984). One of these enzymes has been purifed further and identified as a DNA helicase (Seki et al., 1988). In contrast to most other helicases described, this enzyme is able to hydrolyze all four rNTPs and dATP. For the unwinding reaction no single-stranded tail is required. Double-stranded stretches as long as 140 bp can be unwound in 5' → 3' direction.

Calf thymus tissue is a convenient source for the isolation of replicative enzymes. During the purification procedures of the five cellular DNA polymerases (see below) several different DNA helicases have been found. Each of these helicases tends to purify with a particular DNA polymerase. Whether this reflects any in vivo association or is merely due to simple copurification has to be clarified.

The first homogeneous DNA helicase described from calf thymus tends to stick to DNA polymerase α (Thömmes and Hübscher, 1990b) and has a molecular mass of 47 kDa. The activity is dependent on divalent cations and hydrolysis of ribonucleoside triphosphates. ATP and dATP are the preferred nucleotides but ddATP, CTP and dCTP can also be used. The direction of movement is 3' → 5'. A strand displacement activity was found to copurify with DNA polymerase δ over several chromatographic steps (Downey et al., 1990). Polymerizing and unwinding activities could be separated, and a DNA helicase was identified as the strand displacement activity. This helicase prefers ATP and dATP as nucleotide cofactors but can also use CTP and UTP. Unwinding is in the 5' → 3' direction. Finally, the most recently identified DNA polymerase ε also has its copurifying DNA helicase (G. Siegal and R. A. Bambara, personal communication). In contrast to the two calf thymus helicases described, it requires the presence of a single-stranded DNA binding protein for its action. The direction of unwinding is again 3' → 5'. Altogether these results are a good starting point to analyze in detail the properties and functions of various eukaryotic DNA helicases.

Synthesis of the two daughter strands

The five eukaryotic DNA polymerases

DNA polymerases are the main effectors of polymerization of deoxyribonucleoside triphosphates at the replication fork. Their overall function of performing accurate and fast DNA synthesis is responsible for their complex structure. Five different DNA polymerases have been identified and purified from eukaryotic cells. The polypeptide structure of the yeast enzymes is nearly identical to the one in higher eukaryotes which has lead to a new nomenclature for eukaryotic DNA polymerases (Burgers et al., 1990). In this new classification all eukaryotic polymerases are identified by the Greek letters α, β, γ, δ and ε (Table 2). DNA polymerase α (reviewed by Kaguni and Lehman, 1988; Lehman and Kaguni, 1989) is responsible for important tasks in nuclear DNA replication and appears to replicate the lagging strand of the replication fork (see section entitled Discontinuous replication of the lagging strand, below) to complete the Okazaki fragments (see

Table 2. *The five eukaryotic DNA polymerases*

A revised nomenclature for eukaryotic DNA polmyerases has been proposed adopting the Greek nomenclature for yeast DNA polymerases and introducing a fifth DNA polymerase, DNA polymerase ε (for details see text and Burgers et al., 1990)

DNA polymerase	Biological function	Reference
α	Replication of the lagging strand Completion of Okazaki fragments? Repair of nuclear DNA	So and Downey, 1988 Goulian et al., 1990a Seki et al., 1980
β	Repair of nuclear DNA Completion of Okazaki fragments? Recombination?	Hübscher et al., 1979 S. M. Linn, personal communication Nowak et al., 1990
γ	Replication of mitochondrial DNA	Hübscher et al., 1979
δ	Replication of the leading strand	So and Downey, 1988
ε	Repair of nuclear DNA? Replication of nuclear DNA?	Nishida et al., 1988 Morrison et al., 1990

section on **Processing of Okazaki fragments**, below) and to be involved in nuclear DNA repair. DNA polymerase β (reviewed by Wilson et al., 1988) is the main repair enzyme in the nucleus, might complete Okazaki fragment synthesis after removal of the RNA primer by RNase H and 5′ → 3′ exonuclease (see **Processing of Okazaki fragments**, below) and has recently been suggested to have a role in DNA recombination (Nowak et al., 1990). DNA polymerase γ (reviewed by Fry and Loeb, 1986) is the replicase for the mitochondrial DNA. DNA polymerase δ (reviewed by So and Downey, 1988; Bambara et al., 1990) is involved in nuclear DNA replication and is a candidate for replication of the leading strand of the replication fork (see *Continuous replication of the leading strand*, below). This DNA polymerase was first described in 1976 and was originally distinguished from DNA polymerase α by its 3′ → 5′ proof-reading exonuclease activity (Byrnes et al., 1976). Subsequently all 3′ → 5′-exonuclease-containing DNA polymerases were called DNA polymerase δ. On model homopolymer template/primers [e.g. poly(dA)/oligo(dT)] containing long single-stranded regions, this enzyme requires an auxiliary protein (see *Continuous replication of the leading strand* and *DNA polymerase auxiliary proteins*, below) for processive DNA synthesis (Tan et al., 1986). This auxiliary protein is called 'proliferating cell nuclear antigen' (PCNA; Bravo et al., 1987; Prelich et al., 1987b). Subsequently 3′ → 5′-exonuclease-containing DNA polymerases were isolated from calf thymus (Focher et al., 1988a; 1989), Hela cells (Syvaoja and Linn, 1989) and yeast (Hamatake et al., 1990) which were independent of PCNA for processive DNA synthesis. This enzyme is now called DNA polymerase ε (reviewed by Bambara and Jessee, 1990). Its function might be repair of nuclear DNA (Nishida et al., 1988) but it may also have an important function in DNA replication since its gene is essential in yeast (Morrison et al., 1990).

In the following paragraphs we now focus on the functional aspects of DNA polymerases δ and α at the replication fork, the roles of various DNA polymerase auxiliary proteins and the important question of coupling leading and lagging strand synthesis.

Semidiscontinuous replication of the two DNA strands

Based on the antiparallel arrangement of the two DNA strands and the universal 5′ → 3′ direction of DNA synthesis by every known DNA polymerase, it seems that a semidiscontinuous mode of DNA replication occurs at the fork

(Baker and Kornberg, 1990). One strand is replicated continuously, and called the leading strand, while the other is replicated discontinuously and called the lagging strand. This situation needs two types of DNA polymerase: the enzyme at the leading strand has to be highly processive, while the one at the lagging strand can be poorly processive but must have a capability to bind and retract from the DNA strand frequently (e. g. after completion of each Okazaki fragment of about 150 – 200 bases).

A model originally proposed by Alberts (Sinha et al., 1980) predicted that two DNA polymerases might act coordinately as a dimeric enzyme at the replication fork. According to this model (Fig. 1) one DNA polymerase replicates the leading strand continuously and the other the lagging strand discontinuously. The lagging strand would bend back on itself to form a loop at the replication fork. This would allow the dimeric DNA polymerase to move along both template strands in the 'same' direction without violating the 5′ → 3′ directionality rule. When the lagging-strand DNA polymerase reaches the Okazaki fragment synthesized during the previous round, the freshly synthesized DNA is threaded through the enzyme, allowing the DNA polymerase to recycle to a newly exposed single-stranded region situated further 3′ to the previous priming site for initiation of the next Okazaki fragment. In this way, a dimeric DNA polymerase holoenzyme would guarantee an efficient and coordinated progression of the growing replication fork. Data supporting this model accumulated subsequently in the T4 DNA replication system (Alberts et al., 1983). Biochemical data concerning this asymmetric dimer hypothesis for the *E. coli* DNA polymerase III holoenzyme (McHenry and Johanson, 1984) and for various forms of the DNA polymerase holoenzyme from calf thymus (Hübscher and Ottiger, 1984; Ottiger and Hübscher, 1984) were first published in 1984.

As mentioned above, in prokaryotes this dual DNA polymerase function appears to be performed by an asymmetric dimer of the DNA polymerase III holoenzyme (McHenry, 1988). In *E. coli* the DNA polymerase III core polypeptides are identical but are complexed with a variety of additional subunits (see *DNA polymerase auxiliary proteins*, below). The composition of these subunits provides the respective half of this twin DNA polymerase III with properties suited for replication of the leading or the lagging strand, respectively (McHenry, 1988).

In eukaryotes this asymmetry of the replication complex appears to be based on the action of two different DNA

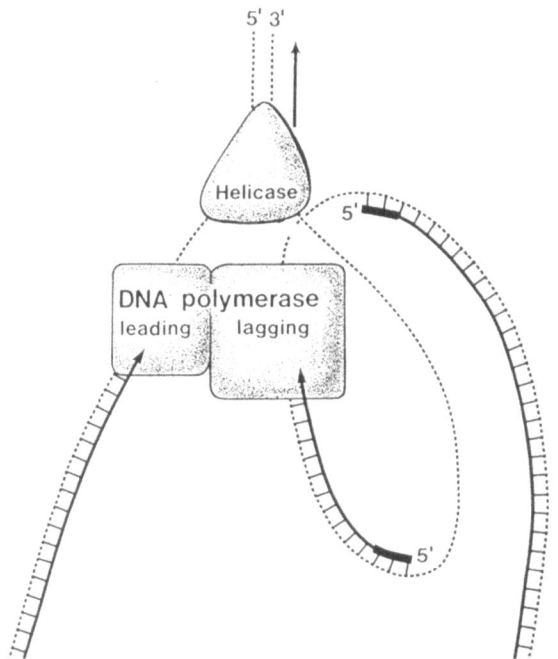

Fig. 1. *The asymmetric DNA replication fork.* Based on explanations given in the text, most organisms require an asymmetric type of DNA replication fork for concurrent replication of both DNA strands. An exception from this general rule is protein-primed replication of bacteriophage φ29 and adenovirus. Parental strands (······), newly replicated DNA (———) primers (▬▬▬)

polymerases, namely DNA polymerase δ and DNA polymerase α (So and Downey, 1988; Burgers, 1989). The genes for both of these enzymes in the yeast *S. cerevisiae* have been cloned and sequenced and both found to be essential for DNA replication (Pizzagalli et al., 1988; Boulet et al., 1989; Sitney et al., 1989). DNA polymerase α contains an intrinsic DNA primase activity in two of its four subunits (Kaguni and Lehman, 1988) and is only moderately processive, suggesting a function in the discontinuous replication of the lagging strand. In contrast, DNA polymerase δ lacks a DNA primase and is very processive in the presence of PCNA (Downey et al., 1988), suggesting a role in the continuous replication of the leading strand.

Continuous replication of the leading strand

As mentioned, a processive DNA polymerase is required for continuous DNA synthesis. An enzyme once bound to a template/primer 3′OH group has to possess the capacity to translocate along the DNA without dissociating from it. There are examples in nature of DNA polymerases consisting of a single polypeptide which can be processive. However, such a minimal strategy of a replicative DNA polymerase as found, for example, in the bacteriophage φ29 (Blanco et al., 1989) is a rare exception, since processive DNA polymerase from most organisms require auxiliary proteins for these functional capabilities. The best characterized example derives from bacteriophage T4. Since the sequences of the DNA polymerases, as well as those of the auxiliary proteins, might be similar to eukaryotes (see *DNA polymerase auxiliary proteins*, below) this prokaryotic system is discussed in some detail here (for an overview see Alberts, 1987). The leading strand DNA polymerase of T4 requires three auxiliary proteins for a

holoenzyme function. The gene products 44/62 and 45 act together as a 'sliding clamp' for the DNA polymerase. By using ATP hydrolysis, they are in a high-energy conformation (Munn and Alberts, 1990a), guide the DNA polymerase to a template/primer and fix it there so that the DNA polymerase can move along a single-stranded DNA in a very processive fashion as a productive holoenzyme (Munn and Alberts, 1990b). The auxiliary proteins can form a specific complex with the gene 32 protein at the primer/template junction (Munn and Alberts, 1990b). Finally, the 41 helicase is required for a maximum rate of leading strand DNA synthesis on a double-stranded DNA (Cha and Alberts, 1990).

In eukaryotes, for a very long period DNA polymerase α was thought to be the main and even the only enzyme involved in DNA replication (Fry and Loeb, 1986). However, this DNA polymerase was never found to be a highly processive enzyme either in its simpler form of only four subunits or in its more complex forms. This view changed when, upon fractionation of the SV40 *in vitro* replication system, it was found that the cell cycle regulated protein proliferating cell nuclear antigen (PCNA, reviewed by Celis et al., 1987; Mathews, 1989) had an important role in DNA replication (Prelich et al., 1987a). At the same time, PCNA was identified as an auxiliary protein of DNA polymerase δ (Prelich et al., 1987b; Bravo et al., 1987).

Although DNA polymerase δ has been known for more than a decade, it was not given the scientific importance that it deserved. This was due to the fact that it could only copy the heteropolymeric DNA poly(dA-dT). PCNA can enable DNA polymerase δ to copy long single-stranded DNA stretches on poly(dA)/oligo(dT) and renders the DNA polymerase processive on this homopolymer (Downey et al., 1988). This finding makes DNA polymerase δ a likely candidate as the enzyme for leading strand synthesis. Two other characteristics stress this conclusion: DNA polymerase δ contains an intrinsic $3′ \rightarrow 5′$ exonuclease which contributes to the fidelity of the enzyme and PCNA is, as mentioned above, highly cell-cycle-regulated and thus able to provide DNA polymerase δ with this property. However, PCNA alone could still not make DNA polymerase δ processive on natural DNA such as, for example, primed M13 DNA (Tsurimoto and Stillman, 1989b). At least two additional proteins which appear to be necessary for processive synthesis on natural DNA. First, a single-stranded DNA-binding protein (called SSB, RF-A or RP-A; see *DNA polymerase auxiliary proteins*, below), and second, replication factor C (RF-C, see *DNA polymerase auxiliary proteins*, below), a multi-subunit protein required for SV40 replication *in vitro* (Tsurimoto and Stillman, 1989a).

Altogether it appears, as we learned earlier from the bacteriophage T4 system, that DNA polymerase δ, once it has the proper composition, can act as a very efficient and processive enzyme with biochemical properties suitable for continuous replication of the leading strand at the fork.

In contrast to the PCNA-dependent DNA polymerase δ, a δ-like DNA polymerase was found that performs processive DNA synthesis in the absence of PCNA (Focher et al., 1988a; 1989; Syvaoja and Linn, 1989; Hamatake et al., 1990). This enzyme is now termed DNA polymerase ε (see above and also Burgers et al., 1990). Originally this enzyme was isolated based on an *in vitro* DNA repair assay (Nishida et al., 1988). Its very high processivity might also suggest a role in DNA replication. DNA polymerase ε has recently been cloned in yeast and was found to be an essential gene for this organism. An involvement in chromosomal DNA replication is therefore very likely (Morrison et al., 1990, see also **Conclusions and perspectives**, below). DNA polymerase ε is found to be a multipolypeptide DNA polymerase as exemplified in calf thymus (Focher et al., 1989; T. Weiser and U. Hübscher, unpublished data), HeLa cell (Syvaoja and Linn, 1989; Syvaoja et al., 1990) and in yeast *S. cerevisiae* (Hamatake et al., 1990).

Discontinuous replication of the lagging strand

The $5' \rightarrow 3'$ synthesis direction of the DNA polymerase indicates discontinuous replication on the lagging strand. The short DNA pieces synthesized on this strand, called Okazaki fragments, have an average length of 150–200 bases in eukaryotic cells (DePamphilis and Wassarman, 1980). Discontinuous synthesis requires frequent initiation by an enzyme called DNA primase and immediate elongation by a DNA polymerase. These RNA-DNA fragments have finally to be processed into a long stretch of DNA which eventually is freed from the RNA primers (for **Processing of the Okazaki fragments**, see below).

The ideal properties to carry out this dual function are found in the DNA polymerase α/primase. Since the first description of DNA polymerase α to be complexed to DNA primase (Conaway and Lehman, 1982), this fact has been shown to be universal in all eukaryotic organisms tested (reviewed by Kaguni and Lehman, 1988). The DNA polymerase α/primase of eukaryotes consists of four subunits. The 160–200-kDa subunit contains DNA polymerizing activity while the 40–50-kDa and 50–60-kDa subunits represent the DNA primase heterodimer. The function of the 65–75-kDa subunit is unknown. An anchoring function between DNA polymerase α and DNA primase has been hypothesized (Campbell, 1986). Cell-cycle-dependent phosphorylation of the human 70-kDa subunit in the G_2/M Phase by the *cdc2* kinase of maturation-promoting factor suggest an important regulatory role of this subunit (T. S.-F. Wang, personal communication).

DNA polymerase α/primase complex does not start primer synthesis at a specific site on single-stranded DNA templates. The results were similar if initiation by DNA primase was measured alone or coupled with DNA synthesis by DNA polymerase α. The position at which primer synthesis was performed was, however, not completely random. The DNA polymerase α/primase is only moderately processive under optimal conditions for DNA synthesis. An average length of 100–200 nucleotides per binding to a primer was found with enzymes from various sources (Kaguni and Lehman, 1988).

It is unlikely that the four-subunit DNA polymerase α/primase reflects the composition of this enzyme at the replication fork. Rapid recycling to start the next primers and interaction with Okazaki-fragment-processing enzymes such as RNase H (see **Processing of the Okazaki fragments**, below) are likely to occur in the cell. Indeed higher-order complexes of DNA polymerase α/primase have been described from calf thymus (Hübscher, 1984; Ottiger and Hübscher, 1984; Ottiger et al., 1987; Hübscher et al., 1987) and HeLa cells (Vishwanatha and Baril, 1986; Malkas et al., 1990) and were, in analogy to the *E. coli* DNA polymerase III, designated as DNA polymerase α holoenzymes.

DNA polymerase auxiliary proteins

From the discussion above, it appears obvious that DNA polymerases δ and α require a helper function to perform their tasks in the replication apparatus. Many proteins have been described that can stimulate DNA polymerases in the test tube or have a tendency to copurify with them (see Spadari et al., 1989). For simplicity we discuss in this chapter three protein factors that have been found to affect DNA elongation in the well characterized SV40 replication system (Challberg and Kelly, 1989; Stillman, 1989). They include proliferating cell nuclear antigen (PCNA), replication factor A (= RF-A, RP-A or SSB) and replication factor C (RF-C). This list might be far from being complete since the SV40 system has also its limitations as a model for the cellular replication fork. For instance this virus codes for its own DNA helicase (Stahl et al., 1986). Accordingly, additional factors for interactions of DNA polymerases with cellular DNA helicases will remain undetected (see also **Separation of the double helix** above).

Proliferating cell nuclear antigen, PCNA. It has been known for a few years that PCNA is a cell-cycle-regulated protein of 36 kDa, whose rate of synthesis correlates directly with the proliferative state of normal cells and tissues (reviewed by Celis et al., 1987; Mathews, 1989). This protein was found to enable calf thymus DNA polymerase δ to work on templates with long stretches of single-stranded DNA (Tan et al., 1986). At the same time, evidence evolved that PCNA is required for *in vitro* replication of SV40 DNA (Prelich et al., 1987a). PCNA was identified in all eukaryotic organisms tested. The yeast PCNA shows 35% amino-acid sequence similarity with its human counterpart (Bauer and Burgers, 1990). The extreme conservation of this protein is also seen at a functional level since the yeast PCNA interacts with mammalian DNA polymerase δ and the mammalian PCNA interacts with yeast DNA polymerase δ (Bauer and Burgers, 1988). PCNA increases specifically the processivity of DNA synthesis by DNA polymerase δ and its productive binding to a template/primer (Burgers, 1988). Homogenous DNA polymerase δ and PCNA alone are inactive on primed single-stranded M13 (M. Gassmann and U. Hübscher, unpublished data). If, however, both RF-A and RF-C are added in optimal amounts (stoichiometric amounts of RF-A and catalytic amounts of RF-C) substantial replication is performed on primed single-stranded DNA (Tsurimoto et al., 1990; Lee and Hurwitz, 1990; Burgers, 1990). In analogy, the UL42 protein of herpes simplex virus 1 influences the processivity of the herpes DNA polymerase (UL30 gene product), suggesting a similar role as PCNA during viral replication (Hernandez and Lehman, 1990).

Replication factor A. Upon fractionation of the SV40 *in vitro* replication system, a three-subunit protein has been purified to homogeneity. Depending on the laboratory, this complex was called replication factor A (RF-A, Fairman and Stillman, 1988), replication protein A (RP-A, Wold and Kelly, 1988) or human single-stranded DNA binding protein (SSB, Wobbe et al., 1987). The protein complex is composed of three subunits, of 70 kDa, 32–34 kDa and 11–14 kDa, and binds preferentially to single-stranded DNA via its 70-kDa subunit (Wold et al., 1989). RF-A appears to have an important role in the generation of the single-stranded region at the SV40 origin prior to initiation of DNA synthesis (Wold et al., 1989). This protein is highly conserved since it has been found to have a similar structure in *S. cerevisiae* (Brill and Stillman, 1989) where its large (70-kDa) subunit is coded for by an essential gene (Heyer et al., 1990). RF-A also has direct effects on DNA polymerases δ and α (Tsurimoto and Stillman, 1989b). First, RF-A can specifically stimulate DNA polymerase α under conditions where heterologous SSBs cannot (Kenny et al., 1989). The stimulatory effect on DNA polymerase δ is more complex. First, DNA polymerase δ requires the presence of at least two other proteins: PCNA (see above) and RF-C (see below) in order to achieve stimulation (Kenny et al., 1989; Tsurimoto and Stillman, 1989b) by RF-A. On the other hand, DNA polymerase δ is much less dependent on RF-A coming from the same organism (Kenny et al., 1989), since the bacteriophage T4 gene 32 protein, *E. coli* SSB and adenovirus DNA binding protein could substitute for the human protein if DNA polymerase δ was first supplemented with PCNA and RF-C. Recent data suggested that the RF-A concentration in the SV40 replication system is critical for the specificity for DNA polymerase α (Tsurimoto et al., 1990). Finally, it appears that the individual subunits have different roles in replication (Kenny et al., 1990). The 32–34-kDa subunit has been cloned and over-expressed (Erdile et al., 1990) and this subunit of yeast and human RP-A is phosphorylated in a cell-cycle-dependent manner (Din et al., 1990).

Replication factor C. The elongation factor RF-C has recently been found in the SV40 system (Tsurimoto and Stillman, 1989a). It is a multipolypeptide protein with subunits of 140, 41 and 37 kDa (Tsurimoto and Stillman, 1989a) or 140, 40, 38, 37 and 36.5 kDa (S.-H. Lee, A. D. Kwong, Z.-Q. Pan and J. Hurwitz, unpublished results). It is likely that this protein is identical to an activator I identified by Lee et al. (1989) and to the C_1C_2 protein factors (Pritchard et al., 1983). Similarly to RF-A, the RF-C complex can stimulate DNA polymerase α alone (Tsurimoto and Stillman, 1989b). To stimulate DNA polymerase δ on primed single-stranded M13 DNA, in addition to RF-C, the other two auxiliary proteins RF-A and PCNA are necessary (Tsurimoto and Stillman, 1989b). RF-C can bind specifically to template/primer DNA and contains an ATPase stimulated by template/primer DNA and by PCNA (Tsurimoto and Stillman, 1990; Lee and Hurwitz, 1990). Using stimulation of DNA polymerase α on primed single-stranded M13 DNA, primer recognition proteins of 36 kDa and 41 kDa have been described recently (Jindal and Vishwanatha, 1990a). The 41-kDa protein has a sequence similar to phosphoglycerate kinase (Jindal and Vishwanatha, 1990b) and the 36-kDa protein to calpactin I heavy chain, a tyrosine kinase substrate (H. K. Jindal, W. G. Chaney, C. W. Anderson, R. G. Davis, and S. K. Vishwanatha, unpublished results). The relationship of these two proteins to part of the RF-C complex remains to be determined. Finally, recent data demonstrated that DNA polymerase ε from yeast *S. cerevisiae* is rendered salt resistant by the combined action of PCNA, RF-A and RF-C (Burgers, 1990) thus suggesting a role for this enzyme in DNA replication.

Coordination of leading- and lagging-strand replication

The crucial question still remains: how are both strands replicated in a coordinated fashion? The model of Alberts a decade ago (Sinha et al., 1980) was essential in the search for two DNA polymerases or one DNA polymerase consisting of two different halves. For instance, DNA polymerase III from *E. coli* can be isolated as an asymmetric dimer with 22 polypeptides (Maki et al., 1988) of which one subunit (the τ subunit) appears to have the capacity to dimerize the DNA polymerase III core (McHenry, 1982; 1988; M. O'Donnell, personal communication). The structural basis of the asymmetric dimer DNA polymerase III may be due to the occurrence of two related polypeptides in the same complex (see e.g. Maki and Kornberg, 1988a; 1988b; Studwell and O'Donnell, 1990; O'Donnell and Studwell, 1990, for recent details of the reconstituted DNA polymerase III holoenzyme structure).

Soon after the discovery that PCNA is required for *in vitro* replication of SV40 DNA (Prelich et al., 1987a), evidence emerged that DNA polymerase δ and α might be responsible for concurrent replication of leading and lagging strand, respectively (reviewed by So and Downey, 1988; Focher et al., 1988b). It appears possible that the DNA polymerase α/primase complex first initiates at the origin of replication (data presented for the SV40 origin, Tsurimoto et al., 1990) that is kept in single-stranded form by RF-A. Afterwards it continues DNA synthesis as the lagging-strand DNA polymerase. Synthesis at the leading strand is initiated on a RNA/DNA primer made by DNA polymerase α which can be used by DNA polymerase δ. PCNA and RF-C have been described to be necessary for the coordination of leading- and lagging-strand DNA replication of SV40 DNA, respectively (Prelich and Stillman, 1988; Tsurimoto and Stillman, 1989a). RF-C has also been proposed as a candidate for a clamp to dimerize DNA polymerase δ and α. The fact that RF-C can stimulate DNA synthesis of both DNA polymerases certainly favors the idea of a clamp factor. Isolation of a DNA polymerase δ/α dimer might provide insight into the interaction of the two DNA polymerases. Table 3 lists the DNA polymerase auxiliary proteins from eukaryotes and compares them to those of *E. coli*, its bacteriophage T4 and to herpes simplex virus 1. PCNA has its counterparts in the β subunit of DNA polymerase III holoenzyme, the gene 45 protein of bacteriophage T4 and possibly the UL42 protein of herpes simplex virus 1. PCNA shares some amino-acid sequence similarity with the T4 protein (Tsurimoto and Stillman, 1990). RF-C as a polypeptide complex of 3–5 subunits might be analogous to the gene 44/62 complex of T4 and to certain combinations of *E. coli* DNA polymerase III subunit complexes (e.g. γ complex, $\tau\delta'$, $\gamma\delta$; O'Donnell and Studwell, 1990). RF-A finally, has single-stranded DNA binding protein counterparts in SSB of *E. coli*, gene 32 of bacteriophage T4 as well as ICP8 of herpes simplex virus 1. Since RF-A has three polypeptides, it might be that its individual subunits have different functions in replication (Kenny et al., 1990).

Processing of the Okazaki fragments

After completion of an Okazaki fragment on the lagging strand by DNA polymerase α/primase, it has to be linked to the previous one. To achieve this three events are required: (a) the removal of the RNA primer made by the primase; (b) the filling of the gap and (c) the sealing of the two Okazaki fragments.

RNases H and $5' \rightarrow 3'$ exonucleases remove the primers. RNases H are enzymes that can hydrolyze RNA from RNA/DNA hybrids (reviewed by Wintersberger, 1990). Even though there is no definitive proof of an involvement of a distinct eukaryotic RNase H in the removal of the Okazaki fragment primers, circumstantial evidence has accumulated in recent years. For example, DNA polymerase α was isolated as a α holoenzyme from calf thymus (Ottiger and Hübscher, 1984) or as a HeLa multiprotein complex (Vishwanatha et al., 1986). These preparations contained RNAse H activity. RNases H purified from yeast (Karwan et al., 1983), *Drosophila melanogaster* (DiFrancesco and Lehman, 1985) and calf thymus (Hagemeier and Grosse, 1989) stimulated their homologous DNA polymerase α/primase. Of these enzymes the *D. melanogaster* RNase H is the best characterized so far. It is a heterodimer of 49-kDa and 39-kDa subunits and can remove RNA primers. When this RNase H was incubated with DNA

Table 3. *The universality of DNA polymerases and their auxiliary proteins*
This table is adapted from Stillman et al. (1990); see text for explanation, abbreviations and citations

Eukaryotes	*Escherichia coli*	Bacteriophage T4	HSV1	Function
DNA polymerase δ	DNA polymerase III	gene 43 protein	UL30	leading strand DNA synthesis
DNA polymerase α	DNA polymerase III	gene 43 protein	UL30	lagging strand DNA synthesis
DNA primase	*dnaG* protein	gene 61 protein	UL8	primer synthesis
PCNA	*dnaN* protein (β subunit)	gene 45 protein	UL42	enhances primer binding of DNA polymerase δ in eukaryotes; stimulates ATPase of RF-C or the DNA polymerase III γ-complex
RF-C complex	various subcomplexes of γ, δ, δ', χ, ψ, τ subunits	gene 44/62 proteins	?	stimulates DNA polmyerase; DNA-dependent ATPase
RF-A complex (= RP-A = SSB)	SSB	gene 32 protein	ICP8	single-stranded DNA binding protein; stimulates DNA polymerase cooperatively

polymerase α/primase, the *in vitro* synthesized DNA chains were shorter and the rate of RNA primer synthesis increased. Accordingly, RNase H appears to form a complex with DNA polymerase α/primase, thus increasing the recycling capacity and therefore also the frequency of chain initiation (DiFrancesco and Lehman, 1985). In the HSV 1 model system, the replicative DNA polymerase contains an intrinsic 5' → 3' exonucleolytic RNase H activity (Crute and Lehman, 1989). This minimal strategy of a virus indicates that the interaction of RNase H and DNA polymerase α/primase, described above, may reflect a true functional relationship between those two activities.

By reconstitution of the SV40 *in vitro* replication system with T antigen and purified proteins from HeLa cells, it was found that covalently closed circular DNA could only be synthesized if an RNase H, a double-strand-specific 5' → 3' exonuclease and a DNA ligase were present (Ishimi et al., 1988). The data suggested that the RNA primers at the 5' end of the Okazaki fragments were completely removed by the 5' → 3' exonuclease but only with the assistance of an RNase H. A similar conclusion can be drawn using a system that is able to convert single-stranded circular DNA to its covalently closed double-stranded form. Upon fractionation in this system, a complex was purified containing DNA polymerase α/primase, a DNA polymerase accessory factor, DNA ligase and RNase H (form H1) as well as a 5' → 3' exonuclease (Goulian et al., 1990a; 1990b). It appears that the combined action of two nucleases is an efficient way to remove primers. Finally, a double-strand-specific 3' → 5' and 5' → 3' exonuclease has been identified in Novikoff hepatoma cells (Mosbaugh and Meyer, 1980). This enzyme, called DNase V, can form a 1:1 complex with DNA polymerase β and displays RNase H activity under these conditions (S. M. Linn, personal communication).

DNA polymerases fill the gaps. The gaps are immediately filled by a DNA polymerase. The close interaction of RNase H and DNA polymerase α described above make perfect sense for their closely related functions. The SV40 *in vitro* replication data suggested that DNA polymerase α alone is sufficient for processing of the Okazaki fragments (Ishimi et al., 1988). This is also confirmed by data from a single-strand to a double-strand conversion assay (Goulian et al., 1990a). An important role for DNA polymerase β is possible but has not been proven yet. Finally, a role for DNA polymerase ε in this reaction cannot be excluded.

DNA ligase joins the fragments. DNA ligases join single-stranded breaks by catalyzing the phosphodiester bond formation and have been found in many eukaryotic tissues (Lasko et al., 1990). Whether these enzymes correspond to the DNA ligase found in *in vitro* systems (Ishimi et al., 1988; Goulian et al., 1990a; 1990b) cannot be decided yet.

Proof-reading of the two DNA strands

Despite the asymmetry of the DNA replication fork, the accuracy for both daughter strands has to be the same. Most cellular DNA polymerases found so far have a similar rate of inaccuracy of DNA synthesis which is $1:10^4-1:10^5$ (Loeb and Reyland, 1987). A 3' → 5' exonuclease activity is part of many DNA polymerases (first described in DNA polymerase I from *E. coli* by Brutlag and Kornberg, 1972) which preferentially excises non-complementary nucleotides from the 3'OH primer terminus. Proof-reading by the 3' → 5' exonuclease enhances the accuracy of DNA synthesis around 100-fold (Kunkel, 1988).

In eukaryotes the first exonuclease-containing DNA polymerase was found in bone marrow cells and called DNA polymerase δ (Byrnes et al., 1976). Recently, a second DNA polymerase with an inherent 3' → 5' exonuclease was found and designated DNA polymerase ε (see above for the five eukaryotic DNA polymerases). DNA polymerase α finally has in most cases no, or no measurable, 3' → 5' exonuclease. But there are a certain number of exceptions: DNA polymerase α holoenzyme (Ottiger et al., 1987) and DNA polymerase multiprotein complexes (Skarnes et al., 1986) contain 3' → 5' exonuclease; the *D. melanogaster* DNA polymerase α 182-kDa catalytic polypeptide possesses a potent 3' → 5' exonuclease but only after separation from the 70-kDa subunit (Cotterill et al., 1987); an immunopurified DNA polymerase α from a lymphoblastoid cell line is able to perform proof-reading (Bialek et al., 1989).

How is proof-reading accomplished at the acting replication fork? A possible scenario might be that the replication complex stalls synthesis as soon as a non-complementary nucleotide has been incorporated (either on the leading or on the lagging strand) and proof-reading has to be carried out before DNA replication can continue by the combined action of DNA polymerases α and δ.

Release of torsional stress

If one imagines the double helix as two pieces of strings wound around each other and fixed at one end; if one now takes the free ends and pulls them apart, the turns of the helix will narrow and finally form what is called a supercoil. This is exactly the situation that is encountered during DNA replication when DNA helicases open up the helix and the replication machinery advances. In that case the DNA is fixed either by forming closed circles or as long linear chromosomes. The supercoils that are created are called positive in the sense that they have the same direction as the double helix itself. To allow further propagation of the replication fork, a swivel is needed to release the torsional stress. The enzymes that can act as such a swivel are called DNA topoisomerases (reviewed by Wang, 1987; Liu, 1989; Richter and Knippers, 1989). By definition, they are able to break and rejoin the phosphodiester backbone of the DNA, thus changing the topological state. Two DNA topoisomerases exist in bacterial and in eukaryotic cells: DNA topoisomerases I breaks one strand of the DNA and DNA topoisomerase II both strands (Liu et al., 1980; Cozzarelli, 1980).

The principal reaction mechanisms of both enzymes are similar (Liu, 1989). The enzyme binds to double-stranded DNA at loosely defined consensus sequences. After breaking the DNA, at or near the binding site, the enzyme becomes covalently attached to the 3′ (DNA topoisomerase I) or 5′ (DNA topoisomerase II) phosphoryl end of the DNA via an O^4-phosphotyrosyl bond. At this stage, the rotation of the DNA can take place. If only one strand is broken due to the action of DNA topoisomerase I, the free end can turn around the intact strand, thus releasing the torsional stress. In the case of DNA topoisomerase II one end of the DNA can rotate freely relative to the other or a double-stranded DNA can pass through the enzyme-bridged gap (decatenation). This reaction mechanism leads to release of supercoiling in steps of one with topoisomerase I or in steps of two with DNA topoisomerase II. In a third step, the energy of the covalent bond between the enzyme and the DNA is used for sealing the previously broken DNA strands. The whole reaction is independent of energy cofactors with DNA topoisomerase I but in the case of DNA topoisomerase II hydrolysis of ATP is required.

Similarly to the replication events described above, our knowledge of DNA topoisomerases during DNA replication results mainly from the study of viral systems and yeast mutants. In addition, specific inhibitors of each DNA topoisomerase have helped to clarify their individual roles (Liu, 1989). For instance, camptothecin is used to inhibit specifically the religation step after cleavage by DNA topoisomerase I. The enzyme remains covalently bound to DNA while the nick in the DNA stays unsealed. Similarly, the quinolones VP16 (etoposide) and VM26 (teniposide) act on DNA topoisomerase II.

Using the SV40 replication system that has been described before, it was found that during initiation in vitro T-ag can unwind only 10−20 bp at the origin of replication before its helicase action is stopped, probably by increasing positive supercoiling of the DNA (Borowiec and Hurwitz, 1988). Addition of topoisomerase releases this topological stress, thus allowing extensive unwinding of the DNA. This task can be performed in vitro either by DNA topoisomerase I or II (Yang et al., 1987). However, the main actor in vivo is most likely DNA topoisomerase I since addition of camptothecin stops replication, as measured by the uptake of [^{3}H]thymidine. The

resulting DNA molecules show a sigma-like structure, indicating a position of DNA topoisomerase I at or near the replication fork (Avemann et al., 1988). In the same way Champoux (1988) has shown that DNA topoisomerase I is preferentially associated with replicative chromatin while it is absent in nonreplicative. Consensus sequences for DNA topoisomerase I cleavage were found at both sides of the SV40 origin of replication (Tsui et al., 1989; Porter and Champoux, 1989) mainly on the template strand for lagging-strand synthesis. Studies on in vitro chromatin assembly also suggest an involvement of DNA topoisomerase I in this process (Almouzni and Méchali, 1988 b).

While DNA topoisomerase II is not essential for fork propagation during SV40 replication, it is absolutely necessary for termination events. In the absence of DNA topoisomerase II late replicative intermediates or catenated dimers accumulate (Yang et al., 1987). Accordingly, if cells are treated with VM26 they are enriched for almost full-length molecules which are connected to each other at a short stretch of unreplicated DNA (Richter et al., 1987; Richter and Strausfeld, 1988).

The second well studied viral system in connection with DNA topoisomerase is adenovirus (Schaack et al., 1990). In contrast to SV40, adenovirus had a long (36-kbp) linear genome, which is replicated from either end by strand displacement synthesis. While some replicative enzymes are encoded by the viral genome, others have to be supplied by the cell. By supplementing viral replication with cellular fractions, factors were identified which are essential for DNA replication. One of these factors, called nuclear factor II, has DNA topoisomerase I activity (Nagata et al., 1983).

Viral replication in vitro can be performed with DNA topoisomerase I only. In contrast to SV40, this enzyme seems to be essential for adenovirus replication and cannot be replaced by DNA topoisomerase II (Schaack et al., 1990). However, inhibition of DNA topoisomerase II stopped replication after approximately one round of replication, indicating a role during late replicative stages as well.

The results gained from the yeast mutants support the findings described so far (Yanagida and Wang, 1987). DNA topoisomerase I is coded for by a nonessential gene, mutants are viable and show no detectable phenotype (Thrash et al., 1985; Uemura et al., 1987). Possibly its activity can be replaced by DNA topoisomerase II. In contrast, cells with a disrupted gene for DNA topoisomerase II are not viable (DiNardo et al., 1984; Goto and Wang, 1985). In the absence of DNA topoisomerase II activity, accumulation of intertwined dimers can be detected, indicating a role for DNA topoisomerase II during strand segregation after replication. These cells are arrested in mitosis due to defects in chromosome condensation and sister chromatid separation.

The conclusive picture gained from these results indicates a role for DNA topoisomerase I as a swivel during advancement of the replication fork. However, it appears that it can be replaced during this process by DNA topoisomerase II. Although there are some data indicating an association of DNA topoisomerase I with the replication fork, it is not certain whether the enzyme is part of the replisome itself. DNA topoisomerase II, on the other hand, most likely acts in vivo before and during decatenation of replicative intermediates.

DNA replication and chromatin

While the basic processes of DNA replication from unwinding to gap filling are similar in all organisms ranging from *E. coli* to man, higher eukaryotes have to encounter one additional problem. The single circular chromosome of bacteria is 'naked' in the cell and can be transcribed and replicated readily. In eukaryotes, however, the DNA is complexed with proteins in a higher-order structure called chromatin (reviewed by van Holde, 1989). The building blocks of chromatin are nucleosomes, which are made up of histones and DNA. Each nucleosome consists of a H3 – H4 tetramer and two H2A – H2B dimers around which 160 – 220 bp of DNA are wound depending on the cell type (van Holde, 1989). In addition, one molecule of H1 is responsible for the compact structure of the chromatin.

The packaging of the DNA in nucleosomes directly influences the gene activity. If the DNA is tightly packed as heterochromatin, no transcription can take place. On the other hand, genes that are actively transcribed have a loose chromatin structure (Pavlovic et al., 1989). Hence arrangement, as well as local modifications, of the nucleosomes are important determinants of the transcriptional abilities of specific cells and these features must be accurately reproduced during cell division.

How this is achieved is still largely a matter of speculation. Two models for the distribution of nucleosomes to the two daughter strands are currently under discussion. On the one hand, if the mode of distribution is dispersive the nucleosomes are distributed randomly to each of the daughter strands. Data confirming this model were given by electron microscopy studies, which showed that nucleosomes were distributed evenly but in small clusters to either strand of the replication fork (Sogo et al., 1986). On the other hand, conservative segregation has been suggested where the leading strand on which DNA synthesis proceeds continuously obtains all the nucleosomes, while the lagging strand has to assemble new ones completely (Handeli et al., 1989).

The structure of the nucleosomal core formed by H2A – H4 is at least partially conserved. While some researchers find a distribution of the intact core to the daughter molecules (Leffak, 1984) others have described a partial breakdown into H2A/H2B dimers and H3/H4 tetramers (Jackson, 1987). Recent data obtained using the T4 replicative proteins and an artificial nucleosome-containing template suggested that the nucleosomes remain bound to the DNA while the DNA is replicated (Bonne-Andrea et al., 1990). In this study the nucleosomes segregated three times as often to the leading strand than to the lagging strand. However, since the system studied had a mixed nature (plasmid DNA, hamster cell histones and the bacteriophage T4 replication apparatus), there may be differences to the situation encountered *in vivo*.

While half of the nucleosomes on the replicated DNA are inherited from the parental strands, the missing ones have to be newly assembled (Svaren and Chalkey, 1990). Several cell-free systems have been developed to study the assembly of chromatin. The current discussion focuses on the issue of whether DNA replication is a prerequisite for nucleosome assembly or not. Lysates from *X. laevis* eggs rapidly and efficiently assemble new chromatin (Laskey et al., 1977). If the DNA input into this system is single-stranded, it is converted to covalently closed double-stranded circles which are packaged into chromatin. However, if the input DNA is already in a double-stranded form, it is assembled into chromatin as well, but with slower kinetics (Almouzni and Méchali, 1988a). ATP and Mg^{2+} are essential for the correct spacing of the nucleosomes (Almouzni and Méchali, 1988b).

The SV40 system also has been used to study chromatin assembly. A nuclear extract that was added to an *in vitro* replication system could promote chromatin assembly. Upon further fractionation, a so-called chromatin assembly factor (CAF-1) was purified, which contributed the ability to form nucleosomes to a cytosolic extract (Smith and Stillman, 1989). This function was completely dependent on the replication of DNA. Using the same approach it was suggested that nucleosome assembly is a two-step pathway of which the first step is replication dependent while the second occurs after replication has been completed (Fotedar and Roberts, 1989).

On the other hand, electron microscopic studies of chromatin assembly on SV40 DNA suggested that replicating and non-replicating DNA are assembled with similar efficiency (Gruss et al., 1990). This is supported by data showing that inhibitors of DNA topoisomerases I and II inhibited DNA replication but did not prevent chromatin assembly (Annunziato, 1989). Banerjee and Cantor (1990) were able to show that it was essentially dependent on the experimental conditions, whether chromatin assembly could take place with human cell extracts. However, chromatin assembly seems to be more rapid and also more accurate in the presence of DNA replication (Gruss et al., 1990).

The structure of newly replicated chromatin is different from 'mature' chromatin. For example, the spacing of the nucleosomes is more compact and less regular than in later stages, suggesting that some maturation steps have to take place. These maturation steps may include deposition of H1 which is the last of the histones to be added to the nucleosomes (Cusick et al., 1983). A model for the maturation of chromatin suggests that nucleosomes are associated randomly on new DNA in an irregular pattern. The mature distribution is finally achieved by sliding of the nucleosomes on the DNA (van Holde, 1989).

Conclusions and perspectives

Many enzymes and proteins are required to replicate the eukaryotic genome faithfully. DNA helicases transiently open the double-helix, the single-stranded DNA can be used by DNA primase to synthesize primers. The leading strand is replicated by DNA polymerase δ and the lagging strand by DNA polymerase α/primase. The Okazaki fragments are processed by at least four different enzymes, i.e. RNase H and $5' \rightarrow 3'$ exonuclease to remove the primers, DNA polymerase α (or possibly β) to fill the gaps and DNA ligase to seal the fragments. The proof-reading $3' \rightarrow 5'$ exonuclease enhances the accuracy of DNA synthesis by DNA polymerases. Topological problems, such as conversions of DNA isomers and segregation of DNA strands, are dealt with by DNA topoisomerases I and II, respectively. Many of these enzymes, especially the DNA polymerases, seem to require auxiliary proteins to be assembled into a functioning replication machine. This machine itself has to perform replication in a structural environment such as chromatin and other higher-order nuclear organization.

Based on the very recent discovery that DNA polymerase ε is a third essential replication DNA polymerase (Morrison et al., 1990), a function of this enzyme during DNA replication was suggested. Accordingly, we propose a prospective model for a eukaryotic replication fork that involves the action of three DNA polymerases (Fig. 2). After unwinding at a given origin of replication, the initiating DNA polymerase (DNA polymerase α/primase) synthesizes primers and short pieces of DNA at the two leading strands (Tsurimoto et al., 1990). Following this initiation synthesis, a switch of DNA poly-

Initiation

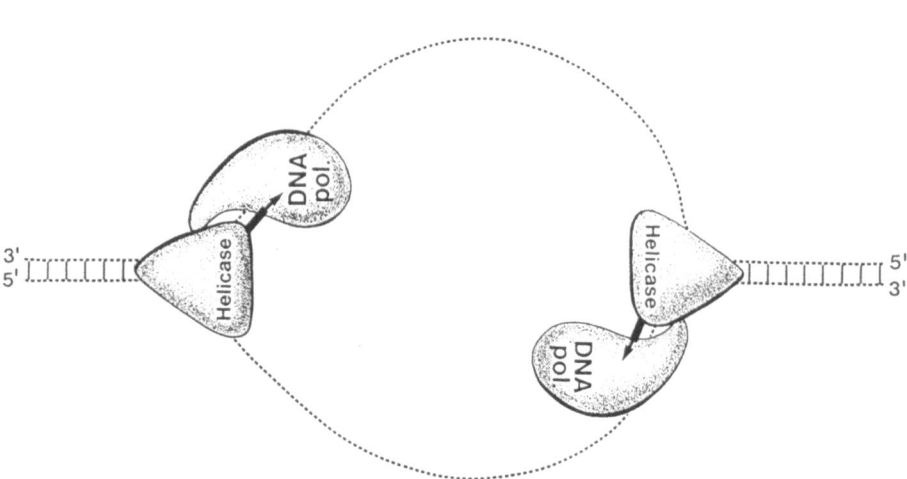

Elongation

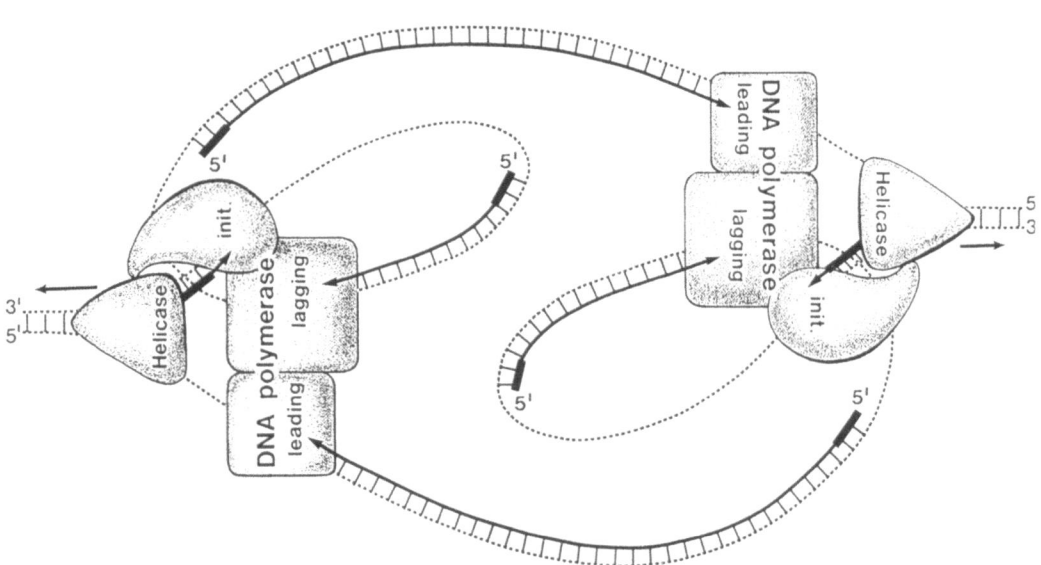

Fig. 2. *Model for eukaryotic DNA replication*. The model is based on the assumption that three DNA polymerases participate in DNA replication. First, an initiating DNA polymerase (DNA polymerase α/primase) synthesizes primers at the two leading strands at a site that has been locally unwound (initiation). This is followed by a switch to the replisome consisting of three DNA polymerases replicating in concert (elongation). A processive DNA polymerase (δ or ε) takes over replication of the leading strand. The initiating DNA polymerase α/primase continues to start and partially synthesize the Okazaki fragments. These are finished by switching to the lagging strand DNA polymerase (δ or ε), which might be complexed to the leading strand enzyme, thus guaranteeing efficient and coordinated replication. The processivity of the lagging strand DNA polymerase can be regulated (by PCNA for δ, or possibly by yet unknown mechanisms for ε). Parental strands (······), newly replicated DNA (——), primers (▬▬)

merases to the advancing replisome occurs where three DNA polymerases might share the replication tasks in the following way: The leading strand is replicated by a processive DNA polymerase (DNA polymerase ε or DNA polymerase δ). At the same time, the initiating DNA polymerase (DNA polymerase α/primase) continues to synthesize primers and short pieces of DNA at the lagging strand. Furthermore, by changing to a third DNA polymerase (DNA polymerase δ or DNA polymerase ε), the Okazaki fragments would be synthesized to completion. It is conceivable that various sub-

sets of auxiliary protein compositions (PCNA, RF-A, RF-C and others) might guide the DNA polymerase δ and ε to their respective functional roles. This model finally could guarantee proof-reading of the two strands since both DNA polymerases δ and ε contain an intrinsic $3' \rightarrow 5'$ exonuclease.

We know only the principle parts of the replication machine. We have to learn now how these different parts interact together to become an efficient engine. Once these parts fit optimally, the engine can be tuned, its speed can be regulated or its accuracy controlled. What enables the engine to start or to stop? We can expect that information of functional details eventually will lead us to a better understanding of two very crucial unsolved problem complexes in medical sciences. First, we will comprehend better the small bandwidth between a normal and a malignant cell and, second, an understanding of differences between cellular and viral replication mechanisms (e.g. HIV) will help to develop more potent anti-viral drugs.

We are indebted to B. M. Alberts, P. M. J. Burgers, M. D. Challberg, J. J. Crute, M. Goulian, J. Hurwitz, D. Lane, I. R. Lehman, T. Lindahl, S. M. Linn, A. Morrison, M. O'Donnell, E. H. A. Poll, B. Stillman, J. K. Vishwanatha, and T. S.-F. Wang for communication of their results prior to publication. We thank P. Hafkemeyer, K. Mendgen, K. A. Neftel and C. P. Schelling for critical comments on the manuscript. The work carried out in the authors laboratory has continuously been supported by the Swiss National Science Foundation and the Kanton of Zürich.

REFERENCES

Abdel-Monem, M., Durwald, H. & Hoffmann-Berling, H. (1976) Eur. J. Biochem. 65, 441—449.

Alberts, B. M. (1987) Proc. Trans. R. Soc. Lond. B317, 395—420.

Alberts, B. M., Berry, J., Bedinger, P., Formosa, T., Jongeneel, C. V. & Kreuzer, K. N. (1983) Cold Spring Harb. Symp. Quant. Biol. 47, 655—668.

Almouzni, G. & Méchali, M. (1988a) EMBO J. 7, 665—672.

Almouzni, G. & Méchali, M. (1988b) EMNO J. 7, 4355—4365.

Annunziato, A. T. (1989) J. Cell Sci. 93, 593—603.

Avemann, K., Knippers, R., Koller, T. & Sogo, J. M. (1988) Mol. Cell. Biol. 8, 3026—3034.

Baker, T. A. & Kornberg, A. (1990) DNA replication, 2nd edn, in the press.

Bambara, R. A. & Jessee, C. B. (1990) Biochim. Biophys. Acta, in the press.

Bambara, R. A., Myers, T. W. & Sabatino, R. D. (1990) in The eukaryotic nucleus: molecular biochemistry and macromolecular assemblies (Strauss, P. & Wilson, S. H., eds) pp. 69—94, Telford Press, Caldwell NJ.

Banerjee, S. & Cantor, C. R. (1990) Mol. Cell. Biol. 10, 2863—2873.

Bauer, G. A. & Burgers, P. M. J. (1988) Proc. Natl Acad. Sci. USA 85, 7506—7510.

Bauer, G. A. & Burgers, P. M. J. (1990) Nucleic Acids Res. 18, 261—265.

Bialek, G., Nasheuer, H. P., Goetz, H. & Grosse, F. (1989) EMBO J. 8, 1833—1839.

Blanco, L., Bernard, A., Làzaro, J. M., Martin, G., Garmendia, C. & Salas, M. (1989) J. Biol. Chem. 264, 8935—8940.

Bonne-Andrea, C., Wong, M. L. & Alberts, B. (1990) Nature 343, 719—726.

Borowiec, J. A. & Hurwitz, J. (1988) EMBO J. 7, 3149—3158.

Boulet, A., Simon, M., Faye, H., Bauer, G. A. & Burgers, P. M. J. (1989) EMBO J. 8, 1849—1854.

Bravo, R., Frank, R., Blundell, P. A. & MacDonald-Bravo, H. (1987) Nature 326, 515—517.

Brill, S. J. & Stillman, B. (1989) Nature 342, 92—95.

Brutlag, D. & Kornberg, A. (1972) J. Biol. Chem. 247, 241—248.

Burgers, P. M. J. (1988) Nucleic Acids Res. 16, 6297—6307.

Burgers, P. M. J. (1989) Progr. Nucleic Acids Res. Mol. Biol. 37, 235—280.

Burgers, P. M. J. (1990) in Eukaryotic DNA polymerases (Engler, J. A. & Wang, T. S.-F., eds) Springer-Verlag, New York, in the press.

Burgers, P. M. J., Bambara, R. A., Campbell, J. L., Chang, L. M. S., Downey, K. M., Hübscher, U., Lee, M. Y. W. T., Linn, S. M., So, A. G. & Spadari, S. (1990) Eur. J. Biochem. 191, 617—618.

Byrnes, J. J., Downey, K. M., Black, V. L. & So, A. G. (1976) Biochemistry 15, 2817—2823.

Campbell, J. L. (1986) Annu. Rev. Biochem. 55, 733—771.

Celis, J. E., Madsen, P., Celis, A., Nielsen, H. V. & Gesser, B. (1987) FEBS Lett. 220, 1—7.

Cha, T.-A. & Alberts, B. M. (1990) Biochemistry 29, 1791—1798.

Challberg, M. D. & Kelly, T. J. (1989) Annu. Rev. Biochem. 58, 671—717.

Champoux, J. J. (1988) J. Virol. 62, 3675—3683.

Cohen, P., van Dam, K., van Deenen, L. L. M., Kennedy, E. P., Radda, G. K. & van der Vliet, P. C. (eds) (1988) Biochim. Biophys. Acta 951.

Conaway, R. C. & Lehman, I. R. (1982) Proc. Natl Acad. Sci. USA 79, 2523—2527.

Cotterill, S. M., Reyland, M. E., Loeb, L. A. & Lehman, I. R. (1987) Proc. Natl Acad. Sci. USA 84, 5635—5639.

Cozzarelli, N. R. (1980) Science 207, 953—960.

Crute, J. J. & Lehman, I. R. (1989) J. Biol. Chem. 264, 19266—19270.

Crute, J. J., Mocarski, E. S. & Lehman, I. R. (1988) Nucleic Acids Res. 16, 6585—6596.

Crute, J. J., Tsurumi, T., Zhu, L., Weller, S. K., Olivo, P. D., Challberg, M. D., Mocarski, E. S. & Lehman, I. R. (1989) Proc. Natl Acad. Sci. USA 86, 2186—2189.

Cusick, M. E., Lee, K. S., DePamphilis, M. E. & Wassarman, P. M. (1983) Biochemistry 22, 3873—3884.

Dean, F. B., Bullock, P., Murakami, Y., Wobbe, C. R., Weissbach, L. & Hurwitz, J. (1987) Proc. Natl Acad. Sci. USA 84, 16—20.

DePamphilis, M. L. & Wassarman, P. M. (1980) Annu. Rev. Biochem. 49, 627—666.

DiFrancesco, R. A. & Lehman, I. R. (1985) J. Biol. Chem. 260, 14764—14770.

Din, S., Brill, S. J., Fairmann, M. P. & Stillman, B. (1990) Genes Dev. 4, 968—977.

DiNardo, S., Voelkel, K. & Sternglanz, R. (1984) Proc. Natl Acad. Sci. USA 81, 2616—2620.

Dodson, M., Dean, M., Bullock, F. B., Echolds, P. & Hurwitz, J. (1987) Science 238, 964—967.

Dodson, M. S., Crute, J. J., Bruckner, R. C. & Lehman, I. R. (1989) J. Biol. Chem. 264, 20835—20838.

Downey, K. M., Tan, C.-K., Andrews, D. M., Li, X. & So, A. G. (1988) Cancer cells, vol. 6, pp. 403—410, Cold Spring Harbor Laboratory Press, Cold Spring Harbor NY.

Downey, K. M., Andrews, D. M., Li, X., Castillo, C., Tan, C.-K. & So, A. G. (1990) in Molecular mechanism in DNA replication and recombination (Richardson, C. C. & Lehman, I. R., eds) pp. 141—152, Alan R. Liss, New York.

Engler, J. A. & Wang, T. S. F. (eds) (1990) Eukaryotic DNA polymerases, Springer-Verlag, in the press.

Erdile, L. F., Wold, M. S. & Kelly, T. J. (1990) J. Biol. Chem. 265, 3177—3182.

Fairman, M. P. & Stillman, B. (1988) EMBO J. 7, 1211—1218.

Focher, F., Spadari, S., Ginelli, B., Hottiger, M., Gassmann, M. & Hübscher, U. (1988a) Nucleic Acids Res. 16, 6279—6295.

Focher, F., Ferrari, E., Spadari, S. & Hübscher, U. (1988b) FEBS Lett. 229, 6—10.

Focher, F., Gassmann, M., Hafkemeyer, P., Ferrari, E., Spadari, S. & Hübscher, U. (1989) Nucleic Acids Res. 17, 1805—1821.

Fotedar, R. & Roberts, G. M. (1989) Proc. Natl Acad. Sci. USA 86, 6459—6463.

Fry, M. & Loeb, L. A. (1986) Animal cell DNA polymerase, CRC Press, Boca Raton, FL.

Goetz, G. S., Dean, F. B., Hurwitz, J. & Matson, S. W. (1988) J. Biol. Chem. 263, 383—392.

Goto, T. & Wang, J. C. (1985) Proc. Natl Acad. Sci. USA 82, 7178—7182.

Goulian, M., Heard, C. J. & Richards, S. H. (1990a) J. Biol. Chem. 265, in the press.

Goulian, M., Bigsby, B. M., Heard, C. J. & Richards, S. H. (1990b) J. Biol. Chem. 265, in the press.

Gruss, C., Guttierez, C., Burhans, W. C., Koller, Th., DePamphilis, M. L. & Sogo, J. M. (1990) EMBO J. 9, 2911—2922.

Hagemeier, A. & Grosse, F. (1989) Eur. J. Biochem. 185, 621—628.

Hamatake, R. K., Hasegawa, H., Clark, A. B., Bebenek, K., Kunkel, T. A. & Sugino, A. (1990) J. Biol. Chem. 265, 4072—4083.

Handeli, S., Klar, A., Meuth, M. & Cedar, H. (1989) Cell 57, 909—920.

Hernandez, T. R. & Lehman, I. R. (1990) J. Biol. Chem. 265, 11127—11132.

‌

Heyer, W.-D., Manachanahalli, R. S. R., Erdile, L. F., Kelly, T. J. & Kolodner, R. D. (1990) *EMBO J. 9*, 2321−2329.

Hübscher, U. (1984) *Trends Biochem. Sci. 4*, 390−393.

Hübscher, U., Kuenzle, C.C. & Spadari, S. (1979) *Proc. Natl Acad. Sci. USA 76*, 2316−2320.

Hübscher, U. & Ottiger, H. P. (1984) *Adv. Exp. Biol. Med. 179*, 321−330.

Hübscher, U., Gassmann, M., Spadari, S., Brown, N. C., Ferrari, E. & Buhk, H. J. (1987) *Phil. Trans. R. Soc. Lond. 317*, 421−428.

Jackson, V. (1987) *Biochemistry 26*, 2315−2325.

Jindal, H. K. & Vishwanatha, J. K. (1990a) *Biochemistry 29*, 4767−4773.

Jindal, H. K. & Vishwanatha, J. K. (1990b) *J. Biol. Chem. 265*, 6540−6543.

Ishimi, Y., Claude, A., Bullock, P. & Hurwitz, J. (1988) *J. Biol. Chem. 263*, 19723−19733.

Kaguni, L. S. & Lehman, I. R. (1988) *Biochim. Biophys. Acta 950*, 87−101.

Karwan, R., Blutsch, H. & Wintersberger, U. (1983) *Biochemistry 22*, 5500−5507.

Kelly, T. J. (1988) *J. Biol. Chem. 263*, 17889−17892.

Kelly, T. J. & Stillman, B. (eds) (1988) *Cancer cells*, vol. 6, Cold Spring Harbor Laboratory Press, Cold Spring Harbor NY.

Kenny, M. K., Lee, S.-H. & Hurwitz, J. (1989) *Proc. Natl Acad. Sci. USA 86*, 9757−9761.

Kenny, M. K., Schlegel, U., Furneaux, H. & Hurwitz, J. (1990) *J. Biol. Chem. 265*, 7693−7700.

Knipe, D. M. (1989) *Adv. Virus Res. 37*, 85−123.

Kornberg, A. (1988) *J. Biol. Chem. 263*, 1−4.

Kornberg, A., Scott, J. F. & Bertsch, L. (1978) *J. Biol. Chem. 253*, 3298−3304.

Kunkel, T. A. (1988) *Cell 53*, 837−840.

Laskey, R. A., Mills, A. D. & Morris, N. R. (1977) *Cell 10*, 237−243.

Lasko, D. D., Tomkinson, A. E. & Lindahl, T. (1990) *Mutation Res.*, in the press.

Lee, S.-H., Kwong, A. D., Ishimi, Y. & Hurwitz, J. (1989) *Proc. Natl Acad. Sci. USA 86*, 4877−4881.

Lee, S.-H. & Hurwitz, J. (1990) *Proc. Natl Acad. Sci. USA 87*, 5672−5676.

Leffak, I. M. (1984) *Nature 307*, 82−85.

Lehman, I. R. & Kaguni, L. S. (1989) *J. Biol. Chem. 264*, 4591−4595.

Liu, L. F. (1989) *Annu. Rev. Biochem. 58*, 351−375.

Liu, L. F., Liu, C. C. & Alberts, B. M. (1980) *Cell 19*, 697−707.

Loeb, L. A. & Reyland, M. E. (1987) in *Nucleic acids and molecular biology* (Eckstein, F. & Lilley, D. M. J., eds) vol. 1, pp. 157−173, Springer-Verlag, Berlin, Heidelberg.

Maki, S. & Kornberg, A. (1988a) *J. Biol. Chem. 263*, 6555−6560.

Maki, S. & Kornberg, A. (1988b) *J. Biol. Chem. 263*, 6561−6569.

Maki, H., Maki, S. & Kornberg, A. (1988) *J. Biol. Chem. 263*, 6570−6578.

Malkas, L. H., Hickey, R. J. & Baril, E. F. (1990) in *The eukaryotic nucleus, molecular structure and macromolecular assemblies* (Strauss, P. & Wilson, S. H., eds) pp. 45−68, Telford Press, Caldwell, NJ.

Mathews, M. B. (1989) in *Growth control during aging* (Wang, E. & Warner, H. R., eds) pp. 89−120, CRC Press, Boca Raton FL.

Matson, S. W. & Kaiser-Rogers, K. A. (1990) *Annu. Rev. Biochem. 59*, 289−329.

Matson, S. W., Tabor, S. & Richardson, C. C. (1983) *J. Biol. Chem. 258*, 14017−14024.

McHenry, C. S. (1982) *J. Biol. Chem. 257*, 2657−2663.

McHenry, C. S. (1988) *Annu. Rev. Biochem. 57*, 519−550.

McHenry, C. S. & Johanson, K. O. (1984) *Adv. Exp. Biol. Med. 179*, 315−319.

Morrison, A., Araki, H., Clark, A. B., Hamatake, R. K. & Sugino, A. (1990) *Cell 62*, 1143−1151.

Mosbaugh, D. W. & Meyer, R. R. (1980) *J. Biol. Chem. 255*, 10239−10247.

Munn, M. M. & Alberts, B. M. (1990a) *J. Biol. Chem. 265*, in the press.

Munn, M. M. & Alberts, B. M. (1990b) *J. Biol. Chem. 265*, in the press.

Nagata, K., Guggenheimer, R. A. & Hurwitz, J. (1983) *Proc. Natl Acad. Sci. USA 80*, 4266−4270.

Nishida, C., Reinhard, P. & Linn, S. (1988) *J. Biol. Chem. 263*, 501−510.

Nowak, R., Woszczynski, M. & Siedlecki, J. A. (1990) *Exp. Cell Res. 191*, 51−56.

O'Donnell, M. & Studwell, P. S. (1990) *J. Biol. Chem. 265*, 1179−1187.

Ottiger, H.-P. & Hübscher, U. (1984) *Proc. Natl Acad. Sci USA 81*, 3993−3997.

Ottiger, H.-P., Frei, P., Hässig, M. & Hübscher, U. (1987) *Nucleic Acids Res. 15*, 4789−4807.

Pavlovic, J., Banz, E. & Parish, R. W. (1989) *Nucleic Acids. Res. 17*, 2315.

Pizzagalli, A., Valsasnini, P., Plevani, P. & Lucchini, G. (1988) *Proc. Natl Acad. Sci. USA 85*, 2771−3776.

Poll, E. H. A. & Benbow, R. M. (1988) *Biochemistry 27*, 8701−8706.

Porter, S. E. & Champoux, J. J. (1989) *Mol. Cell. Biol. 9*, 541−550.

Prelich, G. & Stillman, B. (1988) *Cell 53*, 117−126.

Prelich, G., Kostura, M., Marsahk, D. R., Mathews, M. B. & Stillman, B. (1987a) *Nature 326*, 471−475.

Prelich, G., Tan, C.-K., Kostura, M., Mathews, M. B., So. A. G., Downey, K. M. & Stillman, B. (1987b) *Nature 326*, 517−519.

Pritchard, C. G., Weaver, D. T., Baril, E. F. & DePamphilis, M. L. (1983) *J. Biol. Chem. 258*, 9810−9819.

Richardson, C. C. & Lehman, I. R. (eds) (1990) *Molecular mechanisms in DNA replication and recombination*, vol. 127, Alan R. Liss, New York.

Richter, A. & Knippers, R. (1989) *Life Sci. Adv. 8*, 125−134.

Richter, A. & Strausfeld, U. (1988) *Nucleic Acids. Res. 16*, 10119−10129.

Richter, A., Strausfeld, U. & Knippers, R. (1987) *Nucleic Acids Res. 15*, 3455−3468.

Schaack, J., Schedl, P. & Shenk, T. (1990) *J. Virol. 64*, 78−85.

Scheffner, M., Wessel, R. & Stahl, H. (1989) *Nucleic Acids Res. 17*, 93−106.

Seki, M., Enomoto, T., Yanagisawa, J., Hanoaka, F. & Ui, M. (1988) *Biochemistry 27*, 1766−1771.

Seki, S., Oda, R. & Okashi, M. (1980) *Biochim. Biophys. Acta 610*, 413−420.

Sinha, N. K., Morris, C. F. & Alberts, B. M. (1980) *J. Biol. Chem. 255*, 4290−4303.

Sitney, K. C., Budd, M. E. & Campbell, J. L. (1989) *Cell 56*, 599−605.

Skarnes, W., Bonin, P. & Baril, E. (1986) *J. Biol. Chem. 261*, 6629−6636.

Smith, S. & Stillman, B. (1989) *Cell 58*, 15−25.

So, A. G. & Downey, K. M. (1988) *Biochemistry 27*, 4591−4595.

Sogo, J. M., Stahl, H., Koller, Th. & Knippers, R. (1986) *J. Mol. Biol. 189*, 189−204.

Spadari, S., Montecucco, A., Pedrali-Noy, G., Ciarocchi, G., Focher, F. & Hübscher, U. (1989) *Mutat. Res. 219*, 147−156.

Stahl, H. & Knippers, R. (1987) *Biochim. Biophys. Acta 910*, 1−10.

Stahl, H., Dröge, P., Zentgraf, H. & Knippers, R. (1985) *J. Virol. 54*, 473−482.

Stahl, H., Dröge, P. & Knippers, R. (1986) *EMBO J. 5*, 1939−1944.

Stillman, B. (1989) *Annu. Rev. Cell. Biol. 5*, 197−246.

Stillman, B., Brill, S., Fien, K., Melendy, T., Din, S. & Tsurimoto, T. (1990) in *Eukaryotic DNA polymerases* (Engler, J. A. & Wang, T. S.-F., eds) Springer-Verlag, New York, in the press.

Studwell, P. S. & O'Donnell, M. (1990) *J. Biol. Chem. 265*, 1171−1178.

Sugino, A., Ryu, B. O., Sugino, T., Naumovski, L. & Friedberg, E. C. (1986) *J. Biol. Chem. 261*, 11744−11750.

Sung, P., Prakash, L., Matson, S. W. & Prakash, S. (1987) *Proc. Natl Acad. Sci. USA 84*, 8951−8955.

Svaren, J. & Chalkey, R. (1990) *Trends Genet. 6*, 52−57.

Syvaojo, J. & Linn, S. (1989) *J. Biol. Chem. 264*, 2489−2497.

Syvaojo, J., Siromensaari, S., Nishida, C., Goldsmith, J. S. & Linn, S. (1990) *Proc. Natl Acad. Sci. USA 87*, 6664−6668.

Tan, C.-K., Castillo, C., So, A. G. & Downey, K. M. (1986) *J. Biol. Chem. 261*, 12310−12316.

712

Tarawagi, Y., Enomoto, T., Watanabe, Y., Hanaoka, F. & Yamada, M. (1984) *Biochemistry 23*, 529 – 533.

Thömmes, P. & Hübscher, U. (1990a) *FEBS Lett. 268*, 325 – 328.

Thömmes, P. & Hübscher, U. (1990b) *J. Biol. Chem. 265*, 14347 – 14354.

Thrash, C., Bankiev, A. T., Barrell, B. G. & Sternglanz, R. (1985) *Proc. Natl Acad. Sci. USA 82*, 4374 – 4378.

Tsui, S., Anderson, M. E. & Tegtmeyer, P. (1989) *J. Virol. 63*, 5175 – 5183.

Tsurimoto, T. & Stillman, B. (1989a) *Mol. Cell. Biol. 9*, 609 – 619.

Tsurimoto, T. & Stillman, B. (1989b) *EMBO J. 8*, 3883 – 3889.

Tsurimoto, T. & Stillman, B. (1990) *Proc. Natl Acad. Sci. USA 87*, 1023 – 1027.

Tsurimoto, T., Melendy, T. & Stillman, B. (1990) *Nature 346*, 534 – 539.

Uemura, T., Ohkura, H., Adachi, Y., Morino, K., Shizaki, J. & Yanagida, M. (1987) *Cell 50*, 917 – 925.

van der Vliet, P. C. (1989) *Curr. Opinion Cell. Biol. 1*, 481 – 487.

van Holde, K. E. (1989) *Chromatin*, Springer-Verlag, Heidelberg.

Vishwanatha, J. U. & Baril, E. F. (1986) *J. Biol. Chem. 261*, 6619 – 6628.

Vishwanatha, J. U., Yamaguchi, M., DePamphilis, M. L. & Baril, E. F. (1986) *Nucleic Acids Res. 14*, 7305 – 7323.

Wang, J. C. (1987) *Biochim. Biophys. Acta 909*, 1 – 9.

Wiekowski, M., Schwarz, M. W. & Stahl, H. (1988) *J. Biol. Chem. 263*, 436 – 442.

Wilson, S., Abbotts, J. & Wicken, S. (1988) *Biochim. Biophys. Acta 949*, 149 – 157.

Wintersberger, U. (1990) *Pharmacol. Ther.*, in the press.

Wobbe, C. R., Weissbach, L., Borowiec, J. A., Dean, F. B., Murakami, Y., Bullock, P. & Hurwitz, J. (1987) *Proc. Natl Acad. Sci. USA 84*, 1834 – 1838.

Wold, M. S., Li, J. J. & Kelly, T. J. (1987) *Proc. Natl Acad. Sci. USA 84*, 3643 – 3647.

Wold, M. S. & Kelly, T. (1988) *Proc. Natl Acad. Sci. USA 85*, 2523 – 2527.

Wold, M. S., Weinberg, D. H., Virshup, D. M., Li, J. J. & Kelly, T. J. (1989) *J. Biol. Chem. 264*, 2801 – 2809.

Yanagida, M. & Wang, J. C. (1987) in *Nucleic acids and molecular biology* (Eckstein, F. & Lilley, D. M. J., eds) vol. 1, pp. 196 – 209, Springer-Verlag, Berlin, Heidelberg.

Yang, L., Wold, M. S., Li, J. J., Kelly, T. J. & Liu, L. F. (1987) *Proc. Natl Acad. Sci. USA 84*, 950 – 954.

Subject Index